Lecture Notes in Networks and Systems

Volume 426

The series "Lecture Notes in Networks and Systems" publishes the latest developments in Networks and Systems—quickly, informally and with high quality. Original research reported in proceedings and post-proceedings represents the core of LNNS.

Volumes published in LNNS embrace all aspects and subfields of, as well as new challenges in, Networks and Systems.

The series contains proceedings and edited volumes in systems and networks, spanning the areas of Cyber-Physical Systems, Autonomous Systems, Sensor Networks, Control Systems, Energy Systems, Automotive Systems, Biological Systems, Vehicular Networking and Connected Vehicles, Aerospace Systems, Automation, Manufacturing, Smart Grids, Nonlinear Systems, Power Systems, Robotics, Social Systems, Economic Systems and other. Of particular value to both the contributors and the readership are the short publication timeframe and the world-wide distribution and exposure which enable both a wide and rapid dissemination of research output.

The series covers the theory, applications, and perspectives on the state of the art and future developments relevant to systems and networks, decision making, control, complex processes and related areas, as embedded in the fields of interdisciplinary and applied sciences, engineering, computer science, physics, economics, social, and life sciences, as well as the paradigms and methodologies behind them.

Indexed by SCOPUS, INSPEC, WTI Frankfurt eG, zbMATH, SCImago.

All books published in the series are submitted for consideration in Web of Science.

For proposals from Asia please contact Aninda Bose (aninda.bose@springer.com).

More information about this series at https://link.springer.com/bookseries/15179

Jyotsna Kumar Mandal · Pao-Ann Hsiung ·
Rudra Sankar Dhar
Editors

Topical Drifts in Intelligent Computing

Proceedings of International Conference on Computational Techniques and Applications (ICCTA 2021)

Springer

Editors
Jyotsna Kumar Mandal
Department of Computer Science
and Engineering
University of Kalyani
Kalyani, India

Pao-Ann Hsiung
National Chung Cheng University
Min-Hsiung, Taiwan

Rudra Sankar Dhar
Department of Electronics
and Communication Engineering
NIT Mizoram
Aizawl, India

ISSN 2367-3370 ISSN 2367-3389 (electronic)
Lecture Notes in Networks and Systems
ISBN 978-981-19-0744-9 ISBN 978-981-19-0745-6 (eBook)
https://doi.org/10.1007/978-981-19-0745-6

This Springer imprint is published by the registered company Springer Nature Singapore Pte Ltd.
The registered company address is: 152 Beach Road, #21-01/04 Gateway East, Singapore 189721, Singapore

Preface

The **International Conference on Computational Techniques and Applications (ICCTA 2021)** has been initiated for sharing and preserving the scientific observation and conclusions on the applications of computing in network security, AI, data science, electronics, communication technology, electrical power, control systems, and energy technology. It was supposed to be held physically at Kolkata during 09 and 10 October 2021. But due to the growing pandemic of COVID-19—the coronavirus—the conference was conducted through virtual mode using the Cisco Webex Meetings platform on the same date. It was jointly organized by the Institution of Electronics and Telecommunication Engineers, Kolkata, and Winar Charitable Trust and Electro Inventor, Kolkata. The proceedings was published in the book series "Lecture Notes in Networks and Systems", Springer Nature.

ICCTA 2021 is a multi-institutional effort to exemplify research in the areas of intelligent computing and communication systems including computing, electronics, green energy design, communications, and computers to interact and disseminate information on the latest developments both academically and industrially for computational drifts. This conference constituted an international forum for professionals, industrialists, researchers, academicians, and students from various engineering fields with multi-disciplinary interests. Fifty-nine original contributions were published in the Sixth International Conference on Emerging Applications of Information Technology (EAIT 2020). The technical advancements of image processing, computer vision and pattern recognition, machine learning, data mining, big data and analytics, information security and privacy, wireless and sensor networks, and IoT are reflected in this volume on IoT applications scenarios.

This book contains three parts. These are Computing in Network Security, AI and Data Science, Contemporary Issues in Electronics, and Communication Technology, Intelligent Computing in Electrical Power, Control Systems, and Energy Technology.

The keynotes and technical sessions have been prepared under the conference scope to discuss the challenges, opportunities, and problems of application of computing in various fields.

On behalf of the organizing committee of ICCTA 2021, we would like to express our sincere thanks to the authors for their valuable contributions to this volume. Our sincere gratitude to the reviewers for reviewing the papers.

We express our gratitude for providing us the opportunity to publish the proceedings of ICCTA 2021 in the Lecture Notes in Networks and Systems, Springer Nature.

We also express our sincere gratitude to the authority of IETE Kolkata Centre for conducting this event jointly. Special thanks to the members of EC and OBs of IETE Kolkata Centre.

We hope this book will carry good research information for the researchers and budding engineers.

Kalyani, India Jyotsna Kumar Mandal
Min-Hsiung, Taiwan Dr. Pao-Ann Hsiung
Aizawl, India Dr. Rudra Sankar Dhar

Contents

Editors and Contributors

About the Editors

Jyotsna Kumar Mandal M.Tech. in Computer Science from University of Calcutta in 1987, awarded Ph.D. (Engineering) in Computer Science and Engineering by Jadavpur University in 2000. Working as Professor of Computer Science and Engineering, former Dean, Faculty of Engineering, Technology and Management, KU for two consecutive terms during 2008–2012. Director, IQAC, Kalyani University and Chairman, CIRM, and Placement Cell. Served as Professor Computer Applications, KGEC, as Associate Professor Computer Science, Assistant Professor Computer Science North Bengal University for fifteen years, as lecturer at NERIST, Itanagar for one year. 34 years of teaching and research experience in Coding theory, Data and Network Security and authentication; Remote Sensing and GIS based Applications, Data Compression, Error Correction, Visual Cryptography and Steganography. Awarded 24 Ph.D. Degrees, one submitted and 8 are pursuing. Supervised 03 M.Phil., more than 80 M.Tech.and more than 150 MCA Dissertations. Guest Editor of *MST Journal* (SCI indexed) of Springer. Published more than 450 research articles out of which 190 articles in International Journals. Published 12 books from LAP Germany, IGI Global, Springer etc. Organized more than 50 International Conferences and Corresponding Editors of edited volumes and conference publications of Springer, IEEE and Elsevier etc. and edited more than 50 volumes as volume editor. Received "SikshaRatna" award from Higher Education, Government of West Bengal, India in the year 2018 for outstanding teaching activities. Vidyasagar award from International Society for Science Technology and management in the fifth International Conference on Computing, Communication and Sensor Network. Chapter Patron Award, CSI Kolkata Chapter on 2014. "Bharat Jyoti Award" for meritorious services, outstanding performances and remarkable role in the field of Computer Science and Engineering on 29th August, 2012 from International Friendship Society (IIFS), New Delhi, A. M. Bose Memorial Silver medal and Kali Prasanna Dasgupta Memorial Silver medal from Jadavpur University.

Dr. Pao-Ann Hsiung got his Ph.D. in Electrical Engineering from the National Taiwan University in 1996. He was the Director-General of the Smart Technology Office at the Chiayi City Government, Taiwan. He helped the city to obtain huge funds of around 300M NT$. As a result, Chiayi City was awarded the Top7 Smart City in the World by ICF Forum. He was also the Dean of International Affairs and the CS Department Chair at the National Chung Cheng University, where currently he is a full professor and Director of the Smart Living Technology Research Center. He has published more than 280 papers in top journals and international conferences. His main research interests include smart system design, blockchain, IoT design, distributed security, and smart grid. He is an IET Fellow and has received several international and national accolades such as Excellent Research Awards, etc.

Dr. Rudra Sankar Dhar obtained his B.E. in Electronics and Telecommunication Engineering, from CVRCE, Orissa, M.Phil. in Microelectronics Engineering from University of Newcastle upon Tyne, UK and earned his Ph.D. degree in field of Nanotechnology Engineering from University of Waterloo, Canada. He worked as a Postdoctoral Research Fellow in the area of Nanomaterials technology and Smart Electronics for development of green technology nanodevices at the University of Alberta, Canada. He has also served and worked as teacher and researcher for more than 15 years at some of the reputed Industry and Institutes globally such as Chipworks Inc. Canada, Nanyang Technological University, Singapore and Thapar University, India among few others in the area of nano-microelectronic and optoelectronic technology and devices based on semiconductor, molecular and smart materials for emerging IOT, networking and circuitry applications. Presently, he is working in the Department of Electronics and Communication Engineering at NIT Mizoram since December 2015. He has solely supervised 1 Ph.D. while 10 others are pursuing and many M.Tech. students to completion and executed a few high value sponsored projects in India and abroad. He is the author of one book, and many Scopus indexed international book chapters while he has published more than 60 articles in international peer-reviewed high impact journals and conferences world-wide. He is also a reviewer and editor of many Scopus indexed journals and conferences. His research interests include semiconductor device physics, nanomaterials technology, microelectronic devices, IOT, networking and applications.

Contributors

Md. Abdul Wadud Department of Chemistry, Dhaka Cantonment, BAF Shaheen College, Dhaka, Bangladesh

Saurabh Agarwal School of Computer Science, University of Petroleum and Energy Studies, Dehradun, India

Akshit Aggarwal National Institute of Technology Delhi, New Delhi, Delhi, India

Priyanshu Agrawal National Institute of Technology Delhi, New Delhi, Delhi, India

Gufran Ahmad Department of Electrical Engineering, Dayalbagh Educational Institute, Agra, India

Dia Ali The University of Southern Mississippi, Hattiesburg, MS, USA

Maifuz Ali IIIT Naya Raipur, Naya Raipur, Chhattisgarh, India

Syed Samser Ali Gargi Memorial Institute of Technology, Kolkata, West Bengal, India

Ravshan Aliev Department of Information Systems and Technologies, Tashkent State Transport University, Tashkent, Uzbekistan

Animesh National Institute of Technology Delhi, New Delhi, India

Ravi Kumar Arya Department of Electronics and Communication Engineering, National Institute of Technology Delhi, New Delhi, India;
Department of Electrical and Electronics Engineering, National Institute of Technology, Delhi, India

Anjali Bansal Delhi Technological University, New Delhi, India

Md.Amdadul Bari North South University, Dhaka, Bangladesh

Sujit Barman National Institute of Technology Agartala, Agartala, India

Rituparna Basak University of Engineering and Management, Kolkata, India

Anirban Bhattacharjee National Institute of Technology Agartala, Agartala, India

Soumyendu Bhattacharjee Department of Electronics, Barasat State University, Kolkata, West Bengal, India

Sumanta Bhattacharyya Department of ECE, Greater Kolkata College of Engineering and Management, Kolkata, India

Arindam Biswas School of Mines and Metallurgy, Kazi Nazrul University, Asansol, West Bengal, India

Partha Chakraborty Department of Computer Science and Engineering, Comilla University, Cumilla, Bangladesh

Somenath Chakraborty The University of Southern Mississippi, Hattiesburg, MS, USA

Vinay B. Chandratre Bhabha Atomic Research Centre, Mumbai, India;
Homi Bhabha National Institute, Mumbai, India

Shweta Chauhan Department of Electronics and Communication, National Institute of Technology, Hamirpur, Himachal Pradesh, India

Arvind Choubey Department of Electronics and Communication Engineering, Indian Institute of Information Technology, Bhagalpur, India

Ajay Kumar Choudhary National Institute of Technology Agartala, Agartala, India

Dibyendu Chowdhury Haldia Institute of Technology, Haldia, West Bengal, India

Humayra Afrin Chowdhury Department of Computer Science and Engineering, Comilla University, Cumilla, Bangladesh

Philemon Daniel Department of Electronics and Communication, National Institute of Technology, Hamirpur, Himachal Pradesh, India

Achintya Das Kalyani Government Engineering College, Kalyani, West Bengal, India

Arnab Das Brainware Group of Institutions-S.D.E.T, Kolkata, West Bengal, India

Dipon Das Department of Applied Physics, University of Calcutta, Kolkata, West Bengal, India

Rajdip Das Jadavpur University, Kolkata, India

Rikita Das Department of Electronics and Communication Engineering, RCC Institute of Information Technology, Kolkata, India

Bipa Datta Brainware Group of Institutions-S.D.E.T, Kolkata, West Bengal, India

Anjan Debnath North Eastern Space Applications Centre, Umiam, India

Rajesh Dey Brainware University, Kolkata, West Bengal, India

Arpan Deyasi Department of Electronics and Communication Engineering, RCC Institute of Information Technology, Kolkata, India

Rudra Sankar Dhar Department of Electronics and Communication Engineering, National Institute of Technology Mizoram, Aizawl, Mizoram, India

Ayushi Dogra Electrical Engineering Department, NIT Hamirpur, Hamirpur, H.P., India

Ayush Kumar Dokania National Institute of Technology Delhi, New Delhi, Delhi, India

Anmol Dubey Vellore Institute of Technology, Chennai, India

Amit Dugh Information Technology, Queensland University of Technology, Brisbane, Australia

Ravi Dugh Goergen Institute of Data Science, University of Rochester, Rochester, NY, USA

Joydeep Dutta Department of CS, Kazi Nazrul University, Asansol, West Bengal, India

Kailash Pati Dutta Department of Electronics and Communication Engineering, Cambridge Institute of Technology, Ranchi, Jharkhand, India;
Department of Electronics and Communication Engineering, National Institute of Technology, Durgapur, West Bengal, India

Emenike C. Ejiogu Africa Centre of Excellence for Sustainable Power and Energy Development, University of Nigeria, Nsukka, Nigeria;
Laboratory of Industrial Electronics, Power Devices and New Energy Systems (LIEPNES), University of Nigeria, Nsukka, Nigeria

Md. Ezaz Ahmed CS Department, Saudi Electronic University, Al Madina KSA, Saudi Arabia

Aarthi Ganesan Anna University (MIT Campus), Chennai, India

Sayak Ghosal University of Engineering and Management, Kolkata, India

Subhro Ghosal Department of Electronic Science, Acharya Prafulla Chandra College, Kolkata, West Bengal, India

Kali Krishna Giri University Department of Physics, Ranchi University, Ranchi, India

Abhinav Gola National Institute of Technology Delhi, New Delhi, India

Kunal Goswami National Institute of Technology Delhi, New Delhi, Delhi, India

Siddhant Gudhe Electrical Engineering Department, Maulana Azad National Institute of Technology Bhopal, Bhopal, India

Nisha Gupta Birla Institute of Technology, MESRA, Ranchi, Jharkhand, India

Rashi Gupta School of Computer Science, University of Petroleum and Energy Studies, Dehradun, India

Ritik Ranjan Gupta Department of Electrical and Electronics Engineering, National Institute of Technology Delhi, New Delhi, India

Akshat Gururani Department of Electronics and Communication Engineering, National Institute of Technology Delhi, New Delhi, India

K. Hari Prasad Bhabha Atomic Research Centre, Mumbai, India;
Homi Bhabha National Institute, Mumbai, India

M. N. Islam Khan Nanoscience and Technology Research Laboratory, Atomic Energy Centre, Dhaka, Bangladesh Atomic Energy Commission, Dhaka, Bangladesh

Sanskar Jain National Institute of Technology Delhi, New Delhi, Delhi, India

Vinod Jha KIIT Deemed to be University, Bhubaneswar, Odisha, India

P. Kanungo C. V. Raman Global University, Bhubaneswar, Odisha, India

Rajib Kar National Institute of Technology, Durgapur, West Bengal, India

T. Kar KIIT Deemed to be University, Bhubaneswar, Odisha, India

Menika Karki National Institute of Technology Delhi, New Delhi, Delhi, India

K. A. Khan Department of Physics, Jagannath University, Dhaka, Bangladesh

Kamrul Alam Khan Department of Physics, Jagannath University, Dhaka, Bangladesh

Mohammad Monirujjaman Khan Department of Electrical and Computer Engineering, North South University, Dhaka, Bangladesh

Zuber Khan Department of Electronics and Communication Engineering, Indraprastha Institute of Information Technology, Delhi, India;
Department of Electrical and Electronics Engineering, National Institute of Technology Delhi, New Delhi, India

Kanupriya Khandelwal National Institute of Technology Delhi, Delhi, India

Sumit Kumar Khandelwal Delhi Technological University, New Delhi, India

Sonal Priya Kindo Department of Electronics and Communication Engineering, Cambridge Institute of Technology, Ranchi, Jharkhand, India

Hare Krishna Department of ECE, JUET, Guna, India;
Department of ECE, BIT Mesra, Ranchi, Jharkhand, India

Deo Kumar Department of Electronics and Communication Engineering, BIT Mesra, Ranchi, Jharkhand, India

Jatin Kumar Department of Electronics and Communication Engineering, National Institute of Technology Delhi, New Delhi, India

Ranjeet Kumar Department of Electronics and Communication Engineering, National Institute of Technology, Jamshedpur, India

Sanjeet Kumar Department of Electronics and Communication Engineering, BIT Mesra, Ranchi, Jharkhand, India

Suraj Kumar Cambridge Institute of Technology, Ranchi, Jharkhand, India

Unnathi Utpal Kumar Pathways World School Aravali, Gurugram, India

V. Kumar Electrical Engineering Department, NIT Hamirpur, Hamirpur, Himachal Pradesh, India

Vineet Kumar Electrical Engineering Department, NIT Hamirpur, Hamirpur, Himachal Pradesh, India

Kajal Kumari Department of Electronics and Communication Engineering, Cambridge Institute of Technology, Ranchi, Jharkhand, India

Neha Kumari Cambridge Institute of Technology, Ranchi, Jharkhand, India

Ravita Lamba Department of Electrical Engineering, Malaviya National Institute of Technology, Jaipur, Rajasthan, India

Suman Linda Department of Electronics and Communication Engineering, Cambridge Institute of Technology, Ranchi, Jharkhand, India

Chika Maduabuchi Department of Mechanical Engineering, Federal University of Agriculture, Makurdi, Nigeria

Santosh Kumar Mahto Department of Electronics and Communication Engineering, Indian Institute of Information Technology, Ranchi, India

Ruchika Malhotra Delhi Technological University, New Delhi, India

Sandipan Mallik Department of ECE, NIST Berhampur, Odisha, India

Kumari Mamta Cambridge Institute of Technology, Ranchi, India

Arpita Mandal National Institute of Technology Agartala, Agartala, India

Durbadal Mandal National Institute of Technology, Durgapur, West Bengal, India

Ujjwal Mondal Department of Applied Physics, University of Calcutta, Kolkata, West Bengal, India

Tamasi Moyra National Institute of Technology Agartala, Agartala, India

Moumita Mukherjee Department of Physics, School of Basic and Applied Sciences, Adamas University, Kolkata, West Bengal, India

Sourav Mukhopadhyay Bhabha Atomic Research Centre, Trombay, India; Homi Bhabha National Institute, Anushaktinagar, India

Beddhu Murali The University of Southern Mississippi, Hattiesburg, MS, USA

Lakhindar Murmu IIIT Naya Raipur, Naya Raipur, Chhattisgarh, India

R. Naresh Electrical Engineering Department, NIT Hamirpur, Hamirpur, Himachal Pradesh, India

Farah Nawar Department of Computer Science and Engineering, Comilla University, Cumilla, Bangladesh

Nikhilesh Kumar Neelu UL India Private Limited, Bangalore, Karnataka, India

Ekanshi Pal National Institute of Technology Delhi, New Delhi, Delhi, India

Umesh Pal Jadavpur University, Kolkata, India

Jagdish Chandra Patni School of Computer Science, University of Petroleum and Energy Studies, Dehradun, India

Rimi Paul Aliah University, Kolkata, West Bengal, India

Kholee Phimu Department of Electronic and Communication Engineering, National Institute of Technology, Aizawl, Mizoram, India

Swarna Kamal Pradhan Department of Applied Physics, University of Calcutta, Kolkata, West Bengal, India

P. Prathiksha Anna University (MIT Campus), Chennai, India

Priyadarshini Cambridge Institute of Technology, Ranchi, Jharkhand, India

Srinivasa Nallanthighal Raghava Delhi Technological University, New Delhi, India

Gautam Raina Centre for Energy and Environment, Malaviya National Institute of Technology, Jaipur, Rajasthan, India

Rajdeep Department of Electrical and Electronics Engineering, National Institute of Technology, Delhi, India

B. Rakini Anna University (MIT Campus), Chennai, India

R. S. Rampriya Anna University (MIT Campus), Chennai, India

Salman Rahman Rasel Local Government Engineering Department (LGED), Fulbaria, Mymensingh, Bangladesh

Naved Rehman Department of Electrical and Electronics Engineering, National Institute of Technology, Delhi, India

Suram Rithwik National Institute of Technology Delhi, Delhi, India

Saumya Roy University of Engineering and Management, Kolkata, India

Sabarinathan Cougher Inc, Tokyo, Japan

Pradip Kumar Sadhu Department of Electrical Engineering, Indian Institute of Technology (ISM), Dhanbad, India

Abhishek Saha Department of Electronics and Communication Engineering, National Institute of Technology Mizoram, Aizawl, Mizoram, India

Priyanka Saha Department of Electronics and Communication Engineering, National Institute of Technology Mizoram, Aizawl, Mizoram, India

Tulip Kumar Saha Department of Electronics and Communication Engineering, National Institute of Technology Mizoram, Aizawl, India;
Department of Electronics and Communication Engineering, School of Engineering and Technology, Adamas University, Kolkata, West Bengal, India

Angsuman Sarkar Department of Electronics and Communication Engineering, Kalyani Govt Engg College, Kalyani, India

Debayan Sarkar Department of Electrical Engineering, Indian Institute of Technology (ISM), Dhanbad, India

Kesari Sathvik Vellore Institute of Technology, Chennai, India

Shefali Saxena Department of Electronics and Communication, National Institute of Technology, Hamirpur, Himachal Pradesh, India

Md. Sayed Hossain Center for Research Reactor, Bangladesh Atomic Energy Commission, Dhaka, Bangladesh

Riya Sen Department of ECE, Malviya National Institute of Technology, Jaipur, India

M. Shaiful Islam Department of Physics, Jagannath University, Dhaka, Bangladesh

Deepak Sharma Department of ECE, JUET, Guna, India;
Department of ECE, BIT Mesra, Ranchi, Jharkhand, India

Hitesh Kumar Sharma School of Computer Science, University of Petroleum and Energy Studies, Dehradun, India

Shubham Sharma Centre for Energy and Environment, Malaviya National Institute of Technology, Jaipur, Rajasthan, India

Suraj Sharma IIIT Naya Raipur, Naya Raipur, Chhattisgarh, India

Veena Sharma Electrical Engineering Department, NIT Hamirpur, Hamirpur, Himachal Pradesh, India

Tanay Shubham Department of Electrical and Electronics Engineering, National Institute of Technology Delhi, New Delhi, India

Hiramani Shukla Electrical Engineering Department, Maulana Azad National Institute of Technology Bhopal, Bhopal, India

Ajit Kumar Singh Indian Institute of Information Technology, Ranchi, India

Tanmoy Singha Department of ECE, National Institute of Technology Mizoram, Aizawl, Mizoram, India

Khomdram Jolson Singh Department of Electronic and Communication Engineering, Manipur Institute of Technology, Canchipur, Imphal, Manipur, India

Navaneet Kumar Singh National Institute of Technology, Durgapur, West Bengal, India

Prashant Kumar Singh Department of ECE, University College of Engineering and Technology (UCET), VBU, Hazaribag, Jharkhand, India;
Department of ECE, BIT Mesra, Ranchi, Jharkhand, India

Raj Kumar Singh University Department of Physics, Ranchi University, Ranchi, India

Sachin Singh National Institute of Technology Delhi, New Delhi, India

Shashank K. Singh University College of Engineering and Technology (UCET), VBU, Sindoor, Hazaribag, India

Rashmi Sinha Department of Electronics and Communication Engineering, National Institute of Technology, Jamshedpur, India

Sunanda Sinha Centre for Energy and Environment, Malaviya National Institute of Technology, Jaipur, Rajasthan, India

P. Sonu Department of Computer Science and Applications, Amrita Vishwa Vidyapeetham, Amritapuri, India

Faria Soroni Department of Electrical and Computer Engineering, North South University, Dhaka, Bangladesh

N. Subhashini Vellore Institute of Technology, Chennai, India

R. Suganya Thiagarajar College of Engineering, Madurai, India

Menka Sukhwani Bhabha Atomic Research Centre, Mumbai, India;
Homi Bhabha National Institute, Mumbai, India

Sharath Sunil Department of Computer Science and Applications, Amrita Vishwa Vidyapeetham, Amritapuri, India

Mahmood Tabaddor UL India Private Limited, Bangalore, Karnataka, India

Megha Thomas Electronics Corporation of India Limited, Mumbai, India

Anjini K. Tiwary Department of ECE, JUET, Guna, India;
Department of ECE, Birla Institute of Technology Mesra, Ranchi, Jharkhand, India

Vaibhav Tripathi Vellore Institute of Technology, Chennai, India

Panduranga Vemula Department of Electronics and Communication Engineering, National Institute of Technology Mizoram, Aizawl, Mizoram, India

Promod Verma National Institute of Technology Delhi, Delhi, India

Dickson Warepam Department of Electronic and Communication Engineering, National Institute of Technology, Mizoram, India

Neelam Xalxo Department of Electronics and Communication Engineering, Cambridge Institute of Technology, Ranchi, Jharkhand, India

Menka Yadav Department of ECE, Malviya National Institute of Technology, Jaipur, India

Anil Kumar Yerrola IIIT Naya Raipur, Naya Raipur, Chhattisgarh, India

Potaraju Yugender Department of Electronics and Communication Engineering, National Institute of Technology Mizoram, Aizawl, Mizoram, India

Computing in Network Security, AI and Data Science

Skin Cancer Detection Using Computer Vision

Zuber Khan, Tanay Shubham, and Ravi Kumar Arya

Abstract Diagnosis of skin cancer at an early stage poses a great challenge even in the twenty-first century due to complex and expensive diagnostic techniques currently used for detection. Furthermore, traditional detection techniques are highly dependent on human interpretation. In case of fatal diseases such as melanoma, detection in early stages plays a vital role in determining the probability of getting cured. Several techniques such as dermoscopy, thermography and sonography are used for skin cancer detection, but every technique has its own limitations. Also, it is not feasible for every suspected patient to receive intensive screening by dermatologists. These limitations suggest the need for development of a simpler, cheaper, minimal invasive and accurate methodology independent of human intervention for skin cancer detection. Advancements in various computer vision algorithms have led to their extensive use in the area of bioinformatics. Therefore, this research paper aims to resolve the problem of early detection of skin cancer with a higher accuracy than existing methodologies using computer vision.

Keywords Skin cancer · Computer vision · Dermoscopy

1 Introduction

In recent years, skin cancer has become one of the most dangerous diseases in human beings. According to the World Health Organization, the total fatality cases account to nearly 1 million in 2018. Furthermore, exposure of human beings to ultraviolet rays has increased due to climate change and ozone layer depletion which is one of the major causes in the surge of cases of skin cancer.

Z. Khan (✉) · T. Shubham
Department of Electrical and Electronics Engineering, National Institute of Technology Delhi, New Delhi, India

R. K. Arya
Department of Electronics and Communication Engineering, National Institute of Technology Delhi, New Delhi, India

J. K. Mandal et al. (eds.), *Topical Drifts in Intelligent Computing*, Lecture Notes in Networks and Systems 426, https://doi.org/10.1007/978-981-19-0745-6_1

Skin cancer is broadly classified into basal cell carcinoma, squamous cell carcinoma and melanoma. While basal cell carcinoma is least hazardous among all these types as it does not spread from skin to other parts of the body and early diagnosis and treatment can cure it in most of the cases, the squamous cell carcinoma and melanoma are life threatening in nature. Though, melanoma is less common than other types of skin cancer but its spreading nature in our body poses a greater threat and mostly results in death. Melanoma cases mainly occur on skin, but its origin can be in our eyes, brain or lymph nodes (Fig. 1).

It is necessary to detect skin cancer at early stages as early detection enhances the survival chances of a person during treatment of skin cancer. Staging helps in determining whether the cancer has spread to other body parts or not. As the cancer cells grow deeper into the layer of skin, it can be life threatening.

There are many conventional techniques for detecting skin cancer such as biopsy, Raman spectroscopy, tape stripping and dermoscopy but each has its own limitations and drawbacks. For instance, a high resolution camera and years of experience is required to detect skin cancer using dermoscopy. Similarly, Raman spectroscopy uses complex algorithms and involves a series of processing stages leading to high consumption of time and resources. Furthermore, tape stripping technique needs large number of genetic profiles for differentiation of melanoma and non-melanoma cases, and it might cause irritation to patients too. Biopsy, on the other hand, requires surgical intervention. All of these conventional techniques have high dependency on dermatologist or expert technicians for results which raises the chances of human

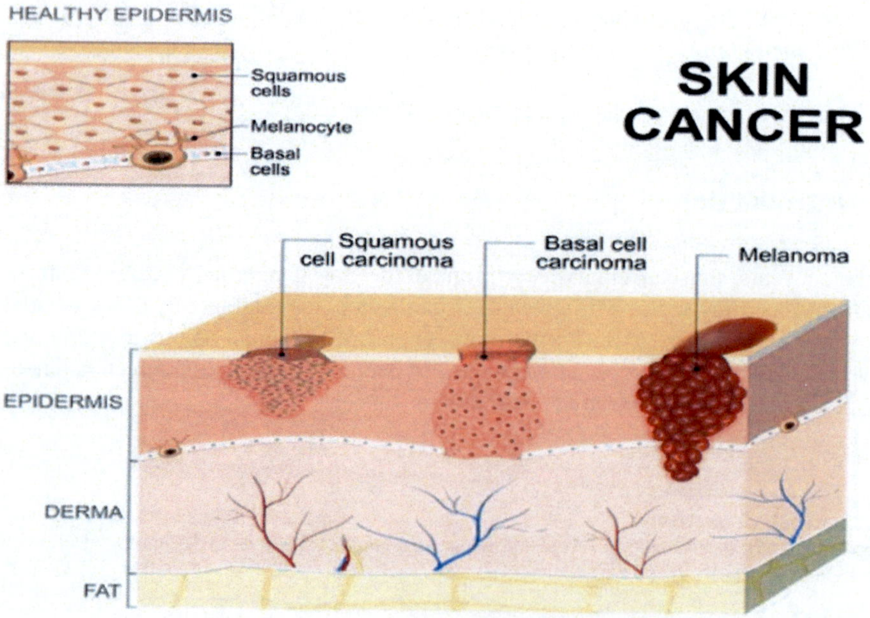

Fig. 1 Illustration of basal cell carcinoma, squamous cell carcinoma and melanoma

error and difficulty in early detection of skin cancer leading to higher mortality rate. Hence, there is a need for accurate and reliable techniques for early detection which will not only help the clinicians but will also prove to be a vital factor in the treatment of skin cancer.

One such advanced method for detection of skin cancer is computer vision. Computers can analyze large amounts of data available from dermoscopic images. When this data is fed into a computer vision algorithm, an appreciably accurate prediction can be made regarding the skin mole being malignant or benign. Therefore, this research work proposes a computer aided non-invasive diagnostic technique for early detection of skin cancer using computer vision.

2 Literature Review

Many researchers have used computer vision techniques in an effort to build precise and most accurate models for skin cancer detection.

Nahata et al. [1] developed multiple convolutional neural network (CNN)-based computer vision models using architectures such as ResNet50, MobileNet, VGG16, Inception V3 and Inception Resnet to classify the type of skin cancer and its early detection. The accuracy of these models ranges from 85% to 91%.

Furthermore, Hameed et al. [2] performed multi-class classification on skin lesion images to categorize them into healthy, acne, eczema, benign and malignant. The proposed methodology got an accuracy of nearly about 86% by using two different algorithms, error correcting output code (ECOC) support vector machine (SVM) and deep convolutional neural network. Traditional machine learning and advanced deep learning algorithms are used to create the classification algorithm. The proposed algorithm is evaluated on 2742 classified images and collected from different sources, but this technique is cumbersome.

Kaur et al. [3] built a model based on CNN with k-fold cross validation on multi-class classification problems defining seven different types of dermoscopic images of lesions. The model performed at 80% accuracy on reaching 150 epochs.

Similarly, Nasr et al. [4] proposed a convolutional neural network-based architecture to predict whether a given clinical image is malignant or benign in nature. Preprocessing of input non-dermoscopic images is done to remove illumination and other artifacts using image processing techniques. The algorithm attained an overall accuracy of 81%.

However, it is provable that an efficient model with higher accuracy can be developed when compared to existing research works in the same domain.

3 Experimentation and Methodology

The world of computer vision is explored via a family of dense convolution neural networks named as EfficientNet with highly fine-tuned hyperparameters using rectified adam [5] optimization leading to the state of art accuracy in image classification. EfficientNet models outperform conventional convolution neural network models in terms of accuracy and efficiency on a variety of scales.

3.1 Dataset

The project utilizes a dataset from Kaggle repository contributed by Fanconi [6] which consists of 3297 medical images of skin moles preprocessed using Ben and Cropping Technique. The dataset is further subdivided into training and testing dataset using manual categorical sorting with the training set consisting of 2637 medical images, while the test dataset consists of 660 images.

3.2 IDE and Tools Used

This project was created in Python using the Jupyter Notebook Integrated Development Environment (IDE). The Jupyter Notebook is a free, open-source web-based software application that helps in developing documents consisting of live code, visual images, mathematical equations and text information. The primary Python libraries used in this project are Numpy, TensorFlow, SciKit-Learn, Keras and Pandas.

3.3 EfficientNetB0 CNN Model and Rectified Adam Optimizer

EfficientNetB0 model. The proposed algorithm is based on the EfficientNetB0 model which is developed by multi-objective neural architecture search that maximizes the neural network's accuracy as well as its floating-point operations. To improve performance, a new baseline network is created using the AutoML MNAS framework to execute a neural architecture search, which optimizes both accuracy and efficiency. The resultant architecture uses mobile inverted bottleneck convolution (MBConv) and MnasNet, then scaling up this baseline network to obtain a family of models named *EfficientNets* (Fig. 2).

Rectified Adam Optimizer. A major disadvantage with conventional optimizers like Adam, RMSProp, etc. is that they have risk of converging into poor local optima

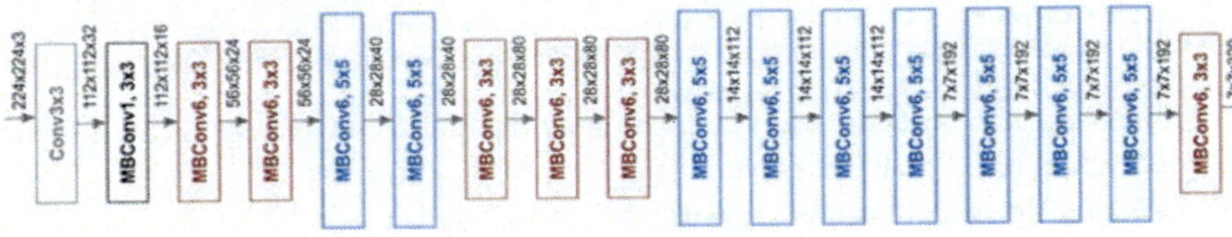

Fig. 2 Efficient net architecture. *Source* Tan et al. [7]

in case they are implemented without a warm-up method. The need to use a warm-up method arises from the fact that the adaptive learning rate optimizers contain too much variation, especially in the early stages of training, and hence make too many leaps based on insufficient training data, resulting in poor local optima.

However, since the degree of warm-up required is unknown and varies from dataset to dataset, an algorithm for dynamic variance reduction is created that results in a rectifier, which allows the adaptive momentum to slowly but steadily build up to full expression as it relates to the underlying variance. According to the divergence of the variance, rectified adam activates or deactivates the adaptive learning rate in real-time. Concluding, it provides a dynamic warm-up with no tunable parameters needed. Aside from that, rectified adam has been demonstrated to be more resistant against learning rate fluctuations and to deliver greater training accuracy and generalization on a range of datasets and inside a variety of AI architectures (Fig. 3).

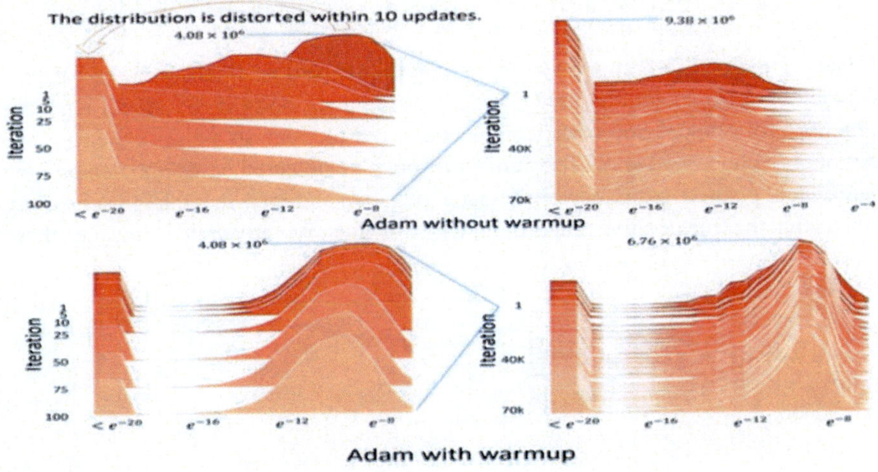

Fig. 3 Performance of Adam optimizer with and without warm-up. *Source* Wright et al. [8]

3.4 Procedure

1. Importing the images from a local machine and converting them into an equivalent array of RGB values using Numpy function.
2. Defining the target variable and one hot encoding it.
3. Performing image augmentation and normalization of RGB values.
4. Defining the build model function (Global Average 2D Pooling, Dropout = 0.5).
5. Defining the rectified adam (R Adam) optimizer with learning rate = 1e − 3, min learning rate = 1e − 7, and warm-up proportion = 0.15.
6. Building the EfficientNetB0 model with 'RAdam' as optimizer, evaluation metrics being 'accuracy' and 'categorical cross entropy' as the loss function.
7. Defining a learning rate reducer and fitting the model on training dataset with a validation split of 20%.
8. Making prediction on test data and evaluation of accuracy with generation of classification report.

4 Results and Observations

Training of the model reveals that the accuracy increases, while the validation loss decreases with the increase in number of epochs. Figure 4 shows the variation of training accuracy and validation accuracy with the number of epochs. It is observed that the accuracy keeps on increasing and becomes roughly constant after the 25th epoch.

The evaluation of a computer vision model should be done on the basis of not only accuracy but also specificity as well as sensitivity especially if the model is meant to be used for clinical purposes. This is due to the fact that in a clinical setting, we are not only concerned about the number of positive detected cases but also on how many actual positive patients were detected negative and vice versa. High sensitivity implies that the true positive rate is high, while the false negative rate is low, i.e., there

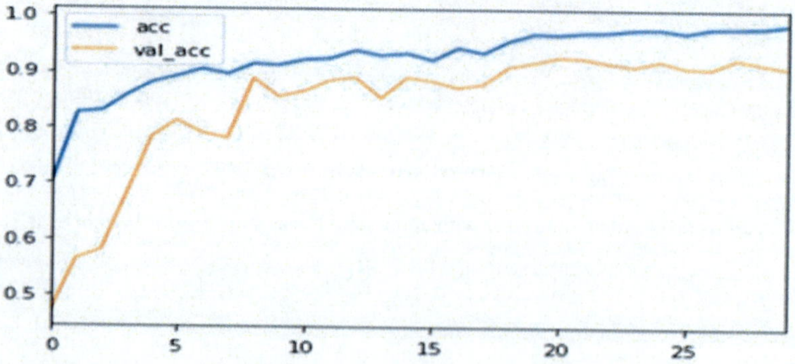

Fig. 4 Accuracy variation with epoch

are very few cancer patients who are missed out from getting detected. Similarly, high specificity implies that true negative cases are very high, while false positive cases are extremely low, i.e., there are very low chances of a healthy individual being falsely detected as a cancer patient.

Region operating characteristic (ROC) curve is one such evaluation metric that depicts trade-off between true-positives and false-positives. The area under the ROC curve reveals whether the model under test is useful for a given problem or not. In general, the value of area under the curve (AUC) greater than 0.7 makes a model acceptable, while a value below 0.5 is considered worthless. AUC value of 0.8–0.9 is considered excellent, while greater than 0.9 is considered as outstanding. Figure 5 shows the ROC curve of our proposed algorithm. Since the area under the curve is greater than 0.9, it shows that our model is outstanding for skin cancer detection. An appreciably high value of precision and recall as shown in Fig. 6 confirms the same.

Table 1 compares our work with prior studies in the same subject. We found from the table that our algorithm provides good accuracy.

Figure 7 shows the actual case vs. the predicted result of our model. It can be seen clearly that the predicted results match with the actual results.

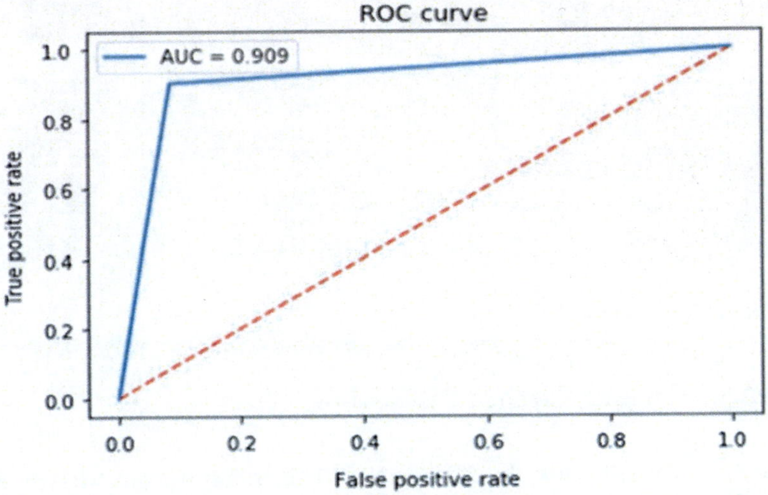

Fig. 5 ROC curve

Fig. 6 Classification report

Classification Report

	precision	recall	f1-score	support
0	0.92	0.91	0.92	360
1	0.90	0.90	0.90	300

Table 1 Comparative analysis

Author	Methodology	Accuracy (%)
Nahata et al.	ResNet50	85
	MobileNet	85
	VGG16	87
	Inception V3	90
	Inception Resnet	91
Hameed et al.	CNN and SVM	86.21
Kaur et al.	CNN	80.93
Esfahani et al.	CNN	81
Khan et al. (our work)	EfficientB0 with rectified Adam optimizer	91.1

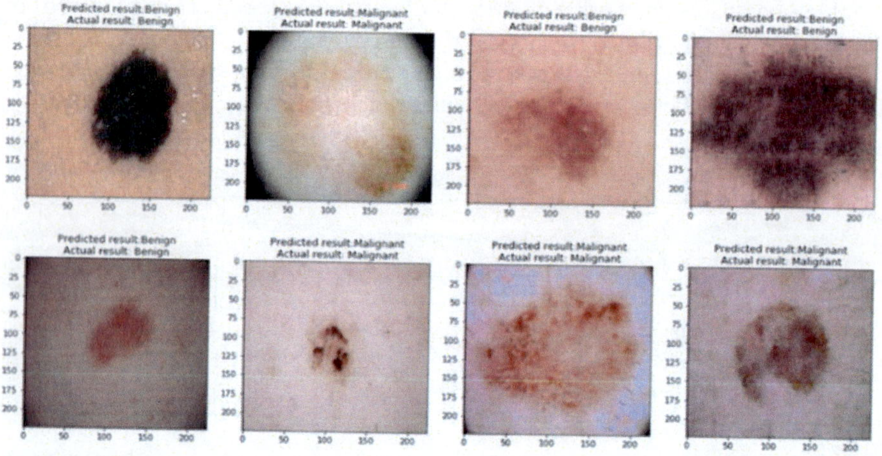

Fig. 7 Predicted versus actual output

5 Conclusion and Further Discussion

Our research establishes that the EfficientNetB0 algorithm when optimized using rectified Adam optimizer performs extremely well for skin cancer detection. Moreover, higher accuracy is achieved in a relatively lesser number of epochs as compared to existing computer vision methodologies implemented for the same goal. Furthermore, it is a non-invasive technique which just requires an image of the skin mole of a concerned patient, and the result is displayed within a few seconds overcoming the limitations of traditional testing procedures. Therefore, it is recommended to use this model in all clinical testing setups wherever feasible after more rigorous training on a larger dataset which will enhance its accuracy to even greater levels.

References

1. Nahata H, Singh SP (2020) Deep learning solutions for skin cancer detection and diagnosis. In: Jain V, Chatterjee J (eds) Machine learning with health care perspective. Learning and analytics in intelligent systems, vol 13. Springer, Cham. https://doi.org/10.1007/978-3-030-40850-3_8
2. Hameed N, Shabut AM, Hossain MA (2018) Multi-class skin diseases classification using deep convolutional neural network and support vector machine. In: 2018 12th international conference on software, knowledge, information management and applications (SKIMA), 2018, pp 1–7. https://doi.org/10.1109/SKIMA.2018.8631525
3. Kaur K, Boddu L, Fiaidhi J (2020) Computer vision for skin cancer detection and diagnosis. TechRxiv. Preprint. https://doi.org/10.36227/techrxiv.12089514.v1
4. Nasr Esfahani E, Samavi S, Karimi N, Soroushmehr SMR, Jafari MH, Ward K, Najarian K (2016) Melanoma detection by analysis of clinical images using convolutional neural network. In: Conference proceedings: annual international conference of the IEEE engineering in medicine and biology society. IEEE Engineering in Medicine and Biology Society. Conference. https://doi.org/10.1109/EMBC.2016.7590963
5. Liu L et al (2019) On the variance of the adaptive learning rate and beyond. arXiv:1908.03265
6. Fanconi C (2021) Skin cancer: malignant vs. benign. Kaggle.Com. https://www.kaggle.com/fanconic/skin-cancer-malignant-vs-benign
7. Tan M, Le QV (2019, May 29) EfficientNet: improving accuracy and efficiency through AutoML and model scaling. Google AI Blog. https://ai.googleblog.com/2019/05/efficientnet-improving-accuracy-and.html
8. Wright L (2019, Aug 20) New state of the art AI optimizer: rectified Adam (RAdam) medium. https://lessw.medium.com/new-state-of-the-art-ai-optimizer-rectified-adam-radam-5d854730807b

A Comparative Study of Machine Learning Algorithms for Anomaly-Based Network Intrusion Detection System

Vaibhav Tripathi, Anmol Dubey, Kesari Sathvik, and N. Subhashini

Abstract Cyber-security has become a major concern with rapid evolution of technology. To counter numerous novel attacks on a regular basis, organizations use intrusion detection systems (IDS). An IDS is often used for monitoring network traffic for detecting any anomaly or data breach. These systems, often called network intrusion detection systems (NIDS) or anomaly-based network intrusion detection systems (A-NIDS), generate alerts when any suspicious activity or anomaly is detected in the network. Machine learning (ML) and deep learning (DL)-based models provide an efficient way to detect these intrusions. Although various algorithms available in these domains can be used for building an A-NIDS which can efficiently detect intrusions, it is important to analyze their performance so as to determine the most suitable model depending on the need of the organization or user. This paper focuses on a comparative study of Naive Bayes classifier, K-neighbors classifier, logistic regression, random forest, gradient boost, SVM and XGBoost algorithms based on their efficiency, accuracy, time complexity and real-time applicability using parameters such as training time, prediction accuracy and confusion matrix. By implementing some of these models and training them on the Knowledge Discovery and Data Mining Tools Competition 1999 (KDD 1999) Dataset, the authors analyzed their performance on various parameters so as to determine which of these algorithms are the most suitable for building an anomaly-based network intrusion detection system. Random forest proved to be the most efficient and robust model with an expeditious computation and therefore suitable for building A-NIDS models for real-time security.

Keywords Machine learning · Network intrusion detection · Cyber-security · Performance evaluation · KDD-99

V. Tripathi · A. Dubey · K. Sathvik · N. Subhashini (✉)
Vellore Institute of Technology, Chennai 600127, India
e-mail: subhashini.n@vit.ac.in

J. K. Mandal et al. (eds.), *Topical Drifts in Intelligent Computing*, Lecture Notes in Networks and Systems 426, https://doi.org/10.1007/978-981-19-0745-6_2

1 Introduction

Intrusion detection systems are software applications that monitor a network for malicious activities. With rapid advent in technology, the attackers also keep developing novel attacks to exploit the systems and servers of organizations in order to steal, manipulate and corrupt their data. One method to detect attacks is using an anomaly-based network intrusion detection system (A-NIDS). These A-NIDS monitor the network using several parameters. If any unusual activity or anomaly is detected in the network, a warning is generated and sent to the administrator who takes necessary actions. Machine learning algorithms can play a pivotal role in detecting these attacks. In the past few years, many ML algorithms have been developed and improvised to detect and categorize network attacks. Using existing data gathered by monitoring networks for identified intrusions, ML algorithms can be trained to detect similar intrusion attempts and even predict new attacks based on any anomaly detected while monitoring the network. Different ML algorithms utilize different approaches to detect patterns and make predictions based on them; hence, it is important to analyze and understand which algorithm is well suited for a particular detection.

The following paper proposes a comparative study of various ML-based algorithms which can be used for building an anomaly-based network intrusion detection system (A-NIDS). The ML algorithm-based prediction models analyzed in the study are Naive Bayes classifier, K-neighbors classifier, logistic regression, random forest, gradient boost, SVM and XGBoost. The authors analyzed these algorithms on parameters of model accuracy, cross-validation accuracy, confusion matrix as well as training time. By training these models on the same datasets and evaluating their performance on the aforementioned parameters, the authors analyzed their performance to determine which of these algorithms is most suitable for building an A-NIDS, in terms of both accuracy as well as time efficiency. The dataset utilized for the proposed study is taken from Knowledge Discovery and Data Mining Tools Competition 1999 (KDD 1999) Dataset.

This paper will walk through the literature survey and their findings in Sect. 2. The dataset utilized for the training and testing of the models, that is, the KDD-1999 dataset, is discussed in Sect. 3 of the paper. The detailed analysis of these A-NIDS models is discussed in Sect. 4 of the paper along with a brief discussion on the findings. Finally, Sect. 5 draws a conclusion of this study and proposes the future work which can be carried out for analyzing and building more robust and efficient ML-based A-NIDS models.

2 Literature Survey

Several ML models have been used to classify traffic in the network on KDD Cup 99 dataset. In [1], the algorithms used are J48 classifier, Naive Bayes [2] and random forest. The results evidently proved that random forest was the best among all as it yielded almost 100% accuracy in detecting the threat. A comparison between a few algorithms, namely logistic linear regression, SVM, gradient boosted decision trees, random forest, neural network and CNN was drawn on UNSW-NB15 dataset to detect the type of cyber-attack on the system [3]. Out of all the algorithms, gradient boosting decision tree did the most accurate prediction with an accuracy of 95%.

One of the popular algorithms used for intrusion detection is XGBoost. In [4], XGBoost is employed on the NSL-KDD dataset. The model predicted the threat with an accuracy of 98.70%. Researchers also tried to propose an intrusion detection system that uses XGBoost algorithm on KDD cup 99 dataset after preprocessing. The results further compared the preprocessed dataset model with the existing algorithms. The XGBoost algorithm-based model yielded the highest accuracy of 96.62% [5].

K-nearest neighbor classification [6] was first implemented in 1951 by Evelyn Fix and Joseph Hodges. This algorithm is simple and easy to implement as it does not involve many parameters and is also capable of adapting itself according to the inputs. It basically works on the principle of distance between two points. The K-nearest neighbor (KNN) algorithm was tested on various datasets for its prediction accuracy in which it yielded an average of 85.15% accuracy [7].

Decision tree classification comes under supervised machine learning algorithm whose structure resembles that of a tree and the concept was first seen in 1963. This algorithm closely mimics the human decision technique and is easier for human interpretation than other algorithms. Decision tree acquired the highest accuracy of 90.9% on KDDTest+ dataset and 83.7% on KDDTest-21 on binary classification and acquired 85.4 and 74.4% on KDDTest+ and KDDTest-21 datasets on multiclass classification when compared with Naive Bayes, logistic regression and random forest [8]. Since decision tree models tend to overfit, the use of random forest [9] is recommended instead for implementation.

Boosting algorithms are generally groups of multiple weak prediction models for increasing accuracy and prediction of a model. Gradient boosting algorithm is a boosting algorithm which combines many decision trees for better results. Such boosting algorithm minimizes the overall error as the new predictions are built based on the errors in previous predictions. Gradient boosted decision trees yielded highest accuracy of 94.8% when compared with logistic regression [10], support vector machine, random forest, neural network and convolution neural network which were trained on UNSW-NB15 dataset [3].

XGBoost or extreme gradient boosting is another decision tree-based ensemble machine learning algorithm that uses the same family of gradient boosting framework and uses it at its core. XGBoost is parallelizable and can harness the power of multi-core computers, making it suitable for big data analysis. The XGBoost model for classification is called XGBClassifier [11]. An anomaly detection framework for IoT devices yielded accuracy of 98.2% with XGBoost, making it suitable for an A-NIDS model implementation [12].

SVM or support machine vector classifier [13] is another algorithm which is widely used for classifications. It is a supervised learning model, and it creates the best line or decision boundary that can segregate n-dimensional space into classes so that we can easily put the new data point in the correct category in the future. Its training on NSL-KDD dataset gave a model accuracy of 99.32% making it another suitable algorithm for the proposed study [14].

These works help identify the models which can be used for A-NIDS implementation, but there is a need to understand their performance against one another to understand which models function the best under similar conditions. Using these works as reference, the authors implemented the aforementioned ML models on the KDD-1999 dataset and analyzed their performance on various parameters as discussed in Sect. 4 to determine the model best suited for A-NIDS implementation.

3 Methodology

The KDD 1999 dataset used in the study gets its name as it was used in 1999 Knowledge Discovery and Data Mining Tools competition. It is a modified version of the 1998 DARPA Intrusion Detection Evaluation dataset which was managed by MIT Lincoln Labs. The dataset is created on the basis of attacks such as denial-of-service (DOS), unauthorized access from remote machines (R2L), unauthorized access to admin privileges (U2R) and probing. The main objective of creating these datasets is to evaluate the performance of various models that can be implemented in an intrusion detection system. The KDD 1999 is a multivariate dataset consisting of 4,898,430 connections and 41 features of each connection on which the connection is classified as normal or anomaly [15]. The 41 features in the dataset can be divided into four categories, namely basic features (Column 1–9), content features (Column 10–22), traffic features (Column 23–31) and host traffic features (Column 32–41). It contains some important features such as connection duration, protocol used, amount of data flowing, access and changes in admin privileges, login type and connection details.

In the proposed study, the authors used this KDD-1999 dataset for training and testing various machine learning-based models. This dataset is partitioned with an 80/20 split, that is, 80% of the data is used for training the models, while 20% of the data is used for validating the models. Training and validating the models on the same dataset simultaneously thus ensures that all models are evaluated equally and the results can be compared without any bias.

4 Results and Discussions

For evaluating the different algorithms, the authors used the KDD-1999 dataset for training as well as testing by splitting the dataset into 80:20 ratio. The training dataset contained 20,153 records with 41 columns while the testing dataset contained 5039 records. For comparing the performance of different models, there were evaluated

Table 1 Model accuracy and cross-validation mean score of models

Models	Model accuracy (%)	Cross-validation mean score (%)
Gaussian Naive Bayes classifier	90.84	90.17
K-neighbors classifier	99.24	99.03
Logistic regression classifier	95.47	95.17
Gradient boosting classifier	97.13	96.60
XGBoost classifier	99.52	99.31
SVM classifier	96.35	95.97
Random forest classifier	99.73	99.68

using parameters such as model accuracy [16], cross-validation accuracy [17], confusion matrix [18], training time and testing time [19]. While model accuracy, cross-validation accuracy and confusion matrix provide an efficient method of assessing the reliability and accuracy of the models, the training time and testing times helps estimate their real-time applicability. On evaluating the models on the aforementioned parameters, the following results are obtained.

4.1 Accuracy of Models

Table 1 describes the model accuracy and cross-validation mean score of all the models. The random forest classifier has the highest model accuracy of 99.73% and cross-validation mean score of 99.68%, followed by XGBoost with a model accuracy of 99.52% and cross-validation mean score of 99.31% and K-neighbors classifier with model accuracy of 99.24% and cross-validation mean score of 99.03%. The Gaussian Naive Bayes classifier has the least model accuracy of 90.84% and the least cross-validation mean score of 90.17%. These results show that among all the models, random forest is the most robust and accurate model for implementing an A-NIDS.

4.2 Confusion Matrix

Table 2 describes the confusion matrix with true positive, false positive, true negative and false negative values. The Gaussian Naive Bayes classifier has the worst performance according to the confusion matrix. It reported 359 false positive and

Table 2 Confusion matrix of models

Models	True positive	False positive	True negative	False negative
Gaussian Naive Bayes classifier	1974	359	2570	136
K-neighbors classifier	2308	25	2682	24
Logistic regression classifier	2179	154	2617	89
Gradient booster classifier	2265	68	2603	103
XGBoost classifier	2316	17	2688	18
SVM classifier	2170	163	2666	40
Random forest classifier	2324	9	2699	7

136 false negative results that sums up to a total of 495 false predictions. The best results are given by random forest classifier, where the false positive is 9 and false negative is 7 followed by XGBoost with only 17 false positives and 18. This means that random forest classifier predicts only 16 instances incorrectly from a total of 5039 records making it the best prediction model in terms of accuracy. Figure 1a is a visualization of true positive and true negative results, that is, the correct predictions from the models. Figure 1b is the visualization of false positives and false negatives of the model, that is, the incorrect predictions from the models. As evident from the graph plots of Fig. 1b, Naive Bayes generated most incorrect predictions indicating a high error percentage, while random forest produced least incorrect predictions, followed by XGBoost, making them the most accurate and robust models.

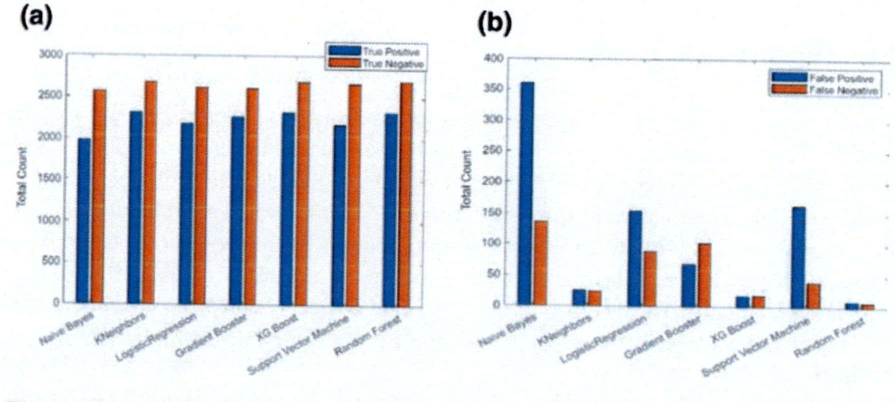

Fig. 1 **a** True positive and true negative instances for the models, **b** false positive and false negative instances for the models

Table 3 Train and prediction time of the models

Models	Training time (in s)	Prediction time (in s)
Gaussian Naive Bayes classifier	1.823	0.218
K-neighbors classifier	31.116	4.725
Logistic regression classifier	9.763	0.248
Gradient booster classifier	11.592	0.159
XGBoost classifier	16.737	0.201
SVM classifier	60.291	2.806
Random forest classifier	11.337	0.294

4.3 *Training and Prediction Time*

Table 3 shows time taken for training and prediction by each model. As mentioned before, the training consisted of the use of a dataset containing 20,153 records, while the testing dataset contained 5039 records. As evident from the data, SVM has the highest training and prediction time summing up to 63.097 s making its real-time implementation impractical. Although Gaussian Naive Bayes and logistic regression offer faster processing, their accuracy is far lower as compared to the other algorithms. Figure 2a is the visualization of training time comparison between the models, while Fig. 2b helps visualize the difference in prediction time during testing.

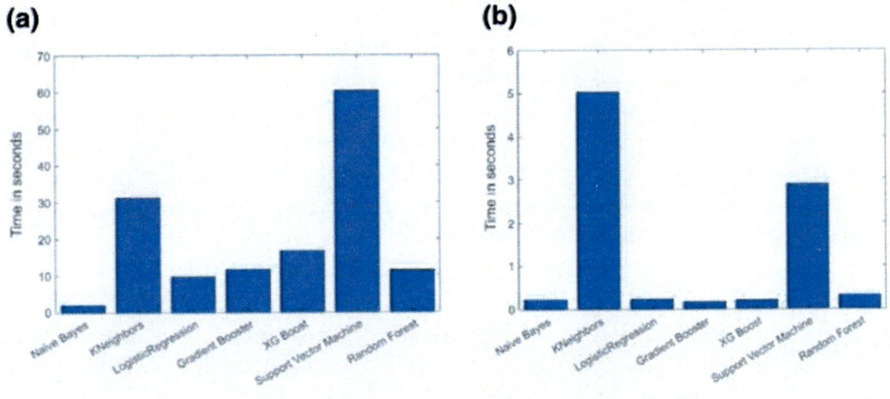

Fig. 2 **a** Training time taken by the models, **b** prediction time taken by the models

Apart from these parameters, the authors also compared the algorithms on the basis of precision, recall, $f1$ score and support. Since these parameters failed to produce a high magnitude of distinction between these models, they were ultimately discarded for the final comparative analysis. Therefore, on analyzing these models using the aforementioned parameters and training and testing them on the same dataset, it is found that K-neighbor and SVM achieved high accuracy but offered a very high runtime, making them unsuitable for real-time network monitoring for detecting intrusions. On the other hand, logistic regression and Naive Bayes had the lowest runtime but produced highly inaccurate results as compared to the other models, making their implementation unsuitable for detecting new intrusion attempts. XGBoost and random forest provided the optimum balance of runtime as well as accuracy. However, random forest with top accuracy and third best overall runtime makes it the best model for implementation as an A-NIDS.

5 Conclusion

From a detailed analysis of these widely used models, the authors observed that even though most algorithms boasted high accuracy, there are some aspects on which they lacked. Models such as Naive Bayes, logistic regression and gradient booster produced a high number of incorrect predictions when compared to the other models. This becomes a significant factor as a single incorrect intrusion detection can cause huge losses to the organizations and can lead to information leak as well as financial losses. The development of new intrusion techniques further diminishes the scope for these algorithms, as they may fail to detect new intrusion attempts. Although K-neighbors classifier and SVM classifier produced highly accurate results, the training as well as evaluation time of these models makes them unsuitable for practical implementation since these A-NIDS need to work in real time with huge datasets to facilitate a secure network connection. XGBoost and random forest had comparable accuracy, but ultimately random forest emerged as the most suitable model for implementing an A-NIDS with the highest accuracy and an optimum overall training and prediction time, far lesser than that of XGBoost making it suitable for real-time detection as well. With regular training of the model on new datasets containing logs of new intrusion attempts, one can even enhance the performance of random forest against new intrusion methods as well. The authors propose the hyper-tuning of these algorithms to make them even more compatible and robust against new intrusion techniques as a scope of future work.

References

1. Aishwarya C, Venkateswaran N, Supriya T, Sreekar M, Sreeja V (2020) Intrusion detection system using KDD cup 99 dataset. Int J Innov Technol Explor Eng (IJITEE) 9(4). ISSN:

2278-3075
2. Kelly A, Johnson MA (2021) Investigating the statistical assumptions of Naïve Bayes classifiers. In: 2021 55th annual conference on information sciences and systems (CISS), pp 1–6. https://doi.org/10.1109/CISS50987.2021.9400215
3. Bhavana D, Kishore Kumar K, Chilakala V, Chithirala HG, Meka TR (2019) A comparison of various machine learning algorithms in designing an intrusion detection system. Int J Sci Technol Res 8(12)
4. Dhaliwal SS, Nahid A-A, Abbas R (2018) Effective intrusion detection system using XGBoost. Information 9(7):149. https://doi.org/10.3390/info9070149
5. Siva Priya MS, Sahu BK, Kumar B, Yadav M (2019) Network intrusion detection system using XG boost. Int J Eng Adv Technol (IJEAT) 9(1). ISSN: 2249-8958
6. Moldagulova A, Sulaiman RB (2017) Using KNN algorithm for classification of textual documents. In: 2017 8th international conference on information technology (ICIT), pp 665–671. https://doi.org/10.1109/ICITECH.2017.8079924
7. Guo G, Wang H, Bell D, Bi Y, Greer K (2004) An kNN model-based approach and its application in text categorization. In: Gelbukh A (eds) Computational linguistics and intelligent text processing. CICLing 2004. Lecture notes in computer science, vol 2945. Springer, Berlin, Heidelberg. https://doi.org/10.1007/978-3-540-24630-5_69
8. Bandyopadhyay S, Chowdhury R, Banerjee P, Dey SD, Saha B (2020) A decision tree based intrusion detection system for identification of malicious web attacks. Preprints 2020, 2020070191. https://doi.org/10.20944/preprints202007.0191.v1
9. Ho TK (1995) Random decision forests. Proceedings of 3rd international conference on document analysis and recognition, vol 1, pp 278–282. https://doi.org/10.1109/ICDAR.1995.598994
10. Zou X, Hu Y, Tian Z, Shen K (2019) Logistic regression model optimization and case analysis. In: 2019 IEEE 7th international conference on computer science and network technology (ICCSNT), pp 135–139. https://doi.org/10.1109/ICCSNT47585.2019.8962457
11. Li H, Cao Y, Li S, Zhao J, Sun Y (2020) XGBoost model and its application to personal credit evaluation. IEEE Intell Syst 35(3):52–61. https://doi.org/10.1109/MIS.2020.2972533
12. Wang X, Lu X (2020) A host-based anomaly detection framework using XGBoost and LSTM for IoT devices. Wirel Commun Mob Comput 8838571:13. https://doi.org/10.1155/2020/8838571
13. Ghosh S, Dasgupta A, Swetapadma A (2019) A study on support vector machine based linear and non-linear pattern classification. Int Conf Intell Sustain Syst (ICISS) 2019:24–28. https://doi.org/10.1109/ISS1.2019.8908018
14. Jha J, Ragha L (2013) Intrusion detection system using support vector machine. IJAIS Proc Int Conf Works Adv Comput ICWAC 3:25–30
15. Choudhary S, Kesswani N (2020) Analysis of KDD-Cup'99, NSL-KDD and UNSW-NB15 datasets using deep learning in IoT. Proc Comput Sci 167:1561–1573. ISSN 1877-0509. https://doi.org/10.1016/j.procs.2020.03.367
16. Yin M, Vaughan JW, Wallach H (2019) Understanding the effect of accuracy on trust in machine learning models. In: Proceedings of the 2019 CHI conference on human factors in computing systems. Association for Computing Machinery, New York, NY, USA, Paper 279, pp 1–12. https://doi.org/10.1145/3290605.3300509
17. Schaffer C (1993) Selecting a classification method by cross-validation. Mach Learn 13:135–143. https://doi.org/10.1007/BF00993106
18. Caelen O (2017) A Bayesian interpretation of the confusion matrix. Ann Math Artif Intell 81:429–450. https://doi.org/10.1007/s10472-017-9564-8
19. Wu Z, Jiang L, Jiang Z, Chen B, Liu K, Xuan Q, Xiang Y (2018) Accurate indoor localization based on CSI and visibility graph. Sensors 18:2549. https://doi.org/10.3390/s18082549

An Effective Approach for Detecting Acute Lymphoblastic Leukemia Using Deep Convolutional Neural Networks

Sharath Sunil and P. Sonu

Abstract In this paper, we present a fully automatic acute lymphoblastic leukemia detection method based on deep neural networks. ALL is a condition affecting the leukocytes. Children tend to be prone to have this melanoma. The fundamental problem with this type of cancer is that, unlike other cancers, it does not create tumors, making it extremely difficult to identify. Prior to automation, manual microscopic testing procedures were used, but they were time-consuming and error-prone. To overcome this issue, many automated systems were introduced, which used machine learning techniques. But, because we are dealing with medical information, we may require better efficiency and accuracy, so as an improvement, our proposed system employed several convolutional neural networks architectures. In the proposed system, five such CNN algorithms were implemented to classify and separate the cancerous and non-cancerous cells. The system accepts the blood cell image from the user and predicts whether the cell contains one/more blasts depending upon the prediction value obtained from the CNN algorithm on the stained cell image. As a result, it significantly reduces the research costs, increases the speed of testing, and can be a lifesaver for millions of cancer patients.

Keywords Acute lymphoblastic leukemia · Deep convolutional neural networks · Medical image processing

1 Introduction

ALL, or acute lymphoblastic leukemia, is a malignancy that primarily affects white blood cells (WBCs). This variety of cancer is induced by a mistake or fault in the DNA of blood cells (Genetic structure). ALL attacks the liver, spleen, and other

S. Sunil · P. Sonu (✉)
Department of Computer Science and Applications, Amrita Vishwa Vidyapeetham, Amritapuri, India
e-mail: sonu@am.amrita.edu

S. Sunil
e-mail: sharathsunil@am.students.amrita.edu

J. K. Mandal et al. (eds.), *Topical Drifts in Intelligent Computing*, Lecture Notes in Networks and Systems 426, https://doi.org/10.1007/978-981-19-0745-6_3

organs after the initial stages. Because ALL does not cause tumors, it is far more difficult to diagnose than other types of cancer. The main risk factors of ALL are exposure to chemical solvents, radiations, electromagnetic fields, etc. To detect the condition, blood cell samples must be manually analyzed, which is a time-consuming and laborious operation [1]. Acute lymphoblastic leukemia, acute myeloid leukemia, chronic myeloid leukemia, and chronic lymphocytic leukemia are some of the many subtypes of leukemia. Acute leukemia tends to spread faster than chronic leukemia, hence more lethal [2, 3]. Automated testing solutions were designed to address the issues of speed and precision in manual testing. Image enhancement and processing, feature extraction, and machine learning-based techniques are used by these automated systems to classify blasts [4]. The main problem with this type of system is that still they do not provide sufficient performance and accuracy needed for medical data classification. Some minor mistakes or miscalculations on the part of the system may cause irreversible consequences. So, to tackle the problem of accuracy and efficiency, convolutional neural networks models are used, which are a class of deep learning models. CNN is used for the classification of image data. The datasets used in the system are ALL-IDB 1 and ALL-IDB 2 (see Fig. 1), which are obtained from Scotti [5], with prior permission to use them for educational and research purposes. The ALL-IDB 1 may be used to test segmentation techniques, as well as classification and image preprocessing approaches. ALL-IDB 2 may be used to evaluate the overall performance of classification models and systems. ALL-IDB2 is a collection of benign and malignant cells' cropped areas of interest.

The deep learning models work similarly to that of human neurons. They are much powerful and accurate as compared to the machine learning counter-parts model. The input layer, hidden layers, and a fully linked output layer all are layers in CNN models. The convolutional layer extracts the features from the image to create feature maps, and the max-pooling layer selects the maximum value from the feature map, to reduce the computational efforts in the upcoming layers. The

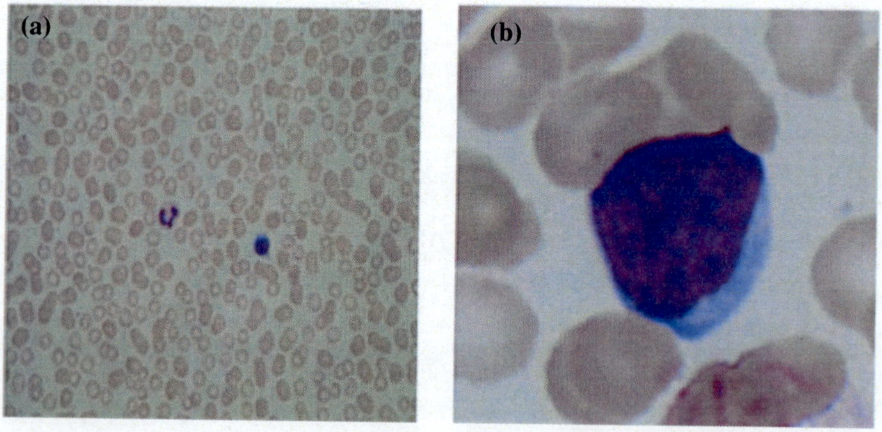

Fig. 1 **a** Blood cell image sample taken from ALL-IDB I, **b** image sample from ALL-IDB II

dense layers otherwise called the actual neural networks perform calculations, after assigning weights and biases, accompanied by ReLU activation. Depending on the type of activation function utilized, such as sigmoid or softmax, the output layer may comprise one or two neurons. The sigmoid function accepts input and outputs a range of 0–1, resulting in an S-shaped graph, whereas the softmax function accepts real numbers as inputs and outputs their associated probabilities [6].

1.1 Background

Leukemia is a group of blood cancers, which origins in the bone marrow, and results in the abnormal multiplication of blood cells. Due to some anomaly or disorder, the bone marrow may produce immature cells that develop into lymphoblast or leukemic WBCs. These defective cells can proliferate and crowd out healthy cells because they are unable to function correctly [7]. There are various risk factors associated with ALL which include:

- Radiation exposure: People who have been exposed to high levels of nuclear or chemical radiation have a high possibility of developing cancer.
- Genetic disorders such as Down's syndrome are associated with an increased risk of ALL.
- People or even children who have undergone chemotherapy or other forms of cancer treatments may have a high chance of developing ALL.

The terms "machine learning" and "artificial intelligence" are frequently interchanged. Machine learning algorithms are built on top of mathematics, computer science concepts, statistics, and probability. With the advancement of computational resources, and improved software and hardware, machine learning has seen exponential growth. Due to morphological differences in the structure of the cells, we can use machine learning algorithms to classify the cells based on their chromatic and morphological features. The feature extraction and image processing have to be done manually. Machine learning algorithms are good solutions for medical science-related problems, but still, they are not reliable as we want to have excellent accuracy and faster development. Deep learning or D.L. is algorithms that are subsets of machine learning algorithms. Neural networks are the main component of these systems; by the name itself, they function in the same way, human neurons work. These neural networks are the successors of the original perceptron models which are linear binary classifiers [8].

Deep neural networks consist of layers, namely input layer, hidden layer and output layer. The input of one layer is obtained from the output of the previous layer. In each neuron, certain weights and biases are applied to get the desired output for the particular input. In all fully connected neural networks, all the nodes of one layer are connected to all the nodes of the next layer. They are powerful, efficient, and perform very well as compared to their predecessors [9]. Some notable deep learning algorithms include:

- Convolutional neural networks.
- Recurrent neural networks.
- Multilayer perceptrons.
- Long short-term memory networks.
- Restricted Boltzmann machines.

Out of the abovementioned algorithms, for medical image analysis and classification, CNN or convolutional neural networks are optimal. In CNN algorithms, the feature extraction process takes automatically via the convolution operation; the maximum value from the pool is selected by the max-pooling operation. Also, it eliminates the need for manual image processing and feature extraction [10]. The process of creating feature maps from images/image data is called convolution operation. For this purpose, a kernel of arbitrary size is chosen, and the convolution operation is applied to get the feature maps; then, to reduce complexity, max-pooling operation is applied to retrieve the maximum value from the feature matrix. Besides that the network is not fully connected to reduce computational complexity. This is what makes it optimal for image classification as compared to other deep learning models. The constraints in deep learning are the need for a very large amount of data and optimal hardware [11, 12].

2 Similar Systems

The system proposed by Shafique and Tehsin [13] used deep learning CNN algorithms, to classify the cancer cells/normal cells. They have employed a pre-trained AlexNet CNN architecture to solve the problem statement. The concept of transfer learning and fine-tuning were implemented in their proposed system. For comparison, they have compared the machine learning models like SVM and KNN. After the application of the image augmentation and training the model on the pretrained AlexNet model, they have attained an accuracy of 99% and a testing accuracy of 96%. It has given insights about pretrained models and image augmentation and overall information about deep learning models particularly CNN. The use of pretrained models significantly increases the accuracy, performance, and speed of development and testing. The ILSVRC or the ImageNet large-scale visual recognition challenge is an annual challenge conducted to evaluate the performance of segmentation and classification CNN models. The main task lies in the identification and segmentation of 15 million images belonging to 22,000 classes (ImageNet). The original AlexNet was the winner of the ILSVRC challenge and was among one of the best models to date.

The system/model proposed by Sajja et al. [14] suggested the use of pretrained CNN models to predict lung cancer. Lung cancer/lung carcinoma is one main cause of cancer-related deaths worldwide. The images obtained from CT scans were used to process, segment, and classify the images. Previous studies and methodologies were put forward which proposed systems/models which gave an accuracy of 67.7%;

later, improvements were made by using transfer learning methodologies which gave an accuracy of 88.41%. When compared to their predecessor, the CNN models removed the overhead of image processing and feature extraction, as it is done automatically in the system. Besides the use of common CNN algorithms, the concept of transfer learning was applied to the dataset. The proposed system was compared with pretrained CNN models, namely AlexNet, ResNet50, and GoogleNet. The AlexNet architecture contains eight layers; GoogLeNet contains 22 hidden layers, and the ResNet50 consists of 50 layers, as the depth went on increasing the training time seem to increase. All the models were capable of classifying 1000 classes and can automatically extract the features. Since the proposed system consisted of numerous layers, it is considered a deep neural network, and with such high complexity, the model may suffer from overfitting. The aforementioned network is built sequentially, but the architecture was changed to sparsely connected networks, in which the model was built upon aggregation of some layers to minimize input channels and reduce convolutions. The accuracies of AlexNet, GoogLeNet, and ResNet50 models were 90, 94, and 96%, while that of the proposed system was 99%, when the models were trained with 90% of the training samples.

The system proposed by Rawat et al. [15] was developed to solve the problem of manual testing, time consumption, and the errors that accompany manual methods. Before their system, some automated systems were developed to solve the problem of manual testing, but they were not much efficient. Image segmentation, feature extraction, and classification are the three modules that make up the suggested system. They have employed machine learning techniques to develop the classification module. Since the features need to be extracted manually, the image-segmentation module and the feature extraction module play an important role in this system. The image is first converted to gray scale; then, 2D statistic filtering is applied after histogram equalization is applied. The aforementioned methods remove the irregularities and equalize the pixel values. To obtain the nucleus and cytoplasm, the image is converted to a binary image by applying Otsu's thresholding, then applied morphological opening to extract the nucleus and subtracted the nucleus to obtain the cytoplasm. For feature extraction, they considered the morphological features like the shape, size, chromatic features, and the area, perimeter, size of nucleus, and vacuoles. The classifier modules consist of three classifiers, which first classified the normal and cancerous cells; then, the L1 and other cell types are accompanied by L2 and L3 cell types. PCA-KNN, PCA-PNN, PCA18 SVM, PCA-SSVM, and PCA-ANFIS are five classifiers that are arranged hierarchically to categorize the cells at each level. The PCA-SVM classifier produced the highest accuracy of 99.2%, compared to 88.4, 96.1, 90.7, and 94.6% for the PCA-KNN, PCA-PNN, PCA-SSVM, and PCA-ANFIS classifiers.

3 The Proposed System

The proposed system uses convolutional neural networks or CNN, which is a class of deep learning algorithms used to classify, segment, process image datasets, and

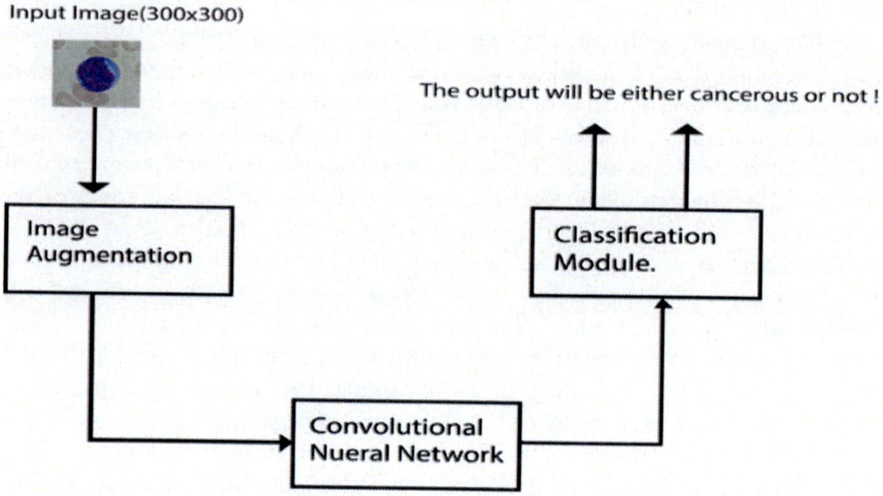

Fig. 2 The high-level architecture of the proposed system, here, the image is resized to 300 × 300 [1]

for other correlated data. The main concept behind the working deep learning algorithms is the layers in these algorithms. The presented system includes several levels, including an input layer, hidden layers, and an output layer. The system's most significant layers are the convolutional and max-pooling layers. Here, we extract the features from the image so that we don't have to extract them manually. The entire system is divided into three modules, namely the image augmentation module, convolutional neural network module, and the classification module. A total of five models has been implemented for the CNN module, namely a custom module, a modified AlexNet module, and three pre-trained models—ResNet50 and VGG. An abstract implementation of the proposed system has been given (see Fig. 2). The blood cell image is first resized according to the CNN algorithm (300 × 300 or 224 × 224 or 227 × 227). Then, the blood cell image is then passed onto the CNN module; in the CNN module, the five CNN algorithms are implemented, after undergoing processing and calculations in the CNN module(s); the values obtained from the output layer of the CNN module will be used to predict whether the cell is cancerous or not. The detailed information regarding individual modules has been given in the modules section.

4 Modules

- Image augmentation module

It is one of the preprocessing tasks to ensure deep learning models with the illusion of a huge amount of data. Our dataset consists of less than 500 microscopic images; due to the small size of the image dataset, there is a chance that the model is very likely to overfit and perform very badly during the testing phase. To solve that issue, image augmentation is performed on the dataset. The entire dataset is divided into 80% training, 10% testing, and 10% validation so that we can later check the training and validation loss for any chance of overfitting. The image augmentation will generate augmented images from the existing dataset so that at each iteration, the model may receive an altered image of the original image. The Keras API provides ImageDataGenerator class, which provides functionalities to perform advanced image augmentation. Advanced image manipulation like mirroring, shearing, zooming, and rotation can be done. We provide the batch size, which in turn determines the augmented images produced or the steps per epochs.

$$\text{Steps Per Epoch} = \text{Training Data}/\text{Batch Size} \tag{1}$$

- CNN Module

In this module, five CNN models are implemented to classify the cancer cells; out of which, three of them are pre-trained models. In conjunction with the sigmoid activation function, the first model is a custom CNN model with five convolutional layers and related max-pooling layers, a fully connected dense neural network layer with 512 neurons, and an output layer with one neuron. The second is a revised version of AlexNet's initial architecture, with an output layer of one neuron accompanied by the sigmoid activation function. Appropriate dropout and batch normalization layers were added to introduce normalization and to prevent the overfitting of the model. Three pre-trained models were implemented, using transfer learning and fine-tuning to compare the performance of the models, which were the VGG16, VGG19, and the ResNet50 CNN models. Transfer learning is the process of reusing a powerful deep learning model, as mentioned earlier which has been previously been trained on very large datasets. This has been widely been popular in computer vision object classification problems. Fine-tuning is a technique used in transfer learning, in which we unfreeze a few of the top layers of a frozen model base and then add a 22 few modified dense fully connected layers. This allows us to fine-tune the base model's higher-order characteristics to make them more efficient for the given use case. The main advantage is that it can increase the training speed and is efficient for smaller datasets (Fig. 3).

- Classification Module

This is the final module (see Fig. 4), where the image is classified based on the prediction value obtained from the CNN model. In this module, the microscopic image is given as an input; after that the image is resized to a specific dimension based on the model that is to be used, for example, 300 × 300 for the custom model, 224 × 224 for VGG, and ResNet, and so on. Then, the image is converted into a NumPy array, and its dimensions are expanded so that it can be directly inputted to the model prediction function. A prediction value will be returned which h will be in the range of 0–1. If the prediction value is less than 0.5, then the image may contain one/blast; otherwise, it is normal.

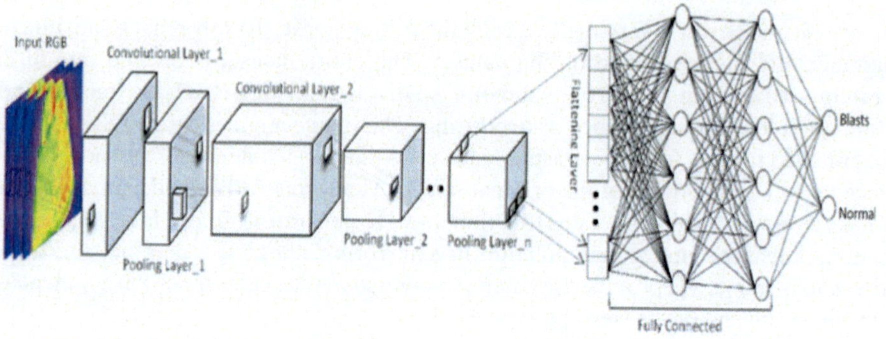

Fig. 3 Architecture of a typical CNN model

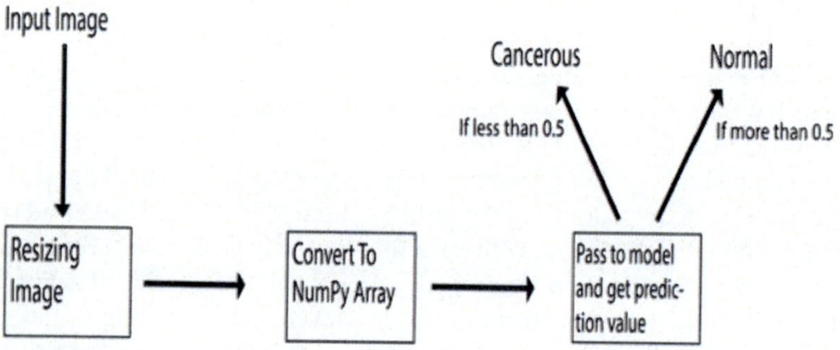

Fig. 4 Working of classification module

5 CNN Models

5.1 Pretrained Models

- VGG

When the original AlexNet was first published in 2012, it proved to be one of the best models. As mentioned earlier, it used dropout and augmentation techniques to address the problems of overfitting and introduce normalization. VGG or OxfordNet (Visual Geometry Group) are a class of networks, which were an improvement toward the original AlexNet; though AlexNet was a capable model, the researchers felt the need for a much advanced and powerful model. The VGG model was runner-up in the annual ILSRVC challenge, while the winner was the GoogLeNet model. The input size of VGG is 224 × 224. VGG uses very narrow receptive fields in its convolutional layers. There are additional 1 × 1 convolution filters that perform a linear change of the input before a ReLU unit is applied. The VGG network consists of three completely linked layers; two of which have 4096 neurons each, and the third has 1000 neurons, allowing it to classify 1000 classes. For our use case, the fully connected layers were removed, and only, the base layers were set to trainable. The VGG16 had a total of 16 layers and 138.30 million parameters, and VGG19 had 19 layers and 143 million parameters.

- ResNet

With the advent of each new model in ILSVRC, the depth and number of parameters always saw a steep rise. But, this increase in layers may cause a problem of vanishing or exploding gradient in which the gradient may seem to be very minute or very large. This phenomenon in turn will significantly increase training and test error rates. ResNet [13] was proposed by Microsoft Research experts in 2015 and is also known as residual network. This architecture introduces the concept of residual network to address the problem of stochastic vanishing/exploding gradient. The aforementioned model uses skip connections. The skip connection bypasses a few stages of training and links directly to the output layer. The main advantage is that any anomaly in the performance is rectified by regularization. The input size of ResNet is 224 × 224. The model is 50 layers deep and has 25 million trainable parameters.

5.2 *Custom Models*

- Custom Model 1

The first CNN is the simplest among all; as mentioned in the previous section, it contains five convolutional and corresponding max-pooling layers. The default input size of this model is 300 × 300. This model was less prone to overfitting due to its simplicity; also, the dense layer has 512 neurons without any dropout layers in between them. This model can also be considered a general-purpose binary classifier. Sigmoid or softmax activation could be used, but it should be noted that one neuron should be used along with sigmoid, while two for softmax.

- Custom Model 2

AlexNet has five convolutional layers, three max-pooling layers, two normalization layers, two fully connected layers, and one softmax layer in its original architecture. The network contained 60 million trainable parameters in total. The input size is 227 × 227 × 3. For our use case, the layer has been modified to classify two classes. Dropout and batch normalization layers have been added to introduce normalization to the input; otherwise, there might be a chance of overfitting. Dropout values were chosen after checking for the chances of overfitting. The output layer was modified to include one neuron in conjunction with a sigmoid activation function.

6 Result Analysis

Five models were implemented as a part of the main module, i.e., the CNN module, namely the custom model with five convolution layers, a modified AlexNet model, pre-trained VGG16 and 19 models, and finally a pre-trained ResNet50 model. The value accuracy and value loss were monitored while training the model, to avoid overfitting. We can detect the overfitting in the model beforehand by analyzing the training loss and validation loss; if the latter is increasing as compared to the former one, then that may be an indication of overfitting.

The testing accuracy is evaluated at first to analyze the performance of the classifiers. But, for the medical data classification, testing accuracy isn't enough. So, to compensate for that we will have to also work out the ROC curve for the binary classifiers. Before evaluating the ROC curve, we have to take care of the following parameters:

- True Positives: the number of cells correctly classified as normal.
- True Negatives: the number of cells correctly classified as cancerous.
- False Positives: the number of cells incorrectly classified as normal also called type I error.
- False Negatives: the number of cells incorrectly classified as cancerous also called type II error.

Using the above values, the accuracy %, true-positivity rate, and the false-positivity rate can be worked out in (Table 1):

$$\text{Accuracy}\,\% = ((\text{TN} + \text{TP}))/((\text{TN} + \text{TP} + \text{FN} + \text{FP})) \tag{2}$$

The ROC or the "Receiver operating characteristics" curve or the area under the curve (AUC) is a measure of how our classification model is performing under various thresholds. It gives an idea of how capable is our model to be able to classify the distinct classes. The ROC curve is plotted with the true-positivity rate (TPR) or recall (Sensitivity) against the false-positivity rate (FPR) or 1-specificity. Usually, ROC curves that come closer to the top left corner may tend to have better performance. The ROC curves are mainly used to judge binary classifiers.

$$\text{Sensitivity or Recall} = (\text{True Positives})/(\text{True Positives} + \text{False Negatives}) \tag{3}$$

Table 1 The accuracies of various CNN algorithms

S. No.	Algorithm	Accuracy %
1	VGG19	96.97
2	VGG16	93.94
3	ResNet50	93.94
4	Custom	93.94
5	AlexNet	90.91

Table 2 The AUC scores of the CNN algorithms

S. No.	Algorithm	AUC score
1	AlexNet	1.000
2	VGG16	0.926
3	Custom	0.867
4	ResNet50	0.824
5	VGG19	0.754

$$\text{Specificity} = (\text{True Negatives})/((\text{True Negatives} + \text{False Positives})) \quad (4)$$

We can obtain the false-positivity rate (FPR) by subtracting specificity from 1. False-positivity rate = 1-Specificity.

$$\text{False Positivity Rate} = \text{False Positives}/(\text{False Positives} + \text{True Negatives}) \quad (5)$$

Usually, when we plot the ROC curve, it is advisable to calculate the "Area under the curve" or AUC which gives an overall idea about the performance of the model in a single measure. For identifying a good model/AUC, we should note the area under the curve and the AUC score. The AUC score is calculated using the FPR and TPR. Suppose, our AUC score is 0.6; then, there is a 60% chance that the model will be able to distinguish between the cancerous and normal cells. The higher the AUC, the better will be the classification capability. Ideally, the AUC should be close to 1, and in the worst case, it is close to 0. Table 2 shows *AUC scores of the models.*

7 Conclusion

In this paper, we present a fully automatic acute lymphoblastic leukemia method based on deep neural networks. Automated systems were introduced in the past to classify cancer cells; they used machine learning and image processing techniques to classify the cancer cells. But, their accuracy and efficiency parameters weren't sufficient enough for medical data classification. To improve the performance and accuracy parameters, CNN models were used, to classify, segment, and process the blast cell images. They are much more powerful and efficient as compared to the machine learning models. Out of the five models which were implemented as part of the CNN module of the proposed system, the VGG19 showed maximum accuracy of 96.97%. The TensorFlow library along with the Keras API was used extensively for the development and testing. The ROC scores of AlexNet were the highest which was 1.000, followed by VGG16 had the highest ROC score of 0.926, custom with 0.867, ResNet with 0.824, and VGG19 with 0.754; we used pre-trained models for fine-tuning and transfer learning, which in turn significantly increases performance and development speed. By making changes in the output layers, we can classify their subtypes of ALL too, since the subtype identification may provide an idea of the severity of the disease. Using such systems reduces the effort and time consumption of doctors and researchers and in turn will be able to provide early treatment for cancer patients.

References

1. Sunil S, Sonu P, Sarath S, Rahul Nath R, Viswan V (2021) An effective approach for classifying acute lymphoblastic leukemia using hybrid hierarchical classifiers. In: Singh M, Tyagi V, Gupta PK, Flusser J, Ören T, Sonawane VR (eds) Advances in computing and data sciences. ICACDS 2021. Communications in computer and information science, vol 1440. Springer, Cham. https://doi.org/10.1007/978-3-030-81462-5_14Srivastava
2. Curesearch for Children's Cancer Research. https://curesearch.org. Last accessed Mar 25, 2021
3. https://homes.di.unimi.it/scotti/all/. Last accessed Jan 31, 2021
4. Al-Jaboriy SS, Sjarif NNA, Chuprat S (2019) Segmentation and detection of acute leukemia using image processing and machine learning techniques: a review. In: Elissa K (ed) Title of paper if known, unpublished
5. Donida Labati R, Piuri V, Scotti F (2011) ALL-IDB: the acute lymphoblastic leukemia image database for image processing. In: Proceedings of the 2011 IEEE international conference on image processing (ICIP 2011), Brussels, Belgium, pp 2045–2048, Sept 11–14, 2011. ISBN: 978-1-4577-1302-6
6. Poorna SS, Reddy M, Kiran R, Akhil N, Kamath S, Mohan L, Anuraj K, Pradeep HS (2020) Computer vision aided study for melanoma detection: a deep learning versus conventional supervised learning approach in advanced computing and intelligent engineering. Advances in intelligent systems and computing
7. The American Society of Hematology. http://www.hematology.org/. Last accessed Mar 25, 2021
8. Erickson BJ, Korfiatis P, Akkus Z, Kline TL. Machine learning for medical imaging
9. Litjens G, Kooi T, Bejnordi BE, Setio AAA, Ciompi F, Ghafoorian M, van der Laak JAWM, van Ginneken B, Sánchez CI (2017) A survey on deep learning in medical image analysis. Med Image Anal 42:60–88. ISSN 1361-8415
10. Bahadure NB, Ray AK, Thethi HP (2017) Image analysis for MRI based brain tumor detection and feature extraction using biologically inspired BWT and SVM. Int J Biomed Imaging 2017:12
11. Gunavathi C, Sivasubramanian K, Keerthika P, Paramasivam C (2020) A review on convolutional neural network-based deep learning methods in gene expression data for disease diagnosis. Mater Today Proc
12. Tamuly S, Jyotsna C, Amudha J (2020) Deep learning model for image classification. In: Advances in intelligent systems and computing, vol 1108. Springer International Publishing, Cham
13. Sarmad Shafique MS, Samabia Tehsin PhD (2018) Acute lymphoblastic leukemia detection and classification of its subtypes using pretrained deep convolutional neural networks. Technol Cancer Res Treat 17:1–7
14. Sajja TK, Devarapalli RM, Kalluri HK (2019) Lung cancer detection based on CT scan images by using deep transfer learning. Traitement du Signal 36(4):339–344. https://doi.org/10.18280/ts.360406
15. Rawat J, Singh A, Bhadauria HS, Virmani J, Devgun JS (2016) Classification of acute lymphoblastic leukemia using hybrid hierarchical classifiers. Multimedia Tools Appl Int J (2016). ISSN 1380–7501
16. http://www.nlm.nih.gov/medlineplus/leukemia.html. Last accessed Jan 31, 2021]

Use of Support Vector Machine to Check Whether Process Metrics are as Good as Static Code Metrics

Ruchika Malhotra and Anjali Bansal

Abstract There can be some faults, while the development of the software. If these faults are not recognized and corrected at any early stage of software development, then it leads to software failure and increase in development cost. So, the most important task is to deal with these software faults. Various models have been proposed to detect the fault proneness of the software using static code metrics. In this paper, we have used support vector machine with poly-kernel and normalized poly-kernel to predict the fault proneness, and both static code metrics and process metrics have been used as independent variables. AUC measure is used as performance evaluation measure to predict the fault proneness. We have also analyzed the results using 3D bar graph and Friedman test. The result of this study shows that in case of both kernels, process metric has positive impact on predicting fault proneness, or we can say that process metric is as good as static code metrics.

Keywords Fault prediction · Process metric · Static code metric

1 Introduction

Software fault prediction is the most widely researched area. Faults can be unintended bugs or errors which can occur during any phase of software development. Software fault prediction (SFP) models use different software metrics to predict the faults in the modules of the software. In software fault prediction models, software metrics work as software quality data. There are many literature reviews published in the field of software fault prediction.

According to Dejaeger et al. [1], as of late, a few researchers directed their study toward another subject of excitement, i.e., the addition of data different from static code variables into defect prediction models such as data on inter-module connections and requirement metrics. The connection to the more ordinarily utilized static code

R. Malhotra · A. Bansal (✉)
Delhi Technological University, New Delhi, India

R. Malhotra
e-mail: ruchikamalhotra@dtu.ac.in

J. K. Mandal et al. (eds.), *Topical Drifts in Intelligent Computing*, Lecture Notes in Networks and Systems 426, https://doi.org/10.1007/978-981-19-0745-6_4

variables remains anyway dubious. Two process metrics were used in this study, but there can be other process metrics as well.

This study shows the requirement to explore the process metrics in the field of fault prediction. This paper focuses on analyzing the predictive performance of process metrics as compared to static code metrics, and we are also analyzing which process metric is effective among all chosen process metrics.

The main contribution of this paper is to compare the results of various kernels such as normalized poly-kernel and poly-kernel. Till now, no one have compared the results of various kernels to check the effectiveness of process metrics on prediction performance.

This paper is organized as follows: Sect. 2 describes the related work. Section 3 describes the proposed work. Section 4 describes the experimental results. Section 5 describes the discussion on results, and Sect. 6 describes conclusion part.

2 Related Work

Many fault prediction models use static code metrics as quality attributes in classifying the modules. Many methods such as logistic regression [2], Naive Bayes [3, 4], random forest, support vector machines, artificial neural network [1], decision tree, k-nearest neighbor have been proposed in the past which shows the interrelationship between fault proneness and static code metrics. A literature review done on 106 paper published between 1991 and 2011 shows that only there are 24% paper in which process metrics were used.

Malhotra et al. [5] analyzed ML techniques to predict fault proneness using Web applications. In this study, they used static code metrics as independent variables. They compare 14 techniques to analyze the connection between object-oriented metrics and fault proneness. The result of this study shows that the ensemble method is the best approach to predict fault proneness in Web applications.

Rahman et al. [6] analyzed 85 versions of 12 projects to check the performance, stasis, and stability of different combinations of metrics. They have used 14 process metrics and 4 classification techniques (J48, Naive Bayes, logistic regression, and support vector machine) to build the prediction model. They concluded that process metrics are more useful than code metrics, and process metrics have low stasis means their value can change from release to release.

Alshehri et al. [7] analyzed three versions of the Eclipse project to compare three different classification techniques. This study's findings focus on different combinations of static code metrics and change metrics, as well as an analysis of the results based on a smaller set of change metrics. They concluded that if we are choosing the G score as a performance evaluation measure, then J48 outperforms logistic regression and Naive Bayes in all cases.

Madeyski and Jureczko [8] conducted an experiment in which they used the same process metrics which we are using in this study. They considered open source and industrial projects in their research. They examined models that contain 1 process

Table 1 Dataset description

Project name	Project version
Ant	1.4, 1.5, 1.6, 1.7
Jedit	4.0, 4.1
Synapse	1.1, 1.2
Xalan	2.5.0, 2.6.0, 2.7.0
Xerces	1.2.0, 1.3.0, 1.4.4

metric and all static code metrics. This study shows that the process metric improves defect prediction and gives a notable contribution, and NDC and NML process metrics improve the defect prediction.

3 Proposed Work

We have used following steps to check whether process metrics are as good as static code metrics and to check which process metric is effective among selected process metrics.

3.1 *Empirical Dataset Collection*

The dataset is collected from five projects. All are java-based projects. Madeyski and Jureczko [6] maintained a metric repository in which all the datasets are available having both static code metrics and process metrics. Table 1 shows the dataset used in this paper. Process metric used in this paper is as follows:

NR Number of revision.
NDC Number of distinct committers.
NML Number of modified lines.
NDPV Number of defects in previous version.

3.2 *Preprocessing Operation*

Datasets from the publicly available repositories may contain some noise, or there can be some missing values which can affect the performance of the generated model, so to avoid this type of problem, some preprocessing is to be done on datasets such as removing unique id field, version field, and class field. In this, we have used one filter "replacing missing values with user constant" available in Weka machine learning tool to fill the missing values if any. There are some fields whose datatype is string;

we change the datatype of these string datatype variable to numeric. We have also used one more filter of Weka tool to convert the bug field data type from "numeric to nominal" as software can be faulty or non-faulty.

3.3 Classification Technique Used

In this paper, we use support vector machine to build the fault prediction model. It is another simple algorithm used for both regression and classification. It is a highly preferred technique as it gives the best accuracy with less computation. It divides the whole dataset into two parts by constructing an N-dimensional hyperplane. The hyperplane is created in such a way that it divides the values of one class of the dependent variable on one side and another class value of the dependent variable on another side. Vectors that are nearer to the hyperplane are called support vectors. Support vector machines have also been used in face recognition, medical diagnosis, text classification [8].

3.4 Performance Evaluation Measure Used

If there is a class imbalance problem in the dataset and we are trying to do classification, then it is a difficult task to choose the correct performance evaluation measure. In such a scenario, AUC is used to estimate the prediction performance. AUC value is the average of all threshold values. AUC closer to 1 shows better prediction performance and closer to 0 shows poor prediction performance. Detailed description of how to calculate AUC values is given in Dejaeger et al. [1].

3.5 Implementation

For implementation, we use Weka machine learning tool. There is a built-in support vector machine classifier available in Weka. There can be some string data type variables due to which there can be some error, so we use filtered classifier which provides the facility of support vector machine with additional functionality. We choose meta-filtered classifier firstly. After this, we choose function. SMO classifier as classifier and filter as unsupervised attribute. String to word vector. All the parameter values are initialized with the default values in Weka, and in SMO, we have used two types of kernels poly-kernel and normalized poly-kernel.

4 Experimental Results

We build and analyze some models based on different combinations of process metrics and static code metric to check the effectiveness of process metrics and to check which process metrics outperform other selected process metrics.

Model-1: Model that solely contains static code metrics.

Model-2: Model that solely contains process metrics.

Model-3: Combined model containing both metrics.

Model-4: Model containing combination of 1 process metric and static code metric.

Model-5: Model containing combination of 2 process metric and static code metrics.

Model-6: Model containing combination of 3 process metric and static code metrics.

- Table 2 shows that in case of poly-kernel, 66.67% cases give better AUC values in combination of both metrics, and in case of normalized poly-kernel, 64.28% cases give better AUC values in combination of both metrics. Among poly-kernel and normalized poly-kernel, poly-kernel gives better prediction results.

Table 2 AUC values of combined model, model solely containing static code metrics and model solely containing process metrics using SVM

	Poly-kernel			Normalized poly-kernel		
Dataset name	Combined	SC	PM	Combined	SC	PM
Ant1.4	0.579	0.500	**0.589**	0.5	0.5	0.5
Ant1.5	**0.616**	0.500	0.553	0.497	0.5	**0.529**
Ant1.6	**0.676**	0.606	0.647	**0.652**	0.608	0.625
Ant1.7	**0.694**	0.620	0.669	**0.686**	0.614	0.664
Jedit4.0	0.735	0.538	**0.748**	0.753	0.536	**0.757**
Jedit4.1	**0.740**	0.609	0.723	**0.701**	0.614	**0.701**
Synapse1.1	**0.671**	0.635	0.499	**0.596**	0.585	0.485
Synapse1.2	0.601	**0.628**	0.546	0.634	**0.651**	0.564
Xalan2.5.0	**0.651**	0.575	0.629	**0.683**	0.620	0.654
Xalan2.6.0	**0.784**	0.695	0.684	**0.791**	0.701	0.711
Xalan2.7.0	**0.981**	0.843	**0.981**	**0.981**	0.879	**0.981**
Xerces 1.2.0	**0.540**	0.500	**0.553**	0.498	**0.5**	0.497
Xerces 1.3.0	**0.594**	0.519	**0.548**	0.531	**0.543**	0.538
Xerces 1.4.4	**0.849**	0.766	**0.679**	**0.851**	0.771	0.677

Bold values shows highest AUC values among all three models for each dataset and for both kernels separately

- If we analyze the results of model 4 on the basis of poly-kernel, it shows that NR metric is more useful in predicting fault proneness, and the result of normalized poly-kernel shows that NML metric is more useful.
- If we analyze the results of model 5 on the basis of poly-kernel, it shows that NR metric combining with NML metric gives better prediction performance, and the result of normalized poly-kernel shows that NML metric combining with NDPV metric is more useful.
- If we analyze the results of model 6 on the basis of poly-kernel, it shows that NR metric combining with NML metric and NDPV metric gives better prediction performance, and the result of normalized poly-kernel shows that NML metric combining with NDPV metric and NDC metric is more useful.

5 Discussion on Results

We analyzed model 1, model 2, and model 3 to check the effectiveness of process metrics on fault proneness using support vector machine. The result of the experimental analysis shows that combined model (model 3) outperforms model solely containing process metrics (model 2) and model solely containing static code metrics (model 1). In this section, we analyze the results using box plots and Friedman test with Nemenyi post-hoc test.

- If we statistically analyze the results of Table 2 using Friedman test at the 0.05 significance level, then results show that there is a rejection of the null hypothesis. Hence, the performance of at least one of the models differs significantly. As the results differ significantly, we will apply a Nemenyi post-hoc test to check the pairwise comparison of the models. The results show that a significant difference exists between the performances of the combined model and model which contain static code metric just. Figure 1 represents the bar graph for Table 2. We can see that in case of both kernels the combined model of process metric and static code metrics outperforms models solely containing process metrics and models solely containing static code metrics, and there is one outlier in combined model and model containing just process metric.
- If we statistically analyze the results of model 4 using Friedman test at the 0.05 significance level, then results show that there is a rejection of the null hypothesis. Hence, the performance of at least one of the pair differs significantly. As the results differ significantly, we will apply a Nemenyi post-hoc test to check the pairwise comparison of the models which contain all static code metric and 1 process metric. The results show that a significant difference exists between the performances of the SC + NR model and SC + NDC model, SC + NR model and SC + NDPV models, and SC + NML model and SC + NDPV model.
- If we statistically analyze the results of model 5 using Friedman test at the 0.05 significance level, then results show that there is a rejection of the null hypothesis. Hence, the performance of at least one of the pairs differs. As the results differ

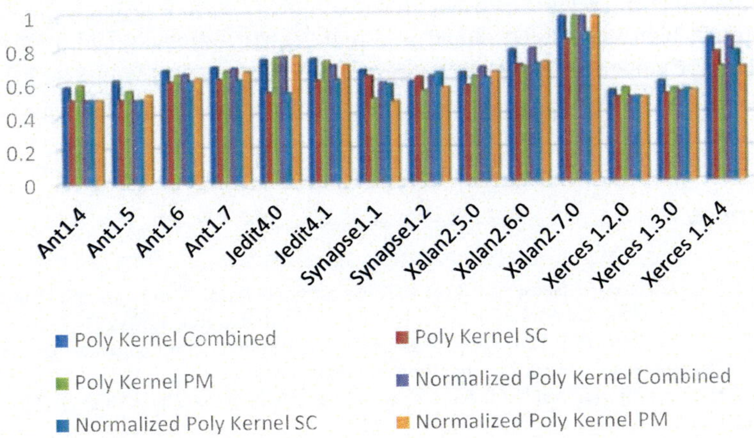

Fig. 1 Bar graph of combined model, model solely containing static code metrics, and model solely containing process metrics using SVM

significantly, we will apply a Nemenyi post-hoc test to check the pairwise comparison of the models which contains 2 process metrics and all static code metrics. The results show that a significant difference exists between the performances of the NR + NML + SC model and NDC + NDPV + SC model and NR + NDPV + SC model and NDC + NDPV + SC model.

- If we statistically analyze the results of model 6 using Friedman test at the 0.05 significance level, then results show that there is a rejection of the null hypothesis. Hence, the performance of at least one of the pairs differs significantly. As the results differ significantly, we will apply a Nemenyi post-hoc test to check the pairwise comparison of the models which contains 3 process metrics and all static code metrics. The results show that a significant difference exists between the performances of the NR + NML + NDPV + SC model and NDC + NML + NDPV + SC model.

6 Conclusion

Various types of models exist and are researched in the field of software fault prediction. Many of them utilize static code metrics to predict faulty modules. Here, we are analyzing the effectiveness of process metrics on fault prediction results using support vector machine. We have analyzed the results based on AUC values, 3D bar graph, and statistical test (Friedman test with Nemenyi post-hoc test). We can conclude that model 3 outperforms model 1 and model 2 in case of both poly-kernel and normalized poly-kernel; in case of model 4, NR metric gives effective outcome in poly-kernel and NML in normalized poly-kernel; in case of model 5, combination of NR, NML metric is effective in poly-kernel, and in case of normalized poly-kernel, combination of NML, NDPV metric is effective, and in case of model 6, combination of NR, NML, NDPV metric is effective in poly-kernel, and combination of NML, NDC, and NDPV metric is effective in normalized poly-kernel. Also, this study shows that poly-kernel gives better prediction results as compare to normalized poly-kernel.

References

1. Dejaeger K, Verbraken T, Baesens B (2013) Toward comprehensible software fault prediction models using Bayesian network classifiers. IEEE Trans Software Eng 39(2):237–257
2. Dreiseitl S, Ohno-Machado L. Logistic regression and artificial neural network classification models: a methodology review. J Biomed Inform 35:352–359
3. Rish I (2001) An empirical study of the naive Bayes classifier. IJCAI Works Empirical Methods Artif Intell 3:41–46
4. Catal C, Sevim U, Diri B. Practical development of an eclipse-based software fault prediction tool using Naive Bayes algorithm. Expert Syst Appl 38(3):2347–2353
5. Malhotra R, Sharma A (2018) Analyzing machine learning techniques for fault prediction using web applications. J Inf Process Syst 14:751–770. https://doi.org/10.3745/JIPS.04.0077
6. Rahman F, Devanbu P (2013) How, and why, process metrics are better. In: 2013 35th international conference on software engineering (ICSE), San Francisco, CA, USA, pp 432–441. https://doi.org/10.1109/ICSE.2013.6606589
7. Alshehri YA, Goseva-Popstojanova K, Dzielski DG, Devine T (2018) Applying machine learning to predict software fault proneness using change metrics, static code metrics, and a combination of them. SoutheastCon 2018, St. Petersburg, FL, USA, pp 1–7. https://doi.org/10.1109/SECON.2018.8478911
8. Madeyski L, Jureczko M (2015) Which process metrics can significantly improve defect prediction models? An empirical study. Software Qual J 23(3):393–422

Social Distancing Using Video Tracking System—an Effort Toward COVID-19

Jagdish Chandra Patni, Saurabh Agarwal, Rashi Gupta, and Hitesh Kumar Sharma

Abstract In the wake of the pandemic that we are all facing these days, we all are advised to maintain some specified social distance from other people in order to keep ourselves safe. The CoVID-19 pandemic started showing its symptoms at the end of 2019 and is still killing thousands of people every day. Although the scientists have been successful in preparing the medicines for it, it is better to take some precautionary measures ourselves only. We are battling from it today as well. So, our tool is just a medium to make this battle a little easier. This will help us to monitor the distance between two objects, here, people, and whether they are at a safe distance from each other or not. This can also be helpful for the officials if they have to keep an eye on everyone and if they are following the proper guidelines to prevent COVID-19 from spreading. This will help us to detect the objects, in this case, people, and track their movements. Anyone can track whether people are maintaining a proper distance from each other or not. We are using three algorithms, object detection, object tracking, and distance measure algorithm, mainly to detect the objects, then track them, and then to analyze the distance between them.

Keywords Object detection · Object tracking · Distance measure · COVID-19 · CNN

1 Introduction

COVID-19 has established its terror all around the globe. Every other country is trying to fight its consequences in its own ways [1]. Every country has its own set of rules and regulations or guidelines that every respective resident has to follow. This pandemic has spread to more than 188 nations all throughout the planet. According to WHO, around 38,394,169 COVID-19 positive cases as on October 15, 2020 and 1,089,047 passing's all throughout the planet [2]. Numerous nations were in the method of lockdown for seemingly forever to keep this infection from spreading; individuals had to wear covers and to keep a particular social separation from others.

J. C. Patni (✉) · S. Agarwal · R. Gupta · H. K. Sharma
School of Computer Science, University of Petroleum and Energy Studies, Dehradun, India

J. K. Mandal et al. (eds.), *Topical Drifts in Intelligent Computing*, Lecture Notes in Networks and Systems 426, https://doi.org/10.1007/978-981-19-0745-6_5

For certain months, even the fundamental food items were not even accessible in the business sectors. This virus created havoc worldwide even is a dangerous threat till now. Some countries have developed their own vaccines to stop coronavirus from spreading, but new mutants of this virus have also started spreading now. Till now, it is extremely hard to anticipate the length and spread of this pandemic.

The circumstance is getting more terrible as time passes. Fever or loss of taste or feeling of smell are some normal side effects of coronavirus [3]. Specialists in China tracked down that 99% of individuals tainted with this infection have shown indications of high temperature. It has been accounted for and broke down that every one of the tainted nations which applied the lockdown for individuals of their nations have shown a decrease in the number of COVID-19 positive cases and the quantity of mortalities.

Vaccines can help us to protect from coronavirus, but till the time, not everyone is vaccinated, and we don't have the proper resources to fight every new mutant of this virus; it is better to take every other precautionary measure that we can from our sides.

Artificial intelligence and ICT tools can be the best way to develop the ways to maintain and ensure the social distancing [4, 5].

Authors target relieving the spread of coronavirus in networks and saving the existences of individuals from one side of the planet to the other [6]. They carried out three calculations through which we can accomplish our last objective of social separating inside individuals.

This application works on three different algorithms [7].

1.1 Object Detection Methods

This becomes more important and place a jumping box around the articles and partner the right article's classification [8]. Profound learning is a compelling strategy to get the object location. It comprises four phases. It begins with bringing the pictures into the info layer; at that point, it removes the provincial proposition; after that it processes the highlights by CNN; lastly, it characterizes these highlights, see Fig. 1 utilizes particular pursuit calculations to produce locale recommendations. It requires some investment as it would need to characterize the districts per picture. R-CNN can't be executed continuously object discovery as each picture need 47 s to process. R-CNN can't be prepared at one time. Maybe, needs to prepare each part freely [9] (Fig. 2).

Another method is fast R-CNN for the local proposition calculation, introduced by a similar creator of the R-CNN model [10]. This model improved the downsides of R-CNN to fabricate a quicker article identification calculation.

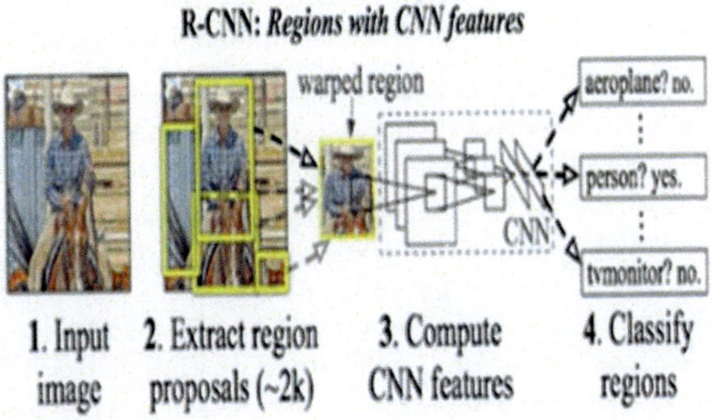

Fig. 1 R-CNN detector [9]

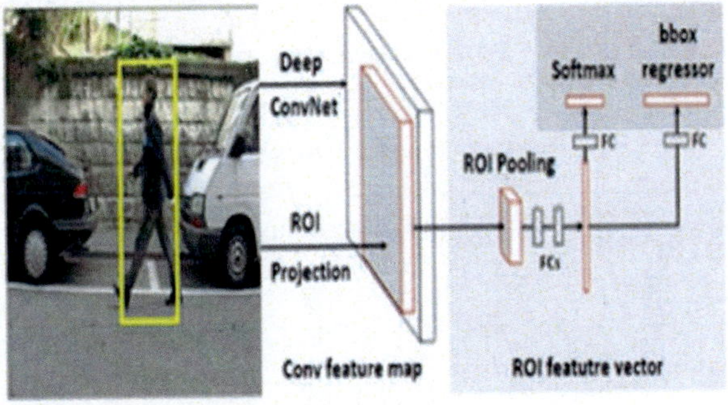

Fig. 2 Fast R-CNN detector [9]

1.2 Overview of Object Tracking Algorithm

With this, we are estimating the target object based on the previous information with us. Most of the researchers suggested two types of object tracking algorithms are single and multiple object tracking [11].

This is not only restricted to 2D grouping information and can be applied to 3D areas as we.

Numerous tracking algorithms (MOT), likewise called multi-target tracking, are a PC vision task that expects to investigate recordings to distinguish and follow objects having a place with at least one classification, like walkers, vehicles, creatures and lifeless things, with no earlier information about the targets. While in single object tracking (SOT), the presence of the objective is known deduced; in MOT, an identification step is important to recognize the objectives that can leave or enter the scene.

1.3 Overview of Distance Measure

Discovering closeness among information focuses is a significant errand when we attempt to discover groups from a given information. Solo learning calculations like K-implies have faith in the hypothesis—"closer the focuses more comparative they are" as there is no unequivocal estimation for similitude [12].

In this way, the bunching, the order of perceptions into gatherings, needs a few techniques for figuring the distance or the (dis)similarity between each pair of perceptions. The consequence of this calculation gives a thought regarding the divergence among the perceptions (why uniqueness rather than comparability? have an independent perspective…) which is called distance framework [13].

1.4 YOLO (You Only Look Once) Model

The YOLO structure [14] (You Only Look Once) manages object recognition in an alternate way. It takes the whole picture in a solitary occurrence and predicts the bouncing box arranges and class probabilities for these containers. The greatest benefit of utilizing YOLO is its speed that can quickly large number of cashing per second.

This shows the best results during object identification in R-CNN calculations. In the impending areas, we will find out about various methods utilized in the YOLO calculation.

2 Literature Review

After the ascent of the COVID-19 pandemic in late December 2019, social removing is considered to be a most extreme dependable practice to forestall the infectious infection transmission and picked as standard practice on January 23, 2020. For one month, the quantity of cases rises incredibly, with 2000–4000 new affirmed cases announced each day in the main seven day stretch of February 2020. Afterward, there has been an indication of help interestingly for five progressive days up to March 23, 2020, with no new affirmed cases. This is a direct result of the social distance practice started in China and, recently, embraced worldwide to control COVID-19 [15]. The examination uncovered that moderate phases of activity could be took into account avoiding an enormous episode. Up until this point, numerous nations have utilized innovation-based arrangements [16–19] to conquer the pandemic misfortune. A few created nations are utilizing GPS innovation to screen the developments of the tainted and suspected people. Give a study of various arising innovations, including Wi-Fi, Bluetooth, cell phones, and GPS, situating (limitation), PC vision, and profound discovering that can assume a pivotal part in a few common-sense social removing situations. A few specialists use drones and other reconnaissance cameras to identify swarm get-togethers [20].

As of not long ago, specialists have accomplished extensive work for location; some give a shrewd medical services framework to pandemic utilizing the Internet of Medical Things. Examined the social separating impacts on the spread of the COVID-19 episode [21–23]. The investigations reasoned that the early and quick act of social separating could step by step diminish the pinnacle of the infection assault. As we as a whole know, albeit social removing is significant for leveling the disease bend; it is a monetarily horrendous advance [24].

There are different profound learning models, for example, R-CNN, single-shot finder (SSD), and YOLO which are applied in various applications for object discovery [25].

3 Problem Statement

This application will help in monitoring the movement of every citizen. It will help in keeping an eye if people are following the proper guidelines laid by the government in the wake of this pandemic. Some people tend not to adhere to the rules and regulations; this tool can help in monitoring them, and we can find them or at least warn them to maintain proper distance in order to help prevent the spreading of this virus. Even the citizens themselves can detect and track other people and stay away from them to take some precautionary actions themselves only.

3.1 Factors to Maintain Social Distancing

Apply the profound learning object finder to identify individuals in warm pictures or video transfers.

Check the quantity of people that are in the images or video transferred.

Calculate the distance from centroids of bounding boxes of individuals.

Finally, the calculation will choose for protected or perilous social separating dependent on the quantity of people and the deliberate distance between the centroid of bouncing boxes (Fig. 3).

Our work begins with the info picture layer, which contains the information picture of size (220 × 220 × 3) for our identifier. Convolutional layers were utilized to plan the highlights for the pictures. It characterizes the stature and width of the information picture. Cluster standardization is used to regularize the model and dispense with the overfitting issue. ReLU initiation capacities were used to acquaint the non-linearity with the neural organization. Max Pooling layers were utilized to down sample the pictures into pooling locales. Took 2 × 2 size of pooling for a neural organization. ReLU_5 was utilized at element extraction layer that is used to separate the highlights from neural organization layers and afterward give them as contribution to YOLOv2.

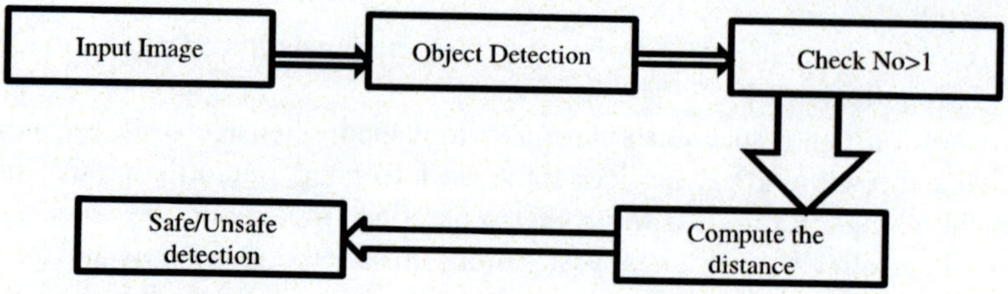

Fig. 3 Steps for people detection and social distancing classification

Its layers were utilized in this finder which builds the YOLOv2 location organization. Its subnetwork comprises of a bunch of layers that incorporate convolutional, clump standardization, ReLU, change, and yield layers. The change layer changes the crude CNN yield into a structure needed to create object identifications. This version utilized and refined the areas of jumping boxes. This version was inspected with a neural organization analyzer and revealed zero mistakes.

3.2 *Object Detection Algorithm*

There are many transfer learning processes for detecting objects like SSD, Mobile net, recurrent neural network (RNN), but the most widely used is YOLO. **YOLO** can detect over 9000 classes, but we are using only the "Person Class" here. After detecting any person, we will put them in boxes. After detection of the Person Class, we will need to track them. For that we will be implementing the second algorithm.

3.3 *Object Tracking Algorithm*

After detecting, we need to track them, so we will assign a new ID to every person, draw a box over them and measure the centroid. To assure that the moving person is the same, we will measure the Euclidean distance from the old centroids to the new centroids, and the closed centroids will denote that both centroids were of the same person. If the same ID does not appear in 50 video frames continuously, then we will re-register that ID in our array (Figs. 4 and 5).

4 Implementation

Use the following code to detect the social distancing (Fig. 6).

The most outstanding imports on Lines 1–3 incorporate our min_conf and nms_thresh capacity to introduce least likelihood to channel powerless discoveries

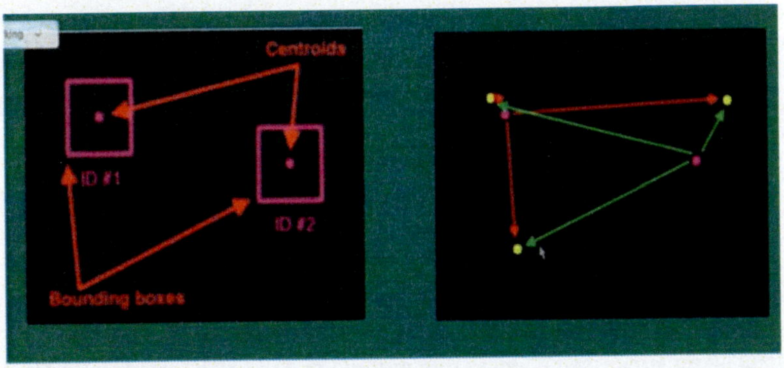

Fig. 4 Object tracking algorithm [7]

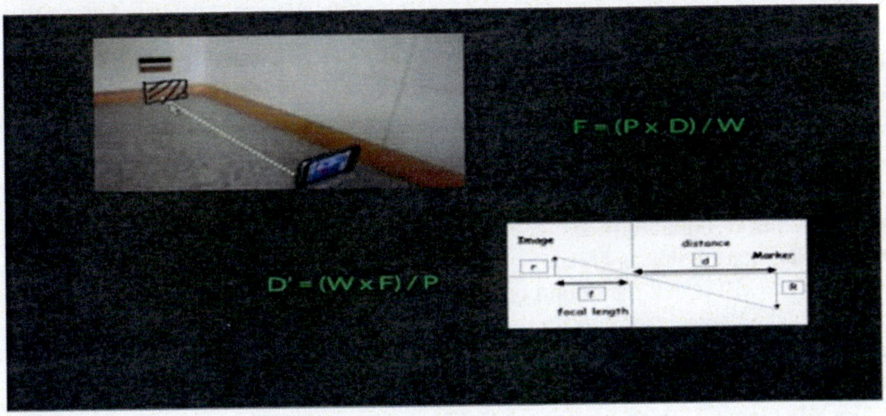

Fig. 5 Distance measure algorithm [7]

```
# initialize minimum probability to filter weak detections along with
# the threshold when applying non-maxima suppression
MIN_CONF = 0.3
NMS_THRESH = 0.3

# define the minimum safe distance (in pixels) that two people can be
# from each other
MIN_DISTANCE = 50
```

Fig. 6 Code to detect the social distancing

alongside the edge while applying non-maxima concealment and the min_distance (characterize the base safe distance (in pixels) that two individuals can be from one another).

We will use the following arguments through command line.

Input: Use to select the video file and in case of no selection, Webcam will be automatically open.

Output: It will display the processed video entered during input process.

Display: It will display the social distancing application and start processing the frames.

4.1 Algorithm

- We use the OpenCV to load YOLO into memory.
- Collect all the processed frames from output layer to get the results and again start the input process to insert the regular input videos.
- Initially, we set the output frame to 0 then start toput the loop for further frames.
- Now, finally, we start the processing of all the received frames to ensure that either we are maintaining the social distance or not.
- Let's assume that two people are there in a frame.
- Compute the Euclidean distance.
- Create a loop on upper part of distance matrix.
- Check for violation of social distancing as per the government regulations. If violation of this, we will add in the violation list.
- Extracting of bound box and center coordinates.
- Initially, mark color as green.
- Now, we will check the bound box in the violation list; if it exist in the list, then mark as red color.
- Make a box that contain the person and its coordinates.
- Found that people are close or not and found total number of violations.
- Finally OpenCV to check the social distancing.

5 Results

Itis time to test the application by our OpenCV social distancing method. We open the terminal and execute the dataset found from Kaggle [26] that we have stored in the local disk.

- Loading YOLO from disk
- Accessing video stream.

Figure 7 showing the number of social distancing violations marked in red color, and where no violations, it is shown by green in color. By this, we can easily manage the violations.

Fig. 7 Output screen of application

6 Conclusion and Future Work

A particular calculation was executed on jumping boxes to recognize protected and dangerous conditions, individually, stamping as represented in colors green for ok condition and red for violation of social distancing. Our approach showed better outcomes for ongoing execution versus other item indicators. Our approach can be executed in a dispersed video reconnaissance framework; in reality, it is an appropriate answer for the specialists to picture the consistency of individuals with social separating and simultaneously screening their internal heat level.

In our study, we have used the fixed image that can be better if we use the adjustment of pictures using cameras that will give more clear idea about an image.

We can input more classes to detect the animals and the non-living things around the environment.

For future scope, we can make our application compatible with mobile devices, tablets, and other devices available.

References

1. Advice for the public on COVID-19—World Health Organization (05 May 2021). www.who.int/emergencies/diseases/novel-coronavirus-2019/advice-for-public
2. Coronavirus disease (COVID-19) dashboard (10 May 2021). https://covid19.who.int
3. Covid map: coronavirus cases, deaths, vaccinations by country (22 May 2021). https://www.bbc.com/news/world-51235105
4. Keniya R, Mehendale N (2020) Real-time social distancing detector using socialdistancingnet-19 deep learning network (Aug 7, 2020). Available at SSRN: https://ssrn.com/abstract=3669311, https://doi.org/10.2139/ssrn.3669311
5. Saponara S, Elhanashi A, Gagliardi A (2021) Implementing a real-time, AI-based, people detection and social distancing measuring system for Covid-19. J Real-Time Image Proc.https://doi.org/10.1007/s11554-021-01070-6
6. Rezaei M, Azarmi M (2020) DeepSOCIAL: social distancing monitoring and infection risk assessment in COVID-19 pandemic. App Sci 10(21):7514. https://doi.org/10.3390/app10217514
7. Georgievski B (2021) Object detection and tracking in 2020 (12 Apr 2021). https://blog.netcetera.com/object-detection-and-tracking-in-2020-f10fb6ff9af3

8. Ahamad H, Zaini N, Latip MFA (2020) Person detection for social distancing and safety violation alert based on segmented ROI. In: 2020 10th IEEE international conference on control system, computing and engineering (ICCSCE), pp 113–118. https://doi.org/10.1109/ICCSCE50387.2020.9204934
9. Girshick R (2015) Fast R-CNN. IEEE Int Conf Comput Vis (ICCV) 2015:1440–1448. https://doi.org/10.1109/ICCV.2015.169
10. Shubham P et al (2019) Real-time object detection using deep learning: a survey. Int Res J Eng Technol 6(10):2395–0056
11. Qian Y, Shi H, Tian M, Yang R, Duan Y (2020) Multiple object tracking for similar, monotonic targets. In: 2020 10th Institute of Electrical and Electronics Engineers international conference on cyber technology in automation, control, and intelligent systems (CYBER), pp 360–363. https://doi.org/10.1109/CYBER50695.2020.9279162
12. Kahale N (2020) On the economic impact of social distancing measures. SSRN Electron J.https://doi.org/10.2139/ssrn.3578415
13. Wu J, Tang B, Bragazzi NL, Nah K, McCarthy Z (2020) Quantifying the role of social distancing, personal protection and case detection in mitigating COVID-19 outbreak in Ontario, Canada. J Math Ind. https://doi.org/10.1186/s13362-020-00083-3
14. Redmon J, Divvala S, Girshick R, Farhadi A. You only look once: unified, real-time object detection. arXiv:1506.02640
15. National Health Commission of the People's Republic of China, 2020 (12 May 2021). http://en.nhc.gov.cn/
16. Yang D, Yurtsever K, Renganathan V, Redmill KA, Özgüner U (2020) A vision-based social distancing and critical density detection system for COVID-19. arXiv:2007.03578
17. Sonbhadra SK et al (2020) Target specific mining of covid-19 scholarly articles using the one-class approach. https://arxiv.org/pdf/2004.11706.pdf
18. Robakowska M et al (2017) The use of drones during mass events. Disaster Emerg Med J 2(3):129–134
19. Punn NS, Sonbhadra SK, Agarwal S (2020) COVID-19 epidemic analysis using machine learning and deep learning algorithms. medRxiv arXiv:2020.04.08.20057679,https://doi.org/10.1101/2020.04.08.20057679
20. Nguyen LH, Drew DA, Graham MS, Joshi AD et al (2020) Risk of COVID-19 among front-line health-care workers and the general community: a prospective cohort study. Lancet Publ Health 5(9):E475–E483
21. Patrick SW, Henkhaus LE, Zickafoose JS, Lovell K, Halvorson A, Loch S, Letterie M, Davis MM (2020) Well-being of parents and children during the COVID-19 pandemic: a national survey. Pediatrics 146(4):e2020016824. https://doi.org/10.1542/peds.2020-016824
22. Chakraborty C, Banerjee A, Garg L, Rodrigues C (2020) Covers comprehensive studies from bioelectronics and biomedical engineering with a prime focus on COVID-19 pandemic. Internet Med Things Smart Healthcare
23. Prem K et al (2020) The effect of control strategies to reduce social mixing on outcomes of the covid19 epidemic in Wuhan, China: a modeling study. Lancet Publ Health
24. Manfredi M, Vezzani R, Calderara S, Cucchiara R (2014) Detection of static groups and crowds gathered in open spaces by texture classification. Pattern Recogn Lett 44:39–48
25. Redmon J, Farhadi A (2017) YOLO9000: better, faster, stronger. IEEE Conf Comput Vis Pattern Recogn (CVPR) 2017:6517–6525. https://doi.org/10.1109/CVPR.2017.690
26. Covid Dataset. https://www.kaggle.com/imdevskp/corona-virus-report (12 July 2021)

A Universal Dependency Treebank for Definitely Endangered Low-Resource Kangri Language

Shweta Chauhan, Shefali Saxena, and Philemon Daniel

Abstract In this paper, we have present the first universal dependency treebank for low resource definitely endangered Himachali Kangri (ISO 639-3xnr) language. The goal of the Kangri Universal Dependency Treebank (KDTB) is to establish a large, syntactically annotated treebank that will aid cross-lingual learning and typological research as well as serve as a significant resource in the development of language technology tools. KDTB data contain syntactic annotation according to dependency-constituency schema, as well as morphological tags. The Kangri UD treebank (KDTB) consists of 2249 tokens and 1108 vocabulary (288 sentences).

Keywords Kangri · Indian languages · Low-resource languages · Universal dependencies

1 Introduction

In the near future, around 40% of all the languages is in danger of extinction. Languages are not just a medium of communication but also a repository of culture and tradition, such as verbal art, songs, storytelling, and rituals. When a language spoken in a community dies out, following generations lose access to a crucial aspect of the culture that is required to fully comprehend it. As a result, language is a vulnerable part of cultural heritage, necessitating its preservation. When it comes to saving such endangered languages, there are two aspects: preservation and revitalization [1]. Significant advancements in major Indian languages have been

S. Chauhan (✉) · S. Saxena · P. Daniel
Department of Electronics and Communication, National Institute of Technology, Hamirpur, Himachal Pradesh 177005, India
e-mail: shweta@nith.ac.in

S. Saxena
e-mail: shefali@nith.ac.in

P. Daniel
e-mail: phildani7@nith.ac.in

J. Mandal et al. (eds.), *Topical Drifts in Intelligent Computing*, Lecture Notes in Networks and Systems 426, https://doi.org/10.1007/978-981-19-0745-6_6

made due to rapid improvements in language technology, while contributions to study in lesser-known/low-resourced dialects have remained negligible.

Low-resource languages are those which have very less or insignificant datasets for applying supervised learning algorithms. The United Nations Educational, Scientific, and Cultural Organization (UNESCO) atlas of global languages in peril lists 2464 languages from throughout the world are in danger [2]. The list is topped by India, which has 197 endangered languages, followed by the United States (191) and Brazil (191). Figure 1 shows that out of 197 languages, 81 is vulnerable, followed by 63 that are certainly endangered, 6 that are highly endangered, and 42 that are critically endangered. Children no longer learn the language as their mother tongue which puts these going extinct languages to endangered languages list. India is a very diverse country in the context of religion, culture, language, art, etc. Himachal Pradesh, an Indian state, has 7 definitely endangered languages, Kangri being the one. Kangri is an Indo-Aryan language variation spoken in Himachal Pradesh's Kangra, Hamirpur, and Una districts. According to the 2011 census, Kangri is spoken by around 1.7 million people in India, of small fraction of fluent speakers [3].

In this paper, we present the creation and development of universal dependencies of Himachal low resource definitely endangered language, Kangri (ISO 639-3xnr) listed in the United Nations Educational, Scientific, and Cultural Organization (UNESCO). UD has been recognized as a promising framework for grammatical annotation that is uniform across languages. There is an open community with over 300 contributors, who have produced 202 treebanks in 114 languages to date.

Section 2 discusses the data and methodology treebank and parser. Section 3 discusses Linguistic Analysis (Morphology and Syntax) of Kangri Dependency Treebank. Section 4 presents Conclusion and Future Work.

2 Data and Methodology

The Kangri dataset conation monolingual and parallel Hindi-Kangri corpora [4], analysis, and speech recognition of under-resourced Kangri dialect with ensemble models presented in [5]. We took 288 Kangri sentences and personally annotated them before releasing them in UD-v2.8. The Bhojpuri have 4,881 annotated tokens

Fig. 1 Categories of Indian endangered language (World Atlas 2017)

in accordance with the annotation scheme of UD [6], and Marathi languages have 3506 tokens for 486 sentences [7].

For part-of-speech categories, morphological features, and dependency connections, the Kangri treebank (KDTB) follows the Universal Dependencies (UD) [8] annotation criteria. We adopted the Hindi tag set wherever possible because Kangri is closely linked to Hindi, and there is already a Hindi treebank in UD. Aside from the universal part-of-speech tag set (UPOS), UD also allows for a language-specific secondary tag set (XPOS), which often continues pre-UD practice. The POS tag sets from the Bureau of Indian Standards (BIS) are used here. This is a tag set for annotating Indian languages in general. The XPOS tags were already present in our incoming data; therefore, we used automatic conversion to get the UPOS tags. We leverage 16 lexical and inflectional properties defined in UD in addition to UPOS. Tables 1 and 2 show the details of the employed morphological features, UPOS tags, and UD relations, as well as their statistics. Kangri statistics in universal dependencies for 288 sentences, 17 POS tags, and 22 DP tags are used.

3 Linguistic Analysis (Morphology and Syntax) of Kangri Dependency Treebank

We have used the UD-v2.8 criteria to annotate KDTB. Some Kangri constructions and their linguistics analysis under UD are listed below [9].

Table 1 UPOS tags statistics used in the KDTB

UPOS tags with description	Total count
Adjective (ADJ)	130
Adposition (ADP)	315
Adverb (ADV)	92
Auxiliary (AUX)	308
Coordinating conjunction (CCONJ)	28
Determiner (DET)	51
Interjection (INTJ)	51
Noun (NOUN)	556
Numeral (NUM)	41
Particle (PART)	95
Pronoun (PRON)	209
Proper noun (PROPN)	89
Punctuation (PUNCT)	311
Subordinating-conjunction (SCONJ)	11
Verb (VERB)	287
Other (X)	0

Table 2 DP tags statistics used in the KDTB

DP tags with description	Total count
Clausal modifier of noun (acl)	0
Adverbial clausal modifier (advcl)	39
Adverbial modifier (advmod)	88
Adjectival modifier of noun (amod)	94
Auxiliary verb (aux)	227
Case marker (case)	300
Coordinating conjunction (cc)	24
Clausal complement (ccomp)	3
Compound (compound)	159
Non-first conjunct (conj)	23
Copula (cop)	66
Unspecified dependency (dep)	93
Determiner (det)	46
Discourse element (discource)	3
Subordinating marker (mark)	30
Nominal modifier of noun (nmod)	212
Nominal subject (nsubj)	225
Numeric modifier (nummod)	31
Direct object (obj)	87
Oblique nominal (obl)	0
Punctuation (punct)	311
Root (root)	289
Open clausal complement (xcomp)	4

3.1 Morphological Structure of Word

A morpheme can be constructed as a minimal word unit in Kangri based syntactic criteria. All the words in the dialect can be analyzed into morphemes. Morphemes are of two kinds free and bound. A morphemes may consist of a base or as an affix. There are two types of affixes in the dialect that is inflectional and derivational; inflectional affixes always serve as a prefixes. A word in Kangri may consist of a single morpheme, or it may consist of two or more morphemes. Words in Kangri can be classified into the following types that is simple complex and compound.

3.2 Word Class

On the basis of morphology and Syntax, two types of word class can be established in Kangri. They are from class words and structure or function words. Class words sets appear in more than one form. Structure words do not form paradigmatic sets.

They form class of themselves and do not appear in more than one form. They serve chiefly to convey grammatical relationship and have functional meaning. But, the form class words are the so-called content words and carry the principal lexical meaning. Each of the form class has an almost unlimited number of words. There are four kinds of form class words in the toilet understudy; they are noun including pronoun, verb, adverb, adjective including numerals in the following we shall study each of these form classes in details.

- **Noun**

Noun in Kangri is either base forms or drive from other classes by employing various morphological processes. All nouns in the dialect inflect for case and number. There are five cases direct, oblique, dative, agentive, and vocative. There are two number: singular and plural and two genders masculine and feminine every noun in country is assigned to one of the genders.

- **Pronoun**

Pronounce in Kangri forms a closed system class. They cannot be derived or extended or invented. Their number is fixed, and there is no choice in their use. However, they share many characteristics with nouns, they can replace noun or noun phrases.

- **Adjective**

Adjective is the class of words which occupy a position immediately before the head in a noun phrase. There are two types of adjectives in Kangra, base adjective and derived adjective. The base adjective contains no derivational affixes. The derivational adjective is drive from the other form class words such as nouns and verbs.

- **Adverb**

Adverbs in Kangra form both an open-class as well as closed-class system. The open-class adverbs are off to types, the base at adverbs which are used without any derivation and the derivative adverbs which are derived from verbs, nouns, post-positions and the base adverbs by affixation, reduplication, and by compounding. The closed-class adverb forms paradigms representing the adverbs of time, place and direction, location, manner and condition indicating proximate, remote, relative, correlative and interrogative forms.

3.3 Syntax

In Kangri, a discourse is made up of sentences which are of different structure and are marked by intonation and terminal contour. In a conversation discourse, a sentence maybe constructed of a single word in addition to information. All single words except postpositions and particles can make a sentence. कुण ?, कुथू ?, फिरी. The sentences in Kangri are of two types in view of their nature of construction. They are (1) elementary and (2) core or kernel. The elementary sentences are obtained from the core or kernel sentences by way of transformation that is they can be expanded from or reduced to the kernel or the basic sentences types. Nominal subject (nsubj) is mostly a noun (राधा) or pronoun (तन्निना, तन्निने) phrase in the nominative case in Kangri. For the purpose of UD, the objects are divided to core objects, obj (कताब) or iobj (मोहने जो) Auxillary verbs mostly enhance the intensity of the verb (तन्निनी राजुए जो कताब दत्ती and तन्निनी राजुए जो The words which participate in the sentence construction are of two kinds that is infected and infected nominal. Verbs are infected and agree with each other. For gender and number a minimum sentence may be composed of a single word form. For example come for come here, or it may be composed of a nominal or a question word sentences composed as such elliptical sentences represent the full form normal sentences the missing element in them become productive from the preceding and the following sentences, for example, गया (left), कुण (who), फेरी (then), तू जान (you know), इक काम कर (do one thing), क्या (what), हुण (right now), तसिजो मना (say no).

3.4 Universal Dependencies Sentence Tags

Table 3 shows the Kangri tags with their annotations. Here, for the first sentence, Sentence 1 = अपर तै वक़्ते दे मताबक तहिनां रुस्सेयाँ रश्तेदारां जो रोटी खुआणे ते मना करी दित्ता। the main verb is करी, and there are 4 noun words.

Figure 2 shows graphically presentation of Kangri sentences with universal dependencies. Here, we can see the annotation "इहनां रुक्खां दे कारण हर रोज़ हवा औंदी है।" has been implemented with SPacy. SpaCy [10] offers a number of linguistic annotations for the understanding of the grammatical structure of a text. This encompasses the many sorts of words, such as parts of speech, as well as how the words are connected to one another. SpaCy is used for natural language processing tasks, and it supports many deep learning models. SpaCy first tokenizes the text; that is, it breaks down the text into individual words and that produces a Doc object.

4 Conclusion and Future Work

This paper describes the creation of the first Kangri dependency treebank using the Universal Dependency annotation scheme. The major goal of this project is to create a dependency treebank for Kangri, which is one of India's low-resourced language. The Kangri treebank currently contains 2249 tokens. The annotation rules employed the annotation procedure, and statistics of the used tags/UD relations

Table 3 Kangri sentence with UD tags and annotation

S. No.	Word	POS tags	DP numbering	DP tags
Sentence 1 = अपर तै वक़्ते दे मताबक तह्निां रुस्सेयाँ रिश्तेदारां जो रोटी खुआणे ते मना करी दत्तिता।				
1	अपर	CCONJ	14	CC
2	तै	ADJ	3	JJ
3	वक़्ते	NOUN	14	NN
4	दे	ADP	3	PSP
5	मताबक	ADP	3	PSP
6	तह्निां	PRON	14	PRP
7	रुस्सेयाँ	ADJ	8	JJ
8	रिश्तेदारां	NOUN	10	NN
9	जो	ADP	8	PSP
10	रोटी	NOUN	11	NN
11	खुआणे	VERB	14	VM
12	ते	ADP	11	PSP
13	मना	NOUN	14	NN
14	करी	VERB	0	VM
15	दत्तिता	AUX	14	VAUX
16	।	PUNCT	14	SYM

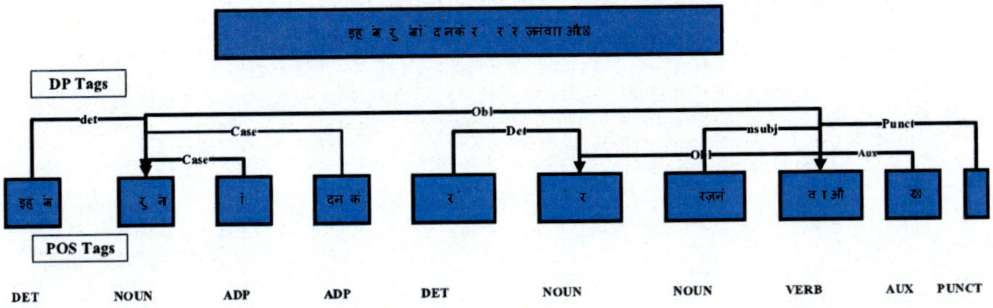

Fig. 2 Graphical presentation of Kangri sentences with universal dependencies

were all discussed in this work. The linguistic analysis of the Kangri treebank was also provided using instances from the language. We intend to expand KDTB sentences and create a Kangri-Hindi treebank in the near future. We will also use a neural model to improve and construct a robust Kangri parser.

Acknowledgments We thank Dr. Karam Singh, Director of Himachal Academy of Arts Culture and Languages, Shimla, Himachal Pradesh, India for their efforts in arranging workshops to collect datasets and to provide annotators. Special thanks to Bhupinder Singh Bhupi and Rajiv kumar Trigarti for preparing Kangri POS and DP tags. We also thank to all the authors of various books specially Praytoosh Guleri, Gautam Vayidhit. We also thank all the language translators writers, Aman Kumar Vishva, Bharti kudailiya, Bhupinder Singh Bhupi, Deepak kulavi, Om parkash Prabhakar, Vijay Puri, Pratibha Sharma, Hari Krishan Murari, Navin Haldwani, Sanju Pal, Suresh

Lata Awasthi, Shakti Singh Rana, Shelly Kiran, Vandana Rana, Vinod Bhavuk, Sonia Dutt, Virender Sharma, Gopal Sharma, Naveen, Manoj Kumar, Durgesh Nandan, Bhagat ram, Kushal, Tej Kumar.

References

1. Zuckermann G (2013) Historical and moral arguments for language reclamation. History and philosophy of the language sciences, p 26
2. Unesco.org (2017) http://www.unesco.org/languages-atlas/, online. Accessed 04 July 2017
3. Wikipedia (2020) Kangri language. https://en.wikipedia.org/wiki/Kangri_language
4. Chauhan S, Saxena S, Daniel P (2021) Monolingual and parallel corpora for Kangri low resource language. arXiv:2103.11596
5. Chauhan S, Jha A, Pant U, Kumar A, Daniel P (2020) analysis and speech recognition of under-resourced kangri dialect with ensemble model. J Crit Rev 7(17):2683–2689
6. Ojha AK, Zeman D (2020, May) Universal Dependency treebanks for low-resource Indian languages: the case of Bhojpuri. In: Proceedings of the WILDRE5–5th workshop on Indian language data: resources and evaluation, pp 33–38
7. Ravishankar V (2017) A universal dependencies treebank for Marathi. In: Proceedings of the 16th international workshop on treebanks and linguistic theories, pp 190–200
8. Universal Dependencies. https://universaldependencies.org/. Last access May 2021
9. Chauhan AR. The Kangri central subsystems by Himachal academy of art, culture and language Shimla
10. Honnibal M, Montani I (2017) spaCy 2: natural language understanding with bloom embeddings, convolutional neural networks and incremental parsing

Analysis of Unsupervised Statistical Machine Translation Using Cross-Lingual Word Embedding for English–Hindi

Shefali Saxena, Shweta Chauhan, and Philemon Daniel

Abstract Machine translation (MT) is automatically converting a piece of text from one language to another. The work provides a comparison and analysis of the phrase-based statistical machine translation (PBSMT) system using monolingual corpora that interpret English text to Hindi. MT quality has improved dramatically over the previous two decades, making it appealing for usage in the translation industry. The importance of MT cannot be overstated. The application of MT systems has been bounded due to the dependency on the bilingual corpus for a range of language pairings. The current paper presents a comparison of two possible cross-lingual word embedding mapping approaches utilizing monolingual datasets for the SMT methodology. When translating from a language to some other language without supervision, inter-lingual word embedding is crucial. We have implemented and compared two cross-lingual mapping approaches are employed in this paper: adversarial training and self-learning methods. The experimental findings for several evaluation methodologies, such as BLEU, METEOR, METEOR-Hi, TER, WER, MER, and NIST, show that the PBSMT approach delivers superior translation using the self-learning method than the adversarial training method for English–Hindi.

Keywords Machine translation · Statistical machine translation · Cross-lingual embedding mapping · Evaluation

S. Saxena (✉) · S. Chauhan · P. Daniel
Department of Electronics and Communication, National Institute of Technology, Hamirpur, Himachal Pradesh, India
e-mail: shefali@nith.ac.in

S. Chauhan
e-mail: shweta@nith.ac.in

P. Daniel
e-mail: phildani7@nith.ac.in

J. K. Mandal et al. (eds.), *Topical Drifts in Intelligent Computing*, Lecture Notes in Networks and Systems 426, https://doi.org/10.1007/978-981-19-0745-6_7

1 Introduction

Machine translation is a branch of artificial intelligence that uses computers to convert a language to another [1]. MT evaluation is an assessment or the comparison of various MT engines' output based on a certain standard.

The traditional SMT methodology is based on massive parallel corpora that are only accessible for a few major languages such as German, English, and French. To remove the dependency on parallel data, the new research line has direct the control to train machine translation systems in a fully unsupervised manner using monolingual corpora. The SMT-based techniques produced a statistical table by cross-lingual mappings, merged it with an n-gram language model, and is then tuned further by iterative back translation [2, 3].

In recent times, much attention has been attracted by cross-lingual word embedding mappings. These approaches employed by training the word embeddings in many languages individually and is then transferred using linear transformations to shared space. Earlier approaches necessitated the use of a training dictionary to determine the initial alignments [4], the adversarial training [5], or self-learning [6] methods have obtained comparable results in developing a model in an unsupervised manner. Cross-lingual embeddings share a common space that can be applied in various machine translation tasks and can also be used to improve language models [7].

Evaluation holds a crucial part in the development of MT models. Word error rate (WER) [8] is the first assessment metric used to determine how well a machine translation system performs on a word-by-word basis. The most popular automatic evaluation metric is the metric bilingual evaluation understudy (BLEU) [9], as it is language independent and easy to implement. This score assesses the n-gram precision (in this work, $n \leq 4$) in comparison to the reference translation, with short translations penalty. It assesses the quality of the translation, with higher BLEU values indicating good quality. Metric for evaluation of translation with explicit ordering (METEOR) [10] score came about to address BLEU's weakness. This metrics are recall oriented. The score calculates and combines the recall and precision with a higher bias toward recall to calculate the harmonic mean. METEOR-Hi [11], another automatic metric, is the extended version of METEOR that includes Hindi-specific features used for evaluating an MT model with Hindi as the target language. The translation error rate (TER) [12] is a character-based automated metric for determining how many edit operations are required to convert the machine-translated output into a human-translated reference. It is more typically used to calculate edit distance, which is used to quantify post-editing work. National Institute of Standard and Technology (NIST) [13] metric is a modification of BLEU, computes specific n-gram's usefulness, i.e., how descriptive an n-gram is in the translated text, by indicating its higher weight on its rareness. Match error rate (MER) shows the probability of the given match is incorrect.

This paper compares the fully unsupervised SMT model with the self-learning and adversarial training cross-embedding technique [2, 3]. We have used the IITB WMT

[14] English–Hindi corpus for training. Different evaluation metrics have been used to compare results, which shows that SMT with self-learning gives better results than adversarial training.

The remaining article is arranged as follows: An overview of the standard SMT system is described in Sect. 2. Section 3 presents the cross-lingual word embedding. The experimental setups are discussed in Sect. 4. Section 5 presents the results and discussion, and finally, Sect. 6 brings out the conclusions and future work.

2 SMT: An Overview

The standard SMT model has three subtasks, namely language model (LM), translation model (TM), and decoding. The TM and LM are built on the Bayesian rule as shown below:

$$t = \arg\max_t P\left(\frac{s}{t}\right) * P(t) \quad (1)$$

In Eq. 1, $P(s/t)$ indicates the probability of the source side sentences for the TM given the target side sentences, while the LM, $P(t)$, estimates the probability of target phrase. And decoding gives the best source sentence machine translations.

The log-linear model as in Eq. (2) is commonly used in the current approach of SMT proposed by [2]. According to it, the translation probability is formulated from the source side s to the target side t as:

$$p\left(t|s; \lambda_1^M\right) = \frac{\exp\left[\sum_{m=1}^{M} \lambda_m h_m(s,t)\right]}{\sum \tilde{t} \exp\left[\sum_{m=1}^{M} \lambda_m h_m\left(s,\tilde{t}\right)\right]} \quad (2)$$

where $h_m(s, t) = \log h_m(s, t)$ denotes the mth feature. In PBSMT [15], the sentence pair is broken down into a series of phrases s_1^K and t_1^L, with K and L as phrases count. Minimum error rate training (MERT) [16] has been used to optimize the feature weights λ_1^M.

3 Cross-Lingual Word Embeddings

The word embedding in one language and the analogous word in some other language are unrelated. As a result, word embeddings are unable to represent themselves in vector space. Word embeddings learned on source and target corpora are connected in a single n-dimensional embedding space in a cross-lingual word embedding. This demonstrates that geometrical symmetry between the organization of words in two distinct languages, or the source and target language embedding space, is always present when large corpora are used. A linear transform can be trained to depict a source embedding space to destination space embeddings. Using an fully unsupervised similarity distribution, the embeddings are rotated continually and the distance between the word pairs is narrowed [6]. As previously noted, obtaining parallel corpora that are aligned is both costly and time-consuming. Recent work in unsupervised machine translation suggests that a competitive system may be developed utilizing just monolingual corpora [2, 3]. The papers [2, 3] offer a unique technique for constructing a statistical machine translation model using monolingual data without any parallel data. The fundamental concept is to learn word embeddings individually and then apply linear transformations to bring them together in a shared space. The phrase table is generated using these embeddings.

3.1 Unsupervised Learning with Adversarial Training

Adversarial learning [5, 17] is used to train a rotation matrix w, whose primary role is to roughly line up the target and source embeddings that has $n * n$ dimensions, where n is the word embedding size. The generator is aiming to mislead the discriminator by making it impossible for it to anticipate the origin from either the source or the target of embedding. Matrix w taught how to perform this method and how to calculate the weights for mapping source to target language embedding's.

3.2 Unsupervised Learning with Similarity Distribution

Without utilizing a seed dictionary, unsupervised learning with similarity distribution creates a cross-lingually mapped terms [6]. The linear transform is learnt in stages, beginning with the normalization of word embeddings and the creation of a seed dictionary based on the similarity distribution of the most common terms.

4 Experiment

4.1 Corpus Statistics

For Hindi and English, the IIT Bombay Monolingual Corpus [14] is utilized. The fast text technique is used to calculate vocabulary size. The English corpus has a value of 61,683, while the Hindi corpus has a value of 60,432.

Table 1 Different evaluation metrics for SMT system using cross-lingual mapping approach

	BLEU	METEOR	METEOR-Hi	MER	WER	NIST	TER
Self-learning	0.043	0.1219	0.1136	0.889	0.1187	0.261	0.1108
Adversarial	0.0364	0.1056	0.980	0.866	0.1009	0.399	0.1283

4.2 Preprocessing

Preprocessing is the initial stage in the MT process. The quality of embedding is decided by the corpus utilized. For English and Hindi corpus, a Moses[1] library and IndicNLP[2] library toolkit has been used respectively.

4.3 SMT Model: Moses

Trained the PBSMT model given by the Moses toolbox [18]. FastAlign is used to align the source language segmented words with the target language segmented words. Symmetrized heuristic and grow-diag-final [1] for the alignments. Used KenLM [19] to train the 5-g LM with modified Kneser–Ney discounting [19]. The Moses decoder (v0.4) [18] was used to decode the data, and the minimum error rate training (MERT) [16] was utilized to optimize the decoder settings. Moses default settings were utilized in all of the experiments.

5 Result and Discussion

The assessment of SMT using two cross-lingual mapping techniques for various evaluation criteria is shown in this section. In addition, several sample phrases are displayed. The assessment for all of the distinct metrics was utilized to execute the SMT model training using the adversarial training and self-learning technique in Table 1.

Figure 1 demonstrates the outcomes of the SMT methodology for monolingual data using self-learning and adversarial training cross-lingual mapping approaches. The BLEU score for the SMT model with the self-learning approach is more significant than with adversarial training, as shown in Fig. 1, and so on. All of the assessment approaches suggest that SMT with self-learning cross-lingual mapping methods produces better outcomes. The work used 272 test sentences, out of which 200 are from the IITB test corpora, and 72 sentences are taken from the WMT test dataset apart from training data.

---

[1] https://github.com/moses-smt/mosesdecoder.git.

[2] https://github.com/anoopkunchukuttan/indic_nlp_library.git.

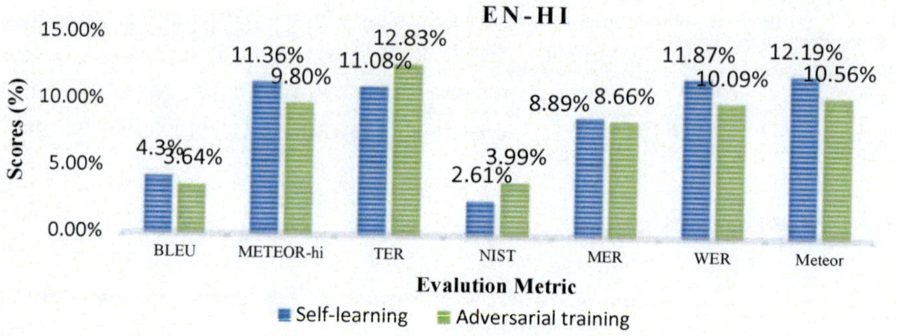

Fig. 1 Evaluation metric of an SMT system for English–Hindi using self-learning and adversarial training techniques

Table 2 Sample test and translated sentence using self-learning and adversarial learning

Test sentences	Translated output using self-learning technique	Translated output using adversarial training technique
Will you eat food?	तुम खाना खाना?	आप खाना खाने स जाए
We are talking about the twenty-first century: servants	हम बातें करें 21वी सदी: मामलें।	हम बात कर के बारे आइ वी सदी लोकसेवक
Seventy percent of my interview was confined to my medical field	सौ प्रतिशत मेरी इंटरव्यू थी किंतु मेरे चिकितसा फील्ड से।	तीस प्रतिशत के मेरे साक्षात्कार था सीमित मेरे लिए मेडिकल फील्ड वेबदुनिय.
Does anybody care that criminals are not being handed over to international justice?	कर सकती है जिससे कि अपराधियों पॉलिसियां को तैनात इटरनेशनल जस्टिस?	कोई देखभाल अपराधी है कि किसी भी तरह से को सौप इटरनेशनल जस्टिस ?
I heard an interesting talk on radio this morning	मैं अपनी रोचक बात रेडियो पर आया था।	मुझे सुना करा मजेदार बात पर रेडियो इस सुबह|

Table 2 illustrates the statistical machine translation output for English and Hindi using cross-lingual mapping approaches such as self-learning and adversarial training. As seen in Table 2, there is a reordering issue with the SMT system when using the adversarial training technique.

6 Conclusion

When translating from a language to other language without supervision, cross-lingual word embedding is crucial. This study compares and analyzes the SMT system utilizing several cross-lingual mappings for English to Hindi. The results of many assessment approaches were also presented, including BLEU, METEOR, METEOR-Hi, TER, MER, WER, and NIST evaluation techniques. Compared to the adversarial training mapping technique, experimental results reveal that SMT trained using the self-learning mapping method works well. We will expand our work in the future by producing cross-lingual word embedding for a variety of languages, primarily morphologically abundant languages, and languages with limited resources. This work will also be extended to improve the translation quality, and the reordering approaches will also be investigated.

References

1. https://en.wikipedia.org/wiki/History_of_machine_translation
2. Artetxe M, Labaka G, Agirre E (2018) Unsupervised statistical machine translation. In: Proceedings of the 2018 conference on empirical methods in natural language processing. Association for Computational Linguistics, Brussels, Belgium, pp 3632–3642
3. Lample G, Conneau A, Denoyer L, Ranzato MA (2018) Unsupervised machine translation using monolingual corpora only. In: Proceedings of the 6th international conference on learning representations (ICLR 2018)
4. Mikolov T, Le QV, Sutskever I (2013) Exploiting similarities among languages for machine translation. arXiv:1309.4168
5. Conneau A, Lample G, Ranzato MA, Denoyer L, Jégou H (2018) Word translation without parallel data. In: Proceedings of the 6th international conference on learning representations (ICLR 2018)
6. Artetxe M, Labaka G, Agirre E (2018) A robust self-learning method for fully unsupervised cross-lingual mappings of word embeddings. In: Proceedings of the 56th annual meeting of the association for computational linguistics, vol 1: long papers, pp 789–798
7. Adams O, Makarucha A, Neubig G, Bird S, Cohn T (2017, Apr) Cross-lingual word embeddings for low-resource language modeling. In: Proceedings of the 15th conference of the European chapter of the association for computational linguistics, vol 1, long papers, pp 937–947
8. Su K-Y, Wu M-W, Chang J-S (1992) A new quantitative quality measure for machine translation systems. In: COLING 1992 volume 2: the 15th international conference on computational linguistics
9. Papineni K et al (2002) Bleu: a method for automatic evaluation of machine translation. In: Proceedings of the 40th annual meeting of the association for computational linguistics
10. Banerjee S, Lavie A (2005) METEOR: an automatic metric for MT evaluation with improved correlation with human judgments. In: Proceedings of the acl workshop on intrinsic and extrinsic evaluation measures for machine translation and/or summarization
11. Gupta A, Venkatapathy S, Sangal R (2010) METEOR-Hindi: automatic MT evaluation metric for Hindi as a target language
12. Snover M et al (2006) A study of translation edit rate with targeted human annotation. In: Proceedings of the 7th conference of the association for machine translation in the Americas: technical papers
13. Przybocki M et al (2008) The NIST 2008 metrics for machine translation challenge overview, methodology, metrics, and results. Mach Transl 23(2):71–103
14. Kunchukuttan A, Mehta P, Bhattacharyya P (2017) The IIT Bombay English-Hindi parallel corpus. arXiv:1710.02855
15. Koehn P, Och FJ, Marcu D (2003) Statistical phrase-based translation. In: Proceedings of the 2003 conference of the North American chapter of the association for computational linguistics on human language technology, vol 1. Association for Computational Linguistics, pp 48–54
16. Och FJ (2003) Minimum error rate training in statistical machine translation. In: Proceedings of the 41st annual meeting on association for computational linguistics, vol 1. Association for Computational Linguistics, pp 160–167

17. Zhang M, Liu Y, Luan H, Sun M (2017, July) Adversarial training for unsupervised bilingual lexicon induction. In: Proceedings of the 55th annual meeting of the association for computational linguistics, vol 1: long papers, pp 1959–1970
18. Koehn P, Hoang H, Birch A, Callison-Burch C, Federico M, Bertoldi N, Cowan B, Shen W, Moran C, Zens R, Dyer C, Bojar O, Constantin A, Herbst E (2007) Moses: open source toolkit for statistical machine translation. In: Proceedings of ACL, pp 177–180
19. Heafield K (2011) KenLM: faster and smaller language model queries. In: Proceedings of the sixth workshop on statistical machine translation, WMT'11, Edinburgh, Scotland, pp 187–197

A Ternary Sentiment Classification of Bangla Text Data using Support Vector Machine and Random Forest Classifier

Partha Chakraborty, Farah Nawar, and Humayra Afrin Chowdhury

Abstract Sentiment analysis refers to the extraction of the underlying sentiment or emotion associated with an opinion. It is a part of the difficult field in the processing of natural languages (NLP) with several applications in the business, education, and political sectors. Unlike other languages, the amount of research performed on opinion mining for Bangla text is very small. There is also a lack of proper NLP tools for Bangla text processing. To bridge this gap, in this manuscript, a classification system has been developed for predicting the polarity of Bangla text data (i.e., positive, neutral, negative) using the two most efficient algorithms SVM and random forest for opinion mining. In this study, we experimented with unigrams, bigrams, and trigrams to illustrate how contextual information affects the overall performance of the classifiers. The dataset we used in this paper is imbalanced, which resembles natural characteristics of opinions in day-to-day life.

Keywords Opinion mining · Classification · Sentiment · *N*-gram · Machine learning

1 Introduction

The tremendous amount of unstructured data that is available online can be used for sentiment analysis. It helps to analyze customer's feedback to improve their products and services or to make decision in political movement or public welfare. There have been very few research carried out in sentiment analysis of Bangla text data in comparison with other languages (e.g., English, Spanish, French). In our paper, we have suggested an Bangla sentiment classification system which works effectively

In our paper, we used both tokenization and stemming of words in combination with *n*-gram (i.e., unigram, bigram, and trigram) for feature selection after removing

P. Chakraborty (✉) · F. Nawar · H. A. Chowdhury
Department of Computer Science and Engineering,Comilla University, Cumilla 3506, Bangladesh
e-mail: partha.chak@cou.ac.bd

J. K. Mandal et al. (eds.), *Topical Drifts in Intelligent Computing*, Lecture Notes in Networks and Systems 426, https://doi.org/10.1007/978-981-19-0745-6_8

the noise and redundancy of the initial raw data through several steps of preprocessing. Tokenization has shown better results than stemming with 68% $F1$-score. N-grams determine the relationship between $N-1$ neighboring words by assigning probabilities to them. It captures contextual information based on the value of n. The higher value of N ensures more contextual information from the sentences. N-gram techniques were used along with the TF-IDF method for vectorization. The processed data was fed into a classification model for evaluation. Our goal is to provide a system of classification that will be able to classify Bangla text into a category of positive, negative, and neutral classes using the two most efficient machine learning algorithms (i.e., SVM and RF) in opinion mining along with n-gram techniques and show a comparative analysis between the classifiers and n-grams.

The remaining sections of the manuscript are divided into four section. Section 2 consists of related works. Section 3 states the overall architecture and methodology of the system. The evaluation of the system and comparisons among the features are showed in sect. 4. Finally, Sect. 5 summarizes the findings and points the work's future directions.

2 Related Work

Over more than 1300 years, the Bengali language has been affected by a diversity of languages and cultures. There are also a few NLP tools for Bengali. Many experts have studied the Bengali language to fill this deficit. Abinash et al. [1] provided a resemblance of results obtained using the classification algorithms of NB and SVM. The algorithms help to decide whether a sentimental assessment is positive or negative. Taher et al. [2] focused their views on Bangla texts by using diverse web-based data. The ML process and the N-gram approach used to categorize Bangla papers were 89% precision achieved using 1–2 grams with SVM. Several works have been done on this work [3, 4]. Lee et al. [5] used the classification of Naive Bayes, SVM and maximum entropy. Pundlik et al. [6] suggested a strategy for categorizing Hindi speech documents into many classes using ontology, combining HSWN and LM classifier. To identify the best combination of user input, Rahman et al. [7] examined the effects of extractive function approaches. Uni, bi, and trigram are used for representations with TF-IDF independently. Tripto et al. [8] created deep learning-based models. They test the model's performance using a fresh corpus of English, Roman, and Bangla feedback from YouTube, and their method achieved 65.97% and 54.24% accuracy in three and five label sentiments, respectively. Related several works have been done in this work [9, 10]. It was developed by Haque et al. [11] to classify feedback using machine learning algorithms where SVM's accuracy is measured at 75.58%., according to a comparison of vectorizers.

3 Methodology

This section provides an overview of the overall system. Different preprocessing procedures reduced noise and redundancy from the underlying raw data. TF-IDF was utilized for vectorization and tokenization. In the training set, SVM and RF classifiers were used to classify the data. A three-category categorization strategy was used to organize the data (positive, neutral, and negative). The model was evaluated based on test results. Our model's system structure is depicted in Fig. 1.

3.1 Data Description

In this research, we used the dataset on Bangla news comments by Ashik et al. [12]. The dataset contains a total of 13802 Bangla text labeled with five categories: positive, slightly positive, neutral, negative, and slightly negative. Each data point was annotated by three different people to obtain three different viewpoints, and the final tag was selected based on the majority's decisions. The dataset has an imbalanced distribution of data at each label. Figure 2 illustrates an example of the dataset. The frequency data for each sentiment label is shown in Fig. 3a.

As the amount of data at each label is small, we put positive, slightly positive classes in the positive category and negative, slightly negative classes in the negative category, keeping the neutral class as it was. Figure 3b shows the percentage of data at each newly annotated label.

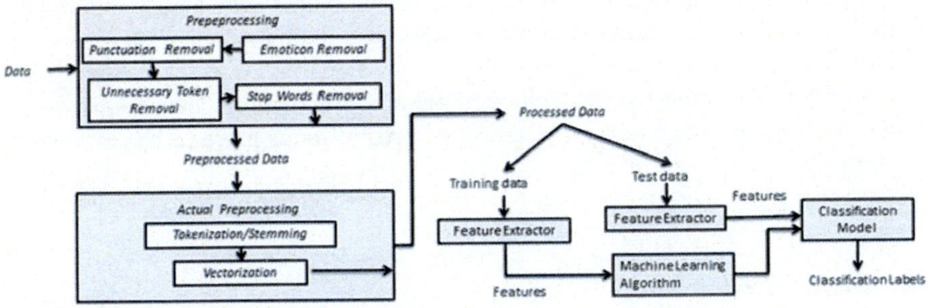

Fig. 1 System structure of the model

Comments	Sentiment
লিখার সময় পারলে সত্য লিখার অভ্যাস শিখুন।	কিছুটা নেতিবাচক
এটা কেন হচ্ছে? সংশ্লিষ্ট সকলের ডিপ্রেশনের ফলে? নাকি সরকার মনোনিত পরিচালনা পর্ষদের ব্যর্থতার কারণে?	কিছুটা নেতিবাচক
আমাদের দেশের স্বাভাবিক অর্থনৈতিক গতিপ্রবাহকে বাধাগ্রস্ত করা হয়েছে। তাই এসব ঘটছে।	নিশ্চিত নেতিবাচক

Fig. 2 Example of Bangla news comments with annotated sentiment label

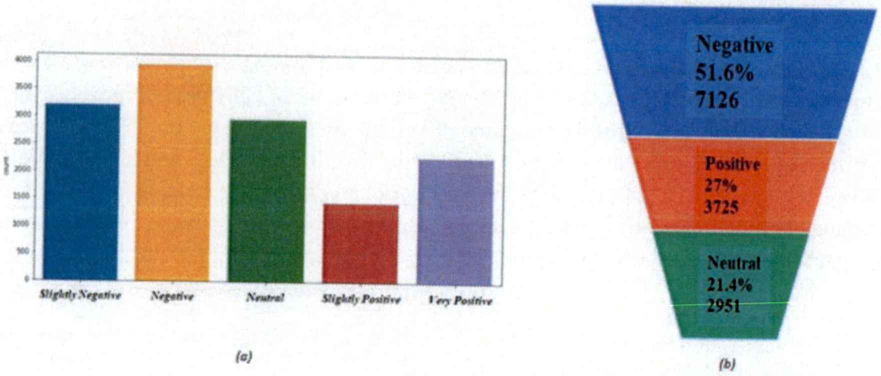

Fig. 3 **a** Amount of data at initial label, **b** percentage of data at newly annotated labels

3.2 Data Preprocessing

The initial raw data comprised of many redundant information and noise that did not offer anything to the sentiment analysis. To remove these, the data goes through several preprocessing steps. The preprocessing steps are given below:

(a) *The removal of emoticon*: Emoticons are basically used to represent different expressions and modes of the face. As in this paper, we are only working with textual information. We removed the emoticons, which do not contribute to our analysis.
(b) *The removal of punctuation marks*: Punctuation marks play a very minor role in determining the sentiment of a sentence. That is why punctuation marks are removed to avoid time and space complexity.
(c) *The removal of stop words*: Conjunctions and prepositions are common stop words. It is important to remove the stopwords to focus more on the important words and reduce complexity.

3.3 Actual Processing

A well-defined series of linguistically significant units was created from the preprocessed texts in this section. Tokenization or stemming of phrases, vectorization of tokens, and n-gram approaches are all involved in this process.

Tokenization and Stemming

In a sentence, a token consists of a sequence of characters that is regarded as a semantic unit. When a sentence is tokenized, the tokens are broken up into smaller chunks of information. For tokenization, we used whitespace as a delimiter. The technique of slicing words in order to return to their base words is known as stemming.

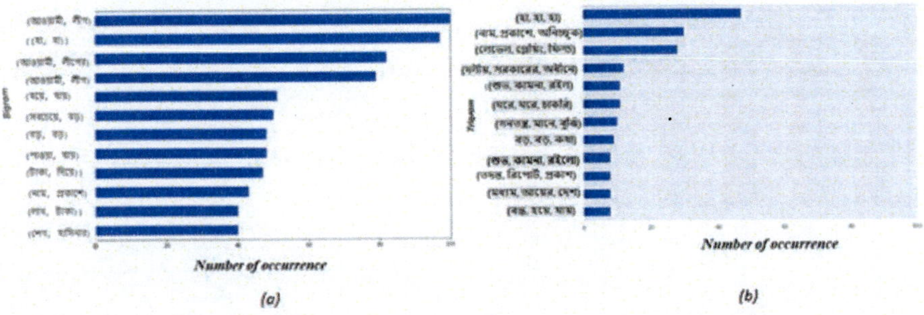

Fig. 4 Twelve most frequently occurring **a** bigram and **b** trigram of the dataset

Because each word derives its meaning from its root, removing affixes has no effect on opinion mining and simplifies calculation.

N-gram techniques We used our data to test various n-gram models in order to find the most useful feature. Based on $n-1$ prior words, n-gram anticipates the possibility of a term emerging. N-gram determines the associations between words and their neighbors and, to some extent, captures the context. Based on the value of n, the N-gram might be of numerous sorts. It is referred to as a unigram when n equals one. Only one word is used at a time by a unigram. The model is called a bigram for $n = 2$ and a trigram for $n = 3$. They, respectively, capture the context of two or three previous words. Figure 4 shows the 12 most frequently occurring bigrams and trigrams in our dataset.

3.4 *Random Forest and Support Vector Machine*

Random Forest Classifier

It enhances the projected accuracy of the dataset by averaging the results of many decision trees on different subsets of a dataset. It creates a classification category rather than a classification and then splits fresh data points based on the classifiers' predictions [13].

Support Vector Machine

SVM is used in machine learning algorithms, in which each node is represented as a data point in n-dimensional space, and every feature value relates to a specific coordinate. Then we classify the data by choosing the hyperplane that clearly divides the two groups [14]. To differentiate positive and negative data, SVMs look for the optimal surface.

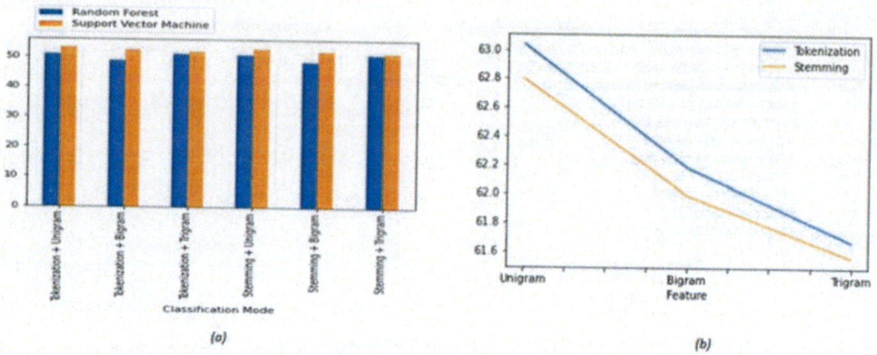

Fig. 5 **a** Comparison of SVM and random forest classifier, **b** comparison of n-grams in SVM

4 Result Analysis and Comparison

As our dataset is not properly balanced, we used four performance metrics to evaluate it accurately. These performance metrics are precision, recall, $F1$-measure, and accuracy. Better performance is associated with higher precision and recall values. Precision gives a measure of how precisely the system captures the correct cases, and recall gives a measure of the minimization of false negatives. The $F1$-measure is the reciprocal of the arithmetic mean of recall and precision. The equations for the performance metrics are given below, true positive and false positive are denoted by (TP) and (FP), true negative and false negative are denoted by (TN) and (FN), and m stands for sample size (TP + FP + FN + TN).

$$\text{Accuray} = \frac{\text{TP} + \text{TN}}{m} \quad \text{Precision} = \frac{\text{TP}}{(\text{TP} + \text{FP})} \quad \text{Recall} = \frac{\text{TP}}{\text{TP} + \text{FN}}$$

$$F1\ \text{measure} = \frac{(2 \times \text{Precision} \times \text{Recall})}{(\text{Precision} + \text{Recall})}$$

Recall, $F1$-measure, and precision scores of each classifier's feature are all shown in Table 1, and the accuracy of the classifiers for each feature is shown in Table 2.

From the performance metrics, we can see that the classifiers predict negative opinions with high perfection (e.g., 96% recall using SVM) where neutral and positive classes are predicted with less perfection. We can say that the imbalance of labeled data has affected the performances highly. In both tokenization and stemming, SVM outperformed random forest. Tokenization has proven to be more effective for opinion mining than stemming. A comparison between the accuracy of SVM and random forest for each feature is given in Fig. 5a.

Table 1 Performance statistics of the proposed model

Features	Support vector machine				Random forest			
Tokenization + unigram		Precision (%)	Recall (%)	*F*1-measure (%)		Precision (%)	Recall (%)	*F*1-measure (%)
	Neg.	54	93	68	Neg.	54	85	66
	Neutral	22	2	3	Neutral	27	8	13
	Pos.	51	17	26	Pos.	43	20	27
Tokenization + bigram		Precision (%)	Recall (%)	*F*1-measure (%)		Precision (%)	Recall (%)	*F*1-measure (%)
	Neg.	53	94	68	Neg.	54	80	64
	Neutral	22	2	4	Neutral	21	8	12
	Pos.	49	13	20	Pos.	41	21	28
Tokenization + trigram		Precision (%)	Recall (%)	*F*1-measure (%)		Precision (%)	Recall (%)	*F*1-measure (%)
	Neg.	52	96	68	Neg.	53	92	67
	Neutral	15	1	2	Neutral	17	3	5
	Pos.	51	7	12	Pos.	45	10	17
Stemming + unigram		Precision (%)	Recall (%)	*F*1-measure (%)		Precision (%)	Recall (%)	*F*1-measure (%)
	Neg.	54	93	68	Neg.	54	84	66
	Neutral	23	2	3	Neutral	25	8	12
	Pos.	50	16	24	Pos.	44	21	28
Stemming + bigram		Precision (%)	Recall (%)	*F*1-measure (%)		Precision (%)	Recall (%)	*F*1-measure (%)
	Neg.	53	93	68	Neg.	53	80	64
	Neutral	25	2	4	Neutral	22	9	13
	Pos.	48	12	19	Pos.	39	21	27
Stemming + trigram		Precision (%)	Recall (%)	*F*1-measure (%)		Precision (%)	Recall (%)	*F*1-measure (%)
	Neg.	52	96	68	Neg.	53	92	67
	Neutral	15	1	2	Neutral	17	3	5
	Pos.	49	6	11	Pos.	44	10	17

We can see that unigram has shown better performance than bigram and trigram, though they capture more contextual information than unigram. We can say that the performance of *n*-gram depends highly on the characteristics of the elements of the dataset and their internal context and relationship among words. The comparison of n-gram techniques in the SVM classifier is shown in Fig. 5b.

Table 2 Performance scores of the proposed model

	Accuracy	
Features	Support vector machine (%)	Random forest (%)
Tokenization + unigram	63.04	60.80
Tokenization + bigram	62.18	58.71
Tokenization + trigram	61.67	60.94
Stemming + unigram	62.80	60.80
Stemming + bigram	61.99	58.44
Stemming + trigram	61.56	60.92

5 Conclusion

SVM and RF are two of the most efficient machine learning algorithms in opinion mining that we used in conjunction with n-gram techniques to create a system for categorizing models for classification of Bangla text based on positive, negative, or neutral classes. Tokenization has showed greater results than stemming with a $F1$-score of 68%. SVM has shown better performance than the RF classifier and the best performance with unigram.

In the future, we want to introduce multi-classes in our system and expand the amount of data in our dataset. We also want to experiment with more preprocessing and feature selection techniques to improve the accuracy.

References

1. Tripathy A, Agrawal A (2015) Rath S (2015) Classification of sentimental reviews using machine learning techniques. Proc Comput Sci 57:821–829
2. Abu Taher SM, Akhter KA, Azharul Hasan KM (2018) N-gram based sentiment mining for bangla text using support vector machine. In: 2018 international conference on Bangla speech and language processing (ICBSLP). IEEE, Sept 2018
3. Ahammad K, Shawon JAB, Chakraborty P, Jahidul Islam M, Islam S (2021) Recognizing bengali sign language gestures for digits in real time using convolutional neural network. Int J Comput Sci Inf Secur (IJCSIS) 19(1):11–19
4. Chakraborty P, Ahmed S, Yousuf MA, Azad A, Alyami SA, Moni MA (2021) A human-robot interaction system calculating visual focus of human's attention level. IEEE Access 9:93409–93421
5. Pang B, Lee L, Vaithyanathan S (2002) Thumbs up? sentiment classification using machine learning techniques. In: Proceedings of the 2002 conference on empirical methods in natural language processing (EMNLP 2002). Association for Computational Linguistics, pp 79–86
6. Pundlik S, Dasare P, Kasbekar P, Gawade A, Gaikwad G, Pundlik P (2016) Multiclass classification and class based sentiment analysis for Hindi language. In: 2016 international conference on advances in computing, communications and informatics (ICACCI). IEEE, pp 512–518
7. Rahman SSMM, Biplob KBMB, Rahman MH, Sarker K, Islam T (2020) An investigation and evaluation of n-gram, TF-IDF and ensemble methods in sentiment classification. In International conference on cyber security and computer science. Springer, pp 391–402

8. Tripto NI, Ali ME (2018) Detecting multilabel sentiment and emotions from bangla youtube comments. In: 2018 international conference on Bangla speech and language processing (ICBSLP). IEEE, pp 1–6
9. Rahman MM, Pramanik MA, Sadik R, Roy M, Chakraborty P (2021) Bangla documents classification using transformer based deep learning models. In: 2020 2nd international conference on sustainable technologies for industry 4.0 (STI). IEEE Publisher, pp 1–5
10. Rahman S, Chakraborty P (2021) Bangla document classification using deep recurrent neural network with bilstm. In: Proceedings of international conference on machine intelligence and data science applications. Springer, pp 507–519
11. Haque F, Manik MMH, Hashem MMA (2019) Opinion mining from Bangla and phonetic bangla reviews using vectorization methods. In: 2019 4th international conference on electrical information and communication technology (EICT). IEEE, pp 1–6
12. Akhter-Uz-Zaman Ashik M, Shovon S, Haque S (2019) Data set for sentiment analysis on Bengali news comments
13. Kominfo Makassar BBPSDMP (2019) Sentiment analysis using random forest algorithm-online social media based
14. Ashis P (2012) Support vector machine–a survey. Int J Emerging Technol Adv Eng 2(8):82–85

Third Wave Prediction Analysis for Kerala in India

Ravi Kumar Arya, Suram Rithwik, Kanupriya Khandelwal, Promod Verma, Ravi Dugh, and Amit Dugh

Abstract COVID-19 is spreading widely across the globe right now. India just went through the second wave of COVID-19 pandemic and has lost more than 425,000 people to this pandemic till date. Most of the other countries have gone through the second wave, with some countries experiencing third and fourth waves. In such difficult times, there is a shortage of resources everywhere. Planning is the need of the hour, and all the countries are expanding their resources, keeping future demands in mind. Some of the states of India, like Kerala, are also expecting imminent danger of the third wave. In this study, we are predicting the arrival and peak of the third wave in Kerala. We also provide the mathematical models and theoretical background to reach such expected dates. Prediction of this type helps to suggest the preparation needed to tackle the upcoming disaster. Governments can prepare themselves so that there is minimal damage to life in future.

Keywords COVID-19 · Coronavirus · Kerala · India · Third wave · Prediction

1 Introduction

COVID-19 was announced as a global pandemic by the World Health Organization (WHO). The "first wave" of COVID-19 slowed down by September–October 2020 after the active patients declining to as low as 10,000; the "second wave" picked up by the middle of March 2021 in India. COVID-19 is currently wreaking havoc in more than 200 countries worldwide. Many states of India, e.g., Kerala, were facing the second wave of this pandemic and are in wait for the third wave.

Many researchers have confirmed that the third wave in India is inevitable, considering the present conditions in India. The results of the SVEIRD model [1] have

R. K. Arya (✉) · S. Rithwik · K. Khandelwal · P. Verma
National Institute of Technology Delhi, Delhi, India

R. Dugh
Goergen Institute of Data Science, University of Rochester, Rochester, NY, USA

A. Dugh
Information Technology, Queensland University of Technology, Brisbane, Australia

J. K. Mandal et al. (eds.), *Topical Drifts in Intelligent Computing*, Lecture Notes in Networks and Systems 426, https://doi.org/10.1007/978-981-19-0745-6_9

concluded that the third wave is inevitable in India, and the only remedy to this is to increase the vaccination rates by five times the current rate, which is practically impossible, given India's population and resources currently in hand. Hence, many researchers have made efforts to build mathematical models for predicting the date of the impending crisis. For example, Rajneesh Bharadwaj and Amit Agarwal [2] employ a logistic model and use regression techniques such as least square fitting to fit the historical curves as an exponential function to predict future scenarios. More complex models have also been employed to get accurate results like the SUTRA model [3].

There have been some studies that have shown that simple models are often better in predicting COVID cases [4, 5]. The goal is to understand a simple trend in the number of new cases versus the number of days plot of several countries. Involving complex mathematical functions to fit a curve may result in overfitting and may not be sensitive to new changes. Hence, efforts have been made to analyze the plot data for several countries and study the possible factors contributing to the third wave in Kerala.

To predict the timing of the third wave in Kerala, we collected data for countries experiencing (or that have experienced) the third wave. The time interval between the second and the third wave for such countries can be instrumental in estimating the general trend of the arrival of the third wave. However, for it to work, the underlying conditions should be the same for the geographical location that has encountered the third wave and the target country (or state) for which prediction is being estimated. We plotted graphs between days and cases in these countries and found gaps between the second wave and third wave to get a better idea about the correlation between the second and third wave. Next, we compared their vaccination rates and also took their population into account. With all this background, we predict when the third wave is expected to arrive in Kerala.

The paper is divided into three different sections. Section 2 deals with the methodology and the mathematical background needed for this work. We will also discuss the results and give insights into the predicted data in this section. In the end, Sect. 3 concludes this work with concluding remarks.

2 Methodology

Multiple factors can contribute to the third wave in Kerala. Most of the states of India are easing lockdowns, while others have already opened public places. All these factors will contribute to the arrival of the third wave.

The prediction of the third wave will depend on several possible factors (predictor variables) such as

1. Number of days between the peaks
2. Vaccination rates
3. Herd immunity.

Next, we elaborate on each possible factor and its effect on predicting the third wave behavior.

2.1 *Number of Days Between the Peaks*

The coronavirus has infected the entire world equally irrespective of the country's economy, medical conditions, etc. To get a better idea about the coronavirus cases in other countries, we show pictorial representations in Figs. 1a, b, and 2. The plots depict the trends of infected people in different countries. Greece, Armenia, Azerbaijan, and Bulgaria have very similar trends as per Fig. 1a. We use the similarity of these trends as the basis for our study. Observing the plot of the number of cases versus days for some countries that have encountered the third wave shows three significant peaks representing the three COVID-19 waves that the countries have experienced. We can study this trend to predict the future action of the virus for nations that are still in earlier phases.

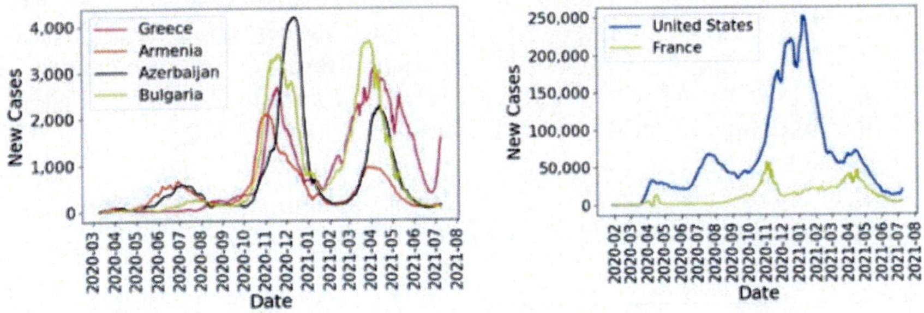

(a) (b)

Fig. 1 COVID-19 trend for **a** Greece, Armenia, Azerbaijan, Bulgaria and **b** the United States and France

Fig. 2 COVID-19 trends for Austria, Canada, Germany, Switzerland, UK, Russia, and Belgium

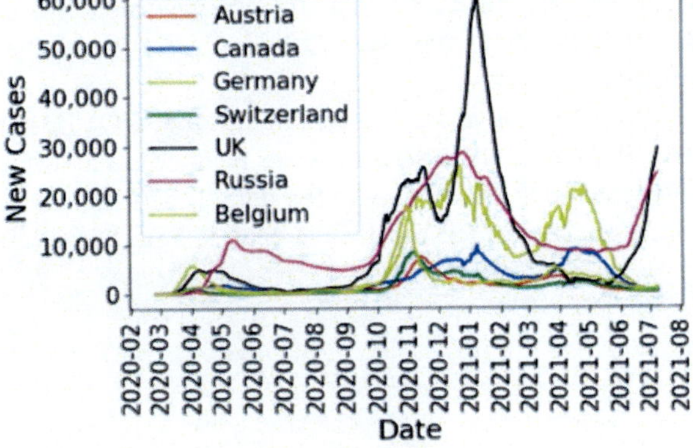

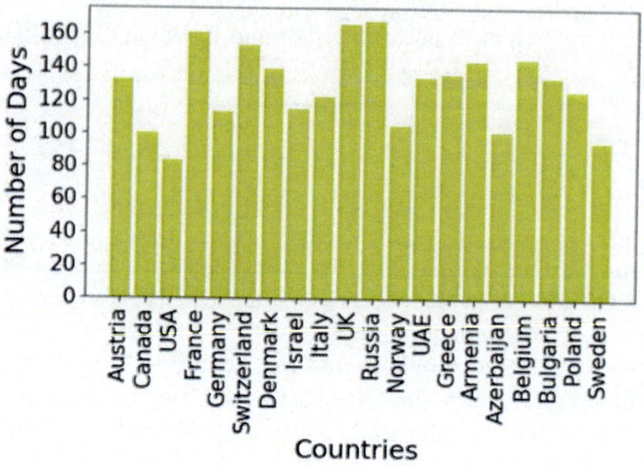

Fig. 3 The number of days between the second wave and third wave peaks for each country

It is difficult to plot and comprehend the trends of different countries in a single plot. So, we show the difference of days between the second and third waves in different countries in Fig. 3. An important observation made here is that the difference of days for all the countries is more or less the same irrespective of each country's lockdowns and government rules. Hence, the bar chart for the number of days taken to reach the third wave from the second wave for various countries can give the correlation that will help our prediction.

In order to remove bias, Eq. (1) is formulated to find the mean interval in days between the second and the third wave for all countries.

$$\text{Mean} = (x_1 + x_2 + x_3 + \ldots)/N \tag{1}$$

where x_1, x_2, x_3 .... are the number of days for each country and N is the total no. of countries. The mean comes out to be 128.6 days which can be rounded up to 129 days on average. In this way, the mean days to reach the third wave peak from the second wave peak are 129 days on average for the concerned countries. We can use this result and the current trend of the coronavirus cases in Kerala and then extrapolate the values to reach the overall trend. Next, we plot the data and discuss the insights (Fig. 4).

From the above study, we can say that COVID-19 will hit its third wave peak in 129 days after its second wave peak. Kerala suffered the second wave peak in early mid-May. According to the data, it was highest on 12th May 2021. Hence, the expected date for the third wave peak is 18th September 2021 for Kerala.

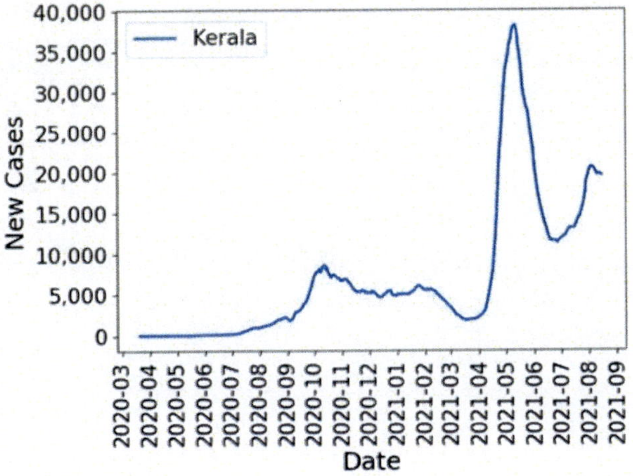

Fig. 4 COVID-19 trend for Kerala

2.2 *Vaccination Rates*

According to the SUTRA model [6], the vaccination rate does not affect the predicted date of the third wave in India unless the rate increases miraculously. The vaccination rate will only affect the number of cases during that period.

Let us look at the vaccination rates [7]. India's vaccination rate as of July 14, 2021 was about 5.6%. Since the vaccination rate is low, this rate can be compared with other similar countries to see if similar peaks can be observed between the first and second and second and third waves. On reviewing the vaccination tracker, two other countries that have nearly 5.6% vaccination rates are Indonesia and Oman. Next, we show the vaccination trends of these countries. In Figs. 5 and 6, we show the trends of new cases and vaccination in Indonesia and Oman, respectively.

India has about 286 doses per thousand population. So, New Zealand is close to India with about 287 doses per thousand population. Figure 7 shows the trends of new cases and vaccination for New Zealand.

From Figs. 5, 6, and 7, it is clear that COVID-19 is spreading in countries where vaccination rates, both as a percentage of the population (case of Indonesia and Oman) and doses per thousand (case of New Zealand), are similar to India. Therefore, from these data, it can be concluded that a third wave appears to be inevitable.

2.3 *Herd Immunity*

To further elaborate on the timing of the third wave, we have to look into the concept of herd immunity. According to John Hopkins [9], depending on the infection spreading rate, herd immunity can be achieved when roughly 50–90% of the population is

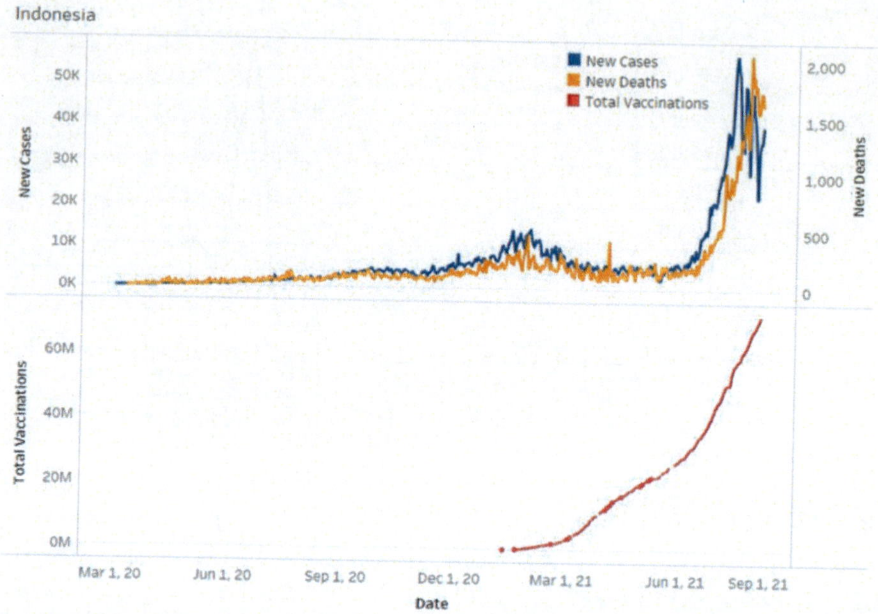

Fig. 5 Indonesia country-level COVID-19 vaccination [8]

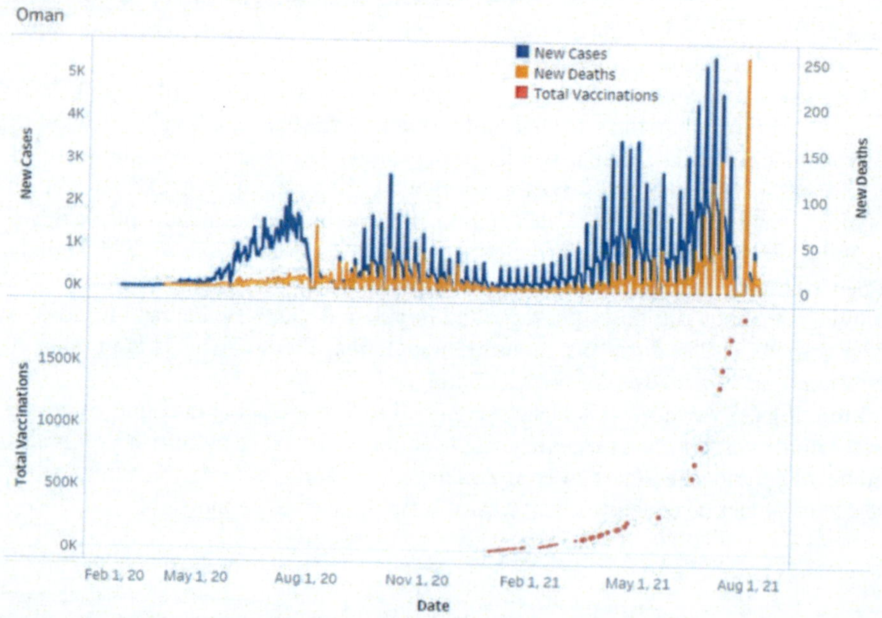

Fig. 6 Oman country-level COVID-19 vaccination [8]

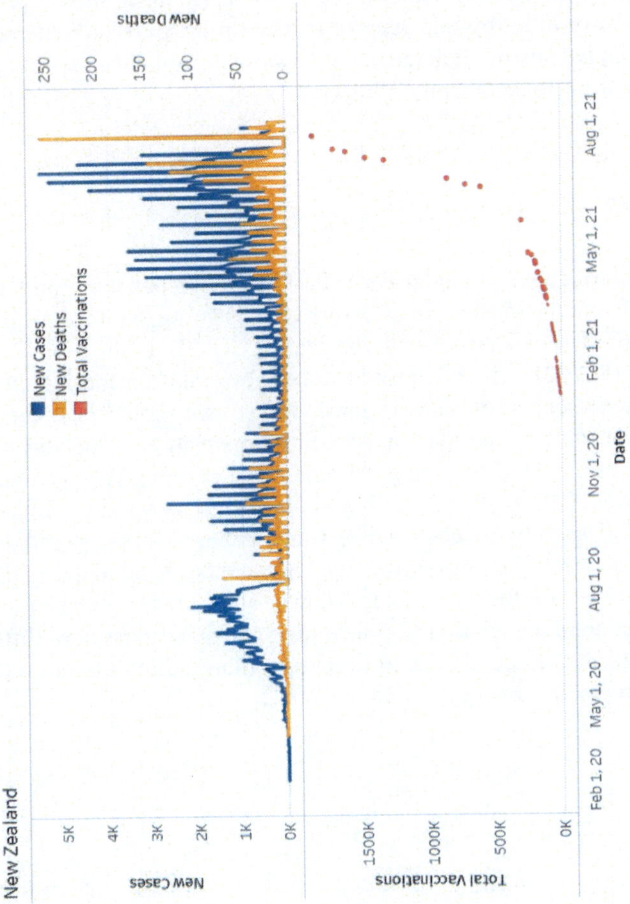

Fig. 7 New Zealand country-level COVID-19 vaccination [8]

vaccinated. Therefore, India is far below the minimum 50% vaccination rate and is susceptible to the third wave of coronavirus. In addition, there is no "magic threshold" that will make the coronavirus go away, especially when new variants (delta being highly virulent) are being detected. Hence, based on our analysis using peak between waves, vaccination rates, and herd immunity, we believe the third wave is likely in Kerala, India.

There have been several other researchers who worked on similar models to predict the third wave date. The IIT Kanpur research predicts that the wave will hit India from July 15 and attain a peak by mid-September [10]. SBI also made similar predictions stating the third wave to hit by mid-August and attain the peak in September [11]. Research by the University of Hyderabad is also similar stating the expected starting date for the third wave to be 4th July 2021 [12].

3 Conclusion

Kerala just went through the second wave of the COVID-19 pandemic and we found from our research that the third wave of coronavirus is going to hit Kerala and other states of India soon. The expected date of the peak of COVID-19 third wave in Kerala is on September 18, 2021, which is proven to be inevitable considering the current herd immunity and vaccination rates in India. Hence, we should plan accordingly to be safe. The government can plan on successful imposition of lockdowns where needed. With the estimated dates, the government can ensure safety by keeping the economy in account. The government should also ramp up vaccine production or procurement so that such future coronavirus waves can be avoided as India needs a very high vaccination rate for herd immunity. On the other hand, citizens of Kerala can prepare themselves for the expected lockdowns and be ready with resources that they cannot access amid the crisis. The time during COVID-19 has been difficult for all. We urge the readers to be careful in these upcoming difficult times and follow instructions laid by their government and the WHO.

References

1. Kannan D, Gurusriram R, Banerjee R, Bhattacharjee S, Varadwaj PK (2021) Will there be a third COVID-19 wave? https://doi.org/10.1101/2021.05.16.21257300
2. Bhardwaj R, Agrawal A (2021) Analysis of second wave of COVID-19 in different countries. Trans Indian Natl Acad Eng https://doi.org/10.1007/s41403-021-00248-5
3. Agrawal M, Kanitkar M, Vidyasagar M (2021) SUTRA: an approach to modelling pandemics with asymptomatic patients, and applications to COVID-19. arXiv preprint arXiv:2101.09158
4. Gola A, Arya RK, Animesh, Dugh R (2020) Review of forecasting models for coronavirus (COVID-19) pandemic in India during country-wise Lockdown. medRxiv preprint. https://doi.org/10.1101/2020.08.03.20167254

5. Gola A, Arya RK, Animesh, Dugh R, Khan Z (2020) Fine-tuned forecasting techniques for COVID-19 prediction in India. medRxiv preprint. https://doi.org/10.1101/2020.08.10.20167247
6. Covid third wave may see half the cases reported during 2nd surge: top expert https://www.ndtv.com/india-news/covid-19-third-wave-may-see-half-the-cases-reported-during-2nd-surge-government-panel-scientist-manindra-agarwal-2478707. Accessed 8 August 2021
7. Covid-19 vaccination tracker https://www.pharmaceutical-technology.com/covid-19-vaccination-tracker. Accessed 8 August 2021
8. Ritchie H, Ortiz-Ospina E, Beltekian D, Mathieu E, Hasell J, Macdonald B, Giattino C, Appel C, Rodés-Guirao L, Roser M (2020) Coronavirus pandemic (COVID-19). https://ourworldindata.org/coronavirus. Accessed on 5 August 2021
9. D'souza G, Dowdy D (2021) What is herd immunity and how can we achieve it with COVID-19?
10. https://www.jhsph.edu/covid-19/articles/achieving-herd-immunity-with-covid19.html Accessed 8 August 2021
11. Ranjan R, Verma M (2021) Along with their team at IIT Kanpur, COVID-19 outbreak in India. https://www.iitk.ac.in/new/covid-19-outbreak-in-india. Accessed 5 August 2021
12. DNA Web Team (2021) COVID-19 alert: third wave may hit India by mid-August, peak in September, says SBI report https://www.dnaindia.com/su/gRNt. Accessed 5 August 2021
13. Akbar S (2021) Third covid wave set in on July 4, says top Hyderabad physicist. http://timesofindia.indiatimes.com/articleshow/84329655.cms. Accessed 8 August 2021

Comparative Study on Sentiment Analysis of Human Speech using DNN and CNN

Sayak Ghosal, Saumya Roy, and Rituparna Basak

Abstract Communication is a way of exchanging thoughts through emotions. In this paper, we have proposed a method where human speech is converted into digital input. The digitized sound is then fed into the proposed models, and the voice of every person is classified into discrete emotional characteristics by its pitch, intensity, timbre, speech rate, and pauses. In the proposed method, we have drawn a comparative study between sentiment analysis of human speech using deep convolutional neural network (CNN) and dense deep neural networks (DNNs). In this method, multiscale area attention is applied in deep CNN as well as dense DNN to obtain emotional characteristics with wide range of granularities and therefore, the classifier can predict a wide range of emotions on a broad scale classification.

Keywords Sentiment analysis · Audio analysis · Deep learning · Neural networks · Emotion detection · Deep neural network (DNN) · Convolutional neural network (CNN)

1 Introduction

Speech is considered to be the most valuable and widely used means of communication. Speech emotion recognition (SER) has wide application-perspectives on psychological assessment, robotics, etc. For example, a doctor treating a patient suffering from depression can keep a track of his patient's development and design a recovery plan according to the patient's speech sentiments. Over the past few years, there has been a consequential development in the field of analyzing the emotions in speech with deep learning. However, deficiencies still exist in SER research due to lack of substantial model accuracy, deficit of useful dataset, and lack of computing resources. In SER, emotion may depict distinct energy patterns in spectrograms with varied granularity of areas. The change in different types of emotions brings changes in energy patterns in spectrograms. Typically, in SER attention, models are

S. Ghosal · S. Roy · R. Basak (✉)
University of Engineering and Management, Kolkata 700156, India
e-mail: rituparna.basak@uem.edu.in

J. K. Mandal et al. (eds.), *Topical Drifts in Intelligent Computing*, Lecture Notes in Networks and Systems 426, https://doi.org/10.1007/978-981-19-0745-6_10

commonly optimized on a fixed scale that may confine the model's capability to deal with diverse areas and granularities. An attention neural network [1] classifier of SER is usually optimized on a fixed attention granularity. In SER, distinct energy patterns are depicted in spectrograms which have varied granularity of areas.

In the proposed method, this constraint is removed by applying multiscale area attention in deep CNN [2] as well as dense DNN to obtain emotional characteristics with a wide range of granularities and therefore, the classifier can predict a wide range of emotions on a broad scale of classification. For example, sentiments such as annoyance and joy have different levels of intensity, so instead of just categorizing the presence or absence of the emotion, level of intensity of the emotion can be identified as well. Hence, a comparative study using both the models is conducted.

Models that have been used for SER previously suffer a problem of sample scarcity. A novel approach has been used to deal with data scarcity. Augmentation of data with addition of stretching, pitch modification, and noise insertion has been used for this reason. This adds variation to the dataset, and it also improves the chances of getting better accuracy. For example, a training data with all anger emotions expressed in higher pitch will now have the same outcome as with an emotion in lower pitch, hence maximizing the chances of identifying anger, when spoken in a lower timbre. Similarly, the sample set might consist of audio of fast speakers and be essentially biased. Stretching helps in removing this bias. Adding noise to the data is proven to be a useful tool for classifying real-time date. To the effectiveness of the proposed method, extensive experiments are carried out on RAVDESS [3], CREMA-D [4], TESS-D [5] dataset.

2 Literature Survey

In the paper [6], Fayek et al. propose a real-time speech emotion recognition system based on end-to-end (E2E) learning. From a one second frame of raw speech spectrograms, the technique of deep neural network is used to study the emotions. A deep hierarchical framework, pragmatic optimization, and data augmentation help in achieving the desired results. Promising results are reported.

In the paper [7], well organized procedure has been provided by author, for implementing SER political debates. The emphasis is laid on manufacturing the outcome and then to prepare visualization of the said results. Two alternative approaches have been considered, such as a classification-oriented approach and a lexicon-oriented approach. In the former universal and domain-oriented sentiment, lexicons are used. Two general methods for implementing domain-oriented lexicons-based approach have also been considered. These are (a) direct generation and (b) adaptation. Direct generation focuses on producing exclusive lexicons depending upon the data labels. Adaptation considers a common and inclusive lexicon-based approach and adjusting it as per necessity to develop it into a non-generic and exclusive symbol of a particular domain. The results obtained from the above discussed approaches were considered and compared with the "classification-based" approach. By observing and analyzing the attitude of the political speakers in the debates, the sentiment mining approaches were compared. Collective labeled speech data were considered, which were of political significance which was extracted from debating transcripts. The outcome of the comparison helped them realize that using sentiment mining, the speakers attitude can be determined conclusively. The proposed debate graph extraction (DGE) framework, in its functioning, effectively extracts the debate graphs from political debate transcripts. They proposed to graphically represent debates with speakers as nodes. In this framework, the speakers are represented as nodes, with nodes having specific labels and links between nodes. These links depend upon the exchange of speeches. The labels on the nodes depended upon the sentiment of the speakers. The attitude of the speaker was then used to classify a link as supporting or non-supporting. If the outcome of both speakers was same, i.e., both positive or both negative, then the link was categorized as supporting or else it was categorized as opposing. Visualization of results was carried out via graphs that represent the essence of the debate, in an abstract manner.

In the paper [8], Tripathi et al. proposed automatic sentiment detection system for natural audio streams. Part of speech tagging and maximum entropy modeling (ME) has been used as the suggested technique to develop a sentiment detection model that was text-based in nature. The number of model boundaries in ME was reduced drastically by an attuning technique while conserving the classification capability. Using decoded automatic speech recognition (ASR) transcripts and the ME sentiment model, sentiments of YouTube videos were able to be determined. As evaluation, they have gathered motivating classification accuracy. According to the results, analysis showed that performance on sentiment analysis on spontaneous speech data is possible in spite of word error rates.

3 Problem Statement

The human speech is the most innate way of expressing oneself. We know emotions play an important role in communication analysis, and the detection of the same is significantly important in today's digital world of remote communication. In text-based classification certain emotions like sarcasm, dual meaning sentences cannot be identified. Tonal qualities of the voice are required to classify the emotions more accurately. An SER system can thus be defined as a collection of methodologies that classifies speech signals to detect embedded emotions. The human speech contains many features different to each individual. If we consider all those features while training the model, then the model will be biased to the training set which is not desired. So, we have considered only the properties common to human voices like loudness, timbre, and quality. Our attempt lies in trying to detect underlying emotions embedded in speech through analysis of the acoustic features of the audio recording.

4 Dataset

Our dataset focusses on RAVDESS, CREMA-D, TESS-D dataset. These three instances of audio datasets used during our analysis and contain the vocal emotional expressions of sentences spoken in a range of varied emotional indexes (joy, grief, rage, agitation, annoyance, and calm). Total of 1440 and 7,440 clips of 115 actors were collected with diverse ethnic background, and it was merged into 8882 files. We have worked only with the audio recordings of the audiovisual data. The sentences are spoken by trained actors belonging to a variety of races and ethnicities (Latino Americans, African, American, Asian, Caucasian). The sentences are classified using one of six different emotions (joy, grief, rage, agitation, annoyance, and calm) and four different emotion levels (low, medium, high, and unspecified). The audio file format used is WAV. To obtain variance in the data, noise is introduced which is stretched and then inserted. Stretching is important for data that have audio cues which are short in nature. To analyze in real-time, noise in an essential component that should not add bias to the results. Therefore, noise is introduced into the training sample.

5 Proposed Solution

Three classes of features can be mainly identified in a speech. These can be classified as lexical features, the visual features, and the acoustic features. For example: the various expressions of the speaker, the terminology used, and properties like vocal quality, pitch, anxiety, noise, energy, etc. (Fig. 1).

Analysis of lingual features requires a script of the speech. However, it will require a processing and extraction of text from speech in order to analyze sentiments from real-time audio. During analyzation of visual features, it requires the access to the video of conversations, and it is not in the scope of this research. Therefore, analysis of the auditory features is done in this work, since analysis of the acoustic features is

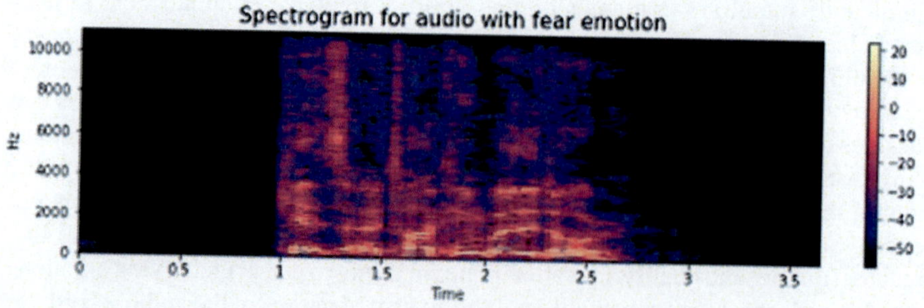

Fig. 1 Spectrogram plotted for audio with fear emotions

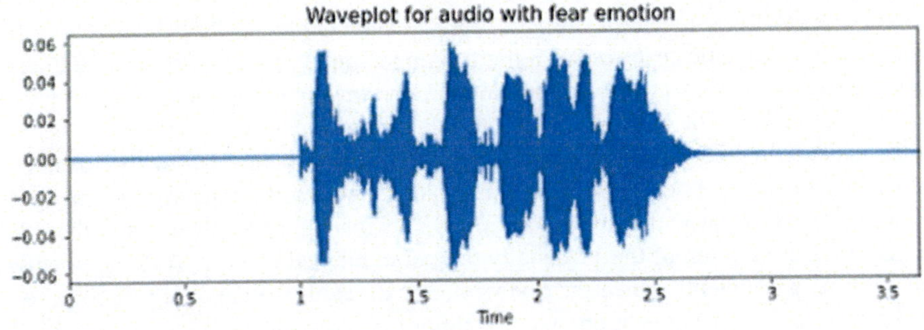

Fig. 2 Various categories of sentiments that our model predicts

possible in real-time. Audio from real-time conversations is extracted and analyzed in order to accomplice the task (Fig. 2).

The representation of emotions can be done in two ways:

- Discrete Classification: Emotions were classified into distinct labels like rage, calmness, neutral, cheerfulness, and joy.
- Dimensional Representation: Representations of emotions with dimensional categories such as activation energy (on a low to high scale), valence (on a negative to positive scale), or/and dominance (on an active to passive scale).

The two mentioned approaches each have their distinct advantages and disadvantages. The dimensional representation approach is an elaborative process, but there is a deficiency of annotated audio data in the dimensional format. The discrete classification is more straightforward and less resource hungry to implement.

In discrete classification approach, emotions are classified on a specified scale for the analysis. Emotions are classified using the trained model and predicted as discrete outputs. This approach is easier to implement and understand and as such has greater outreach, and thus we have focused on this approach.

Dense DNN Architecture:

The different acoustic features from the training sample set are extracted. Features like MFCC, zero crossing rate, chroma STFT, and mel spectrogram have been used to train our model. The MFCCs stand for mel frequency cepstral coefficients. These form cepstral representation for non-linear frequency bands that are distributed according to the mel scale. Zero crossing rate identifies the rate of sign changes of a signal for a particular frame duration. The feature chroma portrays the 12-element representation of the 12 equally tempered pitched classes. These are called the chroma coefficients.

Once the features are extracted, the data are preprocessed and prepared, by noise injection and stretching for audio augmentation. Thereafter, they are fed into the DNN model architecture, for analysis. Our inputs are represented by 128 dimensions, with a dropout of 0.2. We have primarily utilized ReLu activation function to obtain better results, with a final layer having Softmax activation. We have implemented 5 layers.

2D CNN Architecture:

The different acoustic features from the training sample set are extracted. Features like MFCC, zero crossing rate, chroma STFT, and mel spectrogram have been used to train our model. The MFCCs stand for mel frequency cepstral coefficients. These form cepstral representation for non-linear frequency bands that are distributed according to the mel scale. Zero crossing rate identifies the rate of sign changes of a signal for a particular frame duration. The feature chroma portrays the 12-element representation of the 12 equally tempered pitched classes. These are called the chroma coefficients. Once the features are extracted, the data are preprocessed and prepared, by noise injection and stretching for audio augmentation. Thereafter, they are fed into our 2D-CNN followed by a dense DNN model architecture, for analysis. Our inputs are represented by 256 dimensions, with a dropout of 0.3. We have primarily utilized ReLu activation function to obtain better results. We have implemented 4 2D-CNN layers and 2 dense layers.

In Fig. 3, we have taken an audiovisual file as an input for testing. The audio file consists of the voice of an actor who portrays the emotion of anger. The algorithm predicted the emoticon for the emotion portrayed correctly.

The proposed comparative study consists of two separate classification models. The first model is constructed primarily on 2D-CNN, connected to a dense DNN. The second approach that is considered is using only dense DNN. After careful observations of the outputs, the realizations of both the models are portrayed below.

Figure 4 shown above plots the output of the training and testing for 2D-CNN model. In the X-axis, the number of epochs is plotted. In Fig. 4a, training and testing

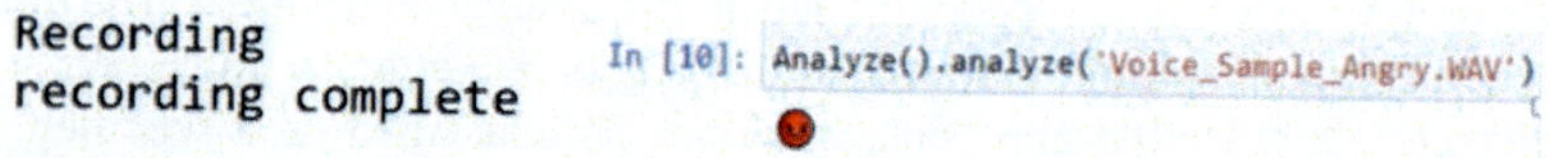

Fig. 3 Analysis of audiovisual data

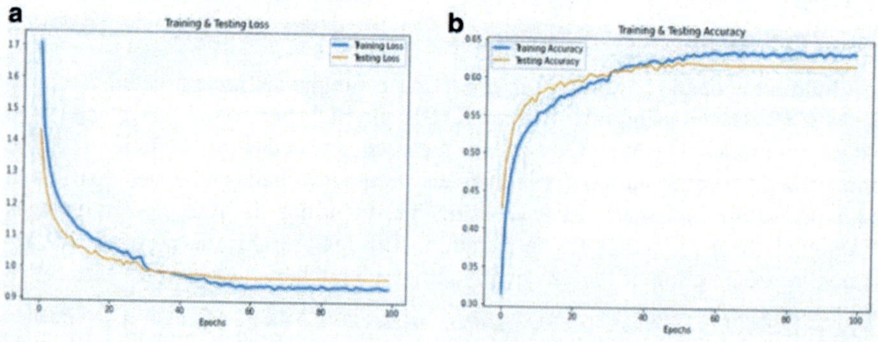

Fig. 4 **a** Training and testing loss 2D-CNN model, **b** training and testing accuracy of 2D-CNN model

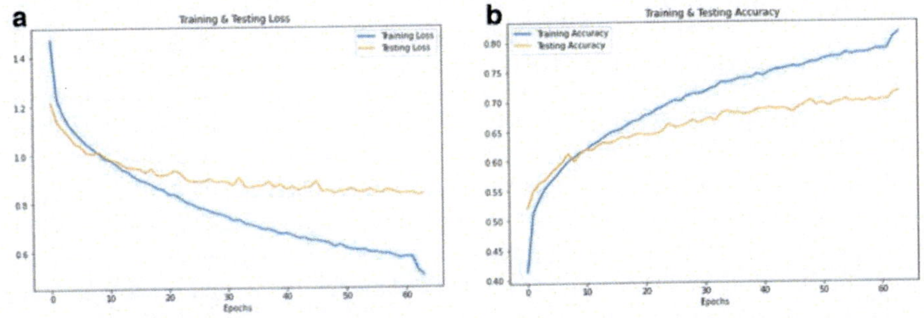

Fig. 5 **a** Training and testing loss dense DNN model, **b** training and testing accuracy of dense DNN model

loss of 2D-CNN model is shown and Fig. 4b training and testing accuracy of 2D-CNN model are shown. On analysis of the graphs, each epoch in both the graphs tend to merge at a point which is desired. After the point of saturation, there is no significant change in the difference between training and testing loss of the model and also training and testing accuracy of the model which shows the strong integrity of the proposed model.

Figure 5 shown above plots the output of the training and testing for the dense DNN model. In the X-axis, number of epochs is plotted.

In Fig. 5a, training and testing loss of dense DNN model are shown and Fig. 5b training and testing accuracy of dense DNN model are shown. On analysis of the graphs, with each epoch in both the graphs, the lines tend to merge at a point. After merging, the lines diverge, which indicates significant variance between the training and the testing loss. With higher number of epochs, the loss decreases since the model refuses to achieve saturation. In case of the accuracy plot, with higher number of epochs, the accuracy increases since the model refuses to achieve saturation.

Experimental Analysis:

The results obtained that of dense DNN model supersedes that of its counterpart, i.e., the model comprising of 2D-CNN connected to dense DNN. The accuracy obtained for the dense DNN model is 71.52% whereas that of its counterpart is 61.89%. The loss incurred in the dense DNN is around 0.83 whereas that of the 2D-CNN model is around 1.0. However, the dense DNN model shows a tendency to overfit on higher epochs, unlike the 2D-CNN model. The advantage of the latter is that it considers lower number of features in each step and operates with a lesser number of coefficients that prevent overfitting.

6 Conclusion

The achieved accuracy in case of dense DNN is of 71.52 and 61.89% in case of 2D-CNN. Though the accuracy of the dense DNN is higher, the model tends to overfit the given sample data. Thus, from the above study, it can be concluded that both the models provide different perspective to the problem of SER and their use is completely dependent on the type of utilization required. The proposed model is able to provide some noteworthy results that can have myriad of applications. Efficient utilization of the audio signals and their tone, pitch, and granularity can help in detection of lies, mimicry, as well as mental state of a person. Furthermore, for exploring broader spectrum like analysis and interpretation, such as analysis of interviews and interrogations, multimodal sentiment analysis should be taken into considerations.

References

1. Yin W, Schütze H, Xiang B, Zhou B (2016) ABCNN: attention-based convolutional neural network for modeling sentence Pairs. In: Transactions of the association for computational linguistics, vol 4. pp 259–272. https://doi.org/10.1162/tacl_a_00097
2. Goecke R, Potamianos G, Neti C (2002) Noisy audio feature enhancement using audio-visual speech data. In: 2002 IEEE international conference on acoustics, speech, and signal processing. pp II-2025–II-2028. https://doi.org/10.1109/ICASSP.2002.5745030
3. Kurpukdee N, Kasuriya S, Chunwijitra V, Wutiwiwatchai C, Lamsrichan P (2017) A study of support vector machines for emotional speech recognition. In: 2017 8th international conference of information and communication technology for embedded systems (IC-ICTES). pp 1–6. https://doi.org/10.1109/ICTEmSys.2017.7958773
4. Abburi H (June 2017) Audio and text based multimodal sentiment analysis using features extracted from selective regions and deep neural networks. International institute of information technology Hyderabad–500032, India
5. Iqbal M. MFCC and machine learning based speech emotion recognition over TESS and IEMOCAP datasets. https://doi.org/10.33897/fujeas.v1i2.32
6. Salah Z (May 2014) Machine learning and sentiment analysis approaches for the analysis of parliamentary debates. ISNI:0000-0004-5365-4219
7. Tripathi S, Kumar A, Ramesh A, Singh C, Yenigalla P. Deep learning based emotion recignition system using peech features and transcription. arXiv:1906.05681 [eess.AS]
8. Lugović S, Dunđer I, Horvat M (30 May–3 June, 2016) Techniques and applications of emotion recognition in speech. In: MIPRO. Opatija, Croat
9. Xu M, Zhang F, Zhang W (2021) Head Fusion: improving the accuracy and robustness of speech emotion recognition on the IEMOCAP and RAVDESS dataset. In: IEEE Access, vol 9. pp 74539–74549, 202. https://doi.org/10.1109/ACCESS.2021.3067460
10. Tai O, Liu Y, Huang J, Yu X, Aljbawi B (1 May 2021) Neural attention frameworks for explainable recommendation. IEEE Trans Knowl Data Eng 33(5):2137–2150. https://doi.org/10.1109/TKDE.2019.2953157
11. Cao H, Cooper DG, Keutmann MK, Gur RC, Nenkova A, Verma R (1 Oct–Dec 2014) CREMA-D: crowd-sourced emotional multimodal actors dataset. IEEE Trans Affect Comput 5(4):377–390. https://doi.org/10.1109/TAFFC.2014.2336244
12. Zheng F, Zhang G, Song Z. Comparison of different implementations of MFCC. J Comput Sci Technol 16:582–589. https://doi.org/10.1007/BF02943243

13. Hermansky H, Cox LA (1991) Perceptual linear predictive (Plp) analysis-rsynthesis technique. In: Final program and paper summaries 1991 IEEE ASSP workshop on applications of signal processing to audio and acoustics. pp. 0_37–0_38. https://doi.org/10.1109/ASPAA.1991.634094
14. Cheong S, Oh SH, Lee S-Y (March 2004) Support vector machines with binary tree architecture for multi-class classification. Neural Inf Proc—Lett Rev 2(3)
15. Jing C, Sha J. An efficient implementation of 2D convolution in CNN. In: IEICE electronics express, vol 14. pp. 20161134–20161134. https://doi.org/10.1587/elex.13.20161134
16. Vetterli M, Herley C (Sept 1992) Wavelets and filter banks: theory and design. IEEE Trans Signal Process 40(9):2207–2232. https://doi.org/10.1109/78.157221
17. Fayek HM, Lech M, Cavedon L (2015) Towards real-time speech emotion recognition using deep neural networks 2015. In: 9th international conference on signal processing and communication systems (ICSPCS). pp 1–5. https://doi.org/10.1109/ICSPCS.2015.7391796

RNN-Based Deep Learning Model for Generating Caption of an Image

Md. Ezaz Ahmed and Hitesh Kumar Sharma

Abstract Image captioning refers to a task wherein an image is seen by a computer to give an output as "what it sees." Image captioning would mean giving an intelligent system an image, say of a dog running across a field to receive a statement back as "dog running across a field" or something similar down its lines. The image caption generator is a deep learning model built using the power of recurrent neural networks (RNNs) which are powered by the long-term and short-term (LTSM) units. The model also uses transfer learning to enable build a better and faster deep learning model. The model is built using the "Flicker8k" dataset and uses the inherent power of a convolutional neural network to predict suitable caption for an image. The seamless integration of computer vision into the model and its high-quality prediction makes it highly usable and in-demand with respect to the current trends in the industry.

Keywords Deep learning · Image caption · Image processing · RNN · LSTM · Machine learning · Artificial intelligence

1 Introduction

The mechanism whereby an intelligent system automatically allots some metadata or associated information in the form of captions, keywords, or vocabulary words to an image is known as image captioning or automated image annotating [1, 2]. This method may be understood as a kind of multiple-class image classification with a humongous number of features—as big as the dictionary size (the vocabulary being used). Generally, image classification and perusal in the form of extricated feature and attribute vectors and the training captioning words are used by machine learning enthusiasts to attempt to automatically apply captions and annotations to

Md. Ezaz Ahmed
CS Department, Saudi Electronic University, Al Madina KSA, Saudi Arabia
e-mail: m.ezaz@seu.edu.sa

H. K. Sharma (✉)
School of Computer Science, University of Petroleum and Energy Studies, Energy Acres, Bidholi, Dehradun, India

J. K. Mandal et al. (eds.), *Topical Drifts in Intelligent Computing*, Lecture Notes in Networks and Systems 426, https://doi.org/10.1007/978-981-19-0745-6_11

new digital image. The first methods learned the interdependence among features and attributes of a digital and training captions, then mechanisms were developed using machine embedded translation to convert the textual vocabulary into the "visual vocabulary." Works following the ventures stated above have incorporated classification techniques, relevance models and so on and so forth. Image captioning, which elucidates the contents observed in a digital image, has received noteworthy amounts of attention in the recent years. Image annotating finds its application in a multitude of scenarios, such as recommendation systems in image-editing software, usage in virtual and voice/media assistants, for image-based indexing, and support of the differently-abled. With the ease of availability of large amounts of data for building the model, neural network-based techniques and models have demonstrated monumental results on image annotating works. The image captioning methods are largely based on recurrent neural networks (RNNs) [3–5], which are generally built over long short-term-memory (LSTM) unit. LSTM [6–9] networks have been considered as the conventional approach for computer vision and vision-vocabulary tasks of image annotating, visual question answering, query generation, and visual dialogs generator. This is because of their superb capability to memorize long-term interdependencies through a memory cell. However, the complex and relatively obscure addressing and overwriting techniques combined with inherently sequential vulcanization of the data in question, and large amounts of storage required due to backpropagation through time (BPTT) [10, 11] creates difficulties during model training phase. Also, in contrast to convolutional neural networks, which are non-sequential models, LSTM units normally demands more meticulous designing. Earlier, LTSM performed better than CNNs on tasks related to computer vision or language-based tasks. Taking inspiration from the successes of convolutional neural network on image generation, image captioning, and other image-based tasks, many models are being trained using CNNs. Building an image annotating model from scratch will require training for a humongous amount of time; hence, we use the concept of transfer learning to build the model. The ResNet50 model is used as a fundamental piece of the entire model. ResNet, which stands for residual network, is a standard neural network used for extensive computer vision works [12–14]. Using the ResNet50 model, the image gets significant annotations and forms a complete caption faster and in a more robust manner using the power of transfer learning. The image captioning looks like image Fig. 1.

2 Methodology

The architecture of the deep learning model is shown in the diagram below Fig. 2. An image is fed to a convolutional neural network, which then feeds and runs a RNN powered by the backbone of LTSM components [15, 16]. A string vector is also given to the network in order to append the most suitable words and form meaningful caption.

Fig. 1 Image captioning example

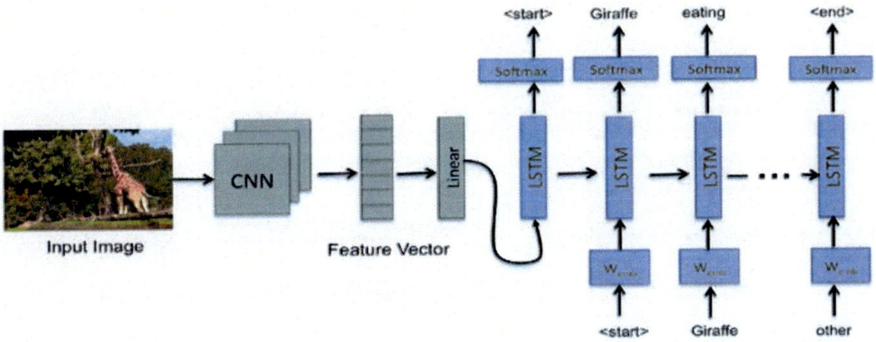

Fig. 2 LSTM network model

The CNN or convolutional neural network may be thought to act like an encoder system. The image given as input to the neural network is used to find hidden attributes, features, and patterns. The last hidden layer of the network is attached to a decoder system [17]. The decoder is a RNN or recurrent neural network [18] which performs the sequence formation and augmentation. Language modeling occurs in the decoder component of the model. The encoder sends the encoded output to the decoder and a vector with the {start} label appended to it [19]. Even if the decoder makes a mistake, it is given a correct input each time, during the training phase (Fig. 3).

3 Model

The model was built in Python using TensorFlow and Keras API. The architecture of the model is shown in the figure given Figure 4.

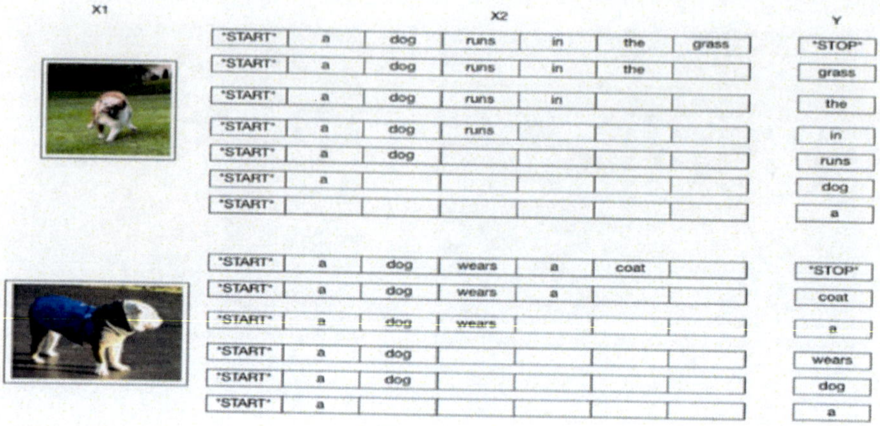

Fig. 3 LSTM workflow

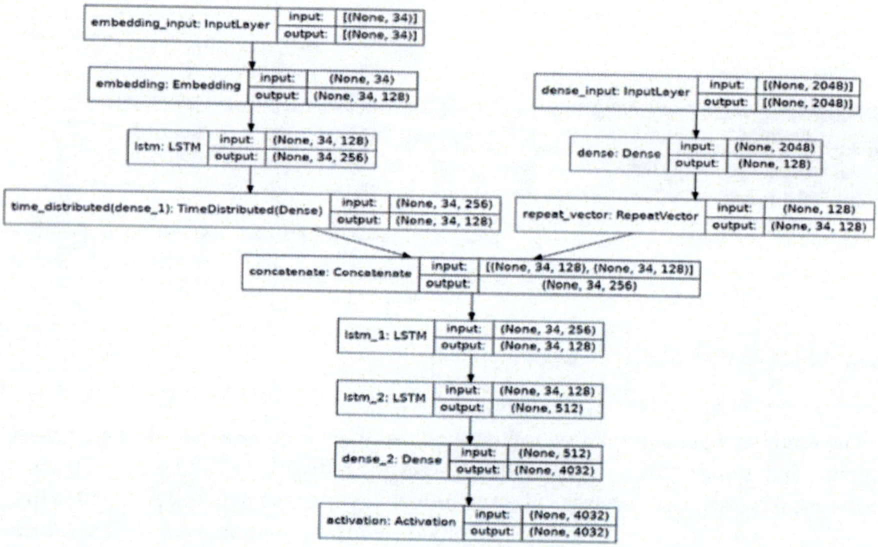

Fig. 4 LSTM model architecture

4 Model Summery

Then, summary of the model is listed below as seen:

Model (proposed): "Seq_model_1."

Model layer	O/P shape	Parameter#
Dense (Dense)	(None, 128)	262,272

(continued)

(continued)

Model layer	O/P shape	Parameter#
Repeat_vector (RepeatVector)	(None, 35, 128)	0

Total params: 262,272
Trainable params: 262,272
Non-trainable params: 0
Model: "Sequential_model_2"

Layer (type)	Output shape	Param#
Embedding (Embedding)	(None, 35, 128)	511,488
LSTM (LSTM)	(None, 35, 256)	394,240
Time_distributed (TimeDistri)	(None, 35, 128)	32,896

Total params: 938,624
Trainable params: 938,624
Parameters (non-trainable): 0
Model (proposed): "Model_1"

Layer (type)	Output shape	Param#
Embedding_input (InputLayer)	[(None, 35)]	0
Dense_input (InputLayer)	[(None, 2048)]	0
Embedding (Embedding)	(None, 35, 128)	511,488
Dense (Dense)	(None, 128)	262,272
LSTM (LSTM)	(None, 35, 256)	394,240
Concatenate (Concatenate)	(None, 35, 256)	0
lstm_1 (LSTM)	(None, 35, 128)	197,120
lstm_2 (LSTM)	(None, 512)	1,312,768
Dense_2 (Dense)	(None, 3996)	2,049,948
Activation (Activation)	(None, 3996)	0

Total no. of parameters: 4,760,732
Parameters (Trainable): 4,760,732
Parameters (non-trainable): 0

5 Results

The results of proposed work are shown in Fig. 5. It represents the confusion matrix of the trained model. In this matrix, rows represent the predicted values and columns represents the actual values. The heat map shows the correctness or correlation

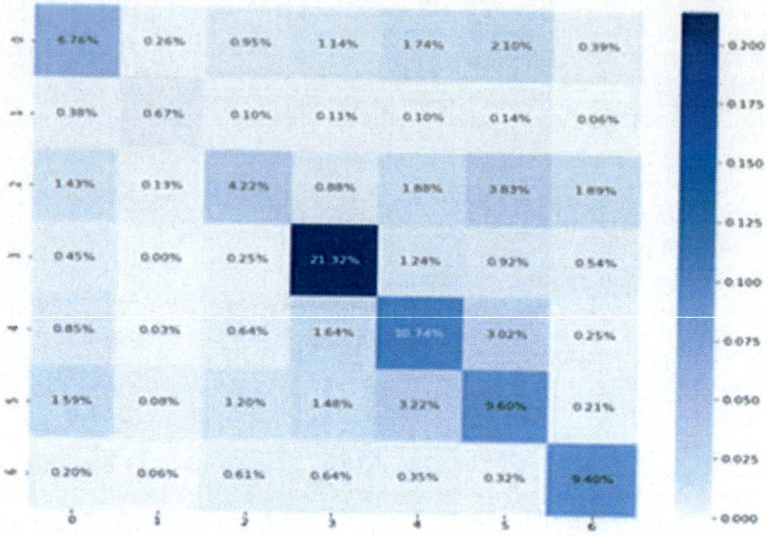

Fig. 5 Heat map of confusion matrix

between actual and predicted values. Confusion matrix is also used for calculation of accuracy, precision, recall, and f-score of proposed mode. As per the following confusion matrix, the model accuracy is approximately 86%. The confusion matrix is captured after passing test data to the model after training the same on given dataset.

The accuracy and loss and error have been shown below in Figs. 6 and 7.

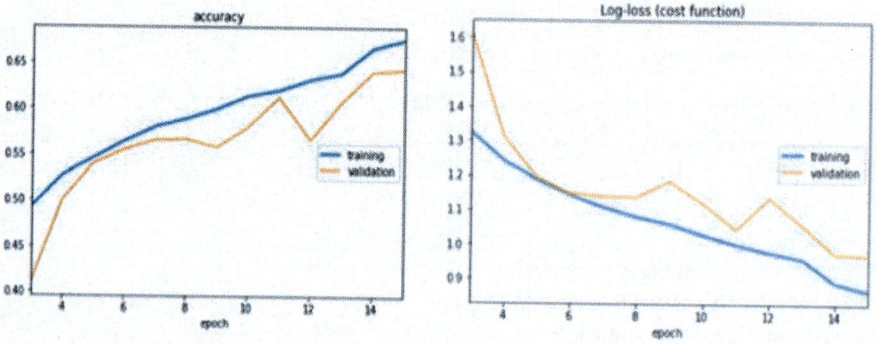

Fig. 6 Accuracy graph and loss graph

Fig. 7 Accurate image captioning examples

5.1 Accuracy

The accuracy of the model was determined using the following table for the epochs = 50 on the ficker8k dataset.

The model reached a loss of 0.8591 and an accuracy of 0.7656 at the end of the 50th run on the training data.

Epoch 1/50 180/180 [==] –454 s 2 s/step – loss: 5.5054 – accuracy: 0.1166
Epoch 2/50 180/180 [==] –453 s 3 s/step – loss: 4.8567 – accuracy: 0.1842
Epoch 3/50 180/180 [==] –442 s 2 s/step – loss: 4.4775 – accuracy: 0.2324
Epoch 4/50 180/180 [==] –444 s 2 s/step – loss: 4.2713 – accuracy: 0.2535
Epoch 5/50 180/180 [==] –444 s 2 s/step – loss: 4.1554 – accuracy: 0.2657
Epoch 6/50 180/180 [==] –443 s 2 s/step – loss: 4.0200 – accuracy: 0.2773
Epoch 7/50 180/180 [==] –443 s 2 s/step – loss: 3.8606 – accuracy: 0.2906
Epoch 8/50 180/180 [==] –454 s 3 s/step – loss: 3.6609 – accuracy: 0.3191
Epoch 9/50 180/180 [==] –448 s 2 s/step – loss: 3.5017 – accuracy: 0.3333
Epoch 10/50 180/180 [==] –445 s 2 s/step – loss: 3.3646 – accuracy: 0.3493
<Continued for 50 runs>
Epoch 40/50 180/180 [==] –455 s 3 s/step – loss: 1.2460 – accuracy: 0.6807
Epoch 41/50 180/180 [==] – 456 s 3 s/step – loss: 1.1988 – accuracy: 0.6927
Epoch 42/50 180/180 [==] – 456 s 3 s/step – loss: 1.1574 – accuracy: 0.7004
Epoch 43/50 180/180 [==] – 464 s 3 s/step – loss: 1.1256 – accuracy: 0.7078
Epoch 44/50 180/180 [==] – 462 s 3 s/step – loss: 1.0749 – accuracy: 0.7204
Epoch 45/50 180/180 [==] – 462 s 3 s/step – loss: 1.0418 – accuracy: 0.7285
Epoch 46/50 180/180 [==] – 459 s 3 s/step – loss: 1.0094 – accuracy: 0.7363
Epoch 47/50 180/180 [==] – 458 s 3 s/step – loss: 0.9731 – accuracy: 0.7449
Epoch 48/50 180/180 [==] – 461 s 3 s/step – loss: 0.9565 – accuracy: 0.7498
Epoch 49/50 180/180 [==] – 471 s 3 s/step – loss: 0.9183 – accuracy: 0.7599
Epoch 50/50 180/180 [==] – 460 s 3 s/step – loss: 0.8951 –
Accuracy: 0.7657

6 Conclusion

A RNN-based deep learning model powered by micro units of LSTM can be used for automated image captioning and annotating. This model has many advantages and can be put to use for extensive purposes. This kind of work will be useful in writing story about a scenery. This can be used to write subtitle of a video. This work can be extended with more deep learning-based model with more layers.

References

1. Anderson P, Fernando B, Johnson M, Gould S (2016) Spice: semantic propositional image caption evaluation. In ECCV
2. Chen X, Zitnick CL. Mind's eye: a recurrent visual representation for image caption generation. In: 2015 IEEE conference on computer vision and pattern recognition (CVPR)
3. Herdade S, Kappeler A, Boakye K, Soares J. Image captioning: transforming objects into words
4. Aneja J, Deshpande A, Schwing AG. Convolutional image captioning
5. Jain U, Zhang Z, Schwing AG. Creativity: generating diverse questions using variational autoencoders. In: Computer vision and pattern recognition
6. Das A, Kottur S, Gupta K, Singh A, Yadav D, Moura JMF, Parikh D, Batra D. Visual dialog
7. Mao J, Xu W, Yang Y, Wang J, Yuille AL. Deep captioning with multimodal recurrent neural networks
8. Mantecón T, del Blanco CR, Jaureguizar F, García N (Oct 2016) Hand gesture recognition using infrared imagery provided by leap motion controller. International conference on advanced concepts for intelligent vision systems, ACIVS 2016. Lecce, Italy, pp. 47–57, 24–27. https://doi.org/10.1007/978-3-319-48680-2_5
9. Patni JC, Sharma HK (2019) Air quality prediction using artificial neural networks. International conference on automation computational and technology management (ICACTM)
10. Khanchi I, et al (2019) Automated framework for real-time sentiment analysis. International conference on next generation computing technologies (NGCT)
11. Kshitiz K et al (2017) Detecting hate speech and insults on social commentary using nlp and machine learning. Int J Eng Technol Sci Res 4(12):279–285
12. Pardeshi V, Sagar S, Murmurwar S, Hage P (2017) Health monitoring systems using IoT and Raspberry Pi—a review. 2017 International conference on innovative mechanisms for industry applications (ICIMIA). Bangalore, pp 134–137
13. Navya K, Dr. Murthy MBR (Sep–Oct 2013) A zigbee based patient health monitoring system. Int J Eng Res Appl 3(5):483–486
14. Mathan Kumar K, Venkatesan RS (2014) A design approach to smart health monitoring using android mobile devices. IEEE international conference on advanced communication control and computing technologies (ICACCCT). pp 1740–1744
15. Alani AA, et al (2018) Hand gesture recognition using an adapted convolutional neural network with data augmentation. 24th IEEE international conference on information management
16. Mukhopadhyay SC (2015) Wearable sensors for human activity monitoring: a review. IEEE Sens J 15(3):1321–1330
17. Djuknic GM, Richton RE (2001) Geolocation and assisted GPS. Computer 34(2):123–125
18. Ahlawat P et al (2020) Sensors based smart healthcare framework using internet of things (IoT). Int J Sci Technol Res 9(2):1228–1234
19. Taneja S, et al (2019) I-doctor: an IoT based self patient's health monitoring system. 2019 international conference on innovative sustainable computational technologies, CISCT 2019

Object Detection in Railway Track using Deep Learning Techniques

R. S. Rampriya, R. Suganya, Sabarinathan, Aarthi Ganesan, P. Prathiksha, and B. Rakini

Abstract Nowadays, deep neural networks are one of the ongoing trends which are having their uses in various kinds of fields. One of the most important applications of neural networks is the object detection framework. Object detection in railways field is the novel one, which includes the detection of obstacles on the railway tracks. But some researchers deployed this model using you only look once (YOLO) and single shot detector (SSD). We found these models to produce a lesser accuracy when compared to that of faster R-CNN. The previously mentioned models consume some time for training. To overcome the existing drawbacks, the upgraded model is being deployed by using faster R-CNN, which comprises two modules, namely regional proposal network (RPN) and fast R-CNN. It helps by detecting the obstacles such as branches, boulders, iron rods, animals, vehicles, and people. So, with the help of both the models (faster R-CNN and YOLO), we can detect the obstacles present on the track. Finally, with the quantitative and qualitative comparisons made on these two models, we chose the best fit model for this purpose. Hence, this provides the novel idea to prevent railway accidents as much as possible.

Keywords Object detection · Deep neural network · Faster R-CNN · YOLO

1 Introduction

Deep learning is the successor of machine learning, which is an AI function that mimics the workings of the human brain in processing data that is used in various fields like object detection, speech recognition, language translation, and decision. The main idea behind this technology is the network of structures. It builds more

R. S. Rampriya (✉) · A. Ganesan · P. Prathiksha · B. Rakini
Anna University (MIT Campus), Chennai, India

R. Suganya
Thiagarajar College of Engineering, Madurai, India
e-mail: rsuganya@tce.edu

Sabarinathan
Cougher Inc, Tokyo, Japan

J. K. Mandal et al. (eds.), *Topical Drifts in Intelligent Computing*, Lecture Notes in Networks and Systems 426, https://doi.org/10.1007/978-981-19-0745-6_12

networks to train the model with unstructured or unlabeled data. They also called these deep neural networks. This kind of framework requires an enormous amount of data to fetch greater accuracy. Hence, inputs were given as huge datasets to the model. One of the most important features of the deep neural networks is to process large numbers of features so that it becomes powerful when dealing with unstructured data. Some of the frequently used algorithms in deep learning are convolutional neural network, long short-term memory networks, stacked auto-encoders, etc. One such application of this framework is the object detection technique.

Object detection plays a significant role in many real-life instances. Its role in railways is quite important for overcoming the challenges that cause many railway accidents. Accidents because of the obstacles on the tracks have become more common nowadays, particularly in rural areas. It also has a tremendous impact on wildlife as most of the accidents happen because of animals crossing the track. All these should be monitored and necessary actions should be taken. For this, we implement object detection on railways that detects the obstacles on tracks such as branches, animals, and people.

The primary purpose of this paper is to find the best fit object detection model to detect the obstacles present on the railway tracks. Here, we consider the two models YOLO and faster R-CNN. With YOLO, it uses a single neural network to process the full image and divides the image into regions and predicts bounding boxes and probabilities of each region whereas faster R-CNN uses region proposal networks to predict the region in which the object is present. This helps to minimize the number of railway accidents or collisions caused by the trains because of the lack of signals.

2 Related Works

They gave a brief introduction to deep learning and CNN in [1]. They have discussed various kinds of object detection, namely generic object detection, salient object detection, face detection, pedestrian detection. But it mainly focuses on typical generic object detection architectures. CNN architectures comprise feature maps and transformations—filtering and pooling. By comparing all these models on various datasets, the efficiency of the models was studied and found the best model for the object detection purpose. This paper has explained the best model with the pedestrian detection application. To do this, the complete process of pedestrian detection starting from dataset creation to computing the evaluation metrics for the results obtained.

In the work described in [2], they compared models SSD and faster R-CNN for object detection. A region proposal network that shares full image convolutional features with the detection network has been introduced in [3]. It focuses on RPN, which is the most important technique of faster R-CNN. This RPN tells the faster R-CNN, where to look at the image to detect the correct one. They did experiments on the MS CoCo and Pascal VOC dataset. Here, they have used 80,000 samples for training and 40,000 for validation. In [4], three major improvements on faster R-CNN algorithms are made, namely feature pyramid structure, region of interest align, usage of soft NMS algorithm (non-maximum suppression)—it sorts all detection boxes based on their detection score, and the one with maximum score is selected while the others are suppressed.

Different neural networks have been shown [5] to achieve classification using the faster R-CNN. For object detection, it exploded the speed of detection as it integrates the process of feature extraction, proposal extraction, and rectification. Experimental results show that its effectiveness comes from the convolutional layers and RPN modules. YOLOv2 model and YOLO9000 used for real-time detection systems for detecting and classifying objects in video records have been used here [6]. They have used GPU to increase speed and processes at 40 frames per second. The computation, processing speed, and efficiency in identifying the objects in the video record have been improved.

The first dedicated dataset for aerial survey of railways has been created by collecting images from Google and frames from YouTube videos, used as dataset for training the CNN to detect the obstacles. This paper [7] uses two versions of the faster R-CNN, i.e., faster R-CNN inception V2 model and faster R-CNN ResNet inception V2 model. Among these models, they took the most efficient one for the application.

3 Proposed Work

There are many sectors in our society like agriculture, transport, pharmaceutical, and many more. Agriculture introduced some techniques like fertility detection and many methods for the welfare of this sector. With pharmaceuticals, market fix modeling is the method introduced using machine learning models to promote the medicines in the markets. Therefore, machine learning and deep learning have invented new technologies for the benefit of humankind and the field. They transported one topic that is left idle because this is the field where these technologies are missing. Among transports, one such area that has to be noted is the railways. This is the sector that is lagging in terms of its technologies. They require technology for monitoring various activities, like monitoring the driver, signaling the driver, proper railway crossings, and many more. One such activity that is very essential is the proper vigilance of railway tracks, because this activity may cost people's lives because of a lack of monitoring of railway tracks. Hence, this project helps the driver to be aware of the objects in the tracks in advance by designing a deep learning model for the early detection of objects on the tracks.

4 Implementation

Implementing the project is carried out on Google Collab with GPU, where both the models are trained and tested on the custom dataset. The dataset creation involves collecting images and annotating them using the "labeling" tool. The proposed work comprises three modules such as data preprocessing, prediction and classification, and comparative study of faster R-CNN and YOLO. A brief explanation of each module is as follows:

4.1 Dataset Description

The dataset that is used is the customized one, where the images were collected from the Web, based on the classes chosen for detection. The custom dataset that is created comprises various classes of images categorized under the labels, namely animal, branch, boulder, iron rod, vehicle, and person. 1050 samples contribute to the training and testing sets for the object detection process. Out of which, the training set comprises 880 samples, and the testing set contains 170 samples. Since there are two models involved in the detection process, the faster R-CNN uses Pascal VOC which gives the annotation details in the ".xml" file, whereas the YOLO uses its own YOLO format and saves its annotation details in the ".txt" file.

4.2 Detection Using Faster R-CNN

It based the implementation on TensorFlow, which is an end-to-end open-source platform for machine learning. Install all the packages like pillow, lxml, Cython, OpenCV-Python, Matplotlib, pandas, etc., using the pip install command. Here, the "pandas" and "OpenCV-Python" packages are used in Python scripts to generate TFRecords. This TFRecord will contain the image info as NumPy arrays and the labels as a string. The model that we used for training is the faster R-CNN inception v2 coco model, downloaded from the TensorFlow detection model zoo [8] repository. In the config file of the downloaded model, changes should be made based on the created train and test datasets, label map, and record files. The hyper-parameters of the model, such as weight values and learning rate, are set as default. It should train the model for at least 60,000 steps and until the loss becomes less than 0.05 (Fig. 4). Once it is done, it will save all the trained models in the respective folder. Now, the last step is to generate the frozen inference graph (.pb file) with which the detection is to be made (Fig. 1). Finally, we can test our model for detecting objects in the input image, which will be annotated with its class name and detection score.

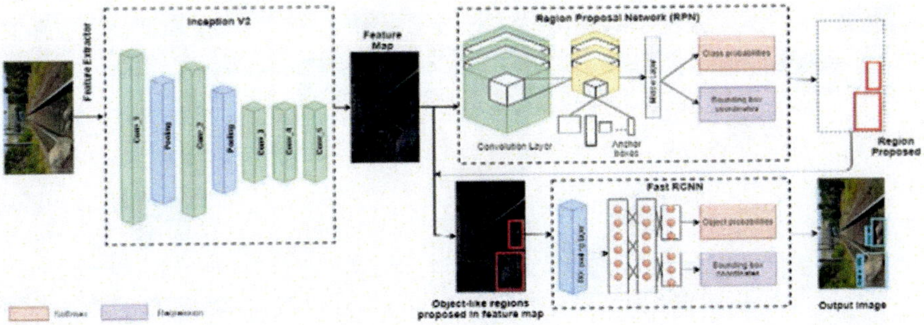

Fig. 1 Architecture diagram of faster R-CNN

4.3 *Detection Using YOLO*

For YOLO, start by cloning the darknet folder [9]. The next step is to create a data file that contains information about the location of the dataset files and the details of the bounding box. Then, split the dataset into train and test text files (80% for training and 20% for testing) which contain the filename of the images. darknet 53. conv.74 is a pre-trained model which should be further trained on the custom dataset. Changes should be made in the config file of YOLO_v3 concerning the training and testing parameters such as batch, subdivisions, and learning rate. It contains 3 YOLO layers, where the number of classes should be changed and in the preceding convolutional layer, change the value of the number of filters used according to the number of classes (Fig. 2). Start the training with the help of the created data file for the custom dataset and by using the darknet function. Once the training is completed, the epoch VS loss graph is plotted (Fig. 4). We then used the trained model for testing. For testing, images are given as inputs and the output image contains the objects detected with the bounding boxes, the classes it belongs to the detection score, and the time taken for the prediction.

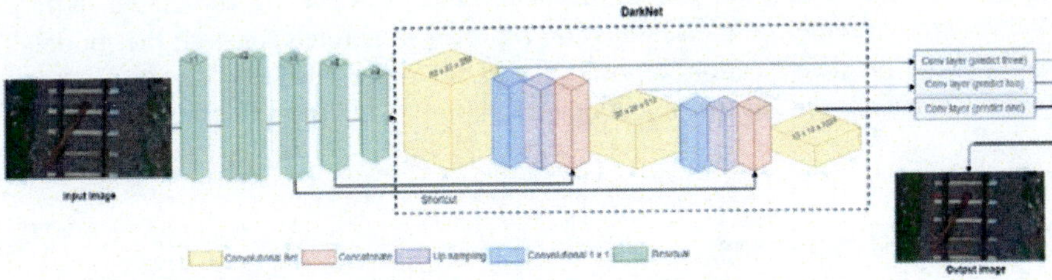

Fig. 2 Architecture diagram of YOLOv3

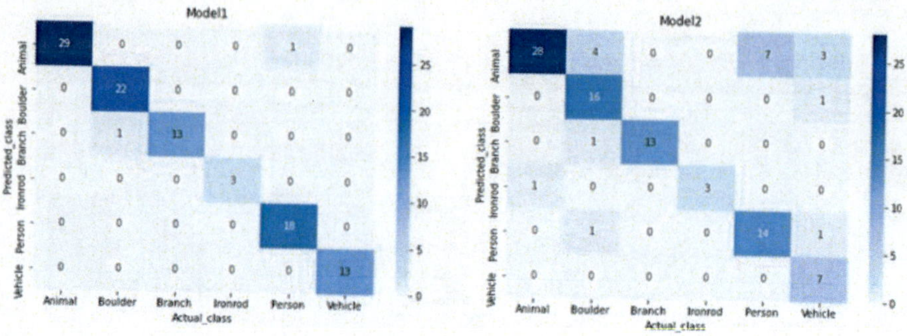

Fig. 3 Confusion matrix of faster R-CNN and YOLO

4.4 Multi-class Classification Testing

Once, the training of faster R-CNN and YOLO is successfully completed, then comes the testing phase, where the evaluation is done by giving an input image to the trained model. Therefore, the results obtained during the testing of both the models should be noted down on separate excel sheets.

To perform the computation for comparison of performance metrics, necessary packages such as pandas and NumPy should be imported. Then, "confusion matrix" is built, for both the models based on the excels created, using "crosstab ()" taken from panda library (Fig. 3). The confusion matrix consists of axis-like, where the horizontal one is the "actual class" and the vertical one corresponds to the "predicted class." So, this confusion matrix helps in finding out various parameters such as "true positive" (TP), "false positive" (FP), "false negative" (FN), and "true negative" (TN). The TP is nothing but the diagonal elements of the confusion matrix, the FP is found by considering all the columns except the values at diagonal, FN is identified by considering all the rows except the ones which have the same class label as the actual class, TN is obtained by summing up all the elements of the confusion matrix and subtracting it from all the above parameters. Once, the above parameters are calculated for each class, the precision, recall, and F1 score are calculated. After the values are computed, the classification report is generated for both the models using the function classification_report(). From this report, the model with the best accuracy will be chosen for the object detection purpose.

5 Result and Analysis

From the comparative study made on faster R-CNN and YOLO, it is evident that faster R-CNN performs better than YOLO in terms of accuracy (faster R-CNN = 98%, YOLO = 81%) and other performance metrics.

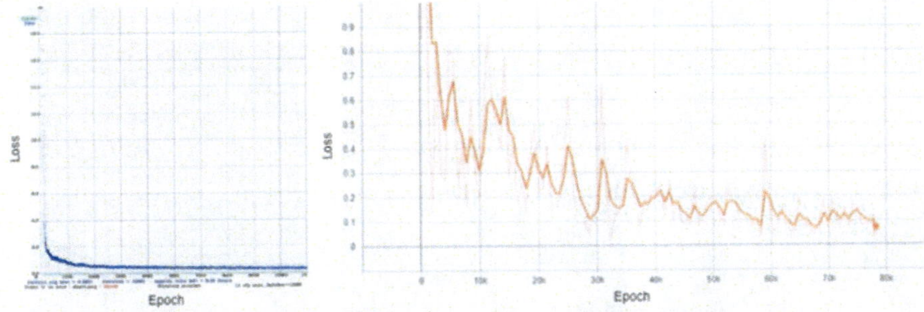

YOLO FASTER R-CNN

Fig. 4 Epoch versus loss graph

During an epoch, the loss function is calculated across every data item and give the quantitative loss measure at the given epoch. But plotting curve across iterations gives the loss on a subset of the entire dataset. So, epoch versus loss graph is plotted for both YOLO and faster R-CNN (Fig. 4).

Precision versus recall graph is plotted for both the model (Fig. 5). The mean average precision (mAP) was calculated for all networks with a pre-defined IoU threshold. Both models have a mAP above 80% on the integrated test set, illustrating that these methods were able to achieve favorable result. Faster R-CNN demonstrated the highest mAP (90.4%) than YOLO. A preliminary analysis suggested that the network inception V2, it performs with fast inference on low computing power, consuming a small amount of memory, playing a fundamental role in the detection accuracy improvement of faster R-CNN.

As output of both the models, we will be obtaining an output image with objects detected along with its class name and the detection score (Fig. 6).

Fig. 5 Precision versus recall graph

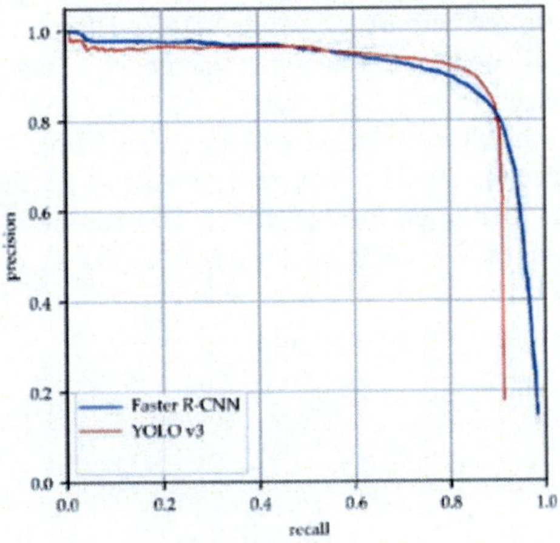

Fig. 6 Prediction of obstacle by faster R-CNN and YOLO

6 Conclusion and Future Work

The project was divided into three phases: The first phase is all about faster R-CNN, where the Pascal VOC dataset is created, augmented, and annotated for the model training. Once the data processing is done, the faster R-CNN model is trained and tested with an input image. The second phase is the implementation of YOLO, where the data processing is similar to faster R-CNN and the model is trained and tested on the dataset. Then, the final phase of the project is the comparison of both the models with the help of test results. Here, different performance metrics are calculated for each model and found that the model, "**faster R-CNN**" is the best one for the object detection on railway tracks.

As future work, we will try to detect objects (obstacles on tracks) in live videos and once the object is detected, an alerting mechanism like alarms can be added to alert the loco pilot. Further, the dataset can be improvised by capturing some real-time images and collecting frames from live videos.

References

1. Zhao Z-Q, Zheng P, Xu ST, Wu X (Nov 2019) Object detection with deep learning: a review. IEEE transactions on neural networks and learning systems 30(11). https://doi.org/10.1109/TNNLS.2018.2876865
2. Galvez, Bandala A, Dadios P, Vicerra P Object detection using convolutional neural networks. TENCON 2018–2018 IEEE region 10 conference. https://doi.org/10.1109/TENCON.2018.8650517
3. Ren S, He K, Sun J (1 June 2017) Faster R-CNN: towards real-time object detection with region proposal network. IEEE Trans Pattern Anal Mach Intell 39:6. https://doi.org/10.1109/TPAMI.2016.2577031
4. Kafetzis D, Fourfouris I, Argyropoulos S, Koutsopoulos I UAV-assisted aerial survey of railways using deep learning. 2020 international conference on unmanned aircraft systems (ICUAS). https://doi.org/10.1109/ICUAS48674.2020.9213928
5. Liu B, Zhao W, Sun Q Study of object detection based on faster R-CNN. 2017 Chinese automation congress (CAC). https://doi.org/10.1109/CAC.2017.8243900
6. Jana AP, Biswas A, Mohana YOLO based detection and classification of objects in video records. 2018 3rd IEEE international conference on recent trends in electronics, information and communication technology (RTEICT). https://doi.org/10.1109/RTEICT42901.2018.9012375
7. Abbas SM, Dr. Singh SN Region-based object detection and classification using faster R-CNN. 2018 4th international conference on computational intelligence and communication technology (CICT). https://doi.org/10.1109/CIACT.2018.8480413
8. https://github.com/tensorflow/models/blob/master/research/object_detection/g3doc/tf1_detection_zoo.md, Tensorflow Model Zoo
9. Darknet https://github.com/AlexeyAB/darknet
10. Pathak AR, Pandey M, Rautaray S Application of deep learning for object detection. Int Conf Comput Intell Data Sci (ICCIDS). https://doi.org/10.1109/ICIS.2017.7960069
11. Fang F, Li L, Zhu H, Lim J-H (22Oct 2019) Combining faster R-CNN and model-driven clustering for elongated object detection. IEEE Trans Image Proc 29. https://doi.org/10.1109/TIP.2019.2947792
12. Liu L, Ouyang W, Wang X, Fieguth P (2020) Deep learning for generic object detection: a survey. 31 October 2019 Int J Comput Vision 128:261–318. https://doi.org/10.1007/s11263-019-01247-4
13. Shah M, Kapdi R Object detection using deep neural networks. 2017 international conference on intelligent computing and control systems (ICICCS). https://doi.org/10.1109/ICCONS.2017.8250570
14. Liu R, Yu Z, Mo D, Cai Y An improved faster RCNN algorithm for object detection in remote sensing images. 2020 39th Chinese control conference (CCC). https://doi.org/10.23919/CCC50068.2020.9189024
15. Mane S, Mangale S Moving object detection and tracking using convolutional neural networks. 2018 Second international conference on intelligent computing and control systems (ICICCS). https://doi.org/10.1109/ICCONS.2018.8662921
16. Wei W. Small object detection based on deep learning. 2020 IEEE international conference on power, intelligent computing and systems (ICPICS). https://doi.org/10.1109/ICPICS50287.2020.9202185

An SVM-Based Approach to Predicting Level of Job Anxiety in Corporate Professionals using Linguistic Markers on Twitter

Unnathi Utpal Kumar

Abstract Work anxiety is linked with decreased job commitment and satisfaction. Yet, work anxiety, like other mental health problems, is not physically diagnosable. The lack of diagnosis and cure of job anxiety leads to lower levels of economic productivity and adds to the mental health epidemic. This study proposes a machine learning model to predict a corporate professional's level of work anxiety using their tweets by identifying linguistic markers associated with work anxiety. The Twitter API was used to create a dataset of over 15,000 corporate professionals. Thousands of tweets were collected from these users over 3 periods of time (May–August 2019, May–August 2020, and January-April 2021). After conducting sentiment and linguistic analysis, tweets from 90 random users (manually labelled for their job anxiety scores according to the job anxiety scale) were used to train/test an SVM regression model. The model achieved an RMSE of 0.2 and an accuracy of 83%. This approach has the potential to enable early detection of work anxiety and alert individuals about their mental health.

Keywords Job anxiety · Social networks · Twitter · Natural language processing

1 Introduction

Job anxiety is a stimulus-related type of anxiety, occurring in the workplace or when thinking of the workplace [8]. Whilst the workplace can provide support [11], it can provoke anxiety too. High standards and expectations can lead to feelings of insufficiency. Social anxiety might be heightened due to monitoring and sanctioning, whilst competition with colleagues might lead to persecution fears [4]. With leadership, greater responsibilities can lead to overburdening, provoking anxiety [5].

Work anxiety is correlated with lower job commitment [1] can cause absenteeism and impact performance [7], further associated with minimized efficiency and increased expenses for businesses, insurances, and public pension funds.

U. U. Kumar (✉)
Pathways World School Aravali, Gurugram, India

J. K. Mandal et al. (eds.), *Topical Drifts in Intelligent Computing*, Lecture Notes in Networks and Systems 426, https://doi.org/10.1007/978-981-19-0745-6_13

Symptoms of work anxiety include typical anxiety symptoms, like blushing, trembling, or palpitations [8]. At present, despite its clinical and economical importance, only 1 questionnaire is available to measure job anxiety, known as the job anxiety scale (JAS) [6]. However, JAS's validity is difficult to assess since it is a self-report scale, such that it consists of shortcomings due to acquiescence and social desirability [9]. Therefore, with no specific ICD codes for work anxieties, it can be difficult to characterize and thus diagnose them dynamically.

In context of these challenges, this study examines and develops a novel solution based on the previous metrics of job anxiety. The research here analyzes social media as a tool to express work-related concerns and thus to predict work anxiety. As people use Twitter to express their thoughts, social media can become a data source for job anxiety. By training an AI-based model on this data, work anxiety can effectively be detected.

This study aims to investigate the following question: can a mathematical model identify and quantify the relationship between a healthy corporate professional's use of language in comparison with that of a corporate professional suffering from job anxiety?

The hypothesis pursued is, if a healthy corporate professional's use of language is compared to that of a corporate professional suffering from job anxiety, then there exists a difference in terms of the linguistic markers, which can be identified by a mathematical model such that the use of these linguistic markers is further heightened in individuals suffering from higher levels of work anxiety.

The main contributions in this paper include

1. A Python script using the Twitter API and Selenium was developed to create a comprehensive dataset with tweets from over 15,000 Twitter users
2. Sentiment analysis of tweets in relation to levels of job anxiety (as expressed linguistically) over 3-month periods in 2019, 2020, and 2021
3. An SVM regression model to predict levels of job anxiety from tweets

The solution proposed in this study, the SVM regression model, combats the problems associated with other ways to measure levels of work anxiety, as it avoids response bias by not requiring a conscious undertaking like a survey and is also specific to job anxiety than anxiety generally. The solution thus has the potential to increase economic productivity and help combat mental health problems.

2 Background Literature

Generally, only some research has been conducted regarding work anxiety, as researchers often adapt different pre-existing instruments or tools. Nonetheless, one empirical way to measure perceived job anxiety is through the job anxiety scale (JAS) [6]. JAS is a self-rating scale to measure job anxiety, containing 5 main dimensions, (i) stimulus-related anxiety and avoidance behaviour, (ii) social anxieties, (iii) health-related anxieties, (iv) cognitions of insufficiency, and (v) job-related worries.

However, the extensivity of the JAS renders it uneconomical and inefficient and places job anxiety in a very specific, unmalleable context. Therefore, to consider job anxiety as a wider construct for this study, a shortened, empirically derived version of the JAS, JAS-15 [10], was considered. The JAS-15 retains the 5 theoretically meaningful dimensions.

Whilst no research has been conducted on language use and work anxiety specifically, the previous research helps identify possible correlations between linguistic structures and anxiety. For example, the use of first-person singular pronouns during negative memory recall is associated with individuals suffering from anxiety [2]. It was also found that the use of unigrams like "withdrawal" and "severe" on Twitter indicated anxiety [3].

Whilst the background literature is useful in defining the study's terms, they either do not consider job anxiety specifically or have not been implemented in a context relevant to social media. This lack of data suggests that further research is needed to determine linguistic markers of work anxiety, along with an efficient and accurate solution to realistically predict work anxiety levels.

3 Data Collection

3.1 Data Collection

A list of Twitter usernames of corporate professionals had to be collated.

First, a list of 10 keywords that corporate professionals might use in their tweets was created, identified based on prior knowledge.

Thereafter, using "search" on Twitter, it was manually checked that, for each keyword, most initial tweets from search results were by corporates. Then, using Web scraping with Python and Selenium (tool automating browsers), each keyword was searched and first 3000–3200 tweets collected. The no. of tweets could not be standardized due to memory overflow issues on Microsoft Edge (driver browser), but no less than 3000 and no more than 3200 tweets were collected from each keyword. The content, username, and name associated with tweets in each keyword were stored in CSVs.

However, not all usernames collected were those of corporate professionals, so irrelevant data had to be removed using filtering on Microsoft Excel. First, all duplicate usernames were removed and usernames containing the words "news" or "job" or links were removed (generally portals and not individuals). Then, going over each keyword file, general elimination trends (Table 1) were identified for each keyword.

Finally, a total of 20,358 usernames was collected.

Next, tweets by these usernames had to be collected for further analysis. To ensure that further algorithms and analyzes were not limited to heightened anxiety arising due to work-from-home during COVID-19, tweets from the identified users were collected over 3 periods of time—May–August 2019, May–August 2020, and

Table 1 Elimination patterns for usernames

Keyword	Raw no. of usernames	Elimination patterns	Final no. of usernames
Company	3142	• Usernames/names with "company" • Tweets using structure of pronoun + company (e.g. "keep me company")	2439
Corporate	3090	• Usernames/names with "corporate"	2059
Job	3082	• "Good job", "great job", "con job"	2112
Manager	3192	• Since "manager" is used in sports contexts, all such tweets were removed • "File manager" and "task manager" • Tweets with "fantastic manager" (news report during data collection)	1753
My Industry	3127	• Tweets containing phrases associated with the entertainment industry	2459
Product	3047	• Tweets for product reviews and promotions ("review" and "check out") • Tweets about skincare products (mostly reviews or promotions) • "Coupon" or "deal"	1728
Recruit	3134	• Since "recruit" is often used in context of college sports recruits, all related tweets were removed	1397
Salary	3032	• "Actor", "cricketer", or "NFL" • Tweets containing "TTS" (trend)	2317
Sales	3057	• Tweets with keywords about music	2032
Software	3134	• "eBay", "Amazon", "Flipkart"	2062

January-April 2021. The Tweepy API (library for communication between Python and Twitter) was used to collect tweets. A script was run for 32 days, and >1 million tweets were collected.

However, not all users had sufficient Twitter activity from all time periods, so minimum 30 tweets were needed per period. Finally, data were retained from 8598 users.

3.2 Pre-Processing

Once the Twitter data were read into Python, pre-processing was conducted. All Unicode characters, emoticons, and links were replaced with an empty string to ensure that these special characters/symbols did not contribute to model training or sentiment analysis.

3.3 Sentiment Analysis

For each user, 3 values were calculated separately for each period. These 3 values were (i) positive sentiment score, (ii) negative sentiment score, and (iii) work anxiety score.

The positive sentiment and negative sentiment scores were calculated using the TextBlob (Python library for common NLP tasks) sentiment analysis algorithm.

For the work anxiety score, some words associated with work anxiety, adapted from JAS-15, were identified, likely to be used by people suffering from work anxiety. Each time a pair of words was used in a tweet, the raw work anxiety score was incremented, and finally, the raw work anxiety score was divided by total number of words by user to obtain final work anxiety score. Whilst this was a primitive approach before a machine learning approach, it helped in understanding the linguistic potential of JAS-15.

1. "always worry" AND "work"
2. "minor matters at workplace" OR "minor matters at work" AND "worry"
3. "workplace" OR "job" OR "work" AND "restricting my capacities for achievement"
4. "suffer" OR "miserable" OR "ill health" OR "bad health" OR "not feeling well" OR "unfairness" OR "avoid" AND "workplace"
5. "work" AND "nervous" OR "my health" OR "suffer" OR "miserable" OR "ill health" OR "bad health" OR "not feeling well" OR "unfairness" OR "avoid"
6. "anxiety" OR "suffer" OR "miserable" OR "ill health" OR "bad health" OR "not feeling well" OR "unfairness" OR "avoid" AND "job"
7. "problem" AND "superiors" OR "boss" OR "colleagues"

   Thereafter, for each period, the Pearson correlation coefficient was calculated between negative sentiment or positive sentiment and work anxiety.

3.4 Data Labelling

After conducting sentiment analysis, data from some users were manually read and labelled on the severity of work anxiety expressed on a scale of 0 to 1, according to JAS-15. The adaptation of JAS-15 for identifying tweets is shown below (Table 2).

Table 2 JAS-15 for Twitter

S. No.	Presence in Tweet/Account
1.	User tweets about missing work purposely or avoiding the workplace
2.	User tweets about haste to leave the workplace
3.	User tweets about specific events from the workplace that cause anxiety
4.	User tweets about problems with colleagues
5.	User tweets about problems with superiors
6.	User tweets about problems with colleagues at least 5 times over two weeks
7.	User tweets about a direct correlation between work and health
8.	User tweets about a direct correlation between workplace and health
9.	User tweets about health and work separately, but constantly
10.	User tweets specifically about work anxiety
11.	User constantly tweets about ambition and growth at the workplace
12.	User tweets about uncertainty at work
13.	User constantly tweets about work, at least five times over two weeks
14.	User tweets about the effect of work on family life
15.	User constantly tweets about work, at least seven times over two weeks

Ground truth train data were collected and labelled from 40 random users, test data were collected and labelled from 15 random users, and data from 20 users (with high work anxiety indexes) were used in creating the token bag (bag of words).

Bag of words contained 50 most common words (no stop words) from those 20 users.

3.5 Model Training

A support vector machine (SVM) algorithm was used for creating the work anxiety regressor, trained on 6 feature vectors. SVM was chosen given the small dataset (with less noise), high prediction accuracy, support for kernels, easy implementation, and memory efficiency. The dimensions of the JAS-15 were used to identify 5 features. The training and testing data also took the same feature vectors into account since they were labelled according to the JAS.

The 6 feature vectors included

Table 3 Correlation between sentiments and work anxiety

	2019	2020	2021
Positive sentiment and work anxiety score	0.045	−0.236	−0.235
Negative sentiment and work anxiety score	0.177	0.454	0.581

1. No. of times workplace mentioned (*stimulus-related anxiety*)
2. No. of times colleagues are mentioned (*social anxiety and cognition of mobbing*)
3. No. of times health is mentioned (in relation to *health- and body-related anxieties*)
4. No. of times ambition is mentioned (in relation to *cognition of insufficiency*)
5. No. of times deadline/pressure is mentioned (in relation to *job-related worrying*)
6. No. of times words from bag of words used

A corpus was created for each feature vector and then values found.

Thereafter, an SVM regressor was trained and tested.

4 Results—Sentiment Analysis

The Pearson correlation coefficients, across the 3 time periods, between scores have been shown in Table 3.

It can be seen that generally, there is a low negative correlation between positive sentiment and work anxiety, whilst there is a moderate positive correlation between negative sentiment and work anxiety. This is consistent with the researchers' expectations as work anxiety is more strongly correlated with negative outcomes.

The correlation between negative sentiment and work anxiety became stronger between 2019 and 2021, which could be due to rising negative sentiment amidst COVID-19 and expected increase in work anxiety due to general increase in uncertainty.

5 Results—SVM Algorithm

The trained SVM regressor achieved a root mean squared error (RMSE) of 0.2 and an accuracy of 83%, which indicates that whilst the algorithm might be successful in predicting work anxiety in a certain range, it needs more data for higher accuracy and lowered error.

6 Discussion

There are several limitations to this study that might have acted as sources of error.

Firstly, the labelling of ground truth and test data was relative, due to the relative nature of the JAS-15 itself, and subject to human error. Furthermore, only 40 users were used in testing and 15 users in training, which is a small sample size. To appropriately train and evaluate, at least 100 users should have been used in testing and training alike.

There were limitations in the sentiment analysis as well. As was seen with username collection, a lot of irrelevant data had been collected. Whilst it was attempted to remove most irrelevant data, it could not be ensured that all of it was removed, and thus, lowered noise in the dataset could not be guaranteed. Furthermore, other algorithms besides TextBlob could have been used, as TextBlob only has an accuracy of ~56% [12].

Future directions could include training the algorithm with more data or implementing an app with a GUI to enable user understanding of work anxiety as well.

7 Conclusions

The hypothesis was supported in that specific linguistic markers (through the dimensions of the JAS) can be used to predict levels of work anxiety, based on JAS-15, as there is a measurable difference in language use between a healthy individual and an individual suffering from some level of work anxiety. Twitter, therefore, serves as a good source of data to predict a corporate professional's level of work anxiety.

References

1. Borg MG, Riding RJ, Falzon JM (2006) Stress in teaching: a study of occupational stress and its determinants, job satisfaction and career commitment among primary schoolteachers. Educ Psychol 59–75
2. Brockmeyer T, Zimmermann J, Kulessa D, Hautzinger M, Bents H, Friederich H-C, Herzog W, Backenstrass M (2015) Me, myself, and I: self-referent word use as an indicator of self-focused attention in relation to depression and anxiety. Front Psychol
3. Choudhury MD, Gamon M, Counts S, Horvitz E (2013) Predicting depression via social media. AAAI
4. Dormann C, Zapf D, Isic A (2002) Emotionale arbeitsanforderungen und ihre konsequenzen bei call center-arbeitsplätzen. Zeitschrift für Arbeits- und Organisationspsychologie 201–215
5. Hobson J, Beach JR (2000) An investigation of the relationship between psychological health and workload among managers. Occup Med 518–522
6. Linden M, Muschalla B, Olbrich D (2008) Die job-angst-skala (JAS). Ein Fragebogen zur Erfassung arbeitsplatzbezogener Ängste. Zeitschrift für Arbeits- und Organisationspsychologie 126–134

7. Muschalla B, Heldmann M, Fay D (2013) The significance of job-anxiety in a working population. Occup Med 415–421
8. Muschalla B, Linden M, Olbrich D (2010) The relationship between job-anxiety and trait-anxiety—a differential diagnostic investigation with the job-anxiety-scale and the state-trait-anxiety-inventory. J Anxiety Disord 366–371
9. Razavi T (2001) Self-report measures: an overview of concerns and limitations of questionnaire use in occupational stress research. University of Southampton School of Management
10. Schmalbach B, Kalkbrenner A, Bassler M, Hinz A, Petrowski K (2020) Psychometric properties of a short version of the job anxiety scale. BMC Med Res Methodol 87
11. Sczesny S, Thau S (2005) Gesundheitsbewertung versus Arbeitszufriedenheit: Der Zusammenhang von Indikatoren des subjektiven Wohlbefindens mit selbstberichteten Fehlzeiten. Zeitschrift für Arbeits- und Organisationspsychologie 17–24
12. ES S (20 July 2021) Sentiment analysis in Python: TextBlob versus vader sentiment versus flair versus building it from scratch. Retrieved from Neptune blog: https://neptune.ai/blog/sentiment-analysis-python-textblob-vs-vader-vs-flair

Markerless Location-Based Augmented Reality Application for Showcasing Deals

Mohammad Monirujjaman Khan and Faria Soroni

Abstract The development and execution of an online mobile app suitable for detecting offers and showing relevant details on a digital augmented reality display via a smart phone's camera are discussed in this paper. The client–server design of this system is two-tiered. Initially, a deal should be recorded on our platform by an entity that is distinct from users. Deal Teal AR uses the camera view to overlay electronic data on areas around you in the current world, based on the position you are facing. Deal Teal AR displays a picture via your phone's camera when you touch it. Deal Teal's database of businesses-restaurants, hotels, places of interest, and much more is often used to generate cards that display information or bargains. The bargains are placed in the current environment and become active, thanks to powerful augmented reality technology like computer vision and object recognition. A virtual terrain modeling platform with deep learning was also employed to increase the application's productivity and construction recognition abilities.

Keywords Augmented reality · Android · Application · GPS · Markerless · Deals

1 Introduction

Dhaka is the most densely populated city in the world. With 19,447 people per square mile, it is a great place for restaurants. It is a city of thousands of restaurants and hundreds opening every day. Dining outside of the house has become a popular focus in recent years, and this has influenced Dhaka's restaurant business. This is attributable to rising incomes as well as improvements in city dwellers' needs and interests. The connection between both the elements that influence restaurant selection and the overall selection choice of restaurant customers was investigated using

M. M. Khan (✉) · F. Soroni
Department of Electrical and Computer Engineering, North South University, Bashundhara RA, Dhaka 1229, Bangladesh
e-mail: monirujjaman.khan@northsouth.edu

F. Soroni
e-mail: faria.soroni@northsouth.edu

J. K. Mandal et al. (eds.), *Topical Drifts in Intelligent Computing*, Lecture Notes in Networks and Systems 426, https://doi.org/10.1007/978-981-19-0745-6_14

both primary and secondary statistics. Customer service, diversity, restaurant registered trademark, and most crucially, the most worth retrieved for the cash a customer would be ready to spend for a food product are all elements that affect customers' requirements, according to the findings [1].

With greater demand in the business, each restaurant is attempting to differentiate itself in order to attract more consumers. The "best bang for the buck" attitude encourages restaurant competitiveness by employing specials and value packages to entice customers from competitors. Restaurants in Dhaka City often used focus on ambiance and quality food to attract customers, but because these characteristics can easily attain a steady stream while remaining competitive, bargains, and discounts were discovered to be the sole recent addition that can provide greater value for the money. However, except from a few Facebook groups that promote the offerings, there are no other options. Platform to showcase the different offers in town. Special occasions are aimed by restaurants to promote new offers, but as they do not reach the customers who are not always actively looking for deals on social media, the offers seldom reach consumers, creating an information gap between the restaurant and its targeted market that needs to be filled up [2–5].

There are four types of augmented reality technology:

- Marker-based AR
- Markerless or location-based AR
- Projection-based AR
- Superimposition-based AR.

The very first two categories—marker-based AR and destination AR—encompass the majority of mobile AR now available. To show virtual information overlaid on specified items, marker-based AR generally uses image recognition capabilities.

AR characteristics that are based on specific place are known as markerless AR, also known as geolocation-based AR. Digital data are projected onto real-world locations, enabling people to engage with it using AR-enabled devices such as smartphones. Every smartphone today has the ability to identify its position. Phones that enable augmented reality apps are rapidly becoming available to customers worldwide, thanks to the advent of Apple's ARKit and Google's ARCore.

Geospatial data defining techniques and augmented reality technology are combined to build geolocation apps. The initial step is to identify geoindicators (also known as "points of interest"). The precise location of the smartphone is obtained by querying the smartphone's appropriate sensors. The internal digital compasses and proximity sensors in smart phones are utilized to evaluate location, together with GPS or beacons. The digital overlays, also known as "augmentations," are virtual overlays that are put on places of interest. Sounds, animations, music, films, and photos are just a few examples. The virtual coverings can be activated and interacted with based on the recipient's physical place. When it comes to designing effective geolocation-based augmented reality apps, precision is crucial. Regardless of the camera position, virtual information must be shown on (portrait, landscape, or tilted). The GPS data utilized to determine the user's exact location must be error-free as well. For the best user experience, all AR features should be adjusted and validated.

Navigation is among the main areas in mobile data processing technology in which augmented reality is widely used, especially in the context of GIS and destination service providers, because it has much the digital objects on the real-world situation and serves as an efficient tool for apps and services provided to the user that are reliant mostly on recipient's or ecosystem's destination [5].

2 Related Work

Feiner et al. detail prior findings on the construction of a destination application interface based on GPS locations and sensors operating on head-mounted monitors. Kahari and Murphy, on the other hand, used a modest smart mobile phone, enhanced by additional sensors, to construct an augmented reality system in 2006. Another augmented reality use is detailed in. Its goal is to offer media data and other information in a real situation while also allowing users to engage with it via augmented reality [6, 7].

Currently, mobile location-based AR has become popular in recent advanced technology. It has been utilized and implemented for developing systems with various purposes. Mobile location-based AR applications are begun to play a significant role in the tourism sector, attributed to the increasing use of phones. Visitors can use these apps to get context-aware data about certain areas, which helps them learn more about the place. Furthermore, mobile location-based AR platforms allow people to discover the environment by layering fresh layers of destination info on their environment and creating a list of their preferred points of attractions (POIs) utilizing this data [8]. Many researches of location-based AR have increasingly been studied and carried out in recent years. The research conducted by Brata and Liang has developed a mobile AR application about the bus stops in Taipei. The developed application based on Android platform called BusAR. BusAR's feature is to locate the nearest bus stop and the route way to get there. A key issue is allocated to a bus stop in augmented reality (POI) [9]. The POI provides information such as the name of the bus station, the distances from the consumer to the bus stop, the path name, the title and kind of the forthcoming bus, and the expected time whenever the bus will reach at the particular bus station.

In addition, this software can guide the user to the closest town by telling him when and how to turn left or right without having to leave the application. Tsai and Shie created a location-based mobile augmented reality application for hot springs vacation in Yilan county as main research method. The AR component and the map component are the implementation of this model that make up this app [10]. The AR method made use of such a POI that displays a two-dimensional digital representation in real time, allowing users to locate hot springs. Whereas the map module utilizes the Google map service to provide the corresponding position and for displaying the hot springs information and route guidance [11, 12].

3 Basic Theory

3.1 Augmented Reality

The term "augmented reality" refers to a technology that allows the user to see the physical environment while also seeing virtual objects projected on it. Textual, pictures, music, video, and location information are examples of virtual objects. The video on a computer, mobile phone, or other electrical gadget captures the true perspective in most cases. AR enhances a user's perspective of and interactions with the actual world by superimposing computer-generated information on the image collected by the camera. Currently, AR mostly has been utilized and carried out in mobile platform environment.

To overcome the problem of the target audience missing out on the offers and deals, we developed an application with an augmented reality feature. The application's AR window allows the consumer to view any nearby deals or offers on their respective locations, even pointing on the levels of the building where the establishment is located in, using 3D pop-ups. Surveys mentioned that more often than not people do not find the time to browse through hundreds of offers on a list to find the perfect one, where locational availability of the offer plays a key role. This application allows the user to point their mobile device at any road or building and easily extract all that information that would otherwise need to be filtered through in multiple steps.

3.2 Website

For a system as data-intensive as ours, we needed an online platform to host as much data as possible so that we can keep our app as light weight as possible. Keeping the amount of contents in the app as minimal as possible allows our app to be responsive and fast. The need to reduce the load on the app from handling data came due to the fact that our AR window itself is very RAM intensive as a result of which, for the app to not be "laggy," we needed to take as much work load off of the app as possible. Our Website hosts all of the available deals and offers along with their location and expiry dates. The deals are further categorized by their type and expiry dates. There are two portals for the Website, an admin-based and a user-based. The user can login; view deals and also add their deals. To add deals, the user pulls up a form where he/she needs to provide all relevant details; see Fig. 1. The user interface of the Website is kept simple and very user-friendly so that everyone can easily navigate through. The user submissions need to be further approved by the admin before they can be aired, maintaining the legitimacy of each and every deal. Figure 2 shows the user login interface. An user can login to the system using user name or Google.

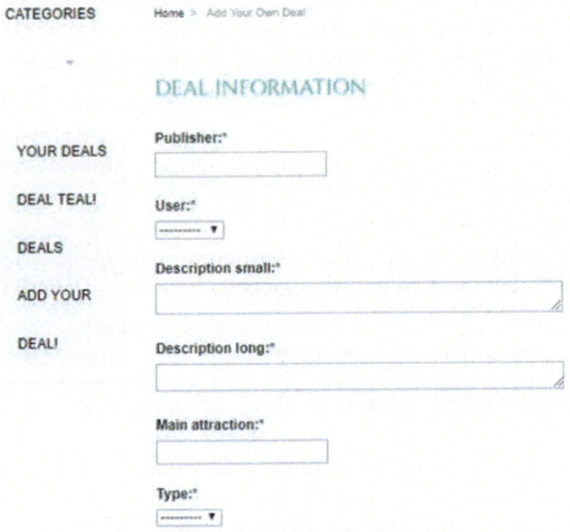

Fig. 1 Adding deal information

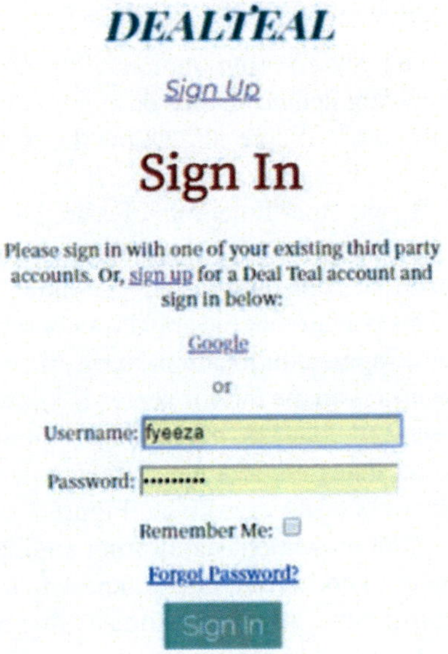

Fig. 2 User login interface

4 Application: AR Core SDK

With numerous benefit of software development kit (SDK), many developers have utilized it to develop numerous applications in diversity of purposes. Nowadays, many augmented reality (AR) SDKs have been carried out by developers for developing various mobile applications. The augmented reality SDKs include Metaio, Vuforia, Wikitude, D'Fusion, etc. The comparison of various augmented reality

SDKs and their features have been studied by. Augmented reality SDK is a set of tools and libraries that provided to developers for developing augmented reality applications. It assists to facilitate many components within the AR application such as object recognition, object tracking, location-based AR, content rendering, and visualization.

Google's augmented reality technology, ARCore, allows users to create their own augmented reality applications. ARCore allows your smartphone to perceive its surroundings, comprehend the universe, and process the information by utilizing various APIs. To allow collaborative augmented reality, several of the APIs are accessible on Android and iOS.

To blend digital environment with the real environment seen through the your cell phone screen, ARCore employs three important capacities:

- Motion tracking enables the phone to comprehend and monitor its location in relation to the rest of the globe.
- The device's environment ability enables it to recognize the size and location of all types of surfaces, including horizontally, perpendicular, and inclined coatings such as the floor, a coffee table, and fences.
- The phone uses brightness prediction to describe the exact illumination conditions and circumstances.

By giving programmers with frameworks and functions either to construct their new augmented reality or upgrade their current ones with just an AR camera view engines, ARCore allows for the technique of evaluating of markerless augmented reality. Image identification and tagging, 3D view generation, video projection, place AR, and other functionalities are all included in this SDK. It incorporates all subsystems, especially POI information and route posting data, within the Internet app. The browser application is a representation of standalone architecture. This architecture gives benefit that the app does not rely on the wide area network (WAN) link. For the development of location-based AR with the ARCore SDK, the object position which overlays on the mobile screen is known through the user position. The user position is received either by the mobile communication network or GPS. The digital compass determines the recipient's position, while the accelerometer sensor determines the device's browser settings. Figure 3 shows AR demo on app.

The movement monitoring engine in the SDK utilizes the smartphone screen to locate good observations, known as characteristics, and monitor how they change across time. ARCore estimates that both position and orientation of the smartphone because it travels across space using a mix of the displacement of such locations and information from the device's sensing devices. ARCore could recognize level objects, such as a desk or the ground, and calculate the mean illumination in the region around that in addition to recognizing critical locations. These features work together to allow ARCore to develop its own understanding of the world. Because of ARCore's awareness of the actual world, you may position items, comments, and other data that blends in with the surroundings. You may put a sleeping cat in the middle of your coffee table or comment a picture the with architect's biography.

(a) (b)

Fig. 3 Augmented reality demo on app

You can walk around rather than observe such things from every aspect with object detection, but even if you turn around rather than leave the building, the cat, or comment will be exactly where you found this when you return. Figure 4 depicts the deals showing on maps (zoomed in, deal pop-ups focused). Figure 5 shows the augmented reality window showing deal pop-ups.

5 System Implementation

The mobile-based augmented reality application developed in this work utilizes coordinate data. The coordinate data are an important component in this research because these data are used to determine the location of tourist attraction and culinary places. These data are retrieved from Google map. This system is implemented on Android powered mobile devices. The system coding development is conducted on Android Studio IDE. The augmented reality SDK utilized in this research is ARCore by Google.

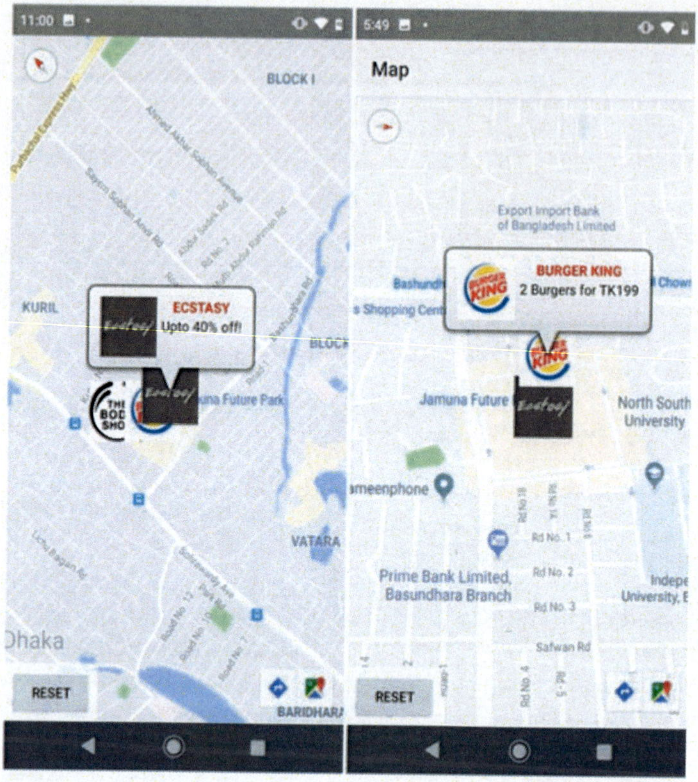

Fig. 4 Augmented reality demo on app

6 Conclusion

According to the previous research, AR technology can be used to provide tourist and map assistance. AR incorporates show maps, and virtual locations may improve users' interactions with computers and global approach. This article proposes an interactive online smartphone application that uses a phone's camera to present bargains in a virtual world. The results of the experiments demonstrate that app recognized the buildings and showcased relevant deals and information. Due to smartphone GPS systems having an accuracy deficiency of around 15 degrees, the models and information tend to move slightly around on-screen. We will be further study how to improve the accuracy of the GPS signals and develop a calibration method to make the augmented reality view run more efficiently and accurately on lower-end smart phones.

Fig. 5 Augmented reality demo on app

References

1. Dr. Islam N (2018) Factors influencing the selection of restaurant for dining in Dhaka City of Bangladesh
2. Cheng J, Chen K, Chen W (2017) Comparison of marker-based AR and markerless AR: a case study on indoor decoration system. https://doi.org/10.24928/JC3-2017/0231
3. Implementation of markerless augmented reality technology based on android to introduction Lontara in marine society
4. Genc Y, Riedel S, Souvannavong F, Akinlar C, Navab N (2002) Marker-less tracking for AR: a learning–based approach. Proceedings. International symposium on mixed and augmented reality. Darmstadt, Germany pp 295–304
5. Prakhar S. Geo-location based augmented reality application, graduate, school of computing science and engineering. Galgotias University, U.P, India
6. Sudarshan SK (2018) Augmented reality in mobile devices. San José State University
7. Rodriguez J, Huang C-Y (Dec 2017) An emerging study in augmented reality and geographical information system. Int J Comput Theor Eng 9(6)

8. Deshmukh SS, et al. 3D object tracking and manipulation in augmented reality, Padmabhushan Vasantdada Patil institute of technology. Pune, Maharashtra, India
9. Gaikwad D, Chikane A, Kulkarni S, Nhavkar A. Professor, computer engineering, NBN Sinhgad school of engineering, Pune, 2,3,4 computer engineering, NBN Sinhgad school of engineering, Pune, augmented reality based platform to share virtual worlds
10. Çolak O. Anadolu university, Assist. Prof. Dr. Lokman Yünlü, Mehmet Akif University, A rewiev on augmented reallity and virtual reality in engineering education
11. Badouch A, Krit S-D, Kabrane M, Karimi K. Augmented reality services implemented within smart cities, based on an internet of things infrastructure, concepts and challenges: an overview
12. Khan H, Soroni F, Mahmood SJS, Mannan N, Khan MM (May 2021)Education system for Bangladesh using augmented reality, virtual reality and artificial intelligence. IEEE world AI IoT congress, 10th–13th. Seattle, USA

Development of a Web-Based Corona Emergency Portal

Mohammad Monirujjaman Khan and Md.Amdadul Bari

Abstract As the present world is devastated by the massive outbreak of coronavirus (COVID-19), Bangladesh is experiencing significant damage in all sectors, especially as a developing country. This paper presents the development of a Web-based corona emergency portal, which will help ordinary people find out all the necessary emergency information in this regard. The Web portal comprises the COVID-19 tracker, corona e-commerce, and blood plasma bank, which are developed with React.js, Node.js and Firebase. The COVID-19 tracker is a much-needed option for this pandemic where daily updates on COVID-19 will be posted. The blood plasma bank aims to serve people with emergency blood plasma, which is very rare to find in critical situations. The developed Web-based system is user-friendly and efficient. Under these ongoing circumstances, this portal aims to be a one-stop service for all types of corona-related services for Bangladesh during this unexpected period of the COVID-19 pandemic.

Keywords Web portal · Corona emergency portal · Blood plasma bank

1 Introduction

Currently, the world is facing a lot of difficulties because of the pandemic. COVID-19 has affected 210 countries around the world and killed around 2 million people. In Bangladesh, it is transmitted at a high rate of 20.37% and doubles the patients within nearly 2 weeks [1]. In this challenging time, people need something that can help them online. Corona emergency portal is a one-stop online service portal where people can find everything they need regarding COVID-19. This Web portal has COVID-19 tracker, blood plasma bank, and corona e-commerce services. The Website will help people to take precautionary steps to fight against COVID-19. As

M. M. Khan (✉) · M. Bari
North South University, Bashundhara, Dhaka 1229, Bangladesh
e-mail: monirujjaman.khan@northsouth.edu

M. Bari
e-mail: amdadul.bari@northsouth.edu

J. K. Mandal et al. (eds.), *Topical Drifts in Intelligent Computing*, Lecture Notes in Networks and Systems 426, https://doi.org/10.1007/978-981-19-0745-6_15

we know, the situation is getting worse because people are facing trouble gathering information about COVID-19 [2]. Even if they find a way to gather information about COVID-19, there is a high chance that the information is not valid enough. Different Websites provide different types of data. And there is no one-stop solution to the disease. In particular, people living in rural areas do not know the proper necessary steps to fight the pandemic. Even if someone knows the extremity of the disease, they cannot find proper support to get tested by the authorities [2]. They do not get the correct contact or information about getting tested. As they do not get tested, they do not know when or how to take the necessary steps. On the other hand, COVID-19 is spread by contacting other people [3]. So, people have to make sure they do not get in contact with other people. And this was the reason the whole world, including Bangladesh, was under lockdown. So, getting out of the house to buy the necessary things to fight with COVID-19 is highly risky for the affected person and the people surrounding him. Next, people who are COVID-19 positive need blood plasma most of the time. Actually, we can see that blood plasma is needed on an emergency basis. But they do not find the blood plasma at the right moment, mostly because it is very hard to find the proper media and information resources about blood plasma. This emergency portal will help to solve all the problems together. This emergency portal provides the correct COVID-19 update with the COVID-19 tracker. The COVID-19 tracker has a real-time COVID-19 update. We believe the whole corona emergency portal will greatly help the people of our country. It will make the fight with coronavirus a bit easier in this disastrous situation.

In Sect. 1, the introduction is provided. In Sects. 2 and 3, the existing works and the methodology are provided, respectively. In Sect. 4, the results are provided. And in the last section, which is Sect. 5, the conclusion of the whole project is provided.

2 Existing Work

Corona-related emergency services are not present in abundance at the moment, but there are a few notable media which do provide such similar services. Below mentioned are some of the notable services. Prava well-being is an Internet entry within the nation which gives virtual clinic administrations [4]. It began from cardiology to brain research and numerous more [5]. They have the benefit of an online specialist interview through earlier arrangements, and for this benefit, they charge an interview charge of Tk.800 [5]. They too give online booking for crown test collecting. Shohojoddha is a government-run online portal where the service of plasma donation is carried on [6, 7]. At present, this is the only portal which provides the service of a blood plasma bank. To track the outbreak of corona, there are a lot of Web portals that are present at the moment, but among them, corona tracker is a notable one [8]. Here, people can track the corona affected cases and treated cases as well as the death cases of a specific region. The portal also provides updates about the travel notices that are currently happening around the world [9]. This helps people to know when to travel, how to travel, and what measures to follow while traveling to a country. This portal even collaborated with the World Health Organization (WHO) to provide basic information related to corona to the general public [10].

We have seen some other works of different organizations. But the main thing that makes this project different than other projects is this is a one-stop service. We can find all the things that are important to fight the COVID-19 in one Website. We have the blood plasma bank, the COVID-19 tracker, and most importantly, the corona e-commerce service. And that too altogether in a Website.

3 Methodology

In this section, all the important methods and materials are given, including the Web-based system, the block diagrams, the use case analysis, software design, targeted audience, unique features, the technologies used, and the tools that are used to make this Web portal.

3.1 Web-Based System

We have developed a Web-based system which will help people to get emergency services regarding COVID-19. This Website will help people on a virtual platform. This system will be a one-stop service for people who roam around mindlessly in search of actual emergency support and treatment. Visiting this Website will provide people with several features, like the COVID-19 tracker, blood plasma bank, and corona e-commerce service. Figure 1 shows the whole structure of the blood plasma bank. Here, we have 3 types of accounts such as the controller, the donor, and the seeker. The donor will register with the needed information through the Website. Blood group and contact information will be needed as the donors' information. From the given information and the blood plasma list, the seeker will search through the list of our donors and will place a request on the system for his required group of blood. The system will then pass the request to the controller. The controller will show the request to the donors. And if any of the donors accepts the request, he will notify the controller that he is available and will be able to donate. Receiving the notification from the controller will then notify the seeker that his requested blood plasma is available and will provide him with the contact number of the donor.

The seeker will then contact the donor. And the last thing that the donor will have to do is to update his status to donate so that he will not have to receive any more requests to donate plasma.

Figure 2 shows the whole block diagram of the corona e-commerce site. Here, users have to register on the Website to create an account. And after that, they can login to the Website with their information. After logging into the Website, they can browse the products and add to the cart whatever they need. And then, they have to confirm the orders. They can also check their order from the view orders section.

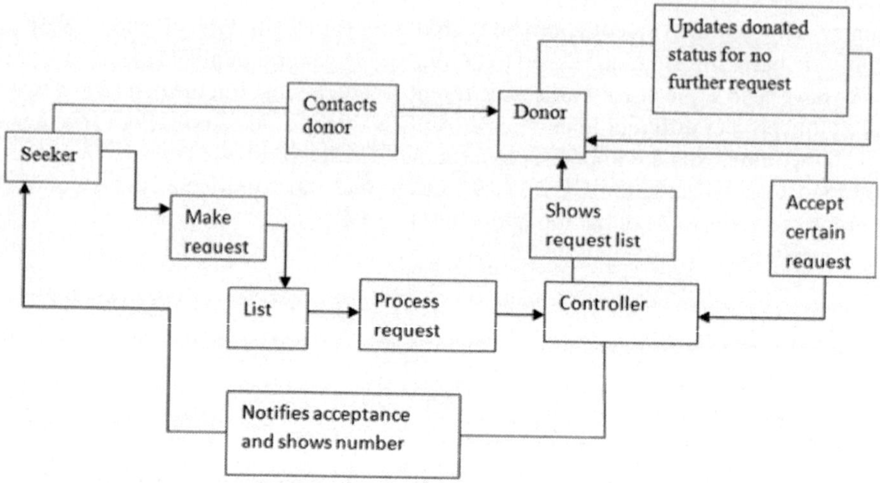

Fig. 1 Block diagram of the blood plasma system

Fig. 2 Block diagram of corona e-commerce site

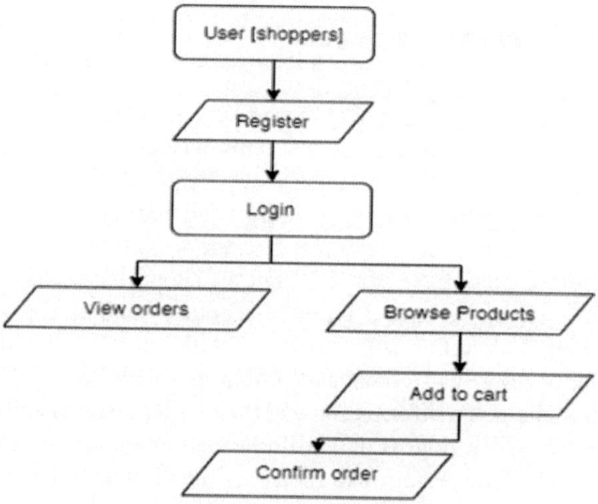

3.2 *Targeted Audience*

The targeted audience for this project is the entire population of Bangladesh. Anyone who is seeking emergency information regarding COVID-19 and has an Internet connection will be able to enjoy the facilities of this Website. More specific targeted users are those who are suspicious of being affected by corona and those who are mildly affected by corona. People outside the city area will benefit from the service to a great extent. The goal of this project is to provide proper support to people in this time of emergency without any confusion and difficulties.

3.3 Unique Features

This Web-based system is an important innovation which is very much needed at the moment. There are some Websites which provide the above-mentioned services, but there is no Website which provides all the services on a single platform. Hospital bed vacancy is one of the features which is very unique at the present moment, since no other Website provides this feature. Another unique feature is the facility of the nearest corona treating hospitals. And so, overall, the project is very innovative.

3.4 Technologies Used

For developing the Web application for this system, NodeJS is used as a backend framework and for the frontend, ReactJS is used. The ReactJS frontend app is deployed over the Firebase cloud server. For the database, NoSQL is used, and for its management, MongoDB is used. It is a cloud platform where managing NoSQL databases is very easy and seamless. For the backend, the Heroku cloud server is used for deployment. For connecting the frontend with the backend, React hooks and API ports are used.

3.5 Software Design

A Website system has been developed which will help people to get emergency services regarding COVID-19 over a virtual platform. This system solves the problem for those people who roam around mindlessly in search of actual emergency support and treatment. The system has four dimensions and each of them has its own type of user. There is a patient user type who can appoint for a corona test, can appoint for doctor consultation, request for plasma, and can also book available hospital beds. Then, there is a doctor user type who can select requested patients from a list and then consult the patients via a live session over video calling or text messaging. The doctors are also able to see patients' histories. Then, there is a hospital user type where an admin from the hospital will be able to upload the hospital information and will also be able to update bed vacancies. Finally, there is a donor user type who is a registered donor for blood plasma donation. All of these user types and facilities are visible after successful registration and verification.

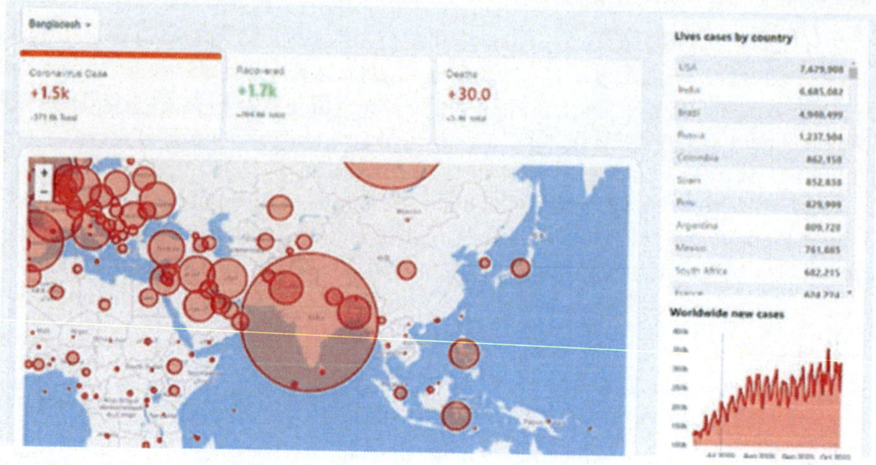

Fig. 3 COVID-19 tracker

3.6 *Web Application*

The landing page of the Website was developed with ReactJS, and the backend is powered by NodeJS. After visiting the Website, users are first able to see the services that are offered by the portal. The landing page is designed in such a way that users can easily identify what the portal offers and only with a simple click on the buttons will the users be able to avail those services. Figure 3 shows the COVID-19 tracker on the Website. This is the first service that the Website offers. It is the tracking of the live update of COVID-19 cases. The live update showcases confirmed cases, recovered cases, and death cases all around the world. The cases can be further analyzed according to different countries. And on the right side, the live update shows a table which showcases the total number of cases by country and the last 120 day update of the cases. The live tracker utilizes the data published by the Website disease.sh. This Website provides disease data for many diseases, and we are using the data provided for COVID-19 via their API endpoints. The portal offers the service of finding hospitals near the user and all the available hospitals that serve corona patients. Users would be able to contact the hospital through the Website and can also book hospital beds that are vacant.

4 Results and Analysis

Figure 4 shows the user's landing page, visible to a user who is not logged on. In the middle is a sidebar that shows the options for displaying the homepage itself and the links to other options. Users can login from this page and also register on the Website if they are not already registered. Figure 5 shows the e-commerce page where users

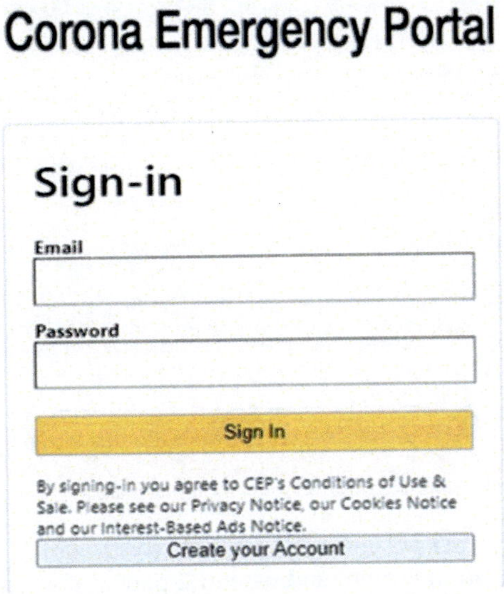

Fig. 4 Landing page of the Website

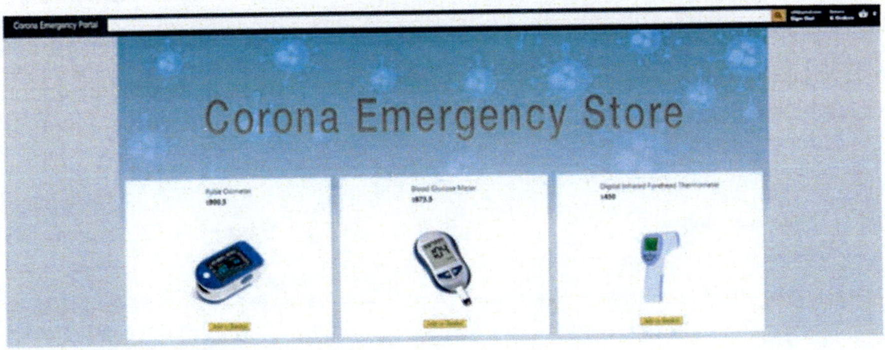

Fig. 5 E-commerce product page

can view their desired products and can add them to the cart. On this page, all the products that can be bought from the Website are shown. People can buy whatever they need from the Website after they register on the Website.

Figures 6a, b show the blood plasma bank page. On this page, people can register themselves as donors of blood plasma. They can register by filling in the information that is needed for the proper management of this donor service. This important section of the Website can help the donors get connected with the people who need blood plasma. It also shows the blood plasma bank seekers' registration page. Just like the donors, the seekers also have to register themselves on the Website. But they will register as a seeker on the seeker's registration page. Blood plasma seekers also have to provide some necessary information to be registered as a blood plasma seeker. The

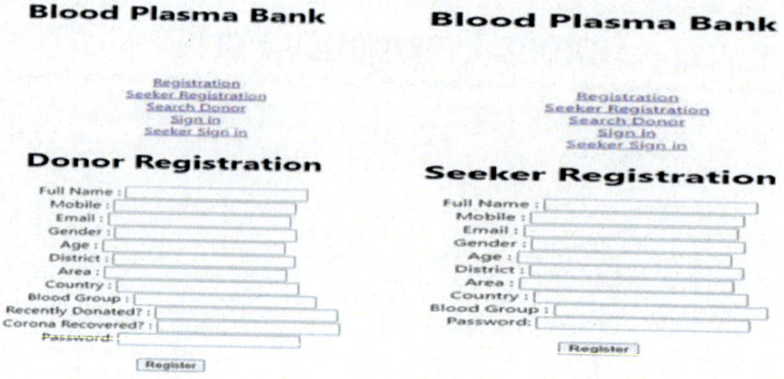

(a) (b)

Fig. 6 Blood plasma bank donor registration and blood plasma bank seeker registration

blood plasma bank is an important and essential part of this Website. Both donors and seekers can find their respective information and blood by browsing the Website.

Figure 7 shows the plasma bank blood donor searching option. This page will come up when a seeker is in need of blood plasma and he searches on our Website. The donor will have to simply fill in the two options. One is the district, and the other is the blood group. These two simple informations are needed to search for the available blood plasma on the Website. From there, users can request the donor for the blood and can get their basic information according to their needs. Figure 8 shows donor search result.

Fig. 7 Donor searching

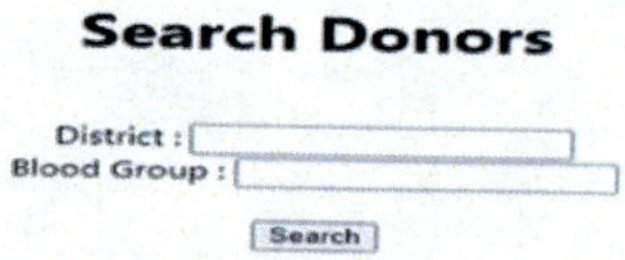

Fig. 8 Donor search result

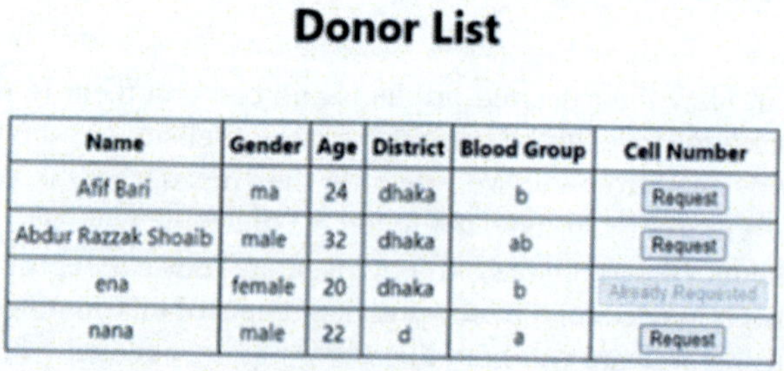

5 Conclusions

The world is facing a lot of difficulties in this time of corona outbreak, and as a result, traditional method to provide service is not sufficient at all. People are trying to stay home as much as possible and those who are moving for treatments are trying to get a rapid treatment solution. This Web portal enables people to get this innovative solution at one platform. Users can find all the things they need in this pandemic from this one Website. They can find the blood plasma that is needed in an emergency basis, they can find the all the necessary corona products in this Website and they can get the daily corona update of the world from this Website. The corona-related information section alone is worth as same as many other Websites that exist at this time. And in future, we have a vision to add video conference, doctors appointment, and some more features. This system will be a very innovative and essential element for the people of Bangladesh.

References

1. Corona tracker, Bangladesh overview. https://www.coronatracker.com/country/bangladesh/. Last accessed 22 Nov 2020
2. Deutsche Welle. Coronavirus: Bangladesh hospitals deal with shortage of doctors. https://www.dw.com/en/coronavirus-bangladesh-hospitals-deal-with-shortage-of-doctors/av-53109402. Last accessed 23 Nov 2020
3. The Daily Star. Coronavirus: where and how to get help in Dhaka, https://www.thedailystar.net/toggle/news/coronavirus-where-and-how-get-help-dhaka-1920405. Last accessed 26 June 2020
4. Disease.sh, Disease.sh. https://disease.sh/. Last accessed 26 June 2020
5. Firebase, Firebase. https://firebase.google.com/. Last accessed 26 June 2020
6. Praava health, medical services. https://praavahealth.com/visitingspecialists. Last accessed 28 June 2016
7. Worldometers, Population of Bangldesh. https://www.worldometers.info/world-population/bangladesh-population/. Last accessed 28 June 2016
8. Heroku. https://www.heroku.com/. Last accessed 26 June 2020
9. Corona Tracker, Prevention. https://www.coronatracker.com/prevention. Last accessed 28 June 2016
10. Shohojoddha, Shohojoddha. http://www.shohojoddha.com/. Last accessed 30 June 2016

Coronavirus Detection Using Computer Vision

Zuber Khan, Tanay Shubham, Naved Rehman, Rajdeep, and Ravi Kumar Arya

Abstract The coronavirus disease (COVID-19) also known as SARS-Cov-2 has largely impacted the entire globe physically, economically, and psychologically. The detection of the virus in early stages is extremely crucial for faster recovery in patients and curbing its spread as its nature is highly contagious. Although several techniques are present today for the detection of coronavirus, they are laborious in nature, costly, require experts from medical science, and the accuracy is questionable in some of the traditional methods. This brings the need to search for a faster and reliable technique. Computer vision produced remarkable results in predicting the onset of various diseases, and the use of machine learning in healthcare has increased tremendously owing to the fast speed and high accuracy of results with minimal human intervention. Hence, this research paper aims to develop a computer vision-based artificial intelligence model that can predict the occurrence of coronavirus using electron microscopic images of the samples. In order to achieve the goal, YOLO v3 object detection algorithm using non-maxima suppression is used to classify whether a particular sample has coronavirus or not. It is proved that the proposed algorithm works faster than existing methodologies with considerably higher accuracy for detection of coronavirus.

Keywords COVID-19 · Computer vision · YOLO object detection · Convolution neural network

1 Introduction

COVID-19, a contagious disease whose first case was found in Wuhan, China, in December 2019 [1] nearly brought the world to halt for several months and has

Z. Khan (✉)
Department of Electronics and Communication Engineering, Indraprastha Institute of Information Technology, Delhi, India
e-mail: zuber21181@iiitd.ac.in

T. Shubham · N. Rehman · Rajdeep · R. K. Arya
Department of Electrical and Electronics Engineering, National Institute of Technology, Delhi, India

J. K. Mandal et al. (eds.), *Topical Drifts in Intelligent Computing*, Lecture Notes in Networks and Systems 426, https://doi.org/10.1007/978-981-19-0745-6_16

changed the meaning of normal lifestyle at least for several upcoming years. This airborne virus has the potential to spread primarily through either the drops of saliva or discharge from the nose of the infected person [2]. The severity of transmission can be realized from the fact that majority of the infected patients are either asymptomatic or have mild to moderate flu-like symptoms [1]. The pandemic's global spread necessitated an immediate commitment from the whole human community. As a consequence of the increasing number of illnesses and deaths, governments throughout the world imposed restrictions on people's activities, movements, and access to products and services, as well as requiring individuals to work from home in order to curb the spreading of virus. Though in an effort to slow down the spread of the virus, researchers are taking big advances, but there is still a need for some sophisticated procedures that are useful in this situation.

Although several techniques are present today for the detection of coronavirus, they are not only time-consuming in nature but also become ineffective in curbing the spread of disease when the number of patients are astonishingly high as the testing capacity per day is limited. Furthermore, the accuracy of some traditional methods in detecting true coronavirus cases is questionable. For instance, antigen tests performed using a nasal or throat swab produce high false-negative rates, with some evidence suggesting rates as high as 42% [3]. Furthermore, the false-negative rate in case of RT-PCR technique varies from 2 to 29% [4]. This brings the need to search for a faster and reliable technique that can produce highly accurate results in fraction of seconds, and this is where computer vision comes into play.

Therefore, this research paper proposes to develop a computer-aided diagnostic system that can help in faster diagnosis of coronavirus as compared to traditional testing techniques. The facilities for human healthcare have proven limited for this abrupt outbreak. Thus, application of computer vision in detection of coronavirus could be best suited in this situation of pandemic. During the course of this pandemic, different techniques were proposed by various researchers employing machine learning for coronavirus detection, but they have certain limitations. This research paper not only overcomes all such demerits but also establishes that a higher accuracy can be obtained with the help of YOLO object detection algorithm proposed by Redmon et al. [5], if employed for the same cause on electronic microscopic images of coronavirus.

2 Literature Review

Machine learning (ML) and image processing techniques are well-known in the field of medical science for predicting diseases. Many researchers have used several machine learning, image processing, and other antibody techniques in effort to build precise and most accurate models for coronavirus detection.

Hosny et al. [6] came up with the introduction of a new ML method for the detection of COVID-19 virus. In their research, they used the chest X-ray images and implemented a visual diagnosis method on the X-ray dataset. They used a fractional moment (i.e., FrMEMs) and a modified version of Manta Ray Foraging Optimization (MRFO), modified by using a differential evolution (DE). Their model generated a set of solutions and used k-nearest neighbor algorithm (k-NN) classifiers to compute the fitness value for each of the solutions in order to decide whether a given chest X-ray image is a normal case or a COVID-19.

Furthermore, Zouch et al. [7], proposed a simple deep learning but effective convolutional neural network (CNN)-based method for classifying COVID-19, pneumonia, and no-finding chest X-ray images. The batch normalization operation was used for input normalization, and it reduced training time and increased model stability.

Recently, Rahman et al. [8], proposed rapid coronavirus identification via computer-aided diagnostics. Six distinct deep learning pre-trained CNN models—ResNet18, ResNet50, Res-Net101, InceptionV3, DenseNet201, and ChexNet—were used to create a unique form of U-Net architecture for lung segmentation from X-ray images.

However, a major problem in the above-mentioned techniques is the reliance on chest X-ray images for computation and detection purposes. It is a well-known fact that excessive X-ray radiations are harmful for the human body since there is a possibility of development of malignant tumors after several years of such exposure. This puts a limit on the number of times an individual can undergo X-ray scan within a short interval of time. Further, it is neither feasible nor practical to perform a chest X-ray of every individual suspected of COVID-19, specifically when the infection is growing exponentially and thousands of samples need to be tested every day. Also, not every country in the world has sufficient X-ray testing labs and well-qualified image processing experts to arrange for a clean dataset required to be fed into the ML models.

Lastly, Susanna et al. [9] with the aid of clinical samples of patients with severe acute respiratory syndrome evaluated the performance of two sandwich enzyme-linked immunosorbent assays (ELISAs) for detection of SARS-CoV nucleocapsid protein with the quantitative real-time polymerase chain reaction (qRT-PCR) and investigated the correlation between their results. This approach can identify recombinant virus antigens and proteins using antisera, but its drawback is the time-consuming antisera synthesis, which can only be done in specialist facilities.

The proposed methodology in this research paper neither requires X-ray images (thereby preventing the exposure to harmful radiation) nor any specialized laboratory to detect the coronavirus. Rather, it produces results within a fraction of second from the electron microscopic image of the sample, detecting coronavirus with higher accuracy than all the above methods.

3 Experimentation and Methodology

The world of computer vision is explored via 53 layered deep convolutional neural network architecture embedded in YOLOv3 model proposed by Redmon et al. [10], pre-trained on CoCo dataset and leading to the state-of-the-art, real-time unified object detection system. In the YOLOv3 model, only a single neural network categorizes the image into multiple regions leading to formation of bounding boxes around the objects present in the image. Then, the classification is done based on probabilistic distribution of each bounding box. The model is 1000 times faster than R-CNN employed for the same purpose.

3.1 Dataset

The proposed work utilizes dataset from OpenAire repository which consists of 126 electron microscopic coronavirus images of ultrathin (60–70 nm) plastic sections through extracellular SARS-CoV particles in Vero cell cultures contributed by Michael et al. [11]. 12-bit images were recorded at 0.64 nm pixel size with 1376 × 1032 pixel dimensions and saved in 16-bit TIFF format and then subsequently converted in JPEG format for use. This dataset is employed for training the YOLOv3 model. For testing purposes, random microscopic images are taken from Google Images.

3.2 IDE & Tools Used

This project was created in Python using the Jupyter Notebook Integrated Development Environment (IDE). The Jupyter Notebook is a free, open-source Web-based software application that helps in developing documents consisting of live code, visual images, mathematical equations, and text information. The primary Python libraries used in this project are Numpy and OpenCV. For training the YOLOv3 model on microscopic coronavirus images, Google Colaboratory is used to access fast computing GPU as hardware accelerator for the TensorFlow framework. For the purpose of graphically labeling the training images, LabelImg is used.

3.3 Procedure

The proposed work is based on YOLOv3 network architecture (Fig. 1) for object detection mechanism. In order to train the YOLOv3 model, its framework is cloned from the GitHub website. In its repository, the GPU, CUDDN, and OpenCV are made functional by assigning confirmatory value. The pre-trained Darknet weights already trained on large datasets are used.

Dataset Preparation. The images are associated with the coordinates and class of the object in a label file having the format of .txt. A graphical user interface (GUI)

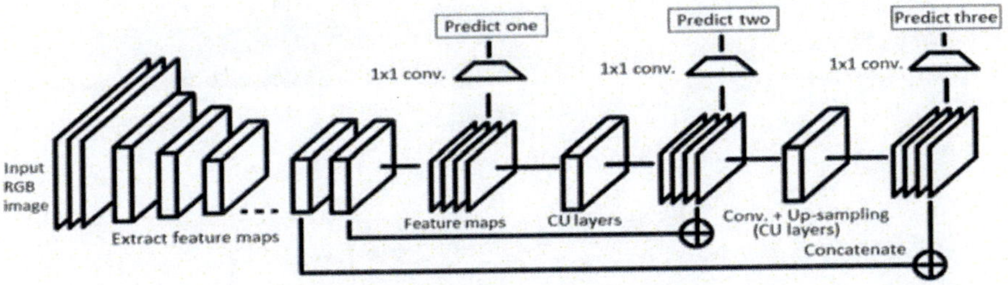

Fig. 1 YOLO v3 architecture (*Source* Punn et al. [12])

named LabelImg is used for generation of the label files. It is a graphical annotation tool and compatible with the model giving satisfactory results.

Splitting of the Dataset, Initializing the Training Path. To avoid over-fitting and mitigate discrepancies in evaluation of model, dataset is split into training set and testing set. We have split it into 20% test dataset and 80% training dataset.

Training in Google Colaboratory. As the complexity of the YOLOv3 model is high and due to its multi-scale detection, the GPU which is provided in personal computers gives unsatisfactory results and mostly proves to be futile for training purposes. So the model is trained on Google Colaboratory using following steps:

- Synchronizing the Darknet folder containing framework, processed dataset, and pre-training weights of the model.
- Installing Dos2Unix to remove Windows Formatting and converting all the files to Unix encoding.
- Compiling the Darknet framework and testing it with sample object detection.
- Making a backup folder in Google Drive to save training weights of customized model.
- Training the model on the COVID-19 training dataset.

Prediction Using Spyder Notebook. Forward method is implemented for extracting feature maps from an image in iteration till every feature map is selected, and the output layer is reached. In the selection process, the model proposes many bounding boxes for a feature map and for optimum performance. Non-maxima suppression (NMS) (Hosang et al. [13]) is used which considers only the maximum-valued bounding box of the feature map while rejecting all other proposals. The prediction for a random electron microscopic image of a coronavirus sample is made in the console and the output is displayed.

4 Results and Observations

The console output displays whether the image has coronavirus, at what location along with percentage accuracy. One such result is shown in Fig. 2. One can clearly

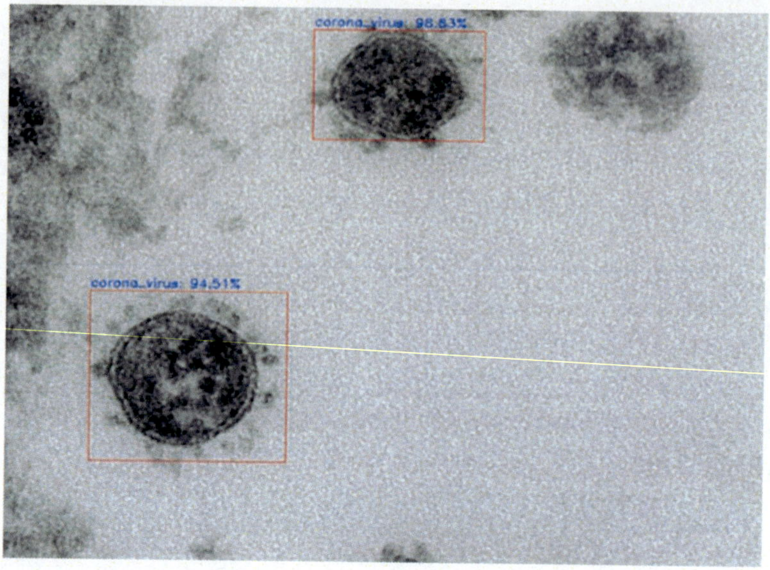

Fig. 2 COVID-19 positive case

observe from the figure that the model produces bounding box around the coronavirus particle along with the percentage accuracy of detection on top of it. This shows that the algorithm is successful in capturing that the sample has coronavirus. If there is no COVID-19 particle in a sample, the resultant image will not contain any bounding boxes.

Table 1 compares our work with previous studies in the same domain.

Clearly, the table shows that our research has higher accuracy than other methodologies. YOLOv3 which is a 53-layered convolution neural network processes the input images and detects the patterns among them. Based on the training of YOLOv3 on 126 coronavirus images dataset, the prediction on incoming image is made. The training itself took 15 days to complete with the rate of around 5 epochs per hour. The training weights were saved after every 100th epoch, and the variation of validation loss with the number of epochs is as shown in Fig. 3.

Table 1 Comparative analysis

Author	Methodology	Accuracy (%)
Hosny et al.	K-NN classifier on chest X-ray images	96.06
Zouch et al.	CNN classifier on chest X-ray images	91.34
Rahman T. et al.	U-Net architecture on chest X-ray images	95.11
Khan et. al (our work)	YOLO v3 on electron microscopic images	98.83

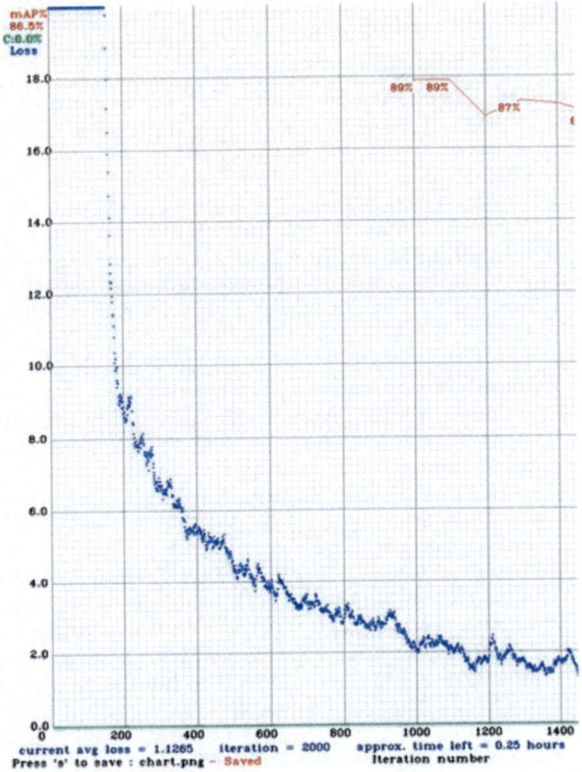

Fig. 3 Validation loss versus epochs

The figure illustrates the variation of validation loss with the number of iterations during the training of the YOLOv3 model. It is observed that the validation loss decreases significantly with the increment in number of epochs and becomes substantially constant after the 1300th epoch.

5 Conclusion and Further Discussion

From the current research, we conclude that the YOLOv3 architecture performs appreciably well for COVID-19 detection using electron microscopic images. The model training saturates within 1300 epochs for the given dataset, and the testing accuracy is as high as 99%. Therefore, its use is suggested for practical purposes in all clinical testing setups wherever feasible after more rigorous training on a larger dataset, i.e., minimum 5000 images for more accurate prediction in practical implementation. It is advocated to observe the performance of single-shot detector (SSD) and spatial pyramid pooling (SPP-net) for the same purpose.

References

1. Kumar R, Nagpal S, Kaushik S et al (2020) COVID-19 diagnostic approaches: different roads to the same destination. VirusDis. 31:97–105. https://doi.org/10.1007/s13337-020-00599-7
2. Alpdagtas S, Ilhan E, Uysal E, Sengor M, Ustundag CB, Gunduz O (2020) Evaluation of current diagnostic methods for COVID-19. APL Bioeng 4:041506. https://doi.org/10.1063/5.0021554
3. Dinnes J, Deeks JJ, Berhane S, Taylor M, Adriano A, Davenport C, Dittrich S, Emperador D, Takwoingi Y, Cunningham J, Beese S, Domen J, Dretzke J, Ferrante di Ruffano L, Harris IM, Price MJ, Taylor-Phillips S, Hooft L, Leeflang MMG, McInnes MDF, Spijker R, Van den Bruel A (2021) Rapid, point-of-care antigen and molecular-based tests for diagnosis of SARS-CoV-2 infection. Cochrane Database Syst Rev (3). Art. No.: CD013705. https://doi.org/10.1002/14651858.CD013705.pub2
4. Watson J, Whiting PF, Brush JE (2020) Interpreting a covid-19 test result BMJ 2020; 369:m1808. https://doi.org/10.1136/bmj.m1808
5. Redmon J, Divvala S, Girshick R, Farhadi A (2016) You only look once: unified, real-time object detection. In: Proceedings of the IEEE conference on computer vision and pattern recognition. pp 779–788
6. Elaziz MA, Hosny KM, Salah A, Darwish MM, Lu S, Sahlol AT (2020) New machine learning method for image-based diagnosis of COVID-19. PLoS ONE 15(6):e0235187. https://doi.org/10.1371/journal.pone.0235187
7. Echtioui A, Zouch W, Ghorbel M, Mhiri C, Hamam H (2020) Detection methods of COVID-19. SLAS Technol 25(6):566–572. https://doi.org/10.1177/2472630320962002
8. Rahman T, Khandakar A, Qiblawey Y, Tahir A, Kiranyaz S, Abul Kashem SB, Islam MT, Al Maadeed S, Zughaier SM, Khan MS, Chowdhury M (2021) Exploring the effect of image enhancement techniques on COVID-19 detection using chest X-rays images. Comput Biol Med 104319. Advance online publication. https://doi.org/10.1016/j.compbiomed.2021.104319
9. Lau S, Che X, Yuen K, Wong B, Cheng V, Woo G et al (2005) SARS coronavirus detection methods. Emerg Infect Dis 11(7):1108–1111. https://doi.org/10.3201/eid1107.041045
10. Laue M, Kauter A, Hoffmann T, Michel J, Nitsche A (14 Aug 2020) Electron microscopy of SARS-CoV particles—dataset 01, openAire. https://doi.org/10.5281/zenodo.3985098
11. Redmon J, Farhadi A (2018) Yolov3: an incremental improvement. arXiv preprint arXiv:1804.02767
12. Punn NS, Sonbhadra SK, Agarwal S, Rai G (2020) Monitoring COVID-19 social distancing with person detection and tracking via fine-tuned YOLO v3 and Deepsort techniques. arXiv preprint arXiv:2005.01385
13. Hosang J, Benenson R, Schiele B (2017) Learning non-maximum suppression. In: Proceedings of the IEEE conference on computer vision and pattern recognition. pp 4507–4515

A Novel Distributed Database Architectural Model for Mobile Cloud Computing

Somenath Chakraborty, Dia Ali, and Beddhu Murali

Abstract Cloud computing is the way by which we connect to servers, large systems into a distributed secure manner without worrying about local memory limits. Here, in this paper, we proposed a Novel distributed database architectural model for mobile cloud computing (NDDAMMCC). Accelerating the exponential growth of wireless technologies and Internet which are following Nielsen's Law of Internet Bandwidth, we are in the new era of cloud computing. In the recent technological era, smart mobile devices play a big role in all sort of day-by-day human needs. The applicability is so huge that the number of apps install on a mobile system becomes a hazard due to local memory limitations for mobile phone users and demands an alternative approach to solve this local memory problems. Mobile cloud computing (MCC) is the ultimate mechanism to this issue, and our model presents a promising path in this new kind of cloud computing technology.

Keywords Cloud computing · Novel distributed database architectural model for mobile cloud computing (NDDAMMCC) · Nielsen's law of Internet bandwidth

1 Introduction

The cloud computing is the way by which we connect servers, large systems into a distributed secure manner without worrying about local memory limits. Initially, we only focused on desktop computing applications to run on cloud systems, but in the recent time, smartphones become so powerful and evident part of our daily life, that we are now capable of performing any kind of computing applications by using only our smartphones. But smartphones faces new challenges especially due

S. Chakraborty (✉) · D. Ali · B. Murali
The University of Southern Mississippi, Hattiesburg, MS 39406, USA
e-mail: somenath.chakraborty@usm.edu

D. Ali
e-mail: dia.ali@usm.edu

B. Murali
e-mail: beddhu.murali@usm.edu

J. K. Mandal et al. (eds.), *Topical Drifts in Intelligent Computing*, Lecture Notes in Networks and Systems 426, https://doi.org/10.1007/978-981-19-0745-6_17

Table 1 Mobile cloud computing applications

Class	Usages and properties	Examples
Private	Very secure applications; all the hardware not shared and own by the owner of the private cloud	Mobile bank applications, server login applications etc.
Public	Anybody can freely access this kind of applications; it is shared on open Internet so anybody can access this services, less secure, hardware, and software not own by the owner of public cloud. It provides cost efficiency and promotes agile methodologies	Any kind of free applications which are deployed on open internet
Hybrid	Integration of private and public both layers	Any MCC applications or CloudOps which consist of free and purchased features of content and services, like credit check systems

to the memory usage, processing infrastructure shortage due to the local smartphone RAMs, and other mobile memory management limitations. Here, this paper mainly focused on presenting distributed database architectural model for mobile cloud computing. With the advancement of Internet speed, Internet bandwidth, efficient dynamic channel allocation for wireless communication [1], Internet of things (IoT) revolution, and the technological development of smartphone, we have much flexibility to develop mobile architectural model that can run on the cloud using mobile cloud computing (MCC) applications. MCC deployments mainly categorized into three classes of applications. The following Table 1 shows the listings.

2 Literature Review

According to NIST SP 800–145 [2, 3], any cloud computing architectural model should adhere at least these properties, like on-demand self-services, ubiquitous network access, resource pooling, rapid elasticity, and pay-per use.

They [2] also represent the standardization of three service model of cloud computing. Starting with the most important one software as a service or SaaS model, a third-party company or business, hosts application software for the end user, other components, like hardware, are maintained by the provider company. The next one is infrastructure as a service or IaaS model. A third-party provider company hosts the software, hardware, services, and other aspects of the system. An organization that wants to maximize costs in the company and needs the scaling would choose this cloud. Another important one is platform as a service or PaaS model, where the third-party company provides the deployment platforms for the end user. It supports cloud versions of application development, deployment of different services, and hosting on cloud platform. The following Fig. 1 portrays the visualization for these

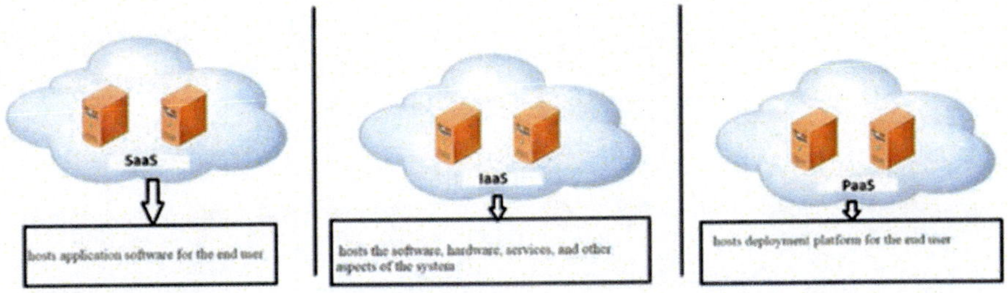

Fig. 1 The general overview of three service model

three types of cloud model. Zhang et al. [4] addresses the issues and challenges faced by the cloud computing infrastructure development and recognize significant development in this emerging field.

Huang and Wu [5] describe in their book chapter, how different kinds of mobile cloud computing framework works and the basic implementation of their architecture, development challenges, and basic properties.

The forum [6] formulates the definition of MCC as a simplest way to support infrastructure, data storage, and data processing should be performed from the outside of the mobile devices. They should be from the cloud and also broaden the services not only the smart phone users, but a host of application supports to its subscribers as well. Aepona [7], shows the potential of mobile cloud computing with the high processing infrastructure support and huge fast processing memory support. Ruay Shiung Chang et al. present a comprehensive description of MCC issues, challenges, and provides the insights of the needs in present times, where there is a demand created for different kinds of cloud applications. Though all those papers provide insightful information and show the pathways for new kinds of cloud support systems, but many are mostly using traditional infrastructure and sometime lacks crucial efficiency and limitations in client demands.

Here, in this paper, we proposed an MCC model which can leverages the advancement of distributed computing through sharing and processing data through distributed data sharing support. Prasad et al. [8] describe resource distributing issues. It also describes the comprehensive analysis of recent research in this domain.

3 Proposed Methodology

Our NDDAMMCC is illustrated in Fig. 2. The architecture supports the advanced distributed cloud supports which enable global agility and robustness to the overall architectural system. In this architectural model, there are basically four paradigm of infrastructure embedded into a complete system model. It is shown in Fig. 3. Different kinds of mobile devices get the dynamic signals [1] in a portable moving environment, where the access point (AP's) establish and support the wireless services by routing

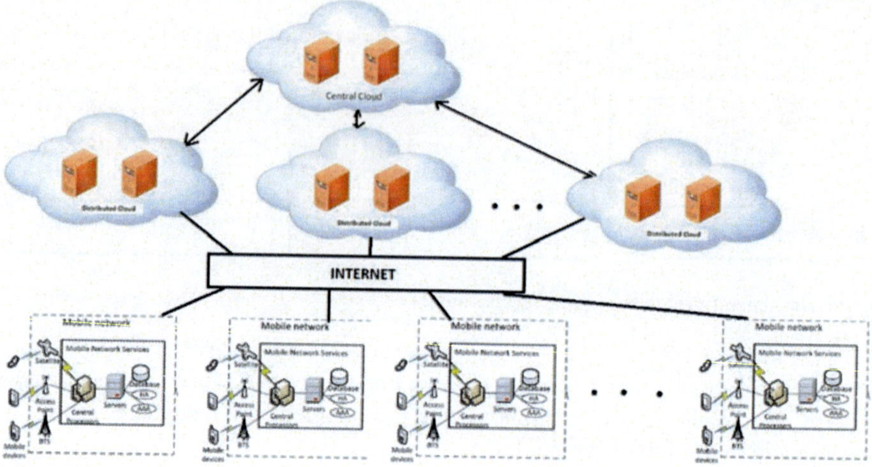

Fig. 2 A novel distributed database architectural model for mobile cloud computing (NDDAMMCC)

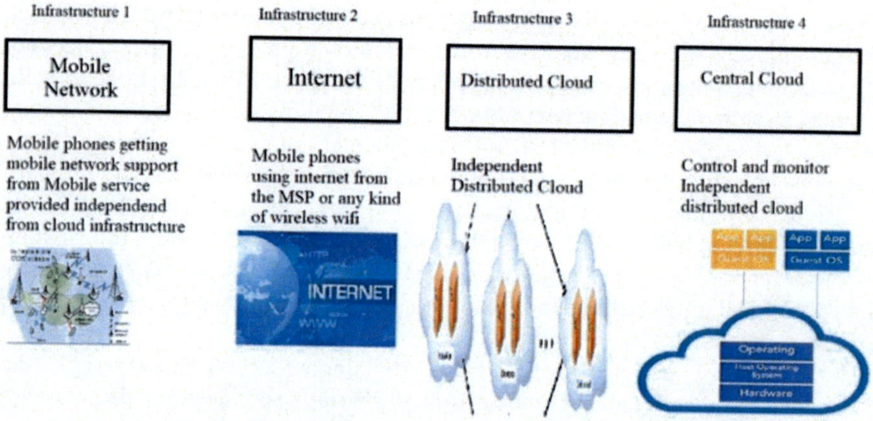

Fig. 3 Four types of infrastructure support for our novel distributed database architectural model for mobile cloud computing (NDDAMMCC)

the signals coming from the mobile base stations (BTS). A set of mobile networks connected through the distributed cloud infrastructure which are capable of providing all kinds of cloud services as well as distributed high speed memory and application processing support and other platforms support. These distributed clouds coupled in an independent architecture, but for better security control, these distributed system integrated with the central cloud systems. Through the mobile networks are shown in an integrated format in Fig. 2, they are fully independent and their other kinds of mobile services do not rely on cloud infrastructure support. The distributed cloud also do not provide hardware or software support to the mobile service providers. So, individually any mobile user can subscribe these distributed cloud services though the Internet and get all those enhanced experiences and facilities of mobile cloud computing (MCC).

The novel distributed database architectural model for mobile cloud computing (NDDAMMCC) algorithm is presented in Algorithm 1.

Algorithm 1: The Novel Distributed Database Architectural Model for Mobile Cloud Computing (NDDAMMCC)

1. Start
2. Input: hostList, ApplicationList Output: Allocation of Distributed shared resource to ApplicationList.
3. DistributedCloud.Orderby.ResourceReq()
4. For each ResourceReq() in ApplicationList Do
5. If ResourceReq() < DistributedCloud_Capacity()
6. Allocate Demand resource from the current distributed cloud
7. Else ResourceReq != Satisfied
8. search for nearest_distriburd cloud n1,n2,n3…upto nn
9. Allocate Demand resource = Distributed_cloud(any of n1, n2, n3…nn)
10. Do: Max_allocation >= Small_ApplicationList
11. Host_allocation with Accesspointsupport(mode=Dynamic)
12. Resource.Add = Resource_Required(mode= dynamic)
13. Use Resource until ApplicationList finished
14. End

4 Experimental Simulations

This distributed cloud are interconnected through distributed database support systems and monitor a distributed log which is auto-updated and protects all kinds of vulnerabilities and attracts from the outside world. We use CloudSim simulator [9], and the simulation results are presented with the following terminologies described in that [9] CloudSim simulator paper (Table 2 and Fig. 4).

5 Conclusion

The main advantages of this model are that if any system fails, then also it provides uninterrupted services in a distributed manner. This proactive model prevents failures of the system as it is auto-managed in a dynamically space-resource by the central log from its distributed child nodes.

Table 2 Analysis of results based on a novel distributed database architectural model for mobile cloud computing (NDDAMMCC)

Distributed cloud details (VMs)	Capacity (Dynamically varying) using space shared	Estimated finish time (in milliseconds)	Total processing capacity of cloud host
12 tasks in 3 VMs	26	239.24	16
23 tasks in 8 VMs	24	568.74	18
39 tasks in 12 VMs	28	698.45	16

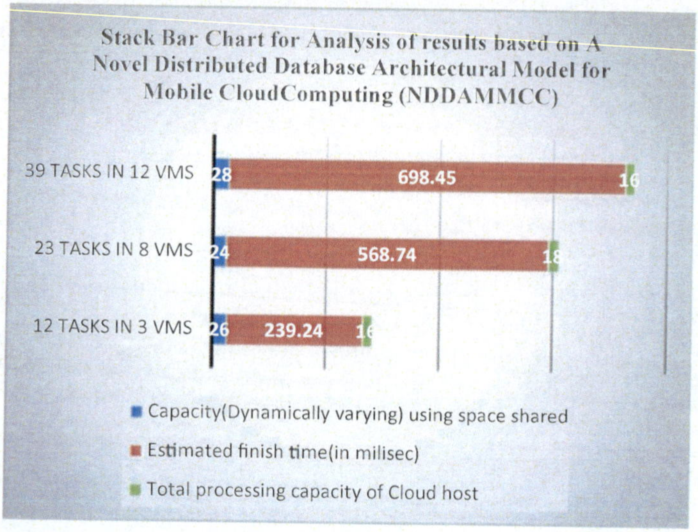

Fig. 4 Stack bar chart for analysis of results based on a novel distributed database architectural model for mobile cloud computing (NDDAMMCC)

References

1. Chakraborty S, Saha B, Bandyopadhyay SK (Sep 2016) Dynamic channel allocation in IEEE 802.11 networks. Int J Comput Appl 149(1):36–38
2. Simmon E (2018) Evaluation of cloud computing services based on NIST SP 800–145, special publication (NIST SP), national institute of standards and technology, Gaithersburg, MD, [online], https://doi.org/10.6028/NIST.SP.500-322 (Accessed 30 May 2021)
3. Badger L, Bernstein D, Bohn R, de Vaulx F, Hogan M, Iorga M, Mao J, Messina J, Mills K, Simmon E, Sokol A, Tong J, Whiteside F, Leaf D (2014) US government cloud computing technology roadmap—volumes I and II, NIST SP 500-293. NIST. p 40 (vol I), 98 (vol II)
4. Zhang Q, Cheng L, Boutaba R (2010) Cloud computing: state-of-the-art and research challenges. J Int Serv Appl 1(1):7–18
5. Huang D, Wu H (2018) Mobile cloud computing service framework. Mobile Cloud Comput 91–114. https://doi.org/10.1016/b978-0-12-809641-3.00006-5
6. Mobile cloud computing forum. http://www.mobilecloudcomputingforum.com/
7. White Paper (2010) Mobile cloud computing solution brief. AEPONA

8. Rajendra Prasad M, Lakshman Naik R, Bapuji V (April 2013) Cloud computing: research issues and implications. Int J Cloud Comput Serv Sci (IJ-CLOSER) 2(2):133–139. ,ISSN: 2089-3337
9. Calheiros RN, Ranjan R, Beloglazov A, De Rose CA, Buyya R (2011) CloudSim: a toolkit for modeling and simulation of cloud computing environments and evaluation of resource provisioning algorithms. Software: Pract Experience 41(1):23–50. https://doi.org/10.1002/spe.995
10. Chang RS, Gao J, Gruhn V, He J, Roussos G, Tsai W-T (2013) Mobile cloud computing research—issues, challenges and needs. Proceedings of IEEE seventh international symposium on service-oriented system engineering. San Francisco, USA
11. Chakraborty S, Murali B (2021) Investigate the correlation of breast cancer dataset using different clustering technique. arXiv preprint arXiv:2109.01538
12. Chakraborty S. Category identification technique by a semantic feature generation algorithm. In: Deep learning for internet of things infrastructure. CRC Press, pp 129–144
13. Chakraborty S, Murali B (2021) A novel medical prognosis system for breast cancer. 2021 international conference on advanced computing applications (ICACA)
14. Chakraborty S, Zhang C (2020) Survival prediction model of renal transplantation using deep neural network. 2020 IEEE 1st international conference for convergence in engineering (ICCE). pp 180–183. https://doi.org/10.1109/ICCE50343.2020.9290695
15. Salama AS (2015) A Swarm Intelligence Based Model for Mobile Cloud Computing. Int J Inf Technol Comput Sci (IJITCS) 7(2):28–34. https://doi.org/10.5815/ijitcs.2015.02.04

Contemporary Issues in Electronics, and Communication Technology

Dual-Band Stop Filter with Controllable Stop-Bands Based on Defect in Shunt Radial Stub

Hare Krishna, Prashant Kumar Singh, Deepak Sharma, and Anjini K. Tiwary

Abstract In this article, a novel planar, compact, and wideband dual band-stop filter (DBSF) with controllable stop-bands are proposed. The first band-stop region is generated by using shunt connected radial shaped resonator to microstrip line. However, the novelty is achieved by generating another band-stop region by means of defect created within the radial stub on the same plane. This defect is created by etching some pattern from shunt connected radial stub, which results in compact filter configuration. The two rejection bands can be independently tuned by varying the dimensional parameters of radial stub and the defected region inside the radial stub, respectively. Further, cascading of the defected radial stub is done in order to achieve improved pass-band region between two stop-bands as well as stop-band regions. The proposed DBSF is designed, fabricated, and tested for WLAN bands at 2.4 GHz and 5.4 GHz. The proposed filter shows good in-band and out-of-band response up to 10 GHz. The experimental results agree well with simulation results, thus validating the proposed design.

Keywords Defected radial stub · Filter · Folded resonator · WLAN

1 Introduction

Modern wireless/microwave communication systems demand low-cost, compact, and multiband systems. These demands are also dictated on the filters, being the integral part of the communication system. The band-stop filter (BSF) [1] rejects the unwanted or spurious signals and permits the desired frequency signals. Dual

H. Krishna (✉) · D. Sharma · A. K. Tiwary
Department of ECE, JUET, Madhya Pradesh, Guna 473226, India

A. K. Tiwary
e-mail: aktiwary@bitmesra.ac.in

P. K. Singh
Department of ECE, UCET, VBU, Hazaribag, Jharkhand 825301, India

H. Krishna · P. K. Singh · D. Sharma · A. K. Tiwary
Department of ECE, BIT Mesra, Ranchi, Jharkhand 835215, India

J. K. Mandal et al. (eds.), *Topical Drifts in Intelligent Computing*, Lecture Notes in Networks and Systems 426, https://doi.org/10.1007/978-981-19-0745-6_18

band-stop filters (DBSFs) have applications in high-power amplifiers and mixers to suppress double sideband spectrum [2]. Various design methods using stepped impedance resonator (SIR), open stubs, open coupled lines, defected ground structure (DGS), etc., for realizing DBSFs are available in literature. SIRs are utilized to realize DBSFs in compact size [3]. However, for designing filter with frequency ratio (f_{r2}/f_{r1}) less than 2, SIRs are not a good selection. The concept of transversal signal interaction and open circuited stubs is employed to achieve compact DBSFs with wide stop-band characteristics [4]. The open stubs and transversal signal interaction concept offer low-frequency ratio but shows structural complexity. A novel compact DBSF [5] is proposed using coupled line stub, which in turn results in narrow stop-bands. A compact DBSF is achieved by means of unequal split ring resonator [6] and embedded structure [7]. Semi-lumped circuits are also a good method to realize the DBSF. A microstrip structure with lumped capacitor [8] is used to design DBSF to realize wide rejection bandwidth along with compactness. Employing lumped elements increases the component and assembling expenditure. The defected ground structure [9] concept is also included to achieve the miniaturization and enhance the performance of DBSF; however, it increases the radiation due to etched ground plane. These problems like radiation, complexity, and stop-band bandwidth, cost, and frequency ratio can be minimized by using the planar radial configuration. It is evident that radial shaped microstrip resonators have inherent wide stop-band feature. A compact DBSF [10] is proposed by embedding a radial stub inside the slotted microstrip radial resonator. Even slotted structure is already presented; though, this paper extends the parametric study of this resonator incorporating a novel radial folded resonator. The novel structure formed by combining these two resonators can be seen as a novel defected radial stub, and the advantage of using the novel folded resonator is the more space within the slot. This in turn gives more design flexibility to choose the upper resonant frequency near to lower resonant frequency by increasing the length of folded resonator. This is not the case with embedded radial stub [10], as there is not much space for further size expansion of inner radial stub.

This paper proposes a novel compact DBSF, for center frequencies 2.4 GHz and 5.4 GHz. The novelty of the work is in terms of novel defect pattern within conventional radial stub, which allows for controlling two stop-bands independently during design. The defect is created by etching out some metallic portion from radial stub on the same plane. This in turn reduces the fabrication complexity as compared to DGS (where alignment between top and ground structures is of utmost importance). Additionally, no extra resonators are attached to the circuit for generating second band-stop region; which may result in increased circuit size. Here, the additional stop-band regime is created by etching metallic portion from the stub; and hence, the metallic loss will decrease as well as compactness will be achieved. The defected radial stub configuration may be seen as the combination of slotted microstrip radial stub and folded open stub resonator in radial fashion. First section explains the shunt connected slotted microstrip radial configuration. Next section presents the novel defected radial stub resonator as DBSF. Further, circuit analysis for unit cell is detailed followed by proposed cascaded DBSF configuration, experimental results, and conclusion.

2 Slotted Microstrip Radial Resonator

In this work, method of moments-based IE3D full wave simulator by mentor graphics is used for simulation of all the microstrip configurations. The substrate used for the design is economical glass epoxy FR-4. The thickness, dielectric constant, and loss tangent of the substrate are 1.56 mm, 4.4, and 0.016, respectively.

The radial shaped resonator illustrated in Fig. 1a has advantages of compact size and good resonance at well specified point in wide frequency band. The lumped equivalent of radial stub is composed of an inductor in series with capacitor. The input impedance (Z_{in}), inductance (L_{rs}), capacitance (C_{rs}), and resonant frequency (f_r) of the radial stub can be calculated as [11]:

$$Z_{\text{in}} \cong -j\frac{120\pi\beta h}{\theta_r\sqrt{\varepsilon_{\text{eff}}}}\left(\ln\frac{r_i}{r_o}+\frac{1}{2}+\frac{2}{\beta r_o^2}\right) \tag{1}$$

$$L_r = \frac{120\pi h}{\theta_r c}\left[\ln\frac{r_o}{r_i}-\frac{1}{2}\right] \tag{2}$$

$$C_r = \frac{\theta_r r_o^2 \varepsilon_{\text{eff}}}{240\pi hc} \tag{3}$$

$$f_r = \frac{1}{2\pi\sqrt{L_{rs}C_{rs}}} \tag{4}$$

The parameters ε_{eff}, c, β, h, and θ_r used in above equations depict effective dielectric constant, speed of light, phase constant, thickness of dielectric, and radial angle, respectively.

The characteristics of the conventional radial stub filter can further be enhanced to provide more compact structure and suppressed higher mode resonance in frequency band of interest. In order to realize a compact structure along with improved characteristics, in the first step, a radial slot is etched out from the conventional radial stub,

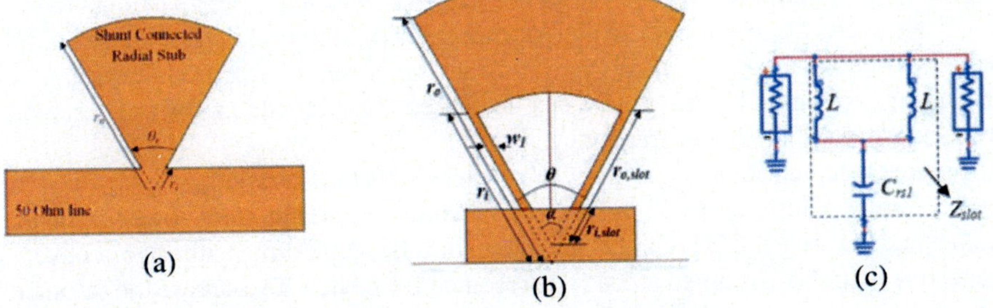

Fig. 1 **a** Shunt connected conventional radial stub resonator, **b** slotted microstrip radial stub filter, **c** lumped equivalent model slotted stub [10]

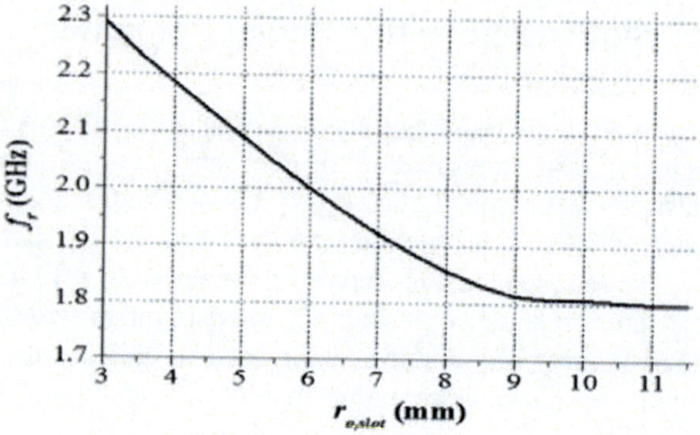

Fig. 2 Shift in the resonant frequency with the variation in outer radius of the radial slot in slotted microstrip radial filter

as presented in Fig. 1b. Figure 1c illustrates its lumped equivalent circuit model. Comparing the microstrip and lumped model, it is clear that the thin high impedance microstrip line (shunt connection with 50 Ω line) resembles inductance, L; and the thick radial microstrip section connected next to the thin line depicts capacitor, C_{rs1}. To investigate the effect of slot, first conventional radial filter is designed at 2.4 GHz resonant frequency and then the frequency shift can be analyzed with the variation of outer radius ($r_{o,\text{ slot}}$) and radial angle (α) of the slot.

The change in resonant frequency (f_r) with the change in outer radius ($r_{o,\text{ slot}}$) of radial slot is shown in Fig. 2. To investigate the effect, other structural parameters shown in Fig. 1b are kept constant. The $r_{o,\text{ slot}}$ is increased from 3 to 11.5 mm, keeping $r_{i,\text{ slot}}$, w_1, r_i, r_o, θ, and α as 2.3 mm, 0.58 mm, 3.5 mm, 16 mm, 60°, and 60°, respectively. The characteristic depicts that as the $r_{o,\text{ slot}}$ is increased, the f_r decreases because of significant increase in inductance as compared to decrease in capacitance due to slot. Equation (4) depicts that the overall increase in inductance-capacitance product results in the shift of resonant frequency toward the lower frequency range.

For $r_{o,\text{ slot}} \geq 8.5$ mm, the resonant frequency is almost constant, i.e., 1.8 GHz, which shows that increase in inductance and decrease in capacitance are in proportion so as to provide the same inductance-capacitance product. To achieve the compactness, lower value of $r_{o,\text{ slot}} = 8.5$ mm is chosen and slot angle (α) is further changed keeping all other parameters same. The change in resonant frequency with the change in slot angle is given in Fig. 3. This shows that as α will increase, f_r will decrease. It is clear from the Fig. 1b–c, while varying the slot angle, the capacitance (C_{rs1}) remains constant as ($r_o - r_i$) and θ are constant; however, inductance increases with increasing slot angle as the width (w_1) decreases. This results in shift of resonant frequency toward the lower frequency.

Figure 3 depicts that by etching the slot with parameters $r_{i,\text{ slot}}$, $r_{o,\text{ slot}}$, and α as 2.3 mm, 8.5 mm, and 60°, respectively, within the radial stub; lower resonant frequency, that is 1.8 GHz, can be achieved with the same dimension as of conventional one, which is designed at 2.4 GHz resonant frequency. To increase the resonant frequency from 1.8 GHz to 2.4 GHz, the overall size of slotted microstrip radial filter needs to be decreased. This is achieved by reducing the outer radius (r_o) of the slotted

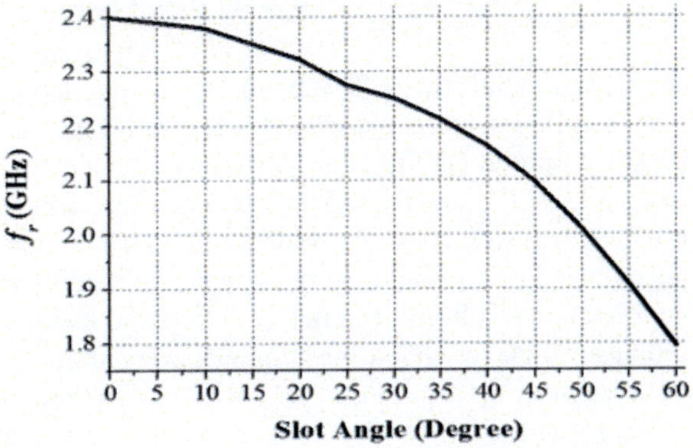

Fig. 3 Shift in the resonant frequency with the variation in slot angle (α) in slotted microstrip radial filter

microstrip radial stub, which in turn reduces the capacitance (C_{rs1}) and thus resonant frequency move toward required frequency 2.4 GHz. Thus, the slot introduces compactness in comparison with the conventional radial filter.

Now, it is clear that the use of slot provides miniaturization of microstrip structure. Additionally, the higher order harmonic can also be suppressed by tuning the parameters of slotted structure, as depicted in Fig. 4. It shows comparison of the transmission characteristics of the radial filter without and with slot for 2.4 GHz resonant frequency, keeping parameters r_i, θ, w_1, α, and $r_{i,\,\text{slot}}$ constant. It is clear from the Fig. 4 that by introducing the slot, the higher mode resonance (in between 7 and 10 GHz frequency band) also gets suppressed in slotted microstrip radial filter and a better pass-band response is achieved. This is because of the fact that the slot itself is equivalent to a resonator and choosing the dimension of the resonator appropriately, a transmission pole can be forced to occur at the point of location of the harmonic; thus, nullifying the effect of harmonic. Hence, the slotted microstrip radial stub not only offers compact structure but also suppresses the harmonic at the specific point in the frequency band.

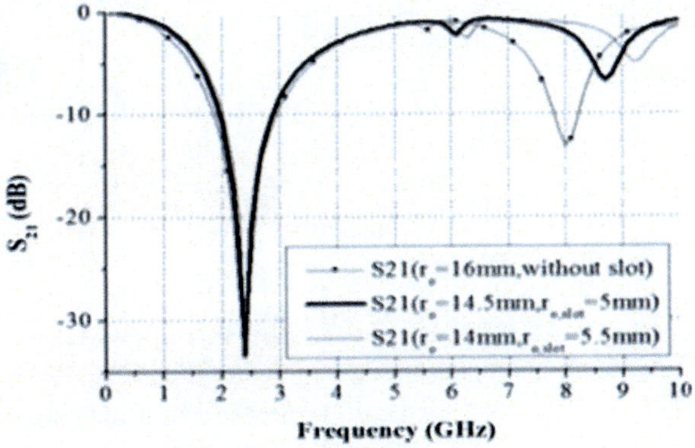

Fig. 4 Transmission characteristics of radial filter with and without radial slot for 2.4 GHz resonant frequency for $r_i = 3.5$ mm, $\theta = 60°$, $w_1 = 0.58$ mm, $\alpha = 60°$, and $r_{i,\,\text{slot}} = 2.3$ mm

3 Proposed Defected Radial Stub

The slotted radial stub is developed by creating defect in radial stub by etching out radial shaped structure as depicted in Fig. 1b. Here, a novel defected radial stub (unit cell for proposed DBSF) is proposed, with new etching pattern as shown in Fig. 5, for $f_{r1} = 2.4$ GHz and $f_{r2} = 5.4$ GHz center frequencies. This novel defected pattern provides dual stop-band performance and also compactness. As depicted in Fig. 5, this defected radial stub may be seen as the combination of slotted radial stub and modified open stub (folded open stub in radial fashion). Obviously, the slotting and folding are done to achieve the miniaturized configuration. Moreover, these two stubs are independently responsible for two stop-band regions of DBSF.

The slotted radial stub has been already discussed in last section. In the proposed configuration, slotted stub is designed for 2.4 GHz. The dimensional parameters $r_{i,\,\mathrm{slot}}$, $r_{o,\,\mathrm{slot}}$, w_1, r_i, r_o, and $\theta = \alpha$ of miniaturized slotted stub for the above said frequency are obtained as 2.3 mm, 8.5 mm, 0.58 mm, 3.5 mm, 13.4 mm, and 60°, respectively. In modified open stub, a is the width of stub, l_{os} is the length of stub, g is the gap between annular sectors, β is the sector angle, r_{i1} is inner radii, and r_{o1} is the outer radii. The resonant frequency of the modified open stub depends on the length of the stub, which is given by

$$f_{r2} = \frac{c}{\lambda} = \frac{c}{4l_{OS}} \tag{5}$$

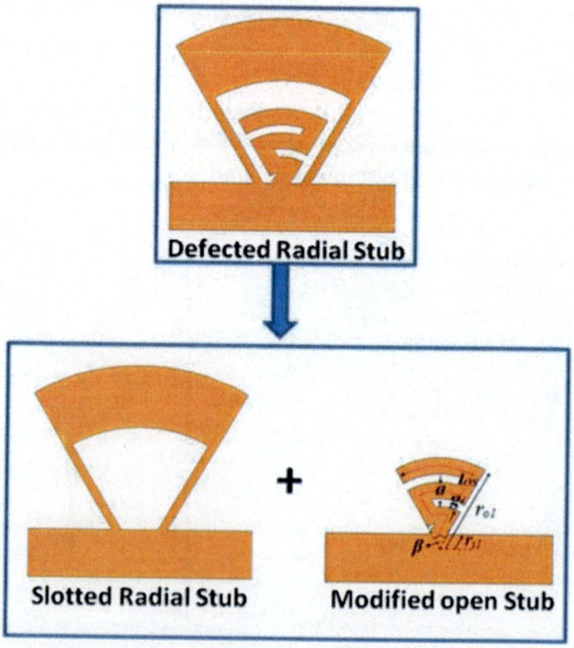

Fig. 5 Proposed defected radial stub resonator as unit cell for DBSF

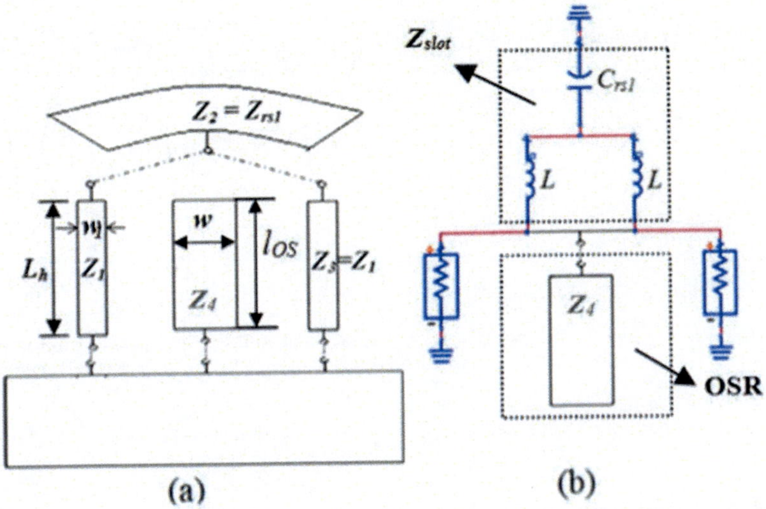

Fig. 6 **a** Basic model of proposed defected radial stub, **b** its equivalent circuit

$$i.e. \quad l_{OS} = (r_{o1} - r_{i1} - 3a) + \pi(r_{o1} - g - 1.5a) \tag{6}$$

The basic model and equivalent circuit of the proposed stub are illustrated in Fig. 6. Here, two arms, namely slotted microstrip radial section with impedance Z_{slot} and modified open stub resonator (OSR) with impedance Z_4, are responsible for two stop-band regions at f_{r1} and f_{r2}, respectively. The slotted microstrip radial section comprises of high impedance lines Z_1 and Z_3 (represent inductors, L) and low impedance line Z_2 (represents capacitor, C_{rs1}). The design part is performed in two parts. First, the compact slotted microstrip radial stub is designed for f_{r1} using (4), which depends on the inductance (L) and capacitance (C_{rs1}). Further, the OSR is designed using (5) for f_{r2}.

The variation of f_{r1}, is presented in Sect. 2, depends on the parameters of slotted stub. However, the position of second resonant frequency, f_{r2}, can be separately controlled by changing the length of modified open stub, which is verified by the simulation results as presented in Fig. 7. The simulation is done by keeping dimensions of slotted stub as obtained for f_{r1} = 2.4 GHz. Figure 7 depicts that the first resonating frequency is undisturbed with the variation in the length of OSR, which shows the independency, and hence validates independently controllable stop-bands.

4 Proposed DBSF

The proposed DBSF is designed, fabricated, and tested for WLAN bands at 2.4 GHz and 5.4 GHz. This DBSF comprises of two cascaded defected radial stubs as depicted

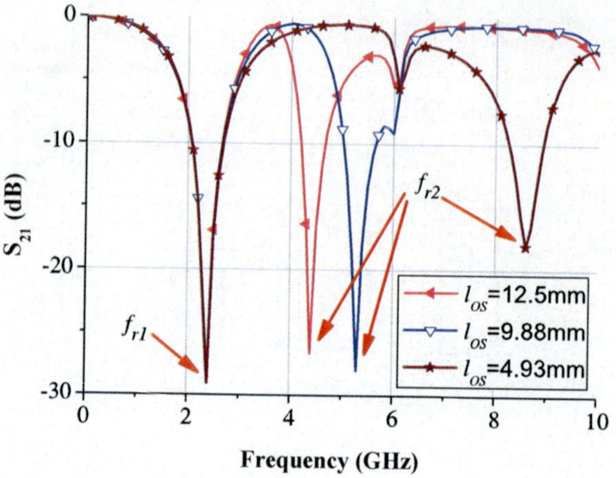

Fig. 7 Simulation result for different lengths of modified OSR with $r_{o1} = 5.5$ mm, $r_{i1} = 1$ mm, $\beta = 60°$, $a = 1$ mm, $g = 0.5$ mm

in Fig. 8. To design proposed DBSF configuration; initially, the dimensions of modified open stub resonator are obtained for 5.4 GHz. Further, the dimension of slotted radial stub should be calculated for 2.4 GHz. This will give the overall dimension of the unit cell, i.e., $x_1 \times y_1 = 13.4\ \text{mm} \times 13.4\ \text{mm} = 179.56\ \text{mm}^2$, as shown in Fig. 8a. The optimized dimension of defected radial stub (unit cell) for abovementioned frequencies is as $r_{i,\,\text{slot}} = 2.3$ mm, $r_{o,\,\text{slot}} = 8.5$ mm, $\theta = \alpha = \beta = 60°$, $r_i = 3.5$ mm, $r_o = 13.4$ mm, $w_1 = 0.58$ mm, $r_{o1} = 5.5$ mm, $r_{i1} = 1$ mm, $a = 1$ mm, $g = 0.5$ mm, and $l_{OS} = 9.6$ mm. Even, this unit cell provides dual band-stop region; though, two unit cells are cascaded, as shown in Fig. 8b, to achieve the better selectivity (in pass-band between two stop-bands), high rejection level, wide stop-band bandwidth, and improved shape factor. The shape factor defines sharpness of the transition between pass-band and stop-band regions, which can be calculated from

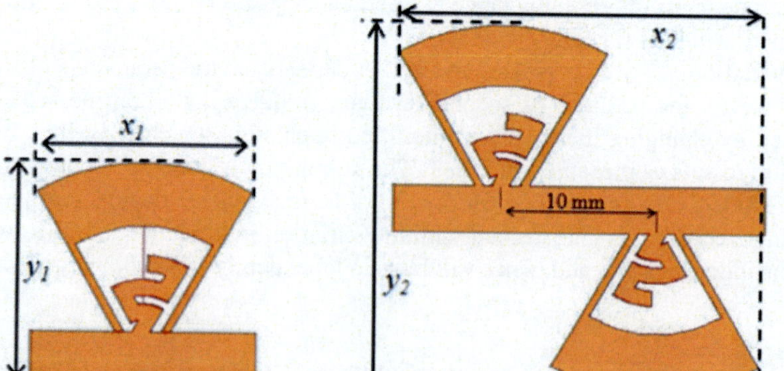

Fig. 8 **a** Defected radial stub (unit cell), **b** proposed DBSF (cascaded defected radial stubs)

insertion loss (S21) graph. This is the ratio of −3 and −20 dB bandwidths; and its minimum and ideal value are 1. The improvement in performance can be easily seen by looking over the comparative S-parameter characteristics in Fig. 9. Table 1 illustrates the performance comparison between the proposed defected radial stub resonator and the proposed DBSF configuration, which also signifies the overall improvement in proposed DBSF. The two similar defected radial stubs are cascaded at a distance of 10 mm. The overall dimension of the proposed DBSF is $x_2 \times y_2 = 23.4\ \text{mm} \times 23.8\ \text{mm} = 556.92\ \text{mm}^2$.

After designing the proposed DBSF, it is finally fabricated and tested. The characteristics of proposed DBSF are measured using vector network analyzer (VNA). Figure 10 illustrates the comparative graph for experimental results and simulation results with the fabricated prototype. Both the results are in good agreement, which verify the proposed DBSF configuration. The characteristic of the proposed DBSF is finally compared with the characteristics of the DBSF configurations available in the literature. The results are tabulated and compared in Table 2, which shows the superiority of defected radial stub and proposed DBSF in terms of wider bandwidth, compact size, and higher rejection level.

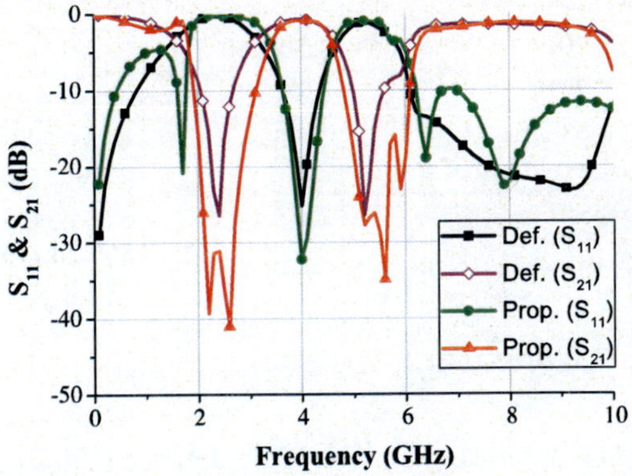

Fig. 9 S-parameter characteristics of defected radial stub and proposed DBSF

Table 1 Comparison table for proposed unit cell and proposed DBSF

Parameters	Defected radial stub	Proposed DBSF
Center frequency (f_{r1}/f_{r2}) (GHz)	2.4/5.4	2.4/5.4
Peak attenuation in stop-band (dB)	−26.5/−27.1	−40/−34.9
−3 dB fractional bandwidth (FBW) (%)	67.9/27.4	69.6/33
−15 dB FBW (%)	13.75/6.29	40/20.74
Shape factor	8.57/8.7	2.06/1.89
Size (mm^2)	179.56	556.92

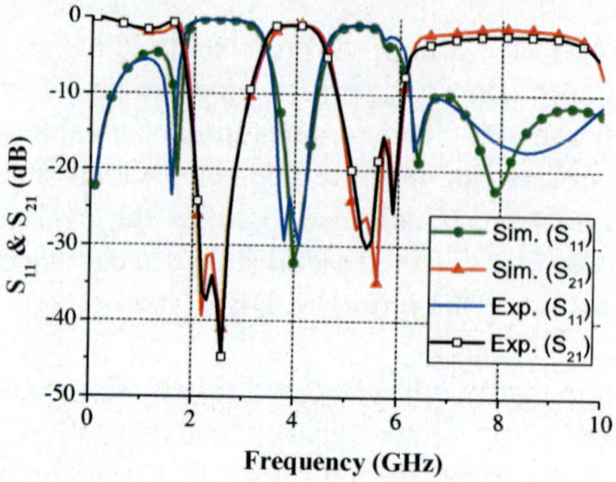

Fig. 10 Simulated and experimental result of proposed DBSF with fabricated prototype

Table 2 Table captions should be placed above the tables

[Reference No.] Year	f_{r1}/f_{r2} (GHz)	FBW (%)	S_{21} (dB)	Size (mm^2)
[2] 2019	2.4/3.75	20.8/21.6 (10 dB)	> 17/ > 17	15.2 × 20
[3] 2019	2.34/7.81	33.2/7.9 (3 dB)	> 45/ > 18	6.2 × 16.8
[4] 2013	2.6/5.3	61/26.8 (10 dB)	> 10/ > 10	42 × 27
[5] 2017	1.50/2.40	23.7/14.6 (3 dB)	> 30/ > 30	39.72 × 18.26
[9] 2019	3.35/5.47	3/ 12.34 (3 dB)	> 35/ > 35	75.8 × 17.5
Proposed work **Defected radial stub:**	**2.4/5.4**	**67.9/27.4 (3 dB) 25/16.3 (10 dB)**	**> 25/ > 20**	**13.4 x 13.4**
DBSF:	**2.4/5.4**	**69.6/33 (3 dB) 50/22.1 (10 dB)**	**> 30/ > 25**	**23.4 x 23.8**

5 Conclusion

Three microstrip structures are detailed in this paper. Initially, exhaustive parametric study of slotted microstrip radial stub is presented, which shows the advantages like wide stop-band, suppression of second harmonic, and compact structure. Further, novel defected radial stub resonator is investigated. This configuration consists of a modified open stub resonator (folded stub in radial manner) and slotted microstrip radial stub; and the structure is formed by only etching some portion from the conventional radial stub resonator. In both slotted and defected radial stub configuration, some metallic portion is etched out from the conventional radial stub; hence, defected radial stub also exhibits the abovementioned advantages of slotted stub. Additionally, defected radial depicts dual band-stop nature having independently controllable stop-bands. One stop-band depends on dimensional parameters of slotted stub; however, other stop-band can be separately controlled by parameter variation of modified open stub. After proper microstrip and circuit analysis, the defected radial stub resonator is designed and simulated for 2.4 and 5.4 GHz center frequencies. The proposed resonator shows wide stop-band with 3 dB fractional bandwidth of 67.9 and 27.4%, shape factor near 8, and peak attenuation above 25 dB. The overall size of the proposed resonator is only 179.56 mm^2.

Finally, this paper presents a novel and compact DBSF for 2.4 and 5.4 GHz frequencies by cascading proposed defected radial stubs in order to achieve improved shape factor, rejection level, and fractional bandwidth. The complete filter structure is analyzed, simulated, fabricated, and tested for operation at two stop-bands around 2.4 GHz and 5.4 GHz. The simulation shows the 3 dB fractional bandwidth of 69.6 and 33%, shape factor near 2, and peak attenuation above 35 dB. The presented compact DBSF provides good pass-band response too and is suitable for wireless applications.

References

1. Matthaei G, Young L, Jones EMT (1980) Microwave filter, impedance-matching networks and coupling structures. Artech House, Norwood
2. Radonić V, Birgermajer S, Cselyuszka N, Crnojević-Bengin V (2019) Compact dual-band bandstop filter based on coupled open ended Hilbert resonators. J Electromagn Waves Appl 33(10):1318–1328
3. Dzhumamuhambetov J, Abykanova B, Gorur A (2019) A novel dual-band microstrip bandstop filter based on stepped impedance hairpin resonators. Prog Electromagn Res Lett 84:139–146
4. Feng WJ, Che WQ, Shi SY, Xue Q (2013) Compact dual-wideband bandstop filters based on open-coupled lines and transversal signal-interaction concepts. IET Microw Antennas Propag 7(2):92–97
5. Feng WJ, Hong ML, Che WQ, Xue Q (2017) Dual-band microstrip bandstop filter with multiple transmission poles using coupled lines. IEEE Microw Wirel Compon Lett 27(3):236–238
6. Castillo-Araníbar P, Garca-Lampérez A, Llorente-Romano S, SegoviaVargas D (2019) Dual-band band-stop microstrip filter with controllable bands based on unequal split ring resonators. IET Microw Antennas Propag 13(12):2119–2128
7. Yang S (2016) A compact dual-band bandstop filter having one spurline and two embedded open stubs. J Electr Syst Inf Tech 3(2):314–319
8. Chiou H-K, Tai C-F (2009) Dual-band microstrip bandstop filter using dual-mode loop resonator. Electron Lett 45(10):507–509
9. Praludi T, Maulana YY, Yuliasih ER, Kurniadi DP, Taryana Y (2019) Dual-band bandstop filter microstrip using T-DMS and U-DGS. In: International conference on radar, antenna, microwave, electronics, and telecommunications (ICRAMET). Indonesia, pp 28–32
10. Singh PK, Tiwary AK (2015) Novel compact dual bandstop filter using radial stub. Microw Rev 21(1):17–22
11. Kwon H, Lim H, Kang B (2007) Design of 6–18 GHz wideband phase shifters using radial stubs. IEEE Microw Wirel Compon Lett 17(3):205–207

Calibration Techniques in ASIC and FPGA Based Time-to-Digital Converters

K. Hari Prasad and Vinay B. Chandratre

Abstract The high-resolution time-to-digital converters (TDC) are currently being implemented in ASIC and FPGA technologies. The methods to implement TDC in ASIC and FPGA technologies are: delay line, Vernier oscillator, and multi-phase clock methods. The TDC implementation has challenges due to spread of delays, delay mismatches, unpredictable place, and route (P&R) delays. The calibration is a crucial aspect to realize high-resolution and robust TDC under process, voltage, and temperature (PVT) variations. This paper describes various time interval measurement methods and their calibration techniques. The unique calibration methodology developed using fewer resources for multi-channel TDCs is also described in this paper. The calibration techniques can be used across technologies of implementation. The TDC using Vernier oscillator method in 0.35 μm CMOS technology having least significant bit (LSB) of 127 ps and Xilinx Spartan-3 FPGA having LSB of 110 ps have been implemented. The delay line method having LSB of 72 ps is implemented in Spartan-6 FPGA.

Keywords ASIC · FPGA · Time-to-digital conversion (TDC) · Calibration

1 Introduction

The high-resolution TDCs are required in many applications such as high energy and nuclear physics experiments [1], time-of-flight mass spectroscopy [2], laser range finding, ultrasonic measurements, and tomography experiments [3]. The FPGAs are present in the data acquisition systems (DAQ) of most of these experiments. The integration of the TDC with the rest of the experimental parameters in FPGA leads to reduction in size and power consumption of the DAQ. Time interval methods

K. Hari Prasad (✉) · V. B. Chandratre (✉)
Bhabha Atomic Research Centre, Mumbai 400085, India
e-mail: khari@barc.gov.in

V. B. Chandratre
e-mail: vbc@barc.gov.in

Homi Bhabha National Institute, Mumbai 400094, India

J. K. Mandal et al. (eds.), *Topical Drifts in Intelligent Computing*, Lecture Notes in Networks and Systems 426, https://doi.org/10.1007/978-981-19-0745-6_19

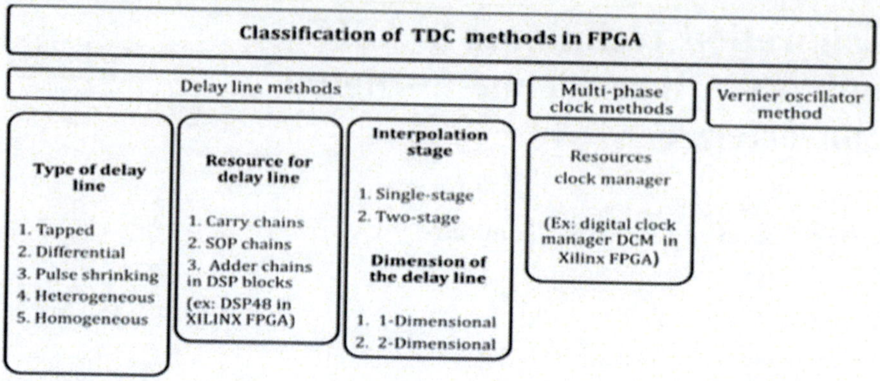

Fig. 1 List of time interval measurement methods and resources in FPGA

like delay line methods [4], Vernier oscillator methods [5, 6], and multi-phase clock methods are being used to realize high-resolution TDC. These methods can be implemented in either ASIC or FPGA technologies. The delay line methods in ASIC technologies are implemented using delay locked loops (DLL). In the FPGA-based delay line methods, dedicated resources like carry chains, sum-of-product (SOP) chains, and adder structures in digital signal processing (DSP) blocks are being used. The brief list of these methods and available resources in FPGA is given in Fig. 1. These delay elements have non-uniform delays due to lack of direct control over delays, delay mismatches, and unpredictable P and R in FPGA. The delays can be measured and compensated for PVT variations using calibration techniques. The various online or offline calibration methods like bin averaging and bin-to-bin estimation is presented in detail in Sect. 2.3. The calibration techniques of delay line methods can also be applied to multi-phase clock methods.

In the Vernier oscillator method, the accuracy, and precision of time measurement depends on accurate calibration of the oscillator time periods. The method to calibrate the oscillator time periods with an accuracy of a few picoseconds, along with results, is described in Sect. 3.

2 The Calibration Schemes in Delay Line Methods

2.1 Introduction to Delay Line Methods

In the simplest TDCs, the time (T) between START and STOP is determined by counting the integer number of clock periods between them. The T is given by clock time period (T_O) times the number of counts. In this method, the time resolution is limited to T_O. To obtain the resolution beyond T_O, various delay line methods like tapped delay line [7], differential delay line [8], heterogeneous and homogeneous [3]

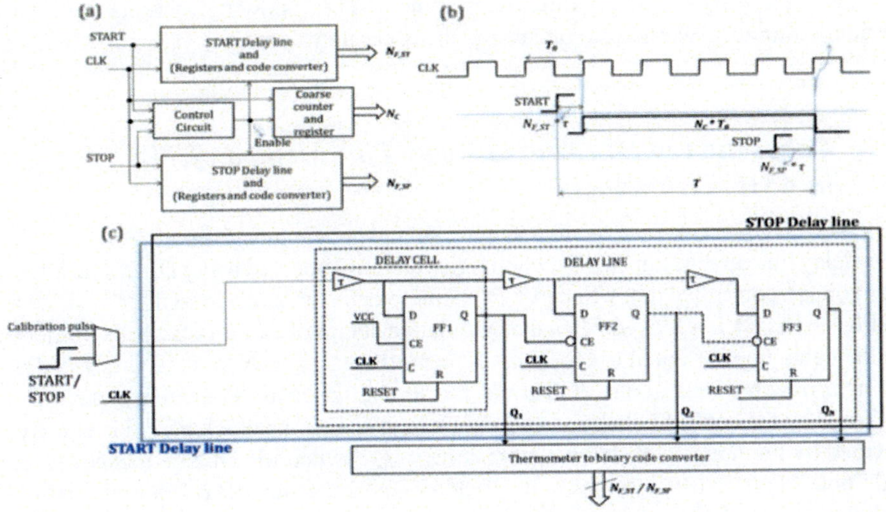

Fig. 2 **a** Block diagram, **b** timing diagram of delay line-based TDC, and **c** delay line schematics

delay line methods are being implemented in ASIC and FPGA technologies using resources like DLLs [4], carry chains [9], and DSP blocks [9]. The delay line methods share the similar design principle illustrated in Fig. 2, irrespective of the type of the delay line, interpolation stages [10], and dimensions of the delay line.

In this method, T is measured by counting the number (N_C) of clock periods (T_0) between them. The fine time within T_0 is measured using the start and stop delay lines as shown in Fig. 2. The coarse counter data and the fine-delay line data are latched on to the corresponding registers on the enable signal generated by the control circuit. The fine-delay line data are converted to binary code using code converters. The detailed explanation of coarse counter, control circuit, and code converters aspects can be found in [11]. If N_{F_ST} is the number of delay cells (buffers) fired between the START, and next rising edge of the clock in start delay line; N_{F_SP} is the number of delay cells fired between STOP and next rising edge of the clock in the stop delay line; τ is the buffer delay of the delay cell, also termed as the least significant bit of the measurement (τ_{LSB}), then T is given by:

$$T = (N_C) * T_0 + (N_{F\_ST} - N_{F\_SP}) * \tau_{LSB} \quad (1)$$

The final time T to be measured can be obtained using Eq. (1). The variables in this equation are count values (N_C and N_F); T_0 and τ_{LSB}. The counter values are first readout. The second variable T_0 is the reference clock period, generally derived from an accurate, stable crystal oscillator. Further, in order to estimate the T_0 with picoseconds accuracy the calibration technique described in [5] can be used. The third variable τ_{LSB} is the crucial factor, which decides the accuracy, precision, and linearity of the time interval measurements. It is often termed as bin width, bin size, or least significant bit of the time measurement. The following sections describe various calibration methods to determine the accurate values of τ_{LSB}.

2.2 Calibration of DLL-Based Delay Line Method of TDC in ASIC Technologies

The delay line methods in ASIC technologies are implemented using DLL. The DLL has voltage-controlled delay line (VCDL), phase detector, and a loop filter as shown in Fig. 3. The DLL is locked when the cumulative delays of all the delay cells (buffer) in the delay line are equal to clock (CLK) period, i.e., $T_0 = N \times \tau$, where N is the number of delay cells in the VCDL and τ is the buffer delay. The buffer delays are adjusted by the control voltage in a feedback loop so that the clock period is equally divided by the number of delay elements $\left(\tau = \frac{T_0}{N}\right)$, when the DLL is locked. As a side note, there exists a very small spread of instantaneous delays due to control voltage noise and phase comparator jitter.

In this method, the delay lines of the TDC are designed using delay cells, whose delays can be adjusted by the control voltage. The control voltage of the DLL is applied to control the delays of the delay cells in the delay line. The principle of time interval measurement method is the same as that explained in Sect. 2.1 (Ref. Figure 2b). The time measurement equation is given in Eq. (1). The variables in Eq. (1) are N_C,N_{F_ST}, and N_{F_SP} which are TDC readout, T_0 the reference clock

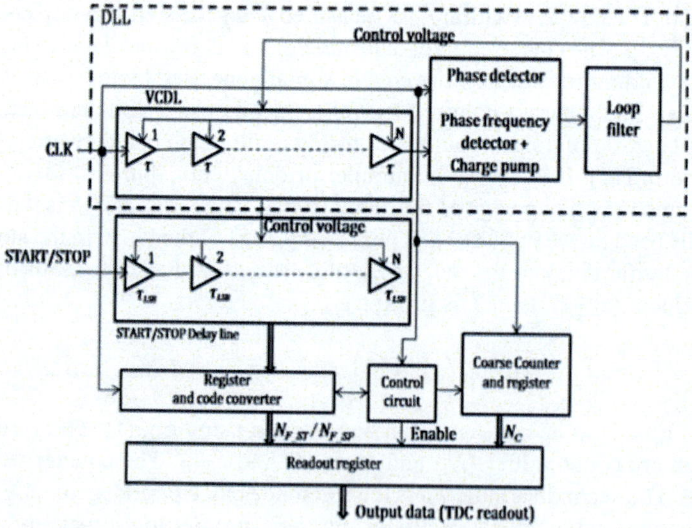

Fig. 3 Typical DLL-based TDC architecture in ASIC technology

period; and τ_{LSB}. In this DLL method, the control voltage that is given from DLL ensures that $\tau_{LSB} = \frac{T_0}{N}$, irrespective of the PVT variations.

The TDC using this DLL method is implemented in 0.35 μm CMOS technology [4]. A current-balanced logic (CBL) delay cell-based DLL has been developed. The reference clock period T_0 is 10 ns. The number of delay cells used is 74. The delay of CBL delay cells τ_{LSB} is 135 ps.

This method provides the most ideal calibration because the delays of the delay cells of the DLL are always (continuously) locked to the reference clock time period by adjusting the control voltage. The τ_{LSB} is automatically calibrated, irrespective of the PVT variations provided the DLL is locked. However, this method is not suitable to implement in FPGA technologies. Even though the DLLs are present in modern FPGAs, the delay cells of the DLL are not accessible to the developer.

2.3 *Calibration of Delay Line Methods of TDCs in FPGA Technologies*

In the FPGA technology, the delay lines are designed using carry chains, sum-of-product (SOP) chains, and adder structures in DSP blocks. In this delay line structure, the buffer delays (τ) are non-uniform (i.e., τ is not identical across the delay line). It is required to calibrate the delay line in order to compensate for these non-uniformities. The non-uniformities are characterized using the following set of equations. In Eq. (1), it is assumed that the buffer delay τ_{LSB} is identical in the START and STOP delay lines, but because of the process variations [12], delay mismatches, the buffer delays across the line are non-uniform. If τ_{LSB_ST} and τ_{LSB_SP} are the buffer delay in START and STOP delay lines, respectively, then Eq. (1) can be modified as

$$T = (N_C) * T_0 + \left(N_{F\_ST} * \tau_{LSB\_ST}\right) - \left(N_{F\_SP} * \tau_{LSB\_SP}\right) \tag{2}$$

The unknown parameters in Eq. (2) that needs to be determined using calibration are τ_{LSB_ST} andτ_{LSB_SP}. The calibration methods like double sampling method and average bin width measurement method are used to determine τ_{LSB}.

Double sampling calibration method: In this method, square pulse whose width and duty cycle are the same as that of the reference clock period is made to pass through the delay line (calibration pulse in Fig. 2) whose width is double the actual delay line width. If number of delay cells fired at the two consecutive clock edges are N_2 and N_1, respectively, then the average bin width is given by $\tau_{LSB} = T_o/(N_2 - N_1)$.

The average bin width calibration method using statistical code density test: In this method, the delay line under test is exposed to a number of random pulses (calibration pulse in Fig. 2). The number of fine bins present in one reference clock period (T_0) is observed. As illustrated in Fig. 4, the minimum bin ($N_{\min}$) and maximum bin ($N_{\max}$) that are fired in the delay line, covering one clock period (T_0) over R samples,

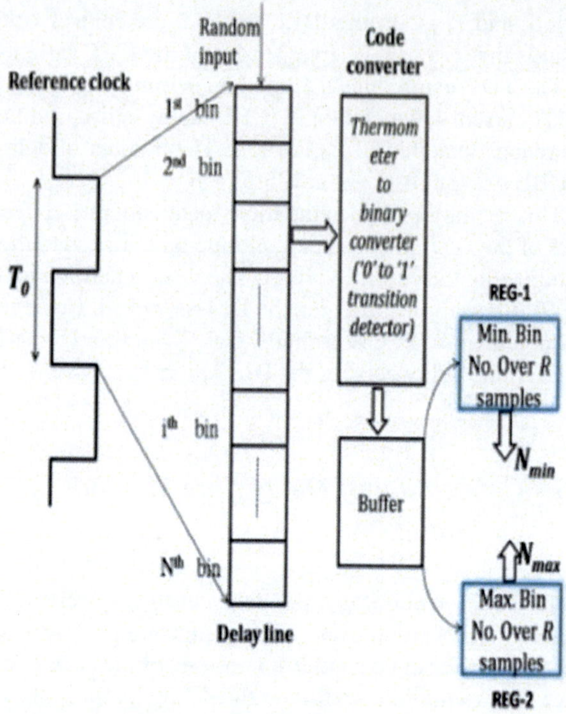

Fig. 4 Average bin width calibration method

are registered in *REG-1* and *REG-2* respectively. The average bin width of the delay line can be formulated as

$$\tau_{LSB} = \frac{T_o}{(N_{\max} - N_{\min} + 1)} \tag{3}$$

If the calibration registers *REG-1* and *REG-2* are updated inside the FPGA, then the calibration can be termed as *online calibration*; if the registers updated in the DAQ software external to the FPGA, then the calibration can be termed as *offline calibration.* This low-resource consuming method is suitable for calibrating TDCs having a large number of channels. The average bin width method is suitable, where the difference between individual bin and the average bin is less than the LSB.

The bin-to-bin calibration method: The calibration methods described above determine the average bin width of the delay line. If the difference between individual bin width and the average bin width is comparable to the LSB, that is if $|\tau_i - \tau_{LSB}| \approx \tau_{LSB}$, indicates that the delay line is highly nonlinear. In this case, it is required to measure the width of each individual bin [3].

If N identical buffers are required to cover one clock period (T_O), then

$$T_0 = N * \tau \tag{4}$$

and in the case where $\tau_1 \neq \tau_2 \neq \tau_N$ Eq. (4) can be written as,

$$T_0 = \sum_{k=1}^{N} \tau_k \tag{5}$$

From Eq. (4) and Eq. (5), the fractional time between START and next rising edge of the clock (t_{ST}) STOP and next rising edge of the clock (t_{SP}) can be written as

$$t_{ST} = N_{F\_ST} * \tau_{LSB\_ST} = \sum_{k=1}^{N_{F\_ST}} \tau_k \quad \text{and} \quad t_{SP} = N_{F\_SP} * \tau_{LSB\_SP} = \sum_{k=1}^{N_{F\_SP}} \tau_k \tag{6}$$

Substituting the values of Eq. (6) in Eq. (2) gives the final time T to be measured as

$$T = (N_C) * T_0 + t_{ST} - t_{SP} \tag{7}$$

The objective of this calibration is to determine t_{ST} and t_{SP} of Eq. (7). In this method, a histogram memory is developed. The memory is addressed by the bin number. The content of the address location has the count indicating the occurrence of the bin over R samples. This count represents the weightage of the particular bin. If the count in the ith bin location is C_i, then the width of the ith bin can be formulated as,$w_i = \frac{C_i}{R} T_0$.

The accumulated time in delay line till *the center* of the ith bin can be formulated as

$$t_i = \frac{w_i}{2} + \sum_{k=1}^{i-1} w_k \tag{8}$$

Equation (8) is used to calculate the final time in Eq. (7). The factor of half in Eq. (8) comes from the fact that the bin widths are measured till the center of the bin. For instance, the t_{ST} and t_{SP} value in Eq. (8) can be written as

$$t_{ST} = \frac{w_{N_{F\_ST}}}{2} + \sum_{k=1}^{N_{F\_ST}-1} w_k \quad \text{and} \quad t_{SP} = \frac{w_{N_{F\_SP}}}{2} + \sum_{k=1}^{N_{F\_SP}-1} w_k$$

3 Calibration Scheme of Vernier Method

In the Vernier oscillator method, the time interval between START and STOP pulses is measured using two stable, start-able oscillators [5, 6] having periods $T_1$and T_2, where ($T_2 < T_1$) as illustrated in Fig. 5. The oscillators are designed to have slight differences in periods $\Delta T = T_1 - T_2$, which is the resolution (LSB) of the time measurement. The oscillator-1 (period T_1) and oscillator-2 (period T_2) were made to start by START and STOP respectively, and are counted by counter-1 and counter-2 respectively. Since $T_2 < T_1$ and STOP come after START, both the oscillators eventually coincide or phase crossover occurs, which is detected by the phase detector. The final time T between START and STOP can be formulated as Eq. (9). The count values n_1 and n_2 are the output of TDC. The final time T depends on the accurate calibration of the oscillator periods T_1 and T_2.

$$T = (n_1 - 1)T_1 - (n_2 - 1)T_2 \tag{9}$$

The objective of the calibration in the Vernier oscillator method is to exactly determine the period of oscillators T_1 and T_2 with an accuracy of a few picoseconds. To achieve the picoseconds accuracy, the oscillator clocks are counted for a very long duration T_L as shown in Fig. 6. If K_1 (K_2) is the number of pulses of the oscillator-1 (2) counted for the duration T_L, then the time period T_1 (T_2) of the oscillator-1 (2) is given by $T_1 = T_L/K_1(T_2 = T_L/K_2)$. The value of T_L is chosen sufficiently long

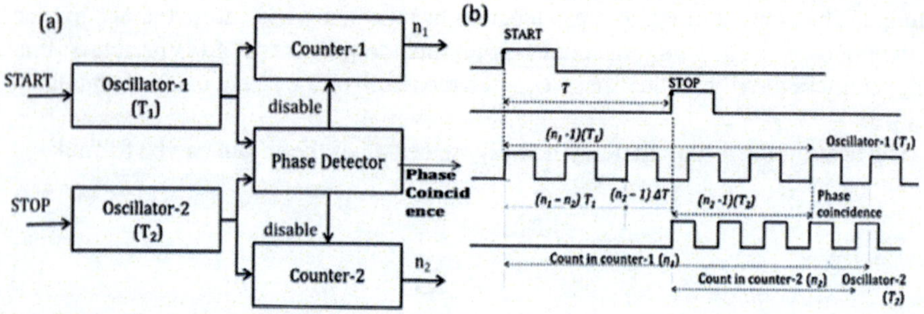

Fig. 5 Design principle of Vernier method a block diagram and b timing diagram

Fig. 6 Calibration scheme in Vernier oscillator method

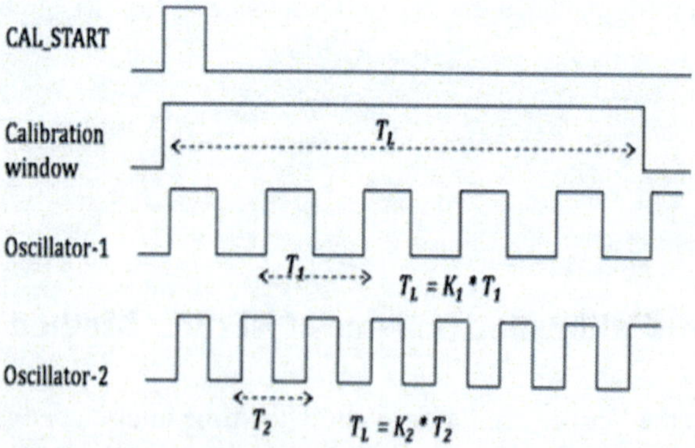

such that an error of ±1 due to quantization in the values of K_1 and K_2 still gives the values of T_1 and T_2 with an accuracy of a few picoseconds. The values of T_1 and T_2 are updated periodically using the calibration in order to mitigate the PVT variations.

4 Implementation Details and Experimental Results

The TDC implementation details and experimental results of various TDCs and calibration schemes implemented in ASIC and FPGA technologies are described in this section.

The TDC-based on the Vernier method is implemented in 0.35 μm commercial CMOS technology [5]. In this method, as shown in Fig. 6 on a calibration start signal (CAL_ START), the oscillator clocks are counted for a large interval (T_L) of 80 μs. The typical value of T_1 is 7.245 ns, T_2 is 7.118 ns, and ΔT is 127 *ps*. The T_1 and T_2 values obtained from the calibration are used to measure the final time T using Eq. (9). The measured precision of the final time T, for the applied inputs of START and STOP having a time difference of 1.52 μs, is 73.6 ps.

The TDC using the Vernier oscillator method was implemented in Spartan-3 FPGA using XORCY [13] elements in the feedback loop of the oscillators. The value of T_L is chosen as 100 μs. The typical value of T_1 is 5.560 ns, T_2 is 5.450 ns, and ΔT is 110 *ps*. The measured precision of the final time T, for the applied inputs of START and STOP having a time difference of 51 ns, is 44.5 ps.

The TDC using tapped delay line method described in Sect. 2.1 is implemented in Spartan-6 FPGA. The 200 MHz reference clock ($T_0 = 5$ ns) is derived inside the FPGA using the on-chip clock manager from an input clock of 40 MHz. The delay line covering one period of T_0 is implemented using CARRY4 [14] elements. From the initial simulation results, a delay line of 80-delay cells is designed. The TDC is calibrated using the average bin width measurement method. This method is chosen for calibration because it is less resource consuming. The statistical code density test is performed by applying the random inputs from the P400 delay generator. The $N_{\max}$ and $N_{\min}$ values in Eq. (3) are measured over a large number of samples ($R =$ 100 k) for each delay line. The measured values of $N_{\max}$ and $N_{\min}$ for the start and stop delay lines are 69 and 1; this results in τ_{LSB_ST} and τ_{LSB_SP} of 72.4 ps. A time interval T of 29 ns between START and STOP is applied to TDC using the delay generator. The output of the TDC calculated using Eq. (2) over 10,000 samples is obtained, and a histogram is plotted as shown in Fig. 7. The measured values of the mean and standard deviation (precision (σ)) are 29.72 ns and 39.28 ps, respectively.

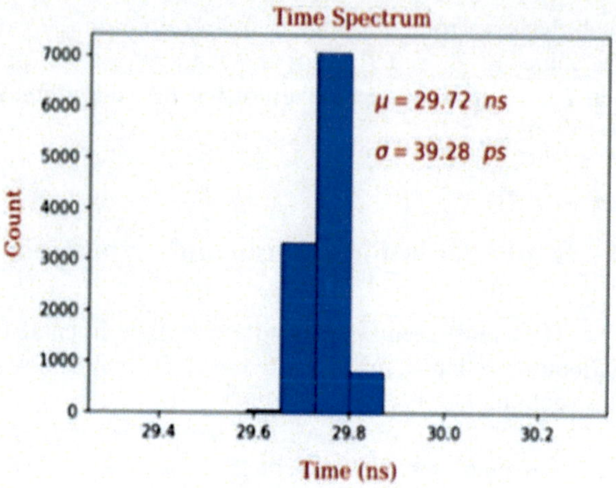

Fig. 7 Time spectrum of the TDC for the applied input of 29 ns between START and STOP

5 Conclusion

Various time interval measurement methods, calibration techniques, and experimental results are described in detail in this paper. The calibration is a crucial aspect which determines the resolution and precision of the TDC. The delay line methods in ASIC technology use DLLs for the calibration of the delay cells. In FPGA-based calibration methods, there exists a trade-off between the average bin width method and bin-to-bin calibration methods. The average bin width method can be applied when the variation in the bin widths is less than the ideal bin width ($|\tau_i - \tau_{LSB}| < \tau_{LSB}$). This method is an ideal solution for the multi-channel TDCs because of its minimal resource consumption. The complex and high-resource consuming bin-to-bin calibration can be applied when the bin width variation is comparable to the ideal bin width ($|\tau_i - \tau_{LSB}| \approx \tau_{LSB}$). The TDC calibration techniques presented in this paper can be used across different processes and FPGA technologies.

The TDC using Vernier oscillator method in 0.35 μm CMOS technology and Xilinx Spartan-3 FPGA have been implemented. The calibration techniques result in a TDC resolution (LSB) of 127 ps in CMOS technology and 110 ps in Spartan-3 FPGA. The delay line-based TDC with a low-resource consuming calibration technique, suitable for TDCs with a large number of channels, was implemented in Spartan-6 FPGA. This technique results in a TDC resolution (LSB) of 72 ps.

References

1. ORTEC AN-42: principles and applications of timing spectroscopy
2. HP Handbook: time-of-flight mass spectrometry
3. Won JY, Kwon SI et al (2016) Dual-phase tapped-delay-line time-to-digital converter with on-the-fly calibration implemented in 40 nm FPGA. IEEE Trans Biomed Circ Syst 10(1):231–242

4. Saxena P, et al (2017) Design of multi-hit time-to-digital converter using current balanced logic delay element. IJEEE 5(2)
5. Hari Prasad K, et al (2014) A four channel time-to-digital converter ASIC with in-built calibration and SPI interface. Nucl Instrum Methods Phys Res Sect. A 737
6. Hari Prasad K, et al (2021) A versatile multi-hit, multi-channel vernier time-to-digital converter ASIC. Nucl Instrum Methods Phys Res Sect. A 990
7. Chaberski D (2016) Time-to-digital-converter based on multiple-tapped-delay-line. Measurement 89:87–96
8. Kalisz J et al (1997) Field-programmable-gate-array-based time-to-digital converter with 200-ps resolution. IEEE Trans Inst Meas 46(1):51–55
9. Qin X, et al (2020) A high resolution time-to-digital converter based on a carry-chain and DSP48E1 adders in a 28-nm field-programmable-gate-array. RSI 91(2)
10. Szplet R, et al 2009 A 45 ps time digitizer with a two-phase clock and dual-edge two-stage interpolation in a field programmable gate array device. IOP Publishing 20(2)
11. Kalisz J (2004) Review of methods for time interval measurements with picosecond resolution. Metrologia 41(17)
12. Henzler S. Time-to-digital converters, 1st edn. Springer
13. Xilinx user guide-ug331: spartan-3 generation FPGA user guide, vol 1.8 (2011)
14. Xilinx user guide-ug615: spartan-6 libraries guide for HDL designs, vol 11.4 (2009)

A High-Speed CMOS Frontend Readout ASIC for Multi-Channel Muon Detectors

Menka Sukhwani, Vinay B. Chandratre, Megha Thomas, and K. Hari Prasad

Abstract A prototype high-speed frontend readout ASIC, designed in 180 nm CMOS process for tracking and precision time-tagging applications in high energy physics experiments, is presented. This ASIC comprises four readout channels, each consisting of a three-stage voltage amplifier, an on-chip analog cable driver and a comparator with LVDS driver. The amplifier and comparator are AC coupled externally. In this ASIC, potential distribution method (PDM) is used to design the high-speed amplifier, cable driver, and comparator stages. This method has proven to be an efficient way of optimizing the target specifications trade-offs. The ASIC exhibited a total voltage gain of ~ 71 and maximum output swing of ~ 600 mV across 50 Ω load for both the input polarities with power consumption of ~ 20 mW/channel. The timing precision of the overall FEE channel is measured to be ~ 530 ps RMS with comparator overdrive of around three times the threshold voltage.

Keywords Frontend electronics · Resistive plate chamber detector · Potential distribution method

1 Introduction

In high energy physics experiments, large area multi-channel detectors, like resistive plate chamber (RPC) and drift tube detectors, are used for muon trigger generation through event tracking and precision time-tagging. These detectors are also suitable for neutrino studies and muon tomography systems. The Iron Calorimeter (ICAL) experiment of the India-based Neutrino Observatory (INO) will also use ~ 28,800

M. Sukhwani (✉) · V. B. Chandratre · K. Hari Prasad
Bhabha Atomic Research Centre, Mumbai 400085, India
e-mail: menkat@barc.gov.in

Homi Bhabha National Institute, Mumbai 400094, India

M. Thomas
Electronics Corporation of India Limited, Mumbai 400028, India
e-mail: meghathomas@ecil.co.in

J. K. Mandal et al. (eds.), *Topical Drifts in Intelligent Computing*, Lecture Notes in Networks and Systems 426, https://doi.org/10.1007/978-981-19-0745-6_20

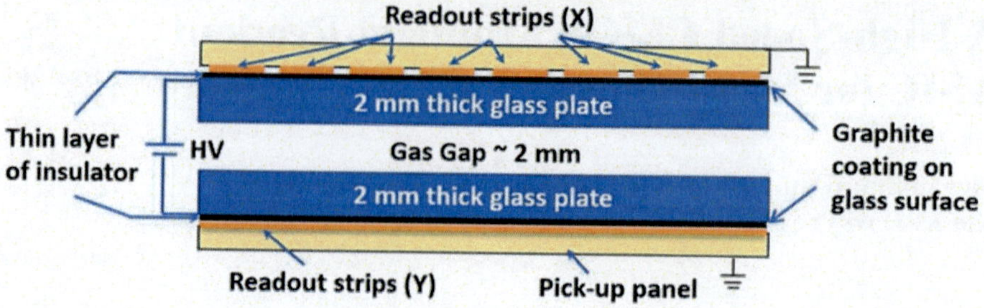

Fig. 1 Schematic representation of the RPC detector

single-gap RPC detectors to study atmospheric muon neutrinos [1]. In this experiment, each RPC detector is of size 2 m × 2 m with 64 X and 64 Y orthogonal readout strips on both sides of the detector, as shown in Fig. 1. Each readout strip exhibits ~ 120 pF capacitance and acts as a strip-line with ~ 50 Ω characteristic impedance.

When a muon particle passes through the gas gap of the RPC detector, it produces charge carriers which move in presence of the applied high electric field. This movement of charge carriers inside the gas gap induces a current signal on the X and Y pick-up strips in opposite polarity. The typical signal of an RPC detector, being operated in the avalanche mode as in the INO-ICAL experiment, has a rise time of ~ 1 – 2 ns with total signal charge in the range of ~ 0.1 pC to a few pC. The measurement requirements of the INO-ICAL experiment involve hit-pattern latching for muon tracking and precision time-tagging of the valid events to identify the direction of incident neutrino.

These detectors require a high-speed, low power, multi-channel amplifier, and leading-edge comparator-based frontend electronics (FEE) to meet the measurement requirements. Furthermore, in the INO-ICAL experiment, access to the amplified detector channels is also required for detector health monitoring, which is not provided in previously reported single-gap RPC FEE solutions [2, 3].

A prototype quad voltage amplifier and leading-edge comparator ASIC is developed in 180 nm CMOS process to meet the novel readout requirements of the single-gap avalanche mode RPC detectors of the INO-ICAL experiment. This FEE ASIC provides an indigenous, low-cost, and low-power solution to cater to the long-term requirements of the experiment. The potential distribution method (PDM) [4, 5] is used for design of the amplifier, cable driver, and comparator stages in this ASIC, which has efficiently allowed optimization of design trade-offs. The detail design aspects and lab test results of the quad FEE ASIC are presented in the following sections.

2 Design of Quad FEE ASIC

The prototype quad FEE ASIC comprises four channels of high-speed amplifier and leading-edge comparator that can be used either individually as stand-alone amplifier and comparator or can be coupled externally to form a complete FEE channel. The external coupling through a bypass capacitor allows a stable overdrive at the comparator input in presence of random DC offset at the amplifier output due to process variations and device mismatch. A single channel configuration of the ASIC is shown in Fig. 2 with details given below.

2.1 *High-Speed Amplifier Channel*

The amplifier channel is built with three stages of single-ended, closed loop voltage amplifiers with DC offset adjustment provided in the first two stages. These amplifier stages are followed by an on-chip analog 50 Ω cable driver.

The critical design aspect of this FEE ASIC was to obtain the overall amplifier bandwidth of ~ 350 MHz, as required to match the minimum detector signal rise time of ~ 1 ns, while achieving the required gain with minimum power consumption (< 30 mW/FEE channel). In the previous versions of INO-ICAL RPC FEE ASIC [6, 7], designed in 0.35 μm CMOS process, single stage transimpedance, and voltage amplifiers were used to build the amplifier channels. However, in this FEE ASIC, owing to the lower intrinsic transistor gain in 180 nm CMOS process [8], a rail-to-rail two-stage amplifier with class AB output buffer [9], as shown in Fig. 3, is used to implement all the amplifier stages and analog cable driver. The design approach followed for optimum and efficient design of these rail-to-rail amplifier stages is described below.

Design Methodology

Initial design using PDM in open loop DC analysis: In short channel CMOS technologies, the g_m/I_D design methodology is typically followed to optimize the analog

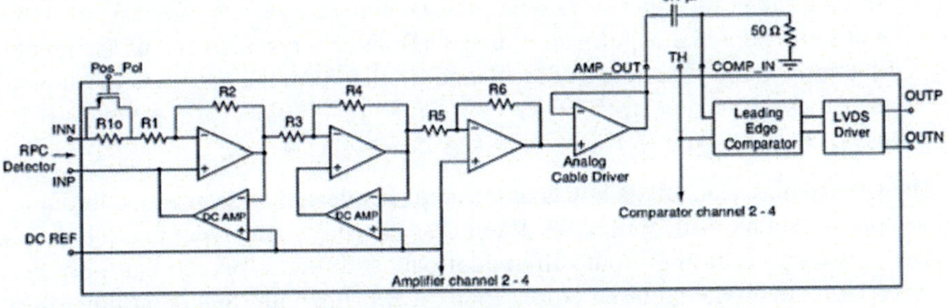

Fig. 2 Single channel schematic of the quad FEE ASIC

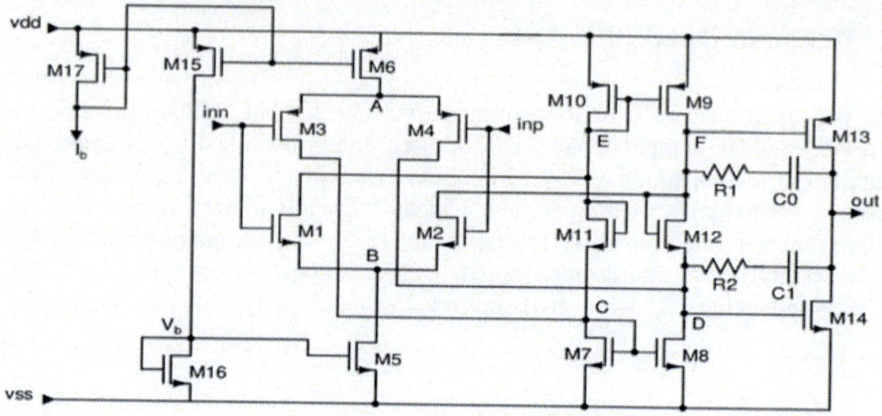

Fig. 3 Schematic of the wide swing amplifier with class AB output stage

designs. However, it was observed that the potential distribution method (PDM) quickly yields the preliminary design through open loop DC analysis and final design can be obtained through minimal optimization iterations. In this method, DC voltage biases at all the amplifier nodes were first estimated based upon following considerations,

(a) Input and output nodes were fixed at $V_{\mathrm{cm}} = \frac{(Vdd-Vss)}{2}$. The bulk of all the transistors were connected to their respective supplies. Further, an overdrive of 5% of (Vdd − Vss), i.e., ~ 90 mV, was considered for all the transistors as in [4, 5], except for output transistors (M13, M14). For these transistors, an additional overdrive of ~ 200 mV was taken to ensure linearity at larger swings.

(b) Gate voltages of the bias transistors (NMOS: M5, M16 and PMOS: M6, M15, M17) were fixed as $V_{GN-\mathrm{Bias}} = Vss + V_{\mathrm{thn}} + 0.05 * (Vdd - Vss)$ and $V_{GP-\mathrm{Bias}} = Vdd - V_{\mathrm{thp}} - 0.05 * (Vdd - Vss)$, respectively.

(c) The common source nodes (A and B) of the input differential pairs were estimated at $V_{SN-\mathrm{Diff}} = V_{\mathrm{cm}} - V'_{\mathrm{thn}} - 0.05 * (Vdd - Vss)$ and $V_{SP-\mathrm{Diff}} = V_{\mathrm{cm}} + V'_{\mathrm{thp}} + 0.05 * (Vdd - Vss)$. Here, V'_{thn} and V'_{thp} represent the threshold voltages of NMOS and PMOS input devices with the given body bias.

(d) The gate and drain nodes (C and E) of diode connected loads were assumed to be at same potential as the output nodes (D and F), respectively, of the input differential pairs. As these nodes (D and F) also form the gate bias of output stage, these were kept at, $V_E = V_F \approx Vdd - V_{\mathrm{thp}} - 0.05 * (Vdd - Vss) - 0.2$ and $V_C = V_D \approx Vss + V_{\mathrm{thn}} + 0.05 * (Vdd - Vss) + 0.2$.

The voltage biases on all the nodes of the amplifier design as determined through above considerations, with Vdd = 0.9 V and Vss = −0.9 V, are given in Table 1.

The bias currents in the input differential pair and output branch can only be finalized based upon the required closed loop 3-dB bandwidth (that determines the amplifier rise time) and stability margins, respectively, in presence of circuit parasitic

Table 1 Node voltages (in V) of amplifier design in Fig. 2 based on PDM

Devices	V_{th}	V_G	V_S	V_D	Devices	V_{th}	V_G	V_S	V_D
M1, M2	0.54	0	0.63	0.1	M3, M4	0.58	0	0.67	−0.13
M5	0.48	−0.33	−0.9	−0.63	M6	0.51	0.3	0.9	0.67
M7, M8	0.48	−0.13	−0.9	−0.13	M9, M10	0.51	0.1	0.9	0.1
M11, M12	0.54	0.1	0.1	−0.13	M13	0.51	0.1	0.9	0
M14	0.48	−0.13	−0.9	0	M15	0.51	0.3	0.9	−0.33
M16	0.48	0.33	0.9	0.63	M17	0.51	0.3	0.9	0.3

while minimizing the power consumption. Initially, these currents were estimated at ~ 100 μA. Then, widths of all the transistors (with L = 0.18 μm) were obtained through DC sweep simulation of single transistor circuit [5] keeping voltage biases at all the terminals as decided previously. Here, a testbench was developed to directly provide the width of the transistor for the given drain current without any graphical analysis, allowing easy iterations. Back annotation of these MOS widths in the amplifier design yielded the designed voltage biases at all the nodes along with the branch currents. The values of miller compensation capacitor and lead compensation resistance were estimated in open loop AC analysis to yield a single-pole response. This base amplifier design exhibited an open loop DC gain of ~ 55 dB with unity gain bandwidth of ~ 2.44 GHz. In the open loop design, the trade-off between the amplifier bandwidth and power consumption can be optimized by careful choice of the bias current.

Design optimization of the closed loop, cascaded amplifiers: Three base amplifiers were DC coupled in closed loop configuration to achieve overall channel gain of ~ 70, as shown in Fig. 2. Now, in order to achieve a rise time of ~ 1 ns, a 3-dB bandwidth of ~ 350 MHz [10] would be required for overall channel. Therefore, closed loop 3-dB bandwidth of each amplifier stage (assuming similar values for all the stages) in the N-stage channel can be estimated as [10],

$$f_{\text{amp}} = \sqrt{N} \times f_{\text{channel}} \tag{1}$$

In this FEE ASIC, the amplifier channel comprises four stages including the analog cable driver, therefore, individual amplifier should have a 3-dB bandwidth of greater than 700 MHz to safely accommodate the post layout parasitic load also. Furthermore, a minimum phase margin of 65° needs to be ensured in closed loop configuration with expected load from the subsequent stages. Therefore, the values of the bias currents, miller compensation capacitors, and lead compensation resistors for each amplifier stage were required to be optimized simultaneously in the closed loop, cascaded configuration to achieve the required stability margins, overall bandwidth, and power consumption. The simulated AC response of all the four stages is shown in Fig. 4.

Analog cable driver: In this FEE design, the analog cable driver consumes the maximum power as it is required to provide a maximum voltage swing of ~ 600 mV

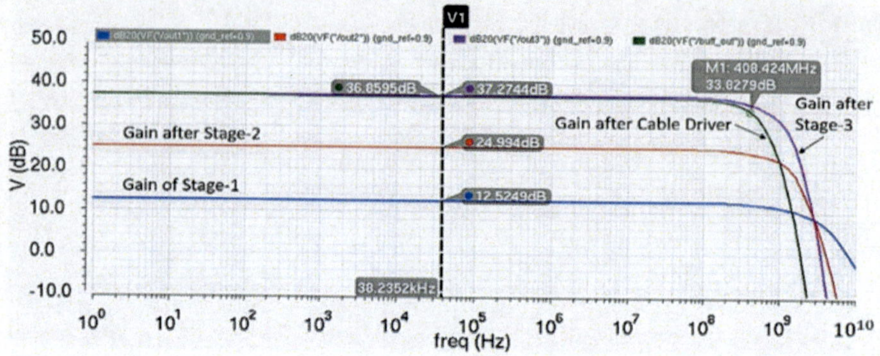

Fig. 4 Simulated AC response of four stages of the amplifier channel

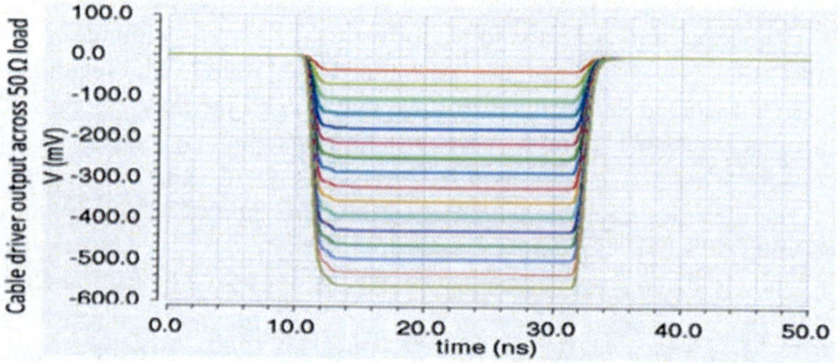

Fig. 5 Simulated transient response of the amplifier channel through analog cable driver across 50 Ω load for input range of 0.5 mV to 8 mV

across external 50 Ω load, as shown in Fig. 5, maintaining the amplifier rise time in presence of parasitic load from two I/O pads and bond wires (amplifier output and comparator input I/Os). Therefore, it was designed to provide only sinking current, i.e., only negative output through NMOS output device leading to lower quiescent power consumption.

However, as the detector provides single-ended complementary outputs from the top and bottom readout channels, the opposite polarity output would require an inversion to obtain negative only output for both the input polarities. A gain trimming option is, therefore, provided in the first amplifier stage in order to equalize the total channel gain in both inverting and non-inverting configurations, as shown in Fig. 2. The analog cable driver was also designed using PDM. The final design details of the amplifier and cable driver stages are given in Table 2.

Table 2 Design details of the amplifier and cable driver stages of the FEE ASIC (L = 0.18 μm)

Devices	Size	Quiescent current	Devices	Size	Quiescent current
M1, M2	5.81 μm	80 μA	M3, M4	18.53 μm	80 μA
M5	16.2 μm	160 μA	M6	50.6 μm	160 μA
M7, M8	1.3 μm	80 μA	M9, M10	3.9 μm	80 μA
M11	0.5 μm	0	M12	0.5 μm	0
M13: Amplifier M13: Cable driver	9.5 μm 20 μm	220 μA 1142 μA	M14: Amplifier M14: Cable driver	3 μm 100 μm	220 μA 1142 μA
M15	17.42 μm	80 μA	M16	7.31 μm	80 μA
M17	21.6 μm	80 μA	C1, C2: Amplifier	45 fF	
R1, R2: Amplifier R1, R2: Driver	5 kΩ 3 kΩ		C1, C2: Driver	120 fF	

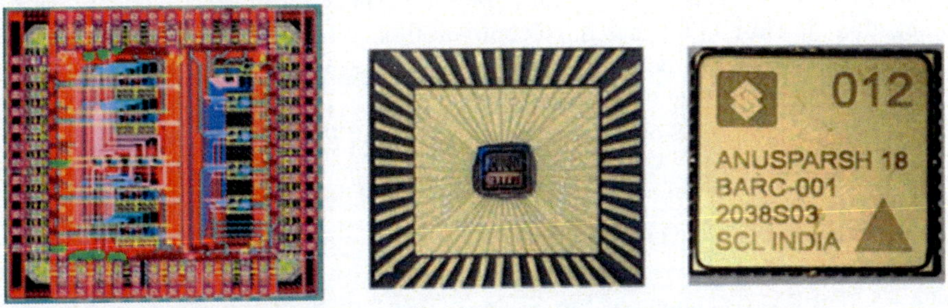

Fig. 6 Die picture of the quad FEE ASIC and packaged ASIC in CLCC-48 package

2.2 *Comparator and Layout Design*

The design of leading-edge comparator [11] was ported to 180 nm CMOS process using PDM. It comprised a pre-amplifier followed by cross-coupled latch and output buffer. An on-chip LVDS driver [12] provided the comparator output on external 100 Ω load. The layout of the FEE ASIC was designed in full custom manner, occupying an area of 2 mm × 2 mm. The ASIC is packaged in CLCC-48 package, as shown in Fig. 6.

3 Experimental Results

The quad FEE ASIC is tested in the laboratory for amplifier gain and linearity along with LVDS output compatibility with the standard receiver and timing precision. The amplifier output, for an input voltage pulse of amplitude ~ 5 mV and rise time ~2.2 ns, is shown in Fig. 7a. The voltage gain of the amplifier channel was measured

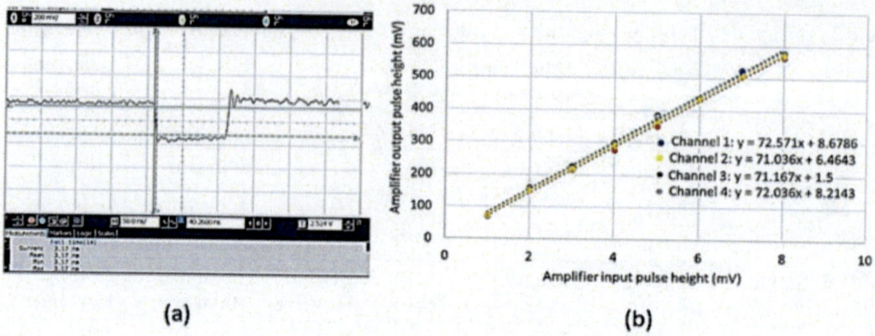

Fig. 7 Experimental results **a** amplifier output profile, **b** transfer characteristics of the four amplifier channels

to be ~ 71, as obtained from its transfer characteristics shown in Fig. 7b. The values of the miller compensation capacitors and lead compensation resistors that were finalized in the closed loop, cascaded configuration considering the actual on-board parasitic, were found to be optimum as amplifier output exhibited good transient stability without significant overshoots and ringing.

The comparator LVDS output was acquired by an FPGA TDC-based data acquisition module, and timing precision was measured by plotting the LVDS width spectrum for comparator overdrive of around three times the threshold voltage (~ 50 mV). The comparator exhibited an intrinsic time precision of ~ 50 ps RMS and time precision of the entire FEE channel, including the amplifier and comparator, was measured to be ~ 530 ps RMS, as shown in Fig. 8a and b, respectively. The power consumption of the FEE ASIC was measured to be ~ 20 mW/channel.

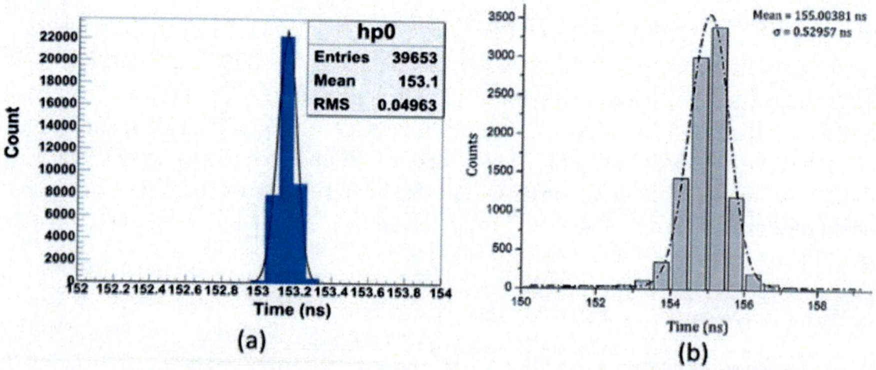

Fig. 8 LVDS pulse width spectrum for **a** intrinsic time precision of the comparator, **b** for time precision measurement of the overall FEE channel including amplifier and comparator

4 Conclusion

A prototype quad FEE ASIC is developed, in 180 nm CMOS process, for readout of large area, multi-channel muon detectors like RPC and drift tube detectors. The ASIC comprises four independent channels of high-speed amplifiers and comparators that are coupled externally. The amplifier and comparator stages were efficiently designed using potential distribution method (PDM). This method quickly leads to an initial design and allows easy iterations of the bias currents and respective MOS aspect ratios to achieve the required specifications. The FEE ASIC has achieved the desired specifications with amplifier channel voltage gain of ~ 71 and overall timing precision of ~ 530 ps RMS with power consumption of ~ 20 mW/channel.

References

1. Kumar A et al (2017) Physics potential of the ICAL detector at the India-based Neutrino observatory (INO). Pramana—J Phys 88:79
2. Manen S, et al (2013) FEERIC, a very-front-end ASIC for the ALICE muon trigger resistive plate chambers. In IEEE nuclear science symposium and medical imaging conference. Seoul, pp 1–4
3. Abbrescia M et al (2000) New developments on front-end electronics for the CMS resistive plate chambers. Nucl Instr Meth A 456:143–149
4. Todani R, Mal AK (2012) Design of CMOS opamp using potential distribution method. In: 2012 IEEE international conference on circuits and systems (ICCAS). Kuala Lumpur, Malaysia, pp 184–189
5. Ghosh NN, Todani R, Chaudhuri C, Mal AK (2013) Simplified design method for fully differential gain-boosted folded cascade OTA. In: 2013 IEEE conference on information and communication technologies. Thuckalay, India, pp 943–948
6. Sukhwani M et al (2013) Regulated cascode preamplifier-based front-end readout ASIC "ANUSPARSH" for resistive plate chamber detector. In: Proceedings of international conference on VLSI, communication, advanced devices, signals and systems and networking (VCASAN-2013). Bangalore, India, pp 71–77
7. Chandratre VB, et al (2015) ANUSPARSH-II frontend ASIC for avalanche mode of RPC detector using regulated cascode trans-impedance amplifier. In: Proceedings of the DAE-BRNS symposium on nuclear physics 60. Mumbai, India, pp 928–929
8. Murmann B, Boser BE (2004) Digitally assisted pipeline ADCs: theory and implementation. Springer, US
9. Atkin EV, Vinogradov SM (2017) Approach to the design of monitoring buffer for read-out ASICs. J Phys: Conf Ser 798
10. Hollister AL (2007) Wideband amplifier design. SciTech Publishing Inc
11. Sukhwani M et al (2020) A high speed BiCMOS comparator ASIC with voltage adjustable hysteresis. Nuclear Inst Methods Phys Res A 980:164503
12. Tajalli YA (2009) A power-efficient LVDS driver circuit in 0.18um CMOS technology. IEEE J Solid-State Circ 44(2):538–548

A High-Gain Photo Sensor in 0.35 μm HV CMOS Process

Sourav Mukhopadhyay and Vinay B. Chandratre

Abstract This paper presents the development of a high-gain photo sensor in 0.35 μm commercial high voltage (HV) CMOS technology. The photo sensor consists of an array of avalanche diodes (APD), each operating in the Geiger-Muller (GM) region together with a passive quenching network, connected in parallel. The primary motivation behind this development is the possibility of monolithic integration of photo sensors and readout electronics in the commercial CMOS process, which is an attractive solution in many areas. Various promising photo sensor structures specific to CMOS technology are implemented, incorporated in a test vehicle (TV), and fabricated in HV CMOS technology. The TCAD simulation, physical design, electrical, and optical characterization of the photo sensor structure that shows the best result among the various structures are presented herewith in this paper.

Keywords High-gain photo sensor · SPAD in HV CMOS · Array of GM-APD

1 Introduction

The detection of weak optical signals down to few photons or even a single photon [1] is challenging but critical for a wide range of scientific and technical applications. This requirement saw the advent of a versatile device called silicon photomultiplier (SiPM) [2], which finds its wide application in various fields, including optical communication [3], astrophysics [4], nuclear radiation-based diagnostic instruments [5], home-land security [6], and high-energy physics [7]. Properties like insensitivity to magnetic fields, high-gain, high photon detection efficiency (PDE), robustness,

S. Mukhopadhyay · V. B. Chandratre (✉)
Bhabha Atomic Research Centre, Trombay Mumbai-400085, India
e-mail: vbc@barc.gov.in

S. Mukhopadhyay
e-mail: souravm@barc.gov.in

Homi Bhabha National Institute, Anushaktinagar Mumbai-400094, India

J. K. Mandal et al. (eds.), *Topical Drifts in Intelligent Computing*, Lecture Notes in Networks and Systems 426, https://doi.org/10.1007/978-981-19-0745-6_21

lower operation voltage, good timing properties make SiPMs an attractive alternative to substitute photomultiplier tube (PMT) in many areas. Furthermore, the "multi micro-cells (self-quenched GM-APD) in parallel like" structure of SiPM converts the digital nature of the individual micro-cells to analog nature collectively as long as the numbers of incident photons on the detector's active area per unit time are less than the number of pixels.

The design of self-quenched GM-APD, i.e., single photon avalanche diode (SPAD), in commercial CMOS technology [8] has been a growing trend over the years. The commercial CMOS process imparts many design challenges by ruling out any scope for modification in the process steps that are primarily optimized for readout electronics. However, the prospect of integrating the sensor and its readout electronics on the same wafer offers an edge over a hybrid system as the parasitic introduced by the external connections bring about system performance degradation in terms of lower sensitivity and speed.

One of the significant challenges in designing SPAD in sub-micron CMOS technology is to avoid the premature perimeter edge breakdown (PEB) [9] or the soft breakdown. There are reported techniques in the literature to mitigate PEB in commercial CMOS processes such as P-well guard ring-based structure [10], N-well guard ring-based structure [11], substrate guard ring-based structures [12], and deep P-tub/N-tub-based structure [13].

In this work, we explore the development feasibility of SPAD, using such various design ideas in a commercial HV CMOS process, by designing an ASIC (test vehicle) incorporating different promising structures. In addition, a novel design technique specific to the standard CMOS process is implemented to enhance the optical sensitivity. Nine different potential designs of arrays of SPAD with various pixel structures, guard ring techniques, and quench resistor layout methods are implemented in the ASIC. The best-performing structure among the nine will be discussed in this paper.

2 Design of High-Gain Photo Sensor

The P-well guard ring-based SPAD [10] structure provides the most promising results among the nine different structures. Hence, this paper will only deal with the development of P-well guard ring-based micro-cell structure from here onwards.

2.1 *Design of a Micro-Cell*

Figure 1 shows the cross-section of the micro-cell, based on the P+/DN-well (deep N-well) active junction.

To avoid premature edge breakdown (PEB), P-well-based guard ring has been implemented. Standard 0.35 μm HV CMOS technology provides these additional P-well and DN-well to facilitate the design of the potential SPAD compared to standard

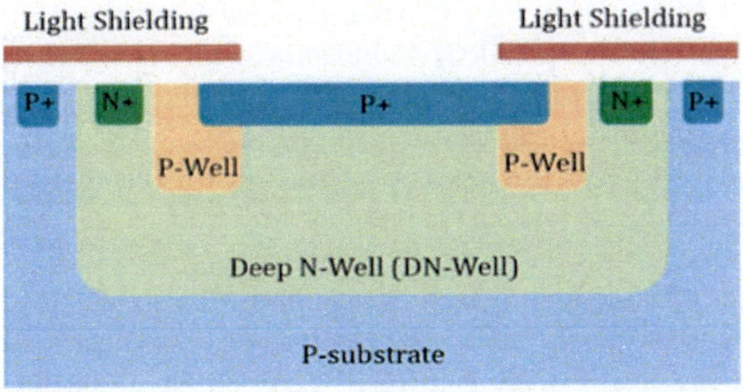

Fig. 1 Cross-section view of a single micro-cell (representative figure, drawn not to scale)

N-well CMOS technology. The active area of the individual pixel is 30 μm × 30 μm. The guard rings, contact for deep N-well (cathode) & substrate, and inter-pixels routings in metal layers reduce the fill factor to a great extent which affects the PDE of the sensor. Laying out the high-value passive quenching resistor degrades the fill factor even further. The non-active area is shielded with one of the available metal layers to avoid light absorption, as shown in Fig. 1.

2.2 Layout of a Micro-Cell

A 16 × 16 array of similar micro-cells as shown above with individual quench resistor based on hypo-poly silicon layer is laid out. The total active area of the high-gain photo sensor is ~1 mm^2 with a fill factor of ~24%. Different techniques were used to implement a high value of quench resistor (~hundreds of kilo ohm) in the smallest possible area without degrading the fill factor too much. Figure 2 shows the layout of different micro-cells with two different ways of quench resistor layout.

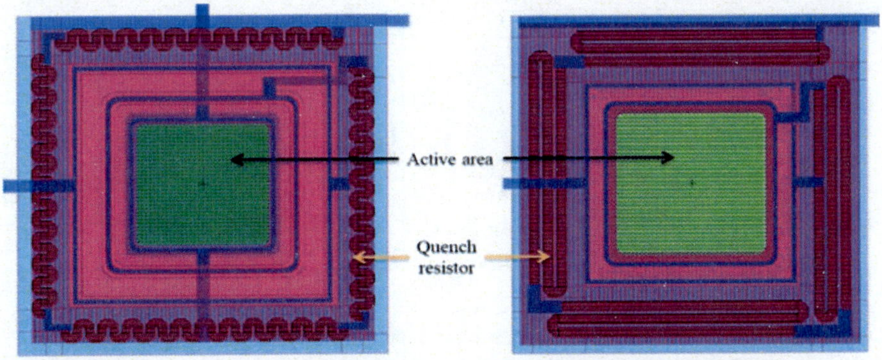

Fig. 2 Physical designs of two micro-cells with different quench resistor layout techniques

As the standard CMOS technology is not optimized for making a photo sensor, some of the design rule check (DRC) and electrical rule check (ERC) rules have been intentionally violated while making the layout of the high-gain photo sensor. Similarly, the stacking of layers from the surface (metals, oxide, and passivation layers) in a standard CMOS process is not tailored to develop an optical device intended to detect a single or few tens of photons. So, in this design, we have utilized the "PAD" layer (the protection opening layer) on top of the device's active area to improve the device's feeble light detection capability. To validate this idea, we have made two designs based on the same micro-cell as discussed earlier, only with the difference that in one design, there is a "PAD" layer on top of the active area while in the other design, it is absent (Comparative optical performance between these two variants are provided in section four).

2.3 TCAD Simulation of a Micro-Cell

Technology computer aided design (TCAD) simulations were carried out before fabricating the design to understand the device's electrical and optical behavior. Due to the limitation of the number of grid points in the device simulator, only one single micro-cell was simulated. Physical models (Boltzman carrier statistics; concentration-dependent and field-dependent mobility; Shockley–Read–Hall and Auger recombination; Selberherr impact ionization) were used to predict the device behavior accurately. Figure 3a shows the electric field profile inside the micro-cell corresponding to Fig. 1 with a peak electric field of 5.5×10^5 V/cm at the center of the active area, while the electric field at the guard rings is considerably less. It indicates that the PEB phenomenon is avoided with this design of the micro-cell. In the 1D electric field profile view (Fig. 3b), by taking a cut-line in the center of the simulated micro-cell, a sharp profile of a high electric field can be seen with a

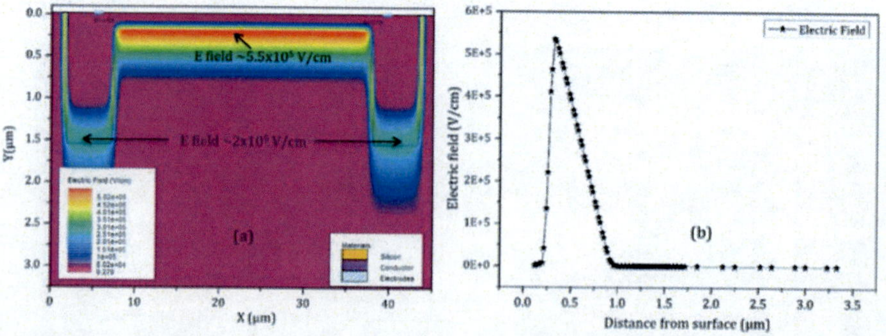

Fig. 3 **a** 2D electric field profile showing max electric field at the center and comparatively lower electric field in the guard rings indicating higher breakdown voltage for the guard rings, **b** 1D electric field distribution along with the depth of the micro-cell

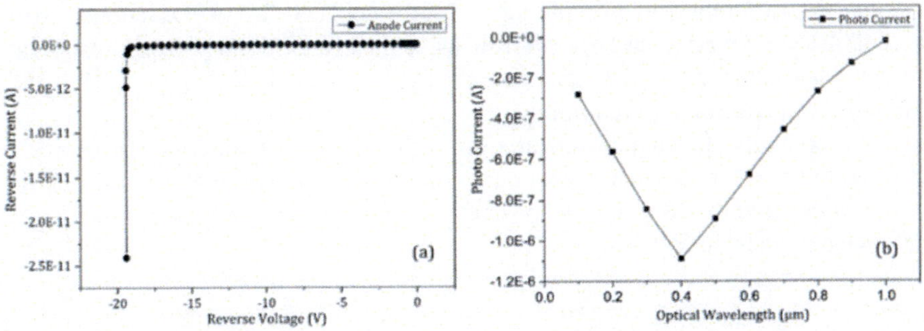

Fig. 4 **a** Reverse I-V characteristics, **b** spectral response of the simulated micro-cell

peak electric field at around 0.25 μm depth from the surface, signifying the junction between P+ and DN-well.

Figure 4a shows the reverse I-V characteristics of the simulated micro-cell with a sharp breakdown voltage of ~18.5 V. The spectral response of the GM-APD micro-cell could be seen in Fig. 4b with a peak sensitivity at ~400 nm wavelength region of the visible light spectrum.

3 Fabrication and Packaging

The design of high-gain photo sensors was fabricated using 0.35 μm HV CMOS technology and packaged in a CLCC-44 package with taped lid. Figure 5a shows the complete layout view of the test vehicle ASIC with a die area of 4.3 mm ×

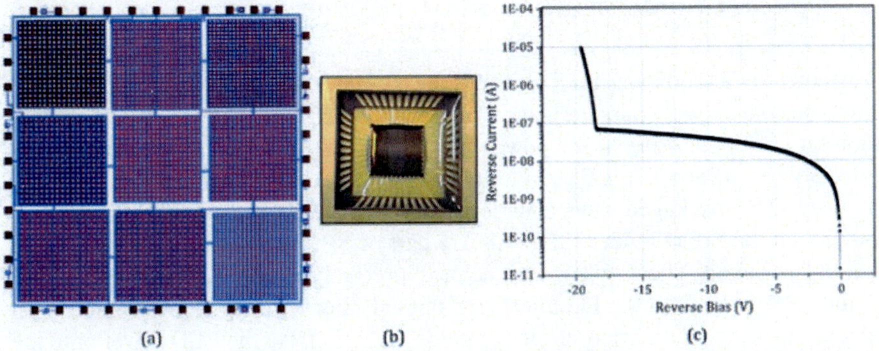

Fig. 5 **a** Complete layout view of the test vehicle ASIC, **b** the packaged device in open lid condition with the bond wires protected with epoxy. The center column could be seen with whitish shade, as in those three sensors the "PAD" layer is present on the active area, **c** reverse I-V characteristics of the fabricated 16 × 16 array of self-quenched GM-APD based on P+/DN-well junction (30 μm × 30 μm) with P-well guard ring. A current compliance limit of 10 μA was set during the measurement

4.3 mm. Nine different structures of arrays of self-quenched GM-APDs (connected in parallel) in a 3 × 3 matrix have been laid out using the nearest bond-pads with minimal routing. The foundry given bond-pads were modified to supply higher voltage into the terminals of the pixels. Figure 5b shows the packaged ASIC with open lid condition. A 99.9% optically transparent (ensured no light loss before entering the device's active area) epoxy and resin hardener combination was applied on the surface of the bare die and cured for 48 h. This step is necessary in order to protect the bond wires and device surface during testing.

4 Device Characterization

Testing of the high-gain photo sensor has been carried out in two steps. Initially, electrical characterization has been carried out with the unpackaged die supplied by the foundry. Subsequently, optical characterization was done using the packaged devices.

4.1 Electrical Characterization

During electrical characterization, current versus reverse bias voltage (I-V) measurement (shown in Fig. 5c) has been done using semi-automatic wafer prober from Cascade, semiconductor parametric analyzer B1500A from Keysight technologies and high-current high-power source measure unit (SMU).

4.2 Optical Characterization and Test Setup

A customized dark box was built to install the optical test bench with tabletop x–y–z mounts for the optical characterization of the TV ASIC. A CAEN SiPM characterization kit [14], consisting of the power supply and amplification unit (SP5600), the device holder (SP5650C), LED driver (SP601), and desktop digitizers (DT5720A) was used. The packaged device under test (DUT) with removed metal lid was mounted on the SiPM holder (SP5650C), a part of SP5600 which is stacked on top of an x–y–z mount. The x–y–z mount helps to precisely align the small area of DUT (~1 mm × 1 mm) with the FC interfaced optical fiber with a 1 mm^2 cross-section that was flashed on the device surface to guide the light from the LED driver onto the DUT. Figure 6 shows the detailed test setup. The ultra-fast LED driver emits pulses at ~405 nm [15]. The optical response of the DUT, amplified and further processed through SP5600, was sent to a high bandwidth oscilloscope.

Figure 7 highlights the optical response of the high-gain photo sensor structure with a violet LED. While Fig. 7a shows the response with a train of LED pulse with

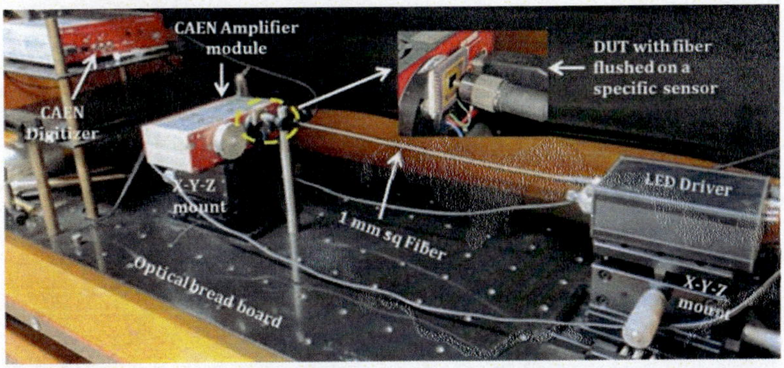

Fig. 6 Test setup to characterize the fabricated and packaged high-gain photo sensors inside a dark box with the CAEN SiPM characterization kit. A zoomed view of the optical fiber flushing on the DUT is shown in the inset

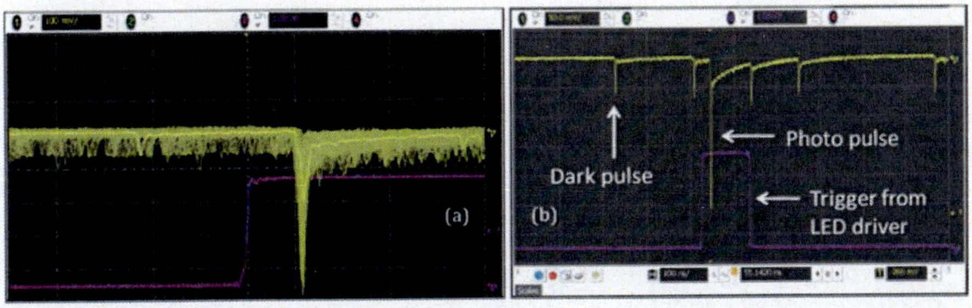

Fig. 7 **a** Hysteresis plot of the spectral response with subsequently increasing light intensity (changing the LED dial) on the oscilloscope, **b** photo pulse corresponding to a Geiger discharge in single mode (oscilloscope) including dark current pulse along with possible second-order effects. The purple track, used as a trigger, corresponds to the synchronization signal from the LED driver

increasing intensity in a hysteresis plot, Fig. 7b shows the optical response in a single mode with a LED trigger. The fall time part of the transient response can be seen to have two components; initially, a fast decay followed by a slow decay. The fast component of the fall time is may be due to the parasitic capacitance [16] of the large quench resistor in the commercial CMOS process.

The optical characterization was also carried out with a yellow LED (~590 nm), but the device showed better optical sensitivity with a violet LED (~405 nm). This behavior is consistent to what we achieve in the simulation (Fig. 4b). Likewise, in line with the value of breakdown voltage achieved in the simulation (~18.5 V), the measured breakdown voltage of the fabricated device is ~18.38 V.

Table 1 Comparative performance among the devices of this work and Hamamatsu MPPC

DUT	Micro-cell size, array size, and effective area	Gain setting in SP5600	Onset of optical response[1]	Transient response		Dark count rate[2]
				Rise time	Fall time	
Hamamatsu MPPC (S10362-11-100C)	100 μm × 100 μm, 10 × 10 and ~1 mm × 1 mm	36 dB	2.3	2 ns	100 ns with single time constant	~250 kHz/mm^2
Type I: P + /DN-well with P-well guard ring-based device (this paper)	30 μm × 30 μm, 16 × 16 and ~1 mm × 1 mm	36 dB	3.3	2 ns	100 ns with fast and slow component	~500–600 kHz/mm^2
Type II: P + /DN-well with P-well guard ring-based device with PAD layer (this paper)	30 μm × 30 μm, 16 × 16 and ~1 mm × 1 mm	36 dB	2.9	2 ns	100 ns with fast and slow component	~500–600 kHz/mm^2

Table 1 provides a comparative performance study carried out among the devices reported in this paper and a commercial MPPC from Hamamatsu using the same test setup and environment as described above.

5 Conclusion

Nine different potential SPAD structures suitable for standard CMOS technology were implemented inside a test vehicle and fabricated using a 0.35 μm HV CMOS process. The design and characterization of the best-performing structure, P+/DN-well-based active junction with P-well guard ring, have been presented in this paper. A 16 × 16 array of such micro-cells with passive quenching resistors has been designed with a 1 mm^2 active area and 24% fill factor. In one of the designs, a "PAD" layer was placed on top of the active area to reduce the material stacking above the active region. The design with "PAD" indeed was shown to have better optical sensitivity than the design containing no "PAD" layer. Vis-à-vis comparison with Hamamatsu MPPC was carried out. Apart from the slightly inferior optical sensitivity, the transient response of the high-gain photo sensor, reported in this paper, contains a fast decay

[1] w.r.t to the LED (~ 405 nm with 5 ns pulse width) driver dial setting [15]. With an external trigger frequency of 10 kHz, dial 6.7 roughly corresponds to 1000 photons.

[2] At 0.5 p.e threshold level and 1 V overvoltage.

component, possibly because of the parasitic capacitance of the large quench resistor. Moreover, it exhibits a higher dark count rate (DCR) than the Hamamatsu MPPC due to the higher doping concentration of sub-micron CMOS technology.

In the next version, few additional steps like surface engineering measures (such as the use of ARC layer) to improve the non-ideal material stacking on top of the active area in the commercial CMOS process, use of optically opaque trench isolation (such as STI) among the pixels are planned to improve the optical sensitivity and reduce the second-order effects (after-pulse, cross-talk).

References

1. Chunnilall CJ et al (2014) Metrology of single-photon sources and detectors: a review. Opt Eng 53(8):081910
2. Acerbi F, Gundacker S (2019) Understanding and simulating SiPMs. Nucl Instrum Methods Phys Res A 926:16–35
3. Ahmed Z et al (2020) A SiPM-based VLC receiver for gigabit communication using OOK modulation. IEEE Photonics Technol Lett 32(6):317–320
4. Link J et al (2019) Silicon photomultiplier use in particle astrophysics and heliophysics missions. ICRC 2019 Proc Sci 358:1–7
5. Lecoq P, Gundacker S (2021) SiPM applications in positron emission tomography: toward ultimate PET time-of-flight resolution. Eur Phys J Plus 136:292
6. Park HM, et al (2013) Design of a silicon photomultiplier based compact radiation detector for homeland security screening. 3rd International conference on ANIMMA. pp 1–5
7. Simon F (2019) Silicon photomultipliers in particle and nuclear physics. Nucl Inst Methods Phys Res A 926:85–100
8. Fisher EMD (2017) Single-photon avalanche diodes in CMOS technologies for optical communications. https://doi.org/10.5772/intechopen.68935
9. Kamrani E, et al (2013) Efficient premature edge breakdown prevention in SiAPD fabrication using the standard CMOS process. Semicond Sci Technol 28(1–6):045008
10. Villa F, et al (2014) Analog SiPM in planar CMOS technology. 44th European solid state device research conference. pp 294–297
11. Karami MA, et al (2010) A new single-photon avalanche diode in 90 nm standard CMOS technology. Optics Express 18(21):22158
12. Rochas A, et al (2002) Low-noise silicon avalanche photodiodes fabricated in conventional CMOS technologies. IEEE Trans. Electron Dev 49(3)
13. Xiao Z, Pantic D, Popovic RS (2007) A new single photon avalanche diode in CMOS high-voltage technology. Transducers 2007:1365–1368
14. Arosio V et al (2015) Development of a silicon photomultiplier toolkit for science and education. JINST 10:C07012
15. DS2477, SP5601 led driver—data sheet, CAEN, Rev 2 (30 June 2011)
16. Rochas A (2003) Single photon avalanche diodes in CMOS technology, a Ph.D. thesis

Design of High-Gain Antenna Incorporated with Left-Handed Material for Satellite Applications

Sujit Barman, Ajay Kumar Choudhary, Tamasi Moyra, Anirban Bhattacharjee, Anjan Debnath, and Arpita Mandal

Abstract This paper proposes a simple low-profile high-gain wideband antenna for Ku-band satellite applications that utilize a block of left-handed metamaterial for gain enhancement over conventional patch antenna. This antenna is designed and simulated using Ansys HFSS. It was found that performances of the simple patch antenna are improved by inserting left-handed material superstrate. The proposed antenna performs in Ku-band which is applicable to satellite applications. The proposed antenna resonates in 11.97 GHz. The left-handed material is realized by incorporating a layer of array of metallic rings with the conventional rectangular patch antenna.

Keywords Conventional antenna · Metamaterial · Circular rings · Gain · Left-handed material · Gain · Permeability · Permittivity · Satellite

1 Introduction

The left-handed material (LHM) is a class of artificial material which is utilized along with normal right-handed material (RHM) for improving profile and performance of conventional antennas, like size reduction, higher gain, wider bandwidth and higher directivity, etc. Victor Veselago in 1968 introduced the concept of left-handed material which was hypothesized to exhibit a negative refractive index, due to simultaneously negative effective magnetic permeability and electric permittivity [1]. According to him, due to the counter progressive direction of phase (actual wave motion) and group velocity (direction of energy propagation) of wave motion through such media, it can support backward wave propagation. As the basic constitutive

S. Barman (✉) · A. K. Choudhary · T. Moyra · A. Bhattacharjee · A. Mandal
National Institute of Technology Agartala, Agartala, India

A. Bhattacharjee
e-mail: anirbanb.ece@nita.ac.in

A. Debnath
North Eastern Space Applications Centre, Umiam, India
e-mail: anjan.debnath@nesac.gov.in

J. K. Mandal et al. (eds.), *Topical Drifts in Intelligent Computing*, Lecture Notes in Networks and Systems 426, https://doi.org/10.1007/978-981-19-0745-6_22

Table 1 Comparison between proposed antenna and other reference antenna

References	Size (mm^3)	Operating frequency (GHz)	Array	Gain (dB)
[4]	32.5 × 16 × 1.6	2.4	3 × 4	1.2
[5]	117 × 127 × 1.6	2.4	8 × 7	4.22
[6]	60 × 50 × 1.6	5.26	7 × 7	6.04
[7]	19.5 × 16.6 × 1.56	7.32	3 × 3	7.32
[8]	61.25 × 61.25 × 1.6	2.5	5 × 5	7.92
[9]	9 × 55 × 1.6	5.7	9 × 3	8.3
Proposed antenna	17.22 × 14.78 × 1.6	11.97	5 × 5	8.04

properties required for such media is simultaneously negative effective permittivity and permeability, which are not observed in nature directly, these were known as metamaterial (beyond normal material). John Pendry in 1999 developed a structure of arrays of split ring resonators which had the property of effective negative permeability [2]. His structure exhibited the elusive property due to strong magnetic coupling between the ring structures of SRR. In 2000, D.R smith implemented the first artificial structure of left-handed material [3] as a 3D prism of SRRs arranged in regular fashion. Periodic structure and strong magnetic coupling of SRR blocks coupled with effective negative permittivity due to plasma-like conductor density yielded the desired property of negative refractive index in a band of microwave frequencies. In this paper, a 5 × 5 array of three circular rings is placed on a layer which is above the conventional antenna. This modified antenna increases the gain from 6.4 to 8.04 dB and bandwidth from 0.84 to 1.1 GHz. The rings exhibit the properties of left-handed material. In Table 1, some references have been mentioned on similar work to show comparative analysis with current work. Some of these designs suffer from low gain, large size, lower bandwidth, etc.

2 Design

2.1 Conventional Antenna Design

Figure 1 shows the conventional rectangular patch antenna fed by a microstrip transmission line. This antenna consists of a metallic ground layer, substrate and metallic patch. The ground is at bottom, and patch is at the top of substrate. Ground and patch are made up of copper, and substrate is made up of FR-4 of relative permittivity 4.4. The dimensions of this antenna are in Table 2. The thickness of the substrate is 1.6 mm. The patch is fed by a transmission microstrip line whose characteristic impedance is 50 Ω. The dimensions are calculated using the standard formula [7, 10].

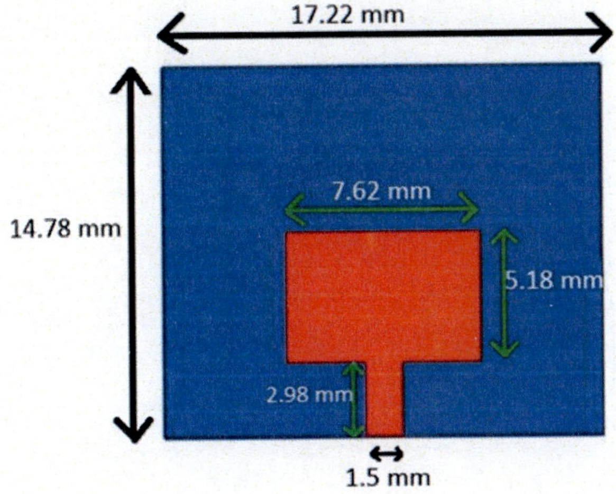

Fig. 1 Conventional antenna

Table 2 Value of all parameters

Parameters	Dimensions (in mm)
W_s	17.22
L_s	14.78
W	7.62
L	5.18
TL	2.98

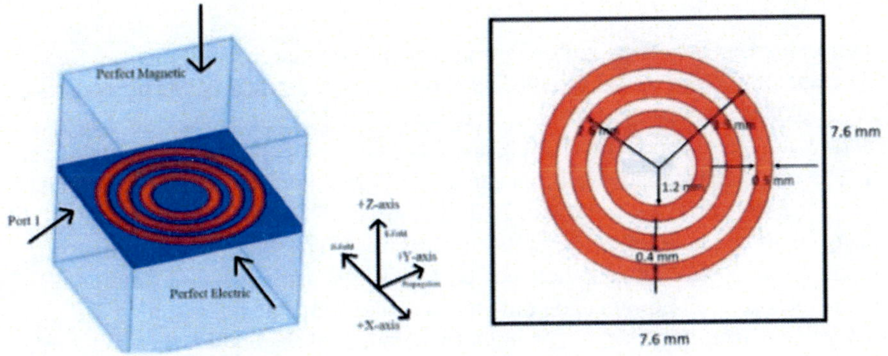

(a) (b)

Fig. 2 **a** Unit cell boundary condition. **b** Dimension of unit cell

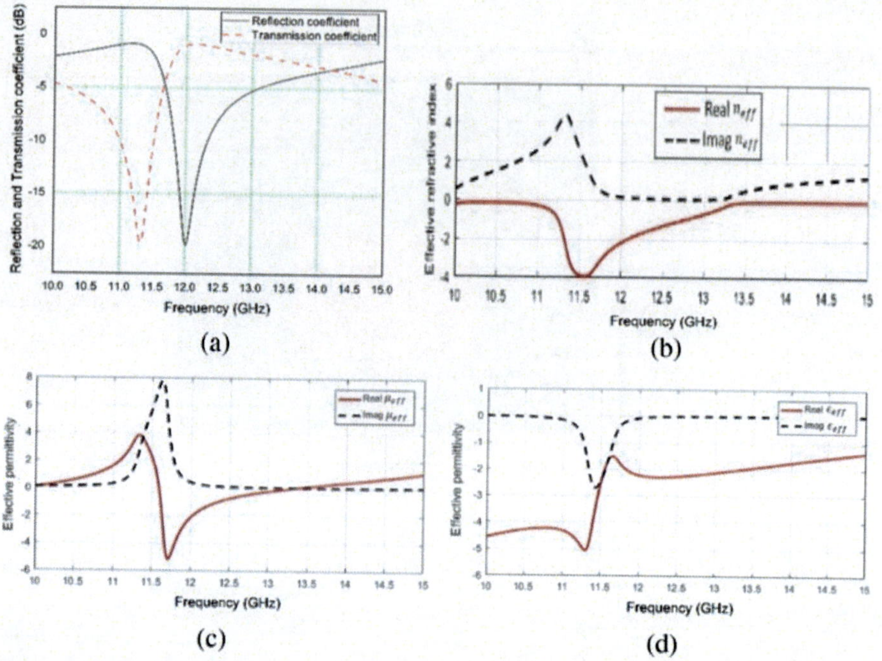

Fig. 3 **a** Reflection and transmission coefficient. **b** Effective refractive index. **c** Effective permeability. **d** Effective permittivity

2.2 Unit Cell Design

A unit cell of LHM along with electromagnetic boundary conditions set for the purpose of simulation in Ansys HFSS is shown in Fig. 2a. The boundary condition along X-axis is given as perfect electric conductor (PEC) and along Z-axis is given as perfect magnetic conductor (PMC), and feed port is assigned to Y-axis of the unit cell. The electromagnetic wave propagates in +Y-axis, while polarization of E-field is along +Z-axis and polarization of H-field is along −X-axis. The dimension of all the rings is shown in Fig. 2b. The unit cell has 3 rings. The width of all circular rings is equal, i.e., 0.5 mm. The gap between each ring is 0.4 mm. The substrate is FR-4 and has thickness of 1.6 mm. The dimension of the substrate is 7.6 mm × 7.6 mm, and relative permittivity is 4.4. The outer most radius of the biggest ring is 3.5 mm, and the inner most radius of the smallest ring is 1.2 mm. The rings are assigned copper material.

2.3 Metamaterial Properties Extraction from Unit Cell

Figure 3a shows the unit cell reflection and transmission coefficient. The unit cell is resonating at 11.97 GHz. S_{11} and S_{21} ($=S_{12}$, since symmetrical) are crossing each other at 11.65 and 14.3 GHz. The effective permeability and effective permittivity are also negative in this range as plotted in Fig. 3c, d. The effective refractive index is also negative in Fig. 3b. These properties show the existence of metamaterial, i.e., left-handed material (LHM). The effective refractive index, effective permeability and effective permittivity are derived from equations illustrated in paper [11].

2.4 Metamaterial Antenna Design

The proposed metamaterial antenna is having a superstrate of FR-4 of relative permittivity of 4.4 at a height of 16 mm above the patch of the conventional antenna. The dimension of this layer is 17.22 mm × 14.78 mm. The composite structure of unit cells has the properties of left-handed material and improves the performance of the patch antenna. The circular-shaped rings are also insensitive to polarization of incident electromagnetic wave. The superstrate suppresses the surface waves for its negative μ properties, and hence, there is a improved in gain. The metasurface also starts acting like a planar surface lens and causes convergence of radiation pattern in broadside direction, thus results into reduction of HPBW and causes gain enhancement. The proposed antenna is shown in Fig. 4.

Fig. 4 Proposed antenna

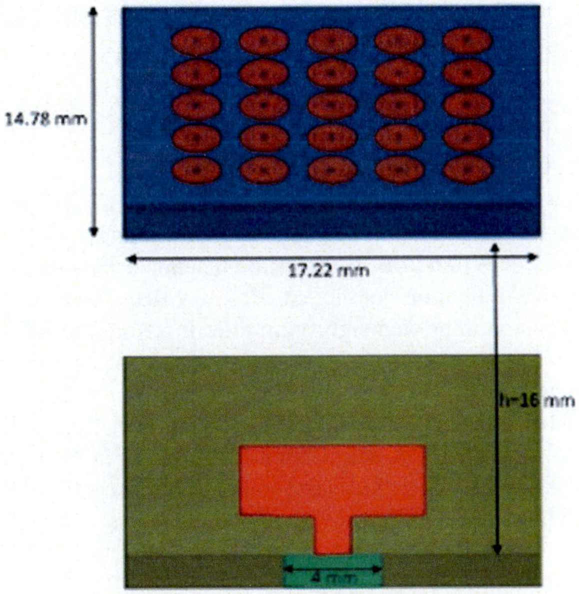

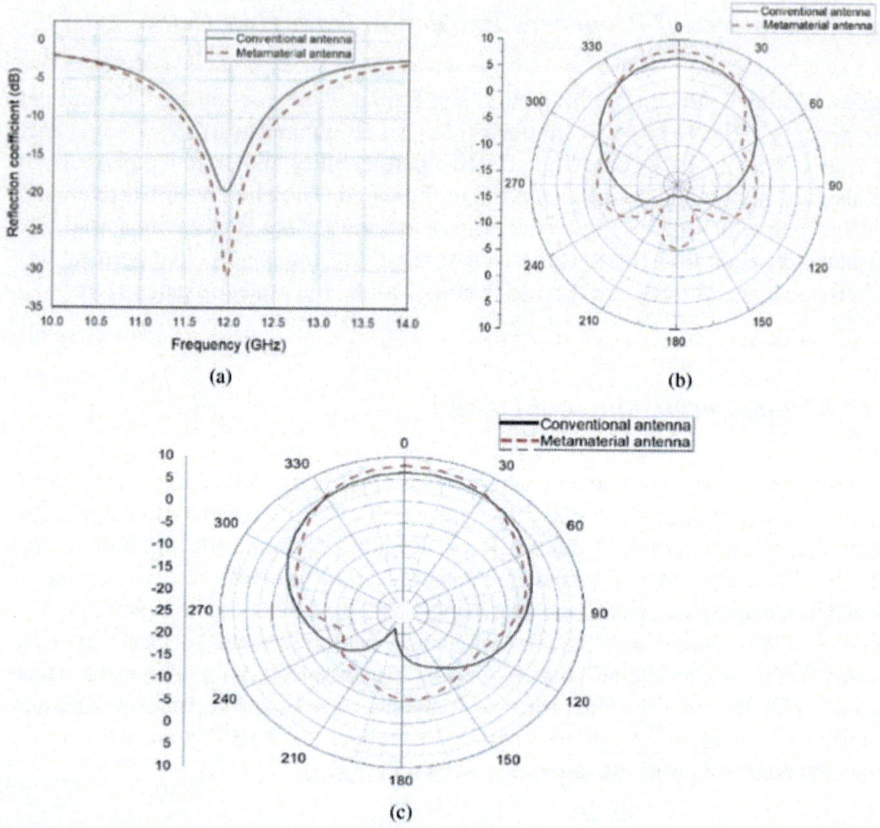

Fig. 5 **a** Return loss comparison of conventional and metamaterial antennas. **b** Radiation pattern at E-plane. **c** Radiation pattern at H-plane

3 Simulated Results

Figure 5a shows the reflection coefficient of the conventional and metamaterial antenna. The conventional antenna is resonating at 11.95 GHz and has reflection coefficient of −19.61 dB, whereas the metamaterial antenna is resonating at 11.97 GHz and has reflection coefficient of −30.9 dB. The reflection coefficient of proposed antenna is improved signifying after inserting the left-handed material. The bandwidth is also wider for the metamaterial antenna than the conventional antenna. Figure 5b, c shows the comparative radiation pattern in E-plane and H-plane. The proposed antenna has gain of 8.04 dB and, i.e., higher than the conventional antenna gain, i.e., 6.4 dB. A comparison is also given in the following Table 3. The −10 dB impedance bandwidth of conventional antenna is 0.84 GHz (7%), while that of MTM antenna is 1.15 GHz (9.6%), a 36.9% increase.

Table 3 Comparison of results of antenna without and with superstrate

Parameters	Antenna without superstrate	Antenna with superstrate
Resonating frequency	11.95 GHz	11.97 GHz
Reflection coefficient	−19.61 dB	−30.9 dB
Bandwidth	0.84 GHz	1.15 GHz
Gain	6.4 dB	8.04 dB
Directivity	7.37 dB	8.85 dB
VSWR	1.23	1.05

4 Conclusion

It is observed from the paper that performance of the proposed antenna is improved in terms of gain, bandwidth and directivity in frequency range of satellite applications. The gain of the proposed antenna becomes 8.04 dB from 6.4 dB, and bandwidth of the antenna becomes 1.15 GHz from 0.84 GHz. All the parameters of proposed antenna are improved. So, it can be used for satellite applications.

References

1. Veselago VGF (1968) The electrodynamics of substances with simultaneous negative values of ε and μ. Sov Phys Ups 10(4):509–514
2. Pendry JBF, Holden AJS, Robbins DJT, Stewart WJF (1998) Low frequency plasmons for thin-wire structure. J Phys Condensed Matter 10(22):4785–809
3. Smith DRF, Padilla WJS, Vier DCT, Nemat-Nasser SCF, Schultz SF (2000) Composite medium with simultaneous negative permeability and permittivity. Phys Rev Lett 84(18):4184–4187
4. Samir Salem Al-Bawri F, Mohd Faizal Jamlos S (2018) Design of low-profile patch antenna incorporated with double negative metamaterial structure. IEEE Int RF Microwave Conf Proc
5. Majid HAF, Rahim MKAS, Masri TT (2009) Microstrip antenna's gain enhancement using left-handed metamaterial structure. Progress Electromagnetics Res 8:235–247
6. Rohan Deshmukh F, Dushyant Marathe S, Kulat KDT (2019) Microstrip patch antenna gain enhancement using near-zero index metamaterial superstrate (NZIM lens). In: 10th International conference on computing, communication and networking technologies, IIT- Kanpur (July 6–8, 2019)
7. Paswan P, Mishra V, Patel PN, Dwivedi S (2014) Performance enhancement of coaxial feed microstrip patch antenna using left-handed metamaterial cover. In: Students' conference on electrical, electronics and computer science, IEEE
8. Saravanan MF, Beslin Geo VS, Umarani SMT (2018) Gain enhancement of patch antenna integrated with metamaterial inspired superstrate. J Electr Syst Inform Technol

9. Shengyuan Luo F, Yingsong Li S, Yinfeng Xia T, Liang Zhang F (2019) A low mutual coupling antenna array with gain enhancement using metamaterial loading and neutralization line structure. Aces J 34(3):411–418
10. Liton Chandra Paul F, Nahid Sultan S (2013) Design, simulation and performance analysis of a line feed rectangular micro-strip patch antenna. Int J Eng Sci Emer Technol 4(2):117–126
11. Mahamed Lashab F, Chemss Eddine S, Naeem Ahmed Jan T, Benabdelaziz FFRA, Abdalhameed F, Marl Child S (2014) CPW-fed antenna based on metamaterial for broadband application. In: Loughborough antenna and propag. Conferences (LAPC), pp 144–147, UK

A Novel and Efficient CNN Architecture for Detection and Classification of ECG Arrhythmia

Abhinav Gola, Animesh, Ravi Kumar Arya, and Sachin Singh

Abstract Cardiac arrhythmia is a result of irregular beating of the heart and occurs when the electrical impulses in our heart fail to send signals regularly. Depending on the type, arrhythmia can cause fainting or dizziness and in some cases, even heart failure which can be fatal. Early and quick detection of arrhythmia can be quite beneficial in many situations, and thus, there has been a surge in research on using artificial intelligence to counter this problem. Until recently, researchers have been using traditional machine learning algorithms to process ECG signals and classify them into different types of arrhythmias. In this manuscript, we propose a novel convolutional neural network (CNN) architecture to classify ECG signals after converting them into two-dimensional images. Our process mirrors the procedure used by medical practitioners in the real world to detect arrhythmia. We also implemented three popular CNNs—DenseNet 169, ResNet-50, and MobileNet V1 to evaluate and compare our proposed model against them. Our model demonstrated better performance than these networks on our evaluation metrics and will be useful for future tasks of ECG arrhythmia classification.

Keywords Cardiac arrhythmia · Machine learning · Convolutional neural network · 2D ECG

1 Introduction

Cardiovascular diseases (CVDs) have been the leading reason for death worldwide for the last 25 years. CVDs are responsible for 28% of deaths in India, according to a report from the Lancet Global Health [1]. Cardiovascular diseases (CVDs) overtook cancer as the leading cause of death in India in this century. This epidemiological transition is primarily due to an increase in the occurrence of cardiovascular diseases in India. In 2016, the approximate prevalence of CVDs in India was 54.5 million

A. Gola (✉) · Animesh · R. K. Arya · S. Singh
National Institute of Technology Delhi, New Delhi, India

S. Singh
e-mail: sachinsingh@nitdelhi.ac.in

J. K. Mandal et al. (eds.), *Topical Drifts in Intelligent Computing*, Lecture Notes in Networks and Systems 426, https://doi.org/10.1007/978-981-19-0745-6_23

people. In India, one of every four deaths is now caused by CVDs, with ischemic heart disease and stroke accounting for more than 80% of this strain. Heart arrhythmia is often caused by irregular electrical conduction of cardiac tissue. Premature ventricular contractions (PVCs) for a long period may potentially progress to ventricular tachycardia (VT) or ventricular fibrillation (VF), which can result in heart failure. As a result, it is important to track heart rates regularly to control and avoid CVDs.

Analyzing electrocardiogram (ECG) signals, which consist of three main components: P wave, QRS complex, and T wave, is the most common method for diagnosing arrhythmias. It aids in the diagnosis of heart disease by monitoring the heart's rhythm and electrical pulse. Heartbeat classification is important in ECG research for assessing the efficacy of arrhythmia diagnosis. Since manually categorizing heartbeats in long-term ECG recordings are time-consuming, the use of computerized algorithms provides critical diagnostic assistance. Significant preparation is needed to understand these dynamic cardiac arrhythmias associated with ECG characteristics consistently.

With the growth and development of AI technologies, many machine learning approaches are being utilized in ECG signal feature detection to solve problems related to vast quantities of ECG signal feature data. Several techniques for classifying ECG arrhythmias have been discussed in the literature. Guler et al. [3] proposed a feed-forward neural network as a classifier and wavelet transformation for feature extraction, both of which used the Lavenberg-Marquard algorithm in training while [8] discussed an optimization-based deep convolutional neural network (CNN).

Hannun et al. [2] have suggested a 1D CNN classifier that used more and broader data than Kiranyaz's CNN model. Although using a bigger ECG dataset, the arrhythmia detection rate is significantly lower than that reported in the literature. Many of the following drawbacks can be seen in many historically written articles for ECG arrhythmia classification: (1) decent results without cross-validation on carefully picked ECG recordings, (2) ECG pulse loss in noise filtering and feature extraction methods, (3) a small number of ECG arrhythmia categories for classification, (4) relatively low classification performance to use in the field. This manuscript shows and discusses how arrhythmia can be classified into seven categories using deep 2D CNN with grayscale ECG images.

The remaining of this manuscript is structured in the following manner. Section 2 delves into the specific methodologies utilized to classify ECG arrhythmias, such as ECG data preprocessing and the use of a CNN classifier. Next, Sect. 3 deals with the evaluation and experimental results of ECG arrhythmia classification. Section 4 concludes the paper with the concluding remarks.

2 Methodology

In this paper, we have classified 7 types of ECG arrhythmia signals through the application of computer vision methods. Particularly, CNNs were used to understand the classification of two-dimensional ECG signal images. This was possible because

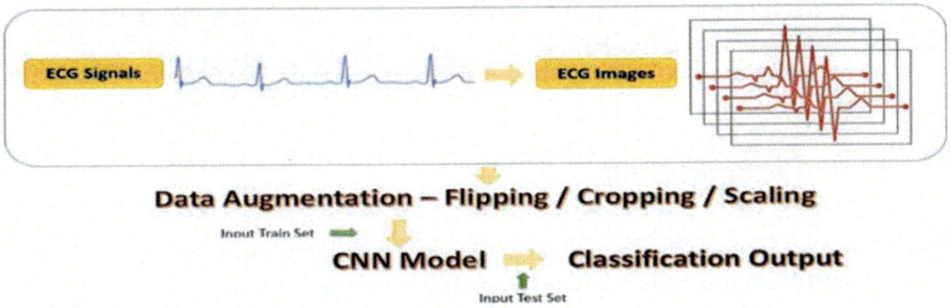

Fig. 1 Summarization of classification process

of the transformation of ECG signals from the MIT-BIH and the PTB databases [11, 12] into 2D images during the data preprocessing phase. Various data augmentation techniques were applied to these images to increase our input data sample. A summary of the implementation of the processes is shown in Fig. 1.

2.1 *Pre-processing of ECG Data*

The 2-dimensional CNN models were chosen for classification instead of signal processing as their performance is independent of the amount of noise in the ECG signals. This allows us to disregard the power line interference, electromyography noise, motion artifacts, line wander among other types of noises without using advanced noise reduction techniques such as wavelet transform or window-based filtering techniques.

Signals for the MIT-BIH dataset were transformed into individual images by first plotting them against time and then using a moving window to obtain images at regular intervals. These images were then converted into 128 × 128 grayscale images using a short Python snippet. This process resulted in 75,000 images classified into 8 types—7 classes of arrhythmia and one normal heartbeat.

2.2 *Data Augmentation*

To achieve high performance on complex tasks, CNN models are largely reliant on the amount and the diversity of the training data. Data augmentation helps with accommodating both these issues coupled with improving the class imbalance problem in the MIT-BIH dataset. Also, according to [13], networks trained with just data augmentation more easily adapt to different architectures and are less complex.

We used geometric transformations—rotation, scaling, flipping, cropping, and color space transformations—Aussian noise injection and color casting to augment the images. Each of the types of ECG beats was then resized back into 128×128 images after undergoing geometric scaling and cropping.

2.3 *Kernel Initialization*

Weight initialization is paramount as using large weights initially can cause the output of activation functions in the model's layers to explode during a forward pass. Similarly, if the weights are initialized to be too small, outputs will vanish while we run through the neural network. This makes the loss gradients to be inefficacious when running the network backward leading to issues with the convergence of the model.

As we will be using the ReLU activation function instead of the hyperbolic tanh function, Xavier initialization can cause our activation outputs to completely vanish. Thus, we used Kaiming's initialization to set our weights.

2.4 *Activation Function*

Activation functions decide which information should be passed on forward through the network by helping in the calculation of error during the backpropagation process. They are also used for introducing non-linearity into the networks and therefore, the sigmoid function. The tanh function and the ReLU function have been used popularly by machine learning researchers as their activation functions.

Convolutional neural networks usually employ ReLU and its other types as the activation functions for their hidden layers. Using ReLU can sometimes lead to missing out on some information as it converts all the negative values to zero, thus deactivating some nodes. Parametric ReLU solves this problem by assigning a small, non-zero gradient when the unit is not active.

2.5 *Regularization*

Batch normalization was originally motivated by the internal covariate shift. ReLU is not zero centered. Hence, initialization and input distribution may not be normalized. Therefore, the input distribution shifts over time. In deeper nets, you even get an amplified effect. As a result, the layers constantly have to adapt and this leads to slow learning. Batch normalization helps with this adaptation by calculating the central tendencies of a layer and then scaling it to adjust for the slow learning.

Shibani and Dimitris showed in their research that batch normalization is effective even if you introduce an internal covariate shift after the batch normalization layer again [17]. Therefore, we have used batch normalization for each block of our model.

Dropout is another method of regularization that randomly drops or ignores some layer outputs. This creates new layers with different numbers of nodes at each iteration which helps to train a neural network with different architectures simultaneously.

2.6 Cost and Optimizer Function

Cost function or loss function conveys the error between the forecasted and the actual output, and various optimizers are utilized to reduce the cost function by updating the parameters accordingly. We made use of the Adam optimization algorithm for our proposed CNN model. It requires very little memory and is computationally efficient to implement.

2.7 Optimized CNN Classifier Architecture

Using the above-recorded hyperparameters, our model's architecture was designed keeping efficiency and speed in mind. It is specifically designed for situations where there is a trade-off between computational power and speed. The architecture is shown in Fig. 2.

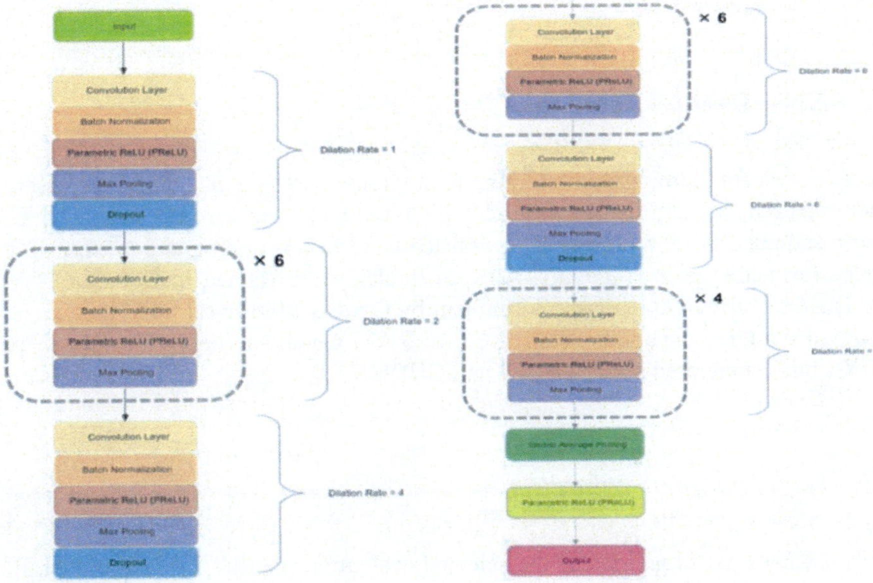

Fig. 2 Proposed CNN architecture

3 Experiments and Result

We implemented three popular convolutional neural network models to benchmark our proposed model with the other CNNs. These are MobileNet V1, DenseNet, and state-of-the-art EfficientNet V2. These networks along with the proposed model were trained and tested on the same MIT-BIH and PTB dataset and evaluated across the same metrics to achieve consistency.

3.1 Data Acquisition

The "PTB diagnostic ECG database" and the "MIT-BIH arrhythmia database" were used to obtain the ECG arrhythmia recordings for this work. The "MIT-BIH arrhythmia archive" is a publicly accessible dataset that contains standard investigative data for heart arrhythmia diagnosis.

There are 48 half-hour ECG recordings in the archive, obtained from 47 individuals. At 360 samples per second, the ECG recording is sampled. In the "MIT-BIH database," there are 110,000 ECG beats with 15 various forms of arrhythmia.

In the PTB diagnostic ECG database, there are 549 records in the archive from 290 different individuals (aged 17–87, average age 57.2).

This paper's experiment aims to validate the success of famous CNN models as well as past ECG arrhythmia classification works. We included regular beat (NOR) and seven forms of ECG arrhythmias from the "MIT-BIH database" and "PTB diagnostic ECG database."

3.2 Experimental Setup and Training

We used transfer learning by first training all the networks on an extensive dataset. Then, we used the feature maps acquired by this model for our specific task of recognizing and classifying 8 classes. This assisted us in faster training and maximizing our performance more efficiently. All four models were trained in Google Colab that is a free Jupyter notebook environment by Google entirely set up in the cloud. It provides a GPU capability with 12 GB of RAM which can be used to run deep CNNs. It is accelerated by CUDA 9.2 and CUDNN 6.0.

3.3 Performance Evaluation

Table 1 shows the properties of the selected evaluation metrics for each model. It can be observed from the table that three models except the MobileNet have a high

Table 1 Performance evaluation

Model	Best activation function	Accuracy	Sensitivity
Mobile Net v2	LeakyRelu	94.5	96.93
DenseNet 169	Softmax	98.6	97.56
ResNet-50	ReLU	98.9	98.34
Proposed model	PReLU	99.4	99.38

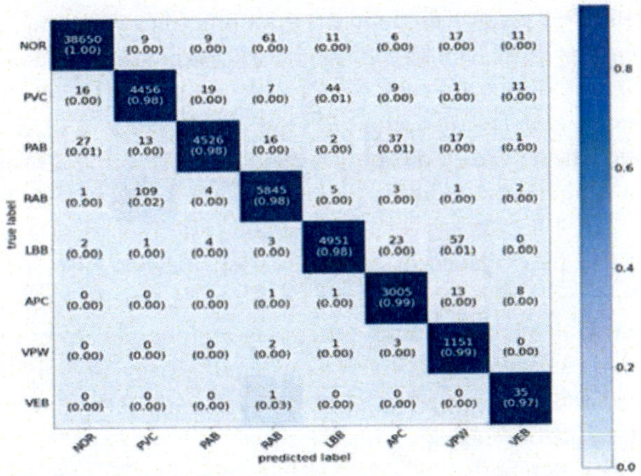

Fig. 3 Confusion matrix of proposed architecture with the precision of all classes

accuracy rate (98.6% for the DenseNet versus 98.9% for ResNet-50 versus 99.88% for the proposed model). We observe that our proposed model outperforms all the other three in the comparison of sensitivity values which are 99.38% for our network and around 97% for the other two models except ResNet-50 which showed around 98.3% sensitivity.

We can also observe from the table that the proposed model obtained excellent precision values for each class of the dataset, the lowest value being 97% for the VEB type while classifying the classes—NOR, APC, VPW correctly for 99% of iterations.

The other four types were also detected precisely about 98% of the time. Each of the networks gave the best performance when paired with a different activation function. MobileNet preferred LeakyReLU, ResNet-50 preferred ReLU while DenseNet provided the best output with Softmax. For our proposed model, we experimented and concluded with parametric ReLU as the optimum activation function (Fig. 3).

3.4 Comparison with the Existing Techniques

The authors of [15] proposed a novel CNN architecture with accuracy and sensitivity values on the MIT-BIH dataset of 99.05 and 97.85%.

Yu et al. [16] used feed-forward neural networks to classify the same 8 classes as ours and they got an accuracy rate of 98.71% and they did not record the recall values. Thus, the proposed model has shown better performance than the previous techniques applied for the same task.

4 Discussion and Conclusion

Four convolutional neural network pipelines were implemented and analyzed for the purpose of ECG arrhythmia classification—MobileNet V1, DenseNet 169, ResNet-50, and our proposed CNN model. We started by discussing the data preprocessing and augmentation techniques we employed and then moved onto the architecture and parameters of our proposed model. We briefly discussed the other three models as well and then laid out the metrics which were used for evaluating these models.

All 4 networks were trained and tested on the MIT-BIH and PTB datasets. Due to the high values of the evaluation metrics from our model, we think that this study will be helpful to take as a baseline for ECG arrhythmia classification tasks. It was designed keeping efficiency and weight in mind and thus could be advantageous when used in situations where there is a bargain of computational power. This research works as a guideline for systems that want to use deep learning for medical classification tasks.

Subsequently, we would like to try to explore some novel CNN architectures that can improve upon the results presented in this research article by experimenting with different activation functions, changing the position of each layer in the architecture, and working with different depth lengths of a network.

References

1. India State-Level Disease Burden Initiative CVD Collaborators (2018) The changing patterns of cardiovascular diseases and their risk factors in the states of India: the global burden of disease study 1990–2016. Lancet Glob Health 6:e1339–e1351. https://doi.org/10.1016/S2214-109X(18)30407-8
2. Park J, Lee K, Kang K (2013) Arrhythmia detection from heartbeat using a k-nearest neighbour classifier. In: IEEE international conference on bioinformatics and biomedicine, pp 15–22
3. Güler İ, Übeylı ED (2005) ECG beat classifier designed by combined neural network model. Pattern Recogn 38(2):199208
4. Kiranyaz S, Ince T, Gabbouj M (2016) Real-time patient-specific ECG classification by 1-D convolutional neural networks. IEEE Trans Biomed Eng 63(3):664–675
5. Zhai X, Tin, C (2018) Automated ECG classification using dual heartbeat coupling based on convolutional neural network. IEEE Access 6:27465–27472
6. Jiang W, Kong SG (2007) Block-based neural networks for personalized ECG signal classification. IEEE Trans Neural Netw 18:1750–1761. https://doi.org/10.1109/TNN.2007.900239

7. January CT, Wann LS, Alpert JS, Calkins H, Cigarroa JE, Cleveland JC, Conti JB, Ellinor PT, Ezekowitz MD, Field ME, Murray KT, Sacco RL, Stevenson WG, Tchou PJ, Tracy CM, Yancy CW (2014) 2014AHA/ACC/HRS guideline for the management of patients with atrial fibrillation: a report of the American College of Cardiology/American Heart Association task force on practice guidelines and the Heart Rhythm Society. J Am Coll Cardiol 64(21):e1–e76
8. Atal DK, Singh M (2020) Arrhythmia classification with ECG signals based on the optimization-enabled deep convolutional neural network. Comp Methods Prog Biomed 196:105607. https://doi.org/10.1016/j.cmpb.2020.105607
9. Osowski S, Hoai LT, Markiewicz T (2004) Support vector machine-based expert system for reliable heartbeat recognition. IEEE Trans Biomed Eng 51(4):582–589
10. Martis RJ, Acharya UR, Lim CM, Mandana K, Ray AK, Chakraborty C (2013) Application of higher order cumulant features for cardiac health diagnosis using ECG signals. Int J Neural Syst 23:1350014. https://doi.org/10.1142/S0129065713500147
11. Moody GB, Mark RG (2001) The impact of the MIT-BIH arrhythmia database. IEEE Eng Med Biol Mag 20(3):45–50
12. Bousseljot R, Kreiseler D, Schnabel A (1995) Nutzung der EKG-Signaldatenbank CARDIODAT der PTB über das Internet. Biomedizinische Technik, Band 40, Ergänzungsband 1:317
13. Hernández-García A, König P (2018) Further advantages of data augmentation on convolutional neural networks. In: International conference on artificial neural networks. Springer, Cham
14. Jun TJ, Nguyen HM, Kang D, Kim D, Kim D, Kim Y-H (2018) ECG arrhythmia classification using a 2-D convolutional neural network. arXiv e-prints
15. Yu SN, Chou KT (2008) Integration of independent component analysis and neural networks for ECG beat classification. Expert Syst Appl 34(4):2841–2846
16. Huang G, Liu Z, van der Maaten L (2018) Densely connected convolutional networks. Expert Syst Appl 34(4):2841–2846
17. Santurkar S, Tsipras D, Ilyas A et al (2018) How does batch normalization help optimization? In: arXiv e-prints, arXiv:1805.11604, arXiv:1805.11604. arXiv: 1805.11604 [stat.ML]
18. Pal A, Srivastva R, Singh YN (2021) CardioNet: an efficient ECG arrhythmia classification system using transfer learning. Big Data Res, 100271, ISSN 2214–5796. https://doi.org/10.1016/j.bdr.2021.100271
19. Sharma K., Masood S (2021) Deep learning-based non-invasive fetal cardiac arrhythmia detection. In: Choudhary A, Agrawal AP, Logeswaran R, Unhelkar B (eds) Applications of artificial intelligence and machine learning. Lecture Notes in electrical engineering, vol 778. Springer, Singapore. https://doi.org/10.1007/978-981-16-3067-5_38

A Uniquely Packed 2.4 GHz ISM Band Microstrip Antenna for Bluetooth Devices

Ranjeet Kumar, Rashmi Sinha, Arvind Choubey, and Santosh Kumar Mahto

Abstract Custom designed microstrip antennas are currently the preferred choice of creative antenna designers because of its record low better linear gain and efficiency, high bandwidth, favorable radiation pattern and easy to manufacture as they can be printed into the circuit board directly. This paper proposes a uniquely packed microstrip antenna of dimension 39.5 × 31 mm^2 for Bluetooth-enabled devices which have myriad of applications ranging from healthcare, Internet of things and entertainment. The antenna structure is simple and consists of concentric elliptical strip-shaped patch designed on an inexpensive FR4 substrate. The dielectric constant and thickness of substrate is 4.4 and 1.6 mm, respectively, to match the microstrip material dimension so that devices can be used over a broad temperature range. The presented antenna operates at 2.4 GHz resonances frequency, transmits and receives linearly polarized radiation which are omnidirectional in nature, provides stable gain covering 2.38–2.42 GHz. VSWR of 2 is obtained signifying a good impedance matching with 50Ω microstrip feedline and SMA connector. Design and analysis is done by HFSS 19.0 simulation software. Measurements of parameters of fabricated antenna are in conformity with simulation values. Equivalent circuit analysis result is presented to along with experimental and simulated results to authorize the need for Bluetooth-enabled devices.

Keywords Bluetooth · Microstrip antenna · HFSS · Return loss

R. Kumar (✉) · R. Sinha
Department of Electronics and Communication Engineering, National Institute of Technology, Jamshedpur, India

R. Sinha
e-mail: rsinha.ece@nitjsr.ac.in

A. Choubey
Department of Electronics and Communication Engineering, Indian Institute of Information Technology, Bhagalpur, India
e-mail: achoubey.ece@nitjsr.ac.in

S. K. Mahto
Department of Electronics and Communication Engineering, Indian Institute of Information Technology, Ranchi, India
e-mail: skumar@iiitranchi.ac.in

J. K. Mandal et al. (eds.), *Topical Drifts in Intelligent Computing*, Lecture Notes in Networks and Systems 426, https://doi.org/10.1007/978-981-19-0745-6_24

1 Introduction

Wireless technology has shown tremendous growth over the years and is being effectively adopted by telecommunication networks, business installations, medical world and homes throughout the world to cut down on the expenditure and incompatibility imposed by introducing cables into building or as a connection between various equipment. Bluetooth is an ideal wireless communication technology that has taken over the connected device market and is being extensively used in numerous types of portable devices and wearables. They include fitness trackers, pulse oximeters, smart watches to glucose monitors, cell phones, pagers, communicators, audio peripheral (headsets, speakers, stereo receivers) data peripherals, personal digital assistants, microphone ovens, garage door openers and embedded applications. Bluetooth utilizes 2.4 GHz Industrial Scientific Medical (ISM) spectrum and around 20 MHz bandwidth to support legacy devices and Wi-Fi standard 802.11 b or 802.11 g. These frequencies do not need to have a Federal Communication Commissions (FCC) license to operate the devices. It is one of the first technologies that meets the requirement of interoperability, low-power operation, secured transmission of data and compatibility with other electronics devices and Internet services, very wide adaptation and low cost. With aforementioned capabilities that are mismatched by any other medium, Bluetooth is poised to set a new standard to benefit mankind in every walk of life. Recently, it is the go-to choice for unique solutions for medical health care, reopening of business, public venues and fulfilling academic aspirations during the COVID-19 pandemic. Bluetooth-enabled devices have been around for more than 20 years but is still grieved with issues related to connectivity, battery life and interference. In order to coexist with the entire assemble of wireless devices that share 2.4 GHz ISM band, Bluetooth keeps on hopping between frequency channels. Designing a cost-effective and compact antenna for Bluetooth devices exhibiting good and reliable performance has been a challenge to antenna design fraternity. There has been a requirement for more antenna area and volume on the device as allocated frequency bands are getting narrower due to heavy congestion but consumers want handy, portable devices. It is the antenna that has to endure the burden for size constraint along with high performance measure. Any radiating structure that can be easily embedded on the printed circuit board of portable Bluetooth devices and is capable of resonating at 2.4 GHz with bandwidth more than 100 MHz and efficiency greater than 50% can be considered as a Bluetooth antenna. Creative antenna designers have presented plenty of options for Bluetooth devices. Numerous methods and structures have been developed in yesteryears to improve the gain and impedance bandwidth of the planar antenna which covers multiple frequency bands along with the narrow and ultra-wideband (UWB) bandwidth [1–3]. In [4], a low-profile F antenna is built on the edge of ground plane of the headset for Bluetooth applications which operate as an internal antenna. A compact monopole antenna consists of arc and omega shape strip, and partial ground plane is designed for Bluetooth, WLAN and WiMAX in [5]. In [6], a horizontal and vertical slot is subtracted from ground plane to get enhanced wideband circularly polarized monopole antenna. In [7], authors have presented a proximity coupled feed multi-band microstrip antenna for operating in the long-term evolution (LTE), Bluetooth, WiMAX and WLAN bands. The frequency-reconfigurable antenna is designed on both sides of the substrate and is fed by a microstrip line for Bluetooth, WiMAX and WLAN applications in [8]. A compact CPW-fed pentagon-shaped radiating patch antenna with an asymmetrical slot antenna has presented for communication system in [9]. A microstrip antenna is a lucrative option for Bluetooth devices as they are characterized by low cost, easy to design and develop with the help of simulation tools, ease of production, better gain and efficiency, compact size, omnidirectional radiation pattern, flexible and can be easily placed on the PCB of the device. This paper presents a uniquely packed low-profile microstrip antenna designed on an inexpensive flexible Flame retardant (FR4) substrate for Bluetooth devices. The radiating patch has a scaled elliptical strip grid-shaped geometry fed by 50Ω feedline. The microstrip dimensions are matched with dielectric constant (ε_r) of 4.4 of the substrate to get the correct resonance frequency. A consistent omnidirectional radiation pattern, good bandwidth and return loss is exhibited by the antenna. The size of antenna is $39.5 \times 31\ \text{mm}^2$ which can be easily mounted on PCB module of Bluetooth devices. It can transmit and receive linearly polarized radiations which are omnidirectional in nature, provides stable gain covering 2.38–2.42 GHz. VSWR of 2 is obtained signifying a good impedance matching with 50Ω microstrip feedline and SMA connector.

2 Antenna Configuration and Implementation

Proposed monopole microstrip antenna is evolved in four distinct steps as depicted in Fig. 1. The patch consists of a circular chunk surrounded by concentric circular strips having different radii $R1$, $R2$, $R3$ and $R4$, separated by a gap 's' as shown in Fig. 1a. In the second step, the above configuration is scaled by 1.4, resulting in elliptical-shaped radiating elements on which a rectangular slot of width 'a' is etched out, to get a unique geometry, as depicted in Fig. 1b. In the third step, a rectangular shape partial ground plane of dimension $l_g \times W$ is as depicted in Fig. 1c. Finally, a 'T'-shaped microstrip feedline is used to complete the antenna structure, as shown in Fig. 1d. This antenna configuration of size $39.5 \times 31\ \text{mm}^2$ is designed on FR-4 epoxy substrate having material permittivity of 4.4, tangent loss of 0.02 and height of 1.6 mm. The optimized designed parameter of this antenna is given in Table 1. Its resonance frequency is calculated by Eq. (2) [1].

$$\varepsilon_{\text{reff}} = \frac{\epsilon_r + 1}{2} \tag{1}$$

$$f_r = \frac{V}{2L_{\text{eff}}\sqrt{\varepsilon_{\text{reff}}}} \tag{2}$$

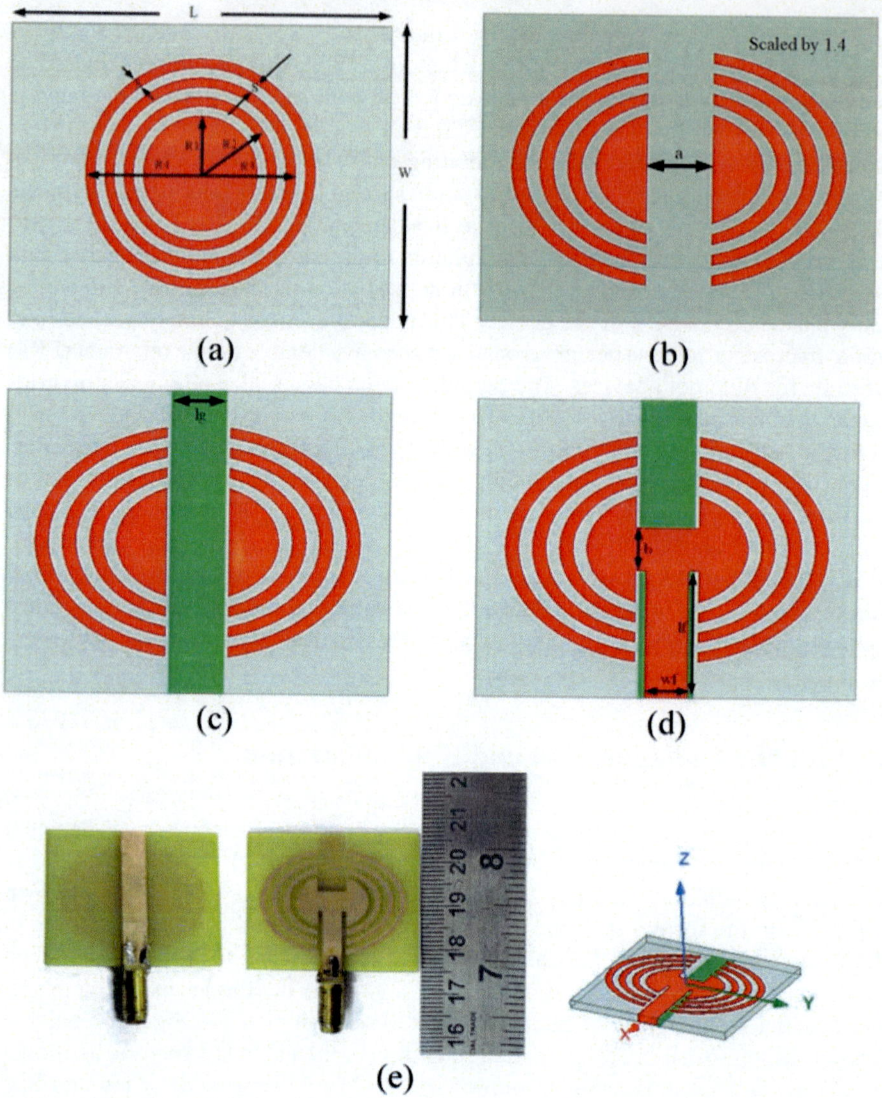

Fig. 1 **a–d** Evolution of the antenna geometry, and **e** fabricated antenna (rear view and front view)

where $\varepsilon_{\text{reff}}$, L_{eff}, f_r and v denote effective dielectric constant (ϵ_r), effective length, effective width, resonance frequency and velocity of electromagnetic waves in the free space, respectively. A prototype of the proposed antenna is captured Fig. 1e.

Table 1 The optimized dimension of the Bluetooth antenna

Parameter	Optimized dimension (mm)
L	39.5
W	31
l_g	5.64
w_f	4.5
l_f	13.25
$R1, R2, R3, R4$	6, 8, 10, 12
a, b, s	6.58, 4.5, 1

3 Circuit Analysis Bluetooth Antenna

The reflection coefficient characteristics antenna behaves as band pass filters. Hence, transmission line model is used to develop lumped circuit model by the help of basic circuit element RLC. The conducting strips (matching transmission feedline, radiating patch, partial ground plane) are considered as combination of surface resistance (R_s) and inductance (L_s). The upper and lower conducting strip are separated by FR4 dielectric substrate, which provides capacitance (C) and conductance (G). These parameters are calculated by Eqs. (3–6) [10–12].

$$R_s = \frac{1}{W_f \sigma_{\text{cond}}\, \delta} \tag{3}$$

$$L_s = \frac{1}{W_f \sigma_{\text{cond}}\, \delta\omega} \tag{4}$$

$$\delta = \sqrt{\frac{2}{\omega\sigma_{\text{cond}}\, \mu_0}} \tag{5}$$

$$C = \varepsilon_{\text{reff}} 2.85 \frac{1}{\ln\left\{1 + \frac{1}{2}\left(\frac{8h}{W_{\text{eff}}}\right)\left[\left(\left(\frac{8h}{W_{\text{eff}}}\right) + \sqrt{\left(\frac{8h}{W_{\text{eff}}}\right)^2 + \pi^2}\right)\right]\right\}} \tag{6}$$

where $W_{\text{eff}} = W_f + \frac{t}{\pi}\ln\left\{\frac{4e}{\sqrt{\left(\frac{t}{h}\right)^2 + \left[\frac{1}{\pi\left(\frac{W_f}{t} + 1.10\right)}\right]^2}}\right\}$.

Where t, σ_{cond}, ω, δ, R_s, L_s, C, W_{eff}, $\varepsilon_{\text{reff}}$ and μ_0 have its own meaning. After considering all the assumptions, the circuit model of this antenna is exhibited in Fig. 2a, b. The ADS is used to simulate and optimize value of the equivalent circuit components given in Table 2, and its $S11$ characteristics are incorporated in Figs. 3, 4, 5 and 6a.

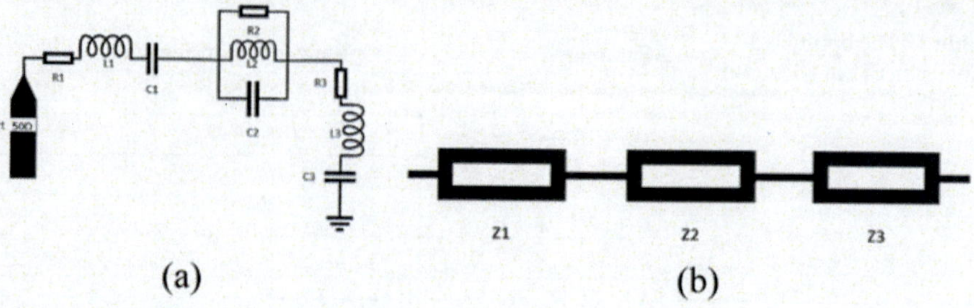

Fig. 2 **a** Basic circuit model and **b** reduced form

Table 2 Optimal values of the circuit parameters

Resistor value (mΩ)	Inductor value (nH)	Capacitor value (pF)
$R1$ 4.320	$L1$ 5.6224	$C1$ 24.959
$R2$ 654.377	$L2$ 1.4610	$C2$ 3.34350
$R3$ 6.432	$L3$ 6.1589	$C3$ 38.9251

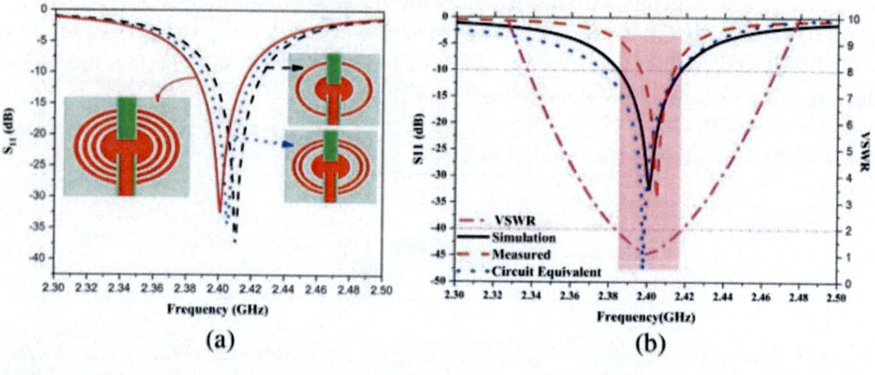

Fig. 3 S_{11} characteristics of **a** different geometry, and **b** proposed antenna

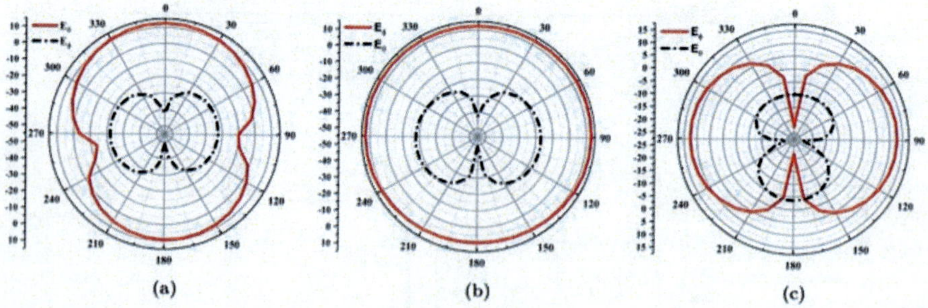

Fig. 4 The simulated radiation pattern at resonance frequency

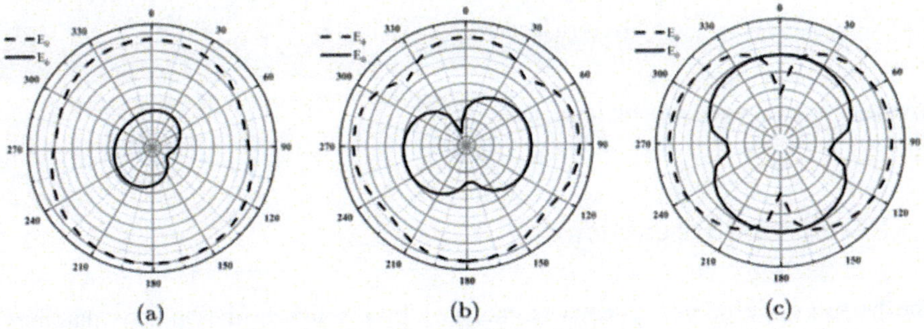

Fig. 5 The measured radiation pattern at resonance frequency

Fig. 6 The gain versus frequency

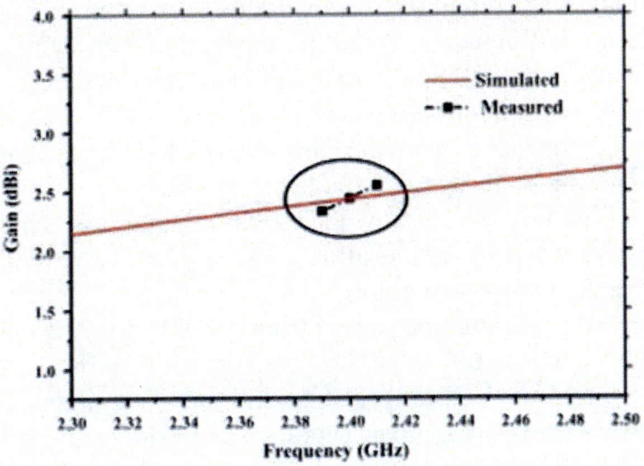

$$Z1 = R1 + j\omega L1 + \frac{1}{j\omega C1}, \frac{1}{Z2} = \frac{1}{R2} + \frac{1}{j\omega L2} + j\omega C2,$$
$$Z3 = R3 + j\omega L3 + \frac{1}{j\omega C3}$$

$$Z_{\text{in}} = Z1 + Z2 + Z3 \tag{7}$$

Now, using Eq. (7), different circuit parameters can be calculated [13] as follows.

$$\text{Reflection coefficient}(\tau) = \frac{Z_{\text{in}} - Z_0}{Z_{\text{in}} + Z_0} \tag{7}$$

$$\text{VSWR} = \frac{1 + |\tau|}{1 - |\tau|} \tag{8}$$

$$\text{Return loss(RL)} = 20\log|\tau| \quad (9)$$

where Z_0 is the characteristic impedance (50Ω).

4 Results and Discussion

The performance of this antenna is analyzed by discussing the antenna parameters. The commercial full-wave electromagnetic (FEM) Ansys Design HFSS 19.0 software is used for simulation of the designed antenna. The $S11$ characteristics of different geometries involved in designing the final antenna are displayed in Fig. 3a. It is observed that the bandwidth increases and shifts toward left to get the desired resonant frequency in the ultimately evolved configuration. Figure 3b exhibits a comparison of return loss (S_{11}) characteristics obtained by simulation, measurement and circuit equivalent model and VSWR, maintaining <2, which signifies good impedance matching characteristics. The impedance bandwidth of 600 MHz is achieved, with resonant frequency 2.4 GHz.

Figures 4 and 5 show the radiation pattern of the antenna. Figure 4a represents both E_Φ and E_θ the elevation plane (*xz*-plane) at $\Phi = 0°$. From the plot it can be seen that E_Φ component uniformly varies between 2 and −5 dB from $\theta = 0°$ to 60°. The variation of radiation pattern from $\theta = 60°$ to 90° is −5 to −20 dB. From $\theta = 90°$ to 120° and $\theta = 60°$ to 90°, the variation of pattern is −20 dB to −7.5 dB and 7.5 dB to −1.15 dB, respectively. The E_Φ and E_θ components are durable and have 5 dB difference in the angular sections. Thus, principally, these planes have a very uniform field. The radiation pattern in elevation plane (yz-plane) at $\Phi = 90°$ is depicted in Fig. 4b. The E_Φ component varies around 0 dB insignificantly throughout the whole angular region. The E_θ component, is not dominated, only adds on where E_Φ and E_θ component is lacking. The radiation pattern in the azimuth plane (xy-plane) at $\theta = 90°$ is shown in Fig. 4c. The E_Φ component, between $\Phi = 30°–150°$ and $\Phi = 200°–340°$, varies between 2.4 and −5 dB in this angular region. Signal strength drops to 30 dB approximately in the remaining sector of the plane. Thus, the data from the radiation pattern reveals that the angular coverage for the component E_Φ and E_θ is favorable for the proposed Bluetooth antenna.

The simulated and measured radiation antenna gain values have good agreement at discrete frequency points as depicted in Fig. 6. Proposed antenna attains maximum gain of 2.43 dBi at 2.4 GHz and has approximately constant average gain of 2.2 dBi in the desired bandwidth. This is almost a good measure of performance for Bluetooth devices.

5 Conclusion

This paper presents a uniquely packed simple and low-cost compact microstrip antenna that can be easily embedded on the PCB module for Bluetooth devices. The antenna operates in 2.4 GHz ISM band spectrum and exhibits performance favorable for Bluetooth-enabled devices. The simulated, measured and equivalent circuit model results are compared and found in conformity with each other. A little difference in the simulated, measured and circuit equivalent model results can be attributed to the measured and fabricated limitations.

References

1. Balanis CA (1982) Antenna theory analysis and design. Willey, New York
2. Kumar G, Ray KP (2003) Broadband microstrip antennas. Artech house
3. Anguera J, Andújar A, Huynh MC, Orlenius C, Picher C, Puente C (2013) Advances in antenna technology for wireless handheld devices. Int J Antennas Propag
4. Wong KL, Hsu MR, Li WY, Su SW, Chen A (2007) Study of the Bluetooth headset antenna with the user's head. Microw Opt Technol Lett 49(1):19–23
5. Naidu PV, Kumar R (2014) Design of a compact ACS-fed dual band antenna for blue tooth/WLAN and WiMAX applications. Prog Electromag Res 55:63–72
6. Zhang L, Jiao YC, Ni T (2015) A planar monopole antenna with wideband circularly polarized characteristic. Electromagnetics 35(4):227–239
7. Rohini S, Sharanya M et al (2017) Proximity coupled microstrip antenna for Bluetooth, WIMAX and WLAN applications. Int J Commun Comput Technol 5(2):4852
8. Li T, Zhai H, Wang X, Li L, Liang C (2014) Frequency-reconfigurable bow-tie antenna for Bluetooth, WiMAX, and WLAN applications. IEEE Antennas Wirel Propag Lett 14:171–174
9. Gorai A, Roy B, Mahanti G (2019) A compact CPW-fed wideband slot antenna with reflector for wireless communication. Electromagnetics 39(8):556–570
10. Doddipalli S, Kothari A (2018) Compact UWB antenna with integrated triple notch bands for WBAN applications. IEEE Access 7:183–190
11. Nella A, Gandhi A (2018) Lumped equivalent models of narrowband antennas and isolation enhancement in a three antennas system. Radio Eng 27(3):647
12. Wang Y, Li J, Ran LX (2008) An equivalent circuit modeling method for ultra-wideband antennas. Prog Electromag Res 82:433–445
13. Singh P (2014) Rectangular notch loaded dual band annular ring patch antenna. J Microw Optoelectron Electromag Appl 13(1):85–96

Investigating Electromagnetic Bandgap for Nano-fishnet Structure with Elliptical Void Embedded Inside Triangular Lattice

Arpan Deyasi, Rikita Das, and Angsuman Sarkar

Abstract Tunable electromagnetic bandgap (EBG) variation can be observed when nano-fishnet with elliptical void structure is embedded inside triangular lattice. Being a double-negative metamaterial, it offers unique variation of mid-band frequency with increasing fill factor, which is monotonically increasing, when variation is limited within feasible mechanical limit. Highest EBG is observed at 46% fill factor, and corresponding field patterns are obtained computed inside first Brillouin zone. Only a magnetic polarized bandgap is generated, whereas a quasi-electrical polarized bandgap is not even found. Results are critically important for photonic filter design using metamaterials.

Keywords Nano-fishnet · Elliptical void · Electromagnetic bandgap · Mid-band frequency · Triangular lattice

1 Introduction

The future optical-integrated circuit has a major dependence on advent and applications of photonic crystal, after its initial discovery [1], and later, wide amount of research with it in the last decade. Based on the principle of Bragg's reflection, its applicability is now covering the area of design of transmitter [2], receiver [3], sensor [4], fiber [5], information processing [6], switch [7] and secured low-loss transmission [8]. Though initial works consist of conventional SiO_2–air or similar type of conventional oxide stacks [9], but recently, introduction of metamaterials greatly revolutionize the concept as it facilitates to integrate the circuit with planar microstrip antenna or frequency-selective surfaces [10]. Addition of nonlinear effects make the results more accurate.

A. Deyasi (✉) · R. Das
Department of Electronics and Communication Engineering, RCC Institute of Information Technology, Kolkata 700015, India

A. Sarkar
Department of Electronics and Communication Engineering, Kalyani Govt Engg College, Kalyani 741235, India
e-mail: angsumansarkar@ieee.org

J. K. Mandal et al. (eds.), *Topical Drifts in Intelligent Computing*, Lecture Notes in Networks and Systems 426, https://doi.org/10.1007/978-981-19-0745-6_25

Metamaterial is the prime material today used for making planar high-frequency integrated circuits [11] owing to its crucial advantage of designing and implementing planar antennas [12]. This feature really enables to scientists and researchers to develop optoelectronic, and, later, photonic integrated circuits embedded with microstrip antenna [13], where research on photonic crystal (PhC) just brightens the future prospects. The most important feature of PhC-based device is filter [14, 15], where selection/rejection of electromagnetic spectrum critically depends on dimensional parameters [16] and material constants [17]. The basic structure of photonic crystal is formed as a highly ordered infinite number of parallel alternating layers, consisting materials of different dielectric constant. Such arrangement gives them their significant properties capable of manipulating electromagnetic wave and identical photonic band gap structure for TE and TM mode. To facilitate the tunable characteristics of the filter, metamaterial–air combination is utilized for PhC design, where already literatures show performance improvement [18, 19] than conventional structure.

The present paper deals with characterization of nano-fishnet (elliptical void)–air combination embedded inside a triangular lattice, and its photonic bandgap and corresponding mid-band frequency variations are investigated. 3D field patterns are also made at the optimized structure, where max electromagnetic bandgap is attained. Section 2 deals with a brief outline of the structure, whereas Sect. 3 analyzes the simulated findings. Paper is concluded in Sect. 4.

2 Structural Description

The present structure as considered in this simulation is the nano-fishnet structure with elliptical void, which is filled up by air. A schematic description is given in Fig. 1.

The structure is an artificial one; magnitudes of refractive indices are −4 (nano-fishnet) and 1.0008 (air). For theoretical calculation, it is assumed that all the voids are almost equivalent.

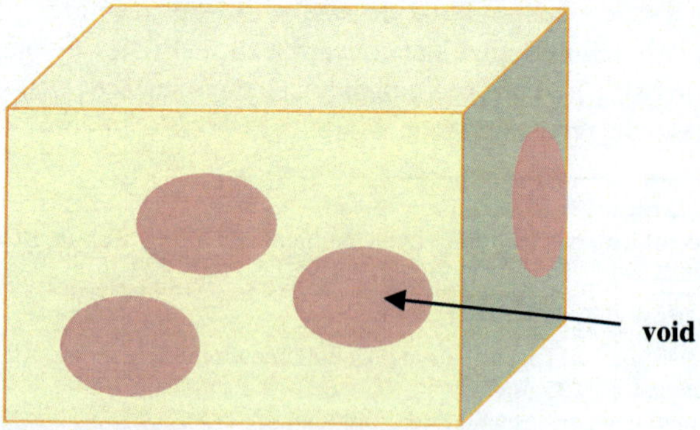

Fig. 1 Schematic nano-fishnet structure (with elliptical void)

The design, as mentioned, in this structure is already available a few years back [20]; however, novelty lies in the fact that fill factor variation is not considered at that time where analyzing TE and TM wave propagations. This is the advantage of this manuscript.

3 Result and Analysis

Based on the simulated findings, it can be stated that the structure will produce complete electromagnetic bandgap for magnetic polarization (evident from Fig. 2a),

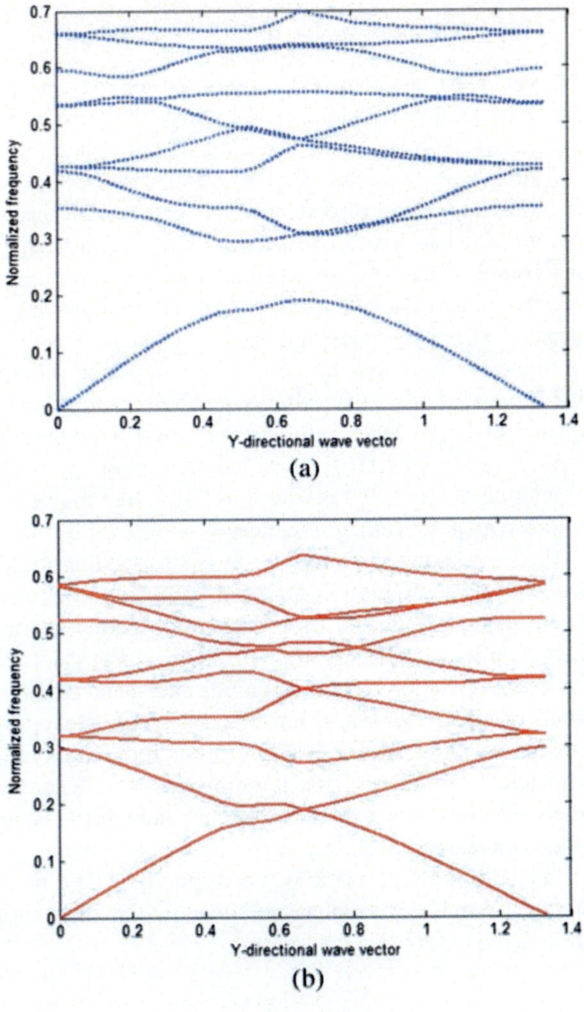

Fig. 2 **a** Electromagnetic bandgap for magnetic polarization, **b** no electromagnetic bandgap for electric polarization

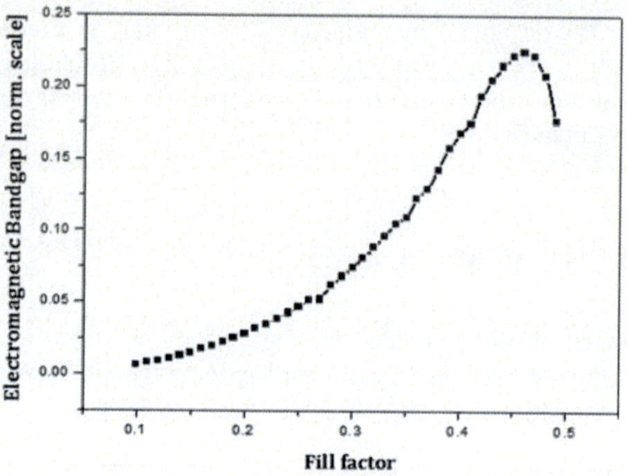

Fig. 3 Electromagnetic bandgap variation with fill factor

but even quasi-bandgap cannot be generated for electric polarization (reflected from Fig. 2b). All the simulations are considered inside first Brillouin zone, which is completely analogous to the semiconductor structure.

The difference occurs for two different modes of propagation is simply due to the relative position of displacement vector with the wave-vector. That is why for this proposed structure, all the results are considered for TM mode only. But only one complete bandgap is obtained within the first ten levels, which means it is a large stop-band filter. Also, magnitude of the passband is comparatively large, which means the filter is a wideband one.

Figure 3 shows the variation of bandgap with fill factor. From the plot, it is seen that bandgap increases with increase in fill factor, attains a peak value at 0.46, and then again decreases. Henceforth, it can be concluded that maximum bandgap formation is possible for 46% fill factor. One point may be noted in this context that limit of simulation sets to 49%, as over it, the structure may become mechanically weak and will lose its practical significance.

Figure 4 shows the variation of corresponding mid-band frequency. It has been observed from simulation that mid-band frequency monotonically increases, even when bandgap reduces. In this context, it may be noted the maximum bandgap formation is possible for 46% fill factor, as evident from Fig. 3. However, Fig. 4 shows the variation of corresponding mid-band frequency monotonically increases when bandgap reduces, even after fill factor cross 46%. This is quite interesting as when electromagnetic bandgap starts reducing, but the position for the bandgap with respect to energy axis is going upwards. Therefore, magnitude of mid-band frequency, which is nothing but the average of the bandgap value (lower and upper levels) increases.

For both the polarizations, corresponding field distributions are calculated and plotted. It may be emphasized that all the distributions are calculated at 0.45 normalized frequency.

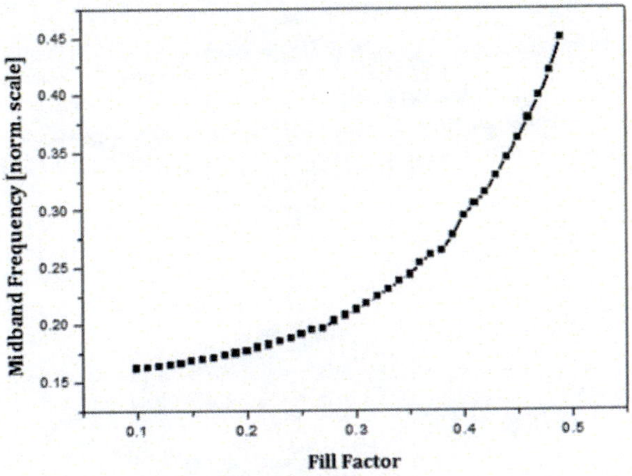

Fig. 4 Mid-band frequency variation with fill factor

Figure 5 shows the field variations at lowermost value of 'kz,' whereas Fig. 6 depicts it for highest magnitude of 'kz.' The intensity variations are clearly evident from the colors, where red signifies highest magnitude. A vis-à-vis comparative study between Figs. 5 and 6 suggests that intensity of electric field decreases for higher value of 'kz,' whereas reverse variation is seen for magnetic field. The colors shown in both Figs. 5 and 6 show the variation of magnitude.

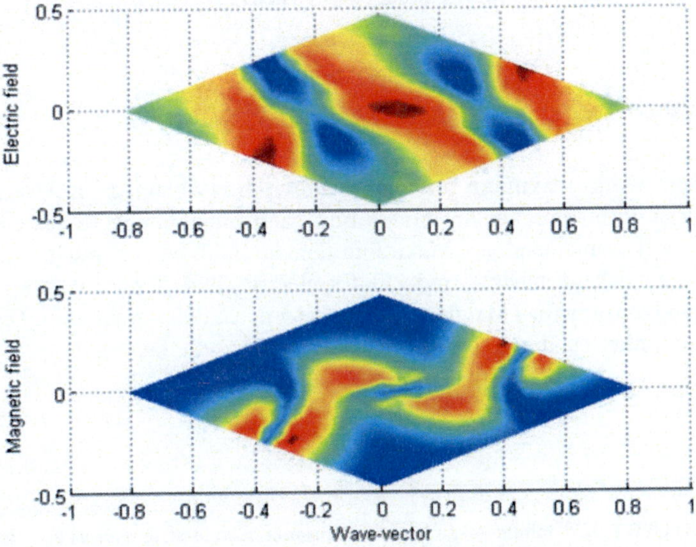

Fig. 5 Electric and magnetic field distributions at 0.45 normalized frequency for minimum value of 'kz'

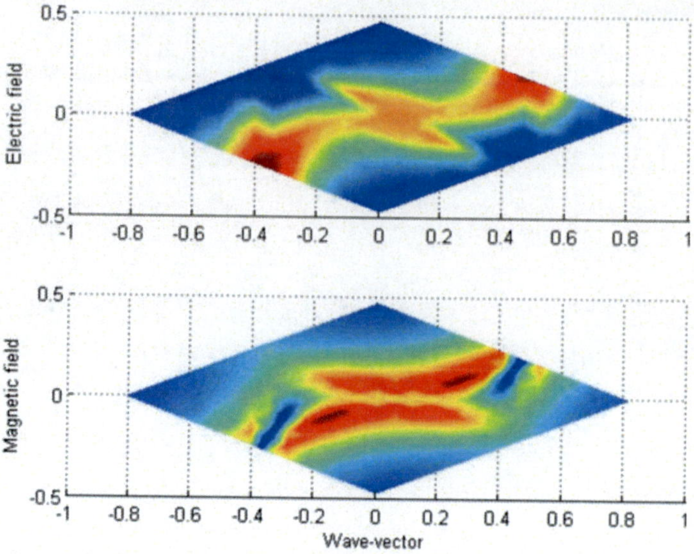

Fig. 6 Electric and magnetic field distributions at 0.45 normalized frequency for maximum value of 'kz'

As per the color distribution, it is seen that in both the plots, intensity decreases from the center for both the fields. However, for electric field, the distribution is along horizontal axis, whereas for magnetic field, it is perpendicular to that. This is just because of the orientation of displacement vector.

4 Conclusion

In the present work, maximum photonic bandwidth is obtained for TM mode of propagation at 46% ripple factor, though mid-band frequency increases even after decreasing of photonic bandgap. Maximum field strength enhances with increasing 'kz' for magnetic field, whereas reduces for electric field. This structure provides complete bandgap only for TM mode, as evident form bandgap simulation. Therefore, best wideband filter can be designed at 46% value of ripple factor.

References

1. Yablonovitch E (1987) Inhibited spontaneous emission in solid-state physics and electronics. Phys Rev Lett 58:2059–2061
2. Xu Y, Lin J, Dubé-Demers R, LaRochelle S, Rusch L, Shi W (2018) A single-laser flexible-grid WDM silicon photonic transmitter using microring modulators. In: Optical fiber

communication conference, OSA technical digest, p W1I.3

3. Nozaki K, Matsuo S, Fujii T, Takeda K, Kuramochi E, Shinya A, Notomi M (2017) Forward-biased photonic crystal photodetector towards amplifier-free bias-free receiver. In: Conference on lasers and electro-optics, OSA Technical Digest, p STh4N.1
4. Lu TW, Wu CC, Lee PT (2018) 1D photonic crystal strain sensors. ACS Photon 5(7):2767–2772
5. Gerosa RM, Vianna PG, Domingues SH, de Matos CJS (2019) Reduced graphene oxide coated photonic crystal fiber for all-fiber laser mode locking. In: Conference on lasers and electro-optics: OSA technical digest, p STu4L.2
6. Norman JC, Jung D, Wan Y, Bowers JE (2018) Perspective: the future of quantum dot photonic integrated circuits. APL Photon 3:030901
7. Jung B-M (2018) X-band microwave photonic filter using switch-based fiber-optic delay lines. Curr Opt Photon 2:34–38
8. Wan X, Wang Z, Sun B, Zhang Z (2020) Low dispersion and confinement loss photonic crystal fiber for orbital angular momentum mode transmission. Opt Quant Electron 52:289
9. Karmakar A, Roy I, Deyasi A, Deb A, Kundu A (2013) Suitability of semiconductor heterostructure over SiO_2-air composition for one-dimensional photonic crystal based bandpass filter. Int J Electron Commun Comput Technol 3(3):403–406
10. Jha KR, Singh G (2012) Analysis and design of terahertz microstrip antenna on photonic bandgap material. J Comput Electron 11(4):364–373
11. Caroline E, Johnson W, Xavier SC, Kabilan A (2016) Compact photonic crystal integrated circuit for all-optical logic operation. IET Optoelectron 10(4):142–147
12. Gonzalo R, Nagore G, deMaagt P (2003) Simulated and measured performance of a patch antenna on a 2-dimensional photonic crystals substrate. Prog Electromag Res 41:257–269
13. Kumar DS, Prithika P, Caroline BE (2019) Investigating the performance of microstrip patch antenna with photonic crystal on different substrate. In: IEEE international conference on system, computation, automation and networking
14. Deyasi A, Sarkar A (2021) Performance estimation of defected ternary photonic crystal-based bandpass filter beyond 100 THz for all-optical circuit. In: Biswas A, Banerjee A, Acharyya A, Inokawa H (eds) Emerging trends in terahertz engineering and system technologies. Springer, Singapore
15. Deyasi A, Sarkar A (2020) THz bandpass filter design using metamaterial-based defected 1d photonic crystal structure. In: Biswas A., Banerjee A, Acharyya A, Inokawa H, Roy J (eds) Emerging trends in terahertz solid-state physics and devices. Springer, Singapore
16. Foteinopoulou S (2012) Photonic crystals as metamaterials. Phys B 407(20):4056–4061
17. Deyasi A, Sarkar A (2019) Computing optical bandwidth of bandpass filter using metamaterial-based defected 1D PhC. In: AIP conference proceedings, vol 2072, p 020003
18. Lotfi ES, Jamshidi-Ghaleh K, Moslemi F, Masalehdan H (2010) Comparison of photonic crystal narrow filters with metamaterials and dielectric defects. Euro Phys J D 60(2):369
19. Deyasi A, Dey U, Das S, De S, Sarkar A (2020) Computing photonic bandgap from dispersion relation for tm mode propagation inside metamaterial-based 1D PhC. Micro Nanosyst 12:201–208
20. Cao T, Cryan MJ (2012) Optical trapping using double negative index fishnet metamaterial. Prog Electromag Res 129:1–4

Dual-Element CPW-Fed MIMO Antenna for ISM Band Application

Ajit Kumar Singh, Santosh Kumar Mahto, and Rashmi Sinha

Abstract This article presents a dual-port circular patch CPW (coplanar waveguide)-fed multiple-input-multiple-output (MIMO) antenna for ISM band (5.8 GHz) applications. The antenna achieves an impedance bandwidth of 1.64 GHz (5.22–6.86 GHz). The optimized dimension of the MIMO antenna is 30 mm × 16 mm. The MIMO structure is obtained by putting the antenna elements orthogonally and fed independently. The matching and isolation of the MIMO antenna are improved by using a rectangular stub associated with the feed and ground plane. The individual antenna has gain and radiation efficiency of 2.52 dBi and 92%, respectively. The antenna has a stable radiation characteristic at 5.8 GHz and co- and cross-polarization are also studied. The performance characteristics of the proposed antenna are dissected as far as the envelope correlation coefficient (ECC), diversity gain (DG), mean effective gain (MEG), total active reflection coefficient (TARC), isolation between the ports, and the values are 0.28, 9.90 dB, ±3 dB, −7 dB, 12 dB, respectively.

Keywords CPW-fed · MIMO · Isolation · ECC · TARC

1 Introduction

The current scenario for wireless communication is to achieve a higher data rate, capacity, low latency, and resolution. It has been shown in [1–3] that to improve the information throughput in a multipath environment for ISM/LTE/5G operations, the MIMO antenna system should be adopted. Because of compact volume in the mobile terminal and to achieve better performance for the MIMO antenna system.

A. K. Singh (✉) · S. K. Mahto
Indian Institute of Information Technology, Ranchi, India

S. K. Mahto
e-mail: skumar@iiitranchi.ac.in

R. Sinha
National Institute of Technology, Jamshedpur, India
e-mail: rsinha.ece@nitjsr.ac.in

J. K. Mandal et al. (eds.), *Topical Drifts in Intelligent Computing*, Lecture Notes in Networks and Systems 426, https://doi.org/10.1007/978-981-19-0745-6_26

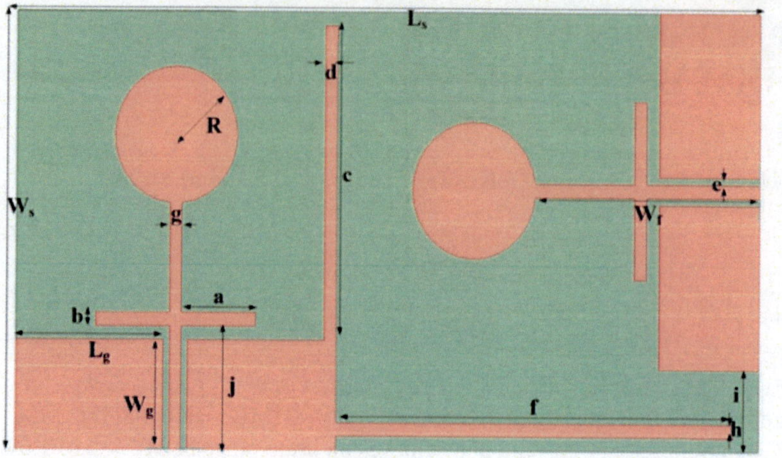

Fig. 1 CPW-fed MIMO antenna

As the number of antennas increases at the receiver, the reliability of the receiving system improves; however, space is a central issue. Likewise, the impact of isolation decreases as the distance between antenna elements diminishes, which influences the diversity characteristics of the antenna [4, 5].

Coplanar waveguide (CPW) taking care of draws in more consideration in light of more bandwidth with low dispersion and radiation spillage as compared to other feed lines [6–8].

In literature, several MIMO antennas have been proposed for ISM/WLAN applications [7–10]. In [7], a monopole printed antenna was studied for ISM band with a dimension of $44 \times 44 \times 0.5$ mm^3. A compact quad element antenna was proposed for body area networks with isolation of −25 dB and ECC of 0.02 in [8]. In [9], a compact MIMO was proposed for ISM band application (5.8 GHz) with pattern and polarization diversity. The compact antenna was proposed with a dimension of $30 \times 25 \times 1.524$ mm^3. In [10], a dual-element CPW-fed MIMO antenna was discussed for WLAN applications with an antenna dimension of $50 \times 50 \times 1.59$ mm^3.

In this article, 1×2 MIMO antenna with a dimension of $30 \times 16 \times 1.6$ mm^{3}is proposed in the frequency range 5.22–6.86 GHz by using a CPW fed along with rectangular stubs. The average isolation is 12 dB in the operating band from 5.22 to 6.86 GHz with ECC < 0.28, the peak gain 2.52 dBi, and the average radiation efficiency of 92%. Simulation is performed using commercially available software High Frequency Structure Simulator (HFSS version 2021R1).

2 Design

The proposed antenna has two circular radiators, as depicted in Fig. 1. It has an overall dimension of 30 × 16 × 1.6 mm^3 and is designed on an FR4 substrate having a dielectric constant of 4.4, and a loss tangent of 0.02. In this case, each element is orthogonally coupled with the others to utilize the polarization diversity. Matching and isolation can be improved by incorporating rectangular stubs in the ground plane and feed, which act as an open-circuit stub, and its dimension is tuned properly to compensate for the input impedance offered by the antenna as the inductance and capacitance value depends on it (Table 1).

Table 1 The optimized parameters of CPW-fed antenna

Representation	Size (mm)	Representation	Size (mm)	Representation	Size (mm)
L_S	30	W_f	9	e	0.25
W_S	16	a	3	f	16
L_g	6	b	0.5	g	0.5
W_g	4	c	11.5	h	0.5
R	2.5	d	0.5	i	3
j	4.5				

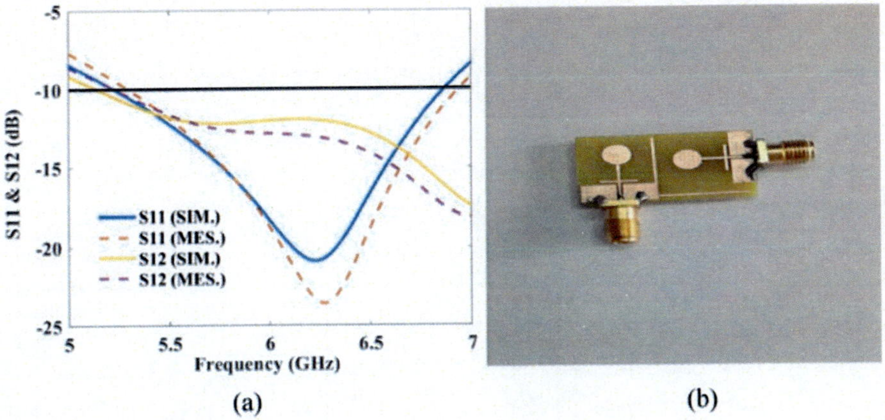

Fig. 2 **a** $S11$ and $S12$, **b** fabricated antenna

3 Results and Discussion

Simulated and experimental, Matching ($S11$) and isolation ($S12$) of the antenna, as represented in Fig. 2a are less than −10 dB, −12 dB, respectively, throughout the entire bandwidth. The fabricated antenna is shown in Fig. 2b.

The MIMO antenna has an average peak gain of 2.42 dBi in the frequency ranging from 5.22 to 6.86 GHz as displayed in Fig. 3a.

Figure 3b represents the radiation efficiency at all ports of the proposed antenna. The average radiation efficiency is 92.1% (−0.90 dB), throughout the entire bandwidth.

Figure 4 represents the 2D patterns of the antenna at 5.8 GHz and observed that at Ø = 0°, designs are similar to dipole and almost omnidirectional at Ø = 90°.

Figure 5 represents the far field patterns of MIMO antenna at 5.8 GHz frequency in x–z and y–z plane.

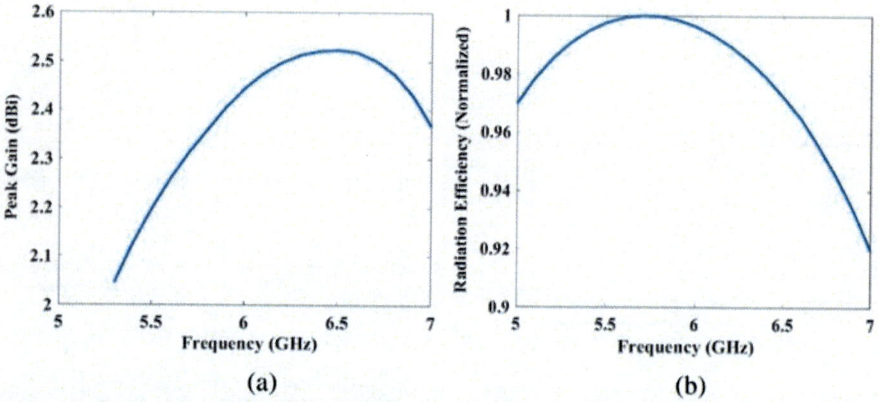

Fig. 3 **a** Peak gain, **b** radiation efficiency

Fig. 4 E and H plane at a 5.8 GHz, frequency for Ø = 0° and 90°

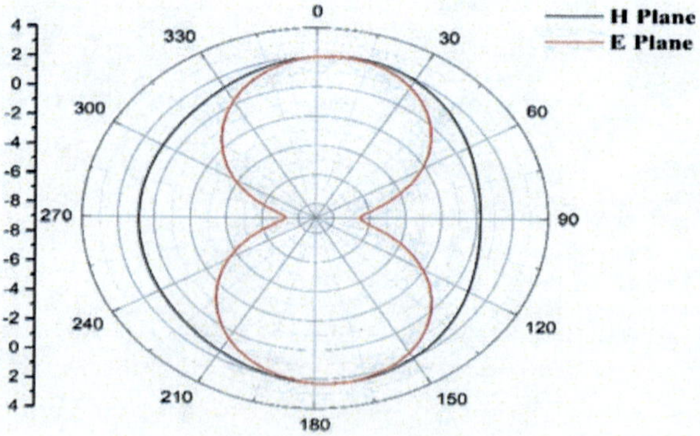

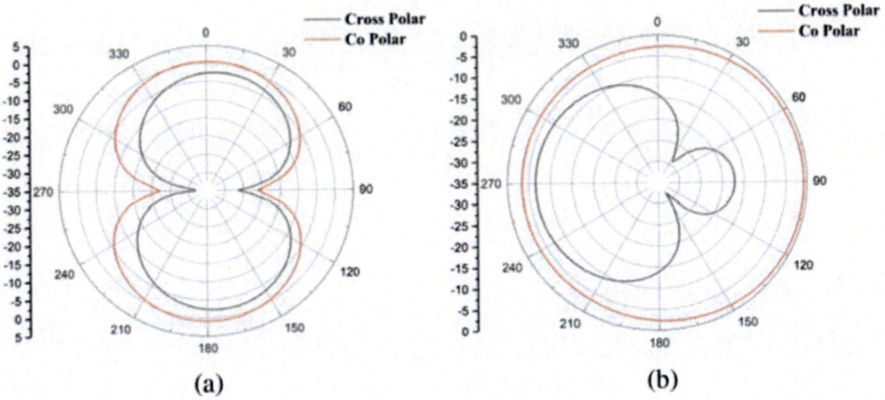

Fig. 5 Co- and cross-polarization, **a** *ZX* plane, **b** *YZ* plane

Fig. 6 Surface current density at 5.8 GHz

Surface currents flow in Fig. 6 from the stimulated port-1 to other ports is the fundamental reason behind the isolation between close antenna elements. At a time, only port-1 is excited and port-2 is matched.

4 MIMO Diversity Characteristics

The ECC which considers the correlations between the two antennas and determined using (1) [11]:

$$|\rho_e(i, j, N)| = \frac{\sum_{n=1}^{N} S_{i,n}^{*} S_{n,j}}{\prod_{k(=i,j)} \left[\sum_{n=1}^{N} S_{i,n}^{*} S_{n,k} \right]} \quad (1)$$

For ideal case, ECC is zero; however, experimental value should be ≤ 0.5.

DG of the MIMO antenna is calculated as:

$$\text{DG} = 10\sqrt{1 - \text{ECC}^2} \quad (2)$$

DG should be near to 10 dB.

The calculated values of ECC and DG are less than 0.28 and 9.90 dB, respectively, as displayed in Figs. 7a, b that show the good diversity performance.

In dual-port antenna, TARC is also an important parameter that gives the relation between radiated and received power.

For dual-element TARC can be calculated using (4) [11].

$$\text{TARC} = \frac{\sqrt{(S_{11} + S_{12})^2 + (S_{21} + S_{22})^2}}{\sqrt{4}} \quad (3)$$

Ideally, TARC should be <0 dB. The simulated value of the TARC is less than ≤ 7 dB in the frequency ranging from 5.22 to 6.86 GHz as depicted in Fig. 7b.

MEG is the ratio of mean received power to mean input power. For two elements, MEG can be determined using (5) and (6) [11] and shown in Fig. 8.

$$\text{MEG}_i = 0.5\left[1 - \sum_{j=1}^{N} |S_{ij}|^2\right] \leq 3\,\text{dB} \quad (4)$$

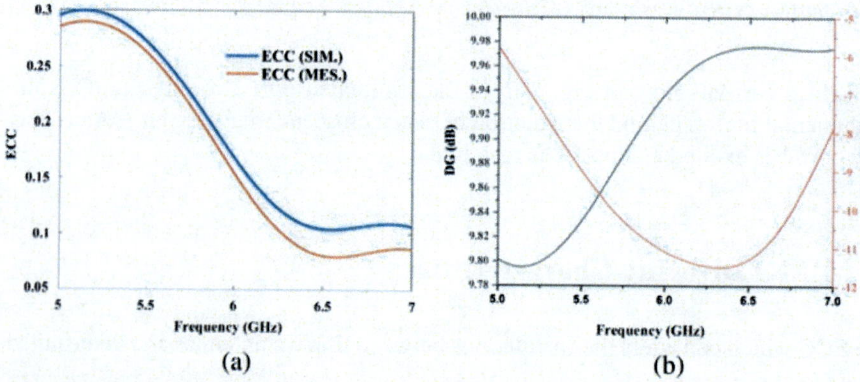

Fig. 7 **a** ECC, **b** DG and TARC

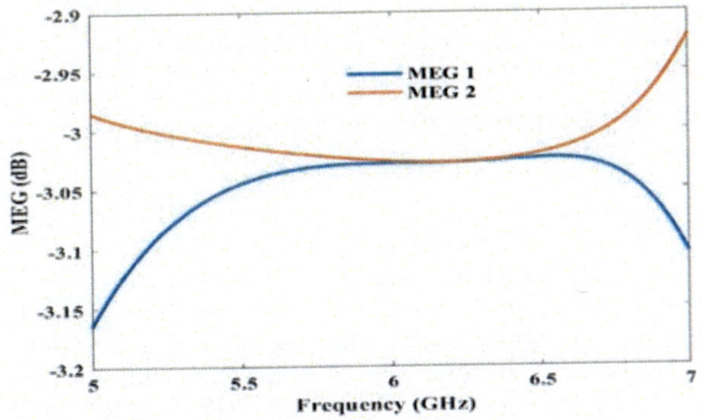

Fig. 8 MEG

Table 2 Comparative chart of proposed MIMO antenna

References	Dimensions (mm^3)	Frequency range (GHz)	Isolation (dB)	ECC
[9]	44 × 44 × 0.5	2.35–2.5	−10	NA
[10]	26 × 26 × 0.8	2.4	−25	0.02
[12]	30 × 25 × 1.524	5.8	−13	NA
[13]	50 × 50 × 1.59	2.25–3.15, 4.89–5.95	−15	0.01
[P]	30 × 16 × 1.6	5.22–6.86	−12	0.28

$$\mathrm{MEG}_j = 0.5\left[1 - \sum_{i=1}^{N} |S_{ij}|^2\right] \leq 3\,\mathrm{dB} \tag{5}$$

in which, i, j denote antenna elements 1 and 2, separately.

The proposed MIMO antenna is compared with other published work in Table 2. This MIMO antenna has a size of 30 × 16 × 1.6 mm^3, which is compact as contrasted to other antennas. The measured value of ECC is 0.28 which is good as depicted in Table 2, except [10, 13]. The MIMO antenna achieved isolation of 12 dB, which is better than [9], however, less than [10, 12, 13]. The MIMO antenna achieves an average radiation efficiency of 92% which is better than other reported antennas.

5 Conclusion

A CPW-fed dual-element MIMO antenna with an isolation of 12 dB is presented for ISM band application. The performance of the proposed design has been analyzed and discussed in terms of its various antenna characteristics parameters like impedance bandwidth, surface current distribution, reflection coefficient, gain, efficiency, and radiation characteristics. It also offers high DG of 9.90 dB and satisfactory TARC $\leq$ 7 dB. Therefore, the proposed MIMO antenna is suitable for ISM bands viably.

References

1. Karaboikis MP, Papamichael VC, Tsachtsiris GF, Soras CF, Makios VT (2008) Integrating compact printed antennas onto small diversity/MIMO terminals. IEEE Trans Antennas Propag 56(7):2067–2078
2. Oh J, Sarabandi K (2014) Compact, low profile, common aperture polarization, and pattern diversity antennas. IEEE Trans Antennas Propag 62(2):569–576
3. Sun L, Huang W, Sun B, Sun Q, Fan J (2014) Two-port pattern diversity antenna for 3G and 4G MIMO indoor applications. IEEE Antennas Wirel Propag Lett 13:1573–1576
4. Dietrich CB, Dietze K, Nealy JR, Stutzman WL (2001) Spatial, polarization, and pattern diversity for wireless handheld terminals. IEEE Trans Antennas Propag 49(9):1271–1281
5. Thaysen J, Jakobsen KB (2006) Envelope correlation in (N, N) MIMO antenna array from scattering parameters. Microw Opt Technol Lett 48(5):832–834
6. Bakariya PS, Dwari S, Sarkar M, Mandal MK (2015) Proximity-coupled microstrip antenna for bluetooth, WiMAX, and WLAN applications. IEEE Antennas Wirel Propag Lett 14:755–758
7. Diallo A, Luxey C, Le Thuc P, Staraj R, Kossiavas G (2008) Enhanced two-antenna structures for universal mobile telecommunications system diversity terminals. IET Microwaves Antennas Propag 2(1):93–101
8. Sonkki M, Antonino-Daviu E, Cabedo-Fabres M, Ferrando-Bataller M, Salonen ET (2012) Improved planar wideband antenna element and its usage in a mobile MIMO system. IEEE Antennas Wirel Propag Lett 11:826–829
9. Mathialagan S (2017) Design of CPW-fed tapered MIMO antenna for ISM band applications. Int J Microw Wirel Technol 9(1):227–230
10. Elfergani I, Iqbal A, Zebiri C, Basir A, Rodriguez J, Sajedin M, Ullah S (2020) Low-profile and closely spaced four-element MIMO antenna for wireless body area networks. Electronics 9(2):258
11. Singh AK, Mahto SK, Sinha R (2021) Compact super-wideband MIMO antenna with improved isolation for wireless communications. Frequenz
12. Malik J, Patnaik A, Kartikeyan MV (2014) Novel printed MIMO antenna with pattern and polarization diversity. IEEE Antennas Wirel Propag Lett 14:739–742
13. Dou Y, Chen Z, Bai J, Cai Q, Liu G (2021) Two-Port CPW-Fed dual-band MIMO antenna for IEEE 802.11 a/b/g applications. Int J Antennas Propag

Investigation of Extended Gate-On-Source and Charge-Plasma-Based Gate-All-Around TFET for Improved Analog Performance

Navaneet Kumar Singh, Rajib Kar, Durbadal Mandal, and Dibyendu Chowdhury

Abstract In this paper, extended gate and gate-stack-based charge-plasma gate-all-around TFET (EG-GS-CP-GAA) is designed for the first time. The device characteristics of EG-GS-CP-GAA are compared with the conventional device (CP-GAA). The proposed device shows lower I_{OFF}, improved I_{ON} and I_{ON}/I_{OFF}. The dual-gate materials in the proposed structure improve the drain current ratio. The presented device holds a subthreshold slope within the Boltzmann limit (<60 mV/decade). The linearity parameters, g_{m2} and g_{m3}, have been compared between both the devices. Various analog parameters, like cut-off-frequency (f_T), transconductance factor (TGF) (g_m/I_d) and transconductance frequency product (TFP) (TGF * f_T), have also been compared. The linearity and analog parameters are improved in EG-GS-CP-GAA when compared with CP-GAA. To understand the device physics, electric field, potential and carrier concentrations have been analysed. Thus, the proposed device has better electrical, linearity and analog performance than the conventional structure.

Keywords Charge plasma · Gate stack · Dual material · GAA · Extended gate

1 Introduction

Low power, multi-functional and high-density devices are the most demanding features of the recent semiconductor industry. To facilitate high-density and portable devices, the design of low-dimension MOSFETs is obligatory [1, 2]. Continuous

N. K. Singh (✉) · R. Kar · D. Mandal
National Institute of Technology, Durgapur, West Bengal 713209, India

R. Kar
e-mail: rajib.kar@ece.nitdgp.ac.in

D. Mandal
e-mail: durbadal.mondal@ece.nitdgp.ac.in

D. Chowdhury
Haldia Institute of Technology, Haldia, West Bengal 721657, India

J. K. Mandal et al. (eds.), *Topical Drifts in Intelligent Computing*, Lecture Notes in Networks and Systems 426, https://doi.org/10.1007/978-981-19-0745-6_27

downscaling is a key principle of the growing semiconductor industry. Moore's law predicts the exponential rate of MOSFET scaling. But the scaling below the submicron region or below 10 nm, surprising power dissipation and leakage current are experienced [3]. In MOSFETs, incapability of lowering the subthreshold slope (SS) below the Boltzmann limit (<60 mV/decade) does not result in lower power supply voltage proportionally with rapid downscaling of the device [4]. To engulf the demerits of the conventional MOSFETs, numerous kinds of transistor architecture have been appraised. Thus, tunnel-field effect transistors (TFETs) which have exceptional switching ability, low voltage operation and high energy efficiency are established [5, 6]. The quantum tunnelling mechanism is the basis of the operation of TFET, which differs from the conventional MOSFETs. Despite many advantages in TFET, it has the drawback of low ON-current (I_{ON}) due to its band-to-band tunnelling (BTBT) mechanism in TFETs. To improve the I_{ON} of TFETs, multi-gate architecture, hetero-gate, hetero-junction, III–V group semiconductor material can be used. Among the above-mentioned structures, GAA nanowire is most appropriate for MOSFETs scaling. The gate-all-around (GAA) nanowire bears excellent channel controllability [7, 8]. Appropriate work functions of source, drain and gate electrode are chosen to create electron/hole plasma in the device. In undoped intrinsic silicon, doping of silicon material can be achieved in the N + drain and source region by using a technique called the charge-plasma. According to Hueting et al. and Sahu et al., the condition for creating electron charge plasma is that the work function of drain/source must be less than that of silicon film [$\varphi_m < \chi_{\mathrm{Si}} + (E_G/2)$] [6]. The electron affinity of the silicon film is $\chi_{\mathrm{Si}} = 4.17$ eV, and E_G is the energy bandgap of silicon material. Another condition for creating charge plasma is that substrate body thickness must be less than Debye's length [9–11]. In charge plasma dual-material (CP-DM) and charge plasma gate stack dual-material (CP-GS-DM), the lateral and top metal electrodes are used at the source/drain side for creating uniform electron charge plasma [12]. The extended gate on the source covers more area of gate electrode for creating adequate charge plasma in the device. To decipher the problem of higher leakage in conventional MOSFET, many efforts have been done with the use of "high-K dielectrics" as gate insulators. The gate dielectric can be used as a stack of SiO_2 and high-k oxide. The gate stack produces a large gate capacitance that improves the ON-current in the device. Furthermore, dual-material (DM) gate structures offer improved drain current (I_d), transconductance (g_m) and reduced SCEs when it is compared with single-material gate (SMG). When DM gate structure is added to charge plasma-based TFET, it improves the analog parameters as compared with single-gate material (SGM) CP-GAA-TFET. Along with dual-material (DM) gate, the spacers with high-*k* dielectric increase the performance of CP-GAA-TFET [12].

In this paper, extended gate and gate stack-based charge plasma GAA TFET (EG-GS-CP-GAA) is designed. The charge plasma method is used for doping by choosing the appropriate work function of the drain and source electrode metals. This paper is split into four sections. The device designing and simulation methodology is shown in Sect. 2. Electrical parameters, linearity and analog parameters of both the devices are calculated, analysed and discussed in Sect. 3. Section 4 incorporates the summary of the presented work.

2 Device Structure and Design Methodology

The two-dimensional structures of CP-GAA and EG-GS-CP-GAA are shown in Figs. 1a, b, respectively. The drain/source length and channel length in both the devices are the same and are equal to 30 nm. The silicon film thickness (diameter of nanowire) and each spacer length in both devices are 10 nm and 5 nm, respectively. Both the devices use the charge plasma technique to create internal doping by choosing the source/drain electrode work functions of 5.4 eV and 4.7 eV, respectively. In the proposed EG-GS-CP-GAA, dual-gate electrodes (M1 and M2) with work functions of 4.92 eV and 4.42 eV, respectively, are used.

Whereas, in the CP-GAA gate electrode, the work function of 4.42 eV is used. To diminish leakage current in the proposed structure, a technique called gate stacking (GS) is used [12]. In the EG-GS-CP-GAA gate stack of SiO_2 and HfO_2 with individual oxide thickness of 0.5 and 1.5 nm, and the combined thickness of $t_{SiO_2} + t_{Hfo_2} = 2$ is used. In CP-GAA, an oxide (SiO_2) thickness of 2 nm is used. The length of the extended gate on the source side included in EG-GS-CP-GAA is 10 nm. The other device parameters of CP-GAA and EG-GS-CP-GAA are shown in Table 1. The transfer characteristics of CP-GAA and EG-GS-CP-GAA are shown in Fig. 2 in both logarithmic and linear scales. During simulations, drain voltage is kept

Fig. 1 Two-dimensional structures of **a** CP-GAA and **b** EG-GS-CP-GAA at V_{ds} = 1 V and V_{gs} = 1.2 V

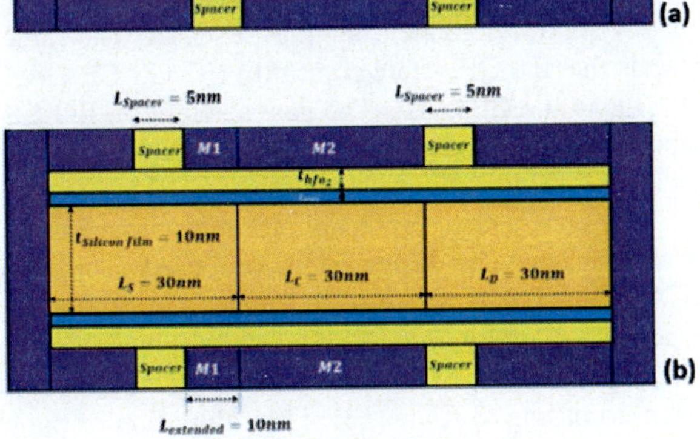

Table 1 Device dimensions of CP-GAA and EG-GS-CP-GAA

Parameters	CP-GAA	EG-GS-CP-GAA	Descriptions
L_g (nm)	30	30	Channel length
t_{ox} (nm)	2	$t_{SiO_2} + t_{HfO_2} = 2$	Oxide thickness
φ_S(eV)	5.4	5.4	Source work function
φ_D(eV)	4.7	4.7	Drain work function
φ_G(eV)	4.42	4.92, 4.42	Gate work function
N_{si} (/cm^3)	Intrinsic	Intrinsic	Silicon film doping
L_{spacer}(nm)	5	5	Spacer length

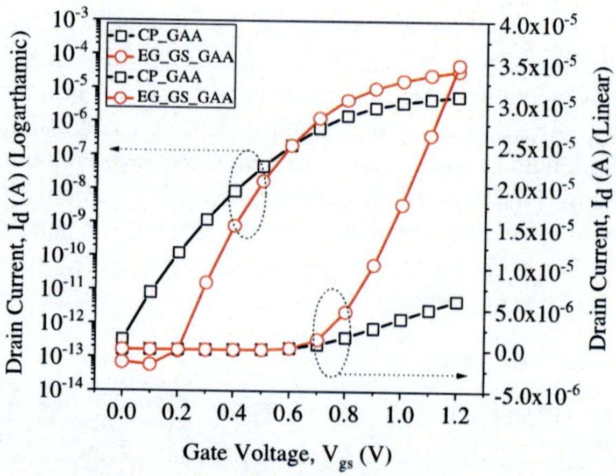

Fig. 2 Comparison of transfer characteristics between CP-GAA and EG-GS-CP-GAA at V_{ds} = 1 V and V_{gs} = 1.2 V

constant at $V_{ds} = 1$ V, and gate voltage, V_{gs} is varied from 0 to 1.2 V. From the linear graph shown in Fig. 2, it can be seen that ON-current in EG-GS-CP-GAA is approximately six times greater than the CP-GAA. The same plot in the logarithmic scale reveals that the OFF-state current in EG-GS-CP-GAA is approximately four times lower than the CP-GAA. The gate stacking in EG-GS-CP-GAA increases the gate capacitance which results in an increase in the drain current in addition to lower the leakage due to the insertion of high-*k* dielectric in gate stacking. The extended gate on the source is also responsible for high-drive current due to the creation of better charge plasma. The design of the proposed device is done on the Silvaco ATLAS TCAD tool. The various models included in simulations are field and concentration-dependent mobility (*fldmob* and *conmob*), *auger*, *srh* for recombination and *bgn* for bandgap narrowing. The band-to-band tunnelling model employed in simulation is given according to Kane [13]. The calibration of the proposed EG-GS-CP-GAA is

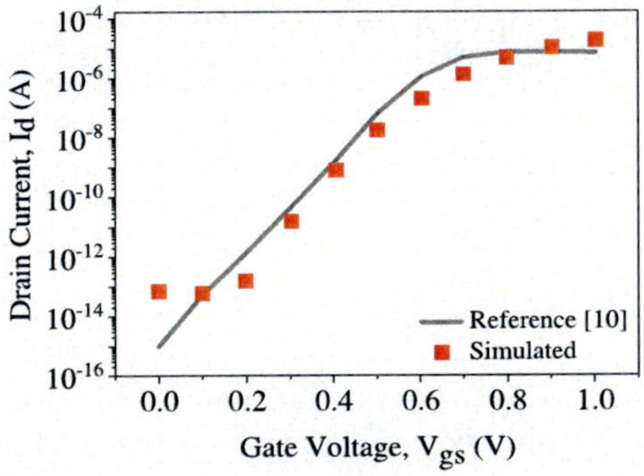

Fig. 3 Calibration of the proposed EG-GS-CP-GAA by reproducing the result of published data from [10]

done by reproducing the drain characteristics similar to the published data [10]. The calibration process shown in Fig. 3 is confirmed by considering similar models as in reference published data.

3 Results and Discussions

In this work, to understand the device physics clearly, electric field, potential and carrier concentration have been analysed. Figure 4 shows the electric field and centre potential plot of the devices, CP-GAA and EG-GS-CP-GAA along the device length. The plot of the electric field in Fig. 4a demonstrates that a higher peak of the electric field is observed in EG-GS-CP-GAA when compared with CP-GAA. The peak electric field is centred about the source–channel junction in CP-GAA; however, it ensues before the source–channel junction in EG-GS-CP-GAA due to the extended

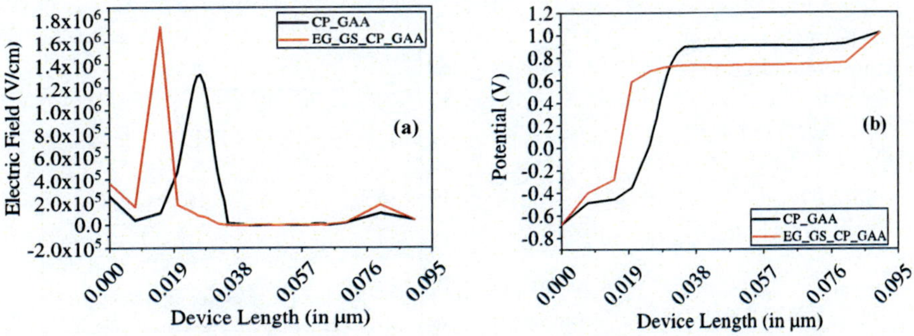

Fig. 4 Plot of **a** electric field and **b** potential in CP-GAA and EG-GS-CP-GAA at $V_{ds} = 1$ V and $V_{gs} = 1.2$ V

Table 2 Electrical parameters comparisons of CP-GAA and EG-GS-CP-GAA

Electrical parameters	CP-GAA	EG-GS-CP-GAA
I_{ON} (A)	5.99×10^{-6}	3.47×10^{-5}
I_{OFF} (A)	3.18×10^{-13}	7.11×10^{-14}
I_{ON}/I_{OFF}	1.88×10^{7}	4.88×10^{8}
V_T (V)	0.55	0.57
SS (mV/decade)	71.08	49.58

gate on the source. The variation of centre potential along the length of the device is displayed in Fig. 4b. The plot shows a higher centre potential in EG-GS-CP-GAA before the source–channel junction and a higher centre potential in CP-GAA after the source–channel junction. The electric field can be defined as the negation of potential gradient. The comparison of the performance of CP-GAA and EG-GS-CP-GAA in terms of electrical parameters such as I_{ON}, I_{OFF}, I_{ON}/I_{OFF}, subthreshold slope (SS) and threshold voltage (V_T) is shown in Table 2. The improved I_{ON} and I_{OFF} are observed in EG-GS-CP-GAA. The current ratio I_{ON}/I_{OFF} is found approximately 25 times larger in the proposed device than that of CP-GAA. The steeper subthreshold slope (<60 mV/decade) in the proposed structure makes the device more acceptable for switching applications when compared to its counterpart [4]. Figure 5 shows a comparative analysis of carrier concentration (electron and hole) for ON condition ($V_{gs} = 1.2$ V, $V_{ds} = 1$ V) along the horizontal length of the device between CP-GAA and EG-GS-CP-GAA. From the plot, it is palpable that the required carrier concentration profile is attained in both the structures by using a charge plasma doping technique similar to conventional doped TFET. The electron and hole concentration in EG-GS-CP-GAA is higher than the CP-GAA that results in improved drain characteristics. The derivative of I_d with V_{gs} at constant V_{ds} is termed the transconductance (g_m). The SI unit of transconductance is the Siemens (S), where 1 S is equivalent to 1 A/V [12]. The transconductance shows the capability of the transistor to translate a smaller gate voltage into the desired drain current. Figure 6 shows the variation of transconductance, g_m with the gate voltage (V_{gs}) at $V_{ds} = 1$ V. The plot shows higher

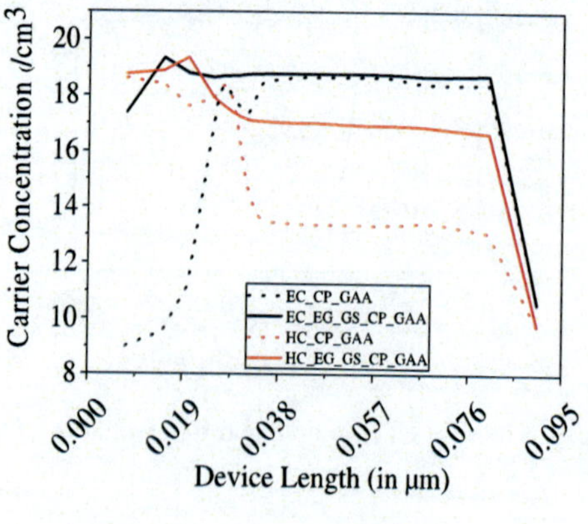

Fig. 5 Plot of electron and hole concentrations in CP-GAA and EG-GS-CP-GAA at $V_{ds} = 1$ V and $V_{gs} = 1.2$ V

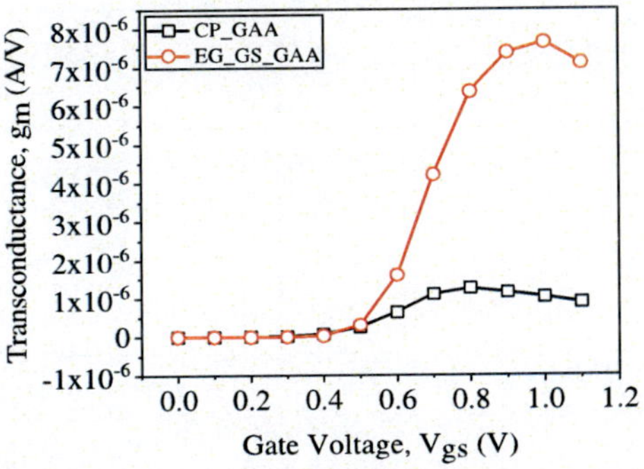

Fig. 6 Plot of transconductance in CP-GAA and EG-GS-CP-GAA at $V_{ds} = 1$ V and $V_{gs} = 1.2$ V

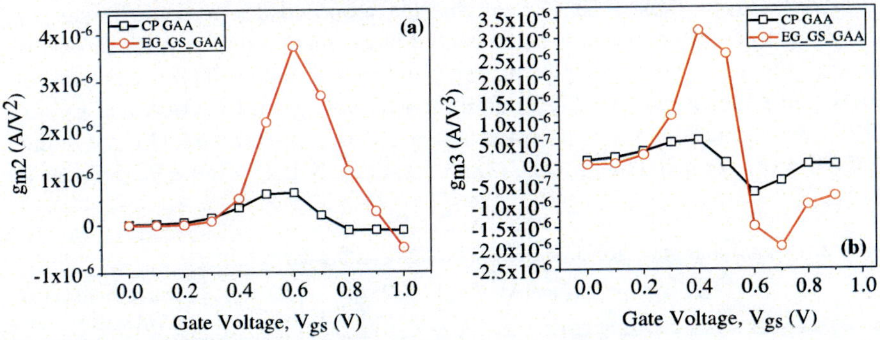

Fig. 7 Plot of **a** g_{m2} and **b** g_{m3} in CP-GAA and EG-GS-CP-GAA at $V_{ds} = 1$ V and $V_{gs} = 1.2$ V

transconductance values in EG-GS-CP-GAA when compared with CP-GAA. The reason behind this is the larger values of the drain current of EG-GS-CP-GAA over CP-GAA. Figure 7 shows a comparative analysis of the higher-order transconductance coefficient for CP-GAA and EG-GS-CP-GAA. The second- and third-order transconductance coefficients, e.g. g_{m2} and g_{m3} are depicted in Figs. 7a, b, respectively. The expressions for g_{m2} and g_{m3} are given as $g_{m_2} = \frac{\partial^2 I_d}{\partial V_{gs}^2}$ and $g_{m_3} = \frac{\partial^3 I_d}{\partial V_{gs}^3}$, respectively [14, 15]. Figure 7a shows that g_{m2} achieves lower values in CP-GAA compared to EG-GS-CP-GAA. To achieve better linearity, g_{m3} value must be lower. From Fig. 7b, it is clear that for extremely low and extremely high values of V_{gs}, the EG-GS-CP-GAA possesses lower values of g_{m3}, whereas for moderate values of V_{gs}, the CP-GAA shows lower values of g_{m3}. Figure 8 demonstrates the variation of cut-off frequency, f_T with the gate to source voltage (V_{gs}) at $V_{ds} = 1$ V. It is defined as $f_T = g_m/2\pi C_{gg}$ [14]. In this work, C_{gg} is approximated as C_{gs}. The structure having lower C_{gg} and higher g_m is favourable for higher cut-off frequency. From the plot, it can be observed that EG-GS-CP-GAA has a higher cut-off frequency than the CP-GAA. This may occur owing to the dependency of f_T over g_m, which is more for EG-GS-CP-GAA.

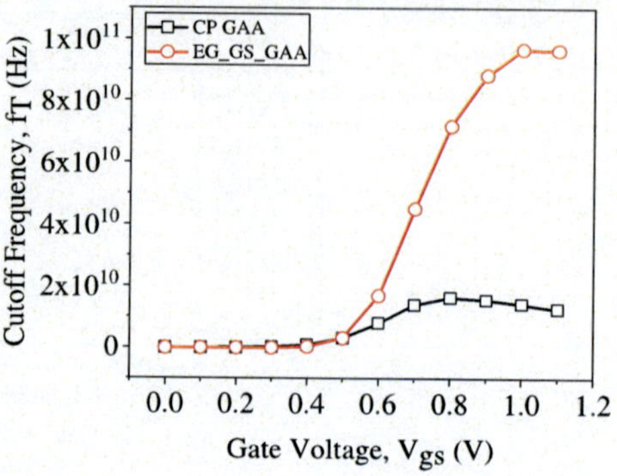

Fig. 8 Plot of unity gain cut-off frequency, f_T for CP-GAA and EG-GS-CP-GAA at $V_{ds} = 1$ V and $V_{gs} = 1.2$ V

Comparison of results of proposed structure with the existing devices is shown in Table 3. From Table 3, it is clear that the proposed device has a higher I_{ON}, higher current ratio and lower subthreshold slope than above-mentioned existing devices. Figure 9 shows the plot of transconductance factor (TGF) and transconductance frequency product (TFP) with the gate voltage in CP-GAA and EG-GS-CP-GAA structures at $V_{ds} = 1$ V and $V_{gs} = 1.2$ V. The TGF and TFP are defined as TGF $= g_m/I_d$ and TFP $= (g_m f_T)/I_d$, respectively [14]. Figure 9a portrays the

Table 3 Comparisons of device characteristics within existing devices

	I_{ON} (A)	I_{OFF} (A)	I_{ON}/I_{OFF}	V_{th} (V)	Subthreshold slope (mV/decade)
Proposed work	3.47×10^{-5}	7.11×10^{-14}	4.88×10^{08}	0.57	49.58
[16]	2.14×10^{-7}	4.49×10^{-14}	4.77×10^{06}	0.90	50.10
[17]	2.40×10^{-6}	2.40×10^{-12}	1.00×10^{06}	0.50	120.00
[18]	5.00×10^{-7}	1.00×10^{-13}	5.00×10^{06}	0.52	12.90

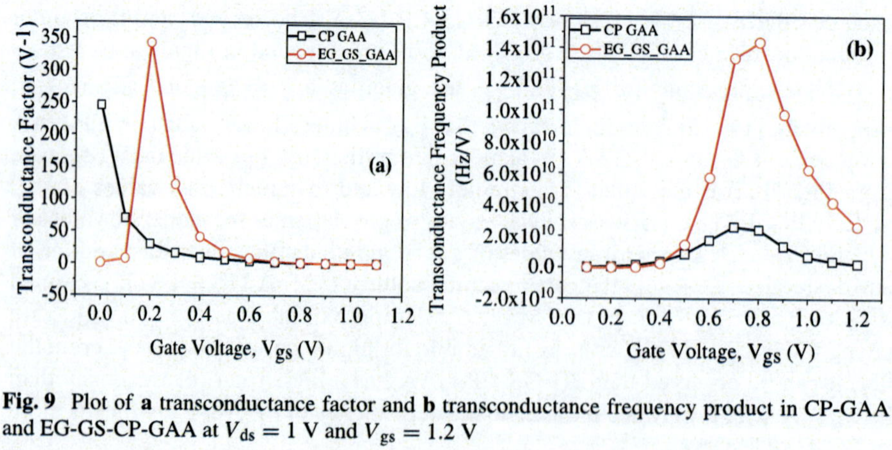

Fig. 9 Plot of **a** transconductance factor and **b** transconductance frequency product in CP-GAA and EG-GS-CP-GAA at $V_{ds} = 1$ V and $V_{gs} = 1.2$ V

plot of the transconductance factor with V_{gs} for both the structures. The g_m/I_d ratio determines how efficiently the current can be used to attain a specific magnitude of transconductance. The benefit of a larger transconductance (g_m)-to-drain current (I_d) ratio is the design of low-voltage operation circuits. From the investigation, the structure EG-GS-CP-GAA shows a higher transconductance. According to the TGF expression, it is clear that higher g_m will yield higher TGF. Hence, EG-GS-CP-GAA gives rise to higher values of TGF when compared to CP-GAA. Furthermore, to analyse the comparative analysis on transconductance frequency product (TFP), Fig. 9b illustrates the plot of TFP of CP-GAA and EG-GS-CP-GAA with gate-to-source variations. The plot shows that a higher TFP is found in EG-GS-CP-GAA as compared with CP-GAA. Its role is essential for the design of moderate to high-speed circuits and signifies the trade-off amidst the power and bandwidth for optimization.

4 Conclusions

In this paper, the linearity and analog parameters have been studied for conventional CP-GAA and proposed EG-GS-CP-GAA structures. The results obtained from the EG-GS-CP-GAA structure ensure that the device can be utilized efficiently for linearity and analog applications. The improvement in analog performance ensured with larger g_m, TGF, and TFP and f_T. The proposed device also shows improved electrical parameters such as I_{ON}, I_{OFF}, I_{ON}/I_{OFF} and SS when compared to CP-GAA. Thus, the EG-GS-CP-GAA can be considered as an improved device than its counterpart.

References

1. Saurabh S, Kumar MJ (2016) Fundamentals of tunnel field-effect transistors, 1st edn. CRC Press, Boca Raton
2. Wang PF (2004) Complementary tunneling transistor for low power application. Solid-State Electron 48:2281–2286
3. Schaller RR (1997) Moore's law: past, present and future. IEEE Spectr 34(6):52–59
4. Krishnamohan T (2008) Double-gate strained-Ge heterostructure tunnelling FET (TFET) with record high drive currents and 60 mV/dec subthreshold slope. In: Proceedings of IEEE international electron devices meeting. San Francisco, CA, USA, pp 1–3
5. Mamidala MK (2016) Tunnel field-effect transistors (TFET): modelling and simulation. Wiley, West Sussex
6. Hueting RJE (2008) The charge plasma P-N diode. IEEE Electron Dev Lett 29(12):1367–1368
7. Gupta AK (2019) Design and investigation of a novel charge plasma-based core-shell ring-TFET: analog and linearity analysis. IEEE Trans Electron Dev 66(8):3506–3512
8. Kumar S (2016) Analysis of I_{ON} and ambipolar current for dual material gate-drain overlapped DG-TFET. J Nanoelectron Optoelectronic 11(3):323–333
9. Trivedi N (2017) Charge plasma technique based dopingless accumulation mode junctionless cylindrical surrounding gate MOSFET: analog performance improvement. Appl Phys A 123:1–7

10. Choi WY (2007) Tunneling field-effect transistors (TFETs) with subthreshold swing (SS) less than 60 mV/dec. IEEE Electron Dev Lett 28(8):743–745
11. Hueting RJE (2008) Charge plasma P-N diode. IEEE Electron Dev Lett 29:1367–1368
12. Singh S (2018) Gate-all-around charge plasma-based dual-material gate-stack nanowire FET for enhanced analog performance. IEEE Trans Electron Dev 65:3026–3032
13. Silvaco ATLAS user manual, 2016
14. Singh NK (2021) Design of $Si_{0.45}Ge_{0.55}$-based core–shell-type dual-material dual-gate nanotube TFET with source pocket technique. Appl Phys A, 1–12, Article no. 233
15. Singh NK (2021) Simulation and analysis of ZnO-based extended-gate gate-stack junctionless NWFET for hydrogen gas detection. Appl Phys A 290:1–10
16. Yadav S (2018) A new structure of electrically doped TFET for improving electronic characteristics. Appl Phys A 124:1–9
17. Moselund KE (2012) InAs–Si nanowire heterojunction tunnel FETs. IEEE Electron Dev Lett 33:1453–1455
18. Asai H (2019) Steep switcing less than 15 mV dec−1 in silicon-on-insulator tunnel FETs by a trimmed-gate structure. Jpn J Appl Phys 58, Art. no. SBBA16

Efficient Optimization Technique for Analysing the Performance of Bifacial Solar Cells Using Fuzzy Logic

Kholee Phimu, Khomdram Jolson Singh, and Rudra Sankar Dhar

Abstract Due to numerous economic innovations and population growth, the electricity demand steadily grew. Meeting the need for electricity is a problem for producing energy focused solely on fossil fuels which eventually lead to numerous environmental issues, such as carbon footprints. Alternative sources of electricity can be used to satisfy the need for power users worldwide. This research study focused on providing a solution to the problem of tracking maximum power points in the solar cell using fuzzy logic. Considering short-circuit current I_{sc} 7.34 A with open-circuit voltage V_{oc} 0.6 V and irradiance used for measurement I_{r0} is 1000 with quality factor N 1.5; when used for the modeling of a bifacial solar cell, the efficiency of the system was found to be ranging from 90 to 97% only because the fuzzy-based logic controller is used. Also, the average duty cycle of the system 0.5 is achieved. The models have been checked for validation and linked to multiple models to create an optimal power model using fuzzy logic.

Keywords Fuzzy logic · Photovoltaic module · MATLAB

1 Introduction

The photovoltaic (PV) industry's key focus areas for lowering the cost of solar power generation are improving performance, upscaling production sizes, solar cells and modules, and low-cost products. More sophisticated manufacturing processes are necessary to further increase the efficiency of solar cells. As a result, the complexity and expense of the gadget are growing, leading to decreased revenues [1]. Bifacial PV modules can create more energy by converting solar power to electricity on both sides of the panel. For the reduction of PV power costs, bifacial modules are suitable

K. Phimu · R. S. Dhar (✉)
Department of Electronic and Communication Engineering, National Institute of Technology, Aizawl, Mizoram 796012, India
e-mail: rdhar@uwaterloo.ca

K. J. Singh
Department of Electronic and Communication Engineering, Manipur Institute of Technology, Canchipur, Imphal, Manipur 795003, India

J. K. Mandal et al. (eds.), *Topical Drifts in Intelligent Computing*, Lecture Notes in Networks and Systems 426, https://doi.org/10.1007/978-981-19-0745-6_28

since they can convert electricity from each end of the panel. Bifacial modules can improve their power density, reducing area-related expenses, such as land, cabling, planting, and so on [2].

In terms of their qualities, renewable energy sources are in high demand or insignificant emissions and the introduction of energy shortages in this sense. The researchers now base their attention on alternative energy sources, like solar, hydro, and wind power. The Sun is the main solar energy resource and is easy to reach worldwide [3]. Owing to the growth of industry and human consumption, energy consumption is increased.

In the current scenario, various strategies for designing high-power trackers are accessible in terms of cost, speed, performance, and hardware use in a wide range of applications [4]. Photovoltaic bifacial solar cells have been a popular topic for a while now, with predictions that they will replace non-renewable sources like fossil fuels in the coming year. PVs must compete with other energy sources, such as fossil fuels, in terms of cost per kilowatt-hour, in order to achieve this level of improvement. PV module efficacy is now just 12–26%, which is very low in the current situation, for transforming solar irradiance into electricity [5]. GaAs solar cell performance is 29%, whereas silicon solar cell efficiency is just 12–14%. In addition, performance can be reduced as a result of decreasing solar separation, photovoltaic temperature, or charging conditions. To gain the greatest possible power from the PV module operation, it is necessary. To this end, the maximum power point tracker (MPPT) controller is necessary. Variation in the solar cells I–V characteristics is regulated by MPPT. MPPT's aim is to monitor the operating point variation, where maximum power is supplied to increase the performance of the PV module [6].

2 Bifacial Solar Cell

By its route, inclination, height, location features, and solar area in the atmosphere, along with the radiation on the front side of the episode as seen in Fig. 1. Bifacial PV uses environmental, mirror, and direct radiation which reaches the module's effective rear side. The intensity rates from the rear are therefore sensitive to the albedo and environmental conditions of the surrounding soil, the setup module format, and climatic data [7].

The sun-oriented module absorbs diffuse radiation emitted by the shadow, but instant and drawn-out radiation is replicated in the unshaded zone, impacting the

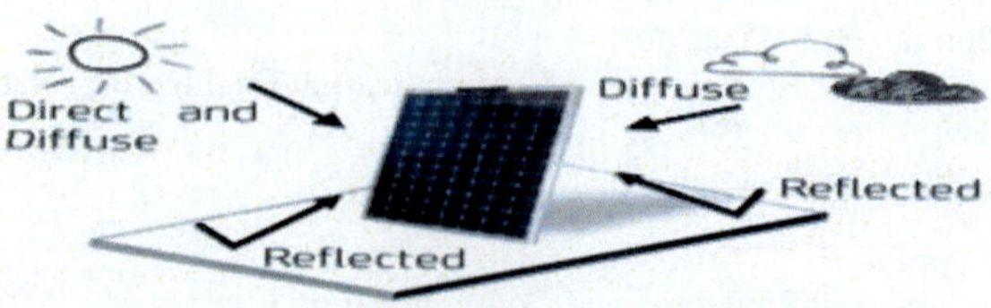

Fig. 1 Schematic of bifacial module

back of the bifacial solar cell module [8]. The evaluation of the bifacial increase is more conspicuous when we consider PV stand [9], given the various variables involved, above them but also because of the pressing spacing of the shadows created by the montage, the exorbitant concealing of the adjacent solar cell modules, and the impedance of replicated energy radiation. The elements mark the bifacial solar cell module PV power plant installation design more responsive than conventional ones that use mono facial modules [10, 11].

3 Mathematical Modeling

The Sun-based cells of this unique bifacial electrical model have been proposed, manufactured, and an attempt has been made to forecast their capacity. The vast majority of the generated models believe that it is possible to that a Sun-powered bifacial solar cell module may be communicated as twin mono-facial solar cells in equality, communicated by a connection made up of double diodes [6]. In Fig. 2, the electricity map is introduced [6]. The main aim of a diode is to connect in parallel to a current source (freewheeling diode) in order to avoid damaging some nearby components sensitive to high voltage and thereby providing safety to the model. Actually, for simulation purpose to realize bifacial solar cell, we show two current sources connected in parallel, one for front and other for rear cell, but as a whole represent the bifacial solar cell, we consider its equivalent model consisting of both.

There is a practical difference in the intensity of light falling on the rear surface of bifacial solar cell which is somewhat lesser than the front surface and depends upon the various factors like albedo, shade, and view factor. To electrically define bifacial PV units, Singh et al. [6] designed a technique for all enhancing situations to electric identify the bifacial solar cell. The model is a mono-facial cell with a single diode, and the electrical boundaries supplied as input to the model are acquired for both sides of the cell from I–V bends produced separately at Standard Testing Conditions (STC). Furthermore, the reactance amid the dissimilar sides is not careful, which may lead to slight comparisons amid investigational knowledge effects. The various parameters like short-circuit current and open voltage from both sides of a bifacial solar cell are given as I_{sc} front and I_{sc} back, V_{oc} front and V_{oc} back.

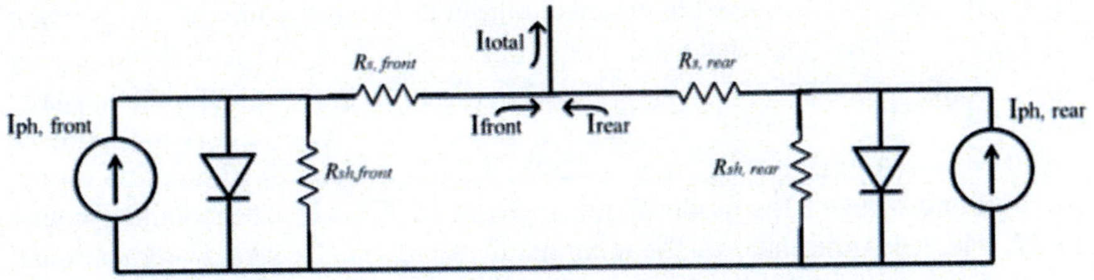

Fig. 2 Solar cell module

By considering the bifacial module operating from both ends, the importance of these parameters for varying amounts of radiation assumes a direct reaction, as it is distinct and certainly has no front or back impact on the cell side. To evaluate these thresholds, the model also represents the proportion of back and front irradiation as a percentage of total irradiance as follows [6]:

$$\chi\,(\text{Ratio of Irradiance}) = \frac{G_{\text{rear}}}{G_{\text{front}}} \tag{1}$$

where G_{rear} is irradiances due to incident radiation on the rear surface and G_{front} is irradiances due to fall of radiation on front. The typical value χ is 0.9 (typical standard value) [6]. Bifacial solar cell modules' power can be defined as follows:

$$P_{\text{bifacial}} = I_{\text{sc,bi}} V_{\text{oc,bi}} \text{FF}_{\text{bi}} \tag{2}$$

With $I_{\text{sc,bi}}$ the short-circuit current for bifacial modules is

$$I_{\text{sc,bi}} = R_{\text{SC}} I_{\text{SC,front}} = \left(1 + \chi \frac{I_{\text{SC,rear}}}{I_{\text{SC,front}}}\right) I_{\text{SC,front}} \tag{3}$$

The bifacial open-circuit voltage is shown below

$$V_{\text{oc,bi}} = V_{\text{oc,front}} + \frac{\left(V_{\text{oc,rear}} - V_{\text{oc,front}}\right)\ln(R_{\text{sc}})}{\ln\left(I_{\text{sc,rear}} - I_{\text{sc,front}}\right)} \tag{4}$$

And, also the bifacial fill factor FF_{bi} is shown below.

$$\text{FF}_{\text{bi}} = \text{pFF} - R_{\text{sc}} \frac{V_{\text{oc,front}}}{V_{\text{oc,bi}}} (\text{pFF} - \text{FF}_{\text{front}}) \tag{5}$$

where pFF = pseudo fill factor (FF) and its value can be calculated by Eq. 6.

$$\text{pFF} = \frac{\left(I_{\text{sc,rear}}/I_{\text{sc,front}}\right)\text{FF}_{\text{front}} - \left(V_{\text{oc,rear}}/V_{\text{oc,front}}\right)\text{FF}_{\text{rear}}}{\left(I_{\text{sc,rear}}/I_{\text{sc,front}}\right) - \left(V_{\text{oc,rear}}/V_{\text{oc,front}}\right)} \tag{6}$$

This approach has been used to estimate output power to within 1% of observed values for a variety of irradiance situations [12]. Considering a substrate resistivity of around 1 cm^2, a cross-section area of 125 $\times$ 125 mm^2 p-type, and a thickness = 200 μm. Furthermore, for typical values V_{oc}, rear = 0.028 V, V_{oc}, front = 0.0099 V, I_{sc}, front = 340 A/cm^2, and I_{sc}, rear = 280 A/cm^2 acquired by simulation on the tanner tool and utilized for model fuzzy logic in MATLAB. The obtained power is 9 μW, which is quite near to the theoretical value (Eq. 2). The aforementioned parameters are utilized for mathematical modeling of the module based on their availability.

4 Energy Conversion

"η" is the energy conversion efficiency of a solar cell, and it is defined as the proportion of incoming light energy that eventually turns into electricity.

$$\eta = \frac{P_M}{A_C E} \tag{7}$$

where E is the input light radiation (W/m^2), the highest energy point at the STC, A_C is solar cell surface (m^2), and P_M is maximum power point. The current impact because of diode law is illustrated below, taking into account the effect of normal dark current.

$$I = I_o\left[\exp\left(\frac{qV}{nKT}\right) - 1\right] - I_L \tag{8}$$

where I_L denotes light-produced current, q denotes charge, V denotes voltage, K denotes Boltzmann constant, and T denotes standard temperature. I_L, is the current at zero voltage in ideal condition, I_{sc}.

$$V_{oc} = \frac{nKT}{q}\ln\left(\frac{I_L}{I_o} - 1\right) \tag{9}$$

The FF is the ratio of peak power (P_M) to open-circuit voltage (V_{oc}) and short-circuit current (I_{sc}):

$$\mathrm{FF} = \frac{P_M}{V_{oc} \cdot I_{sc}} = \frac{\eta A_C E}{V_{oc} \cdot I_{sc}} \tag{10}$$

Preliminary averaged results are shown in Table 1.

The rear-to-front efficiency ratio in percentage is 86, representing a 9% gain over traditional models.

Table 1 Showing parameters of different value

	FF (%)	I_{sc} (μA/cm^2)	V_{oc} (V)	η (%)
Front side	76.3	340	0.0099	17.3
Rear side	74.8	280	0.028	15

5 Fuzzy Logic System and Membership Function

During the reasoning process, a fuzzy logic deals with imprecision, vagueness, uncertainty, partial truth, and so on. It is used to assist in deciding for incomplete and imprecise information as shown in Fig. 3.

The nature of real-world problems is complex and imprecise, whether in a decision-making process or in a single case. A real-life situation-based model based on fuzzy logic can be built to tackle the ambiguity issue [13, 14]. Fuzzy logic is a branch of artificial intelligence that deals with reasoning systems. These logics are employed in machine learning to imitate human cognitive patterns and decision-making abilities [15]. By adjusting rules that are a combination of expert inputs and desired outputs, fuzzy logic completes a variety of tasks [16]. The structure of a fuzzy logic system is shown below in Fig. 4.

To acquire output of various power components, a fuzzy logic controller (FLC)-based MPPT for PV bifacial is designed and simulated using MATLAB in this research work. Various rules are designed to optimize the operation of the controller and placed in the rule base. According to the input parameter, the particular rule is provided to the inference engine and optimum output is generated as shown in Fig. 4. The membership function (MF) [17] is a graphical representation of the level of participation of individual inputs. The MF assigns a specific value to the inputs that also functions as overlap among the inputs. The degree of membership function has a range of 0–1.

Fig. 3 Fuzzy logic controller: block diagram

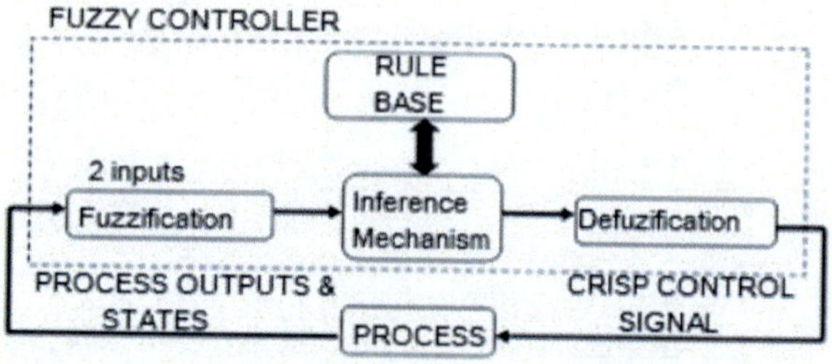

Fig. 4 Fuzzy logic system structure

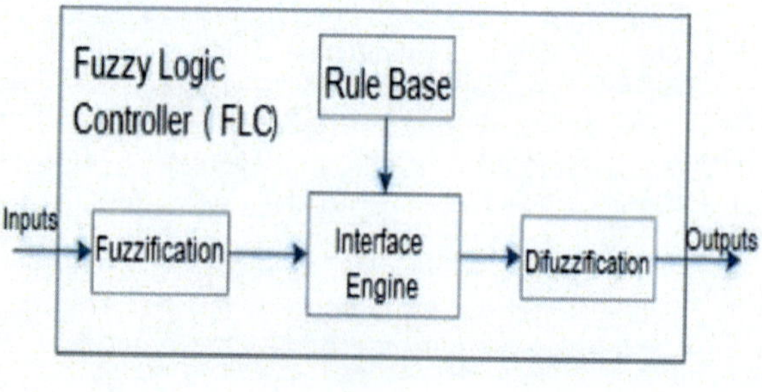

6 Result

We created a FLC-based MPPT for a PV bifacial cell and simulated it by MATLAB to acquire output of various power components, which we used in our study. For the proposed system, the fuzzy rules have been derived and verified using MATLAB Simulink. Rules for fuzzy logic controller are shown in Fig. 5.

Where MD represents the duty cycle of the fuzzy logic-based control system. As shown in Fig. 6, we used MATLAB Simulink software to develop the fuzzy logic algorithm-based controller.

MATLAB Simulink software is used to create fuzzy logic, the 3-D representation of results obtained from rule base as shown in Fig. 7. Also, a fuzzy logic system is seen in Fig. 8. The relationship between current in ampere (A), a voltage in volt (V), and duty cycle is also shown in Table 2.

The implementation of the above model has been done using MATLAB Simulink software at various changes of irradiance. Different readings of input and output power at solar irradiance, viz. (1000, 800, 600, 400, 200) W/sq m, were taken to check the efficiency of the converter and the functioning of the fuzzy controller.

1. If (Current is low) and (Voltage is low) then (MD is high) (1)
2. If (Current is low) and (Voltage is medium) then (MD is high) (1)
3. If (Current is low) and (Voltage is high) then (MD is medium) (1)
4. If (Current is medium) and (Voltage is low) then (MD is high) (1)
5. If (Current is medium) and (Voltage is medium) then (MD is high) (1)
6. If (Current is medium) and (Voltage is high) then (MD is medium) (1)
7. If (Current is high) and (Voltage is low) then (MD is medium) (1)
8. If (Current is high) and (Voltage is medium) then (MD is medium) (1)
9. If (Current is high) and (Voltage is high) then (MD is medium) (1)

Fig. 5 Rules for FLC

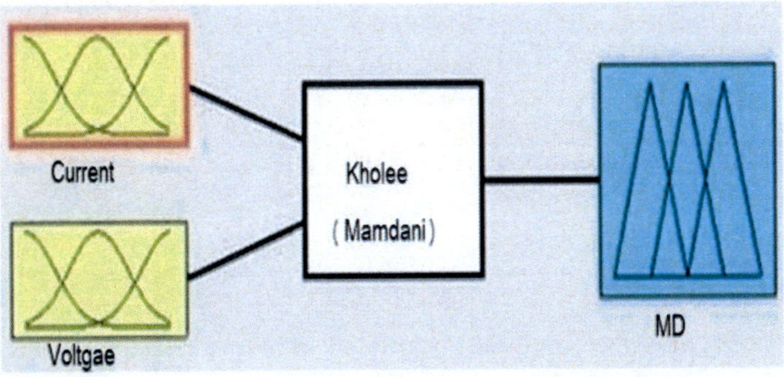

Fig. 6 Fuzzy logic designer

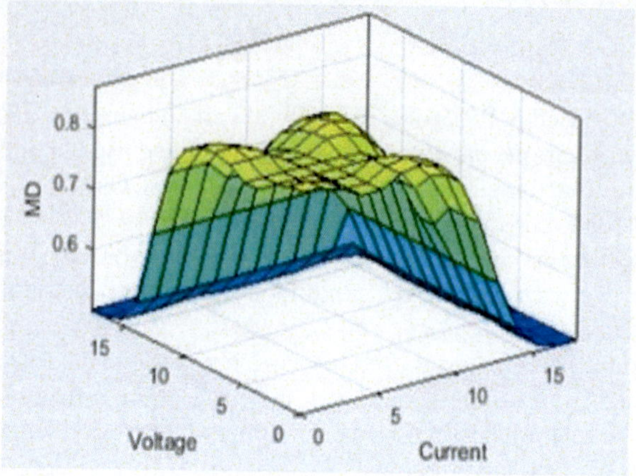

Fig. 7 Graphical representations for fuzzy-based controller rules

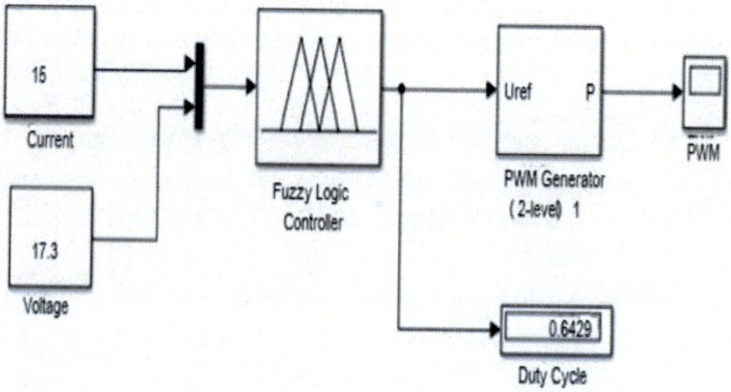

Fig. 8 Fuzzy logic system

Table 2 Relation between current voltage and duty cycle

I	V	DC
0.5	12	0.856
3.0	15.0	0.767
5.0	18.0	0.629
10.0	17.0	0.683
10.0	25.0	0.500
10.0	17.33	0.520

Table 3 Fuzzy-based MPPT controller simulation results

Irradiance (watts/sq m)	P_{in} (Watts)	P_{out} (Watts)	$\eta = \frac{P_{out}}{P_{in}}$ (%)	DC
1000	257.7	252	$\frac{252}{257.7} = 0.98$	0.65
800	209.7	200.9	$\frac{200.9}{209.9} = 0.96$	0.65
600	150.7	142.7	$\frac{142.7}{150.9} = 0.95$	0.59
400	98.98	92.5	$\frac{92.5}{98.98} = 0.92$	0.49
200	48.86	44.4	$\frac{44.4}{48.86} = 0.91$	0.35

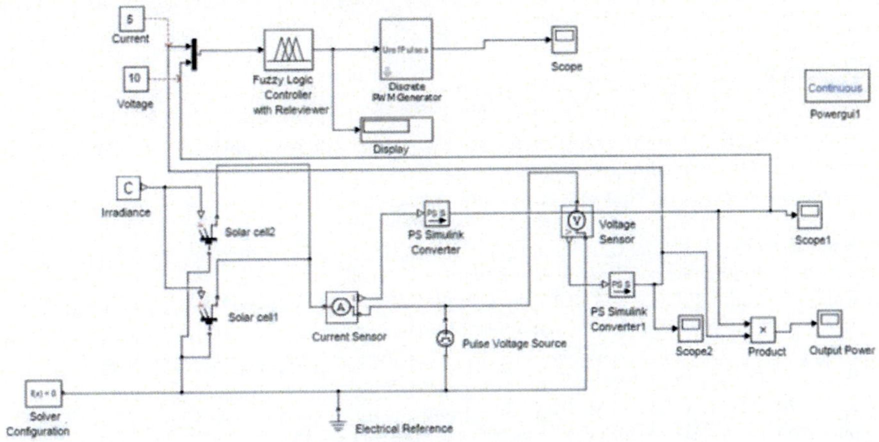

Fig. 9 Simulink model of complete system

On the same irradiation settings, the duty cycle has been observed. Table 3 shows fuzzy-based MPPT controller simulation results.

MATLAB Simulink software was used to assemble and simulate the entire system for the different values of solar radiation is shown in Fig. 9.

7 Conclusion

Bifacial solar cell-based has a nonlinear characteristics photovoltaic system which are affected by climate condition, and in these characteristics, maximum available power of PV is obtained from an operating point. The artificial intelligence-based maximum power point tracking approach for attaining the maximum power point is the fuzzy logic controller. Defining the logical rule and particular range of membership function in this method has a big impact on getting the best outcomes. In this study, a unique fuzzy logic-based approach for tracking optimal power is proposed. Various subsystems and components have been modeled and evaluated in our study. Thus, the photovoltaic module with bifacial solar cell has been modeled using Simulink. After combining the converter model and the photovoltaic module with FLC, we developed a complete MPPT model and utilized it to match the fuzzy logic controller's rules and membership functions. According to our findings, the photovoltaic facial solar cell model is practically correct and can be utilized to model solar panels with the help of applicable information found in the manufacturer's datasheets. The simulation findings also show that an FLC-based MPPT can extract a significant amount of additional energy from a bifacial solar cell PV module. These findings indicated that PV power systems could enhance their energy efficiency. Improvements in energy efficiency parameters are, in the end, a cost-cutting measure.

References

1. Harder NP (2012) Simplified cost analysis of n-type high-efficiency solar cells. In: Proceedings of the 21st nPV workshop, Amsterdam
2. Hazel R (2009) A novel high-efficiency rear-contact solar cell with bifacial sensitivity. High-eff. Low-cost Photovoltaics. Springer Ser Opt Sci 140:65–93
3. Kabir E, Kumar P, Kumar S, Adelodun AA, Kim KH (2018) Solar energy: potential and future prospects. Renew Sustain Energy Rev 82:894
4. Ruelas J, Munoz F, Lucero B, Palomares J (2019) PV tracking design methodology based on an orientation efficiency chart. Appl Sci 9(894):1–15
5. King DL, Boyson WE, Kratochvil JA (2014) Photovoltaic array performance model. Sandia Rep. No. 2004–3535, vol 8, pp 1–19
6. Singh JP, Aberle AG, Walsh TM (2014) Electrical characterization method for bifacial photovoltaic modules. Sol Energy Mater Sol Cells 127:136–142
7. Chiodetti M, Binesti D, Mousel S, Lutun KE (2016) Enhancing bifacial PV modelling with ray-tracing. In: 6th PV performance modeling collaborative workshop, Oct 2016
8. Johnson J, Yoon D, Baghzouz Y (2012) Modeling and analysis of a bifacial grid-connected photovoltaic system. IEEE Power Energy Soc Gen Meet, 1–6
9. Design guide for bifacial solar modules design guide for bifacial solar modules. Prism Solar Technologies, Inc.
10. Ishikawa N, Nishiyama S (2016) World first large scale 1.25 MW bifacial PV power plant on snowy area in Japan. 3rd Bifacial PV workshop, Japan
11. Joge T, Eguchi Y, Imazu Y, Araki I, Uematsu T, Matsukuma K (2004) Basic application technologies of bifacial photovoltaic solar modules. Electr Eng Japan (English Transl. Denki Gakkai Ronbunshi) 149(3):32–42
12. Calculating the additional energy yield of bifacial solar modules. Solar World. 2015
13. Singh H, Gupta MM, Meitzler T, Hou ZG, Garg KK (2013) Real-life applications of fuzzy logic. Adv Fuzzy Syst. https://doi.org/10.1155/2013/581879
14. Whig P, Ahmed N (2017) Fuzzy logic implementation of photocatalytic sensor. Int B=Rob Auto J 2(3)
15. Norihiro K (2020) Sequential fuzzy description logic: reasoning for fuzzy knowledge bases with sequential information. In: IEEE 50th international symposium on multiple-valued logic (ISMVL)
16. Pekaslan D, Garibaldi JM (2020) ADONiS-adaptive online non singleton fuzzy logic systems. IEEE Logic Syst IEEE Trans Fuzzy Syst 28(10)
17. Omone OM, Aggrey E, Takaes M, Kozlovszky M (2020) Implementation of a fuzzy logic progression for alcohol addicts using fuzzy control system (FCS). In: IEEE 18th international symposium on intelligent systems and informatics (SISY)

Nanowire GaSb Infrared Solar Cell

Dickson Warepam, Khomdram Jolson Singh, and Rudra Sankar Dhar

Abstract The 3D geometry structure along with the nanopillar array photovoltaics gives unique properties for effective solar cells comparing to the ordinary solar cell. It is found that with the use of GaSb, the infrared spectrum can be absorbed by the nanopillars having large interface area. A combination of AlGaSb and GaSb is analyzed for designing the infrared solar cell. Our simulation results indicate that GaSb-based infrared solar cell has higher efficiency as compared to that of Si-based nanowire solar cell. The comparison between the GaSb and Si nanowires is done regarding the I–V characteristics, efficiency, short-circuit current and open-circuit voltage. GaSb works in the infrared region of the solar spectrum and is simulated using TCAD. The efficiency of the infrared solar cell increases from 5.6 to 12.7% when the illumination increases from 1 to 100 suns. Charge carriers are collected along the radial structures as indicated by short-circuit current scaling. Simulation of the solar cell gives an open-circuit voltage, V_{oc} of 0.34 V under 100 suns, a short-circuit current (I_{sc}) of 0.31 nA under 1 sun and a fill factor of 71.0%. The results are verified with already published experimental data.

Keywords Photovoltaic · Nanowires (NWs) · Technology Computer-Aided Design (TCAD)

1 Introduction

Nano-structure that is being used for the infrared solar cell is advantageous regarding cost and size, yielding better efficiency for the solar cell [1–9]. These nanowires and

D. Warepam (✉) · R. S. Dhar
Department of Electronic and Communication Engineering, National Institute of Technology, Mizoram 796012, India

R. S. Dhar
e-mail: rdhar@uwaterloo.ca

K. J. Singh
Department of Electronic & Communication Engineering, Manipur Institute of Technology, Canchipur, Imphal, Manipur 795003, India

J. K. Mandal et al. (eds.), *Topical Drifts in Intelligent Computing*, Lecture Notes in Networks and Systems 426, https://doi.org/10.1007/978-981-19-0745-6_29

nanorods are able to accumulate charge carriers more efficiently for various solar cells [4–6]. Absorption of photons for converting it to charge carriers effectively is used in third-generation solar cells [7]. Since we are dealing with infrared solar cells, solar cells should be able to operate in various temperatures [3–6, 10]. Further investigation has shown that nanorod arrays have that advantages for the optimal light absorption and photon conversion as compared to the conventional solar cell with single-crystal bulk semiconductors when the same materials is used [8, 9]. It is to be noted that a combination of nanosturctures and single crystal bulk semiconductor-based solar cells leads to decrease in efficiency and poor stability [1].

In this paper, we are implementing a 3D coaxial infrared solar cell having three layers, namely positive type, intrinsic type and negative type layers. The three layers mentioned above will be forming the (p-i-n) layout. Recently, Si-nanowires-based solar cells are of major concern in the research area, and our paper is compared to that of the experimental data obtained for the Si-nanowires-based solar cells [11]. Si is vastly used for making solar cells owing to its properties such as abundant availability and developed manufacturing method. However, its indirect band gap gives rise to certain limitation as it is unable to harness maximum solar energy due to phonon emission. The power conversion efficiency has increased within these last five years, and in this article, we will be modeling a nanowire-based solar cell which will work mostly in the infrared region. An efficiency of 5.6% can be obtained for GaSb nanowire solar cell under 1 sun. With the use of nanopillars, we are able to obtain large interface area and cascading of nanopillars will increase the overall efficiency of the solar cell. It can act as an alternative renewable energy fuel for exploring outer space as GaSb works in the infrared region. GaSb solar cell can be used further for understating the working of electron and holes generation optically for energy conversion and artificial photosynthesis [12]. Coaxial GaSb nanowire solar cell can be used for detecting low light level photons [13]. Nanowire technology is used in various photonic application [14, 15].

From Fig. 1, we are able to observe the arrangement of the positive type, intrinsic and negative type which make up the 3D coaxial solar cell [16, 17]. N-type AlGaSb crowned the n-type GaSb and p-type GaSb. AlGaSb reduces the band gap with rise in temperature and short-circuit current decrease in a nonlinear trend [18]. GaSb has

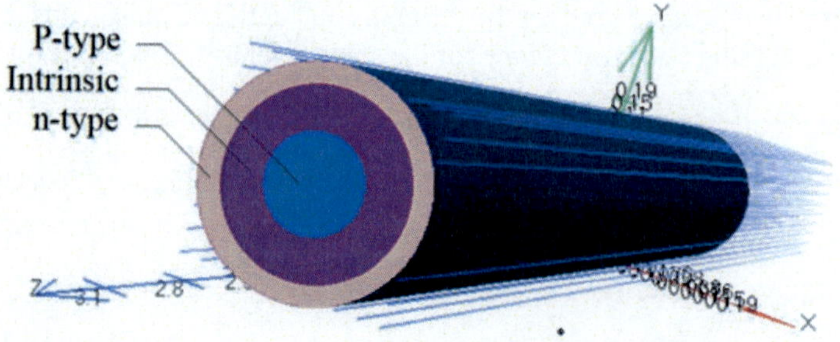

Fig. 1 Simulation scheme of a coaxial nanowire

some desirable properties when working in higher temperature as photo-generated current increases with the decrement in band gap [19]. This type of structure made with GaSb and AlGaSb is beneficial as charge separation takes place radially instead of axial direction which in turn gives advantages to the photon-induced carriers to reach the p-i-n junction easily as compared to the bulk crystal structure [8]. When the 3D co-axial infrared solar cell is illuminated, regular charge distribution occurs radially as the voltage across the shell is reduced due to highly conductive n-shell [8]. Nanowire is advantageous over thin-film solar cells regarding reduced reflection, more light absorption, band gap flexibility, cost of manufacturing and least problem with defects [20].

2 Photovoltaic Simulation

2.1 Nanowire Modeling

The device structure depicted in Fig. 1 is simulated using ATLAS in cylindrical coordinate system with the outer n-type radial size of 190 nm. The intrinsic type has a maximum radius of 160 nm and that of p-type having a maximum of 80 nm. Thus, the overall thickness of the nanowire is 380 nm, and it is similar to the nanowire give in structure of the paper with smaller length [2].

Drift diffusion model is being used for calculating the carrier movement, considering the smaller length of the nanowire [13, 14]. It takes less amount of time for simulation if the nanowire is of smaller length. We use Shockley–Read–Hall recombination and concentration-dependent mobility for evaluating the recombination and mobility. We need to check whether quantum effect is present during our simulation. Poisson-Schrodinger equations self-consistently are used for quantum effect check, and later, we found it to be very small. Self-consistent calculation is confirmed with the first bound state having a position of 1 meV just above the conduction band when infinite potential well approximation is implemented. The dark current–voltage relation was attained initially, after which the structure is illuminated with AM 1.5 Global solar irradiance. Coaxial nanowires parameters such as potential and the band energy were obtained self-consistently. The spectrum of the solar illumination using multi-spectral photogeneration in the simulator is given in Fig. 2.

I_{sc} is set to pico Ampere range by scaling the power of the solar spectrum by 1:5 $\times 10^{-3}$. The generation rate is calculated from the equation as below:

$$G = \eta_o \frac{P\lambda}{hc} \alpha e^{-\alpha y}$$

where G is the photogeneration rate, P is the cumulative effects, transmissions and loss due to absorption over the ray path, y is the relative distance for the given ray, h is the Planck's constant, λ is the wavelength, c is the speed of light and α is the

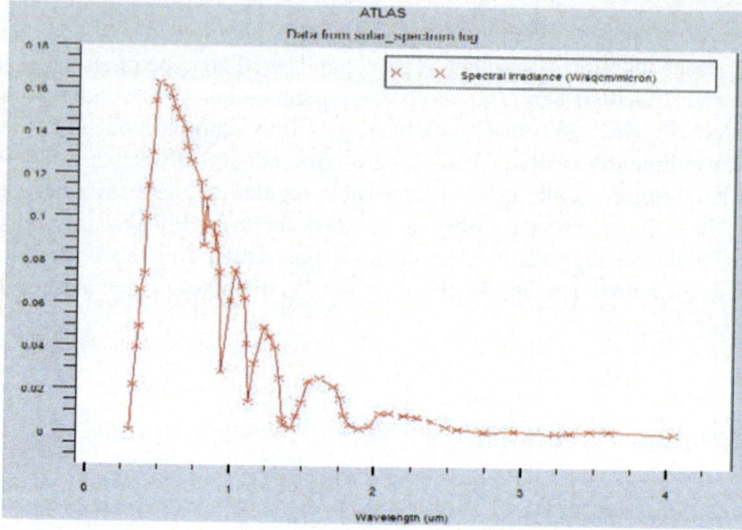

Fig. 2 AM1.5G spectrum used to illuminate the model

absorption coefficient calculated for each set of (n, k) value. Absorption coefficient is given by the total number of photons absorbed by a specific wavelength (λ) falling on it. This particular property is constant for the given material. The absorption coefficient in terms of the extinction coefficient is given by:

$$\alpha(\lambda) = \frac{4\pi k}{\lambda} 10^7 \text{cm}^{-1}$$

where λ is in Nano meter.

The simulated nanowire structure is being illuminated with an air mass of 1.5 global and various profiles can be obtained as shown in Figs. 3, 4, 5 and 6. Photogeneration rates can be obtained by considering the different legends expressed in the log of the electron–hole pair generation rate with respect to the color-coded display. From the legions, it is evident that more photo generation rates are occurring at upper solar layer as it has the advantages to absorb photons earlier than the inside layer.

2.2 *TCAD Simulation*

Charges are produced in a nanowire when photons fall on it. The charges are pairs of electrons and holes which has different movement paths radially. Electrons flow outward from the center, whereas holes flow towards the center of the nanowire. With the movement of the carriers, it can produce power in the range of a few nano Watt. In order for the output power to drive external circuit, we need to either increase

Fig. 3 Electric field profile

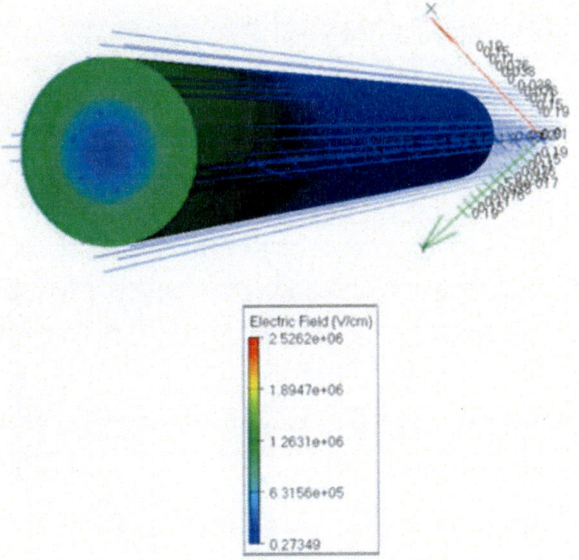

Fig. 4 Potential-generated profile

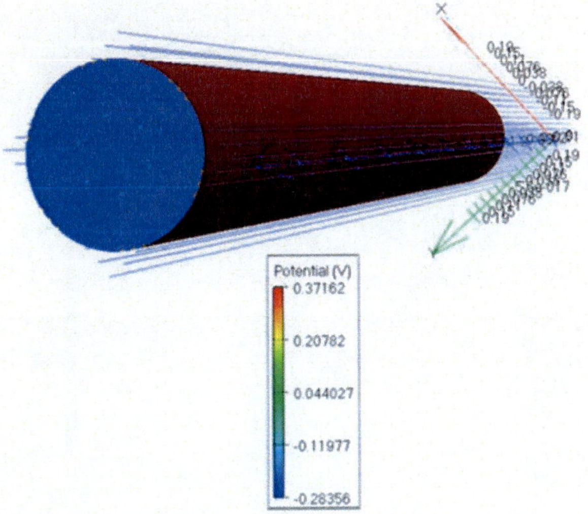

the light intensity or cascade various nanowires. Practically, the illumination is one directional, and the photogeneration rate along the radial direction has been plotted. For the nanowire to absorb the photon in all the direction, simulation is performed with all round illumination using three dimension simulation (Table 1).

Fig. 5 Net doping profile

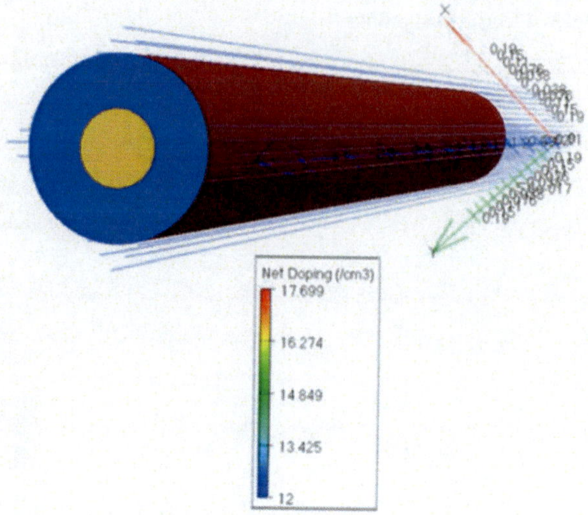

Fig. 6 Recombination rate profile

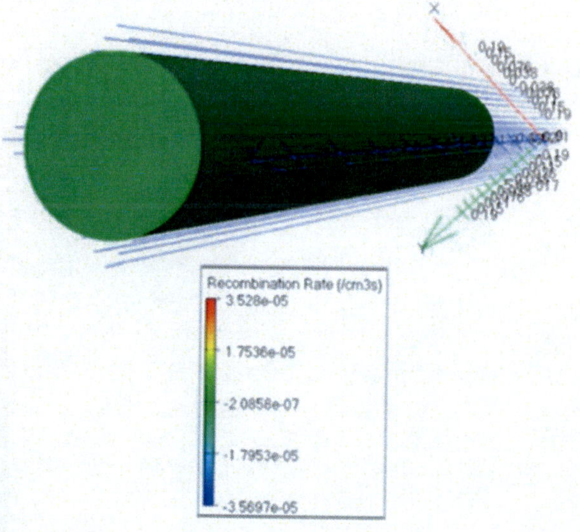

3 Result and Discussion

The nanowire is being illuminated with air mass 1.5 G and various photovoltaic properties of p-i-n coaxial GaSb are obtained as shown in Fig. 7. Even though the solar cell is mainly focused on the infrared spectrum, we were able to obtain an open-circuit voltage, V_{oc} of 0.34 V under 100 suns, a short-circuit current (I_{sc}) of 0.31 nA under 1 sun and a fill factor of 71.0%. A maximum efficiency of 5.6% can be obtained as the output of the nanowire under 1 sun.

Table 1 Standard major parameters used in our model

Material	GaSb
Band gap [eV] @300°K	0.72
Permittivity	14.4
Affinity [eV]	4.06
Epsilon	15.7
e − mobility [$cm^2/V \times s$]	4000
h + mobility [$cm^2/V \times s$]	1400
e − density of states NC [cm^{-3}]	5.68E + 18
h + density of states NV [cm^{-3}]	2.95E+18
Lifetime (el)[ns]	1
Lifetime (ho)[ns]	1
ni (per cc)	3.66E+12
Vsatn (cm/s)	1.00E+08
Vsatp (cm/s)	1.00E+08

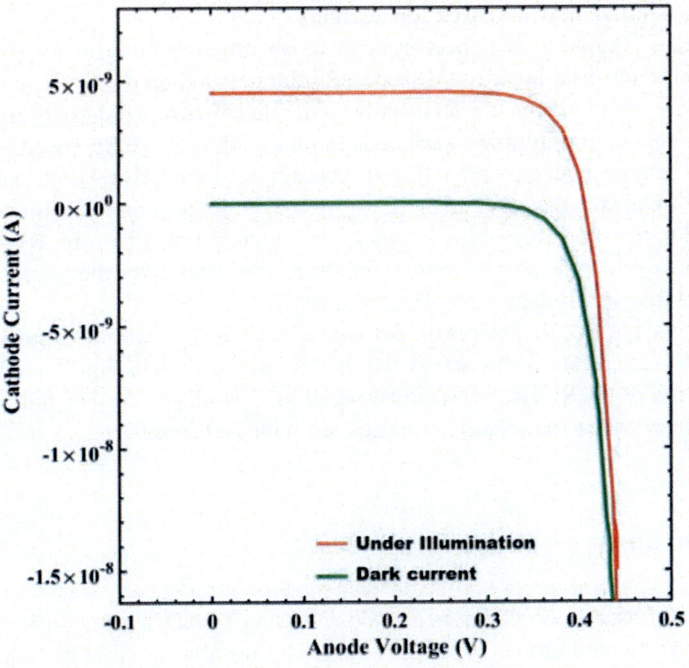

Fig. 7 I–V characteristic under AM1.5G illumination and dark current

Table 2 Output parameters obtained from the simulated model

P-I-N material	Sun	Jsc (mA/cm^3)	Vm (V)	Im (A)	Eff. %
GaSb	1	3.09849e−11	0.22	2.87615e−11	5.6
GaSb	100	4.53582e−09	0.34	4.24268e−09	12.7
Si	1	2.27140e−011	0.42	2.1484e−011	6.4
Si	100	2.28628e−09	0.53	2.16278e−09	8.8

V_{oc} remains independent on the length of the p-i-n Si nanowire but I_{sc} has linear scaling along the Si nanowire length. This short-circuit current scaling indicates that induced charge carriers are uniformly collected along the radial structures and metal contact's light scattering does not contribute to the I_{sc} (Table 2).

The simulated model can be further improved so that the upper and lower bound efficiency can be increased with the usage of vertical integration or multi-layer stacking [8]. Photo-generated carriers have systematic increment when the light is incident on the nanowire with linear increment of open-circuit voltages and logarithmic increment of short-circuit current [21]. When the model is illuminated with multiple suns, efficiency increases considerably.

The infrared region of the spectrum can be absorbed more effectively with the introduction of intrinsic layer in GaSb-based solar cell. When the GaSb is based on p-i-n structure, more charge carriers can be generated as compared to normal diode-based solar cell, as it minimizes surface recombination [16]. GaSb-based solar cell can convert heat generated by the infrared spectrum to electrical power in an efficient way [10]. We can note that most of the energy from the solar radiation is found within the infrared region at wavelengths longer than 1 micron, but silicon-based solar cell responds to wavelengths shorter than 1.1 micron [10]. This gives the advantages for GaSb nanowire over Si nanowire in infrared region.

In Fig. 7, we are able to observe the i–v characteristics under illumination and dark current. Cathode current of the simulated model has 0.04 nA in Fig. 7 as compared to the 0.02 nA of the Si nanowire model when same voltages are applied. Cathode current increase when the model is illuminated with 100 suns.

4 Conclusions

GaSb-based infrared solar cell has been modeled using TCAD, and results are verified after running the simulation. With this work, we are able to simulate and model infrared solar cell, without actual product which gives a better option for testing different model. The details of the model can be analyzed with the 3D structure. When the number of suns illuminating increases, GaSb nanowire performs better than Si nanowire. The output power can be increased in future work by connecting co-axial nanowires together to form nanopillar array PV system. In the nano photovoltaic study, the methodology used in this work and the precise details that can be obtained from the 3D structure can be beneficial for further research and application. It is found from other literature that such a model is able to provide power in the range of nano Watt for nanoelectronics devices.

References

1. Lewis NS (2007) Toward cost-effective solar energy use. Science 315:798–801
2. Lewis NS, Crabtree G (2005) Basic research needs for solar energy utilization. Report of the basic energy sciences workshop on solar energy utilization. US Department of Energy, Washington DC
3. Gratzel M (2001) Photoelectrochemical cells. Nature 414:338–344
4. Huynh WU, Dittmer JJ, Alivisatos AP (2002) Hybrid nanorod-polymer solar cells. Science 295:2425–2427
5. Law M, Greene LE, Johnson JC, Saykally R, Yang P (2005) Nanowire dyesensitized solar cells. Nat Mater 4:455–459
6. Baxter JB, Aydil ES (2005) Nanowire-based dye-sensitized solar cells. Appl Phys Lett 86:053114
7. Luque A, Marti A, Nozik AJ (2007) Solar cells based on quantum dots: multiple exciton generation and intermediate bands. MRS Bull 32:236–241
8. Kayes BM, Atwater HA, Lewis NS (2005) Comparison of the device physics principles of planar and radial p-n junction nanorod solar cells. J Appl Phys 97:114302
9. Zhang Y, Wang LW, Mascarenhas A (2007) Quantum coaxial cables. Nano Lett 7:1264–1269
10. Fraas LM, Huang HX, Ye SZ, Hui S, Avery J, Ballantyne R (1997) Low cost high power GaSB photovoltaic cells. AIP Conf Proc 401:33
11. Singh KJ, Leiphrakpam C, Singh NP, Singh NB, Sarkar SK (2014) 3d Single Gaas Co-axial nanowire solar cell for nanopillar-array photovoltaic device. Int J Comput Sci Appl (IJCSA) 4(3), June 2014
12. Process and Device, 2D and 3D Physics-based Simulation. SILVACO International www.silvaco.com
13. Hayden O, Agarwal R, Lieber CM (2006) Nanoscale avalanche photodiodes for highly sensitive and spatially resolved photon detection. Nat Mater 5:352–356
14. Karpov VG, Cooray MLC, Shvydka D (2006) Physics of ultrathin photovoltaics. Appl Phys Lett 89:163518 (2006)
15. Shah AV et al (2004) Thin-film silicon solar cell technology. Prog Photovolt Res Appl 12:113–142
16. Maji1 B, Chattopadhyay R (2021) Design and optimization of high efficient GaSb homo-junction solar cell using GaSb intrinsic layer. Springer, 03 Jan 2021
17. Kret J, Tournet J, Parola S, Martinez F, Chemisana D, Morin R, de la Mata M, Fernández-Delgado N, Khan AA, Molina SI, Rouillard Y, Tournié E, Cuminal Y (2021) Investigation of AlInAsSb/GaSb tandem cells—a first step towards GaSb-based multi-junction solar cells. Solar Energy Mater Solar Cells 219(2021):110795, 29 Sept 2020
18. Vadiee E, Fang Y, Zhang C, Fischer AM, Williams1 JJ, Renteria EM, Balakrishnan G, Honsberg CB (2018) Temperature dependence of GaSb and AlGaSb solar cells.: current applied physics, 6 Mar 2018
19. Sahoo GS, Mishra GP (2019) Efficient use of low-bandgap GaAs/GaSb to convert more than 50% of solar radiation into electrical energy: a numerical approach. J Electron Mater 48(1)
20. Garnett EC, Brongersma ML, Cui Y, McGehee MD (2011) Nanowire solar cells. Ann Rev Mater Res 41:269–295
21. Luque A, Hegedus S (2003) Handbook of photovoltaic science and engineering. Wiley, Chichester

Triple-Strap EMI Suppression with Frequency-Selective Exterior

Rajdip Das and Umesh Pal

Abstract Over a century, electromagnetic shielding (EMI) has been restricted to a very small frequency. The curiosity comes from the necessity to safeguard the radio receiver's circuits and equipment from EMI and the radiated field. The development of wireless, and satellite technologies possess potential health hazards directly or indirectly. Therefore, it is necessary to find solutions to effectively isolate interference. The major aim is to keep all of the notch filter's properties, while ventilation needs to compel the enclosure to pass a specific spectrum of EMI. This research paper describes the plan and manufacture of a super slim adaptable EMI safeguard equipped for dismissing a couple of bothersome frequentness. The outline starts with a plan of a small part, which allows us to effectively calculate the ring's initial geometric measurements, and then uses full-wave electromagnetic modeling to correct the measurements and the intended frequency response's end. Various EMI shielding geometric patterns with concentric rings are studied and discussed. Based on these findings, a screen-printed radical-thin and malleable EMI shield were created. This paper demonstrates a strong relationship between measurement and simulation.

Keywords Frequency selective surfaces (FSS) · Print screening · Triple-strap EMI

1 Introduction

As the connections grow at an exponential rate, the quantity of camp stations is likely to grow in the next few years, in the direction of achieving better service coverage. Therefore, this communication through wireless in a conjunction with other sensitive electronic gadgets is likely to generate an EMI [1]. Like a shield, it draws people's attention to themselves. Architectural shielding is becoming increasingly important in protecting sensitive electronics from potential EMI hazards.

Traditional shielding cabinets are bulky and impose structural stress on existing structures [2], thereby shielding them. An EMI protector that is ultra-lightweight, highly malleable, and can be attached to the walls of existing structures will be

R. Das (✉) · U. Pal
Jadavpur University, Kolkata 700032, India

J. K. Mandal et al. (eds.), *Topical Drifts in Intelligent Computing*, Lecture Notes in Networks and Systems 426, https://doi.org/10.1007/978-981-19-0745-6_30

appealing [3]. In addition to the weight issues, traditional metal housings lack frequency-selective shielding (FSS) [4–7], they effectively block electromagnetic waves of all frequencies.

The FSS design allows for blank regions on the screen, resulting in another sensible feature sight lucidity. It should be observed that cascaded FSS is usually used to obtain a frequency-selective multi-band surface [8–10]. There have been many works on the frequency selection structure of various applications, and we will not repeat them here [11–15]. The cascading nature of the design results in a relatively thick screen. The hinge design is an excellent choice to maintain the design level. EMI shielding uses printed periodic components to provide a frequency selection function, which may put away undesirable frequencies without interfering with other mobile assistance. Due to the possibility of mass production of coils, screen printing technology is used to make EMI boards. A screen-printed prototype was made, and its multi-band suppression capability was proved through experiments.

The paper is arranged as follows: The sketching of several band stop EMI shields are presented in Sect. 2 which consists of building a full-wave structure and followed by the analysis of full-wave model geometrical parameters in Sects. 2.1, and 2.2, respectively, the experimental analysis is discussed in Sect. 3. Finally, the conclusion and future works are reviewed.

2 Sketching of Several Band Stop EMI Shield

FSS is a periodic structure made up of one- or two-dimensional conductive components or holes that allow the filter to function in the presence of electromagnetic wave radiation. It has a complete transmission or reflection near the resonance frequency when emitted with electromagnetic waves. This FSS spatial filter characteristic is employed in the construction of an FSS screen [16]. These filter actions are determined by the geometry of the component.

A ring structure is proposed to provide multi-band suppression capabilities in a single-layer structure because the usage of concentric rings with various radii can create numerous resonances. Figure 1 shows the peak and part views of the band cut-out EMI shield. FSS is made by placing conductive strips in an orderly manner. The screen is utilized as a size band limit channel, and along these lines, the best format is the fix sketch. This permits the attenuator to go about as an indented channel at the ideal sound recurrence. The concentric rings would be the finest sketch to satisfy the additional aims of enabling optical transmission.

Since it consists of empty rings instead of spots, it is ideal for applications that require a high level of optical clarity. The conducting rings are made up of sticky silver paste. The conductivity of a round-shaped entirely covered accompanied by a silver paste may be measured using a tetrad-point probe. A round shape completely plating with adhesive silver has a conduction of 2.6×10^6 S/m and a thickness of 10 μm after post profile measuring. The relative permittivity of the 0.1 mm wide substrate

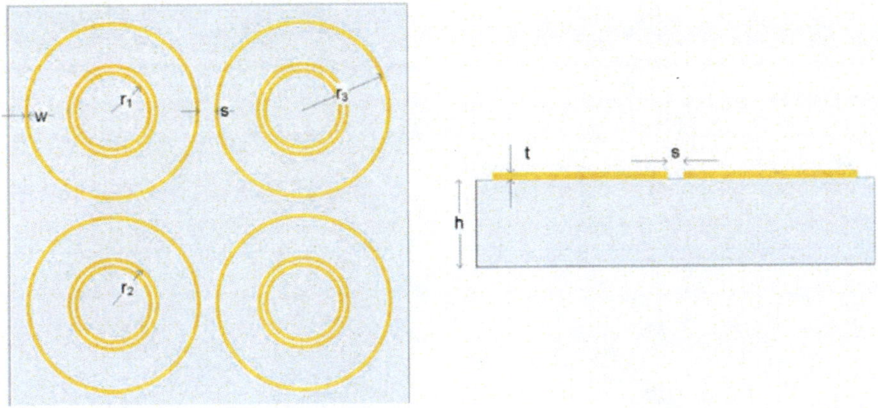

Fig. 1 Peak and part views of band stop EMI safeguard

is 2.3. For improved accuracy, the voltaic characteristics of the two substrates and the silver paste are included in the full-wave simulation prototype [17, 18].

2.1 Building a Full-Wave Structure

The equivalent circuit model yields the initial geometric dimensions of the ring construction, commercial 3D EM software can be used to model the full-wavelength 3D structure, taking into account the required boundary conditions. The unit cell consists of two waveguide ports arranged in the middle of the ($-z$, $+z$) course, which is used as an excitation source and an obtaining source. Under time-domain conditions, the ideal charged wall is to set down in the middle of the ($-x$, $+x$) course, and the ideal barrier is to set down in the middle of the ($-y$, $+y$) course [19, 20]. Figure 2 exhibits a three-dimensional ingredient cell replica with multiple concentric rings and parameters and con-centric rings with geometric dimensions.

Teflon is utilized as the dielectric substrate in this case. The substrate has a thickness of 0.1 mm. Two air boxes are formed around the substrate. The air boxes have a height of 20 mm. The FSS structure is now formed on the substrate. PEC is used to make the ring constructions. On the substrate, there are three concentric circles. The innermost circle has a radius of 19.7 mm and a broadness of 3.51 mm. The semi-diameter of the subsequent circle is 26 mm, the corresponding width is 25.5 mm, the semi-diameter of the outermost ring is 50 mm, and the corresponding broadness is 3 mm.

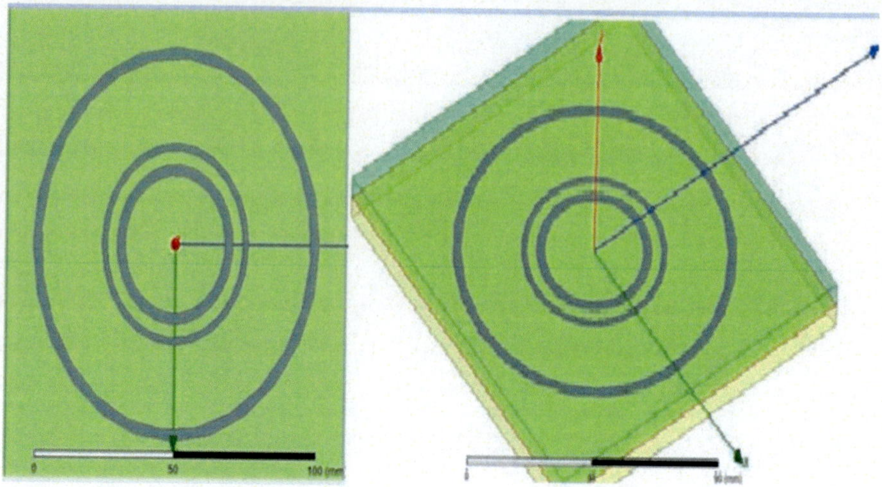

Fig. 2 The three-dimensional model of a unit cell with multiple concentric rings and parameters and concentric rings with geometric dimensions

2.2 Analysis of Full-Wave Model Geometrical Parameters

The reflectance $S_{(1,1)}$ and gain $S_{(1,2)}$ are measured from 0 to 3.50 GHz, with three shielded frequency bands. Figure 3 denotes the simulation results of the three-band EMI suppression shield.

Figure 4 shows the estimated or simulated transmission conditions for various radio stations at full-wavelength. It can be noticed in other words, the estimation result formed on the corresponding circuit, architecture is very similar to the result of the full-wavelength simulation using the simulation tool, except that the change in the resonance frequency is very small. However, the equivalent circuit model

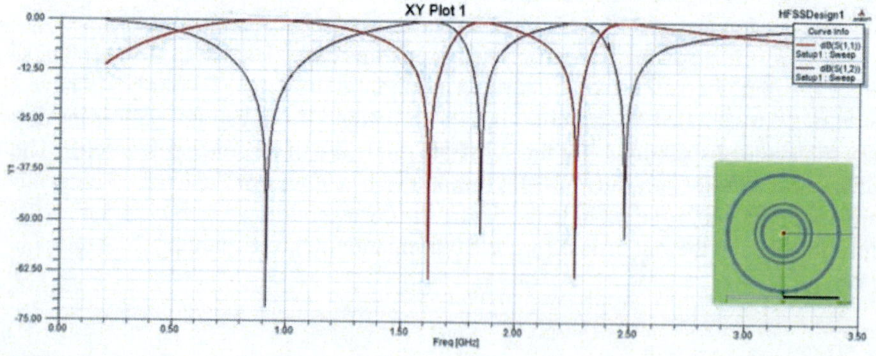

Fig. 3 Experimental and simulation results of three-band EMI suppression screen

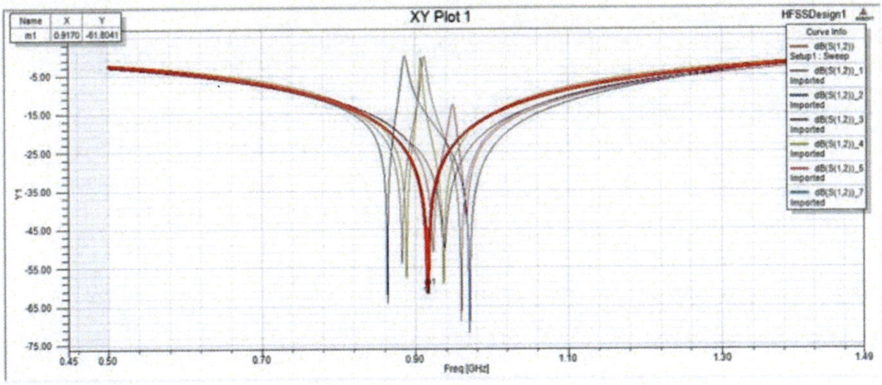

Fig. 4 The estimated or simulated transmission conditions for various radio stations at the full-wavelength [30, 35, 40, 45, and 50 mm]

(ECM) method is very effective in rapid design synthesis that does not require a lot of calculations, so it can perform parameter analysis more effectively than full-wave modeling.

3 Experimental Results

The 3-channel EMI shielding is based on the results of previous research. According to previous studies, it has been observed that the radius of the ring is roughly about 19.4, 25.6, and 50 mm, discretely, to achieve screens with resonance frequencies of 918 MHz, 1.848 GHz, and 2.48 GHz. Based on initial results, further adjustments are made to achieve the required size of all resonances performed in a single design (Table 1).

Figure 6 demonstrates the imitated S_{21} of the beginning and closing blueprint of the three-band EMI screen. Make prototypes based on the final geometric dimensions. Figure 5 shows the manufactured EMI tape. The guide is 450 mm × 450 mm. Figure 7 shows the experimental setup used to measure the shielding efficiency of EMI ribbon connectors [21]. A pair of duplex sheets oscillate guide antennas (18 GHz) are used for measurement. The horn antenna is used for transmission and reception. Plane waves pass through the screen. There is a 350 mm x 350 mm [22] hole on the top of the metal box. The EM safeguard is situated between the sending receiving wire and the getting radio wire. This helps reduce the diffraction accuracy of surrounding waves. The interval betwixt every antenna and the EMI bandwidth is kept at 300 mm with a hollow wooden box. The 8 GHz vector web browser is used to retrieve the S variable from the shared configuration.

The shielding efficiency for this computation system may be calculated as follows:

$$SE_1 = S_{21\,\text{without shield}} - S_{21\,\text{with shield}} \tag{1}$$

Table 1 The introductory and closing geometric proportions of the three-way concentric circles of three orbital discharges

Parameter	Initial dimension (mm)	Final dimension (mm)	Description
w	2.00	2 0.00	Breadth of a conductor
r_1	19.5	21.5	The inmost radius of the circle
r_2	25.0	26.0	The inmost radius of the circle
r_3	51.0	49.5	The inmost radius of the circle
s	9.99	10.0	Range betwixt adjacent circles
h	0.01	0.1	Substrate broadness
t	0.001	0.01	Conductor broadness

Fig. 5 Photo of flexible and transparent three-band EMI shielding

where $S_{21withoutshield}$ is the insertion depletion evaluated without shielding, while the shielded $S_{21withshield}$ is the insertion depletion shielding accompanied by. Figure 8 shows the protecting proficiency of the last plan got through measurement. Contrasting the outcomes, it very well may be seen that the reenactment results are the same as the test results. Experiments have shown that the passband EMI shield can provide an S_{E1} of 10.1–25.1 dB at the required resonant recurrence, and the amplitude was observed no later than the t the corresponding resonance frequency S_{E1}.

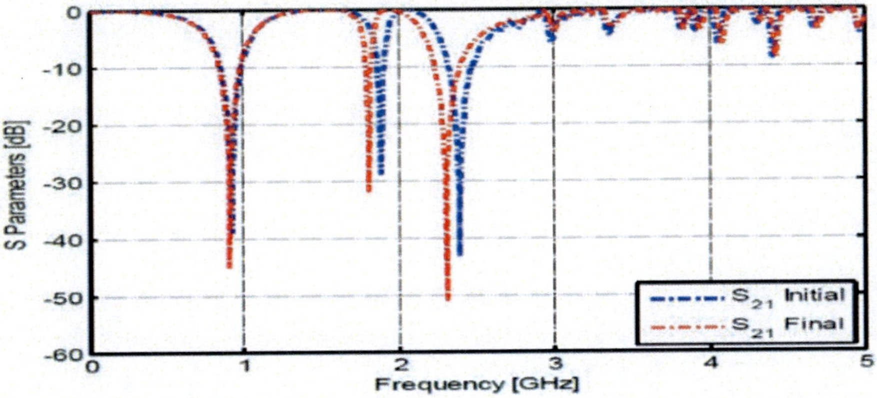

Fig. 6 Tri-band EMI shielding simulation results from the preliminary and final draft

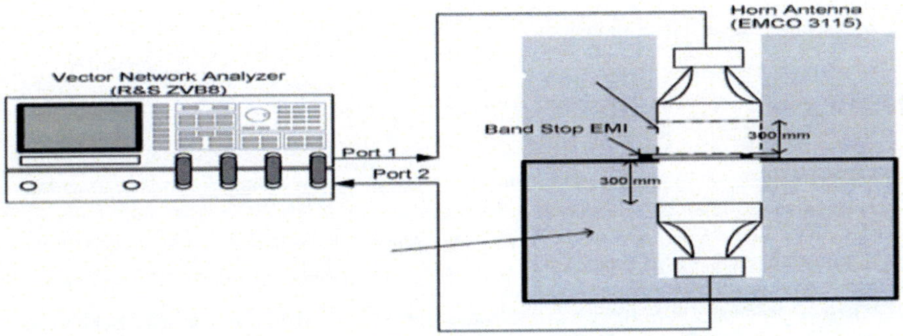

Fig. 7 Demonstration system for identifying the protecting productiveness of the screen

4 Conclusion

This research paper shows that a multi-channel EMI shield with conductive concentric rings can be screen-printed on a distinct surface, making it super light and exceptionally adaptable. The initial blueprint of the EMI safeguard lay off handily made utilizing the successful equivalent circuit. Extra centering and tweaking can be accomplished through full-frequency reenactment. Taking the three-frequency EMI shielding as an example, it shows a good match. This sort of EMI safeguarding can be handily joined to the dividers of existing structures to smother undesirable recurrence without influencing the necessary recurrence groups.

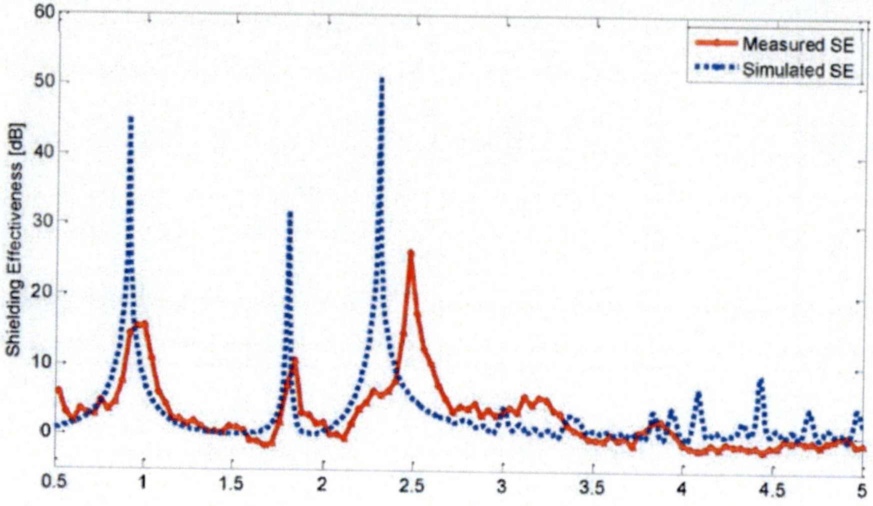

Fig. 8 Experimental and final simulation outcome of a final guide

References

1. Unal E, Gokcen A, Kutlu Y (2006) Effective electromagnetic shielding. IEEE Microw Mag 7:48–54. https://doi.org/10.1109/MMW.2006.1663989
2. Hemming LH (1992) Architectural electromagnetic shielding handbook: a design and specification guide. IEEE Press.
3. Roll-to-roll, flexible, and multi-functional. www.simtech.a-star.edu.sg
4. Wireless products: FSS (frequency selective surface)-based electromagnetic interference shielding film. http://www.mitsubishicable.co.jp [online]
5. Stefanelli R, Thinchero D (2009) Scattering analysis of frequency selective shields for electromagnetic indoor isolation. Microw Opt Technol Lett 51(II):2758–2762
6. Stefanelli R, Thinchero D (2010) Reduction of electromagnetic interference by means of frequency selective devices. In: IEEE 17th international conference on telecommunications, pp 239–243
7. Wang LB, See KY, Zhang JW, Salam B, Lu ACW (2011) Ultrathin and flexible screen-printed metasurfaces for EMI shielding applications. IEEE Trans Electromag Compat 53(3):700–705
8. Vardaxoglou JC (1997) Frequency selective surfaces: analysis and design. Wiley, New York
9. Munk BA (2000) Frequency selective surface: design and theory. Wiley, New York. https://doi.org/10.1002/0471723770
10. Celozzi S, Araneo R, Lovat G (2008) Electromagnetic shielding. Wiley-Interscience, Hoboken.https://doi.org/10.1002/9780470268483
11. Mittra R, Chan CH, Cwik T (1988) Techniques for analyzing frequency selective surfaces—a review. IEEE Proc 76:1593–1615
12. Koleck T, Diez H, Bolomey JC (1997) Techniques for analysing finite frequency selective surfaces. In: Antennas and propagation conference, pp 161–165.https://doi.org/10.1049/cp:19970229
13. Savia SB, Parker EA (2003) Equivalent circuit model for superdense linear dipole FSS. IEEE Proc Microw Antennas Propag 150:37–42
14. Cai Y, Mias C (2006) Time and frequency domain modelling of a normally incident plane wave at a periodic structure. In: The 6th international conference on computational electromagnetics (CEM), pp 1–2

15. Fernandez L, Garcia E, Castro D, Segovia D (2005) Tool to design frequency selective surfaces using an equivalent circuit model. Microw Technol Lett 47:464–467
16. Marcuvitz N (1986) Waveguide handbook: Peter Penguins Pte Lte
17. Wang LB, See KY, Chang WY, Lu CW, Ng ST (2010) Electromagnetic shielding analysis of printed flexible meshed screens. In: Asia-Pacific symposium on electromagnetic compatibility, pp 965–968
18. Wang LB, See KY, Zhang W, Lu ACW, Ng ST (2010) Full-wave modeling and analysis of screen-printed EMI shield. In: 2010 Asia Pacific microwave conference, pp 1348–1351
19. CST Studio Suite 2006 Application Note (2006) Periodic Arrays: FSS/PBG
20. Ko W, Mittra R (1993) Implementation of Floquet boundary condition in FDTD for FSS analysis. In: IEEE AP-S International on Symposium Digest, pp 14–17
21. Marvin AC, Dawson, Flintoft D, Dawson IF (2009) A method for the measurement of shielding effectiveness of planar samples requiring no sample edge preparation or contact. IEEE Trans Electromag Compat 51(2):255–262
22. Mur G (1981) Absorbing boundary conditions for the finite-difference approximation of the time-domain electromagnetic- field equations. IEEE Trans Electromag Compat, EMC-23, 377–382. https://doi.org/10.1109/TEMC.1981.303970

Design Analysis of Uniformly Weighted Circular Planar Antenna Array Using Efficient Meta-heuristic Algorithm in MATLAB

Kailash Pati Dutta, Sonal Priya Kindo, Neelam Xalxo, Suman Linda, and Kajal Kumari

Abstract Circular planar array antenna with minimized peak side lobe levels is desired in many advanced wireless applications. In this paper, an innovative strategy is applied to the designing problem of the planar array antenna having concentric rings that are uniformly weighted with circular aperture using a human intelligence-based meta-heuristic algorithm, namely teaching–learning-based optimization. The objective of this work is to design the concentric circular array antenna with constraints like number of array elements as well as radius of each ring and then finally present an analysis of the same using different examples. Four cases with different number of rings are presented considering different constraints individually. Firstly, concentric circular array antenna with 5 and 6 rings is discussed with optimized ring radii. Secondly, with optimized number of array elements, 7 and 8 rings are taken into account. Results are superior in the example having rings in larger number. The statistical data for every design are also presented to showcase the effectiveness of the simulation approach. The result comparison of the proposed work with state of the art further confirms the effectiveness of the proposed design.

Keywords Circular planar array antenna · Teaching–learning-based optimization algorithm · Side lobe levels · Evolutionary algorithms

1 Introduction

Concentric circular array antenna (CCAA) [1] has multiple circular rings with a common centre. Each ring in CCAA has different number of array elements. Regardless of there being many different types of antenna array, CCAA has an important role to play in wireless communication [1, 2]. CCAA has numerous advantages over

K. P. Dutta · S. P. Kindo (✉) · N. Xalxo · S. Linda · K. Kumari
Department of Electronics and Communication Engineering, Cambridge Institute of Technology, Ranchi, Jharkhand 835103, India

K. P. Dutta
Department of Electronics and Communication Engineering, National Institute of Technology, Durgapur, West Bengal 713209, India

J. K. Mandal et al. (eds.), *Topical Drifts in Intelligent Computing*, Lecture Notes in Networks and Systems 426, https://doi.org/10.1007/978-981-19-0745-6_31

other antenna. It has insufficient edge element, thus it is less sensitive to mutual coupling. Also, the main lobe of CCAA is capable of radiating isotropically and provide 360o azimuth scanning feature without any variation in beam pattern. It offers the radiation pattern that covers the whole space [1, 3].

In spite of its numerous advantages, it has one drawback. Though having a large directivity, they have side lobe levels with higher value which is not desirable [3, 4]. We customize CCAA in order to achieve better performance with respect to minimized value of the side lobes. To solve antenna design problem by optimizing inter-element spacing with fixed radii, optimizing using radii with fixed inter-element spacing and optimizing inter-element spacing and ring radii [6, 7]. Customizing CCAA not only targets in reduction of SLL, but also decreases the manufacturing cost and weight of antenna arrays. For these reasons, CCAA has been applied extensively to a variety of application over last 42 years [1–7].

Readers can refer to a number of synthesis techniques for planar antenna array that are available in the literature [6–20]. Hybrid approach (HA) [7], moth flame optimization (MFO) [8], genetic algorithm (GA) [9], global and local real-coded genetic algorithms [10], chaotic bee colony algorithms (CBCAs) [11], Gaussian tapering window technique [12], seeker optimization algorithm (SOA) [13], modified teaching–learning-based optimization (MTLBO) [14], QPSO [15], chaotic adaptive invasive weed optimization (CAIWO) [16, 19], and symbiotic organism search (SOS) [20] algorithm are few of the distinguished methods with notable contributions in this area.

In a nutshell, there exists a rich literature that addresses the solving of the synthesis problem of planar array antenna. However, as per the no free lunch theorem by Wolpert and Macready [21], any evolutionary/meta-heuristic algorithm is not sufficient to solve all types of the problems related to the optimization. Hence, meta-heuristics with proven performance and computational efficiency are always welcoming towards applying to some specific problem.

In this article, we propose the application of well-established evolutionary algorithm specifically teaching–learning-based optimization (TLBO) [8] in the designing analysis of the CCAA that yields optimized side lobe levels. The impact of the constraints like number of array elements in each ring and inter-ring radii is analyzed with four different examples that include 5, 6, 7 and 8 rings separately with constraints. Unlike other optimization algorithms, TLBO is a easy to use tool that has no additional tuning parameter other than the common algorithm specific free optimization parameter [8, 17] like number of runs, population size, etc. Hence, users need not to bother paying special attention for appropriate tuning of any algorithm specific control. This is not only attributed for a faster convergence rate but also simplifies the entire simulation process, thereby reducing the operational complexities. Due to its effectiveness, TLBO has been used to address optimization problems related to diverse engineering problems [8, 14, 17, 18].

The rest of our paper is structured as with the following sections. Section 2 discusses CCAA geometry with problem formulation, Sect. 3 gives the detailed discussion on the TLBO algorithm, and Sect. 4 presents the results of the simulation and their analysis. The conclusions were presented in the Sect. 5 followed by the references.

2 CCAA Geometry and Problem Formulation

Uniformly, weighted concentric circular antenna array (CCAA) yields radiation beam that is capable of covering the whole space with 360o azimuthal scanning ability. Its elements are organized in different concentric circular rings that have common-centre, different radii and unlike number of antenna array elements in each circle. They are supplied with current excitations that are of uniform value for each of the antenna elements. The CCAA array geometrical layout is shown in Fig. 1.

Considering the central array element feed, the array factor symbolized by AFactor$(\theta, \emptyset, I)$ of the CCAA geometry lying on the x-y plane is expressed by (1).

$$\text{AFactor}(\theta, \emptyset, I) = 1 + \sum_{m=1}^{M} \sum_{n=1}^{N_m} I_{mn} e^{j[\text{kR}_m \sin\theta \cos(\emptyset - \emptyset_{mn}) + \alpha_{mn}]} \tag{1}$$

where M symbolizes total rings in the CCAA, N_m represents total array elements (isotropic in nature) in the mth ring, d_m is separation between each array element in the mth ring. $\text{R}_m = N_m d_m / 2\pi$. This represents the radius of mth ring, $\emptyset_{mn}$ is angular-position of nth element of the mth ring such that$1 \leq n \leq N_m$. θ and $\emptyset$ are polar and

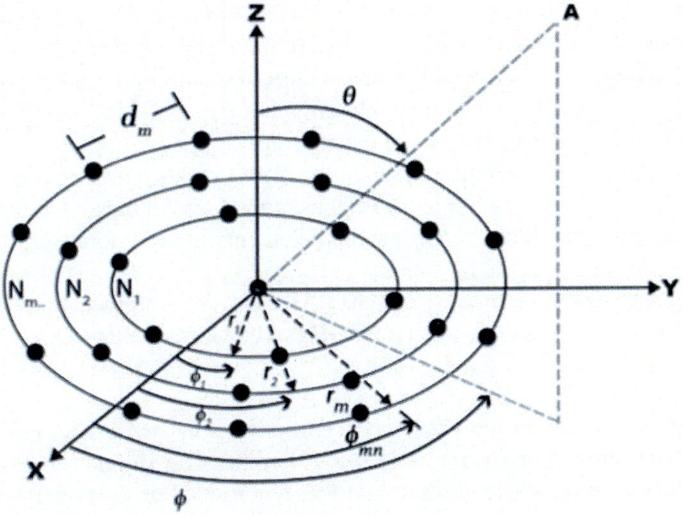

Fig.1 Concentric circular planar array geometry layout presented in the x–y plane

azimuthal angles, respectively. K gives the wave-number which equals $2\pi/\lambda$, whereas λ represents wave length in operation, j is complex number (imaginary unit). α_{mn} is phase of individual array element, and I_{mn} is current amplitude excitation of the mnth array elements of that is set to uniform but unity value. Uniform phase excitation of 0o is associated with each element.

We have designed the objective (or cost or fitness) function in such a style that the side lobe levels (SLL) get optimized applying appropriate searching of the inter-ring radii, and the number of the antenna array elements present in each concentric circle.

3 Teaching–Learning-Based Optimization (TLBO) Algorithm

TLBO [17, 20] was proposed by R. V. Rao, V. J. Savsani and D. Vakharis in 2011. TLBO is a novel optimization technique applied in diverse engineering applications. It has basically two phases such asthe teacher's phase and the learner's phase. The readers may refer papers [17] and [20] for the theoretical details and implementation steps of TLBO.

4 Simulation Results

The bundles of elements are considered as isotropic antenna. Simulation is done using four cases. In case 1 and case 2, we try to optimize the ring radii (R_{m}) using 5 and 6 concentric rings, respectively. And in case 3 and 4, we have attempted optimizing the array elements in each particular ring using 7 and 8 concentric rings, respectively. These are examples only and one is free to consider any number of rings as per choice. We have used TLBO in MATLAB for the simulation purpose. TLBO does not have any tuning parameter, it has only common parameters to design antenna array like population density number of generations [20].

As per the results obtained after simulation, TLBO yields better results than the results available in the existing literature as far as suppression of the SLL is concerned with respect to total number of array antenna elements in each concentric rings as constraints.

TLBO algorithm is implemented in MATLAB software with the help of a laptop having 8 GB RAM, i5 processor with 2.20 GHz of clock frequency.

Case 1. Optimized Ring Radii with 5 Rings

This is the first case where we used 5 concentric rings and with ring radii as the constraints. The simulations were done using 4 trials. The simulation results are shown in Table 1 from where we see that the synthesized array is very bulky in nature with a large number of array elements counting up to 492. This array will also

Table 1 Simulation results and their comparison with [7, 9]

Algorithm	Array type	N_R	N	Aperture (λ)	SLL (dB)	Std. deviation	Worst (dB)
HA [7]	Opt R_m	8	201	4.98	–29.03	–	–
TLBO (Case 1)	Opt R_m	5	492	13.4	–19.98	0.11	–19.69
TLBO (Case 2)	Opt R_m	6	147	3.86	–23.03	0.08	–22.83
GA [9]	Opt N_m	9	183	4.5	–25.58	–	–
TLBO (Case 3)	Opt N_m	7	115	3.5	–31.28	0.5	–30.05
TLBO (Case 4)	Opt N_m	8	184	4	–34.26	1.25	–31.07

be larger in size as the array aperture is as large as 13.41λ. Due to large number of array elements, there will be existence of the grating lobes. Moreover, the costing and the maintenance of such a bulky array will be difficult and thus not desirable. More importantly, the SLL in this case is found as –19.98 dB which is infact very high and against the objective of this work.

Case 2. Optimized Ring Radii with 6 Rings

In this case, also the common control parameters are same as that in case 1. The simulation results are summarized in Table 1 given which clearly indicates that the related SLL found is –23.03 dB. This result is better than the previous case (–19.98 dB). The worst value found here is–22.83 dB. That means, the worst value seen here is even better than the best value of case 1. The aperture size of the array is found to be 3.86λ which is also far better and smaller than found in case 1 (13.41λ).

Case 3. Optimized Number of Array Antenna Elements with 7 Rings

This case demonstrates the value of minimized SLL as –31.28 dB with a maximum ring-aperture size of 3.5λ. These results are showcased in Table 1 from where readers can easily see the worst value obtained in this case is –30.05 dB. As far as individual ring is concerned, it may be stated from Table 2 that the first ring contains 31 elements, second ring yields 20 elements, third, fourth, fifth and sixth rings hold 13, 15, 18 and 8 array elements, respectively. The last ring that is seventh ring yields 10 number of array elements. In other words, we can say that to obtain this design, a total number of only 115 elements were used. The standard deviation obtained from simulation results in this case is 0.50. The graph of the mean of best fitness value vs fitness function evaluations is presented in Fig. 2a. Also, Fig. 2b gives the normalized power pattern for 7 rings CCAA obtained after the MATLAB simulation.

Case 4. Optimized Number of Array Antenna Elements with 8 Rings

This is the fourth case where best results are obtained in terms of SLL. The results obtained are shown in Tables 1 and 2. The maximum minimized SLL that we found

Table 2 Distribution of array elements in each ring and inter-elemental separation

Array type	N_{total}	N	Circle number							
			Ring 1	Ring 2	Ring 3	Ring 4	Ring 5	Ring 6	Ring 7	Ring 8
Opt R_m	492	N_m	44	75	90	114	169	–	–	–
		R_m	3.50	5.96	7.15	9.07	13.4	–	–	–
Opt R_m	147	N_m	5	12	18	27	36	49	–	–
		R_m	0.36	0.92	1.42	2.12	2.89	3.86	–	–
Opt N_m	115	N_m	31	20	13	15	18	8	10	–
		R_m	0.5	1	1.5	2	2.5	3	3.5	–
Opt N_m	184	N_m	14	20	29	30	30	27	16	18
		R_m	0.5	1	1.5	2	2.5	3	3.5	4

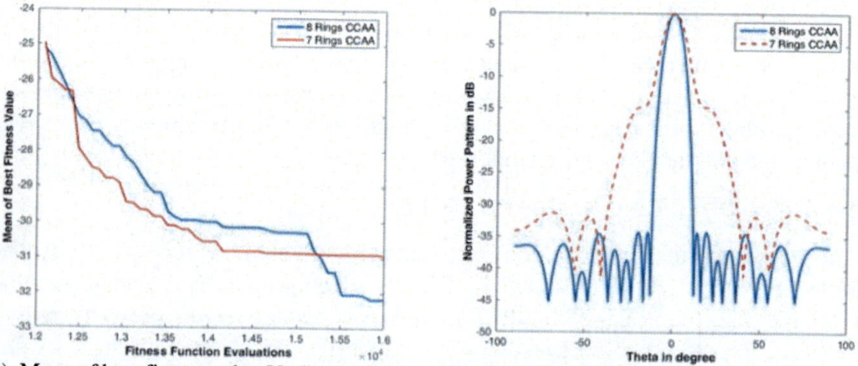

(a) Mean of best fitness value Vs fitness function evaluation in case 3 and case 4

(b) Normalised radiation pattern obtained in case 3 and case 4

Fig. 2 Comparative performance of results obtained in case 3 and case 4

here is–34.23 dB. The optimal number of total array elements is found to be 184. The worst value obtained is –31.07 dB where as the standard deviation is 1.25. If we talk about the ring wise distribution of array elements, the first ring has the lowest number of array antenna elements (14) whereas the fourth and fifth ring have equal number of array elements (30 each) both counting to the maximum value. The second, third, sixth and seventh ring yield 20, 29, 27 and 16 array elements, respectively. The 8th ring has a total of 18 antenna elements. The array synthesis results of all cases are summarized in Table 1 whereas Table 2 presents the ring radii, total number of array elements and ring wise array element distribution. The mean of best fitness value versus fitness function evaluations is presented in Fig. 2a. Also, Fig. 2b gives the normalized power pattern.

In general, it is observed that case 4 yields best result in terms of SLL suppression which is better than case 3, case 2 and case 1 with values of 2.98 dB, 11.24 dB and 14.28 dB, respectively. However, case 3 gives lowest number of array elements (115) with lowest aperture of 3.5λ. As compared to case4, case 3 saves 69 array elements. As a result, it can be said that though case 3 and case 4 are demonstrated to achieve peak SLL with optimized antenna array elements, however, both these cases establish a trade-off between SLL versus optimal number of array elements.

In other words, if we work on the higher SLL, the number of array elements increases. And if we work on decreasing the later one in order to reduce the designing cost, the SLL is increased. Initially, in paper [7], we observed hybrid approach was used to optimize ring radii. The ring radii are optimized using HA with 8 rings having 201 elements and give –29.03 dB of peak SLL. Similarly, in paper [9], ring radii were optimized using GA with 6 rings having 201 elements and give –25 dB peak SLL. We approached with TLBO algorithm for same constraints. And, as a consequence, our optimal results obtained in both case 3 and case 4 are better. As compared to [7], case 3 and case 4 give 2.25 dB and 5.23 dB lower SLL, respectively. Similarly, as compared to [9], the SLL achieved in cases 3 and 4 are better with dB values of 5.7 and 8.68, respectively. Apart from this, this work is different from [17] as well in a way that in [17], 9 rings are considered to achieve the optimal results; however, in the present work, we have used a maximum of 8 rings only. Here, better algorithmic specific settings have yielded to preferred results. This will not only save the cost of the array but also reduce the design complexities, bulkiness and weight of the antenna system when subjected to real-time implementation. Figure 2 displays the comparative performance of results obtained in cases 3 and 4.

5 Conclusions

The MATLAB-based simulation using TLBO as an efficient meta-heuristic algorithm was successfully conducted. Four cases with different number of rings were considered in which the third and the fourth cases with total number of array elements as constraints were found to be more effective. The fourth case outperformed all other cases in terms of SLL suppression, however, case 3 obtained best results with maximum savings on the part of array antenna elements. Using the proposed approach, this work overtakes the prior arts [7, 9]. TLBO being an algorithm having no tuning parameter makes the computation easier, more simple but yet efficient. This approach may be implemented for different other array geometries.

References

1. Ballanis A (1997) Antenna theory analysis and design, 2nd edn. Willey Inc., New York
2. Sotirios K, Christos K, Mittra R (2016) Evolutionary algorithms applied to antennas and propagation: emerging trends and applications. Int J Antennas Propag 2016:1–12
3. Magill E, Wheeler H (1966) Wide-angle impedance matching of a planar array antenna by a dielectric sheet. IEEE Trans Antennas Propag 14(1):49–53

4. Hsu SH, Ren YJ, Chang K (2009) A dual-polarized planar-array antenna for s-band and x-band airborne applications. IEEE Antennas Propag Mag 51(4):70–78
5. Keizer WPMN (2007) Fast low-sidelobe synthesis for large planar array antennas utilizing successive fast fourier transforms of the array factor. IEEE Trans Antennas Propag 55(3):715–722
6. Stearns C, Stewart A (1965) An investigation of concentric ring antennas with low sidelobes. IEEE Trans Antennas Propag 3(6):856–863
7. Guo Q, Chen C, Jiang Y (2017) An effective approach for the synthesis of uniform amplitude concentric ring arrays. IEEE Antennas Wirel Propag Lett 16:2558–2561
8. Dutta KP, Mahanti GK (2020) Meta-heuristic optimization algorithms for simultaneous optimization of sidelobe level and directivity of uniformly excited concentric ring array antennas. Int J Microw Wirel Technol 12(2):183–192
9. Haupt RL (2008) Optimized element spacing for low sidelobe concentric ring arrays. IEEE Trans Antennas Propag 56(1):266–268
10. Martínez CG, Lozano M, Herrera F, Molina D, Sánchez AM (2008) Global and local real-coded genetic algorithms based on parent-centric crossover operators. Eur J Oper Res 185(3):1088–1113
11. Alatas B (2010) Chaotic bee colony algorithms for global numerical optimization. Expert Syst Appl 37:5682–5687
12. Dessouky MI, Sharshar HA, Albagory YA (2007) Optimum normalized-Gaussian tapering window for side lobe reduction in uniform concentric circular arrays. Prog Electromagnet Res 69:35–46
13. Dhali M, Sindhu C, Mandal D, Ram G, Kar R (2018) Thinned concentric circular array antenna synthesis using seeker optimization algorithm. Int Electr Eng Congress (iEECON) 2018:1–4. https://doi.org/10.1109/IEECON.2018.8712326
14. Zailei L, He X, Chen X, Luo X, Li X (2015) Synthesis of thinned concentric circular antenna arrays using modified TLBO algorithm. Int J Antennas Propag 2015:1–9
15. Dutta KP, Sundi S, Khalkho SS, Tirkey J, Hembrom KK (2020) Minimization of sidelobe level of uniformly excited concentric circular array antenna using evolutionary technique. In: Raj JS, Iliyasu AM, Bestak R, Baig ZA (eds) Innovative data communication technologies and application. Lecture notes on data engineering and communications technologies, vol 59. Springer
16. Wu H, Liu C, Xie X (2014) Pattern synthesis of planar nonuniform circular antenna arrays using a chaotic adaptive invasive weed optimization algorithm. Math Probl Eng. https://doi.org/10.1155/2014/575860
17. Dutta KP, Mahanti GK Synthesizing uniform concentric circular array antenna for minimized sidelobes using teaching–learning-based optimization algorithm. In: Hassanien AE, Bhattacharyya S, Chakrabati S, Bhattacharya A, Dutta S (eds) Emerging technologies in data mining and information security. In: Advances in intelligent systems and computing, vol 1286. Springer. https://doi.org/10.1007/978-981-15-9927-9_25
18. Chen X, Xu B, Yu K, Du W (2018) Teaching–learning-based optimization with learning enthusiasm mechanism and its application in chemical engineering. J Appl Math. https://doi.org/10.1155/2018/1806947
19. Goto N, Tsunoda Y (1977) Sidelobe reduction of circular arrays with a constant excitation amplitude. IEEE Trans Antennas Propag 25(6):896–898
20. Dutta KP, Mahanti GK (2020) Evolutionary algorithms for optimal synthesis of thinned multiple concentric circular array antenna with constraints. Int J Electron 107(10):1649–1662
21. Wolpert DH, Macready WG (1997) No free lunch theorems for optimization. IEEE Trans Evolutionary Comput 1(1):67–82

Design of Novel Radial Folded Microstrip Patch Antenna for WiMAX Application

Prashant Kumar Singh, Shashank K. Singh, Anjini K. Tiwary, Gufran Ahmad, Sandipan Mallik, and Syed Samser Ali

Abstract In recent years, the study of microstrip patch antennas (MPA) has made great progress because of its advantages in terms of weight, volume, cost, fabrication, and dimension. This paper presents a novel radial folded microstrip patch antenna, which operates at 3.55 GHz frequency. The microstrip design composed radial folded resonator with partial ground plane. The folded structure is used in order to achieve the compactness, and partial ground is used to suppress the high order harmonic. The proposed antenna shows mono narrowband behavior with the –10 dB bandwidth of 480 MHz (3.3–3.78 GHz), which is one of the 5G bands used for Worldwide Interoperability for Microwave Access (WiMAX) application. The obtained results through simulation depict return loss below –30 dB, VSWR below 0.6, and peak gain of 5.1 dB at operating frequency of 3.55 GHz. These results are obtained through HFSS software.

Keywords Microstrip antenna · Radial folded resonator · WiMAX

1 Introduction

The rapid development in mobile communication and emergence of newer technologies require design of antenna with smaller size, higher gain, lower power loss, high

P. K. Singh (✉) · S. K. Singh
University College of Engineering and Technology (UCET), VBU, Sindoor, Hazaribag, India

A. K. Tiwary
Department of ECE, Birla Institute of Technology, Mesra, Ranchi, Jharkhand 835215, India
e-mail: aktiwary@bitmesra.ac.in

G. Ahmad
Department of Electrical Engineering, Dayalbagh Educational Institute, Agra 282005, India

S. Mallik
Department of ECE, NIST Berhampur, Odisha 760010, India

S. S. Ali
Gargi Memorial Institute of Technology, Kolkata, West Bengal, India

J. K. Mandal et al. (eds.), *Topical Drifts in Intelligent Computing*, Lecture Notes in Networks and Systems 426, https://doi.org/10.1007/978-981-19-0745-6_32

bandwidth, and higher data speed. In order to reach tradeoffs amid these requirements, several antenna configurations are developed [1–3]. The advent of integrated devices in modern communication system popularizes microstrip patch antenna due to its properties like light weight, integration compatibility with passive and active devices, low volume, low cost, and easier fabrication. The microstrip patch antenna includes a metallic radiating patch at the top of dielectric substrate with metallic ground on the bottom of the substrate.

The antenna is vital element in wireless system used to transmit and receive the electromagnetic signals through the air media. Nowadays, its use is not limited to only human communication but is used in variety of applications like microwave imaging [4], e-health [5], satellite communication, IoT (Internet of things), radar, military, cognitive radio, energy harvesting, and so on. The individual applications need the antenna performance as per their requirement. Many researches had been already done in the designing of antenna, and many more are going on due to regular escalation in technology with improved performance and addition of newer technologies within the same volume of the equipment, as one can easily see from 1 to 5G wireless communication. In order to achieve these, a variety of antenna designs with numerous radiations patterns and performance parameters had presented [6]. Numerous patch shapes are also researched to realize miniaturized antenna structure, enhanced bandwidth, multiband operations such as Z-shape [4], Koch Fractal [7], T-shape [8], U-shape [9], spline shape [10], and so on.

There are different methods proposed in literature for wideband and narrowband performance. As specific frequency bands are allotted for different applications like for wireless local area network (WLAN), WiMAX, wireless fidelity (Wi-Fi), Global Positioning System (GPS) etc. So, these systems need multi or mono narrowband antennas, and for easier portability microstrip technology is best choice due to its light weight and easier integration capability. In past, many application-specific mono-band antennas are presented like E-shaped microstrip antenna [11] for WLAN (5–6 GHz) application, metamaterial-based antenna with bowtie stub for WLAN (2.45 GHz) [12], Coplanar waveguide-fed antenna for WiMAX (5.5 GHz) application [13], microstrip rectangular patch antenna [14] with defected ground structure comprise of array of complementary split ring slots in ground plane for WiMAX (2.6 GHz), fractal patch antenna with partial ground [15] for WiMAX (3.5 GHz) application.

This paper illustrates a narrowband microstrip patch antenna (MPA) with partial ground plane for WiMAX application at 3.55 GHz. The microstrip line fed is used with quarter wave transformer for impedance matching. The novelty of the work is in terms of patch configuration which shows compactness due to radial folded configuration and its narrowband performance. HFSS software is used for all the simulations in this work. The achieved bandwidth, VSWR, and realized gain show its suitability for WiMAX.

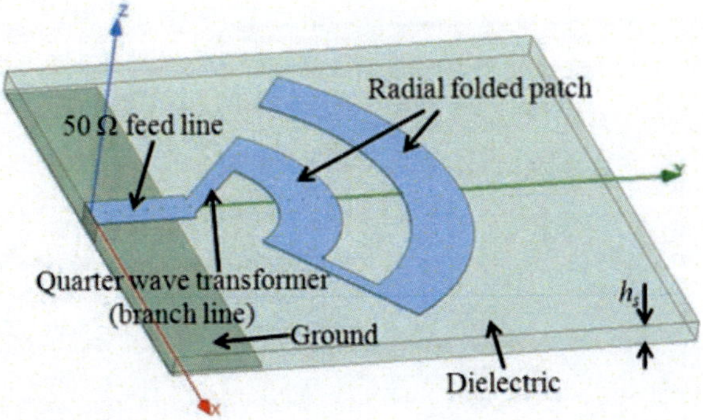

Fig. 1 Three-dimensional view of the proposed radial folded patch antenna

2 Antenna Design Methodology and Configuration

Initially, the dielectric substrate with a particular height and dielectric constant and the desired operating frequency are chosen or known in the typical antenna design procedure. The designed antenna is radial folded microstrip patch. The overall volume of the antenna geometry is $W_s \times L_s \times h_s$, i.e., $40 \times 40 \times 1.6$ mm^3; however, the overall dimension of the radiating patch is only $W_g \times L_p$, i.e., 39×29 mm^2.The proposed antenna system has been simulated on FR4 glass-reinforced epoxy laminate material with relative permittivity, dielectric loss tangent, and height of 4.4, 0.02, and 1.6 mm, respectively. For effective radiation from the antenna, impedance matching must be ensured by the designer in between the radiating patch and feed line for maximum power transfer from the input to the patch and less reflection. Here, a branch line of quarter wavelength is used for the matching purpose. The width of the branch line can be calculated easily after the evaluation of its characteristic impedance $\left(Z_{\lambda/4}\right)$, which can be determined by using (1).

$$Z_{\lambda/4} = \sqrt{Z_o Z_e} \tag{1}$$

Here, Z_o is characteristic impedance of main feed line (50 Ω) and Z_e is input impedance at patch edge. The three-dimensional view of the designed antenna is illustrated in Fig. 1. Figure 2 depicts the top and bottom view of the design with dimensional variables. The optimized dimensional parameters are specified in Table 1.

3 Results

The mono and narrowband antenna presented in this paper use partial ground plane and radiating element as radial folded patch. The simulated results demonstrate the applicability of the proposed antenna for WiMAX band at 3.55 GHz. The operating

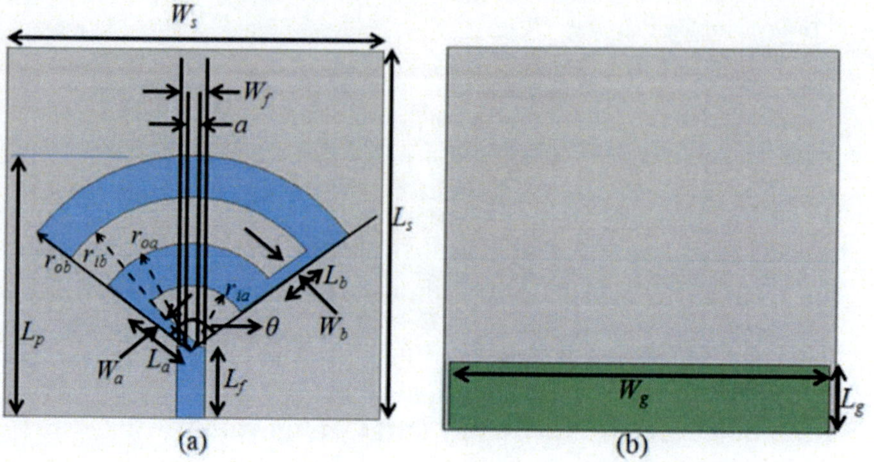

Fig. 2 Two-dimensional view **(a)** radiating patch at top and **(b)** partial ground at bottom

Table 1 Antenna design variables

Variables	Dimension
W_s	40 mm
L_s	40 mm
W_g	39 mm
L_g	7 mm
W_f	3 mm
L_f	7.6 mm
A	2 mm
W_a	1.3 mm
L_a	6.5 mm
r_{ia}	6.5 mm
r_{oa}	11 mm
W_b	1.3 mm
L_b	5 mm
r_{ib}	16 mm
r_{ob}	20.5 mm
L_p	29 mm
Θ	108°

frequency and bandwidth of the antenna are achieved by optimizing the dimensional parameters shown in Table 1. The partial ground concept is utilized in the design to excite the higher mode and enhance the return loss. The effect of the partial ground plane can be easily demonstrated by analyzing the return loss (S11) plot in Fig. 3.

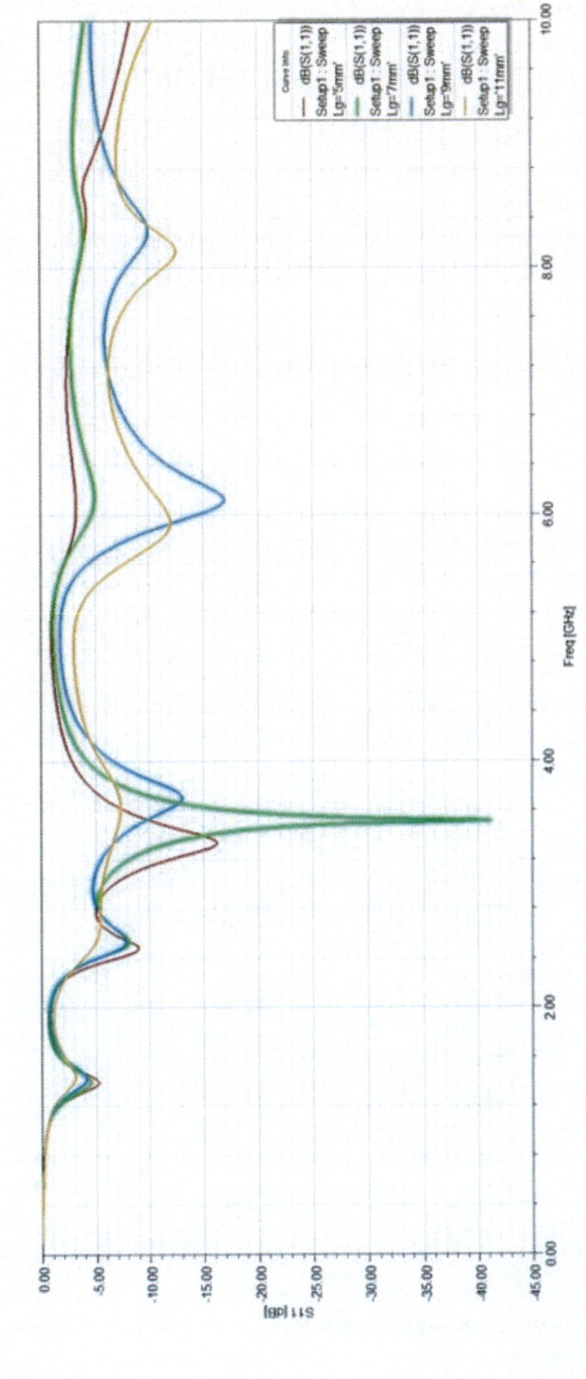

Fig. 3 Simulated return loss (S11) of designed antenna with variation in length of ground conductor (L_g)

The return loss of the final structure with $L_g = 7$ mm, is illustrated in Fig. 4. The designed antenna resonates at 3.55 GHz and shows –10 dB bandwidth of 480 MHz (3.3–3.78 GHz) i.e. 14%. The simulated S-parameter illustrates a return loss at the operating frequency well below –30 dB.

In Fig. 5 VSWR plot is shown, which demonstrates the VSWR level within 6 dB for 3.3–3.78 GHz frequency range; however, the VSWR at 3.55 GHz is below 0.6 dB. Figure 6 illustrates the polar plot of radiation pattern at the resonant frequency 3.55 GHz for 0 and 90 phi angle. The 3-D radiation plot for total gain at 3.3, 3.55 and 3.7 GHz are depicted in Fig. 7, which show peak gain of 5.6, 5.1 and 4.7 dB, respectively. This demonstrates the applicability of the antenna over the bandwidth.

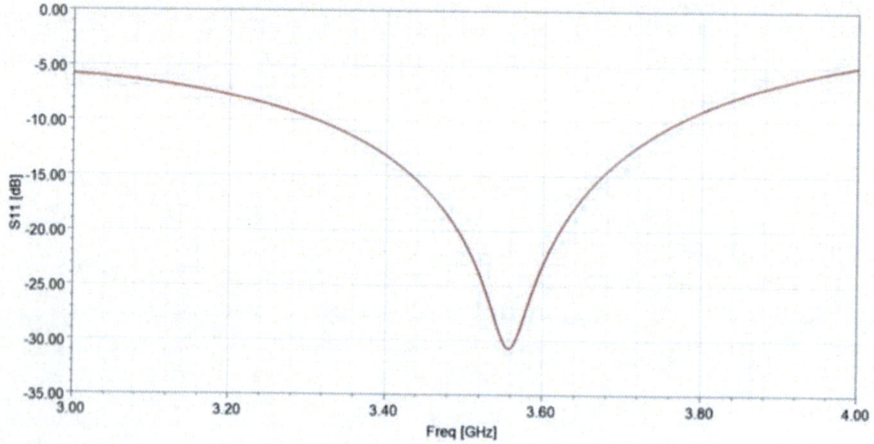

Fig. 4 Simulated S11 plot for proposed antenna configuration ($L_g = 7$ mm)

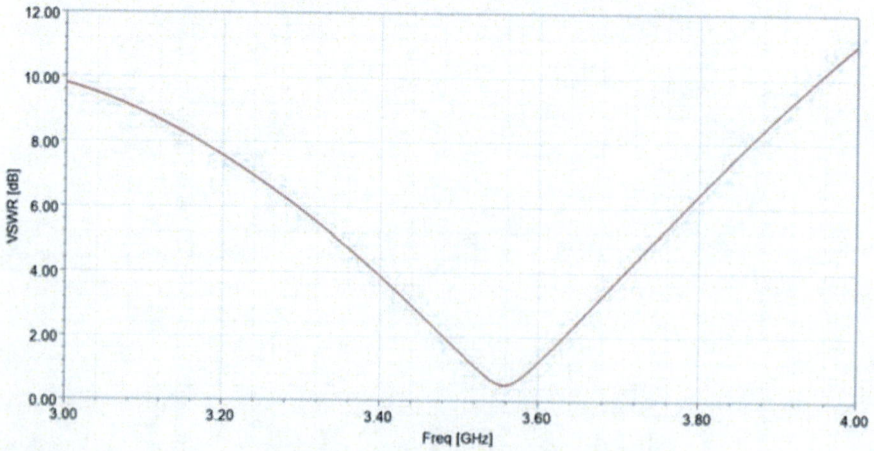

Fig. 5 The simulated VSWR plot of the proposed antenna

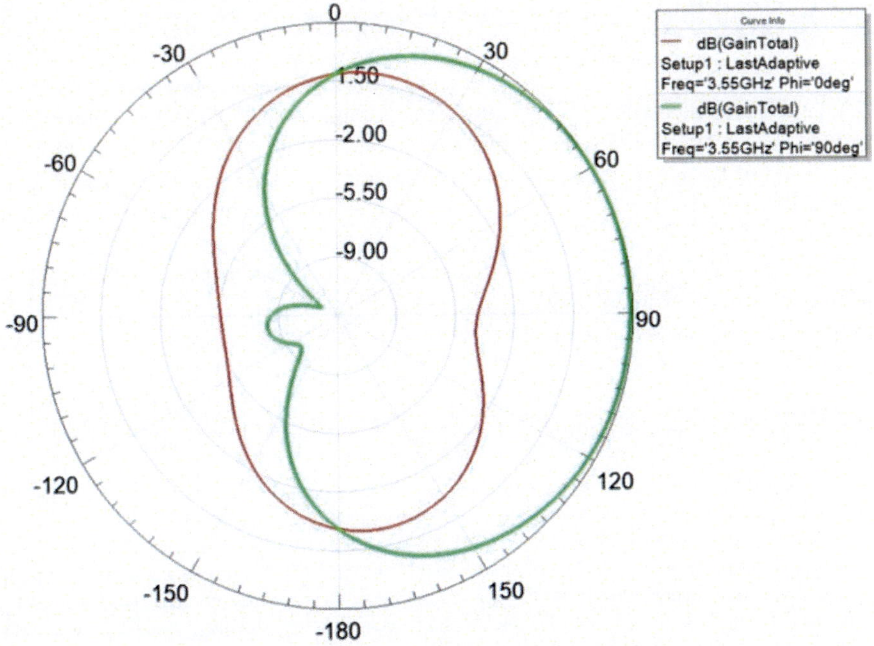

Fig. 6 Polar plot for radiation pattern at 3.55 GHz for phi = 0° and 90°

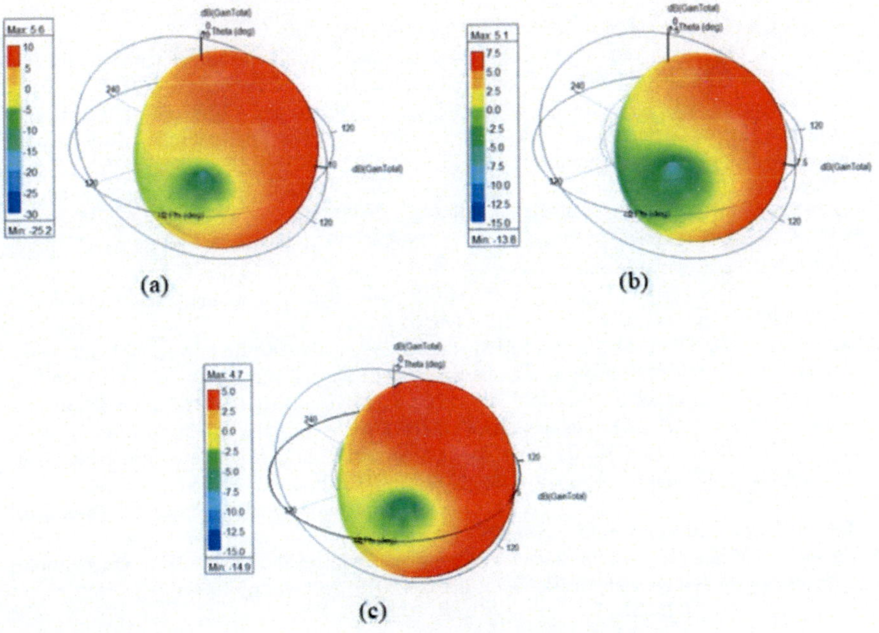

Fig. 7 3-D plot of total gain **a** at 3.3 GHz, **b** at 3.55 **GHz** and **c** at 3.7 GHz

4 Conclusion

In this work, a novel radial folded MPA has been presented. The design of radial folded MPA is done by using HFSS simulator. A radial folded MPA is simulated and it provides 3.55 GHz resonating frequency which is used for WiMAX applications. The folding of the resonator is done in order to achieve miniaturization. The Radial folded MPA gives a gain of 5.1 dB for 3.55 GHz. The VSWR at operating frequency is below 0.6 dB and the −10 dB bandwidth is 480 MHz (3.3–3.78 GHz). This proposed antenna depicts appreciable gain, bandwidth and return loss, which makes the antenna suitable for WiMAX applications.

References

1. Jacobs IS (1963) Fine particles, thin films and exchange anisotropy. Magnetism 1963:271–350
2. Maxwell JC (1873) A treatise on electricity and magnetism. Clarendon press
3. Outerelo DA, Alejos AV, Sanchez MG, Isasa MV (2015) Microstrip antenna for 5G broadband communications: Overview of design issues. In: 2015 IEEE international symposium on antennas and propagation and USNC/URSI national radio science meeting. IEEE pp 2443–2444
4. Ullah S, Ruan C, Sadiq MS, Haq TU, He W (2020) High efficient and ultra wide band monopole antenna for microwave imaging and communication applications. Sensors 20:115
5. Nafis F, Yahyaouy A, Aghoutane B (2019) Ontologies for the classification of cultural heritage data. In: 2019 international conference on wireless technologies, embedded and intelligent systems (WITS). IEEE, pp 1–7
6. Balanis CA (2016) Antenna theory: analysis and design, p 318
7. Tripathi S, Mohan A, Yadav S (2014) Hexagonal fractal ultra-wideband antenna using Koch geometry with bandwidth enhancement. IET Microwaves Antennas Propag 8:1445–1450
8. Paga P, Nagaraj HC, Rukmini TS, Nithin NE (2015) Design and fabrication of a microstrip printed T monopole antenna for ISM application. In: 2015 International conference on microwave, optical and communication engineering (ICMOCE). IEEE, pp 264–267
9. Tiwari RN, Singh P, Kanaujia BK (2020) Asymmetric U-shaped printed monopole antenna embedded with T-shaped strip for bluetooth. WLAN/WiMAX appl Wirel Netw 26:51–61
10. Lizzi L, Azaro R, Oliveri G, Massa A (2011) Printed UWB antenna operating over multiple mobile wireless standards. IEEE Antennas Wirel Propag Lett 10:1429–1432
11. Logan B, Healey J (2005) Robust detection of atrial fibrillation for a long term telemonitoring system. In: Computers in cardiology. IEEE, pp 619–622
12. Mulchandani JD, Gupta D, Sharma SK, Mishra N, Chaudhary RK (2016) Narrow-band electrically small metamaterial-inspired antenna with bowtie-shaped stub for WLAN. In: 2016 11th International conference on industrial and information systems (ICIIS). IEEE, pp 32–36
13. Sahitya V, Kumar SA, Shanmuganantham T (2019) Design of CPW fed antenna for WIMAX applications. In: 2019 TEQIP III sponsored international conference on microwave integrated circuits, photonics and wireless networks (IMICPW). IEEE, pp 95–97
14. Ajay VG, Parvathy AR, Mathew T (2019) Microstrip antenna with DGS based on CSRR array for WiMAX applications. Int J Electr Comput Eng 9:157
15. Sanap K, Labade RP (2020) Fractal patch antenna with exalted bandwidth and harmonic suppression for WiMAX applications. In: ICCCE 2019. Springer pp 143–149

Gain Enhancement of Open-Ended Waveguide with Finite Circular Ground Plane and Slots

Anil Kumar Yerrola, Suraj Sharma, Maifuz Ali, Ravi Kumar Arya, and Lakhindar Murmu

Abstract The gain enhancement of a rectangular open-ended waveguide (OEW) with a finite circular ground plane in the mmWave frequency range is presented in this paper. The circular ground plane, made of dielectric (FR-4) sandwiched by two perfect electric conductor (PEC) sheets, is placed on the open end of a standard WR-28 waveguide. Two optimized slots are etched out on the radiating front side of PEC sheets of the circular ground plane such that the slots expose the FR-4 substrate. In this way, the OEW with a slotted ground plane provides a gain of 14.93 dBi as compared to the gain of 5.74 dBi by a standard WR-28 waveguide. The full-wave simulator, Ansys HFSS, is used to carry out simulations and optimizations of the antenna structure.

Keywords 5G communications · Gain · Ground plane · mmWaves · Open-ended waveguide (OEW) · Perfect electric conductor (PEC)

1 Introduction

The increasing demand for wireless data traffic and wireless connectivity drives the shortage of spectrum in microwave frequency bands. This is urging researchers to explore the new millimeter wave (mmWave) frequencies (30–300 GHz) for future 5G communications with high-speed wireless connectivity for mobile, imaging,

A. K. Yerrola (✉) · S. Sharma · M. Ali · L. Murmu
IIIT Naya Raipur, Naya Raipur, Chhattisgarh, India
e-mail: anil@iiitnr.edu.in

S. Sharma
e-mail: suraj17107@iiitnr.edu.in

M. Ali
e-mail: maifzali@iiitnr.edu.in

L. Murmu
e-mail: lakhindar@iiitnr.edu.in

R. K. Arya
NIT Delhi, New Delhi, India

J. K. Mandal et al. (eds.), *Topical Drifts in Intelligent Computing*, Lecture Notes in Networks and Systems 426, https://doi.org/10.1007/978-981-19-0745-6_33

and multimedia applications [1]. The mmWaves have found applications in intra-vehicular and inter-vehicle applications, aerospace communication, etc. Antennas with small power consumption, ease of fabrication, and a small on-chip area overhead are very much required for future 5G communications. To fill the void left by these requirements, one of the most straightforward solutions is to use a rectangular open-ended waveguide (OEW) as an antenna. The OEW antenna has received wide attention for a long time because of its vast applications in radar and wireless communications due to its ease of design, simple feeding, high-power handling capability, low loss, and ease of fabrication. Most of the published works on OEWs are on the calculation of mutual and self-admittance, reflection coefficients, and radiation properties [2–7]. A significant breakthrough in OEWs was made by Marqués et al. in [8], which showed the transmission of electromagnetic waves through a hollow waveguide, embedded with split-ring resonators (SRRs) structure, within a defined frequency band. In [9], an OEW with a thin slow-wave surface plasmon structure is used to generate an endfire electromagnetic beam and enhances the gain from 7 to 14.8 dBi. Chandra et al. in [10] presented a circularly polarized dual-band OEW antenna, loaded with double complementary split-ring resonator (DCSRR) and superstrate and achieved a maximum gain of 8.68 dBi.

In this paper, first, we improve the gain of OEW by using a finite ground plane made of dielectric (FR-4) sandwiched by two perfect electric conductor (PEC) sheets. Next, the gain is enhanced further by etching two slots on one of the PEC sheets on the radiating side of the OEW that exposes the FR-4 substrate. In this way, the placement of the ground plane with slots enhances gain to 14.93 from 5.74 dBi.

2 Gain of Open-Ended Waveguide and the Effects of Ground Plane

High path loss is the inherent property of the EM wave at mmWave frequencies, and to circumvent this, mmWave antennas need to have high gain. In this work, we took a WR-28 OEW of length 50 mm with a larger dimension, $a = 7.112$ mm and the smaller dimension, $b = 3.556$ mm as a base design (also called design 1 here). The WR-28 OEW has a cut-off frequency of 21.0 GHz [11] and is widely used for 26–40 GHz communication systems. We have considered the WR-28 waveguide for 26 GHz in this work. The coordinate system used in this work is as shown in Fig. 1a. The gain of the simulated OEW at $\theta = 0°$ is 5.74 dBi which is very close to the calculated gain of 6.92 dBi using (1).

$$G = 21.6 f a' \tag{1}$$

where f is the frequency in GHz, and a' is the larger dimension of the rectangular waveguide in meters [2].

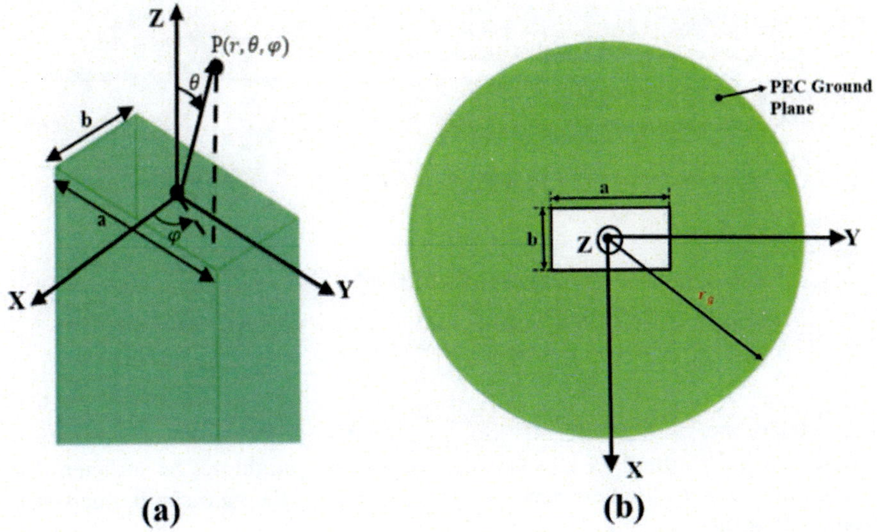

Fig. 1 **(a)** Open-ended waveguide (OEW) with the coordinate system (design 1). **(b)** Front view of OEW with the ground plane (design 2)

Table 1 Substrate used

Name	FR-4
Dielectric constant (ε_r)	4.4
Dielectric loss tangent	0.02
The thickness of the dielectric	1.2 mm

The gain close to 6 dBi is not sufficient for many of today's applications. For high gain, the aperture of the antenna should be large. So, to increase the gain of the OEW, a circular ground plane of radius, r_g, made of dielectric (FR-4) sandwiched by two perfect electric conductor (PEC) sheets, is placed on the open end of the waveguide as shown in Fig. 1b, and the details of the substrate used is shown in Table 1. The gain of this composite made of OEW and ground plane depends upon the radius of the ground plane. To determine how the radius of the ground plane affects the gain of OEW, a systematic study is conducted. We change the radius of the ground plane and track the gain values of the grounded OEW as shown in Fig. 2. The figure shows that when the radius (in terms of λ corresponding to the frequency of 26 GHz) of the ground plane changes, the gain changes in a nonlinear fashion and looks like a damped sinusoidal curve where peaks and nulls appear at regular intervals with respect to the radius of the ground plane.

In summary, if $r_g \approx n\left(\frac{\lambda}{2}\right)$, the gain shows either a peak (high gain) or a null (low gain). For peak, n is an odd integer and for null, n is an even integer. This behavior shows that to get a high gain, r_g should be an odd multiple of half wavelength of operation. For our next design step (design 2), we fix r_g to be 2.68λ (31 mm) which

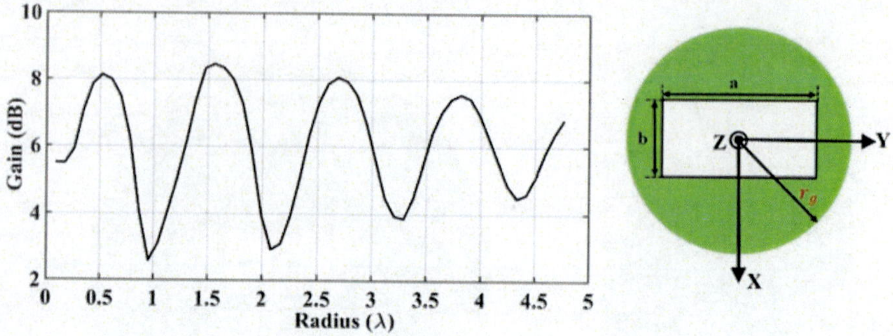

Fig. 2 The gain behavior with variable ground plane radius (r_g)

gives a maximum gain of 8.45 dBi. So, the introduction of the ground plane has increased the gain from 5.74 to 8.45 dBi, but that still might not be sufficient for some applications. In the next section, we explain how the gain can be increased beyond 8.45 dBi.

3 Gain Improvement by Slotting on the Ground Plane

Next, to increase the gain of the OEW antenna with the ground plane, slots are etched on the radiating side of the ground plane which exposes the dielectric substrate as shown in Fig. 3. In Fig. 3a, a single slot is etched (also called design 3) and is shown by blue color. The width of the slot, S_{w1}, is optimized and found to be 5 mm, i.e., 8 mm $\leq r_g \leq$ 13 mm. The introduction of a single slot increases the gain to 12.57 dBi.

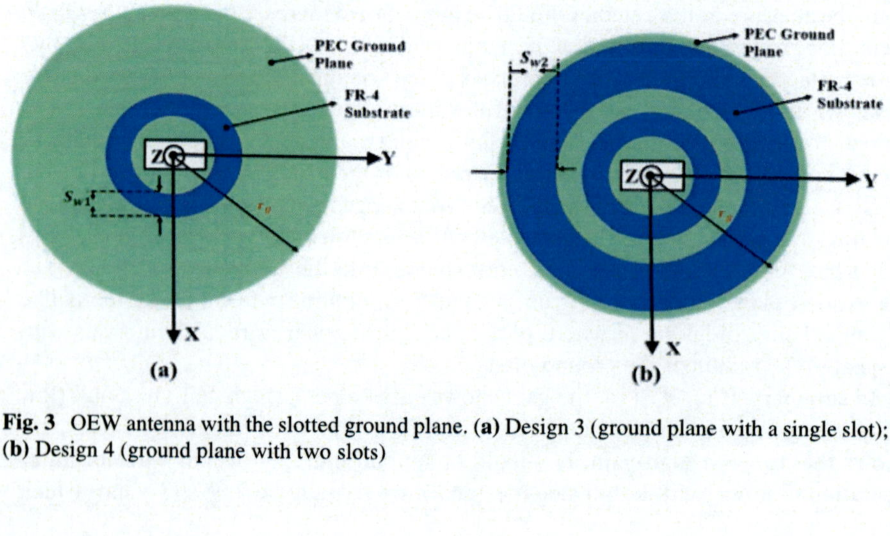

Fig. 3 OEW antenna with the slotted ground plane. (a) Design 3 (ground plane with a single slot); (b) Design 4 (ground plane with two slots)

Next, one more slot of width S_{w2} is added (also called design 4), keeping the first slot as it is, at an optimized position (18 mm $\leq r_g \leq$ 28 mm) as shown in Fig. 3b. The optimized value of S_{w2} is found to be 10 mm, which increases the gain of the antenna to 14.93 dBi.

The simulated gain patterns at E-plane ($\varphi = 0°$) and H-plane ($\varphi = 90°$) for a frequency of 26 GHz of all the designs are shown in Figs. 4 and 5, respectively.

The peak gains from all the designs are also compared and shown as histograms in Fig. 6. Overall, the introduction of the slotted ground plane has enhanced the gain of the OEW antenna by 9.19 dBi. The slots of the different sizes help to increase the gain of the OEW antenna. The slots re-radiate the radiations in such a way that the gain of the antenna is enhanced by constructive interference by these radiations. For completeness, Fig. 7 shows the simulated return loss for the different designs. It is evident from Fig. 7 that all the designs have good return loss in the desired frequency band (24–28 GHz).

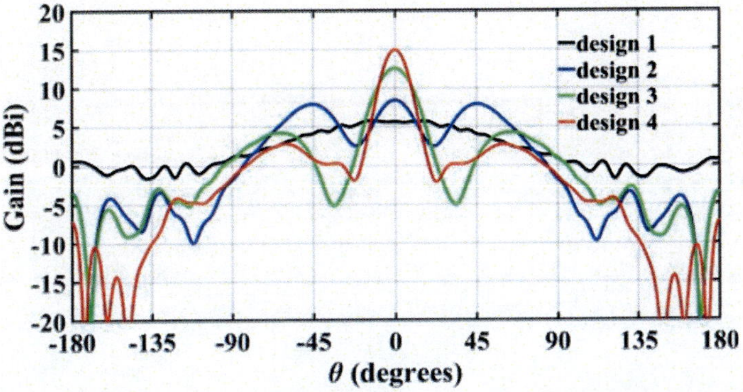

Fig. 4 E-plane gain pattern of OEW (design 1), OEW with the ground plane (design 2), OEW with ground plane and a single slot (design 3), and OEW with a ground plane and two slots (design 4)

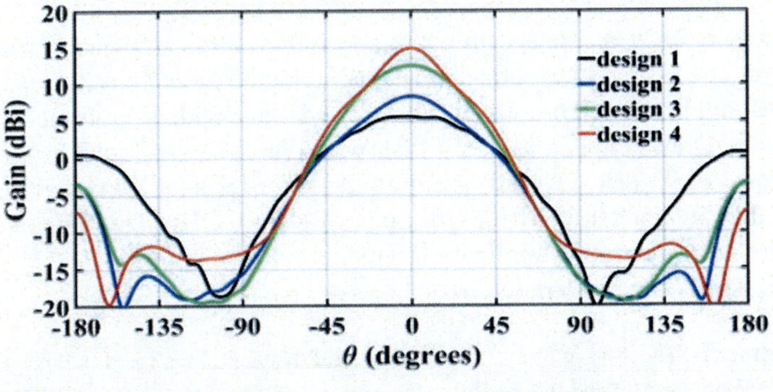

Fig. 5 H-plane gain pattern of OEW (design 1), OEW with the ground plane (design 2), OEW with ground plane and a single slot (design 3), and OEW with a ground plane and two slots (design 4)

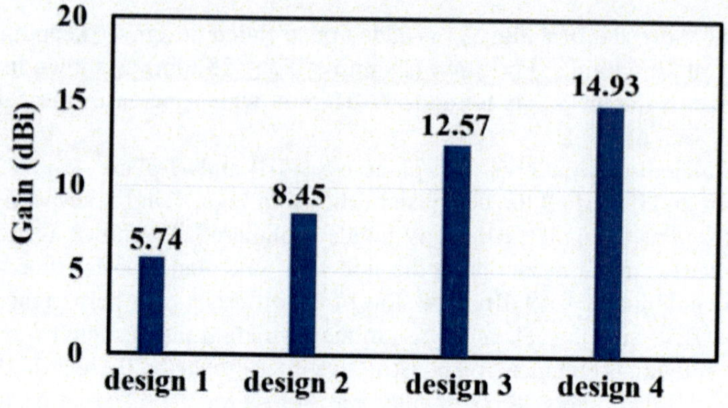

Fig. 6 Gain enhancement with the introduction of different configurations of the ground plane

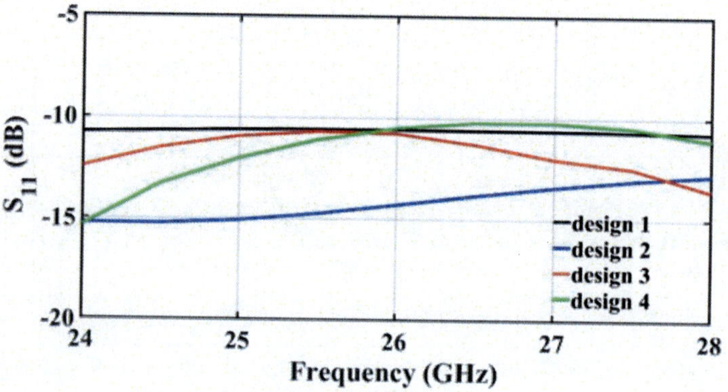

Fig. 7 S-parameters of all the designs

4 Conclusion

In this work, the OEW antenna with a slotted ground plane has been introduced to enhance gain. The systematic approach of introducing the slotted ground plane and its effect on the gain have been presented. The ground plane with a single slot increased the gain of the OEW antenna to 12.57 dBi, while the double slotted ground plane increased it to 14.93 dBi. Overall, the double-slotted ground plane has increased the gain of the OEW antenna by 9.2 dBi operating at 26 GHz. This antenna will be useful for the next-generation applications antennas where high gain and high power handling capability are needed.

Acknowledgements The first author is thankful to IIIT Naya Raipur for providing necessary financial support for carrying out this work.

References

1. Elkashlan M, Duong TQ, Chen H-H (2014) Millimeter-wave communications for 5G: fundamentals: part I [Guest Editorial]. IEEE Commun Mag 52(9):52–54
2. Bhattacharya A, Gupta S, Chakraborty A (1999) Analysis of rectangular waveguides and thick windows as EMI sensors—abstract. J Electromagn Waves Appl 13(7):981–982. https://doi.org/10.1163/156939399X00448
3. Yaghjian A (1984) Approximate formulas for the far field and gain of open-ended rectangular waveguide. IEEE Trans Antennas Propag 32(4):378–384
4. Stewart J, Havrilla M (2006) Electromagnetic characterization of a magnetic material using an open-ended waveguide probe and a rigorous full-wave multimode model. J Electromagn Waves Appl 20:2037–2052
5. Baudrand H, Tao J, Atechian J (1988) Study of radiating properties of open-ended rectangular waveguides. IEEE Trans Antennas Propag 36(8):1071–1077
6. Fred E, Gardiol (1985) Open-ended waveguides: principles and applications. Adv Electron Electron Phys 63:139–187
7. Das S, Chakraborty A (2006) A novel modeling technique to solve a class of rectangular waveguide based circuits and radiators. Prog Electromagn Res 61:231–252
8. Marqués R, Martel J, Mesa F, Medina F (2002) Left-handed-media simulation and transmission of EM waves in subwavelength split ring resonator-loaded metallic waveguides. Phys Rev Lett 89(18):183901
9. Kandwal A, Nie Z, Li J, Liu Y, Liu WY, Liu L, Das R (2019) Bandwidth and gain enhancement of endfire radiating open-ended waveguide using thin surface plasmon structure. Electron 8(5):504
10. Chandra A, Das S (2017) Superstrate and CSRR loaded circularly polarized dual-band open-ended waveguide antenna with improved radiation characteristics and polarization reconfiguration property. IEEE Trans Antennas Propag 65(10):5559–5564
11. Balanis, Constantine A (2015) Antenna theory: analysis and design. Wiley

Illumination Insensitive Video Cut Detection Using Phase Congruency

T. Kar, P. Kanungo, and Vinod Jha

Abstract Shot boundary detection is the first and the most crucial step towards video content management applications including indexing, retrieval and summerisation. In this paper, an abrupt transition detection algorithm has been proposed based on phase congruency feature of the frames. The phase congruency feature is insensitive to illumination variation, change in contrast and scale. Besides this, it captures edges, corners and structural information of the frames. Motivated by this, a PC-based similarity measure is proposed for illumination insensitive video cut detection. The proposed approach is experimentally validated with standard algorithms available in the literature using TRECVid data set and other publicly available videos. The favourable results are in agreement with the proposed model.

Keywords Cut detection · Phase congruency · Illumination insensitive feature

1 Introduction

Development of the multimedia technology has made available plenty of high performance and easy to operate video capturing devices at affordable cost to the common public. As a result, everyday a huge volume of video data is created. Statistically, it is observed that videos from different sources are uploaded, downloaded and viewed in an unimaginable rate. This led to an escalation of the digital video information in the cyber space. This huge volume of video data is also easily available to and accessible by the common public. So, there is a substantial need for efficient management of the video information starting from video indexing, retrieval and classification to summerisation [1]. Hierarchical levels of video structure from bottom to top are frames,

T. Kar (✉) · V. Jha
KIIT Deemed to be University, Bhubaneswar, Odisha, India
e-mail: tkarfet@kiit.ac.in

V. Jha
e-mail: vjhafet@kiit.ac.in

P. Kanungo
C. V. Raman Global University, Bhubaneswar, Odisha, India

J. K. Mandal et al. (eds.), *Topical Drifts in Intelligent Computing*, Lecture Notes in Networks and Systems 426, https://doi.org/10.1007/978-981-19-0745-6_34

shots, scenes and stories. Combination of shots forms a scene, combination of scene forms a story and so forth combined to produce a video. Scene change detection is the first and the most crucial step towards this goal. It divides the video data into semantically related frames sharing the same content. The boundary between two consecutive shots is generally a hard-cut or a gradual transition (GT). Hard-cut also called cut transitions is observed to be the most predominant type of transition in videos than gradual transitions [2].

2 Review of Literature

Many algorithms have been developed in literature for automatic detection of type and location of transitions. However, sudden illumination change such as flashlight, stage effects and high-speed object/camera motion in videos can be misunderstood as shot change leading to false detection in the shot boundary. Sudden change in illumination and high motion scenes are common in fantasy and thriller movies, news reports or sports videos, that are intentionally included in the video to make them attractive for the audience. So it is a crucial step to eliminate the influence of these effects on shot change detection. To ensure this, a suitable feature needs to be extracted followed by formulation of the similarity measure and shot boundary detection(SBD). Many contributions on video SBD can be found in the literature [3–5]. Several features deployed for SBD are pixel intensity [6], histogram [7], edge, SURF and SIFT features [8]. Intensity-based features are found to have higher sensitivity towards large motion and light variation. Histogram feature-based algorithms are comparatively better than intensity feature-based approach, in terms of handling motion and illumination variation. Besides these, various edge and gradient-based features can be found in literature. Edge feature-based approaches are insensitive to small variation in light; however, for videos with large illumination, variations tend to destroy the edge features leading to false transitions. In order to improve efficiency, some researchers combined multiple features [9–11] for transition detection. Many authors proposed shot boundary techniques based on new feature space. LBPHF [12], LBP [13], CSLBP [14] and LDP [15] deployed for shot boundary detection are efficient in handling videos with sudden change in lighting condition. As an improvement in LBP-based method, Chakraborty et al. [16] proposed LTP-based approach in laboratory colour space under high object camera motion for SBD but has lower sensitivity to noise and illumination. In the current work, the advantage of illumination insensitivity feature of phase congruency (PC) is explored to develop a new feature similarity(FS) measure, i.e. PCFS. The proposed PCFS outperforms histogram and LBP-based feature similarity measure. This paper is arranged as follows: Sect. 2 describes the related work. Section 3 introduces image feature extraction using phase congruency, its significance in applying for current problem and the algorithm for development of similarity measure for the transition detection. Section 4 presents the simulation result discussions to support for the effectiveness of the extracted feature and algorithm followed by conclusions in Sect. 5.

3 Phase Congruency

Phase congruency has been developed from the phase information of the signal obtained from the frequency domain representation. It is found from the literature that phase is a crucial parameter for the perception of visual features [17]. Further, as per the evidence given by Morrone [18] indicated that the human visual system strongly responds to the image locations. It perceives features with highly ordered phase information. Thus, a human has a tendency to sketch an image, by precisely through the edges and interest points as perceived in a scene. This highlights the points having highest order in phase in a frequency-based image representation. The locations in the image where the Fourier components have highest phase correlation emphasize the visually differentiable image features such as edges, lines and mach bands. This indicates that more informative features can be captured at points of high value of PC. Thus, the PC model defines features as points in an image with high phase order and uniquely defines the image luminance function. PC value lies between 0 and 1. It is a dimensionless quantity and identified as invariant to scale, illumination and contrast of an image [19, 20]. It allows a wide range of feature types to be detected within a single framework. The PC captures edge, corner, structure or contour information of objects. Gradient-based edge detection operators are sensitive to illumination variations and do not have accurate and consistent localization, which is overcome by PC-based feature. Moreover, PC mimics the response of the human visual perception to contours. Yu et al. [21] showed that it is well capable of distinguishing structural information content of the scene. The points where the Fourier waves at different frequencies have congruent phases capture the visually differentiable features. That is, at points of high phase congruency (PC), highly informative features can be extracted. PC can be defined by the frequency response of the log Gabor filter which is given by the following transfer function

$$LG(w, \theta) = e^{\frac{-(log(\frac{w}{w_0}))^2}{2(log(\frac{k}{w_0}))^2}} e^{\frac{-(\theta-\theta_0)}{2\sigma_\theta^2}} \quad (1)$$

where ω_0: centre frequency of the filter and θ_0: orientation component of the filter.

For an image $f(x, y)$ with M_{so}^{ev} and M_{so}^{od} as the even symmetric and odd symmetric components of the log Gabor filter at scale **s** and orientation **o**, the responses of the two quadrature pair filters are given by $ev_{so}(x, y)$ and $od_{so}(x, y)$, respectively, in (2)

$$[ev_{so}(x, y), od_{so}(x, y)] = [f(x, y) * M_{so}^{ev}, f(x, y) * M_{so}^{od}] \quad (2)$$

The amplitude at scale **s** and orientation **o** is given by (3)

$$A_{so} = \sqrt{ev_{so}^2(x, y) + od_{so}^2(x, y)} \quad (3)$$

Hence, the phase congruency representation of an image $f(x, y)$ in the simplest form without considering weight and noise component is given by (4)

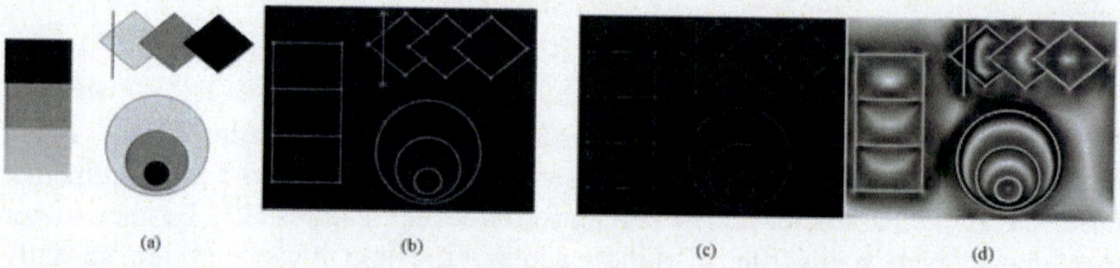

Fig. 1 **a** Hypothetical image having corners and edges, **b** corresponding edge strength image, **c** corresponding corner strength image, **d** corresponding PC map of the image

$$PC(x,y) = \frac{\sum_o \sqrt{(\sum_s ev_{so}(x,y))^2 + (\sum_s od_{so}(x,y))^2}}{\epsilon + \sum_o \sum_s A_{so}(x,y)} \tag{4}$$

ϵ is a small constant for avoiding zero in the denominator of (4). The phase congruency (PC) representation is a frequency-based modelling of visual information. It supposes that, instead of processing visual data spatially, the visual system can do similar processing via phase and amplitude of the individual frequency components in a signal. In PC evaluation, frequency domain processing is achieved through the Fourier transform. Kovesi [22] showed that corners and edges are well detected using PC. The problem of discriminating between abrupt shot transition in videos has not been addressed earlier by using phase congruency features. Kovesi formulated PC via a log Gabor filter function. In contrast to the Gabor function, it maintains zero DC for arbitrarily large bandwidth. Moreover, log Gabor is characterized by extended tail at higher-frequency region preserving the high-frequency details in the image [23]. To show the features captured by PC, we have considered a hypothetical image having prominent edges and corners given in Fig. 1a. The corresponding edge strength, corner strength and the complete PC map are illustrated in Fig. 1b, c and d, respectively. Edge and corners are well captured through PC. Again to illustrate the illumination insensitivity characteristic of the PC feature map, we have considered two consecutive frames, i.e. 165th and 166th frame of the video "Littlemiss sunshine", as shown in Fig. 2a, out of which 166th frame is exposed to the flashlight. The original image and the corresponding PC feature frames of the two consecutive frames are given in Fig. 2a and b, respectively. It is clearly visible from the figure that the PC feature frames of the frame numbers 165 and 166 are very similar and not much affected by illumination variation through flashlight effect and hence can be suitable to develop an illumination insensitive similarity for abrupt transition detection. Due to the mentioned characteristics of the PC feature, we are motivated to use it for shot boundary detection problem. Towards this goal, this paper introduces the use of PC-based feature extraction for representing frame content of the video and for illumination insensitive shot boundary detection.

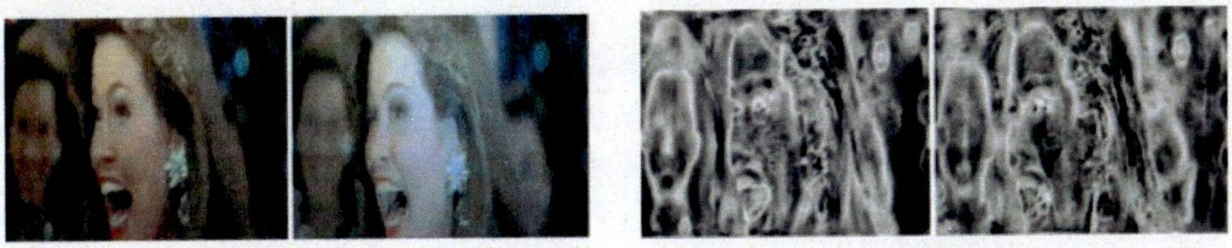

(a) (b)

Fig. 2 **a** 165th and 166th frame of video "LM" , **b** PC feature image of corresponding frames

4 Proposed PC-Based CUT Detection

In this section, the proposed PC-based abrupt transition detection algorithm has been explained. The PC feature-based similarity (PCFS) between consecutive PC feature frames is given by (5).

$$PCFS(t,t+1) = \sum_{k=1}^{\text{Row}} \sum_{l=1}^{Col} |PC_t(k,l) - PC_{t+1}(k,l)| \quad (5)$$

$PCFS(t,t+1)$: is the PC-based feature similarity measure between tth and $(t+1)$th frame. PC_t and PC_{t+1} are the PC feature frames of tth and $(t+1)$th frame, respectively. Row and Col are the number of rows and columns in the image. For AT detection, PC-based similarity is compared with Th as given by Zhang et al. [7].

$$\text{Th} = \mu_s + \beta \times \sigma_s \quad (6)$$

where β is a constant and its value lies between 4 and 8 and μ_s and σ_s are the average and standard deviation of the PC feature-based similarity value.

5 Simulations and Result Discussions

For validation of the proposed PC-based similarity measure, we considered ten videos of different genre such as English movies, Sitcom video, Soccer video, Cartoon video and Documentary videos, consisting of 78,447 frames in total and 437 cuts collected from [24] and Internet. The detailed information of the test videos is given in Table 1. The proposed PCFS is compared with the histogram-based similarity approach (ASHD) [7] and LBP-based similarity approach [14]. Recall (Rec), precision (Pr) and $F1$-measure are used to validate different SBD methods. It is clearly observed from Table 2 that the performance of the proposed PCFS is the highest in terms of average Rec, Pr and $F1$-measures.

Table 1 Ground truth transition details of different genres of test videos

Video ID	Video name	Total frames	Total cut	Source
LM	Little miss sunshine	4665	34	CD
2*B*	2 Brothers	4798	38	CD
BS	Before sunrise	6146	25	CD
TBT	The Big Bang theory	14,261	153	Sitcom
Car	Cartoon	3000	39	Internet
Soc	Soccer	4159	14	UEFA
*a*4	anni 004	3895	13	TRECVid 2001
*a*5	anni 005	11,363	39	TRECVid 2001
*N*57	NAD 57	12,510	42	TRECVid 2001
*N*58	NAD 58	13,650	40	TRECVid 2001
	Total	78,447	437	

Table 2 Performance comparison of histogram, LBP and proposed PCFS methods

Video ID	Histogram [17]			LBP [13]			Proposed PCFS		
	Rec	*Pr*	*F*1	*Rec*	*Pr*	*F*1	*Rec*	*Pr*	*F*1
LM	88.24	90.91	89.55	91.18	91.18	91.18	88.24	88.24	88.24
2*B*	92.11	89.74	90.91	94.74	81.82	87.80	92.11	87.50	89.74
BS	100	100	100	100	100	100	100	100	100
TBT	90.85	95.21	92.98	94.77	92.95	93.85	98.69	94.37	96.49
Soc	92.86	92.86	92.86	92.86	92.86	92.86	92.86	100	96.3
Cart	100	90.70	95.12	94.87	94.87	94.87	87.18	87.18	87.18
a_4	84.62	84.62	84.62	100	100	100	100	100	100
a_5	43.59	42.50	43.04	92.31	85.71	88.89	100	95.12	97.50
*N*57	71.43	71.43	71.43	80.95	75.56	78.16	97.62	85.42	91.11
*N*58	77.50	72.09	74.70	62.5	62.5	62.5	92.50	82.22	87.06
Avg	84.12	83	83.6	90.41	87.74	89.01	94.92	92	93.36

6 Conclusions

In this article, an illumination insensitive phase congruency feature-based abrupt transition detection algorithm has been proposed. Besides illumination insensitivity, this feature is robust against contrast and scale changes as well. The performance of the proposed model is validated on publicly available video and the standard benchmark TRECVid data set. The limitations of using PC are the computational complexity in the evaluation process, setting of too many parameters to suit the application and sensitivity to image noise. Techniques of noise reduction prior to the PC evaluation may further improve the result. In future, the PC-based feature can be integrated with other features for efficient detection of gradual transitions.

References

1. Ejaz N, Mehmood I, Baik SW (2014) Feature aggregation based visual attention model for video summarization. Comput Electrical Eng 40(3):993–1005 April
2. Smeaton AF, Over P, Doherty AR (2010) Video shot boundary detection: seven years of TRECVid activity. Comput. Vis. Image Understand. 114(4):411–418 April
3. Abdulhussain SH, Ramli AR, Saripan MI, Mahmmod BM, Al-Haddad SAR, Jassim WA (2018) Methods and challenges in shot boundary detection: a review. Entropy 20(4)
4. Pal G, Rudrapaul D, Acharjee S, Ray R, Chakraborty S, Dey N (2015) Video shot boundary detection: a review. In: Emerging ICT for bridging the future, proceedings of the 49th annual convention of the computer society of India (CSI), vol 338, pp 119–127
5. SenGupta A, Singh KM, Thounaojam DM, Roy S (2015) Video shot boundary detection: a review. In: IEEE international conference on electrical, computer and communication technologies (ICECCT)
6. Lu Z, Shi Y (2013) Fast video shot boundary detection based on svd and pattern matching. IEEE Trans Image Process 22(12):5136–5145 December
7. Zhang HJ, Kankanhalli A, Smoliar SW (1993) Automatic partitioning of full motion video. Multimed Syst 1(1):10–28 Jan
8. Birinci M, Kiranyaz S (2014) A perceptual scheme for fully automatic video shot boundary detection. Signal Process Image Commun 29(3):410–423 March
9. Lakshmi Priya GG, Domnic S (2014) Walsh-Hadamard transform kernel-based feature vector for shot boundary detection. IEEE Trans Image Process 23(12):5187–5197
10. Chakraborty S, Thounaojam DM (2021) Sbd-duo: a dual stage shot boundary detection technique robust to motion and illumination effect. Multimed Tools Appl 80:3071–3087
11. Duan FF, Meng F (2020) Video shot boundary detection based on feature fusion and clustering technique. IEEE ACCESS 8:214633–214645
12. Singh A, Thounaojam DM, Chakraborty S (2019) A novel automatic shot boundary detection algorithm: robust to illumination and motion effect. Signal Image Video Process
13. Kar T, Kanungo P A texture based method for scene change detection. In: 2015 IEEE power, communication and information technology conference (PCITC), pp 72–77, 15–17 October 2015
14. Kar T, Kanungo P Cut detection using block based centre symmetric local binary pattern. In:2015 international conference on man and machine interfacing (MAMI), pp 1–5, 17–19 December 2015
15. Kar T, Kanungo P Abrupt scene change detection using block based local directional pattern. In: Data management, analytics and innovation proceedings of ICDMAI, vol 2, pp 191–203, January 18–20 2019
16. Chakraborty S, Thounaojam DM, Singh A (2021) A novel bifold-stage shot boundary detection algorithm: invariant to motion and illumination. Visual Comput
17. Xiao Z, Hou Z (2004) Phase based feature detector consistent with human visual system characteristics. Pattern Recogn Lett 25(10):1115–1121 July
18. Morrone MC, Burr DC (1988) Feature detection in human vision: a phase dependent energy model. In: Proceedings of the royal society of London, biological sciences, vol 235 of *B*. The Royal society, pp 221–245
19. Kovesi P (1999) In: Sun T, Ourseli, Adriaansen (eds) Image features from phase congruency. Videre J Comput Vis Res 1:1–26

20. Yu G, Zhao S (2020) A new feature descriptor for multimodal image registration using phase congruency. Sensors
21. Yu J, Sato Y (2015) Structure-preserving image smoothing via phase congruency-aware weighted least square. In: Stam J, Mitra NJ, Xu K (eds) Proceedings Pacific graphics short papers. The Eurographics Association
22. Kovesi P (2003) Phase congruency detects corners and edges. In: Sun T, Ourselin, Adriaansen (eds) Digital image computing: techniques and applications. Sydney, Australia edn, vol 1. CSIRO Publishing, Victoria, pp 309–318
23. Morrone MC, Owens RA (1987) Feature detection from local energy. Pattern Recognit Lett 6(5):303–313 Dec.
24. The open video project. [online] Available at: http://www.open-video.org

A Comparative Study on Label-Free Detection of Biomolecules Using Various Biosensing Techniques

Tulip Kumar Saha, Moumita Mukherjee, and Rudra Sankar Dhar

Abstract This paper presents the results of a comparative analysis of label-free biomolecule detection using different biosensing techniques with conventional SiO_2 and high-K dielectric metal-oxide-semiconductor high-electron-mobility transistor (MOSHEMT)-based biosensors. A MOSHEMT-based biosensor with high-K dielectric material to improve its detection sensitivity and selectivity. The use of high-K material decreases the substantial amount of leakage current due to the effect of quantum tunneling and improve two-dimensional electron gas (2DEG) carrier confinement. Hence, power consumption of the device is reduced and increased the gate capacitance without leakage effects. The numerical modeling is carried out using TCAD Silvaco Atlas. Different performance parameters of high-K MOSHEMT-based biosensors are studied using the simulation and compare with SiO_2 MOSHEMT for the identification of biomolecules without introducing the label. AlGaN/GaN MOSHEMTs with high-K dielectric are excellent candidates for making biosensors.

Keywords Biosensors · High-K dielectric · 2DEG · Nano-cavity · Sensitivity · Label-free detection

T. K. Saha (✉) · R. S. Dhar
Department of Electronics and Communication Engineering, National Institute of Technology Mizoram, Aizawl 796012, India

R. S. Dhar
e-mail: rdhar@uwaterloo.ca

T. K. Saha
Department of Electronics and Communication Engineering, School of Engineering and Technology, Adamas University, Kolkata, West Bengal 700126, India

M. Mukherjee
Department of Physics, School of Basic and Applied Sciences, Adamas University, Kolkata, West Bengal 700126, India

J. K. Mandal et al. (eds.), *Topical Drifts in Intelligent Computing*, Lecture Notes in Networks and Systems 426, https://doi.org/10.1007/978-981-19-0745-6_35

1 Introduction

In COVID-19 pandemic, requirement of biosensors is now put forward to detect viruses, nucleic acids, and proteins. Comprehensive study has been carried out by researchers in worldwide on FET-based biosensors for rapid, accurate, and reliable detection of various biomolecules. FET-based biosensors have been designed for diverse applications such as medical diagnosis and study of biomolecule interaction, food analysis, crime detection, and drug development [1]. Because of the high-electron-mobility of 2DEG at heterointerface, AlGaN/GaN MOSHEMTs have smaller on-resistance and larger switching speed than SiC. AlGaN/GaN MOSHEMT-based biosensor exhibits biosensing application due to the presence of 2DEG at heterointerface [2]. Different biomolecules are easily attached to the nano-cavity which is placed above to AlGaN barrier layer and make significant change in channel properties as a result of this surface charges at AlGaN/GaN vary.

K.H. Chen et al. proposed a change in 2DEG concentration in AlGaN/GaN heterointerface HEMTs are employed in the detection of c-erbB-2 antigen for a breast cancer marker [3]. Choi et al. demonstrated short-channel effects (SCEs) are suppressed, and mobility deterioration is minimized by employing a high-K gate dielectric in a dielectric-modulated field-effect transistor (DMFET)-based biosensor [4]. Kim et al. have invested a MOSFET-based charge sensor for detection of the deoxyribonucleic acid (DNA) [4–6]. Lee et al. proposed nanogap FET to detect an extremely virulent Avian Influenza (AI) virus of influenza type-A [6, 7]. Maesoon et al. reported nanogap-embedded separated double-gate FETs (nanogap-DGFETs) for detection of Avian Influenza (AI) virus [7, 8]. Recently, Seo et al. designed a graphene-based field-effect transistor (FET) biosensors are used to detect SARS-CoV-2 virus [8, 9].

To detect immunoglobulin E (IgE), Maehashi et al. developed label-free protein biosensors based on aptamer-modified carbon nanotube field-effect transistors (CNT-FETs) [10, 11]. Therefore, in this work, a comparative study of biosensors based on AlGaN/GaN MOSHEMT has been used to detect biomolecules using label-free detection techniques. Section 2 describes the in-silico device design. In Sect. 3, we present simulated results including drain and transfer characteristics as well as sensitivity curves for the detection of different biomolecules and finally, our work has been concluded in Sect. 4.

2 In-Silico Device Design

Figure 1a shows a schematic cross section of an AlGaN/GaN MOSHEMT-based biosensors in which SiO_2 is used as conventional dielectric material and the authors have considered HfO_2 as a high-K dielectric material for miniaturization of device. The structure comprises on top of a 2 μm GaN buffer layer, there is a 25 nm $Al_{0.3}Ga_{0.7}N$ barrier layer, the entire structure fabricated on Sapphire substrate.

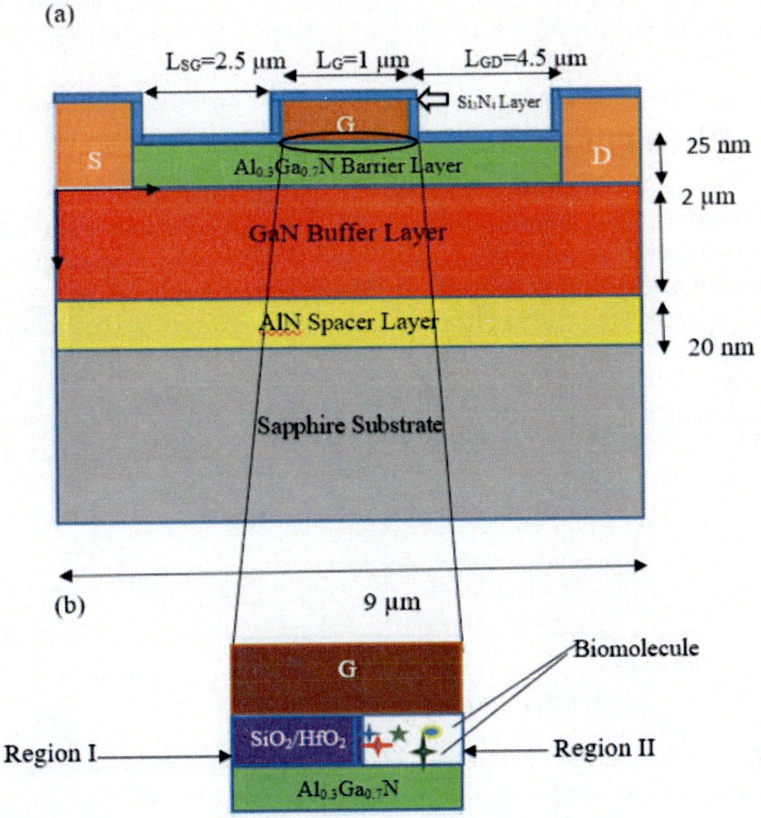

Fig. 1 **a** MOSHEMT-based AlGaN/GaN SiO_2/HfO_2 biosensor structure, **b** enlarge view of nano-cavity with biomolecules

Schottky gate contact is made of Ni/Au metal. Aluminum is used for source/drain ohmic contact. The source to gate length (L_{SG}) is 2.5 μm, gate length (L_G) is 1 μm, and gate to drain length (L_{GD}) is 4.5 μm. The entire region below the gate electrode was first filled with SiO_2 and HfO_2, respectively, and then, a 500 nm long (L_{Cavity}) and 20 nm width (W_{Cavity}) nano-cavity is formed in region-II which is depicted in Fig. 1b. This nano-cavity is formed on the top of AlGaN barrier layer where detection biomolecules is performed. To optimize device performance, the gate is placed close to the source [9, 10, 12].

All the simulations are carried out using Silvaco Atlas and models used are polarization model, cal.strain, albrct.n, Shockley-Read-Hall (SRH) model, field dependent drift velocity (FLDMOB), concentration dependent mobility (CONMOB). The interface charge at AlGaN/GaN is considered as $8.8 \times 1012\ cm^{-2}$.

The lattice mismatch increases with increasing mole fraction so for optimize mole fraction, $x = 0.3$ is considered and the sheet charge density increases as it is dependent on the polarization charge in the channel, which has the greatest impact on sheet charge density.

3 Results and Discussion

The different I_D–V_D curves depicted in Fig. 2a–d are generated using different biomolecules in nano-cavity for $L_{Cavity} = 500$ nm and $W_{Cavity} = 20$ nm with $V_G = -4$ V using HfO2 MOSHEMT and SiO_2 MOSHEMT, respectively. So due to presence of high-K material more 2DEG carrier confinements in AlGaN/GaN heterointerface, we get more drain current using HfO_2 MOSHEMT than SiO_2 MOSHEMT.

The I_D–V_G curves depicted in Fig. 3a–d are generated by introducing various biomolecules into a nano-cavity with $L_{Cavity} = 500$ nm and $W_{Cavity} = 20$ nm with $V_D = 5$ V. The ON current, I_{ON}, becomes higher with lowing the dielectric constant of biomolecules with good sensitivity. Figure 3a–d depicts the variation in drain current for ChO_x, Protein, Streptavidin and Uricase, respectively. The ChO_x exhibits

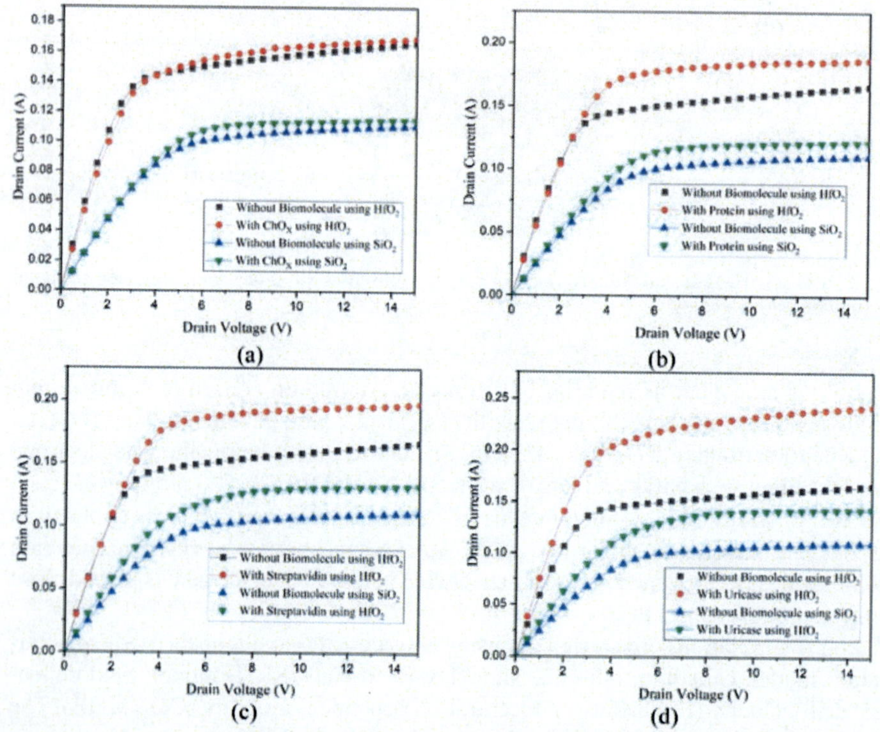

Fig. 2 **a–d** I_D–V_D curves for without cavity and different biomolecules with $V_G = -4$ V using high-K MOSHEMT and SiO_2MOSHEMT [1]

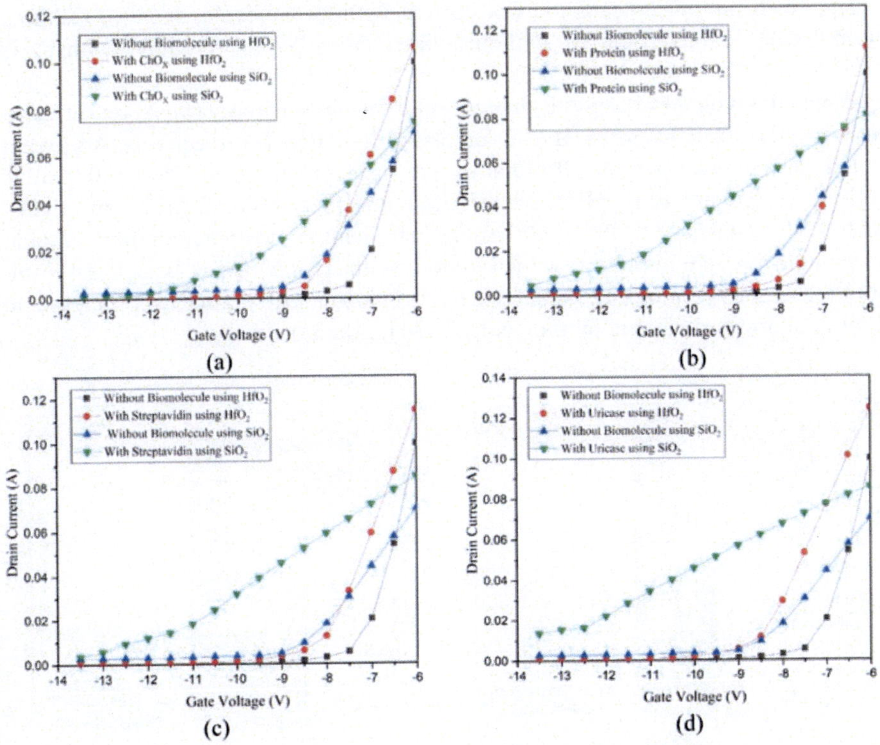

Fig. 3 **a–d** I_D–V_G curves for without cavity and different biomolecules with $V_D = 5$ V using high-K MOSHEMT and SiO_2 MOSHEMT [1]

a lower drain current variation of 6.4 and 4.2 mA at a gate voltage of −6 V for HfO_2 MOSHEMT and SiO_2 MOSHEMT, respectively; while biomolecule with smaller dielectric constant, Uricase provides a greater variation in drain current of 24.4 and 15.4 mA at a gate voltage of −6 V for HfO_2 MOSHEMT and SiO_2 MOSHEMT, respectively.

The different K-value of these biomolecules is shown in Table 1.

$$S_{ION} = \frac{I_{ON}(\text{With Biomolecule}) - I_{ON}(\text{Without Biomolecule})}{I_{ON}(\text{Without Biomolecule})}$$

Table 1 K-value of different biomolecules [1]

Biomolecule	K-value
ChOx	3.3
Protein	2.5
Streptavidin	2.1
Uricase	1.54

I_{ON} (with biomolecule) is the ON current with biomolecule, and I_{ON} (without biomolecule) is the ON current without biomolecules. S_{ION} is measured with V_G is fixed.

Figure 4a–d demonstrates the changes of drain-on-sensitivity (S_{ION}) for different biomolecules for HfO_2 MOSHEMT and SiO_2 MOSHEMT, respectively. A greatest change in sensitivity among all biomolecules of an amount of 0.739 and 0.0853 is noticed for Uricase using HfO_2 MOSHEMT and SiO_2 MOSHEMT, respectively, when L_{cavity} ranges from 400 to 600 nm as increment in cavity length more amount of biomolecules are introduced which interact with AlGaN barrier layer. As a result of interaction of biomolecules, charge density changes with drain current. Figure 4d shows that lower dielectric biomolecules have higher sensitivity.

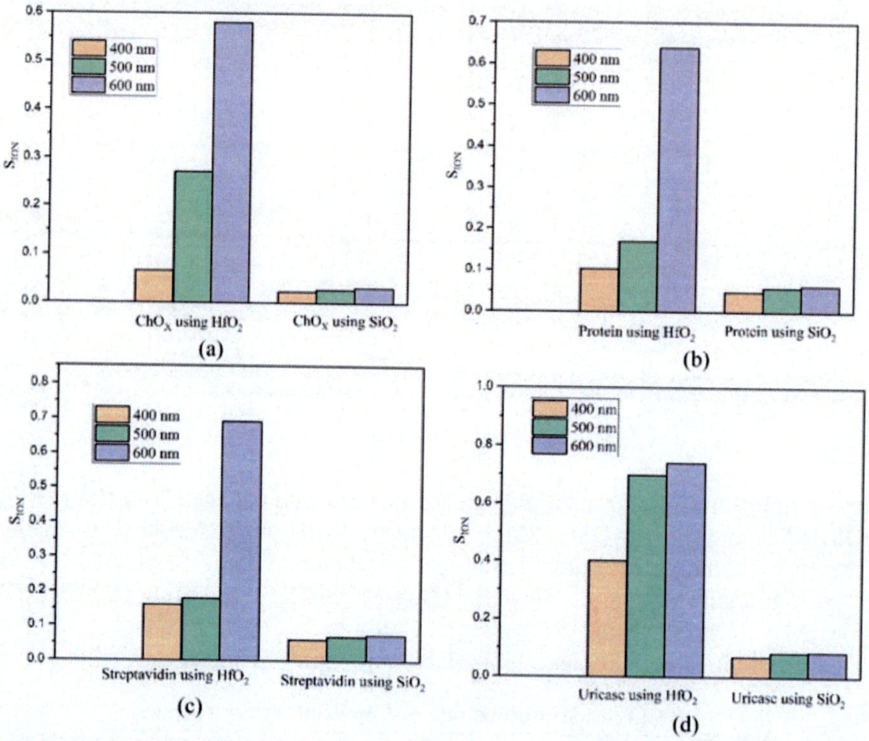

Fig. 4 a–d Drain-on-sensitivity (S_{ION}) for different biomolecules for high-K MOSHEMT and SiO_2 MOSHEMT [1]

4 Conclusion

In this comparative study, high-K MOSHEMT-based biosensors for label-free biosensing improve the sensitivity in compassion with SiO_2 MOSHEMT-based biosensors. By introducing high-K material, the gate to source capacitance is enhanced. As a result, the biomolecules on the surface have more potential changes. In this comparative study, it is observed that the sensitivity is improved by introducing high-K material in MOSHEMT-based biosensors.

Acknowledgements Authors would like to acknowledge Adamas University and NIT Mizoram for providing necessary facilities for conducting research work.

References

1. Pal P, Pratap Y, Gupta M, Kabra S (2018) Modeling and simulation of AlGaN/GaN MOS-HEMT for biosensor applications. IEEE Sens J 19(12):587–593
2. Shaveta HM, Ahmed M, Chaujar R (2020) Rapid detection of biomolecules in a dielectric modulated GaN MOSHEMT. Springer
3. Chen KH, Kang BS, Wang HT, Lele TP, Ren F, Wang YL, Chang CY, Pearton SJ, Dennis DM, Johnson JW, Rajagopal P, Roberts JC, Piner EL, Linthicum KJ (2008) C-erbB-2 sensing using AlGaN/GaN high electron mobility transistors for breast cancer detection. J Appl Phys
4. Choi JM, Han JW, Choi SJ, Choi YK (2010) Analytical modeling of a nanogap-embedded FET for application as a biosensor. IEEE Trans Electron Dev 57(12):3477–3484
5. Kim D, Jeong YT, Park HJ, Shin JK, Choi P, Lee JH, Lim G (2004) An FET-type charge sensor for highly sensitive detection of DNA sequence. Biosens Bioelectron 20(1):69–74
6. Kim C, Jung C, Lee KB, Park HG, Choi YK (2011) Label-free DNA detection with a nanogap embedded complementary metal oxide semiconductor. Nanotechnol 22(13):135502
7. Gu B, Park TJ, Ahn JH, Huang XJ, Lee SY, Choi YK (2009) Nanogap field-effect transistor biosensors for electrical detection of avian influenza. Small 5(21):2407–2412
8. Maesoon AJH, Han JW, Park TJ, Lee SY, Choi YK (2011) Development of a point-of-care testing platform with a nanogap-embedded separated double-gate field effect transistor array and its readout system for detection of avian influenza. IEEE Sens J 11(2):351–360
9. Seo G, Lee G, Lee J-O, Kim BT, Park EC, Kim SI (2020) Rapid detection of COVID-19 causative virus (SARS-CoV-2) in human nasopharyngeal swab specimens using field-effect transistor based biosensor. Am Chem Soc 14(4):5135–5142
10. Dechun G, Kankan Q, Junfeng C, Xiaobin L, Chao Y (2012) A simulation about the influence of the gate-source-drain distance on the AlGaN/GaN HEMT performance at Ka-band, In: Microwave workshop series on millimeter wave wireless technology and applications (IMWS), IEEE MTT-S international, pp 1–4
11. Kenzo M, Katsura T, Kerman K, Takamura Y, Matsumoto K, Tamiya E (2007) Label-free protein biosensor based on aptamer-modified carbon nanotube field-effect transistors. Anal Chem 79(2):782–787
12. Russo S, Di Carlo A (2007) Influence of the source–gate distance on the AlGaN/GaN HEMT performance. IEEE Trans Electron Dev 54(5):1071–1075

Inset Feed Semi-circular Slot Compact Microstrip Antenna for 10 GHz Mobile Applications

Kali Krishna Giri, Raj Kumar Singh, and Kumari Mamta

Abstract In this paper, the effect of dielectric constant value and the thickness of substrate material on the performance of an inset feed semi-circular slot compact microstrip antenna for 10 GHz mobile communication, alike satellite and terrestrial applications, is reported. The performance analysis of the compact antenna achieves a resonance frequency of 9.52 GHz at reflection coefficient of −22.14 dB, voltage standing wave ratio of value 1.72, and a gain 4.9 dB. Y parameter, group delay, and directivity are also part of the analysis. The bending of electric flux lines near at the edge of patch toward ground is fringing effect. Fringing effect plays a key role in the designing of antenna as it is function of radiating patch dimension. Fringing effect on antenna's performance due to varying dielectric constant and substrate thickness has been studied, and result parameters of antenna like S11, VSWR, 2D radiation pattern, and 3D polar plot due to variation of substrate thickness and dielectric constant have been obtained. Fringing field is enhanced due to increasing substrate height and due to decreasing substrate dielectric constants. The design, simulation, result parameters determination, and variations are done on HFSS.

Keywords S11 · VSWR · Fringing effect · Microstrip antenna · HFSS · Directivity

1 Introduction

The next-generation mobile communication requires high operational frequency. Wireless devices and applications call for high-speed data transmission and reception [1–6] with high bandwidth user requirement in mobile communication [7–9]. 5G communication systems exploit the use of millimeter-wave spectrum seeking large bandwidth and high gain. Microstrip patch antennas accommodate the spectrum

K. K. Giri (✉) · R. K. Singh
University Department of Physics, Ranchi University, Ranchi 834008, India

K. Mamta
Cambridge Institute of Technology, Ranchi 835103, India

J. K. Mandal et al. (eds.), *Topical Drifts in Intelligent Computing*, Lecture Notes in Networks and Systems 426, https://doi.org/10.1007/978-981-19-0745-6_36

limitation and high energy limitations. Patch antennas possess low profile, low cost, easy configuration, and small volume, apt for wireless communication.

Microstrip antennas are automatic choice for reconfigurability. Substrate height changing or selecting suitable proper height is one technique to achieve reconfigurability among other techniques. Fringing effect has great impact on the antenna performance. Due to fringing effect, electric field extends the area of patch. In microstrip antenna, patch and ground work as two conductive objects and a dielectric exist between them in the form of substrate, and this makes antenna behaves like a capacitor. The bending of electric flux lines near at the edge of patch toward ground is the *fringing effect* (Fig. 1). Inside the antenna, the flux lines are parallel and uniform, but at the edges of patch, the flux lines are not straight but curve to the ground. The amount of fringing depends on the distance between two conducting objects (patch and ground) which is the thickness of substrate. Fringing effect takes decisive role to the antenna dimension and design. It also plays vital role in shifting the resonant frequency. Keeping these points in mind, thickness of substrate and dielectric constant value is varied independently to achieve the desired fringing fields, keeping other design parameters unchanged.

Due to fringing effect, dielectric constant becomes effective dielectric constant and is given by the following expression:

$$\varepsilon_r^{\text{eff}} = \frac{1}{2}(\varepsilon_r + 1) + \frac{1}{2}(\varepsilon_r - 1)(1 + 12\frac{h}{w})^{-1/2} \tag{1}$$

If dielectric constant and substrate thickness are changed, then effective dielectric constant changes which in turn affect the fringing pattern. Usually, FR4 and Duroid (t_m) are used (not limited to them only) as substrate material for compact patch antenna design with parameters as in Table 1.

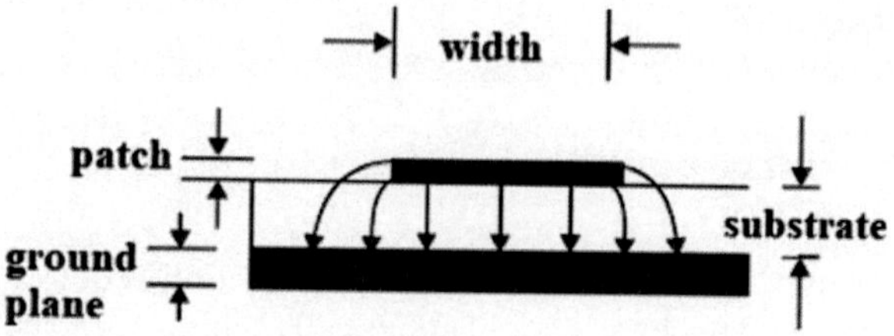

Fig. 1 Fringing effect in an antenna. Electric field lines are bent at the edge of patch

Table 1 Different substrate material in use

Substrate material	Dielectric constant	Cost	Loss tangent
Glass epoxy (FR4)	4.4	Low	0.02
Duroid (t_m)	2.2	Very high	0.0009

2 Methodology

A compact microstrip patch antenna is designed on HFSS to perform at 10 GHz. Optimization in the designing process is done through a circular slot of radius 1.88 mm drawn on a rectangular patch with matching thickness of patch (0.001 mm). This circular area is subtracted from patch, and then, feed-patch is united. Thus, inset feed is designed, and there becomes a semi-circular slot on the patch. Substrate used is glass epoxy (FR4) with dielectric constant 4.4. The patch equation is given by [10, 11].

$$W_{\mathrm{p}} = \frac{v_0}{2f_r}\sqrt{\frac{2}{\varepsilon_r + 1}} \tag{2}$$

The effective dielectric constant follows from Eq. (1). Patch length is given as in Eq. (3) where ΔL represents the extension in patch length arising due to fringing effect

$$L_{\mathrm{p}} = \frac{c}{\sqrt[2]{\varepsilon_r^{\mathrm{eff}}}}\left(\frac{1}{f_r}\right) - 2(\Delta L) \tag{3}$$

The length extension following the fringing effect is governed by Eq. (4)

$$\frac{\Delta L}{h} = 412\frac{\left(\varepsilon_r^{\mathrm{eff}} + 0.3\right)\left(\frac{W_{\mathrm{p}}}{h} + 0.264\right)}{\left(\frac{W_{\mathrm{p}}}{h} + 0.8\right)\left(\varepsilon_r^{\mathrm{eff}} - 0.258\right)} \tag{4}$$

Ground length equation is

$$L_g = 12h + L_{\mathrm{p}} \tag{5}$$

And the ground width equation is given by

$$W_{\mathrm{g}} = 12h + W_{\mathrm{p}} \tag{6}$$

Design parameters for compact microstrip patch antenna are given in Table 2.

With these parameters, a semi-circular slotted compact microstrip antenna is designed at 10 GHz as in Fig. 2. After designing the compact antenna at 10 GHz, the substrate thickness and dielectric constants are varied independently, keeping other design considerations fixed.

Table 2 Design parameters for compact microstrip patch antenna

Parameters	Along X axis (mm)	Along Y axis (mm)	Along Z axis (mm)	Axis
Ground	16.02	18.72		Z
Substrate	16.02	18.72	1.6 (thickness)	Z
Patch	6.42	9.12	0.001	Z
Feed	8.01	3		Z
Port		3	−1.61	X
Radiation box	28	28	14	

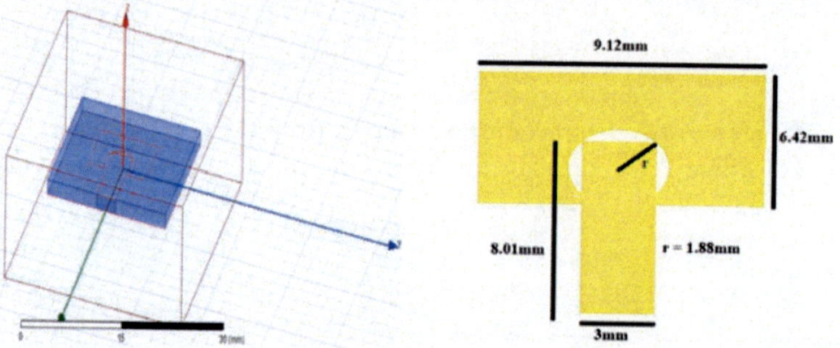

Fig. 2 Left: the proposed semi-circular slotted compact microstrip patch antenna designed to operate at 10 GHz. Right: front view of the patch with dimensional parameters in the XY plane. Patch dimension along X and Y axis being 6.42 and 9.12 mm that of feed being 8.01 and 3 mm. Radius of slot being 1.88 mm

3 Simulation and Results

Antenna design and simulation have been done for three different situations—

A. Substrate dielectric constant being 4.4 and substrate thickness being 1.6 mm.
B. Different substrate materials at a fixed value of substrate thickness (1.6 mm).
C. Different substrate thickness at a fixed value of dielectric constant (4.4).

3.1 Substrate Dielectric Constant is 4.4 and Substrate Thickness is 1.6 mm

S11 parameter and VSWR have been plotted against the range of operational frequency with substrate dielectric constant value of 4.4 and its thickness being 1.6 mm (Figs. 3 and 4). Simulated resonance frequency is 9.52 GHz, and the value of return loss is −22.14 dB at this frequency. VSWR value of 1.7 is achieved at 9.52 GHz which is within the recommended value of 2. Simulated 2D radiation pattern is shown in Fig. 5. 3D polar gain plot (Fig. 6) achieves a gain value of 4.9 dB.

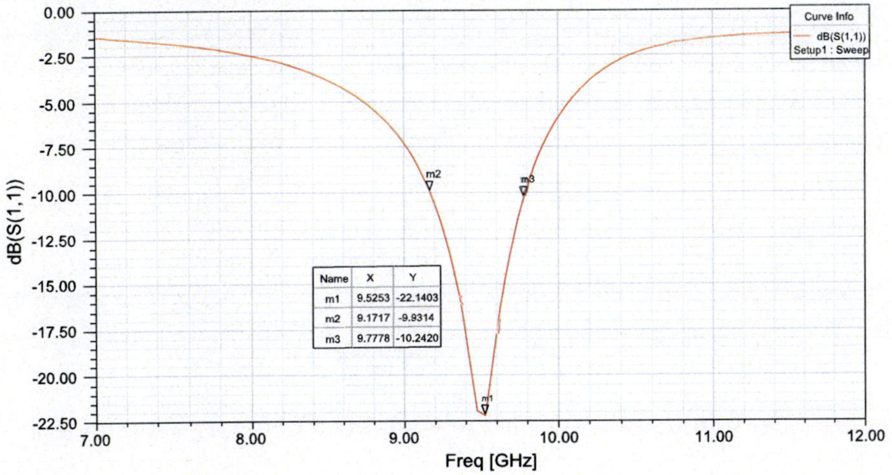

Fig. 3 S11 parameter against the operational frequencies. Simulated resonance frequency is 9.52 GHz, and the value of return loss is −22.14 dB at this frequency

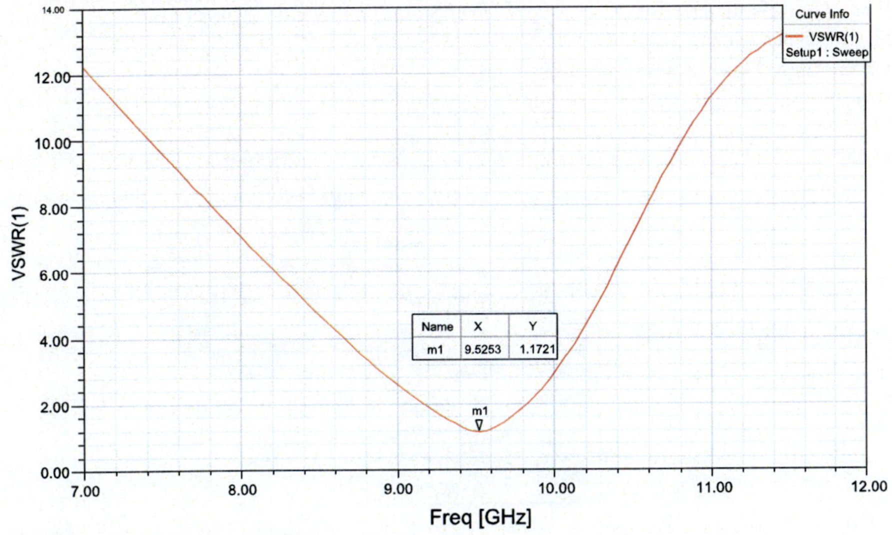

Fig. 4 VSWR versus frequency. VSWR value of 1.7 is achieved at 9.52 GHz which is within the recommended value of 2

Imaginary and real Y parameter are obtained in Figs. 8 and 9. Maximum admittance is achieved at the resonant frequency of 9.52 GHz. Gain versus frequency, directivity against gain, and group delay graphs have been shown in Figs. 10, 11, and 12. Directivity and gain are stable up to the resonant frequency, and the group delay plot has a resonance peak at the resonant frequency 9.52 GHz (Fig. 7).

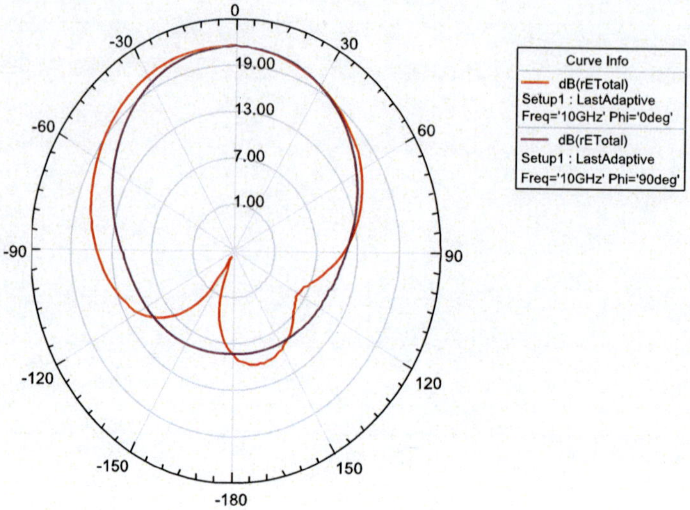

Fig. 5 2D radiation pattern for the designed antenna

Fig. 6 3D polar gain plot achieves a gain value of 4.9 dB

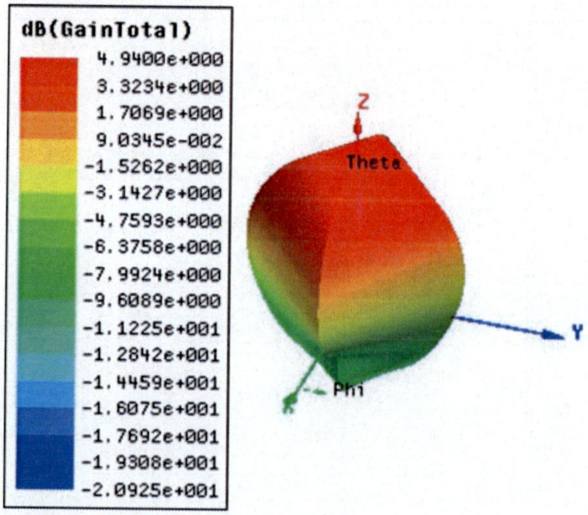

3.2 *Different Substrate Materials at a Fixed Value of Substrate Thickness*

Substrate material is used as an insulator. In this situation, only, substrate materials have been changed in order to have different dielectric constant value, and the substrate thickness has been kept fixed at 1.6 mm, and other design parameters are

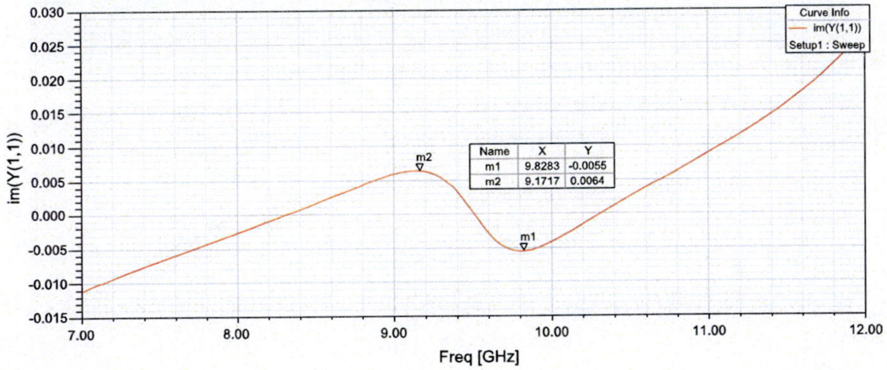

Fig.7 Imaginary Y parameter

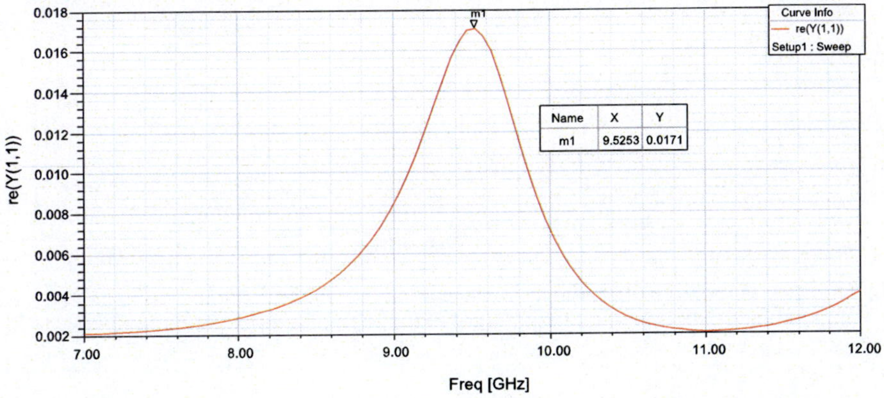

Fig. 8 Real Y parameter

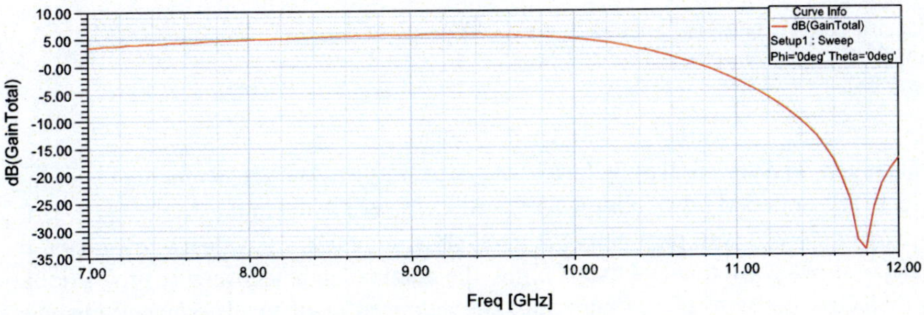

Fig. 9 Gain versus frequency plot for proposed antenna

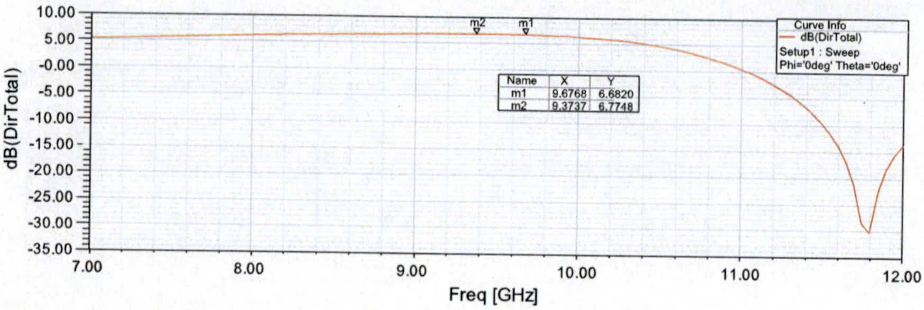

Fig. 10 Directivity versus gain plot

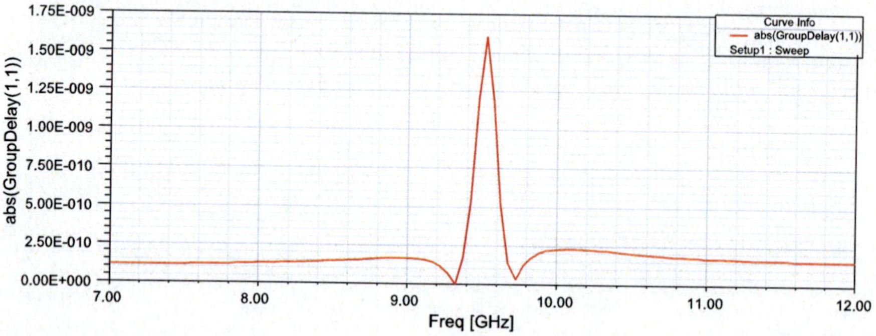

Fig. 11 Group delay graph of the simulated antenna

Fig. 12 Effect of dielectric constant value on the gain. Gain becomes higher at dielectric constant 4–4.4, becomes lower when dielectric constant is larger than 4.4 and again when lower than 4.4

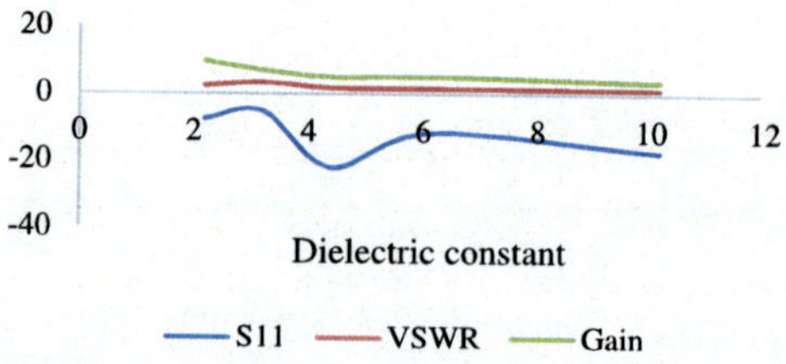

unchanged. The variation of dielectric constant confirms the achievement of fringing effect. The obtained result parameters in this case are listed in the Table 3 [12, 13].

The obtained result parameters at fixed substrate height establish a few concrete ideas. It is observed that at a fixed height, the obtained gain is inversely proportional to dielectric constant of the material. The achieved resonance frequency changes with different dielectric constants though the target resonance frequency is fixed. Here, the whole dimension remains fixed; resonance frequency shifts. The up and down nature of graph of dielectric constant and gain (Fig. 12) is due to mismatching of impedance. For this condition, VSWR is very good and lowest for Rogers RT [7]. Below and beyond this range, there is a downfall trend in the gain of the antenna.

Table 3 Different substrate materials at a fixed thickness of substrate (1.6 mm)

Substrate	Dielectric constant	Resonance frequency (GHz)	S11 (dB)	VSWR	Gain (dB) from 3D polar plot
FR4	4.4	9.52	−22.14	1.72	4.9
Duroid tm	2.2	12.86	−7.68	2.41	9.6
Taconic TLC (t_m)	3.2	11.1	−5.24	3.42	6.96
Rogers RO (300t_m)	6.15	11.49	−11.77	1.69	5.14
Rogers RT	10.2	9.15	−17.51	1.30	3.63

3.3 *Different Substrate Thickness at a Fixed Value of Dielectric Constant (4.4)*

From Table 3, it is observed that the antenna has good performance for dielectric constant 4.4 rather than other values. Keeping this point into consideration, the thickness of the substrate is changed, and the VSWR and gain are studied at resonance frequency of 10 GHz. The variation results are mentioned in Table 4.

Effect of variation in the thickness of substrate on VSWR and gain is shown in Figs. 13 and 14, respectively. VSWR is found to increase much beyond the recommended limit with increasing thickness of substrate. Resonance nature is found in the gain value against the increasing value of substrate thickness. These graphs establish the influence of fringing effect [7, 14, 15]. Substrate thickness in between 1.4 and

Table 4 Gain and VSWR of antenna with dielectric constant 4.4 at different substrate thickness

Substrate thickness (mm)	VSWR	Gain (dB)
1.6	1.72	4.9
1.7	6.5	2.18
1.8	8.5	1.89
1.9	9.77	2.23
2	11.19	2.32
2.1	13.43	2.63
2.2	13.78	2.77

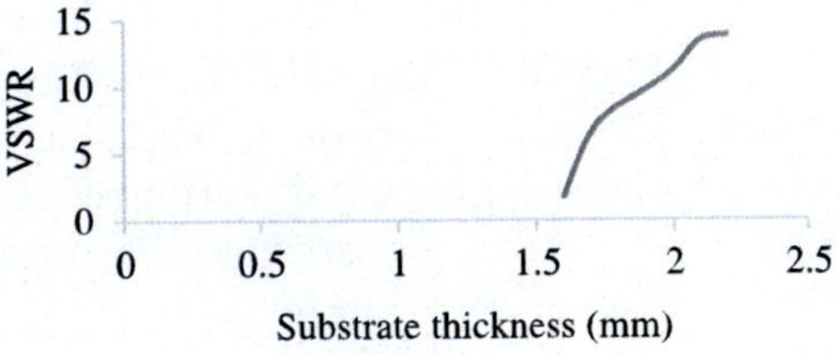

Fig. 13 Substrate thickness versus VSWR. VSWR shows an increasing trend with the increase in the thickness of substrate and goes beyond the recommended limiting value of 2

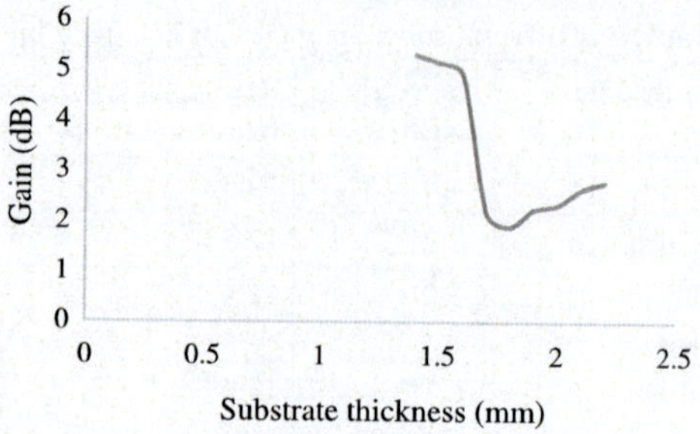

Fig. 14 Substrate thickness against the gain. Resonance type effect can be seen in the graph

1.6 mm, the gain is obtained at nearly the same level. But, it drops sharply apart from 1.6 mm. That is, gain becomes lower with the increasing thickness of the substrate [16–18].

It is inferred that the gain and VSWR depend on the proper choice of substrate thickness. With the increase in substrate thickness, VSWR also increases. That is, impedance mismatching also increases with increasing substrate thickness.

4 Conclusion

In this paper, the effect of dielectric constant value and the thickness of substrate material on the performance of an inset feed semi-circular slot compact microstrip antenna for 10 GHz mobile communications is reported. The performance analysis of the compact antenna achieves a resonance frequency of 9.52 GHz at reflection coefficient of −22.14 dB, voltage standing wave ratio of value 1.72, and a gain 4.9 dB. Y parameter, group delay, directivity are also part of the analysis. Gain and VSWR depend on the proper choice of substrate thickness. For the substrate dielectric constant 4.4 and substrate thickness 1.6 mm, S11 parameter and VSWR have been plotted, and the simulated resonance frequency obtained is 9.52 GHz. The value of return loss is −22.14 dB at this frequency. VSWR value of 1.7 is achieved at 9.52 GHz which is within the recommended value of 2. 3D polar gain plot achieves a gain value of 4.9 dB. Maximum admittance is achieved at the resonant frequency. Gain versus frequency, directivity against gain, and group delay graphs have also been obtained. Directivity and gain are stable up to the resonant frequency, and the group delay plot has a resonance peak at the resonant frequency 9.52 GHz [19].

For Duroid, Taconic, Glass, Rogers at 10 GHz frequency and substrate height 1.6 mm the VSWR and S11 do not show up proper resonance frequency. So, S11 and VSWR cannot be detected properly due to variation of substrate materials. But, gain has been obtained for each substrate used for variation. With the increase in substrate thickness VSWR also increases. That is, impedance mismatching also increases with increasing substrate thickness. Performance of antenna depends on proper selection of substrate material whose dielectric constant value is very important. Substrate thickness takes decisive role for achieving satisfactory result parameters. As radiation depends on fringing field, so on obtaining gains for different substrate thickness confirm that fringing effect is the function of substrate height. Fringing effect impacts on achieving resonance frequency.

References

1. Wang C-X, Haider F, Gao X, You X-H, Yang Y, Yuan D et al (2014) Cellular architecture and key technologies for 5G wireless communication networks. IEEE Commun Mag 52:122–130. https://doi.org/10.1109/MCOM.2014.6736752
2. Chen Z, Zhang YP (2013) FR4 PCB grid array antenna for millimeter-wave 5G mobile communications. In: Microwave workshop series on RF and wireless technologies for biomedical and healthcare applications (IMWS-BIO), 2013 IEEE MTT-S International, pp 1–3. https://doi.org/10.1109/IMWS-BIO.2013.6756214
3. Mamta K, Singh RK (2019) Frequency reconfigurable millimeter wave antenna design for 5G application. Int J Mag IT Eng 4:137–147
4. Mamta K, Singh RK, Sinha NK, Keshri R (2018) Design and development of microstrip patch antenna for mm wave application, vol 476. LNEE, Springer, pp 79–85
5. Andrews JG, Buzzi S, Choi W et al. (2014) What will 5 G be? IEEE J Sel Areas Commun 32(6):1065–1082. https://cupdf.com/document/what-will-5-g-be.html
6. Khan F, Pi Z (2011) Millimeter wave mobile broadband (MMB): unleashing the 3–300 GHz spectrum. In: Proceedings of the 34th IEEE Sarnoff symposium. https://doi.org/10.1109/SARNOF.2011.5876482
7. Haraz OM, Elboushi A, Alshebeili SA, Sebak A-R (2014) Dense dielectric patch array antenna with improved radiation characteristics using EBG ground structure and dielectric superstrate for future 5G cellular networks. IEEE Access 2:909–913. https://doi.org/10.1016/j.promfg.2018.03.071
8. Haraz OM, Elboushi A, Sebak A-R (2014) New dense dielectric patch array antenna for future 5G short-range communications. In: Antenna technology and applied electromagnetics (ANTEM), 16th international symposium, pp 1–4. https://doi.org/10.1109/ANTEM.2014.6887689
9. Giordani M, Mezzavilla M, Zorzi M (2016) Initial access in 5G mm wave cellular networks. IEEE Commun Mag 54(11):40–47. https://doi.org/10.1109/MCOM.2016.1600193CM
10. Singh I et al (2011) Microstrip patch antenna and it's application: a survey. IJCTTA
11. Singh VP, Thakur M (2014) Band width enhancement of microstrip line inset feed patch antenna. IJERT 3
12. Kiran T, Mounisha N, Mythily C, Akhil D, Phani Kumar TVB (2018) Design of microstrip patch antenna for 5g applications 13:14–17
13. Smyth BP, Barth S, Iyer AK (2016) Dual-band microstrip patch antenna using integrated uniplanar metamaterial-based EBGs. IEEE Trans Antennas Propag 64(12):5046–5053. https://doi.org/10.1109/TAP.2016.2618854
14. Islam M, Hasan R, Rahaman M Design and analysis of microstrip patch antenna using different dielectric materials for wimax communication system. IJRCES and IT. https://doi.org/10.3991/IJES.V4I9.5569
15. Chen D, Yu Y, Zhao C, Tian G (2019) A novel compact microstrip antenna with an embedded lamda/4 resonator. https://doi.org/10.1155/2019/2431760
16. Elrashidi A, Elleithy K, Bajwa H (2011) The fringing field and resonance frequency of cylindrical microstrip printed antenna as a function of curvature. EURASIP J Wirel Commun netw 3(2)
17. Guha D, Antar YM Microstrip and printed antennas. Print ISBN: 9780470681923, ISBN:9780470973370, https://doi.org/10.1002/9780470973370

18. Mamta K, Singh RK (2017) Analysis of reconfigurable polarisation antenna as an EMI sensor. Br J Appl Sci Technol 19(5):1–8. https://doi.org/10.9734/BJAST/2017/31798
19. Panda RA, Mishra D, Panda H (2016) Biconvex patch antenna with circular slot for 10 GHz application. In: International conference on signal processing, communication, power and embedded system (SCOPES), pp 1927–1930. https://doi.org/10.1109/SCOPES.2016.7955782

A Comparative Study of GaN and Si-Based SOI FinFET

Abhishek Saha, Rudra Sankar Dhar, and Subhro Ghosal

Abstract Over the last decade, the continuous strife for miniaturization has led to an exponential increase in the power density of semiconductor electronics. The average operating temperature of devices has sharply increased due to the combined effects of high frequency switching and lower surface area per device. A possible way out is the introduction of with high bandgap materials or alloys as an alternative to silicon electronics. Of the different options, GaN with its high bandgap and mobility has emerged as one of the most promising materials. However, use of GaN has primarily been restricted to applications in HEMT and optoelectronic devices. In this paper, a comparative simulation study of GaN and Si field effect device has been presented using SOI FinFET as the device of choice. It has been shown that besides a higher ON current GaN shows a reduced sensitivity of I_{ON}/I_{OFF} ratio to variations in operating temperature justifying its choice in high-power density devices. Effect of an LDD profile has been studied. Moreover, this work reports a higher susceptibility of threshold voltage of GaN-based devices to variations in dielectric. This has promising implications with respect to its use as a bio-detector.

Keywords TG-FinFET · *GaN* channel · Bio-detector

1 Introduction

Over the past couple of decades, silicon technology has dominated the semiconductor industry and the major factors responsible for it, besides easy availability, and excellent oxide properties have been the scalability of Si devices. However, the

A. Saha · R. S. Dhar (✉)
Department of Electronics and Communication Engineering, National Institute of Technology Mizoram, Aizawl, Mizoram 796012, India
e-mail: rdhar@uwaterloo.ca

A. Saha
e-mail: abhishek.ece.phd@nitmz.ac.in

S. Ghosal
Department of Electronic Science, Acharya Prafulla Chandra College, Kolkata, West Bengal, India

J. K. Mandal et al. (eds.), *Topical Drifts in Intelligent Computing*, Lecture Notes in Networks and Systems 426, https://doi.org/10.1007/978-981-19-0745-6_37

aggressive demands on device and circuits parameters which are being set by the wireless communication sector in the microwave and RF region will soon appear to be out of reach of Si-based devices whose parameters are close to their theoretical maxima. Moreover, relatively lower values saturation velocity besides a low carrier mobility precludes silicon as the material of choice for modern high speed switching applications. Over the last decade, the continuous strife for miniaturization has led to an exponential increase in the power density of semiconductor electronics. Besides the obvious bottlenecks due to short-channel effects and leakage current due to tunneling, the average operating temperature of devices has sharply increased due to the combined effects of high frequency switching and lower surface area per device. However, it is very much problematic to use silicon electronics device as a power device because of its temperature constraint. Hence, the focus of device scientists on new compound semiconductors which include group III–V materials, silicon–carbide, and silicon–germanium. In the last two-decade, group III-nitride semiconductors and in particular gallium nitride (GaN), and its associated alloys, have emerged as the most promising nitride semiconductor for commercial applications, especially in the field of optoelectronic and high-power devices due to its inherent material properties. Indeed, GaN has a larger breakdown field (20 MV/cm) [1] than GaAs (4 MV/cm) and Si (3 MV/cm) and also high peak electron velocity [2] of 3 × 107 cm/s as compared to GaAs (2 × 107 cm/s) and Si. The acceptability of GaN technology to the stringent military as well as commercial sector owes itself the superior electron transport properties of GaN. However, widespread acceptability cannot be achieved unless some bottlenecks such as drain current collapse and other reliability problems [3–5] are overcome.

As we move further into the DSM regime, effective control of channel is an area of concern. As per the roadmap provided by ITRS, the effects associated with short channel together with the stringent leakage requirements for devices beyond 32 nm technology node will seriously impede further developments using planar devices. Multigate devices like tri-gate transistor or fin-shaped field effect transistor appear to be one of the most promising devices below the 20 nm technology node, as the effective increase in the inversion layer owing to multiple gates leads to a better channel control. The gate-all-around architecture provided by FinFET suppresses the short-channel effects (SCE) while achieving high trans conductance value and lower subthreshold swing [6–8]. Miniaturization of the components and low energy consumption makes it possible to use them as switching power supplies, amplifiers, frequency converters in electronic equipment [9, 10], sensors [11–13], as well as high frequency switching generators and modulators in medical surgical devices for high frequency and ultrasonic welding of biological tissues.

In this study, the proposed structure is simulated with two different channel materials, i.e., GaN and Si. An LDD doping profile has been chosen. The parameters varied are temperature and gate material, and different figures of merit of relevance, namely I_{ON}/I_{OFF} ratio, threshold voltage, leakage current, saturation current, and subthreshold slope have been calculated from the simulation data.

This paper is organized as follows. In Sect. 2, the simulation model is described. The simulated results of GaN-based n-channel TG-FinFET are analyzed and discussed in Sect. 3, and Sect. 4 discusses the significance of the study.

2 Materials and Methods

2.1 *FinFET*

In this work, a model of GaN-based FinFET structure is considered for the simulated study by Atlas of Silvaco_TCAD device simulator. An SOI FinFET is the chosen structure for simulation. While the larger gate area leads to a better control, the SOI structure helps in the elimination of parasitic capacitance leading to a better high-frequency response. Moreover, due to BOX layer, there is no unwanted leakage paths which are far from the gate and lesser effect of back-gate leading to lower power consumption. A major drawback of SOI structure is self-heating as the silicon oxide substrate prevents heat dissipation. However, the better thermal properties of the chosen material GaN make this structure better suited for it than Si.

The simulated device structure of the n-channel TG (Tri gate) FinFET, used in this study, is shown in Fig. 1. A buried oxide thickness (TBOX) of 50 nm has been used.

The device consists of a Si fin of rectangular cross-sectional wrapped around the channel which terminates in the source and drain regions at both ends. The inversion current is controlled by the polarization charge on gate oxide which forms the interface between the channel and the fin. Thus, the output drain current is a function of gate and drain bias voltages.

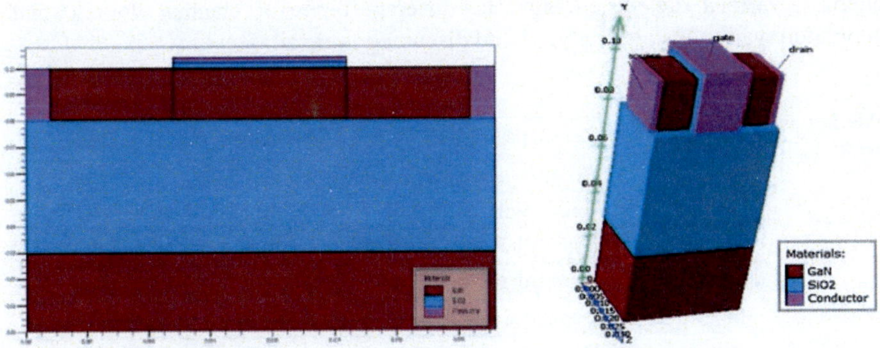

Fig. 1 TG-FinFET structure (2D and 3D view)

2.2 *Simulation Method*

All simulations of this work have been carried out using ATLAS SILVACO TCAD, 3D simulation. Basic parameters of this n-channel TG-FinFET structure are presented in Table 1.

A Gaussian doping profile has been used for the channel region. In this entire simulation procedure, work function of gate electrode is assumed as 4.6 eV. Solution method applied is Gummel Newton. For carrier statistics, instead of the default Boltzmann statistics, we have used Fermi to account for reduced carrier concentrations in heavily doped regions. As this study considers the effects of high temperature, the energy exchange with the surrounding lattice environment through carrier heating and through generation-recombination processes need to be taken into account. To take account of the recombination effects, the leakage currents that owe their origin to thermal generation are simulated by the Shockley–Read–Hall (SRH) model while consrh accounts for concentration dependent lifetimes. Auger is used to account for recombination at high current densities and high-level injection effects. The Auger recombination rate is reduced at higher values of carrier density as the inter-carrier interactions become more significant. Experimental work has shown that in heavy doped regions SOI MOSFET starts behaving like a bipolar transistor as the pn product becomes doping dependent, and bandgap narrowing occurs; BGN accounts for this effect. The quantum mechanical effect of carrier confinement at the gate oxide interface in MOSFETs is accounted for by the correction given by Hansch (Model HANSCHQM).

3 Simulation and Results

In this section, the simulated results of n-channel TG-FinFET with different channel materials, various dielectric materials, different types of channel doping, and temperature variations are analyzed and discussed in detail.

Table 1 Basic parameters used for SOI TG-FinFET

Device parameter	Value	Unit
Length of source and drain	10	Nm
Gate length	14	Nm
Source/drain height	20	Nm
Source/drain width	20	Nm
Doping of source and drain	1×10^{18} (n-type)	Atoms/cc
Doping of substrate and channel	1×10^{15} (p-type)	Atoms/cc
T_{BOX}	50	Nm
T_{sub}	30	Nm
T_{OX}	2	Nm

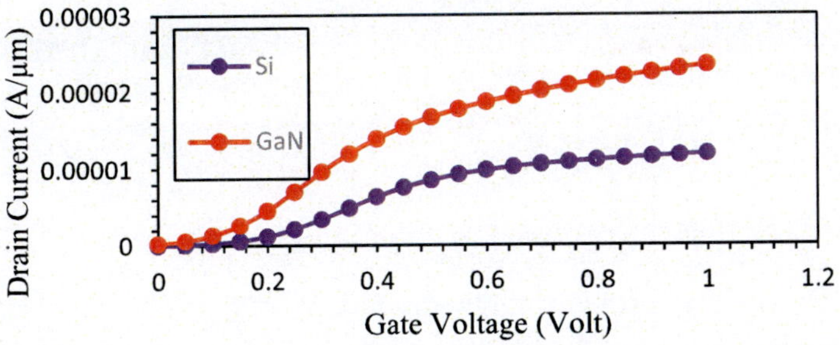

Fig. 2 I_D–V_G curve for n-channel TG-FinFET (uniform doping) with different channel materials

3.1 Variation in Channel Material

Figure 2 shows the variation of drain current with the increment of gate voltage for silicon (Si) and gallium nitride (GaN) channel TG-FinFET. Here, the DC voltage is applied to the gate terminal which is varied from 0 to 1 V with the drain bias fixed at $V_{DS} = 1$ V.

The transfer characteristics show that the higher mobility of GaN translates to a higher ON current than Si. This makes such GaN-based field effect devices a better choice for power-electronics circuits where a higher drive current is required.

3.2 Variation in Drain Doping Profile

Various works have been done on the effects of gradation of doping of field effect devices in the channel region [14–17]. The gradation may be deliberate or an undesired consequence of the implantation process. However, in this study, it was observed that replacement of a uniform doping profile with a Gaussian one in the channel region led to an insignificant change. This could be a consequence of the light doping and tight control of the channel region by the surrounding gate structure. However, the source and drain regions are highly doped, and FinFET devices with LDD implantation have been reported [15] to exhibit better electrostatic characteristics resulting in reduced values of subthreshold slope and DIBL. In this study, also, the effect of an LDD doping profile (Fig. 3) was studied, and the transfer characteristics show a notable improvement for GaN though the effect is less pronounced for the Si channel device (Fig. 4).

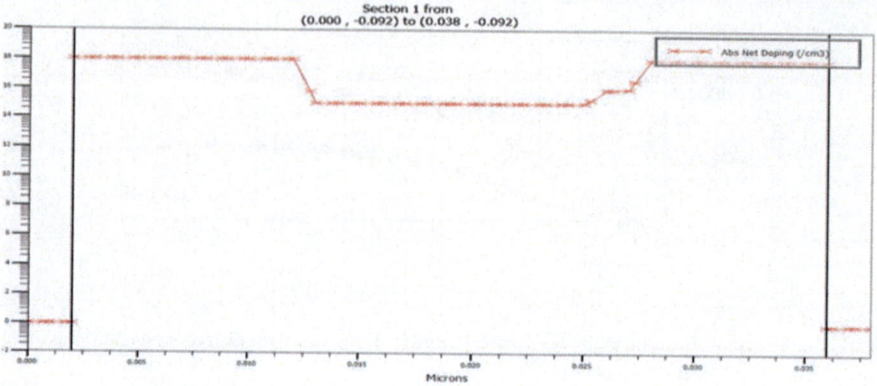

Fig. 3 Doping profile of LDD region

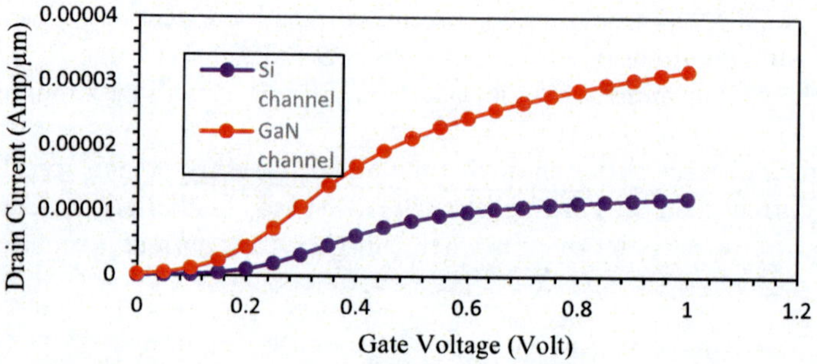

Fig. 4 I_D–V_G curve for n-channel TG-FinFET with LDD region

3.3 Dielectric Material Variation

Simulations were done to study the outcome of gate dielectric material on the output of both GaN and Si-channel devices. It can be seen that in both the cases GaN channel-based field effect device produces more current than silicon channel-based device and significantly reduction in threshold voltage also. The different dielectric materials used were aluminum oxide (Al_2O_3), hafnium oxide (HfO_2), silicon nitride (Si_3N_4), and zirconium dioxide (ZrO_2) in contrast with the silicon dioxide (SiO_2). Figure 5 shows the simulated result.

Extracted parameters from the simulations are given in Table 2. From the comparison shown in the Table 2, it can be noted that with the decreasing value of dielectric constant of different dielectric material the value of the threshold voltage is decreased. The value of maximum saturation drain current I_{DSmax} is also decreased, whereas the value of leakage current is increased. Furthermore, it is observed that the value of threshold voltage for all kind of dielectric material used in GaN channel-based

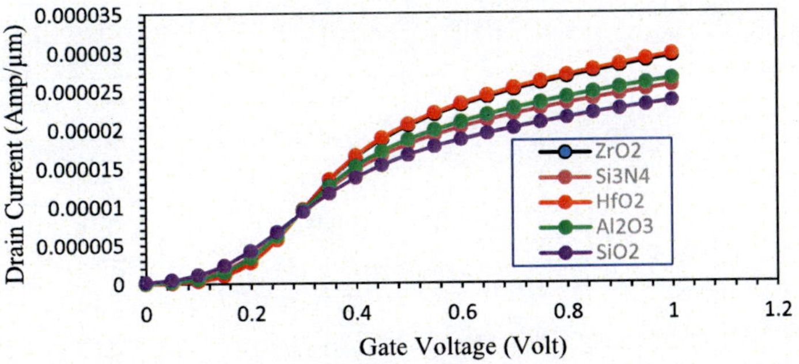

Fig. 5 I_D–V_G characteristics (GaN channel TG-FinFET) for different dielectric materials

Table 2 Extracted parameter value for n-channel TG-FinFET

Channel material	Dielectric material (dielectric constant value)	V_t V	I_{DSmax} ($\times 10^{-6}$) A	Leakage current ($\times 10^{-9}$) A	Sub V_t mV/decade
GaN	HfO_2 (25)	0.127	29.6	29.85	91.29
	ZrO_2 (23)	0.126	29.3	31.36	91.91
	Al_2O_3 (9)	0.108	26.4	68.85	103.31
	Si_3N_4 (7)	0.101	25.5	92.20	108.58
	SiO_2 (3.9)	0.073	23.5	207.2	127.79
Si	HfO_2 (25)	0.204	14.0	2.52	88.74
	ZrO_2 (23)	0.203	13.9	2.69	89.37
	Al_2O_3 (9)	0.181	13.0	7.97	100.77
	Si_3N_4 (7)	0.172	12.8	11.87	105.88
	SiO_2 (3.9)	0.144	12.2	35.4	124.03

TG-FinFET is lower than that of the Si channel-based TG-FinFET. The sensitivity to dielectric material [17] can be explained as follows:

The drain current in the subthreshold region has an exponential dependence of the form exp $\left({}^{v_{gs}}/_{nVT}\right)$ which results a relation for subthreshold voltage

$$S = nVT\ln(10) \text{ where } n = \left(1 + \frac{C_b}{C_g}\right),\ C_b = {}^{\int Si}/_{w_d},\ C_g = {}^{\int_{ox}}\!\Big/_{t_{ox}}$$

Now, as HfO_2 has the higher dielectric constant, there will be an increment in oxide capacitance which translates to a lower value of *n*. Thus, a higher value dielectric constant of gate oxide helps the subthreshold slope to approach its ideal value. The lower leakage current can be accounted for by the decrease in threshold voltage.

A very significant observation is given in Table 3. Table 3 provides the data for percentage change in threshold voltage with variable dielectric material, in this case

Table 3 Percentage change in threshold voltage with dielectric

Channel material	Dielectric material	Percentage change in threshold voltage
GaN	HfO_2 (25)	73.97
	ZrO_2 (23)	72.6
	Al_2O_3 (9)	47.94
	Si_3N_4 (7)	38.36
Si	HfO_2 (25)	41.66
	ZrO_2 (23)	40.97
	Al_2O_3 (9)	25.69
	Si_3N_4 (7)	19.44

GaN taken as the primary material for this study, whereas silicon data are used as a comparison. The percentage change in threshold voltage with dielectric constant is much more for GaN than Si. Over the past decade, a significant amount of work [18–22] has been done on the use of field effect devices as a biosensor. The increased sensitivity of GaN-based device to dielectric variation makes it a promising candidate for use as a detector of biologically relevant substances.

3.4 Temperature Variation

Applicability of group III-nitride-based devices for high-frequency, power, and temperature applications is well documented [22–25]. Therefore, for device optimization, self-heating effect and ambient temperature effect of a device are such a thing that can be studied to establish its suitability as a power electronic device. In this work, the transfer characteristics of GaN FINFET have been studied for temperatures ranging from 300 to 600 °K. From the result plotted in Fig. 6 it can be concluded that

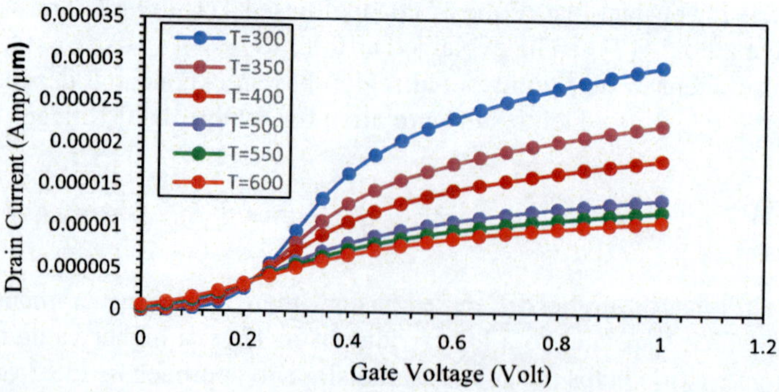

Fig. 6 Temperature variation in transfer characteristics for n-channel (GaN) TG-FinFET

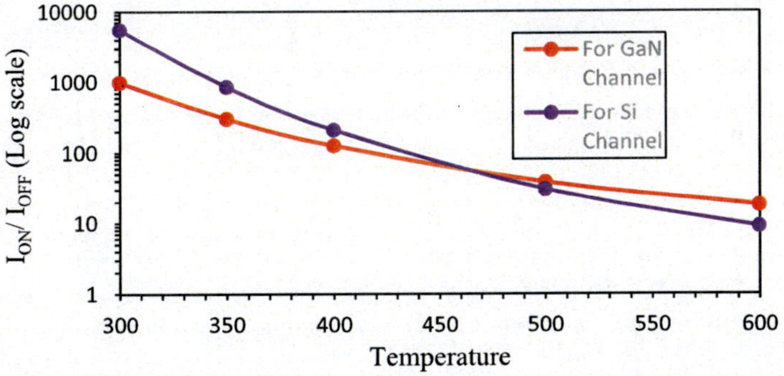

Fig. 7 Variation in I_{ON}/I_{OFF} ratio with temperature for different channel material

with the variation of the temperature the saturation drain current decreases with the increasing temperature. Furthermore, if the ratio of I_{on}/I_{off} is considered, significant phenomena is being noticed. With the increasing temperature, the ratio of I_{on}/I_{off} decreases more sharply in case of silicon channel device compared to GaN channel device. This comparison is plotted in Fig. 7. It may be observed that the sensitivity of I_{on}/I_{off} ratio to temperature is much lower for GaN as compared to Si over the measured temperature range. This indicates the issue of degradation of performance with temperature is much lower for GaN making it a promising candidate for devices with high-power density.

4 Conclusion

A survey of existing scientific reports shows that the existing work on GaN has been limited to HEMT-based devices. The wafer technology of GaN is not mature that coupled with the technologically challenging HEMT structure does not encourage its acceptance as a mass-scale replacement for Si. The objective of this study was to show its suitability as simple field effect device vis-à-vis silicon without taking into consideration the piezoelectric advantage associated with the AlGaN/GaN heterostructure. The SOI n-channel TG-FinFET structure is successfully designed and simulated by the influence of different channel material, lightly doped drain profile, and temperature variation. It has been shown that the GaN channel devices provide a higher ON current and lower threshold voltage though its leakage current and subthreshold performance lag behind silicon. Moreover, the temperature sensitivity of GaN is much lower than Si FinFET indicating its suitability for high-power density devices. Perhaps, the most interesting outcome of this study has been the observation of a significantly enhanced sensitivity of threshold voltage of GaN FINFET to dielectric variation as compared to Si. This has promising implications by way of application as a biosensor. In future, this work will be extended to study the switching performance of such devices as well its sensitivity to variations in structural parameters.

References

1. Tanaka T, Takano K, Mishima T, Kohji Y, Otoki Y, Meguro T (2005) GaN epitaxial wafers for high breakdown voltage RF transistor applications. Hitachi Cable Rev 24:11–14
2. Gangwani P, Pandey S, Haldar S, Gupta M, Gupta RS (2007) Polarization dependent analysis of algan/gan hemt for high power applications. Solid State Electron 51(1):130–135. https://doi.org/10.1016/j.sse.2006.11.002
3. Pengelly RS, Wood SM, Milligan JW (2012) A review of GaN on SiC high electron-mobility power transistors and MMICs. IEEE Trans Microw Theory Tech 60(6):1764
4. Klein PB, Binari SC, Ikossi-Anastasiou K (2001) Investigation of traps producing current collapse in AlGaN/GaN high electron transistors. Electron Lett 37(10):661
5. Meneghesso G, Verzellesi G, Pierobon R (2004) Surface-related drain current dispersion effects in AlGaN–GaN HEMTs. IEEE Trans Electron Dev 51(10):1554
6. Kuhn KJ, Avci U, Cappellani A, Giles MD, Haverty M, Kim S, Kotlyar R, Manipatruni S, Nikonov D, Pawashe C, Radosavljevic M, Rios R, Shankar S, Vedula R, Chau R, Young, I (2012) The ultimate CMOS device and beyond. IEEE Int Electron Devices Meet (IEDM), pp 171–174. https://doi.org/10.1109/IEDM.2012.6479001
7. Lin CH, Kambhampati R, Miller RJ, Hook TB, Bryant A, Haensch W, Oldiges P, Lauer I, Yamashita T, Basker V, Standaert T, Rim K, Leobandung E, Bu H, Khare M (2012) Channel doping impact on FinFETs for 22 nm and beyond. VLSI Technology (VLSIT). Honolulu, pp 15–16. https://doi.org/10.1109/VLSIT.2012.6242438
8. Li Y, Hwang CH (2007) Effect of fin angle on electrical characteristics of nanoscale round-top-gate bulk FinFETs. IEEE Trans Electron Dev 54:3426–3429. https://doi.org/10.1109/TED.2007.908908
9. Kumari A, Lakra BS, Pahuja H, Jambagi SG Design and performance analysis of operational transconductance amplifier using FinFET. In: IEEE conference. https://doi.org/10.1109/CONFLUENCE.2008.8442585
10. Rahin AB, Rahin VB (2016) A low voltage and low power two-stage operational amplifier FinFET transistors. Int Acad J Sci Eng 3(4):80–95
11. Nguyen HV, Kim Y (2012) 32 nm FinFET based 0.7 to 1.1V digital voltage sensor with 50 mV resolution. IEEE Xplore. https://doi.org/10.1109/ICICDT.2012.6232846
12. Verma G, Negi H, Saxena AK (2020) Design and simulation of FinFET based temperature sensor. IEEE Xplore. https://doi.org/10.1109/Icon-CuTE47290.2019.8991440
13. Hashim Y Investigation of FinFET as a temperature nano-sensor based on channel semiconductor type. IOP Conf Ser Mater Sci Eng 226
14. Woo DS, Lee JT, Choi (2013) Electrical characteristics of FinFET with vertically non uniform source/drain profile. IEEE Trans Nanotechnol 1(4):233–237
15. Mohamed YC, Jo M (2013) Junctionless MOSFETs with laterally graded-doping channel for analog/RF applications. J Comput Electron 12(4)
16. Ferhati H, Djeffal F (2018) Graded channel doping junctionless MOSFET: a potential high performance and low power leakage device for nano electronic applications. J Comput Electron 17:129–137
17. Ghosal S, Ganguly M, Ghosh D (2020) A Study on sensitivity of some switching parameters of JLT to structural parameters. Nanosci Nanotechnol Asia 10(4):433–446
18. Rigante S, Lattanzio L, Ionescu AM (2011) FinFET for high sensitivity ion and biological sensing applications. Microelectron Eng 88(8):1864–1866
19. Rahman E, Shadman A, Khosru QDM (2017) Effect of biomolecule position and fill in factor on sensitivity of a dielectric modulated double gate junctionless MOSFET biosensor. Sens Bio-Sens Res 13:49–54
20. Hideshima S, Saito M, Fujita K, Harada Y, Tsuna M, Sekiguchi S, Kuroiwa S, Nakanishi T, Osaka T (2018) Label-free detection of allergens in food via surfactant-induced signal amplification using a field effect transistor-based biosensor. Sens Actuators B Chem 254:1011–1016

21. Singha S, Raja B, Vishvakarma SK (2018) Analytical modeling of split-gate junction-less transistor for a biosensor application. Sens Bio-Sens Res 18:31–36
22. Sadighbayanab D, Hasanzadeh M, Ghafar-Zadeh E (2020) Biosensing based on field-effect transistors (FET): recent progress and challenges. TrAC Trends Anal Chem 133:116067
23. Baliga BJ (2013) Gallium nitride devices for power electronic applications. Semicond Sci Technol 28(7):074011
24. Li X, Taofei Pu, Li L, Ao J-P (2020) Enhanced sensitivity of GaN-based temperature sensor by using the series Schottky barrier diode structure. IEEE Electron Dev Lett 41(4):601–604
25. Kuzmík J, Javorka P, Alam A, Marso M, Heuken M, Kordoˇs P (2002) Determination of channel temperature in AlGaN/GaN HEMTs grown on sapphire and silicon substrates using DC characterization method. IEEE Trans Electron Devices 49(8):1496–1498

Synthesis and Characterization of Zinc Oxide Nanoparticles Using *Catharanthus Roseus* Leaf Extract

K. A. Khan, M. Shaiful Islam, M. N. Islam Khan, and Sumanta Bhattacharyya

Abstract Zinc oxide (ZnO) nanoparticles (NPs) have been successfully synthesized and characterized. Firstly, it has been synthesized the ZnO NPs and then after calcined at 400, 500 and 600 °C for getting better crystalline ZnO NPs. More than 70 components including alcohol, keton, steroid, terpinoid and carotinoide were present in the leaf extract which was confirmed by gas chromatography mass spectrometer (GC–MS) analysis. It has been characterized the synthesized ZnO NPs through XRD, FESEM, EDX, FTIR, DSC, TGA and VSM. The crystalline size of the XRD was around (33.58–25.8) nm. It was shown by the elemental analysis of the ZnO NPs by EDX, and it is shown that the zinc, oxygen, carbon is in the ZnO NPs. The FTIR analysis showed that the capping agents of the NPs contained the functional groups alcohol, alkene, kitone, terpinoied, organic acid. The thermal stability was determine. It was shown that the exothermic peaks ware created. It was also shown that the heat enthalpy was of ZnO NPs were 2326 j/g and 242.7 μVs/mg, respectively. Using TGA, it is shown that the percentage of weight loss was of about 10% for ZnO. It is also found that the ZnO NPs were super paramagnetic in nature with zero coercivity. It is also found that the crystalline size has been changed with calcinations temperature. But the crystallite size increases with calcinations temperature, which supported the results of XRD and VSM.

Keywords ZnNPs · *Catharanthus roseus* leaf extract · FTIR · Characterizations · Crystalline size · Calcination temperature

---

K. A. Khan · M. Shaiful Islam
Department of Physics, Jagannath University, Dhaka 1100, Bangladesh

M. N. Islam Khan
Nanoscience and Technology Research Laboratory, Atomic Energy Centre, Dhaka, Bangladesh
Atomic Energy Commission, Dhaka 1000, Bangladesh

S. Bhattacharyya (✉)
Department of ECE, Greater Kolkata College of Engineering and Management, Kolkata, India

J. K. Mandal et al. (eds.), *Topical Drifts in Intelligent Computing*, Lecture Notes in Networks and Systems 426, https://doi.org/10.1007/978-981-19-0745-6_38

1 Introduction

Green synthesis is basically a chemical process [1–3]. The advantage of green synthesis is cheap equipment and simple experimental set up, easy to fabricate, easy to control, uniform particle size, no need any extra surface functionalizing agent [4–6]. There are various kinds of nano-synthesis methods. The reducing and precipitation are the main chemical reactions involved in the process [7–9]. These are physical, chemical and biological [10–13]. The physical synthesis is ball milling, thermal evaporation, lithography and vapour phase [14, 15]. The chemical synthesis is sol–gel processing, solution-based synthesis and co-precipitation method [16–19]. The biological synthesis is in bacteria, in fungi, yeast and using plant extracts [20–26]. The attention focussed on characterization of the synthesis of ZnO NPs. Many bio molecules in leaf extract such as sterols, carotinoid, terpinoid, alkaloids, alcoholic ketonic and organic acid compounds could be involved in bio reduction, formation and stabilization of ZnO NPs [27–29]. Preparation of ZnO NPs via green synthesis route. Investigation of the properties of the ZnO NPs has been conducted like elemental analysis of leaf extract by GC–MS, crystal structure by XRD [30, 31], surface morphology by SEM and FESEM, functional group by FTIR, elemental composition by EDX, magnetic properties by VSM and thermal properties by DSC/TGA.

2 Methodology

2.1 Materials Used for this Research Work

The material as precursors is zinc acetate dihydrate (Zn(CH3COO)2-2H2O) M.W = 219.50 g/mol. The other materials are solvent leaf extract and distilled water. The synthesis conditions are synthesis temperature = 80^0C, concentration Zn(CH3COO)2-2H_2O = 0.3 M, 200 mL, leaf extract = 200 mL, magnetic starrier speed = 800 rpm. The method of leaf extract preparation and the chemical reactions for ZnO NPs is given by the following:

$$\text{Catharanthus roseus} + H_2O + Zn^{2+}_{(aq)} \xrightarrow{\text{Stirring}} [\text{Cathararanthus roseus}/\ Zn^{2+}] \quad (1)$$

$$[\text{Cathararanthus roseus}/Zn^{2+} \rightarrow [\text{Cathararanthus roseus}/ZnO] \downarrow (s) + H_2O\ (aq) \quad (2)$$

It is shown (from Fig. 1) that the different steps of *Catharanthus roseus* leaf juice preparation. At the starting, it is shown that a *Catharanthus roseus* leaf tree. At the end of the Fig. 1, it was found the *Catharanthus roseus* leaf extract for production of zinc oxide nanoparticles (NPs).

Fig. 1 Different steps of *Catharanthus roseus* leaf juice preparation

It is also shown (from Fig. 2) that the different steps for getting ZnO NPs.

Fig. 2 Different steps for getting ZnO NPs

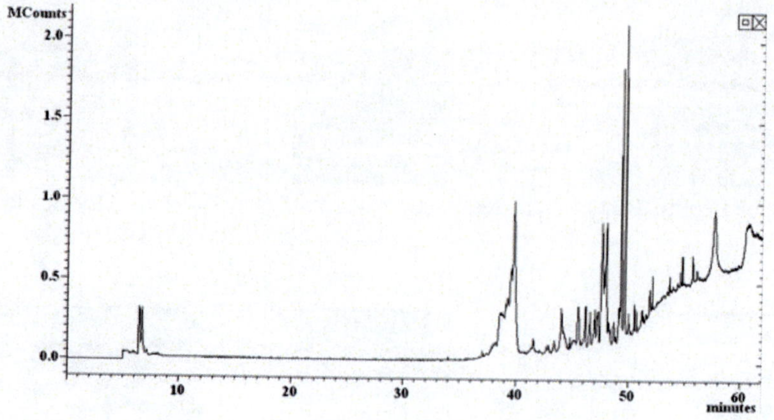

Fig. 3 GC–MS analysis of *Catharanthus roseus* leaf extracts

3 Results and Discussion

3.1 GC–MS Analysis

The GC–MS shows (Fig. 3) more than 65 prominent peaks in the retention time range 5–59 min. The GC–MS analysis of *Catharanthus roseus* reveals the presence of poly-alcohols, carboxylic acids, ketons, sterols, terpnoid, carotenoid, etc.

3.2 XRD Analysis

From Fig. 4, it is shown that the XRD patterns of ZnO nanoparticles. It is found that the reflection peaks becomes sharper and intensity of diffraction peaks increase with increasing calcinations temperature.

3.3 ZnO NPs Using Scherer's Method

From Fig. 4, it is found that the presence of diffraction peaks (100), (002), (101), (102), (110), (103), (200), (112) and (201) for 400 °C at ($2\theta = 31.92°$, 34.60°, 36.33°, 47.77°, 56.62°, 63.00°, 68.07); further calcined at 500 °C ($2\theta = 31.89°$, 34.59°, 36.34°, 47.68°, 56.73°, 63.01°, 68.10°); and finally calcined at 600 °C and $2\theta =$ 31.82°, 34.43°, 36.42°, 47.63°, 56.73°, 62.96°, 66.53°, 68.07°, 69.30), respectively. The average crystallite size of the ZnO NPs for different calcinations temperature was determined by the X-ray line broadening method using the Scherer equation.

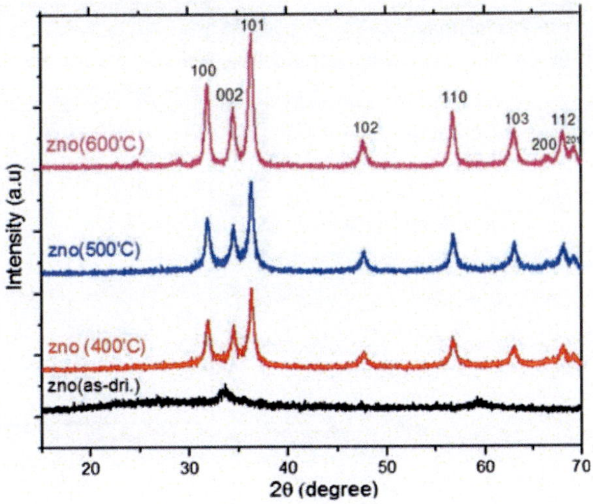

Fig. 4 XRD patterns of ZnO NPs

Table 1 Crystallite size at different temperature

S. no.	Temperature (°C)	Crystallite size D in (nm)
A	400	33.58
B	500	28.98
C	600	25.83

$$D = \frac{k\lambda}{\beta \cos\theta} \tag{3}$$

Here, D = Crystallite size; k = Shape parameter, which is 0.89 for ZnO, λ = X-ray wavelength (0.15406 *nm*), β = Full width at half maximum (FWHM) in radians in the 2θ scale and θ = Bragg angle [5] (Table 1).

The decrease in crystallite size from 35 to 25 nm was found with the variation of calcination temperature [6]. The variation of crystallite size with increase in calcinations temperature is shown in Fig. 5. The decrease in crystallite size at high calcination temperature is indicative of re-structuring process. From this XRD analysis, the structural parameters such as dislocation density and microstrain are also determined. The dislocation density and microstrain of ZnO NPs were calculated by using following formula, respectively.

$$\delta = \frac{1}{D^2} \tag{4}$$

$$\varepsilon = \frac{\beta \cos\theta}{4} \tag{5}$$

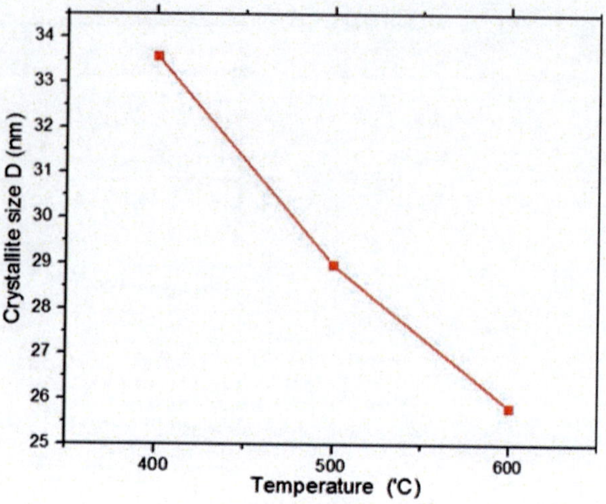

Fig. 5 Crystallites size versus calcinations temperature

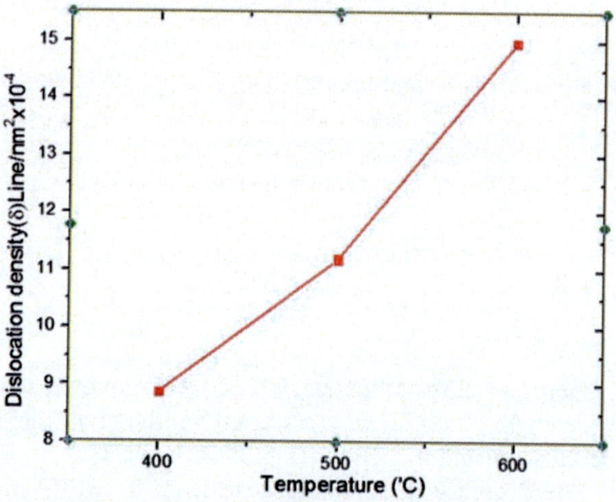

Fig. 6 Dislocation density (δ) versus calcinations temperature

where $D =$ crystallite size which is calculated using Scherrer approximation and β is the FWHM which is obtained from XRD analysis of ZnO NPs. Variation in the structural parameter of ZnO NPs with calcination temperature obtained from XRD analysis is shown in Table 2.

Table 2 Value of different calcination temperature

Temperature (°C)	D (nm)	δ (line/nm^2 × 10^{-3})	ε ($\times 10^{-3}$)
400	33.58	8.87	0.852
500	28.98	11.19	1.015
600	25.83	14.98	1.178

Table 3 Elements versus mass (%)

Elements	Mass (%)
Zinc (Zn)	43.36
Oxygen (O)	29.57
Carbon (C)	26.96

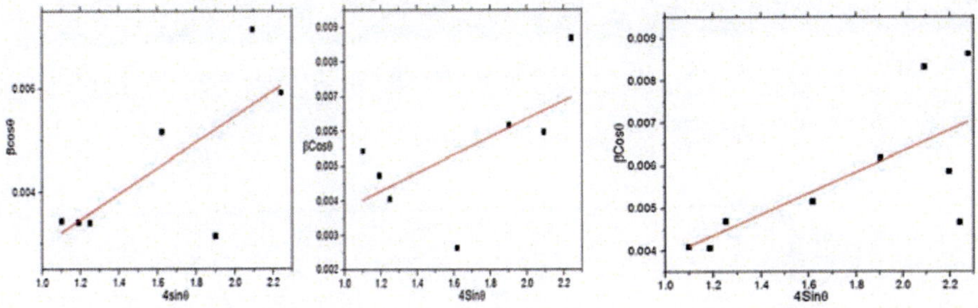

(a) Slope is 2.49 x 10^{-3} for 400^{0}C (b) Slope is 1.17 x 10^{-3} for 500^{0}C (c) Slope is 2.47 x 10^{-3} for 600^{0}C

Fig. 7 Crystallite size and strain of ZnO NPs using Williamson-Hall method

It is shown (from Table 3) that structural parameter for different calcinations temperature. Here, $D =$ Crystallite size, $\delta =$ Dislocation density and $\varepsilon =$ Microstrain.

Figure 5 shows the average crystallite size decreases with increasing calcination temperature. Figure 6 shows the dislocation density increases with increasing calcinations temperature.

3.4 *Crystallite Size and Strain of ZnO NPs Using Williamson-Hall's Method*

It is found (from Fig. 7) that the Williamson-Hall method was employed to separate the peak broadening due to microstrain and due to particle size effect. The ZnO samples calcined at 400, 500 and 600 °C temperatures displayed in the Fig. 7.

3.5 *FESEM Analysis*

Figure 8 shows the shape of the image after analysis for different temperatures are 400 °C, 500 °C and 600 °C, respectively. From Fig. 8a, b, it shows that it is calcined at 400, 500 °C and the particles are in nanosize. It is also shown from Fig. 8c that it is calcined at 600 °C, and it is formed nanoparticles with homogeneous and spherical shape.

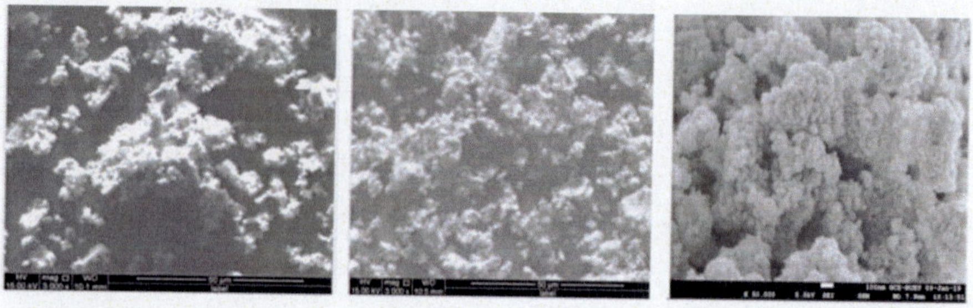

(a) FESEM image calcined at 400°C (b) FESEM image calcined at 500°C (c) FESEM image calcined at 600°C

Fig. 8 FESEM image after analysis

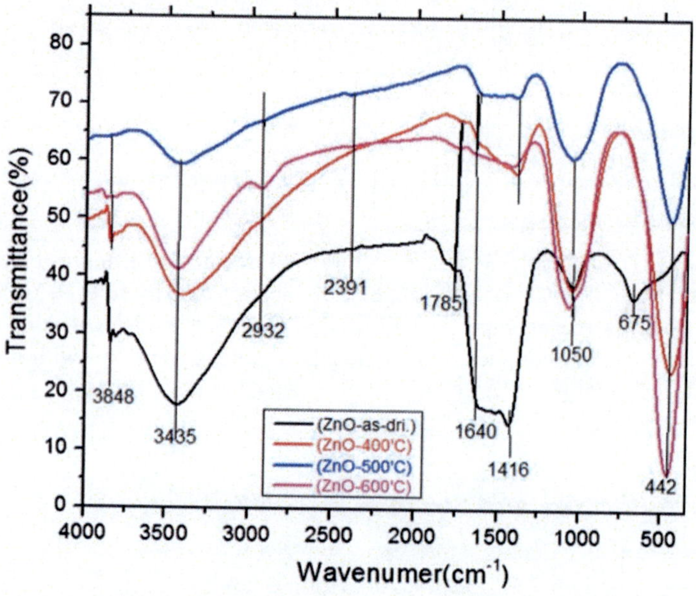

Fig. 9 Functional group analysis of ZnO nanoparticles using FTIR spectroscopy

3.6 *FTIR Analysis*

Fourier transform infrared (FTIR) spectroscopy has been used in order to determine the presence different functional groups on synthesized ZnO NPs. FTIR measurements are also performed in order to identify the presence of bioorganic compounds responsible for reducing, capping and stabilizing the ZnO NPs is in the *C. roseus juice*. In this study, ZnO NPs are synthesized through the reduction of zinc acetate using *C. roseus* leaf extract as a reducing agent, and the presence of reducing agents is confirmed from the FTIR spectra of as dried sample as shown in Fig. 9, where there are two strong peaks at 3848 and 3430 cm^{-1} correspond to N–H stretching

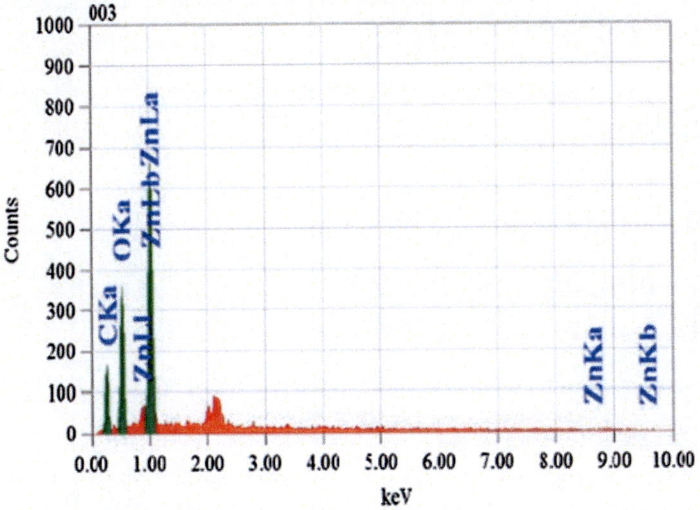

Fig. 10 Elemental analysis at 600 °C: EDX

of amide and –OH stretching vibration (phenolic compounds) are found. There are also some intense peaks are observed at 2932 and 2391 cm^{-1} 2391 correspond to the C–H stretching vibration and C=O stretching vibrations for as dried sample acting as reducing and capping agents [11].

3.7 Elemental Analysis and Magnetization of ZnO NPs Using EDX and VSM Analysis

It is shown in Fig. 10 that the different parts of the components of the ZnO nanoparticles (NPs).

Figure 10 shows the elemental analysis at 600 °C and, Fig. 11 shows the VSM image of ZnO NPs after analysis. It is shown in Fig. 10 that there is no corecivity, and it is also shown in Fig. 11 that ZnO NPs are super paramagnetic, and it is also shown that the saturation magnetization of the ZnO NPs was 0.00182 emu/g (Fig. 12).

3.8 TGA and DSC Analysis

This study will help in synthesis of other metal oxide NPs by the depicted procedure in the future. More diverse plant species will be exploited in the future era using this non-toxic, easy, simple low cost and environmental friendly process. The anti-bacterial activities of these type NPs can be used in medical science for human

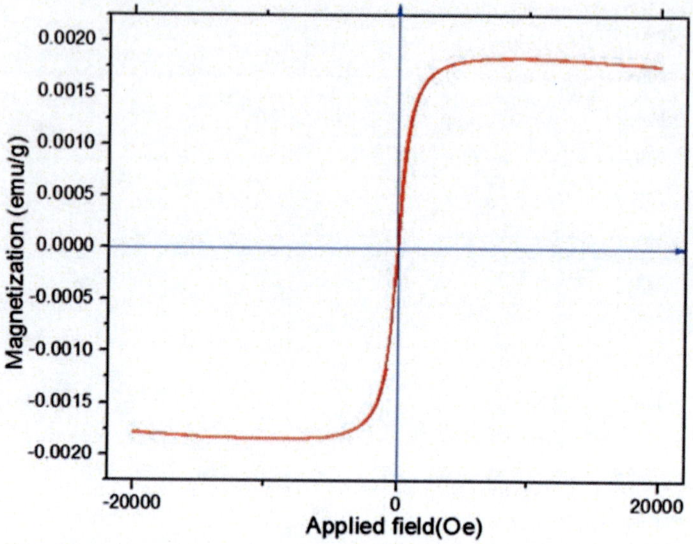

Fig. 11 VSM image of ZnO NPs

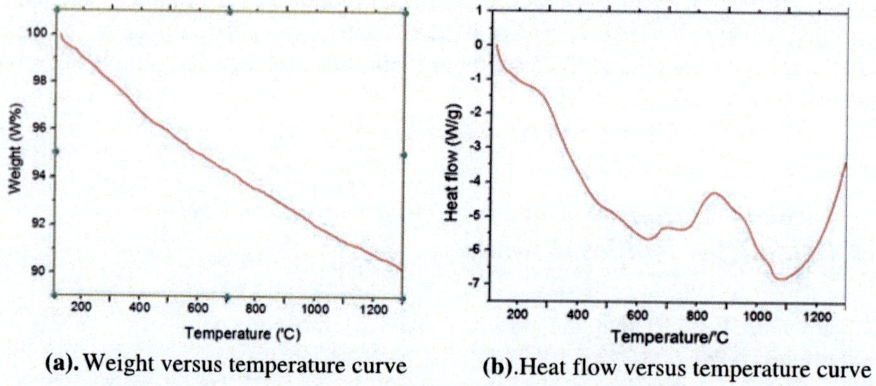

Fig. 12 Thermal analysis of nanoparticles: TGA and DSC

welfare. This ZnO NPs can be used in PKL electrochemical cell for PKL electricity generation.

Acknowledgements Authors are grateful to the Ministry of Education for financing during the research work (Project/User ID: PS2019949).

References

1. Hamid MR (2013) Characterization of a battery cell fueled by bryophyllum pinnatum sap. Int J Sci Eng Res 9(3). ISSN 2229-5518
2. Hamid MR, Yusuf A, Wadud AMA, Rahaman MM (2016) Design and performance test of a prototype of a 12 volt DC battery fueled by bryophyllum pinnatum sap and improvement of its characteristics. Int J Electron Electric Eng 4(5)
3. Pal S, Chaudhuri AKN (1991) Studies on the anti-ulcer activity of a *Bryophyllum pinnatum* leaf extract in experimental animals. J Ethanopharmacol 33:97–102
4. Ying CY, Dayou J (2014) Increasing the energy output from living-plants fuel cells with natural 8(14):20–23
5. Ojewole JAO (2005) Antinociceptive, anti-inflammatory and antidiabetic effects of *Bryophyllum pinnatum* (Crassulaceae) leaf aqueous extract. J Ethnopharmacol 99:13–19
6. Asiedu-Gyekye IJ, Antwi DA, Bugyei KA, Awortwe C (2012) Comparative study of two kalanchoe species: total flavonoid, phenolic contents and antioxidant properties. Afr J Pure Appl Chem 6:65–73
7. Khan KA (1999) Copper oxide coatings for use in a linear solar Fresnel reflecting concentrating collector, Published in the Journal of Elsevier, Renewable Energy, An International Journal, WREN (World Renewable Energy Network), UK, RE: 12.97/859,1998, Publication date 1999/8/1. J Renew Energy 17(4):603–608. Publisher—Pergamon
8. Ruhane TA, Tauhidul Islam M, Rahman MS, Bhuiyah MMH, Islam JMM, Bhuiyah TI, Khan KA, Khan MA (2017) Impact of photo electrode thickness annealing temperature on natural dye sensitized solar cell. Sustain Energy Technol Assess Elsevier. https://doi.org/10.1016/j.seta.2017.01.012
9. Ruhane TA, Tauhidul Islam M, Rahaman MS, Bhuiyan MMH, Islam JMM, Newaz MK, Khan KA, Khan MA (2017) Photo current enhancement of natural dye sensitized solar cell by optimizing dye extraction and its loading period. Optik Int J Light Electron Opt Elsevier
10. Hasan M, Khan KA (2018) Dynamic model of Bryophyllum pinnatum leaf fueled BPL cell: a possible alternate source of electricity at the off-grid region in Bangladesh, microsystem technologies micro- and nano-systems information storage and processing systems, Springer. Microsyst Technol. ISSN 0946-7076. https://doi.org/10.1007/s00542-018-4149-y
11. Khan KA, Hazrat Ali M, Obaydullah AKM, Wadud MA (2019) Production of candle using solar thermal technology, microsystem technologies micro- and nano-systems information storage and processing systems, Springer. Microsyst Technol 25(12). ISSN 0946-7076. https://doi.org/10.1007/s00542-019-04390-7
12. Khan KA, Rasel SA, Ohiduzzaman M (2019) Homemade PKL electricity generation for use in DC fan at remote areas, microsystem technologies micro- and nano-systems information storage and processing systems. Microsyst Technol 25(12). ISSN 0946-7076. https://doi.org/10.1007/s00542-019-04422-2
13. Hasan M, Khan KA (2019) Experimental characterization and identification of cell parameters in a BPL electrochemical device, Springer. SN Appl Sci 1:1008. https://doi.org/10.1007/s42452-019-1045-8
14. Hassan L, Khan KA (2019) A study on harvesting of PKL electricity. Springer J Microsyst Technol (2020) 26(3):1031–1041. https://doi.org/10.1007/s00542-019-04625-7
15. Khan KA, Mamun MA, Ibrahim M, Hasan M, Ohiduzzaman M, Obaydullah AKM, Wadud MA, Shajahan M (2019) PKL electrochemical cell: physics and chemistry. Spr J SN Appl Sci 1:1335. https://doi.org/10.1007/s42452-019-1363-x
16. Hazrat Ali M, Chakma U, Howlader D, Tawhidul Islam M, Khan KA (2019) Studies on performance parameters of a practical transformer for various utilizations, microsystem technologies, Springer, Accepted: 03 Dec 2019. https://doi.org/10.1007/s00542-019-04711-w
17. Khan KA, Hassan L, Obaydullah AKM, Azharul Islam SM, Mamun MA, Akter T, Hasan M, Alam MS, Ibrahim M, Mizanur Rahman M, Shahjahan M (2018) Bioelectricity: a new approach to provide the electrical power from vegetative and fruits at off-grid region. Microsyst Technol Springer. https://doi.org/10.1007/s00542-018-3808-3(0123456789)

18. Khan KA, Bhuyan MS, Mamun MA, Ibrahim M, Hasan L, Wadud MA (2018) Organic electricity from Zn/Cu-PKL electrochemical cell. In: Mandal JK et al (eds) Contemporary advances in innovative and applicable information technology, advances in intelligent systems and computing, vol 812, Chap. 9. © Springer Nature Singapore Pvt. Ltd., pp 75–90
19. Khan KA, Hazrat Ali M, Mamun MA, Mahbubul Haque M, Atique Ullah AKM, Islam Khan MN, Lovelu Hassan AKM, Obaydullah MAW (2020) Bioelectrical characterization and production of nanoparticles (NPs) using PKL extract for electricity generation, Received: 31 July 2018/Accepted: 4 February 2020. Microsyst Technol Spr J. https://doi.org/10.1007/s00542-020-04774-0
20. Khan KA, Rahman A, Rahman MS, Tahsin A, Jubyer KM, Paul S (2016) Performance analysis of electrical parameters of PKL electricity (An experimental analysis on discharge rates, capacity & discharge time, pulse performance and cycle life & deep discharge of PathorKuchi Leaf (PKL) electricity cell). In: Innovative smart grid technologies-Asia (ISGT-Asia), 2016 IEEE. IEEE, pp 540–544
21. Khan MKA, Paul S, Rahman MS, Kundu RK, Hasan MM, Moniruzzaman M, Mamun MA (2016) A study of performance analysis of PKL electricity generation parameters: (An experimental analysis on voltage regulation, capacity and energy efficiency of pathorkuchi leaf (PKL) electricity cell). In: Power India international conference (PIICON), 2016 IEEE 7th. IEEE, pp 1–6
22. Khan MKA, Rahman MS, Das T, Ahmed MN, Saha KN, Paul S (2017) Investigation on parameters performance of Zn/Cu electrodes of PKL, AVL, tomato and lemon juice based electrochemical cells: a comparative study. In: 2015 3rd international conference on electrical information and communication technology (EICT). IEEE, pp 1–6
23. Khan KA, Rahman A, Rahman MS, Tahsin A, Jubyer KM, Paul S (2016) Performance analysis of electrical parameters of PKL electricity (An experimental analysis on discharge rates, capacity & discharge time, pulse performance and cycle life & deep discharge of Pathor Kuchi Leaf (PKL) electricity cell). In: Innovative smart grid technologies-Asia (ISGT-Asia), 2016 IEEE. IEEE, pp 540–544
24. Khan MKA, Paul S, Rahman MS, Kundu RK, Hasan MM, Moniruzzaman M, Mamun MA (2016) A study of performance analysis of PKL electricity generation parameters: (An experimental analysis on voltage regulation, capacity and energy efficiency of pathor kuchi leaf (PKL) electricity cell). In: Power India international conference (PIICON), 2016 IEEE 7th. IEEE, pp 1–6
25. Paul S, Khan KA, Islam KA, Islam B, Reza MA (2012) Modeling of a biomass energy based (BPL) generating power plant and its features in comparison with other generating plants. IPCBEE 44. IACSIT Press. https://doi.org/10.7763/IPCBEE.2012.V44
26. Khan KA, Paul S (2013) A analytical study on electrochemistry for PKL (Pathor Kuchi Leaf) electricity generation system, Publication date 2013/5/21, Conference—Energytech, 2013 IEEE. Publisher, IEEE, pp 1–6
27. Hasan M, Khan KA (2016) Bryophyllum pinnatum leaf fueled cell: an alternate way of supplying electricity at the off-grid areas in Bangladesh. In: Proceedings of 4th international conference on the developments in renewable energy technology [ICDRET 2016], p 01. https://doi.org/10.1109/ICDRET.2016.7421522
28. Hasan M, Khan KA, Mamun MA (2017) An estimation of the extractable electrical energy from bryophyllum pinnatum leaf. Am Int J Res Sci Technol Eng Math (AIJRSTEM) 01(19):100–106
29. Ullah AA, Haque MM, Akter M, Hossain A, Tamanna AN, Hosen MM, Khan MKA et al (2020) Green synthesis of Bryophyllum pinnatum aqueous leaf extract mediated bio-molecule capped dilute ferromagnetic α-MnO2 nanoparticles. Mat Res Exp 7(1):015088
30. Sharma BK (1998) Electrochemistry, Krishna Prakashan Media (P) Ltd., 11
31. Khan MKA, Rahman MS, Das T, Ahmed MN, Saha KN, Paul S (2017) Investigation on parameters performance of Zn/Cu Electrodes of PKL, AVL, tomato and lemon juice based electrochemical cells: a comparative study. In: 2017 3rd international conference on electrical information and communication technology (EICT). IEEE, Khulna, pp 1–6. https://doi.org/10.1109/EICT.2017.8275150

Effects of Tilting ESD Gun on Discharging Current

Nikhilesh Kumar Neelu, Nisha Gupta, and Mahmood Tabaddor

Abstract In this paper, an investigation is carried out to study the impact of tilting the electrostatic discharge (ESD) gun on discharging current waveform which is critical for studying electromagnetic compatibility (EMC) issues due to ESD and ESD compliance of electrical or electronics (E/E) modules. The impact with respect to various angles and ESD voltages are simulated, and results are compared with IEC 61000-4-2. The ESD gun is rotated in the range of $\pm 10°$ in the X and Y axis, and its effects on discharge current are monitored between ESD gun tip and target through ground strap with respect to the different ESD voltage levels. It is found that the angular rotation of the gun affects the discharging current significantly. To test the discharge waveform more accurately, current target is simulated.

Keywords Electrostatic discharge (ESD) · ESD generator · Electromagnetic compatibility (EMC) · Electromagnetic interference (EMI) · Immunity · Numerical modeling · Numerical simulations

1 Introduction

Electrostatic discharge is a known threat to the electronic devices, circuits, and ICs [1–6]. The larger integration of electronics has led to aggravation of the ESD problem in high-speed circuits. The ESD phenomenon often treated as a natural phenomenon occurs when two dissimilar non-conducting materials are brought in contact and then separated. As a result, one material gets positively charged while the other one negative. Both positively and negatively charged materials may charge up to −25 kV

N. K. Neelu (✉) · M. Tabaddor
UL India Private Limited, Bangalore, Karnataka, India
e-mail: nikhilesh.neelu@ul.com

M. Tabaddor
e-mail: mahmood.tabaddor@ul.com

N. Gupta
Birla Institute of Technology, MESRA, Ranchi, Jharkhand, India
e-mail: ngupta@bitmesra.ac.in

J. K. Mandal et al. (eds.), *Topical Drifts in Intelligent Computing*, Lecture Notes in Networks and Systems 426, https://doi.org/10.1007/978-981-19-0745-6_39

and +25 kV ESD voltage, respectively, that can discharge if any conductive material is kept in the vicinity of ESD-charged materials. This leads to the development of high electrostatic voltage which discharges as a current impulse. High voltages that result from ESD might cause high current spikes of from 0.6 to 1 ns rise time and is sufficient to damage any E/E (electrical/electronics) module, ICs, microcontrollers, etc. ESD either causes partial or permanent damage of E/E module on subsystem or system level.

The research related to ESD phenomena can be categorized into three main streams such as ESD simulation and modelling, its effects on modern electronic devices, circuits, and ICs, and finally, the protection schemes to control it. The proposed work is placed under the first category. The ESD gun is nothing but an ESD generator used to create ESD over desired and in the vicinity of electronic devices, circuits, and ICs. In absence of the experimental ESD gun, it is generally modeled using several 3-D simulation tools such as HFSS, CST Microwave studio, etc., to study the behavior of ESD discharge current. The gun can model the ESD in air or contact discharge mode.

The study of tilt effect is necessary because several parts and sections of the automobiles, aircraft, other objects, etc., may not always be accessible directly to the ESD gun, and it becomes difficult for the gun to make a contact with the body without tilting it. As a result, this may lead to the situation where it becomes necessary to investigate the effect of the tilting [1, 2] the ESD gun on discharge current.

In [1], there is some discussion about the sensitivity aspect of the test with respect to relative positions and rotation of the gun, length, and termination of the grounding strap, as well as other factors. However, no simulation or practical data or results are shown to validate the above statements. In the proposed work, the simulations are carried out using CST microwave studio (MWS) to examine the effects of angular displacement of the ESD gun and investigate its effects on generation of ESD current waveform over a metallic sheet.

Model of ESD Simulator.

1.1 3-D Physical Model

To reproduce the human–metal ESD waveform, ESD generators are used. The model presented in [4] is used for simulation in the present work shown in Figs. 1 and 2, according to the standard IEC 61000-4-2 [7, 8]. The model contains dielectric material, metal as PEC and lumped element part to reproduce ESD physical current waveform more accurately. The lumped element combination is taken to reproduce the human–metal ESD current reference waveform shown in Fig. 2 for the contact mode discharge testing.

The geometry of simulated model is shown in Fig. 1 and the schematic model of ESD gun designed in MWS is shown in Fig. 2. The excitation voltage, the ideal voltage source at port-3 is assumed to be +8 kV, with a rise time of 1 ns to simulate slow charging and fast discharging of ESD gun. Port-1 and Port-2 are

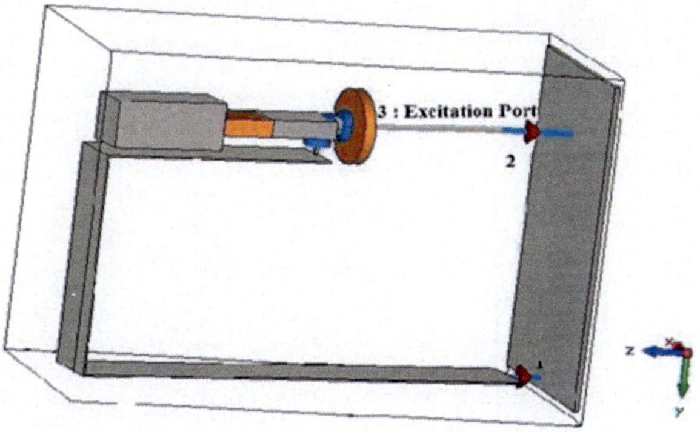

Fig. 1 Full model of ESD simulator in MWS showing ports, strap, discharging tip, and discharge wall simulated in CST-MWS

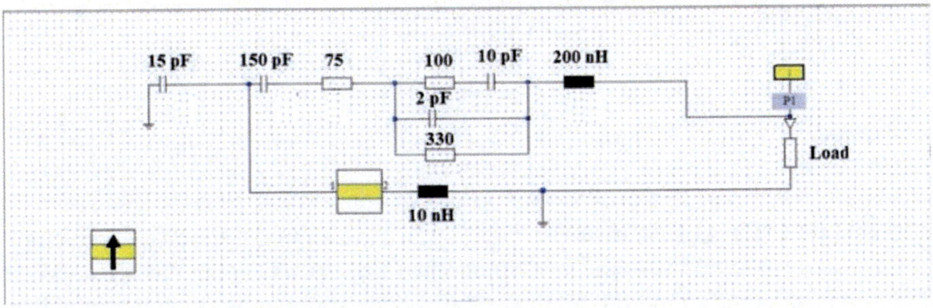

Fig. 2 Equivalent schematic model of ESD simulator in MWS

S-parameter port. The S-parameter ports are terminated into a very low impedance of 1 Ω according to the ESD-target specification detailed in [4, 7]. The simulated model is made to discharge through port 2, which allows to observe time varying o/p current waveform as shown in Fig. 3. 2. The port-3 is a coupling port which couples the discharged current to the gun model through the grounding strap. The model is in "direct application" of static charge to EUT. In the present case, EUT is assumed to be a metallic wall. The "energy storage capacitor" for human–metal ESD is taken as 150 pF, the discharging resistance is 330 Ω, the discharging tip capacitance is as 2 pF, and to couple discharge waveform through coupling strap, and the lumped values are 2 Ω and 1 nH. The details of this combination are described in [4]. The generated waveform is in accordance with the reference discharge current waveform proposed by IEC in IEC 61000-4-2, except for the second peaks, which are mostly due to the ringing effect caused by the reactance of the dielectric material [5, 6, 8]. However, the present IEC standard does not include any specification or guidelines for the consideration of the ringing effects.

Next, an investigation is carried out to study the effects of the angular orientation of the gun on discharging current. The waveform depicting the discharging current when the ESD gun is kept perpendicular to the discharging surface is shown in Fig. 3. In Fig. 5a, b, the angular orientation of ESD gun with respect to vertical axis and

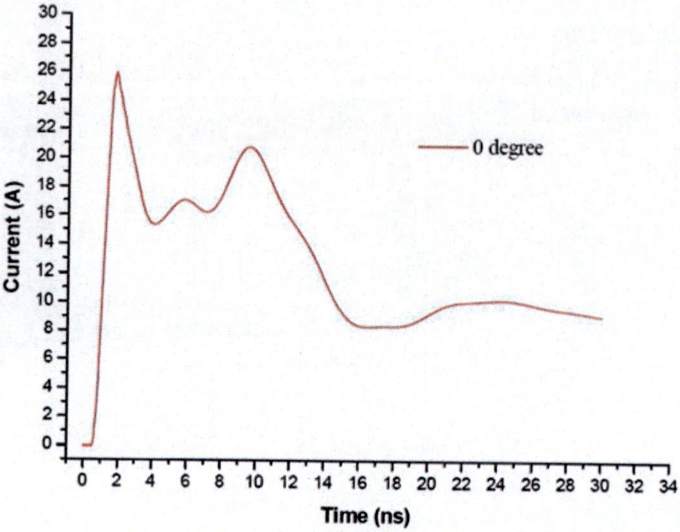

Fig. 3 Current waveform for 0 (zero) degree angular rotation for +8 kV discharging potential

horizontal axis, respectively, is shown, and Figs. 6 and 7 show the simulated discharge current with respect to the angular orientation of ESD gun keeping the discharging tip at a fixed position for all the orientation. It is seen that the rise time lies between 0.6 and 1 ns, though not constant for all the angular orientation, i.e., X-direction and Y-direction rotation of discharge gun.

1.2 3-D Physical Model of Coaxial Current Target of ESD Simulator

To calibrate the ESD generator system for checking the repeatability and performance of current measurement, ESD target is needed [8]. The coaxial current target needs to be attached to an attenuator for the purpose of measurement of the ESD up to 15 kV. For 25 kV measurement, one more attenuator of 20 dB will be required. Inner cross-section coaxial current target is shown in Fig. 4. The target has been designed as per [8].

2 Results

For the 0 (zero) degree angular orientation, the peak of the discharge current is highest compared to all the other orientations and lies in the range as specified in IEC61000-4-2. The discharge current level at 30 ns is 9.14 A, which is below the

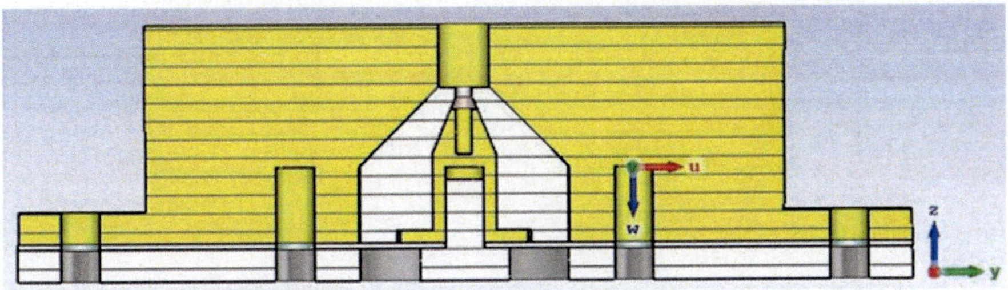

Fig. 4 Inner cross-sectional view of simulated structure of coaxial current target, (dimensions and materials as in IEC-61000-4-2)

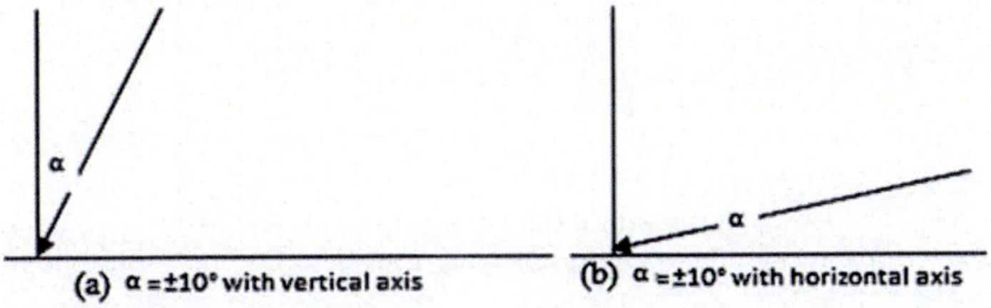

Fig. 5 **a, b** The angular orientation of ESD gun with vertical axis and horizontal axis, respectively

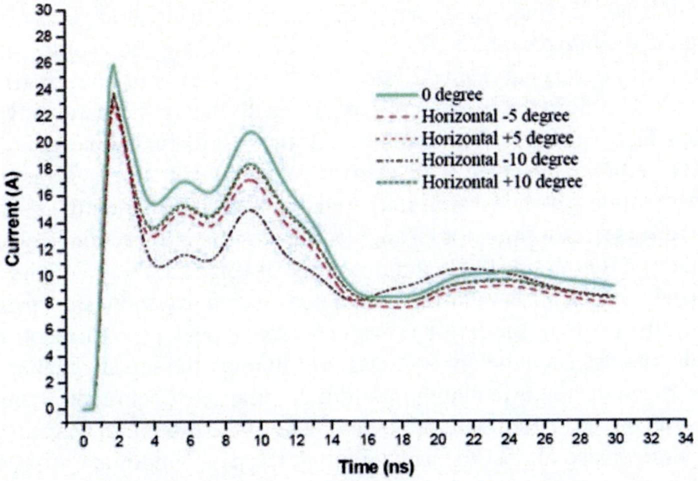

Fig. 6 Discharge current waveform with different angular position varied perpendicular to the discharge surface in X-direction for +8 kV of discharging voltage

specified level and the transient time duration is high, and due to this, the radiated emission frequency band is expected to be high as described in IEC61000-4-2.

While investigating the amplitude of the discharging waveform, it is found that the amplitude of current waveform for all the cases other than the zero-degree case

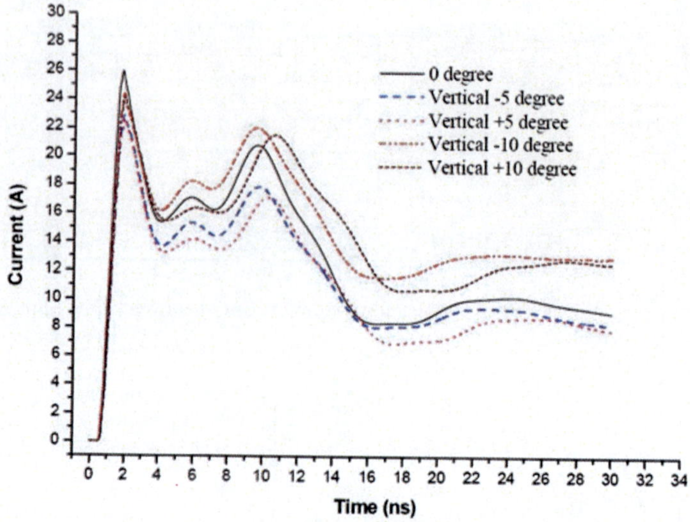

Fig. 7 Discharge current waveform with different angular position varied perpendicular to the discharge surface in Y-direction for +8 kV discharging voltage

are well below the limit 30 ± 15%A as specified in IEC61000-4-2 [8]. Next, the ESD gun is rotated in horizontal X-direction and vertical Y-direction between −10 to + 10°, and results are obtained as shown in Fig. 5. For a different angular rotation of the ESD gun in the horizontal, i.e., X-direction, the peak amplitude of the discharge current waveform is varying from 22.7 to 23.6 A, but in case of angular rotation in vertical, i.e., Y-direction, the peak amplitude of the discharging current waveform is varying from 22.6 to 24.5 A. The symmetry in the results for all the cases such as −5, +5, −10, +10 are observed with an error percentage of about 4% maximum at 30 ns discharge time which is a well in correlation with IEC standard.

The rise time is almost same for all angular positions in X-direction. However, for the Y-direction, it is not same but in defined range of IEC. The second higher current waveform peak may be observed in the simulated discharge current waveform of the gun, which is the result of the ringing effect of the first peak [4, 6]. The other peaks add up to the total transient behavior duration. Although the second peak is highest in case of 0 (zero) degree orientation in X-direction, in Y-direction discharge for − 10° orientation, the peak dominates all other peaks of the discharge waveforms. The occurrence time is also almost the same, though the peak amplitude of second and third peak has major differences for various orientation.

The above observed behavior may be caused by the less contact area upon rotation of the ESD gun. Due to this, the amount of discharge current observes different reactive behavior of the contact area.

The IEC61000-4-2 standard specifies the magnitude of the current to be 16 A ± 30% at 30 ns. However, our simulation result of discharging current amplitude shows a deviation in the range of from 7.9 to 13 A. For 0 (zero) degree orientation, the current

Table 1 Values for different orientation in X-direction for +8 kV discharging voltage

Position in X-direction (Deg)	First peak current of discharge 30 ± 15% (A)	First peak occurrence time (ns)	Rise time 0.8 ± 25% (ns)	Current at 30 ns 16 ± 30% (A)
0	25.937	1.787	0.789	9.139
−5	22.779	1.806	0.803	7.940
+5	22.862	1.800	0.802	8.056
−10	22.990	1.811	0.803	8.231
+10	23.512	1.800	0.814	8.179

Table 2 Values for different orientation in Y-direction for +8 kV discharging voltage

Position in Y-direction (Deg)	First peak current of discharge 30 ± 15% (A)	First peak occurrence time (ns)	Rise time 0.8 ± 25% (ns)	Current at 30 ns 16 ± 30% (A)
0	25.937	1.787	0.789	9.139
−5	22.865	1.881	0.783	8.287
+5	22.602	1.928	0.811	7.949
−10	24.318	1.882	0.820	12.972
+10	24.420	1.902	0.825	12.559

level is in the range of 16 A ± 30%, but results for other orientations fail to qualify for X-direction orientation. The details of the results for various orientations are tabulated in Tables 1 and 2. In Fig. 7, a comparison of the results for 0 (zero) degree orientation and −10° vertical and horizontal orientations are presented (Fig. 8).

3 Conclusions

In all the cases discussed so far, large differences in the current amplitude of second and third peak are observed. These peaks may be due to the discharge gun contact area which decrease with angular orientation. The present standard IEC61000-4-2 [7, 8] does not include *s* effect. Therefore, the study carried out reveals that there is a need to include the effect of tilting of the ESD gun in the present standard.

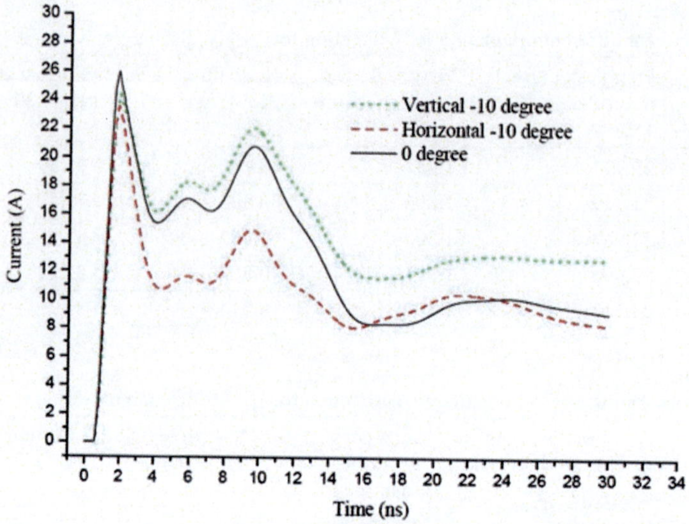

Fig. 8 Comparison of waveform for single orientation of −10° in both X and Y axis

Acknowledgements The authors wish to acknowledge Mr. Md Anjarul Haque and Dr. Prashant Kumar Singh for fruitful discussions and suggestions.

References

1. Zhang J, Beetner G, Moseley R, Herrin S, Pommerenke D (2011) Modelling electromagnetic field coupling from ESD gun to an IC. IEEE Int Symp Electromagn Compatibility (EMC) 553–558
2. Kang TW, Chung YC, Won SH, Kim HT (2002) On the uncertainty in the current waveform measurement of an ESD generator. IEEE Trans Electromagn Compatibility 42(4):405–413
3. Centola F, Pommerenke D, Kai W, Van Doren T, Caniggia S (2003) ESD Excitation model for susceptibility study IEEE Int Symp Electromagn Compat 1:58–63 18-22
4. Caniggia S, Maradei F (2006) Circuit and numerical modeling of electrostatic discharge generators. IEEE Trans Ind Appl 42(6):1350–1357
5. Caniggia S, Maradei F (2007) Numerical prediction and measurement of ESD radiated fields by free-space field sensors. IEEE Trans Electromagn Compatibility 49(3):494–503
6. Chundru R, Pommerenke D, Wang K, Van Doren T, Centola FP, Huang JS (2004) Characterization of human metal ESD reference discharge event and correlation of generator parameters to failure levels—part I: reference event. IEEE Trans Electromagn Compatibility 46(4):498–504
7. Electromagnetic compatibility (EMC)—part 4-2: testing and measurement techniques—electrostatic discharge immunity test (IEC 61000-4-2:2000)
8. Electromagnetic compatibility (EMC)—part 4–2: testing and measurement techniques—electrostatic discharge immunity test (IEC 61000-4-2:2008)

Design and Development of 8T SRAM Cell Using 14 nm FinFET

Panduranga Vemula and Rudra Sankar Dhar

Abstract The main purpose of conventional CMOS is to style SRAM, but the performance of systems is affected due to increase in leakage currents and high-power dissipation. The purpose of memories is to possess short time interval, less power consumption, and low leakage currents so SRAM cells with FinFET preferable. Multi-threshold CMOS and variable threshold voltage CMOS include stacking method and circuit power gating and self-controllable threshold voltage-level techniques. This paper details the advantage of multi-threshold CMOS method in order to develop a FinFET-based SRAM cell, and then, the designed cell is differentiated in terms of active power dissipation. By using predictive technology mode (PTM) and 14 nm technology on cadence, all the simulations are performed.

Keywords FinFET · Static random-access memory · Dynamic power consumption · Energy efficiency · CMOS and MTCMOS

1 Introduction

As process technology is bringing down, the threshold voltage and leakage current changes rise. In the predominant 6T cell, it is very challenging task to find a perfect architecture because of its write stability and its write margin should be considered. At low supply voltage, the 6T cell fall off in write stability. The leakage power is the very high priority considerable parameter due to feature scaling in high-performance memory architectures. Low power and high stability is the major parameters of SRAM designs in the last decade. In this paper, we use 8T SRAM cell to address the various above problems, and also we compare the CMOS-based SRAM cell, conventional 6 T and proposed 8T SRAM cell.

To recompense the requirement for superior storage system, FinFET SRAM is considered as a best method to offer 14 nm-sized transistor design with help of advanced VLSI field. The controlling dependencies on conventional drain and source

P. Vemula · R. S. Dhar (✉)
Department of Electronics and Communication Engineering, National Institute of Technology Mizoram, Aizawl, Mizoram 796012, India

J. K. Mandal et al. (eds.), *Topical Drifts in Intelligent Computing*, Lecture Notes in Networks and Systems 426, https://doi.org/10.1007/978-981-19-0745-6_40

terminal will be lowered by the three-dimensional design of the gate by using this revolutionary technology. Many short channel problems will be faced by the conventional transistor design which can be completely overcome by the present design of FinFET [1]. The changes of arbitrary dopant in conventional MOSFET are removed by using FinFET [2]. CMOS have high supply voltage when compared to FinFET circuits also a reduced number of energy points and product delay points, which results stability in potential due to FinFET [3]. Simultaneously, SRAM suffers with high existence of cache memory in chip region, and also, at the chip power, maximum energy consumption takes place.

An SRAM cell is one of the prominent components for storing a single-bit binary data. Greater than 90% of the chip silicon area is expected to occupy by memory circuits that mainly comprises of cache, which makes the design to be robust without any errors [4]. Because of lowest static power dissipation, the 6T CMOS SRAM is more recognizable in IC industry among different circuit combinations; also, the complementary metal oxide semiconductor offer different noise margins [5]. The SRAM cell consists of two inverters which form a latch and pair of n-type transistors which the transistor gate terminal is connected to the world line and the source terminal with bit line, the other with complement of bit line. For operations such as read and write, the entry transistors should be kept ON. The need of a SRAM cell is to provide high noise margin and more speed, but the main problem is the increase in leakage power [6].

The principle issue with standard SRAM is the using of large extent technologies in the design to provide reduced size with lesser supply voltage, to provide a less variation in cutting voltage and supply voltage [7]. When the design requirement increases, the number of devices with smaller size is to manage larger storage areas, the less difference becomes highly unstable. In this, stability and also power consumption and time required to access SRAM cell govern the design for a SRAM [8]. The power consumption in 6T SRAM will be more due to the transistor switchings in between the on state to off state; in order to minimize the power consumption, we need to go for the various types of methods [9]. The various types of schematics with advanced and enhanced devices are used to reduce the power consumption in the transient state and also during the switching, and the advanced methods are also used to increase the switching speed. The multi-threshold FinFET method is possibly the best alternative to reduce the power consumption and also to improve the switching speed [10].

2 Theory and Circuit Simulation

The occupancy level of memories will be more in upcoming designs. Fact scaling becomes the more difficult and important aspect because of this occupancy of memories. In FinFET device, electrostatic control gate is susceptible as a result of gate control from multiple sides of its fin. Due to multi-gate MOSFET, short channel effects such as V_{th} decreases the width and length of channel, subthreshold degradation and also the drain-induced barrier lowering will be enhanced. To visualize lesser access time and reduced power losses and leakage, currents in FinFET-based SRAM cells are more recognized when compared to SRAM cells with CMOS. Also, FinFET-based SRAM cells are familiar for lower dissipation of power. When considered with traditional 6T SRAM, there is a difference in the shape of 6T SRAM cell using FinFET. FinFET-based SRAM consists of a pair of cross-coupled inverters which act as primary memory element, also a pair of FinFET-based access transistors as shown in Fig. 1. The entry transistor gate terminal and word line are joined together, and also, the source terminal and bit line bar are coupled together. For memory read and memory write tasks, the transistor is turned ON, and while reading and writing, operation of a SRAM cell and kept off in hold. Figure 1 illustrates the design of 6T SRAM cell using FinFET.

While in standby mode, majority of microelectronic frameworks spend lots of time and energy. The energy of DC converter while entering or depart less energy mode should be taken into consideration. On the occasion that the fee of changing state and create a small standby energy nation is satisfactorily small so that the coverage of coming into the low energy state while the gadget is in idle state is probably applied. In specified area the predicted length in the accurate state has to be exactly calculated and taken into consideration in the use of an energy control approach.

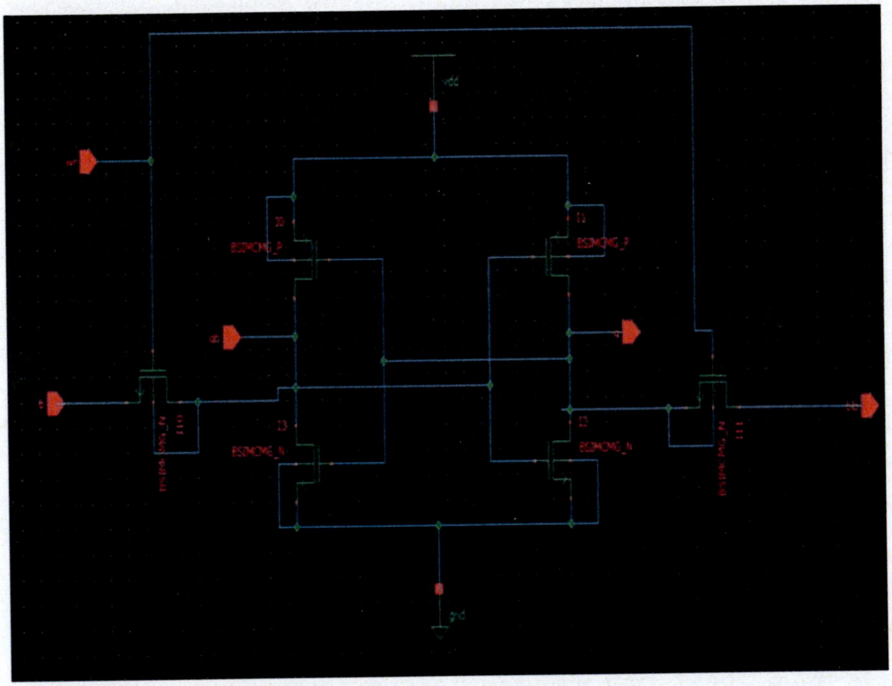

Fig. 1 Schematic of basic SRAM using FinFET

3 Results and Discussion

Multi-threshold CMOS (MTCMOS) will be a shape of CMOS device in the new model which consists of devices with more than one operating voltages preserving in thoughts; the pinnacle intention is to reinforce put off and power. Gate voltage in which an opposite coat is made on the bond among gate oxide layer along with additionally the base material of the device. Low V_{th} gadgets are very a whole lot appropriate to exchange speedy and perform a totally critical function in postpone paths to restrict clock periods. But those gadgets have better static leakage strength. To lower the static leakage strength on non-important paths, high V_{th} gadgets are used; in addition, they lessen the static leakage ten instances while as compared to low V_{th} gadgets.

Sleep transistor is kept on during the active mode keeping the circuit functioning as usual. Within the stand-in condition, the FinFET is biased, which disengage the gate in distinction to the underside. It is to be considered that the leakage currents should be bringing down, and brink voltage of the transistor must be larger. The leakage in the inactive transistor will be high if the above state is not satisfied.

In MTCMOS method, the low V_{th} transistors get separated from the input supply by using high V_{th} sleep transistors at the largest and also the base of digital circuit. The device with the low V_{th} which is employed to design logic. The clock is utilized to control the sleep transistors. In changing mode, high clock signal is generated, resulting in high of both the V_{th} transistor to visualize and provides an imaginary supply with low V_{th} digital circuit. The sleep signal is designed to travel low during standby mode, leading both the high V_{th} transistor are allowed to cut-off and separate supply from low V_{th} circuit. A detailed circuitry of the FinFET-based 8 T SRAM CELL using MTCMOS technique is shown in Fig. 2.

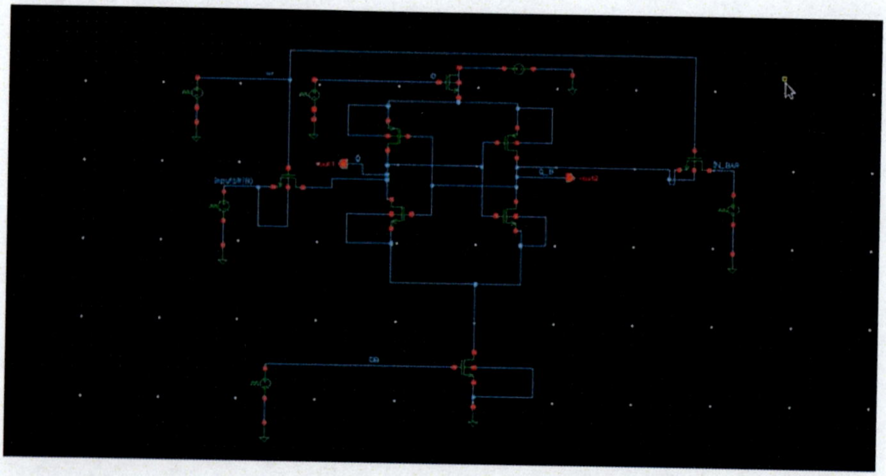

Fig. 2 Detailed circuitry of FinFET-based 8 T SRAM cell with MTCMOS

This method lowers power consumption by utilizing pFET operates, with high V_{th} in the middle the mains and also small V_{th} SRAM memory cell device to disconnect the facility input and the n-FET operates with larger V_{thn} in between ground and less V_{th} SRAM device to disconnect the bottom from least V_{th} SRAM memory cell. During the sleep mode of circuit, the largest V_{th} sleep transistors are kept ON as they create separation of smaller V_{th} transistor from supply potential and ground which in turn decreases threshold leakage current. The main disadvantage with MTCMOS method is the requirement for added making process for larger V_{thp} of pFET and larger V_{thn} of nFET, and the reality is that it is not possible to restore the data by using this method. The newly developed 14 nm FinFET-based 8 T SRAM cell is designed with MTCMOS technique as is illustrated in Fig. 3.

Using this 14 nm technology in cadence tool with MTCMOS, the FinFET-based 8 T SRAM cell simulation is performed. The write operation waveform is acquired for the novel circuitry developed here for 8 T SRAM cell employing the 14 nm technology FinFETs which provides a step ahead for the future as observed in Fig. 4. As detected from Fig. 4, where *Wr* is the write control signal (wordline), controlling the full operation in the SRAM cell, input bit (bitline) is the signal applied at the

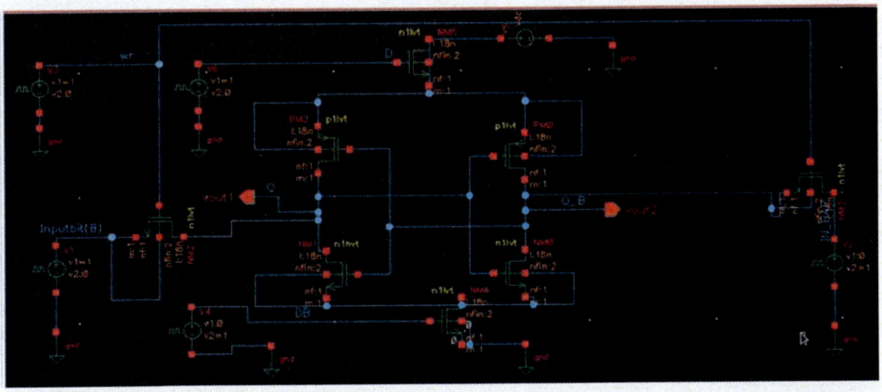

Fig. 3 Illustration of FinFET-based SRAM cell using MTCMOS method

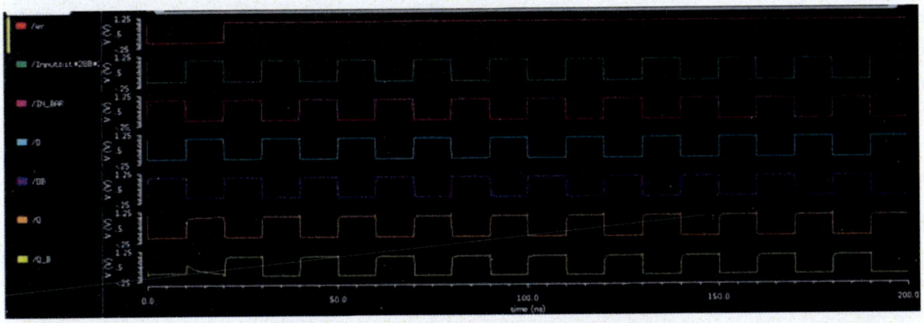

Fig. 4 MTCMOS-based 14 nm node FinFET write operation waveform

Table 1 Power usage of CMOS SRAM cell, 6 T-FinFET-based SRAM cell and advanced 8 T-FinFET-based SRAM cell

Type of technology used	Power consumed (nw)
Active power usage in CMOS SRAM in write operation (45 nm)	240
Power consumed during write operation in FinFET SRAM in 6 T	40.12
Used power during write operation in 8 T SRAM cell-based FinFET using MTCMOS	20.52

input to be stored in the circuit and IN_BAR (bitlinebar) is the complimented bit of the input D is the control signal which is used to control the switching of the nFinFETs, DB is the control signal used to control the switching of pFinFETs while Q is the stored bit available at output, and Q_B is the compliment bit of the stored bit Q. From Fig. 4, the change in the output at Q and Q_B with respect to the change in input, i.e., input bit and IN_BAR is evidently observed so that the digital bit either at one or zero is stored in the memory cell by using back to back connected inverters.

The active power consumption is calculated that is the product of available power and flowing current, while in the MTCMOS SRAM memory device, it was calculated by product of the current flowing through the drain terminal of the pFET device by the power supply, for the other SRAM circuitry shown in Table 1. The active power dissipation for MTCMOS 14 nm technology node FinFET for 8 T SRAM is achieved to be 20.52 nw. The power dissipation for 6 T SRAM circuitry is also calculated along with the active power for 45 nm-based CMOS SRAM are also calculated and are compared as presented in Table 1. From the comparison of the active power dissipation, it is evidently observed that the FinFET with 8 T SRAM using the MTCMOS method lowers dynamic power dissipation especially for the 14 nm technology node for write operation, which is quite advantageous for the circuitry and is beneficial for future VLSI circuit implementation.

4 Conclusion

This work details the design of FinFET SRAM cell with the use of MTCMOS. A pair of back to back connected inverters for static memory cell. Power output from second inverter is coupled to the first inverters input. It also consists of two access transistors. To bit line, the success transistor source terminal is connected. The transistors are activated while read and write operations, and they are kept in off state in hold position. In this process, sleep transistor is coupled with supply and with the low V_{th} device and the V_{SS}. The active power consumption is computed as product of current flowing through the drain of pFET transistor using the power supply. The operation of each design and the simulation results of them are also provided. Using 14 nm technology on cadence software, all the circuits are designed.

References

1. Dhanumjaya SK, Sudha M, Giri Prasad MN, Padmaraju K (2012) Cell stability analysis of conventional 6T dynamic and 8T SRAM cell in 45nm technology. Int J VLSI Design Commun Syst 3(2)
2. Trivedi VP, Fossum JG (2004) Source/drain-doping engineering, for optimal nanoscale FinFET design. In: International SOI conference, pp 192–194
3. Alluri S (2019) High performance SR latch in VLSI circuits using FINFET 18nm technology. J Continua Math Sci ISSN 14(6):329–346
4. Nanda S, Dhar RS Implementation and characterization of 14 nm trigate HOI n-FinFET using strained silicon channel with reduced area on chip. In: 6th international conference for convergence in technology (I2CT), pp. 1–4. IEEE, India (2021)
5. Gupta V, Khandelwal S (2015) Leakage current reduction in finfet based 6T SRAM cell for minimizing power dissipation in nanoscale memories. In: Nirma University international conference on engineering (NUiCONE), 978-1-4799-9991-0/15/©2015 IEEE
6. Kim YB, Kim YB, Lombardi F (2008) New SRAM cell design for low power and high reliability using 32 nm independent gate FinFET technology. In: IEEE international workshop on design and test of nano devices, circuits and systems, 978-0-7695-3379-7/08 © IEEE
7. Saha P, Yugender P, Nanda S, Dhar RS (2021) Geometrical impact on short channel parameters at 14 nm gate length in trigate n-FinFET. In: 1st International conference for women in multifaceted research 2021 (ICWMR 2021)
8. Fan M, Wu Y, Hu VP, Hsieh C, Su P, Chuang C (2011) Comparison of 4T and 6T FinFET SRAM cells for subthreshold operation considering variability—a model-based approach. IEEE Trans Electron Devices 58(3):609–616
9. Ma Y, Zhang L, Liu J (2020) A new 6T SRAM memory cell based on FINFET process. In: IEEE 5th international conference on integrated circuits and microsystems (ICICM), pp 251–254
10. Rahman M, Mahapatra R (2014) Design of a 32 nm independent gate FinFET based SRAM cell with improved noise margin for low power application Int Conf Electron Commun Syst (ICECS) 1–5

A Novel Solar Thermoelectric Generator with Conical Frustum Leg Geometry

Ravita Lamba, Chika Maduabuchi, and Emenike C. Ejiogu

Abstract A new method of improving the overall performance of a solar thermoelectric generator (STEG) using a TEG with frustum leg geometries is presented in this study. Six STEG models are proposed compromising the conventional rectangular (rect), frustum (frust)—rect, di-frust—rect, di-frust—frust, frust and di-frust legs, respectively. A comparison of the volume occupied by the total thermoelectric legs in the diverse models reveals that the models which utilized only frustum legs reduced the volume occupied by rectangular legs in a conventional TEG cell by 42%, thus implying a reduction in fabrication cost as lesser materials will be required for modules comprising the novel leg geometries. Furthermore, it follows that under the same amount of concentrated solar radiation, system 6, which utilized only di-frustum legs, provides a temperature gradient, power output density, and thermodynamic efficiencies (energy and exergy efficiencies) that are 2.1, 3.8, and 1.3 times higher than that of the conventional STEG module, respectively. The findings of this paper will greatly advance the research in optimizing the performance of contemporary STEG systems.

Keywords Solar thermoelectricity · Thermoelectric generator · Thermal engineering · Variable leg geometry · Frustum legs

R. Lamba (✉)
Department of Electrical Engineering, Malaviya National Institute of Technology, Jaipur, Rajasthan 302017, India

C. Maduabuchi
Department of Mechanical Engineering, Federal University of Agriculture, P.M.B. 2373, Makurdi, Nigeria
e-mail: chika.maduabuchi.191341@unn.edu.ng

E. C. Ejiogu
Africa Centre of Excellence for Sustainable Power and Energy Development, University of Nigeria, Nsukka, Nigeria
e-mail: emenike.ejiogu@unn.edu.ng

Laboratory of Industrial Electronics, Power Devices and New Energy Systems (LIEPNES), University of Nigeria, Nsukka, Nigeria

J. K. Mandal et al. (eds.), *Topical Drifts in Intelligent Computing*, Lecture Notes in Networks and Systems 426, https://doi.org/10.1007/978-981-19-0745-6_41

1 Introduction

In a world with a rapid population growth rate and increasing energy demand, the need for alternative energy sources becomes inevitable. More importantly, the health challenges and environmental degradation as a result of the heavy reliance on fossil fuel energy sources suggest that the potentials of renewable energy sources like solar energy need to be fully explored. For instance, a recently published paper [1] revealed that the environmental pollution as a result of burning fossil fuels resulted in the brain damage of babies and young children. These and many more factors have necessitated the search for better means of solar energy-based power generation systems. Two main ways of harnessing solar energy are through light (photovoltaics, PV) or heat (thermal). Thermoelectric generators (TEGs) as devices that convert heat directly to electricity using the Seebeck effect fall under the latter [2–4]. Recently, these devices have gained much research interest due to their relatively low price compared to their PV counterparts, compact size, noiseless and frictionless operation, and requiring little maintenance requirements [5–7]. However, a hindrance to the full realization of these devices is that they are characterized by low thermoelectric (TE) efficiencies [8–10]. Thus, much research is needed on improving the TE efficiency as this will increase its competitive advantage in the world energy market [11–13]. The several means of improving TE efficiency that has been explored over the years include altering TE leg geometry [14–16], utilizing multi-stage TEGs [17–19], incorporating solar concentrators [9, 20, 21], segmentation [22–24], etc.

Recently, some authors have experimentally [25, 26] and theoretically [27, 28] established that altering TE geometry resulted in an increased TE efficiency and overall device performance. However, the analysis was carried out using an assumed temperature variation in the TEG. This is totally inaccurate when applied to a STEG system. This is because in a STEG, the incident solar radiation from the sun is the major determinant of the temperature distribution developed in the hybrid system. Thus, assuming a random temperature is totally unacceptable and inappropriate. Furthermore, amongst the diverse leg geometries studied, frustum legs have never been studied. It is expected that a suitable utilization of a frustum leg in a STEG system will provide the much-expected improvement in TEG performance, thus giving it a higher competitive advantage in the world energy market. Hence, for the first time, this paper unlike the others seeks to fill in the knowledge gap by determining the effect of varying concentrated solar radiation and frustum leg geometries on the performance of a conventional rectangular leg STEG. Furthermore, this paper will also incorporate an exergy analysis of the frustum leg STEG in order to assess a more realistic performance of the hybrid system.

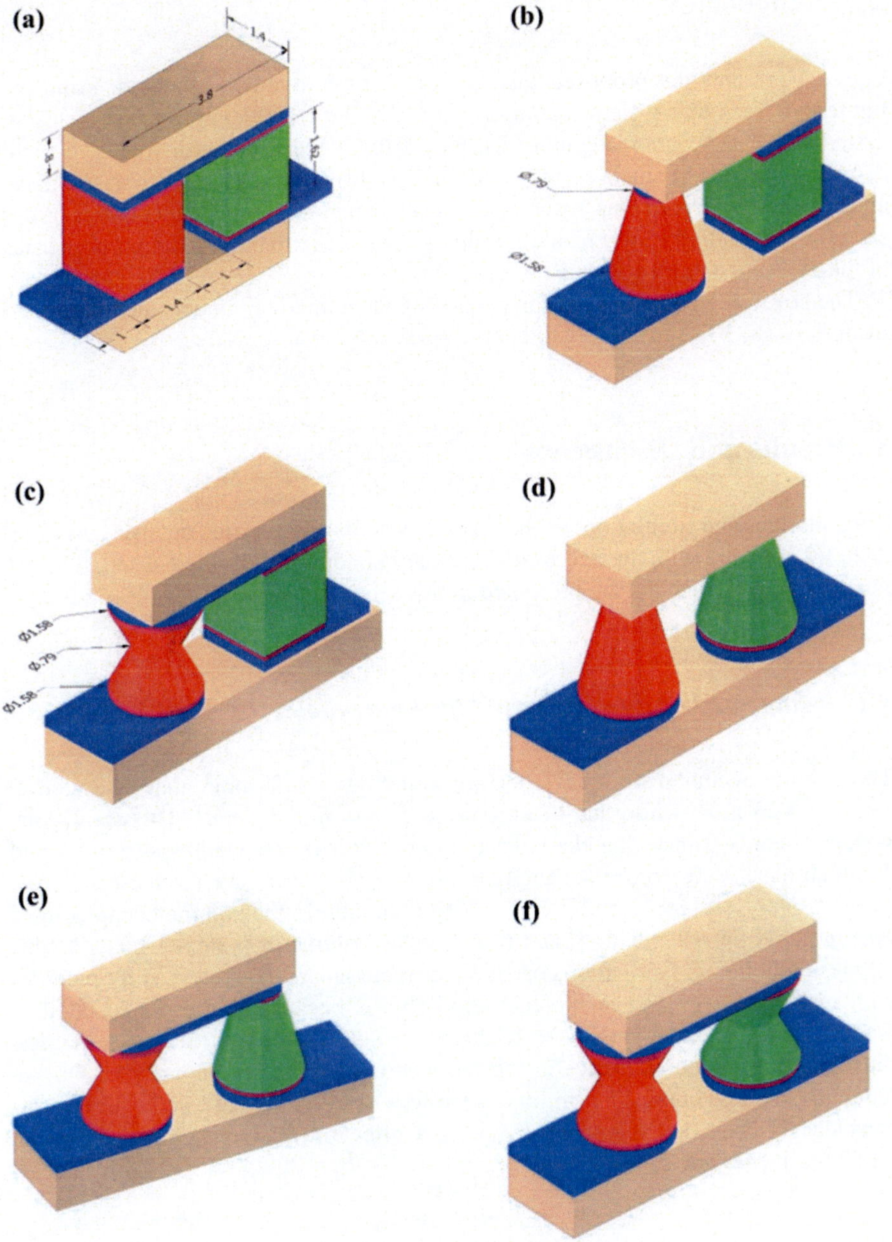

Fig. 1 Proposed systems. **a** System 1, **b** System 2, **c** System 3, **d** System 4, **e** System 5, **f** System 6

2 Methodology

Figure 1 presents the proposed TEG system configurations considered. From the figure, it is seen that system 1 comprises rectangular (rect) leg. On the other hand, systems 2, 3, and 5 are made up of frustum (frust)—rectangular, di-frust—rect and di-frust—frust leg geometries, respectively. Finally, systems 4 and 6 consist of frust and di-frust thermoelements, respectively. The temperature-dependent material properties of unmodified (pure) bismuth telluride and the remaining material properties of the TEG are obtained from Refs. [29, 30].

The governing equations used in evaluating the numerical model can be obtained from Refs. [8, 10].

3 Results and Discussions

Computations and simulations of the various systems were carried out using ANSYS 2020 R2 software with the necessary boundary conditions specified. Then, relevant plots were generated and are presented in this section.

3.1 *Material Volume Occupied by Thermoelectric Legs*

The volume occupied by the TE legs in system 1 is 6.35 nm^3, that of system 2 and 3 is 5.03 nm^3, while that of systems 4, 5, and 6 is 3.7 nm^3. This shows that system 1 (conventional) has the largest volume. This is followed by systems 2 and 3, which have same volumes. Finally, systems 4, 5, and 6 have the least volumes. Quantitatively, systems 2 and 3 give a 21% reduction in the volume occupied by a system 1. While systems 4, 5, and 6 reduce the volume of system 1 by 42%, this implies reduction in fabrication cost as lesser materials will be required for modules with rect—trap legs and rect—frust legs. This is because of the use of additive manufacturing technologies in the fabrication of all thermoelectric leg geometries, thus reducing the complexities in the fabrication of these leg geometries. Further reductions can be obtained with trap—trap legs, trap—frust legs, and di-frust—di-frust legs. This is good as it shows the positive effect of frustum leg geometry on the cost effectiveness of TEGs.

3.2 *Solar Radiation Intensity*

Results on the effect of optical concentration on the performance of STEGs are shown in Fig. 2. Figure 2a reveals a linear relationship between the temperature gradient,

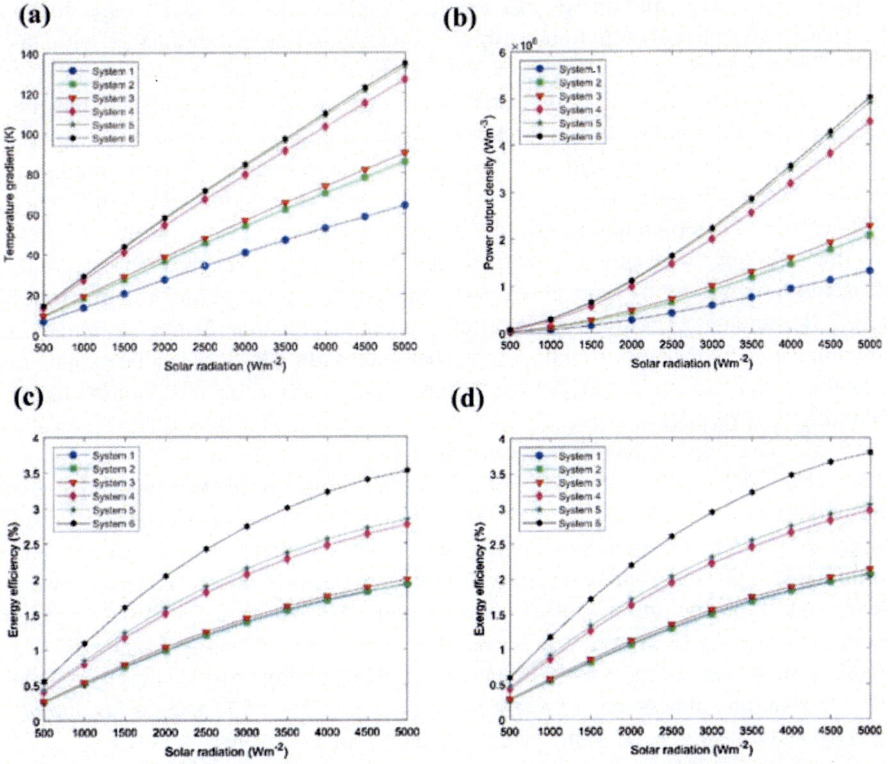

Fig. 2 Effect of solar radiation on **a** Temperature gradient, **b** Power output density, **c** Energy efficiency, **d** Exergy efficiency

ΔT, and the concentrated solar radiation (CSR), with system 6 having the highest ΔT, followed by systems 5, 4, 3, 2, and 1. Also, the plots show a widening gap between ΔT of system 1 and that of system 2 as CSR increases, which becomes even wider as one goes from system 3 to system 4. However, between systems 2 and 3, and between systems 4 and 5, the disparity is not so wide. The disparity becomes even almost negligible between system 5 and system 6. Typically, at CSR of 500 W/m^2 and 5000 $^W/m^2$, respectively, ΔTs of 7 and 64, 9 and 85, 10 and 90, 14 and 127, 14 and 133, and 15 and 135 K were obtained for systems 1, 2, 3, 4, 5, and 6, respectively.

Then, in Fig. 2b, a, slightly exponential rise in the power output density with increasing CSR is noted, with system 6 clearly recording the highest power output densities, followed by systems 5, 4, 3, 2, and 1. Again, a significant difference exists between system 1 and system 2, and between system 4 and system 5, respectively. While a much larger disparity is observed between system 3 and system 4, but very little between system 5 and system 6, and between system 2 and system 3, respectively. Quantites 2, 3, 4, 5, and 6 are 1.3×10^6, 2.1×10^6, 2.3×10^6, 4.5×10^6, 4.9×10^6, and 5×10^6 $^W/m^3$, respectively.

Lastly, the energy and exergy efficiencies are plotted in Fig. 2c, d, respectively. As clearly seen from both figures, both efficiencies rise parabolically with the CSR. Comparatively, system 6 achieves the highest energy and exergy efficiencies, followed by systems 5, 4, 3, 1, and lastly system 2. However, the disparity in the efficiency of system 1 and system 2 can be seen to be negligible. More so, the disparity between energy and exergy efficiencies of system 1 and system 3 is very small. But between systems 3 and 4 as well as between system 5 and system 6, the difference is quite large. The efficiency plots indicate that asymmetry has no significant effect on either rect legs (conventional) or frust legs. Moreover, as one goes from the least efficient (system 2) to the most efficient (system 6), the rate of increase in efficiency with CSR increases. Though, for all systems, it can be seen that the rate of change in efficiency with CSR begins to reduce at higher values of CSR. This can be explained to be due to the fact that the higher the incident insolation on the absorber, the larger the radiative losses are.

The results in Fig. 2c, d closely agree with those in Fig. 2a, b with some exceptions. First, in Fig. 2a, b, system 2 has a reasonably higher-temperature gradients and power output densities, respectively, than system 1. But in Fig. 2c, d, there is no significant difference between the energy and exergy efficiencies, respectively, of system 1 and system 2. This disparity can be traced to the effect of the total volume occupied by the TE legs on the power output, with efficiencies being independent of material volume. Second, in Fig. 2a, b, system 5 and system 6 show slight disparity in temperature gradient and power output density, respectively. While in Fig. 2c, d, there is a sizeable difference in the efficiencies of system 5 and system 6. This difference is because of the varying current generated in system 5 and system 6.

4 Conclusions

The numerical modelling of six STEG models comprising diverse conical frustum and di-frustum leg configurations was carried out in this study. The finite element solver, ANSYS 2020 R2 software, was used in simultaneously solving the coupled field equations of thermoelectricity.

The total volume occupied by the thermoelements in the proposed models was compared, and it was detected that the models which utilized only frustum legs reduced the volume occupied by rectangular legs in a conventional TEG cell by 42%, thus implying a reduction in fabrication cost as lesser materials will be required for modules comprising the former.

Furthermore, the effect of concentrated solar radiation on device performance was determined, and it was shown that for the same amount of solar radiation, system 6, which utilized only di-frustum legs, provided a temperature gradient, power output density, and thermodynamic efficiencies (energy and exergy efficiencies) that were 2.1, 3.8, and 1.3 times higher than that of the conventional STEG module, respectively.

It was also noted that using thermoelectric pairs which comprised of rectangular and conical frustum/di-frustum legs resulted in a rather slight increment in device performance. This implies that maximum device performance is only obtained when thermoelectric pairs consist of only frustum leg geometries.

Conclusively, the results discussed in this paper argue that rectangular leg geometries in conventional STEG systems should be replaced with frustum legs which require lesser material cost and higher overall performance compared to the former. The findings in this paper will greatly advance the research in optimizing the performance of contemporary STEG systems.

Acknowledgements The second and third authors would like to acknowledge and appreciate the support offered by the Africa Centre of Excellence for Sustainable Power and Energy Development (ACE-SPED), University of Nigeria, Nsukka, towards the successful completion of this work.

References

1. Kotcher J, Maibach E, Choi WT (2019) Fossil fuels are harming our brains: Identifying key messages about the health effects of air pollution from fossil fuels. BMC Public Health 19:1–12. https://doi.org/10.1186/s12889-019-7373-1
2. Maduabuchi C, Eke M (2021) Solar electricity generation using a photovoltaic-thermoelectric system operating in Nigeria climate. IOP conference series: Earth and environmental science 730:012029. https://doi.org/10.1088/1755-1315/730/1/012029
3. Maduabuchi C, Lamba R, Njoku H, Eke M, Mgbemene C (2021) Effects of leg geometry and multistaging of thermoelectric modules on the performance of a photovoltaic-thermoelectric system using different photovoltaic cells. Int J Energy Res 45:17888–902. https://doi.org/10.1002/er.6925
4. Maduabuchi C, Ejenakevwe K, Jacobs I, Ndukwe A (2020) Analysis of a two-stage variable leg geometry solar thermoelectric generator. 2nd African Int Conf Ind Eng Oper Manag Harare, Zimbabwe: 2020, p. 1–7
5. Maduabuchi C, Singh S, Ozoegwu C, Njoku H, Eke M (2022) The combined impacts of leg geometry configuration and multi-staging on the exergetic performance of thermoelectric modules in a solar thermoelectric generator. J Energy Resour Technol 144:1–23. https://doi.org/10.1115/1.4051648
6. Maduabuchi CC, Mgbemene CA (2020) Numerical analysis and simulation of a hybrid concentrated thermoelectric module with phase change material. Sustain Eng Ind Technol Conf, Nsukka, Enugu, Nigeria, p. 1–3
7. Maduabuchi CC, Eke MN, Mgbemene CA (2021) Solar power generation using a two-stage X-leg thermoelectric generator with high-temperature materials. Int J Energy Res 45:13163–81. https://doi.org/10.1002/er.6644
8. Maduabuchi CC, Ejenakevwe KA, Mgbemene CA (2021) Performance optimization and thermodynamic analysis of irreversibility in a contemporary solar thermoelectric generator. Renew Energy 168:1189–206. https://doi.org/10.1016/j.renene.2020.12.130
9. Maduabuchi CC, Mgbemene CA (2020) Numerical study of a phase change material integrated solar thermoelectric generator. J Electron Mater 49:5917–36. https://doi.org/10.1007/s11664-020-08331-3
10. Maduabuchi C, Njoku H, Eke M, Mgbemene C, Lamba R, Ibrahim JS (2021) Overall performance optimisation of tapered leg geometry based solar thermoelectric generators under isoflux conditions. J Power Sources 500:229989. https://doi.org/10.1016/j.jpowsour.2021.229989

11. Maduabuchi CC, Mgbemene CA, Ibeagwu OI (2020) Thermally induced delamination of PV-TEG: Implication of leg's joule and thomson heating. J Electron Mater 49:6417–27. https://doi.org/10.1007/s11664-020-08390-6
12. Maduabuchi C, Ejenakevwe K, Ndukwe A, Mgbemene C (2021) High performance solar thermoelectric generator using asymmetrical variable leg geometries. E3S Web Conf 239:00005. https://doi.org/10.1051/e3sconf/202123900005
13. Ibeagwu OI, Eke MN, Maduabuchi CC, Mgbemene CA, Aka TV (2020) Particle overlay obstruction modelling, parametric and output characteristics evaluation of a photovoltaic system. Niger Res J Eng Environ Sci 5:679–93
14. Shittu S, Li G, Zhao X, Ma X, Akhlaghi YG, Ayodele E (2019) High performance and thermal stress analysis of a segmented annular thermoelectric generator. Energy Convers Manag 184:180–93. https://doi.org/10.1016/j.enconman.2019.01.064
15. Lin S, Yu J (2016) Optimization of a trapezoid-type twostage Peltier couples for thermoelectric cooling applications. Int J Refrig 65:103–10. https://doi.org/10.1016/j.ijrefrig.2015.12.007
16. Wang X, Wang H, Su W, Mehmood F, Zhai J, Wang T, et al (2019) Geometric structural design for lead tellurium thermoelectric power generation application. Renew Energy 141:88–95. https://doi.org/10.1016/j.renene.2019.03.128
17. Hadidi A (2017) A novel approach for optimization of electrically serial two-stage thermoelectric refrigeration systems using chemical reaction optimization (CRO) algorithm. Energy 140:170–84. https://doi.org/10.1016/j.energy.2017.08.087
18. Asaadi S, Khalilarya S, Jafarmadar S (2019) A thermodynamic and exergoeconomic numerical study of two-stage annular thermoelectric generator. Appl Therm Eng 156:371–81. https://doi.org/10.1016/j.applthermaleng.2019.04.058
19. Kanimba E, Pearson M, Sharp J, Stokes D, Priya S, Tian Z (2017) A modeling comparison between a two-stage and three-stage cascaded thermoelectric generator. J Power Sources 365:266–72. https://doi.org/10.1016/j.jpowsour.2017.08.091
20. Mgbemene CA, Duffy J, Sun H, Onyegegbu SO (2010) Electricity generation from a compound parabolic concentrator coupled to a thermoelectric module. J Sol Energy Eng 132:0310151–8. https://doi.org/10.1115/1.4001670
21. Ejenakevwe KA, Mgbemene CA, Njoku HO, Ekechukwu OV (2020) Parametric optimization of exergy efficiency in solar thermoelectric generators. J Electron Mater 49:3063–71. https://doi.org/10.1007/s11664-020-08021-0
22. Ruiz-Ortega P, Olivares-Robles M, Enciso-Montes de Oca O (2019) Segmented thermoelectric generator under variable pulsed heat input power. Entropy 21:929. https://doi.org/10.3390/e21100929
23. Cui YJ, Wang BL, Li JE, Wang KF (2020) Performance evaluation and lifetime prediction of a segmented photovoltaicthermoelectric hybrid system. Energy Convers Manag J 211:1–13. https://doi.org/10.1016/j.enconman.2020.112744
24. Zare M, Ramin H, Naemi S, Hosseini R (2016) Exact optimum design of segmented thermoelectric generators exact optimum design of segmented thermoelectric generators. https://doi.org/10.1155/2016/6914735
25. Mohammad Siddique AR, Mahmud S, Van Heyst B (2020) Performance comparison between rectangular and trapezoidal-shaped thermoelectric legs manufactured by a dispenser printing technique. Energy 196:117089. https://doi.org/10.1016/j.energy.2020.117089
26. Fabián-mijangos A, Min G, Alvarez-quintana J (2017) Enhanced performance thermoelectric module having asymmetrical legs. Energy Convers Manag 148:1372–81. https://doi.org/10.1016/j.enconman.2017.06.087
27. Wang R, Meng Z, Luo D, Yu W, Zhou W (2020) A comprehensive study on x-type thermoelectric generator modules. J Electron Mater 49:4343–54. https://doi.org/10.1007/s11664-020-08152-4
28. Lee M-Y, Seo J, Lee H, Garud KS (2020) Power generation, efficiency and thermal stress of thermoelectric module with leg geometry, material, segmentation and two-stage arrangement. Symmetry (Basel) 12:786. https://doi.org/10.3390/sym12050786

29. Xuan X, Ng K, Yap C, Chua H (2002) The maximum temperature difference and polar characteristic of two-stage thermoelectric coolers. Cryogenics (Guildf) 42:273–8. https://doi.org/10.1016/S0011-2275(02)00035-8
30. Shittu S, Li G, Zhao X, Zhou J, Ma X, Akhlaghi YG (2020) Experimental study and exergy analysis of photovoltaic-thermoelectric with flat plate micro-channel heat pipe. Energy Convers Manag 207:112515. https://doi.org/10.1016/j.enconman.2020.112515

Eyes Say It All: Deep fake Detection Method Analysis Using Different Metrics

Ravi Kumar Arya, Priyanshu Agrawal, Akshit Aggarwal, Ayush Kumar Dokania, Ekanshi Pal, Menika Karki, Kunal Goswami, Sanskar Jain, Ravi Dugh, and Amit Dugh

Abstract With the advancement of artificial intelligence (AI) techniques, fake digital content has mushroomed in recent years. The production of very realistic fake content, even by a novice computer literate, has become a cakewalk with the latest deep learning and generative adversarial network (GAN) models. The objectionable content created by the use of deepfake technology is getting exploited by bad elements to tarnish the image of celebrities. With the advancement of digital technologies, it is almost impossible to detect fake content with naked eyes and is posing a big challenge for the public. In this manuscript, we use a method to detect deepfakes by using eye reflection and comparing both eyes in terms of different metrics such as structural similarity index measure (SSIM), intersection over union (IoU), and mean-squared error (MSE). We used Flickr-Faces-HQ (FFHQ) dataset for real human eye images, and the thispersondosentexist.com Web site for GAN-synthesized images. Next, we used automatic detection of the corneal region and compared pixels of both eyes. Despite their simplicity, the methods can achieve AUC values of up to 0.9.

Keywords Deepfake · GANs · Deep learning · GAN image detection

1 Introduction

Recently, it has become very easy to create hyper-realistic videos by the use of face swaps due to the latest technological advancements. Such digital content, also called "deepfakes," is so well created that it can fool normal human eyes. Faceswap, one of the techniques, is a well-known form of deepfake videos which uses one person's face

R. K. Arya (✉) · P. Agrawal · A. Aggarwal · A. K. Dokania · E. Pal · M. Karki · K. Goswami · S. Jain
National Institute of Technology Delhi, New Delhi, Delhi 110040, India

R. Dugh
Goergen Institute of Data Science, University of Rochester, Rochester, NY, USA
e-mail: rdugh@ur.rochester.edu

A. Dugh
Information Technology, Queensland University of Technology, Brisbane, Australia

J. K. Mandal et al. (eds.), *Topical Drifts in Intelligent Computing*, Lecture Notes in Networks and Systems 426, https://doi.org/10.1007/978-981-19-0745-6_42

in place of another person's face. Several of these faceswap deepfakes are made by the use of a generative adversarial network (GAN) [1], including deepfake faceswap [2], faceswap-GAN [3], and FS-GAN [4]. Other techniques use conventional computer graphics programs to create deepfakes. Similarly, face2face [5] and faceswap [6] work on the principle of puppet-master deepfakes where head movements and facial expressions of one person are replaced by another person.

It is only a matter of time before a very easy-to-use piece of software becomes available that will make it possible to do this type of multimedia manipulation with minimal technical skills. When will that happen? Politics is going to get real interesting once it does. It is only a matter of time before a world leader will be calling a controversial comment or video "a deepfake" to unburden themselves from their wrongdoings. Sometimes, a deepfake could be widely shared and accepted to be true before it is flagged as fake [7, 8]. Often, this happens with static images with made-up quotes plastered over them. Due to such repercussions, it is important to tame the beast of deepfakes by detecting them with certainty. This manuscript discusses such a method.

The whole manuscript is divided into different sections. Section 2 gives the background of the deepfake methods, while Sect. 3 gives details of the method used in this work. Next, in Sect. 4, some results from the different metrics are tabulated and compared. In the end, Sect. 5 concludes this research work with some concluding remarks.

2 Deepfake Detection Method Analysis and Different Comparison Metrics

The deepfake images used for this work are GANs synthesized images extracted from https://thispersondosentexist.com (see Fig. 1) which uses the StyleGAN2 model. StyleGAN2 [9] is a generative adversarial network that builds on StyleGAN with several improvements. On the other hand, real images are from the Flickr-Faces-HQ (FFHQ) dataset.

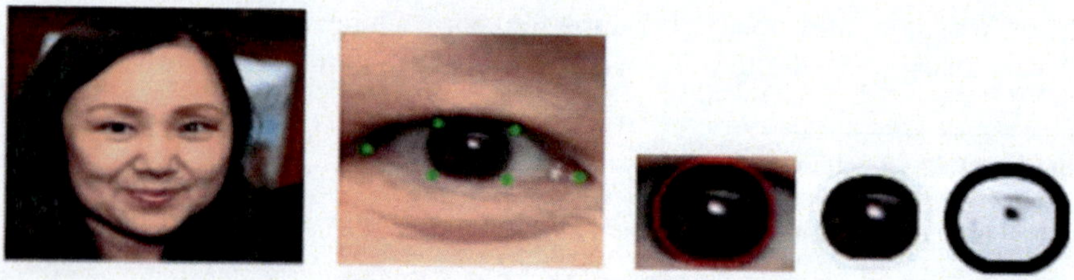

(a) (b) (c) (d) (e)

Fig. 1 Different stages of eye processing

In this study, corneal specular highlights [10] are used for exposing the human eyes. The cornea is a semi-sphere and reflects light well; therefore, the two eyes should have very similar reflective patterns both in shape and color. The real human eye images show the same corneal specular highlights as they result from the same lighting environment. They are co-related by a transform that is determined by: (1) the skeletal part of two eyes, which includes distance between the corneal limbus of both eyes and size (diameter) of the corneal limbus (2) the distance between eye and light source and position of a light source measured relative to camera coordinates, and (3) head orientation plays an important role as it determines the eye pose relative to the camera position. It is helpful to have the same corneal specular highlights for both the eyes during detection for the photographs. To achieve it, the following conditions should be satisfied.

- The two eyes have a front-side pose.
- The eyes should be at a large distance from the source that emits or reflects light.
- The light reflectors or the sources should be visible to both eyes.

In this research work, the method [11] is used, which starts by detecting the face followed by eyes with dlib landmarks. Then, we extract both eyes using those landmarks (see Fig. 1). Further, we use the Canny edge detector followed by the Hough transform to extract the corneal limbus (see Fig. 1c). The convergence of the obtained corneal limbus with the eye region provides us the corneal region. For extracting the corneal specular highlights, an adaptive image thresholding method is applied [12]. In this, we used the fact that specular highlights gravitate to have brighter intensities than the background iris; thus, restricting to only pixel locations above the adaptive threshold gave us the required specular highlights (see Fig. 1e). Then, for comparison in both eyes, we used three different metrics, i.e., (MSE [13], SSIM [14], IoU [15]). Figure 2 shows a flowchart explaining the whole algorithm of the method we used to compare both eyes. The MSE is the square of Euclidean distance [16]. SSIM helps to distinguish two similar images by pointing out the similarities within pixels. IoU, on the other hand, is the area of intersection between two images divided by the area of union between such images.

3 Results and Discussions

The real human eye images are derived from the Flickr-Faces-HQ (FFHQ) dataset [17], while the GAN-synthesized human eye images are the faces generated by StyleGAN2 models (taken from https://thispersondoesnotexist.com). The images are with a resolution of 1024 $\times$ 1024 pixels and a sample size of 40 images (20 real and 20 GAN generated).

Next, we use the face detector and landmark extractor by dlib [18, 19]. We took some sample images [11] and made a table comparing its corneal specular highlights in terms of MSE, SSIM, and IoU. Table 1 shows the analysis for images of GAN-synthesized and real human eyes.

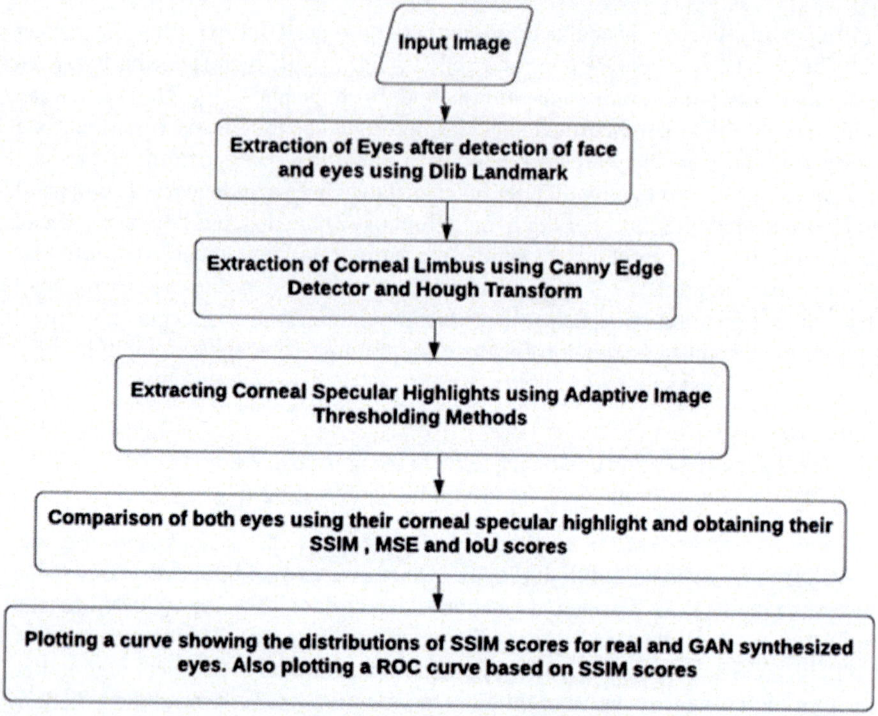

Fig. 2 Flowchart explaining the whole algorithm of comparing both eyes

Table 1 shows different metrics (MSE, SSIM, and IoU) which are calculated by pixel difference of corneal specular highlights of the left eye and right eye, respectively. The image set used contains both real human eyes and GANs-synthesized eyes. The images that have higher values of SSIM and IoU relate to a strong resemblance between both corneal specular highlights, while the lower values signify different relative locations of specular highlight regions or the different geometric shapes of the two eyes. From Table 1, it is evident that the real human eye images have higher values of SSIM and IoU, while the GAN-synthesized images have low values of SSIM and IoU. We also made the ROC [20] curve of SSIM (see Fig. 3), which has an AUC score of 0.9, while the ROC curve for IoU has an AUC of 0.94 (taken directly from [11]). It shows that the IoU and SSIM are somewhat giving the same difference, but IoU is better than SSIM as IoU is a metric used in semantic segmentation. In Fig. 3, the distribution of SSIM scores is shown which compares the detected corneal specular highlights of the GAN-synthesized and the real eyes. The figure also shows the ROC curve for the SSIM scores. However, the MSE is inefficient which we further discuss in the next section.

Table 1 Comparison metrics

Real human eye	Corneal specular highlights (left eye)	Corneal specular highlights (right eye)	MSE	SSIM	IoU
1.			3189.85	0.67	0.7006
2.			2232.51	0.85	0.7522
3.			2097.45	0.77	0.8406
4.			6345.74	0.33	0.5824
5.			1547.88	0.75	0.6842
GAN synthesized eye			**MSE**	**SSIM**	**IoU**
1.			753.56	0.73	0.2535
2.			4399.61	0.44	0.3512
3.			1571.03	0.26	0.2429
4.			4670.88	0.35	0.2500
5.			9563.50	0.26	0.29030

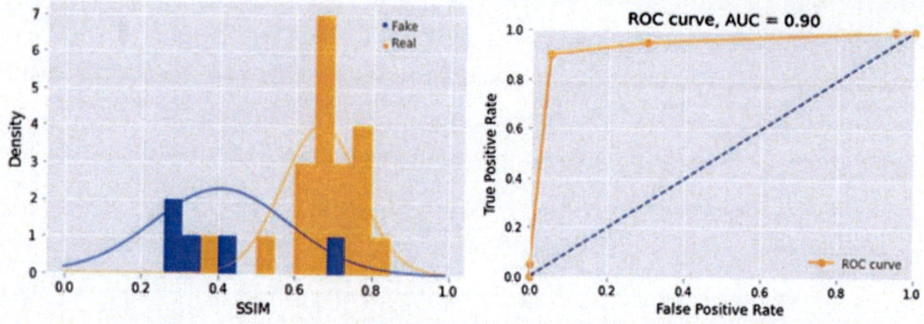

Fig. 3 **a** SSIM score distribution between eyes shown in Table 1; **b** ROC for SSIM scores

4 Limitations and Future Work

Because you have to train two networks from a single backpropagation, GANs are more unstable to train. As a result, picking the correct goals can make a tremendous difference.

However, MSE is also calculated, but it does not show proper results due to some of its disadvantages for image comparison [21]:

- It does not align with the human visual system properly.
- In the human visual system HVS, a decrease in quality leads to errors. But, with MSE, the occurrence of distortions or errors doesn't depend upon the image quality. The errors are entirely unpredictable.
- The images cannot be compared on the basis of MSE values as the image quality, and MSE value do not relate to each other in any sense.

Before concluding this work, we want to inform the readers that there are also positive aspects of the deepfakes. Besides limitations, deepfakes are associated with many advancements and are considered to bring many in the future. Some of them are as follows:

1. The construction of a deep generative model will be a step forward in the field of health care since it will protect patients' privacy throughout therapy, and continuing research should be protected. We will be able to construct a completely fictional population of virtual patients, eliminating the requirement to share real-world patient data.
2. For conducting any experiment on facial figures, we won't be required to show the real face of the people. We could easily use deep generative models to create a fake face and will be able to conduct experiments on it [22].
3. Deepfake technology can be used to make personalized avatars, which are helpful for some applications that use such avatars to show mock clothing and hairstyles.
4. Digital reconstruction and safety of the public: Reconstructing a crime scene falls under the category of forensic science and is also an art that involves

employing inductive and deductive reasoning as well as evidence. Artificial intelligence-generated synthetic media can aid in the reconstruction of a scene involving the interplay of spatial and temporal artifacts.

5 Conclusion

In this work, we compared different metrics for deepfake detection by comparing their pixel differences without considering their geometry and scene. We show that GAN-synthesized faces have inconsistent corneal specular highlights. These can be manually fixed with editing in various photo editing software, but they will not be unique. The point to be added is that in a portrait photograph (real human face), we can have false positives because there are many limitations of this method such as when the light source is very near to the subject or where the source is not visible in both the eyes. In the real human eyes, we observed that although it is a real human eye, we are getting low values of SSIM and IoU signifying that it does not apply to images that do not have specular patterns. We also used MSE, which is quite unpredictable since it poorly correlates with the loss of quality of an image. Since the reconstruction of distorted images with MSE is very unpredictable, the images which appear to be identical according to the human visual system (HVS) may be very different as per MSE and vice versa. The SSIM, on the other hand, uses three criteria, brightness, contrast, and structure, which are very much structured. The SSIM error map depicts the area most affected by noise. So, using SSIM to remake the distorted image is easier. Therefore, we can say IoU and SSIM are effective for deep fake detection, but the IoU will be more accurate than SSIM. This study will be helpful for the researchers to get insights into deepfake detection.

References

1. Goodfellow I, Pouget-Abadie J, Mirza M, Xu B, Warde-Farley D, Ozair S, Courville A, Bengio Y (2020) Generative adversarial networks. Commun ACM 63(11):139–144
2. Deepfakes faceswap. https://github.com/deepfakes/faceswap. Last accessed 25 July 2021
3. Faceswap-GAN. https://github.com/shaoanlu/faceswap-GAN. Last accessed 4 Aug 2021
4. Nirkin Y, Keller Y, Hassner T (2019) Fsgan: Subject agnostic face swapping and reenactment. In: Proceedings of the IEEE/CVF international conference on computer vision, pp 7184–7193
5. Thies J, Zollhofer M, Stamminger M, Theobalt C, Nießner M (2016) Face2face: Real-time face capture and reenactment of rgb videos. In: Proceedings of the IEEE conference on computer vision and pattern recognition, pp 2387–2395
6. Faceswap. https://github.com/MarekKowalski/FaceSwap. Last accessed 25 July 2021
7. Agarwal S, Farid H, Gu Y, He M, Nagano K, Li H (2019) Protecting world leaders against deep fakes. In: CVPR workshops, vol 1
8. Westerlund M (2019) The emergence of deepfake technology: a review. Technol Innov Manage Rev 9(11)
9. Karras T, Laine S, Aittala M, Hellsten J, Lehtinen J, Aila T (2020) Analyzing and improving the image quality of stylegan. In: Proceedings of the IEEE/CVF conference on computer vision and pattern recognition, pp 8110–8119

10. Saboia P, Carvalho T, Rocha A (2011) Eye specular highlights telltales for digital forensics: A machine learning approach. In: 2011 18th IEEE international conference on image processing. IEEE, pp 1937–1940
11. Hu S, Li Y, Lyu S (2021) Exposing GAN-generated faces using inconsistent corneal specular highlights. In: ICASSP 2021–2021 IEEE international conference on acoustics, speech and signal processing (ICASSP). IEEE, pp 2500–2504
12. Yen J-C, Chang F-J, Chang S (1995) A new criterion for automatic multilevel thresholding. IEEE Trans Image Process 4(3):370–378
13. Rao CR (1980) Some comments on the minimum mean square error as a criterion of estimation. Pittsburgh Univ PA Inst for Statistics and Applications
14. Hore A, Ziou D (2010) Image quality metrics: PSNR vs. SSIM. In: 2010 20th international conference on pattern recognition. IEEE, pp 2366–2369
15. Rezatofighi H, Tsoi N, Gwak JY, Sadeghian A, Reid I, Savarese S (2019) Generalized intersection over union: A metric and a loss for bounding box regression. In: Proceedings of the IEEE/CVF conference on computer vision and pattern recognition, pp 658–666
16. Henry J, Montavont N, Busnel Y, Ludinard R, Hrasko I (2021) A geometric approach to noisy EDM resolution in FTM measurements. Computers 10(3):33
17. Karras T, Laine S, Aila T (2019) A style-based generator architecture for generative adversarial networks. In: Proceedings of the IEEE/CVF conference on computer vision and pattern recognition, pp 4401–4410
18. King DE (2009) Dlib-ml: A machine learning toolkit. J Mach Learn Res 10:1755–1758
19. Van der Walt S, Schönberger JL, Nunez-Iglesias J, Boulogne F, Warner JD, Yager N, Gouillart E, Yu T (2014) scikit-image: image processing in Python. PeerJ 2:e453
20. Bosnić Z, Lukowicz P, Coughtrey D, Hoshino R (2006) "n introduction to ROC analysis. Pattern Recogn Lett 27(861):874
21. Gandhi SA, Kulkarni CV (2013) MSE Vs SSIM. Int J Sci Eng Res 4(7). ISSN 2229–5518
22. Korshunova I, Shi W, Dambre J, Theis L (2017) Fast face-swap using convolutional neural networks. In: Proceedings of the IEEE international conference on computer vision, pp 3677–3685

Evolution of Biomedical Implantable Antennas: Requirements, Challenges, Designs, and Applications

Sumit Kumar Khandelwal, Ravi Kumar Arya, and Srinivasa Nallanthighal Raghava

Abstract Today, implantable antennas are very popular among researchers, as they are used to find simple, effective, and real-time solutions for many health-related issues inside the human body. Biomedical implantable devices are receiving great attention to find solutions to different medical conditions. Implantable wireless sensors integrated with implantable antennas have many advantages like early detection of disease and continuous and real-time monitoring of health conditions. Thus, it reduces healthcare costs, improves the quality of life of an individual, and improves the accuracy of diagnostics systems. The implantable sensors are placed inside the human body to measure real-time information of the various body parameters like glucose level, body temperature, blood pressure, and ocular pressure. However, the implantable antennas used for transmitting this information face major challenges such as poor gain and large size. The human tissues due to their frequency-dependent permittivity and highly lossy nature exacerbate the performance of the implantable system. This manuscript covers an overview of the major requirements of implantable antennas, various design aspects, challenges, simulation tools, testing methods, and various applications of biomedical implantable antennas.

Keywords Biomedical implantable devices · Implantable antennas · Biocompatible antenna

S. K. Khandelwal (✉) · S. N. Raghava
Delhi Technological University, New Delhi, India
e-mail: sumit.khandelwal@dtu.ac.in

S. N. Raghava
e-mail: nsraghava@dtu.ac.in

R. K. Arya
National Institute of Technology Delhi, New Delhi, India

J. K. Mandal et al. (eds.), *Topical Drifts in Intelligent Computing*, Lecture Notes in Networks and Systems 426, https://doi.org/10.1007/978-981-19-0745-6_43

1 Introduction

With today's lifestyle that includes more work and less physical activities, people are more vulnerable to various diseases; thus, routine health checkups are important. It becomes even more important for people who already suffer from chronic diseases like diabetes, blood pressure, Alzheimer's, asthma, and cancer. But, due to the hectic lifestyle, people rarely go for regular health checkups. Thus, the current focus of researchers is moving to mitigate this problem. With the evolution of wireless technology and biomedical implantable antennas, real-time monitoring of health conditions is possible. Implantable medical devices (IMDs) are those devices that are inserted inside the human body, and it can be used for several diagnostics and monitoring real-time health conditions of a patient [1]. Typical examples of IMDs include implantable pacemakers, breast cancer detection systems, ocular pressure measurement systems, and real-time blood insulin measurement. Nowadays, biomedical implantable antennas are used in a various healthcare-related applications such as monitoring brain atrophy and lateral ventricle enlargement, [2], breast tumor detection [3], wireless capsule endoscopy [4], contactless measurement of arterial pulses [5], heart failure detection system [6], rectenna-based pacemaker, and many more. Researchers are working on healthcare applications to reduce healthcare costs and improve the accuracy of healthcare systems. The frequency of operation is also very important. The wireless medical telemetry service (WMTS) band, medical device radiocommunications service (MedRadio) band, and industrial, scientific, and medical (ISM) bands are recommended frequency bands for biomedical applications. The MedRadio band is in the frequency ranges of 401–406, 413–419, 426–432, 438–444, and 451–457 MHz [1]. The WMTS band has a frequency range of 1427–1432 MHz, while the frequency range of 2.4–2.5 GHz comes under ISM bands.

In this manuscript, a brief literature review on the requirement, design criteria, biocompatibility issues, different design techniques, and various testing techniques has been covered.

2 Implantable Antenna Requirements

As biomedical implantable antennas are placed inside the body, these antennas have various constraints. There are various requirements that one should touch upon while designing an implantable antenna.

2.1 Antenna Structure

The designed antenna should have a design structure in a way so that it is conformal to the human body.

2.2 Antenna Size

Implantable antennas should be compact as they will be embedded inside the body. Also, compact antennas provide more flexibility. Researchers devote extra time and effort to minimize the size of an implantable antenna. However, reducing antenna dimensions directly affects the electromagnetic performance of an implantable antenna [7].

2.3 Human Safety

Human safety is the most important requirement of an implantable antenna. As implantable antennas radiate power inside the body, thus the analysis of power absorption inside the body is very crucial. The absorption of radiated power by the human body tissues generates heat, which could be dangerous and could affect human health. The parameter which is introduced to measure the absorption of electromagnetic energy by human body tissues due to an implantable device is called specific absorption rate (SAR) [8]. SAR is considered an index parameter by the FCC to control the amount of exposure of the human body tissues to electromagnetic radiation [9]. According to IEEE C95.1-1999, for 1 g human tissue sample, the average SAR distribution must be within the limit of 1.6 W/kg and IEEE C95.1–2005 suggests that for 10 g human tissue; the average SAR must be below 2 W/kg [8].

2.4 Bandwidth

Generally, implantable antennas which have a very high bandwidth are preferred for biomedical applications. The major advantage of wideband antennas is that it operates at a broad range of frequencies and can work in harsh communication environments with low SNRs. Thus, wideband antennas are highly desired for implantable applications.

2.5 Biocompatibility

Biocompatibility is one of the major aspects that need to be considered when it comes to designing implantable antennas [10]. The material used in antennas should not be harmful or toxic to living tissues. The list of biocompatible materials which are used by researchers in designing biocompatible antennas is given in Table 1.

Table 1 List of bio-compatible materials used in implantable antennas

Materials	Relative permittivity
Poly-tetra fluoro ethylene	2.2
Macor	6.03
Alumina	9.3
Zirconia	29
Zirconium dioxide	21

2.6 *Polarization*

Circularly polarized (CP) antennas are recommended to reduce polarization mismatch loss, fading effects due to multipath, and improve data transmission rate for implantable antennas [11].

3 Implantable Antenna Designs

There is continuous research on implantable antennas with different design types. In this section, a brief review of the different design types of implantable antennas is presented. A simple structure and extremely miniaturized implantable antenna which shows the dual-band characteristics in the ISM bands (2.45 and 0.915 GHz) and a biotelemetry device for implantation in the scalp is proposed in [12]. A dual-loop symmetrical antenna is presented in [13] with ultra-wideband performance and operating in the 2.45 GHz ISM band for use in wireless endoscopy applications. A miniaturized single-fed hybrid slot implantable antenna is discussed in [14] for the MICS band (402–405 MHz). Inserting of the different slots in the ground, the effective size reduction of an implantable antenna is possible. A compact, coplanar waveguide-fed, dual-ring slot wideband implantable antenna [15] is presented for the ISM band. The comparison among different antennas and their frequency bands in reference papers is presented in Table 2.

4 Designing of an Implantable Antenna

Designing a compact, flexible, planar, human-safe, high bandwidth, and high gain antenna is a very challenging task. In this section, different techniques to improve the characteristics of implantable antennas will be discussed.

Table 2 Comparison of different implantable antenna structures

References	Design type	Dimension (mm)	Frequency
[16]	Dual folded	25 × 34 × 2.53	ISM, WMTS
[17]	Spiral	14 × 14 × 15	MedRadio
[18]	PIFA	15 × 15 × 2	ISM
[19]	Slot PIFA	19 × 30 × 1.6	MedRadio, ISM
[20]	Hexagonal and T-shaped slots	7 × 7 × 0.2	MedRadio, ISM
[12]	Dual-loop	10 × 15 × 0.25	ISM
[13]	Meander slot	10 × 16 × 1.27	MedRadio
[14]	Dual-ring slot	10 × 10 × 0.4	ISM
[15]	Sigma-shaped monopole	20 × 20 × 1.6	MedRadio

4.1 Antenna Size Reduction Techniques

Implantable antenna size reduction is a very important requirement. Different techniques to reduce the size of the antenna are as follows:

Dielectric material. A material with high dielectric constant (relative permittivity) can be chosen as a substrate or superstrate material. As permittivity of material increases, the capacitance between metallic patches increases. As resonant frequency is given by $1/2\Pi\sqrt{}(LC)$, so increase in capacitance decreases the frequency. Now, for the same resonant frequency, antenna size need to be reduced.

Slotting Techniques. Recently, different techniques have been adopted by research groups to minimize the size of an implantable antenna by introducing slots in the antenna ground plane [21] and creating multiple slots in the patch [22]. For size reduction, a substrate embedded periodic structure is used in the substrate of an implantable antenna [23]. Also, a split ring resonator (SRR) [24] and a combination of complementary SRR and C-shaped slot [25] are utilized for size reduction of an antenna.

Increasing the current path of a radiator. Increasing the current path, length of the antenna can be thought as the most effective and easiest way for reducing size. This increase in current path length reduces the resonance frequency and hence achieves size reduction of an implantable antenna. Current path length can be increased by adopting different design techniques such as curved line, loop, spiral, helical, meandered, and slot antennas [15].

4.2 Wideband Techniques

To provide mobility to the dynamic implantable antenna, wide bandwidth is required. In general, the bandwidth of the antenna can be increased by adopting various methods such as by choosing a low dielectric constant substrate or superstrate material, increasing the thickness of substrate or superstrates, by cutting slots in an antenna, by cutting notches or by using probe feeding.

Bandwidth can also be increased by merging two or more resonant frequency modes. Two π-shaped meandered strips have been used on the PIFA in [26] to increase the bandwidth.

4.3 Gain Enhancement Techniques

There are different types of methods used by researchers to enhance antenna gain. As an implantable antenna operates inside the complex human body environment, its radiation characteristics deteriorate. Thus, for achieving short-range communication with external instruments, antenna gain should be nominally higher than antenna gain in free space.

Increasing the substrate height, making antenna arrays, using EBG superstrates, increasing the effective area of an antenna using techniques like parasitic patches, meandering of edges, using metamaterials, are few methods that may improve the gain of an implantable antenna.

The implantable antenna gain can be improved by introducing printed grid surfaces [27] or by introducing the combination of hemispherical glass lens and parasitic ring [25].

5 Simulation and Testing of Implantable Antenna

5.1 Simulation

Several simulation software tools are used for simulating and designing an implantable antenna for biomedical applications. The commonly used simulation software's are Ansys high-frequency structure simulator (HFSS), CST Microwave Studio, ADS-Momentum, Sonnet Suites, Altair-FEKO, antenna design explorer, etc. Out of these, Ansys HFSS is more popular as it is based on the finite element method (FEM), and it is considered as most accurate for designing implantable antennas.

Ansys HFSS provides readily available human tissue models. The Ansys human body model (HBM) contains geometry as well as material properties that represent the tissues of the human body. It includes a full-body adult male and an adult-female model.

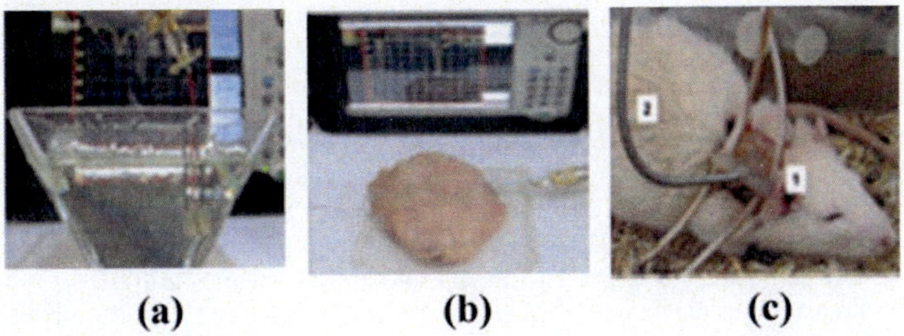

Fig. 1 Different testing techniques: **a** in vitro testing; **b** ex vivo testing; **c** in vivo testing

5.2 *Testing Methods*

Various testing methods are used to study the characteristics and performance of an antenna in the same environment as human tissues. It also helps to estimate SAR values. Following testing methods are mostly employed.

In vitro testing. In this testing technique, the human tissue mimicking liquid phantom is prepared in the laboratory. The liquid phantom electromagnetic characteristics are nearly like human muscle tissue. The fabricated antenna is inserted into a liquid phantom, as shown in Fig. 1a. The recipe for preparing the mimicking liquid phantom [26] is given as deionized water (44.5%), sodium chloride (0.5%), sugar (55%).

Ex vivo testing. In this type of measurement technique, a meat or pork sample is taken, and an antenna is embedded inside the sample for investigation. Minced meat, minced pork, piece of pork, rat skin, Chicken breast, etc. are used in ex-vivo testing. Properties of these meats are very close to human muscles.

In vivo testing. In this type of measurement technique, implantable antennas are placed inside the body of living animals to achieve more realistic scenarios in their validation test.

6 Conclusion

This paper covers the various applications of implantable biomedical antennas. Various constraints for designing implantable antennas are explored in this manuscript. In this manuscript, various methods of decreasing implantable antenna size, increasing the bandwidth and gain have been investigated. Several simulation software's for designing an implantable antenna have been explored. This review in a single manuscript will help readers to hone their skills in biomedical implantable antennas.

References

1. Blanos P (2013)Miniaturization of implantable antennas for medical applications. National Technical University of Athens, School of Electrical & Computer Engineering, Athens
2. Saied IM, Arslan T (2020) Noninvasive wearable RF device towards monitoring brain atrophy and lateral ventricle enlargement. IEEE J Electromagn RF Microwaves Med Biol 4(1):61–68. https://doi.org/10.1109/JERM.2019.2926163
3. Alshaibi MA, Almoneef TS, Attia H, Ramahi OM (2019) Near-field microwave loop array sensor for breast tumor detection. IEEE Sens J 19(24):11867–11872. https://doi.org/10.1109/JSEN.2019.2936993
4. Wang J, Leach M, Lim EG, Wang Z, Pei R, Huang Y (2018) An implantable and conformal antenna for wireless capsule endoscopy. IEEE Antennas Wirel Propag Lett 17(7):1153–1157. https://doi.org/10.1109/LAWP.2018.2836392
5. Johnson JE, Shay O, Kim C, Liao C (2019) Wearable millimeter-wave device for contactless measurement of arterial pulses. IEEE Trans Biomed Circ Syst 13(6):1525–1534. https://doi.org/10.1109/TBCAS.2019.2948581
6. Rezaeieh SA, Abbosh AM (2014) Wideband and unidirectional folded antenna for heart failure detection system. IEEE Antennas Wirel Propag Lett 13:844–847. https://doi.org/10.1109/LAWP.2014.2320495
7. Kiourti A, Nikita KS (2012) A review of implantable patch antennas for biomedical telemetry: challenges and solutions. IEEE Antennas Propag Mag 54(3):210–228
8. Seo Y, Wang ZJ (2021) Measurement and evaluation of specific absorption rate and temperature elevation caused by an artificial hip joint during MRI scanning. Sci Rep 11:1134. https://doi.org/10.1038/s41598-020-80828-7
9. Psathas KA, Kiourti A, Nikita KS (2014) Biocompatibility of implantable antennas: Design and performance considerations. In: The 8th European conference on antennas and propagation (EuCAP 2014), pp 1566–1570. https://doi.org/10.1109/EuCAP.2014.6902083
10. Yang Y-H, Sun B-H, Guo J-L (2019) A Low-Cost, Single-layer, dual circularly polarized antenna for millimeter-wave applications. IEEE Antennas Wirel Propag Lett 18(4):651–655. https://doi.org/10.1109/LAWP.2019.2900301
11. Gabriel C, Gabriel S, Corthout E (1996) The dielectric properties of biological tissues: I. Literature survey. Phys Med Biol 41(11):2231–2249. https://doi.org/10.1088/0031-9155/41/11/001
12. Shang J, Yu Y (2019) An ultrawideband capsule antenna for biomedical applications. IEEE Antennas Wirel Propag Lett 18(12):2548–2551. https://doi.org/10.1109/LAWP.2019.2942842
13. Liu C, Guo Y, Xiao S (2012) A hybrid patch/slot implantable antenna for biotelemetry devices. IEEE Antennas Wirel Propag Lett 11:1646–1649. https://doi.org/10.1109/LAWP.2013.2237879
14. Das S, Mitra D (2018) A compact wideband flexible implantable slot antenna design with enhanced gain. IEEE Trans Antennas Propag 66(8):4309–4314. https://doi.org/10.1109/TAP.2018.2836463
15. Tsai C-L, Chen K-W, Yang C-L (2015) Implantable wideband low-SAR antenna with C-shaped coupled ground. IEEE Antennas Wirel Propag Lett 14:1594–1597
16. Kumar SA, Shanmuganantham T (2012) Implantable CPW fed dual folded dipole antenna for biomedical applications. In: Computing Communication & Networking Technologies (ICCCNT), Coimbatore, India
17. Abadia J, Merli F, Zurcher J-F (2009) 3D-spiral small antenna design and realization for biomedical telemetry in the MICS band. Radioengineering 18(4):1–5
18. Tofighi M (2011) Characterization of biomedical antennas for microwave heating, radiometry, and implant communication applications. In: Wireless and Microwave Technology Conference (WAMICON)
19. Farhad G, Mohan AS (2013) Miniaturized slot PIFA antenna for tripleband implantable biomedical applications. In: IEEE MTTS International

20. Faisal F, Zada M, Ejaz A, Amin Y, Ullah S, Yoo H (2020) A miniaturized dual-band implantable antenna system for medical applications. IEEE Trans Antennas Propag 68(2):1161–1165. https://doi.org/10.1109/TAP.2019.2938591
21. Li R, Xiao S (2014) Compact slotted semi-circular antenna for implantable medical devices. Electron Lett 50(23):1675–1677
22. Hashemi S, Rashed-Mohassel J (2017) Miniaturization of dual band implantable antennas. Microw Opt Technol Lett 59(1):36–40
23. Zhang H, Li L, Liu C, Guo Y-X, Wu S (2017) Miniaturized implantable antenna integrated with split resonate rings for wireless power transfer and data telemetry. Microw Opt Technol Lett 59(3):710–714
24. Liu XY, Wu ZT, Fan Y, Tentzeris EM (2017) A miniaturized CSRR loaded wide-beamwidth circularly polarized implantable antenna for subcutaneous real-time glucose monitoring. IEEE Antennas Wireless Propag Lett 16:577–580
25. Lee C-M, Yo T-C, Huang F-J, Luo C-H (2009) Bandwidth enhance-ment of planar inverted-F antenna for implantable biotelemetry. Microw Opt Technol Lett 51(3):749–752
26. Zhu S, Almari S, AlAmoudi AO, Langley RJ (2013) Implanted antenna efficiency improvement. In: Proceedings of 7th European conference on antennas propagation, Gothenburg, Sweden, pp 3247–3248
27. Alamri S, AlAmoudi A, Langley RJ (2016) Gain enhancement of implanted antenna using lens and parasitic ring. Electron Lett 52(10):800–801

The Impact of Elon Musk Tweets on Bitcoin Price

Ritik Ranjan Gupta, Ravi Kumar Arya, Jatin Kumar, Akshat Gururani, Ravi Dugh, and Amit Dugh

Abstract Bitcoin is based on peer-to-peer technology where there is no need for central authority or banks as intermediaries. The network carries out Bitcoin issuing and transaction management. It is open-source, decentralized by design, nobody owns it or controls Bitcoin, and everyone can participate. Bitcoin challenges the previous payment system through its unique properties and allows compelling use cases that were not feasible in the past. Due to its decentralized properties and lack of price control, many a time, celebrities can influence its value using their social media fan followership. Twitter is becoming common place for celebrities to share their sentiments about Bitcoin. Elon Musk, a juggernaut entrepreneur of this century, is a prominent celebrity who has had a significant influence on Bitcoin and has been instrumental in promoting and criticizing Bitcoin in past months. In this paper, we analyze how Elon Musk's tweets affect Bitcoin prices.

Keywords Twitter · Elon Musk · Bitcoin · Sentiment analysis · Cryptocurrency

R. R. Gupta
Department of Electrical and Electronics Engineering, National Institute of Technology Delhi, New Delhi 110040, India

R. K. Arya (✉) · J. Kumar · A. Gururani
Department of Electronics and Communication Engineering, National Institute of Technology Delhi, New Delhi 110040, India

A. Gururani
e-mail: 171220004@nitdelhi.ac.in

R. Dugh
Goergen Institute of Data Science, University of Rochester, Rochester, NY, USA
e-mail: rdugh@ur.rochester.edu

A. Dugh
Information Technology, Queensland University of Technology, Brisbane, Australia

J. K. Mandal et al. (eds.), *Topical Drifts in Intelligent Computing*, Lecture Notes in Networks and Systems 426, https://doi.org/10.1007/978-981-19-0745-6_44

1 Introduction

Twitter [1] is one of the top leading social media platforms, where the users can craft status messages, called "tweets", to express their opinions. Tweets have a limit of 280 characters. One of the wealthiest entrepreneurs, Elon Musk, has come into the limelight with his tweets about the Bitcoin cryptocurrency [2]. Unlike the common currencies used for commerce and trade, cryptocurrencies are decentralized and are not regulated by any government. Musk's tweets have had a markable impact on Bitcoin's value, and he has even been accused by many on the Internet and by other crypto investors of manipulating the said cryptocurrency.

Bitcoin pricing forecasting has been a feasible topic of study for some time. As a forerunner in blockchain financial changes, Bitcoin [3, 4], plays a critical part in the entire monetary climate of the bitcoin business. As a result, the machine learning and data mining communities have been interested in (I) predicting Bitcoin price changes. (II) share experiences to gain a better understanding of the causes of Bitcoin instability and how to assess related risks in the cryptocurrency space. Many academics analyzed Bitcoin price estimation using social media sentiment analysis and machine learning. We believe that we can not only capture short-term price movements in Bitcoin prices using an appropriate machine learning model but also show that bitcoin price can be impacted by tweets from celebrities (notably, Elon Musk, who has the single largest followers and tweets about Bitcoin).

2 Related Works

Bitcoin is a new technology that is now the most costly cryptocurrency in the world. There are a variety of price prediction models for it. Garcia et al. discovered that rising Bitcoin prices are preceded by rises in opinion polarisation and trading volume [5], and Amjad et al. employed historical time series pricing data for price forecasting and swapping methodologies [6]. By looking at and separating the Twitter feed, Chen and Lazer [7] settled on investing ideas. Iya et al. [8] use a database for remote monitoring to train dividers and approve operational separation. Iya et al. cite Pang et al. influential's study [9], where those analysts set the bar for psychological analysis based on machine learning. Their approach is regarded as one of the most important machine learning projects in the field of doctrinal analysis [9].

Some of the most recent initiatives concentrate on the most popular trade and the use of deep learning methodologies such as RNN data predictions for a series of time-tested networks, feeds that are comparable to job kinds [10]. McNally [11] uses machine learning using neural networks (RNNs) and long short-term memory (LSTM) to predict Bitcoin pricing and compare the results to those obtained using autoregressive integrated average (ARIMA) models. An exogenous model (NARX) was constructed based on [12] multi-layer perceptron (MLP) and non-linear autoregressive models are compared. The authors determined that while MLP did not perform better than the NARX model in price forecasting, it might be utilized to anticipate the stock market. To design and analyze network performance, the authors used the MATLAB network toolkit. The other publication [13] is about daily data for 10-min and 10-min time series. The authors created 30-, 60-, and 120-min series data sets, then performed a GLM / random forest analysis on each set of data. To forecast the price of Bitcoin, these three classes are pooled together. According to [14], the author examines the past research to forecast the stock market in the United States. The author discovered that the prediction network's mean square error (MSE) was equal to the excess return's standard deviation, according to his findings. The author, on the other hand, shows that several basic financial and economic characteristics can anticipate the market excess return.

Rather than directly predicting future prices, the authors of [15] forecast stock trends. It is possible to classify the practice as a pattern. They make both short-term (daily) and long-term projections and discover that the latter has a 79% accuracy rate. The performance of this paper is another intriguing characteristic. The performance testing algorithm evaluates whether to buy, sell, or keep stock based on the expected release.

In this paper, we have explored some of the most effective ways to predict Bitcoin price using the sentiment of tweets. Next, we discuss the methodology used in this work.

3 Methodology

To analyze how Elon Musk is affecting the flow of Bitcoin prices, Elon Musk's tweets were scraped using the Tweepy library in Python. After cleaning the tweets data, the sentiment of each tweet was analyzed as positive or negative and compared the Bitcoin price variations at the same time.

3.1 Twitter Data Collection

The Bitcoin's closing prices were considered from 2014 till June 2021. Also, the volume required to generate the particular cryptocurrency was taken into account, so that the impact on prices with the changes in volume can be studied. The Twitter API was utilized to collect tweets for the sentiment analysis, and the data was then analyzed filtered by incorporating the keywords or hashtags like #BT, #Bitcoin, #Dogecoin, etc. Since the manuscript is about Elon Musk's tweets, all of his tweets related to Bitcoin were collected and the aftermath of them was investigated.

3.2 Twitter Data Preprocessing

Tweets consist of a lot of abbreviations, emoticons, and any unnecessary data such as images and their URLs. So, first, the tweets are processed to represent the rights of the feelings and emotions of the Twitter users. For the initial processing of the tweets, we used the three steps of filtering: Tokenization, removal of the stop words, and the match of the regular expression to remove special characters. Tokenization is the process of breaking down tweets into individual words, depending on the situation, and discarding data, such as emoticons, which are removed. Now, in the next stage, the elimination of the stop words takes place. These are the words that do not express any feelings and emotions. After sharing the tweet, the words such as a, b, c, etc., should be deleted from the dictionary. Finally, the special characters are removed: to take away the special characters, regular expression comparison in Python is carried out, to match a URL, and in the long-term, to be replaced by a URL.

3.3 Twitter Sentiment Analysis

Tweet sentiment analysis is a task in and of itself, and it is highly field-specific. Based on the sentiment present, the tweets were categorized as good, negative, or neutral to make things easier. All of the tweets were carefully reviewed and assigned a score of 1 for positive emotions, 0 for neutral emotions, and -1 for negative emotions. The features retrieved from these annotated tweets are fed into a machine learning model, which is then used to classify other tweets.

3.4 Model Training

Training and prediction are done using long short-term memory (LSTM) architecture. The formation of LSTM is part of repetitive neural networks and is often used in the field of deep learning. LSTM has a feedback loop, which makes it very useful to process the sequence of all the details. Before feeding data to LSTM, the data must be processed in advance (Fig. 1).

4 Machine Learning Methods Used

Following were the machine learning methods used for this work.

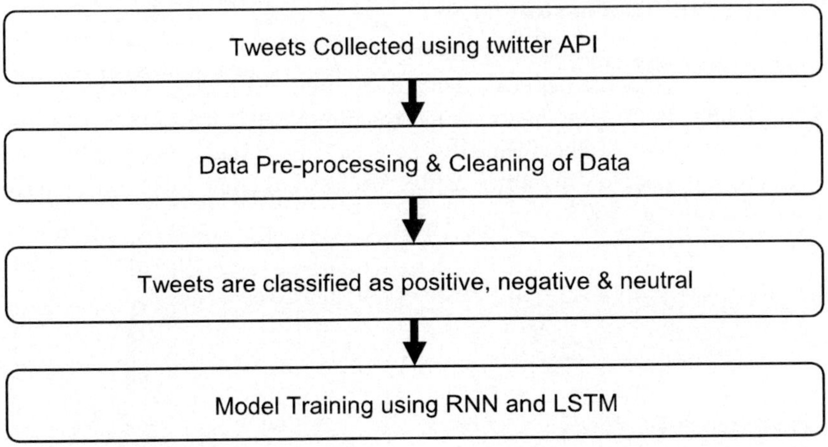

Fig. 1 Architecture

4.1 Recurrent Neural Networks (RNNs)

Artificial neural networks known as recurrent neural networks (RNNs) are widely utilized in speech recognition and natural language processing. RNNs recognize the sequential properties of data and use patterns to forecast the next most likely outcome. Both, deep learning and the models imitating neuron activity in the human brain, use RNNs. They are very useful where context is important for forecasting an outcome, and they differ from other types of artificial neural networks in that they process a series of input that informs the final output through feedback loops. Due to these feedback loops, the information can persist.

4.2 Long Short-Term Memory (LSTM)

The problem of vanishing and exploding gradients in a deep recurrent neural network has been addressed in a variety of ways. One of the most well-known networks is the long short-term memory (LSTM) network. Hochreiter and Schmidhuber created LSTM, an evolution of RNN, to address the aforementioned RNN shortcomings by adding more interactions per module (or cell). In general, an LSTM recurrent unit attempts to "remember" all of the network's prior data while "forgetting" irrelevant data. This is done by establishing a variety of activation function levels known as "gates" for various purposes. Each LSTM recurrent unit is also responsible for a vector known as the inner Cell State, which theoretically describes the information that the previous LSTM recurrent unit selected to be maintained. An extended short-term memory network's basic process is similar to that of a recurrent neural network, with the exception that the interior Cell State is also transmitted along with the Hidden State.

5 Results and Discussion

Before using the methodologies discussed in Sects. 2 and 3, we show how the different trends of Bitcoin are. Figure 2a shows the Bitcoin closed price from 2014 to June 2021. From the figure, it is evident that there have been major fluctuations in the closed price in the past few months. The figure also shows how Bitcoin moved from under a dollar to about $60 k.

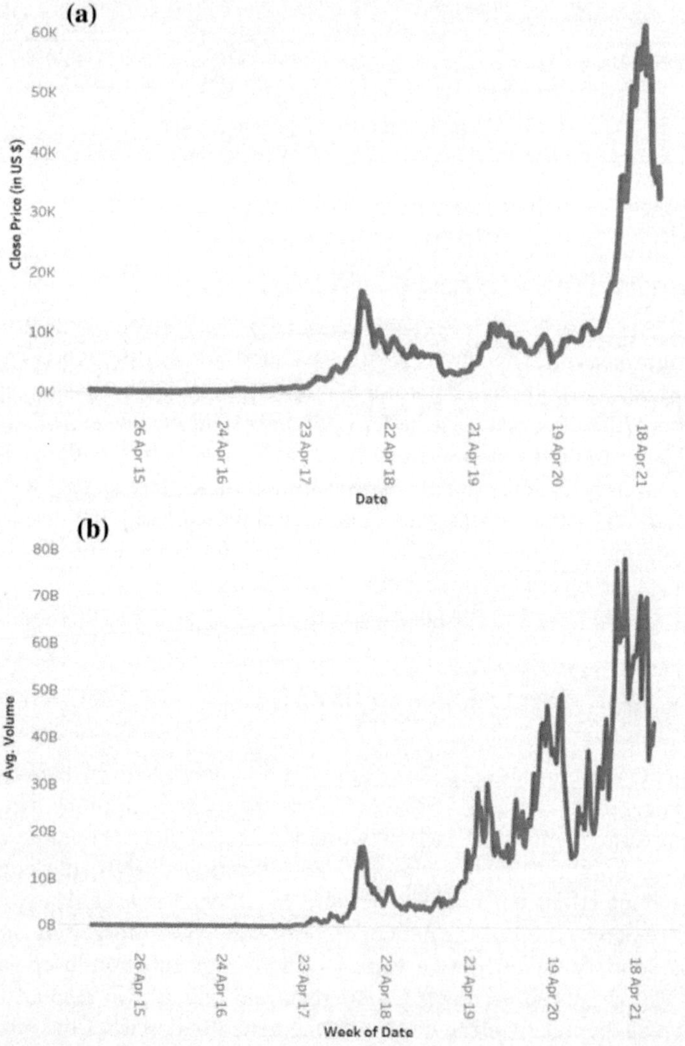

Fig. 2 Bitcoin trend: **a** closed price; **b** volume

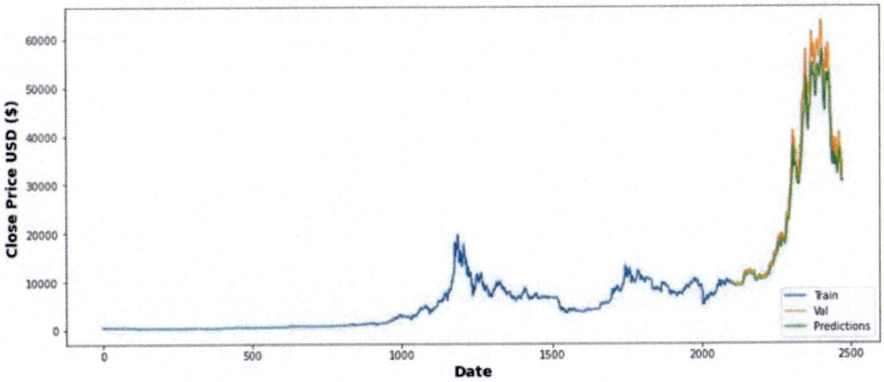

Fig. 3 Bitcoin price prediction using LSTM

Bitcoin volume is also important. Figure 2b shows the Bitcoin volume that went through transactions from 2014 to June 2021. It is evident from Fig. 2b that the volume has increased significantly as the price rose (see Fig. 2a), this indicates that the popularity of Bitcoin is somehow associated with its price.

By using the Bitcoin price in Fig. 2a, it is plausible to predict the future Bitcoin price. In Fig. 3, we show the Bitcoin price prediction using the LSTM model. The blue colored line shows the points considered for training the LSTM model while the Val (orange line) signifies the test values, which are real values, in the dataset. The green-colored line shows the predictions for the future data points after training the model.

Until now, we have relied solely on LSTM to forecast Bitcoin's future price data from Elon Musk's tweets. First, we need to judge if Elon's tweets do have any effect on Bitcoin price. Figure 4 shows the major tweets from Elon Musk and the effect of those tweets in changing the price of Bitcoin. Whenever the tweet had an appreciation for Bitcoin, the Bitcoin price increased, and when the price decreased when Bitcoin was criticized or restricted in any way. Tweets are also tabulated in Table 1 with their sentiments.

6 Conclusions

A method for predicting the Bitcoin price by looking at the sentiment of Elon Musk's tweets was proposed. It was found that the price of Bitcoin is seriously affected by the juggernaut entrepreneur's tweets. Whenever the sentiments are positive like changing the Twitter bio to #bitcoin or indicating the power of Bitcoin by advising people to buy Tesla with it, etc., the value of Bitcoin went up. On the other hand, whenever there were negative indications regarding Bitcoin like refusing to accept Bitcoin or breaking up with Bitcoin, the value of the cryptocurrency deprecates. This shows the

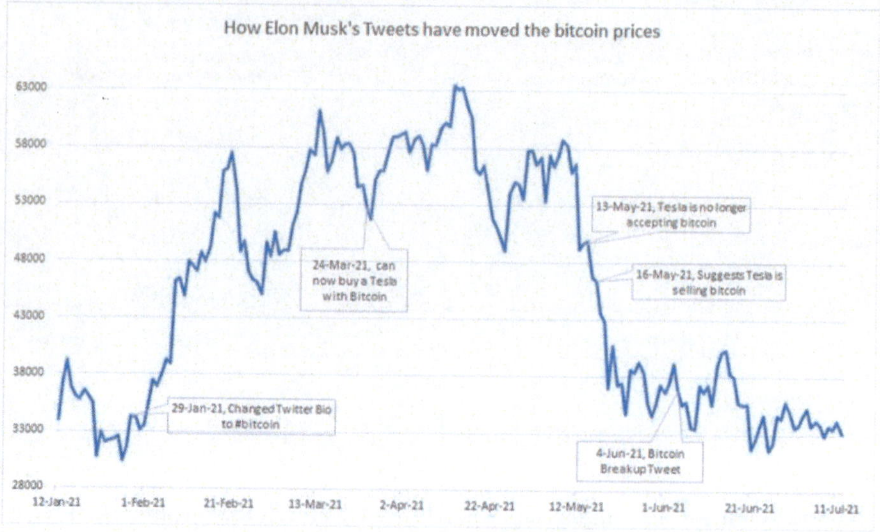

Fig. 4 Elon Musk's tweets and Bitcoin price change

Table 1 Sentiments of Elon Musk Tweets

Date	Tweet	Sentiment
29-01-2021	Changed Twitter Bio to #bitcoin	Positive
24-03-2021	You can now buy a Tesla with Bitcoin	Positive
13-05-2021	Tesla is no longer accepting bitcoin	Negative
16-05-2021	Suggests Tesla is selling bitcoin	Negative
17-05-2021	Tesla has not sold any Bitcoin	Positive
04-06-2021	#Bitcoin breakup	Negative

volatility of this cryptocurrency. Bitcoin is owned by nobody, i.e., no central entity controls it but is controlled by millions of Bitcoin users around the world. In this way, it was found that when some influential personality like Elon Musk affects the way of thinking of millions of users, the value of Bitcoin affects accordingly.

References

1. What is Twitter and why should you use it? (2021) https://esrc.ukri.org/research/impact-toolkit/social-media/twitter/what-is-twitter. Accessed 5 Aug 2021
2. https://bitcoin.org/en. Accessed 5 Aug 2021
3. Top 5 Celebrities Who Love Crypto (2021) https://www.coolwallet.io/celebrities-who-love-cryptocurrency-bitcoin-and-dogecoin. Accessed 5 Aug 2021
4. Vasilev I, Slater D, Spacagna G, Roelants P, Zocca V (2019) Python deep learning: exploring deep learning techniques and neural network architectures with Pytorch, Keras,

and TensorFlow. Packt Publishing Ltd
5. Amjad M, Shah D (2017) Trading bitcoin and online time series prediction. In: NIPS 2016 time series workshop
6. Nakamoto S (2008) Bitcoin: a peer-to-peer electronic cash system
7. Garcia D, Schweitzer F (2015) Social signals and algorithmic trading of Bitcoin. R Soc Open Sci 2(9)
8. Chen R, Lazer M (2011) Sentiment analysis of twitter feed for the prediction of stock market movement. Stanford Comput Sci 229:15
9. Go A, Huang L, Bhayani R (2009) Twitter sentiment classification using distant supervision. Stanford Comput Sci
10. Pang B, Lee L, Vaithyanathan S (2002) Thumbs up: sentiment classification using machine learning techniques. In: ACL-02 Conference on empirical methods in natural language processing, Philadelphia, PA, USA
11. Dixon M, Klabjan D, Bang JH (2017) Classification-based financial markets prediction using deep neural networks. ArXiv
12. McNally S, Roche J, Caton S (2018) Predicting the price of Bitcoin using machine learning. In: 26th Euromicro international conference on parallel, distributed and network-based processing (PDP)
13. Daniela M, Butoi A (2013) Data mining on Romanian stock market using neural networks for price prediction. Informatica Econ 17(3)
14. Shah D, Zhang K (2015) Bayesian regression and Bitcoin. In: 52nd annual Allerton conference on communication, control, and computing (Allerton)
15. Jang H, Lee J (2017) An empirical study on modelling and prediction of Bitcoin Prices with Bayesian neural networks based on Blockchain information. IEEE Early Access Articles

Design of 22 nm Strained Silicon Channel Gate All Around FET Device

Potaraju Yugender and Rudra Sankar Dhar

Abstract As the Scaling of Devices is in turn facing many problems regarding the performance of the device and to keep Moore's law alive a better replacement for a FINFET is a GAAFET. The strained silicon technology has made a great revolution in the last few years makes the device with enhanced mobility and reduces the short channel effects (SCEs) with less leakage current are the main parameters of a semiconductor device. The paper presents a 22 nm GAAFET device of tri-layered structure (s-Si, s-SiGe, s-Si) with a strained silicon channel of (2 nm, 4 nm, 2 nm) thickness, respectively. The biaxial strain-induced will increases the mobility of the charges. The controlling ability of the channel with four gates makes it better performance. The proposed device is compared with a silicon GAAFET using a Silvaco TCAD tool which is superior with a 22% of improvement in the drain current and less leakage current.

Keywords GAAFET · SILVACO TCAD · Strained silicon · Short channel effects

1 Introduction

These days, integrated circuits are present in all aspects of our daily life, starting with a computer and mobile phone which play an important role in our life. This makes the semiconductors industry to achieving great heights and increases the economy. As we are looking forward for higher performance and smaller size of electronic devices. Latest trends predictions is the market size of consumer electronics products will reach to much higher rate by 2024. As we move for the smaller electronic devices many wide range of applications are needed in the market. These qualities are found when we down scale the CMOS processor which requires very less amount of supply voltages. As we go for smaller sized devices which affects the drain voltage in controlling the channel region and gate voltage. The design of the MOSFET is varying frequently and size is decreasing to boost the performance of the integrated

P. Yugender · R. S. Dhar (✉)
Department of Electronics and Communication Engineering, National Institute of Technology Mizoram, Aizawl, Mizoram 796012, India
e-mail: rdhar@uwaterloo.ca

J. K. Mandal et al. (eds.), *Topical Drifts in Intelligent Computing*, Lecture Notes in Networks and Systems 426, https://doi.org/10.1007/978-981-19-0745-6_45

circuit. Since the depletion region width of source and drain and the channel length are same then the Short Channel Effects (SCEs) [1] will occur. "These SCEs are carrier velocity saturation, threshold voltage, drain induced barrier lowering (DIBL), sub-threshold leakage current, etc." Thus to improve the device performance we need to decrease the short channel effects. Many devices were used to lower the short channel effects [2–4].

To overcome these SCE the development started in 1991 when an ultra-thin gate structure of MOSFET was proposed and it was proved to reduce the SCE. This was the first step for the foundation of FinFET structure that was developed later. In this development Prof Chenming Hu role was very important in this period. He said that by changing the gate material [5, 6] and adding more number of gates the physical view of FinFET can be changed and improved to achieve the further down scaling. Double gates, Trigate FinFET are some of the improved devices. Further the channel control was not capable enough due to SCE. Then the inventors started developing a GAAFET structure which improves the performance of the device. So the much improved version of FinFET can be accomplished with the electrostatic control of the gate is Gate all around field effect transistor (GAAFET) [7, 8]. GAAFET is very similar to FinFETs but the conducting channel is enclosed by gate all around with this we get the better control over the gate. When the strained silicon channel is introduced in GAAFET it changes the physical parameters of the silicon channel by stretching the atoms beyond their normal interatomic distance. Therefore with invoked a strain in the channel makes the GAAFET device faster as the charge carries move faster within the channel, i.e. the electrons and the holes [9]. The proposed device in this paper increases the carrier mobility with less threshold voltage therefore the carrier mobility degradation is over come with the strain silicon technology. By implanting the tri-layered channel which increases the channel current and it also has the ability to decrease the SCEs with a minimum parasitic capacitance.

In this paper, a 22 nm GAAFET device is designed with strained silicon channel of tri-layered structure with (s-Si, s-Si-Ge, s-Si) [10]. Strain silicon technology ensures high performance of semiconductors devices which enhances carrier mobility with minimum threshold voltage and reduces the SCEs [11]. The GAAFET structure is designed using the Silvaco TCAD tool [12]. The performances of 22 nm channel GAAFET structure is compared with a 22 nm silicon GAAFET [13] transfer and output characteristics of the devices i.e. I_D versus V_{GS} and I_D versus V_{DS} and I_{on} and I_{off} ratio of the device is also simulated. The simulation results shows the 22 nm GAAFET with strained silicon the gate has more ability to control the channel at 22 nm or even for less gate length and the SCEs are reduced within the acceptable range. The values of ON to OFF current ratio makes the device more power efficient and a very good choice for the semiconductor device.

2 Device Structure

The gate all around FET with strained silicon channel is implemented on 22 nm gate length consisting of tri-layered of (s-Si, s-SiGe, s-Si) materials with a layer thickness of (2, 4, 2)nm using Silivaco TCAD tool [12]. A 22 nm gate length device is implemented and simulated with a strained silicon of three layer channel and is compared with a 22 nm silicon Gate all around FET [13]. The three dimensional view of the 22 nm GAAFET with strained silicon is shown in Fig. 1 and the device parameters are given in Table 1 and the GAAFET structure with Gate all around view of strained layered channel is shown in Fig. 2.

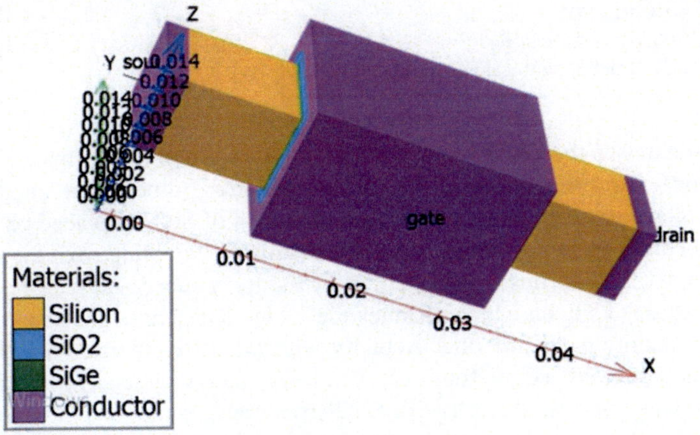

Fig. 1 Three dimensional view of 22 nm GAAFET structure

Table 1 Device parameters

Device parameters	Dimensions
	strained silicon GAAFET
Source length	22 nm
Drain length	10 nm
Channel length	10 nm
Channel thickness (s-Si, s-SiGe, s-Si)	8 nm (2, 4, 2)
Gate oxide thickness	1 nm
Source doping concentration	$3 \times 10^{19}/cm^3$
Drain doping concentration	$5 \times 10^{19}/cm^3$
Channel doping concentration	$1 \times 10^{-15}/cm^3$
Work function of gate contact	4.45 eV
Mole fraction	0.4

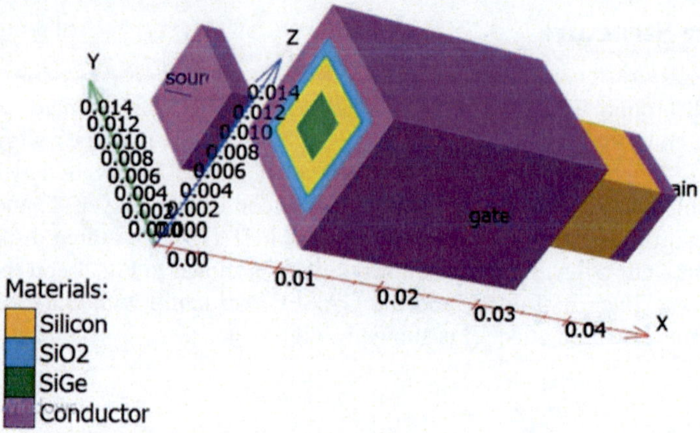

Fig. 2 GAAFET structure with gate all around view

The four gates on the device are located on all the sides of the channel, i.e. front, back, top and bottom of the strained silicon channel and are joined to one another with the same voltage to drive the gates [6]. The thickness of SiO2 is placed on all four sides. This configuration provides a better electrostatic coupling over the strained silicon channel. Due to this floating effect SCEs are reduced. As the presence of tri-layered strained silicon which changes the lattice constant of material and also changes the energy band structure. Actually with inclusion of strain the electrical symmetry is destroyed and the four energy valleys go down in energy two go up in biaxial strain and unaxial strain. The unaxial strain reduces the intervally scattering and effective mass and biaxial strain reduces the scattering. Finally the device has the positive benefits of lower capacitance, faster switching and less leakage current.

3 Results and Simulations

The GAAFET device with strained silicon is simulated using SILVACO TCAD in Atlas [14] of 22 nm channel length for a enhancement of drain current, less leakage current and to reduces the drain-induced barrier lowering. As the insertion of the strain in the channel region the mobility of the charge carries increases due to the quantum confinement. The plot $I_D - V_{GS}$ characteristics in logarithmic scale of the GAAFET with tri-layered strained silicon are compared with silicon GAAFET [13] is shown in Fig. 3. The transfer characteristics were obtained at constant drain voltage of 1v with Y-axis as the drain current and X-axis as the gate voltage. The Ion current and Ioff current are obtained from the linear scale graph of $I_D - V_{GS}$ transfer characteristics from the graph it is observed that the Ion current of the GAAFET with strained silicon is more than the silicon GAAFET and the I_{off} is less than silicon GAAFET. The presented device with the strained silicon in the channel is showing

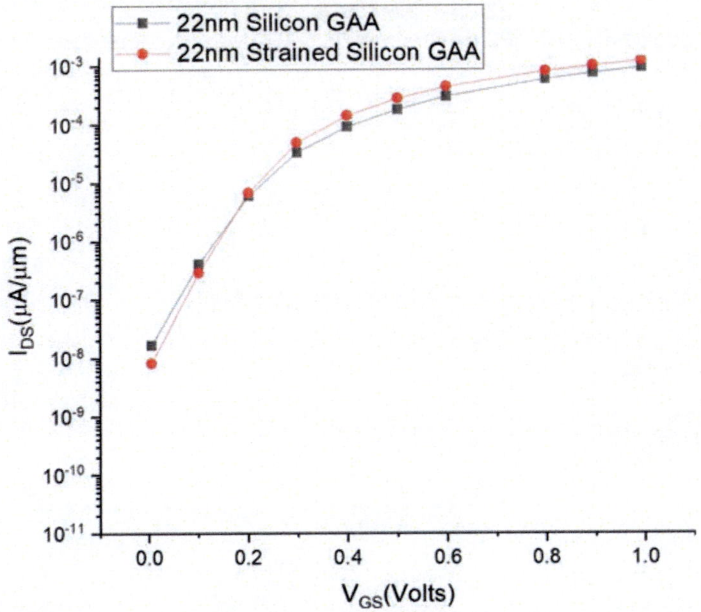

Fig. 3 I_D versus V_{GS} characteristics in logarithmic scale

better results clearly from the graph when compared to the silicon GAAFE [15]. Due to the increase in the mobility it in turn increases the inversion layer and electric field penetration is controlled with all the four gates around the channel which improves the drain current across the channel.

The plot of I_D V_{GS} transfer characteristics shown in Fig. 3 is a logarithmic scale. As in logarithmic scale we can easily find out the Ioff current which is known as leakage current. The I_{off} current is given in mathematical form as Eq. 1.

$$I_{off}(nA) = 100\ \mathrm{W/L}\ 10^{-Vth/ss} \quad (1)$$

In Eq. 1 width of the channel is represented as W, the length of the channel is represented as L, the threshold voltage is given as V_{TH} and the sub-threshold swing as SS [12].

The drain-induced barrier lowering (DIBL) is a short channel effect which is the change in the threshold voltage for a large change in the drain voltage the standard equation is expressed as

$$\mathrm{DIBL\ (mV/V)} = \Delta V_{TH} / \Delta V_{DS} \quad (2)$$

where ΔV_{TH} is the change in threshold voltage and ΔV_{DS} is the change in drain to source voltage [12].

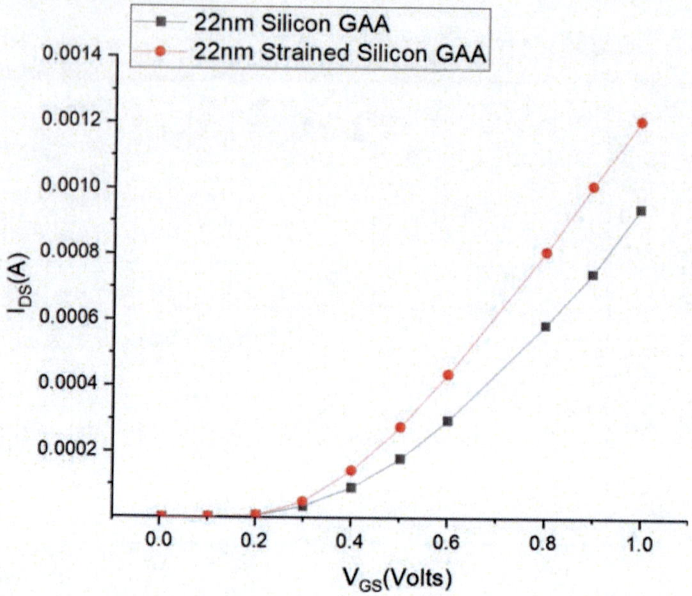

Fig. 4 I_D versus V_{GS} characteristics in linear scale

The linear scale plot of I_D V_{GS} characteristics shown in Fig. 4 from the graph we see the comparison of silicon GAA and the strained silicon GAA we can clearly observe the leakage current Ioff is less and the drain current is more.

The I_D versus V_{DS} characteristics are shown on Fig. 5 with two different gate voltage of 0.3 and 0.6 V which gives the output performance of the device. The I_{ON}/I_{OFF} ratio of silicon GAA and strained silicon GAA is shown in Fig. 6 and from the fig we can say that the strained silicon device is a better device compares to a silicon GAA as the ratio is almost more than 52% with a good performance.

4 Conclusion

In this paper, we developed strained silicon GAAFET with 22 nm gate length using Silvaco TCAD tool. The GAAFET strained silicon device is compared with silicon GAAFET of 22 nm gate length. The gate oxide layer is taken as 2 nm and the length of source and drain is 20 nm. The GAAFET with tri-layered strained silicon shows better gate electrostatic control over the channel compared to a circular GAAFET. From the simulation results the device has better electrical properties in terms of drain current, leakage current and drain-induced barrier lowering (DIBL).The performance of the strained silicon GAAFET has 22% of improvement in the drain current, the leakage current is in the acceptable range and a low drain-induced barrier lowering

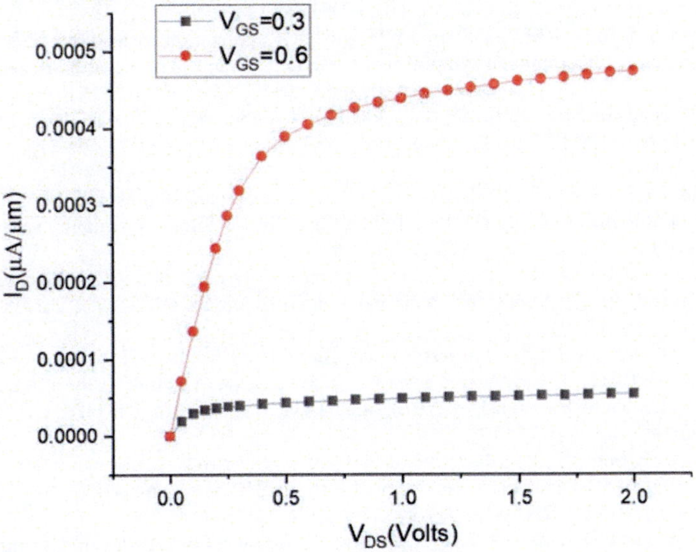

Fig. 5 I_D versus V_{DS} characteristics

Fig. 6 I_{ON}/I_{OFF} ratio of silicon GAA and strained silicon GAA

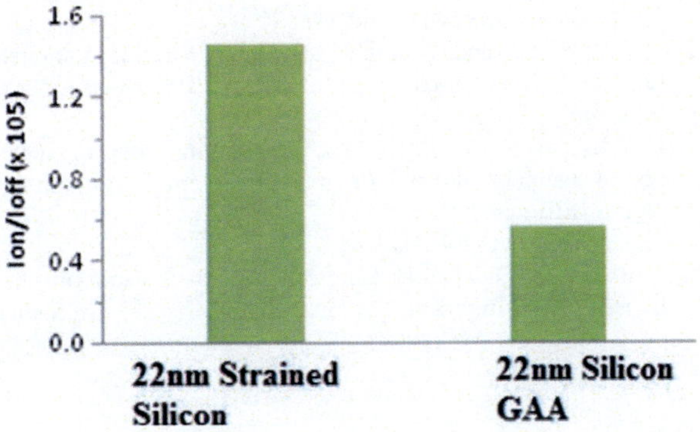

value. Finally, the developed 22 nm strained silicon GAAFET is capable to meet the requirements of the present trends with low power and faster operating device.

References

1. Kang S-M, Leblebici Y, Kim C (2016) CMOS digital integrated circuits. McGraw Hill Education, India Edition
2. Sahu SA, Mohapatra SK, Goswami R (2018) Comparative analysis of double gate TFET and hetero dielectric double gate TFET. In: 2018 International conference on applied electromagnetics, signal processing and communication (AESPC), Bhubaneswar, India, pp 1–4.

https://doi.org/10.1109/AESPC44649.2018.9033293

3. Rajendran S, Lourde RM (2015) FinFETs and their application as load switches in micro mechatronics. In: International symposium on nanoelectronic and information systems. IEEE, Indore, pp. 152–157. https://doi.org/10.1109/iNIS.2015
4. Kumar A, Pattanaik M, Srivastava P, Jha KK (2020) Reduction of drain induced barrier lowering in DM-HD-NA GAAFET for RF applications. IET Circ Dev Syst 14(3):270–275. https://doi.org/10.1049/iet-cds.2019
5. Kumar M, Dubey S, Tiwari PK, Jit S (2013) An analytical model of threshold voltage for short-channel double-material-gate (DMG) strained-Si (s-Si) on Silicon-Germanium-on-Insulator SGOI MOSFET. J Comput Electron 12(1):20–28
6. Kumar K, Dhar RS (2020) Development of 14nm tri-layered (s-Si/s-SiGe/sSi) channel DG-NanoFET. In: IEEE VLSI device, circuit and system conference (VLSI-DCS), IEEE Kolkata, India
7. Yakimets D, Eneman G, Schuddinck P, Bao TH, Bardon MG, Raghavan P, Veloso A, Collaert N, Mercha A, Verkest D, Thean AV, De Meyer K (2015) Vertical GAAFETs for the ultimate CMOS scaling. IEEE Trans Electron Dev 62(5):1433–1439. https://doi.org/10.1109/TED.2015.2414924
8. Connelly D, Zheng P, Liu TK (2017) Channel stress and ballistic performance advantages of gate-all-around FETs and inserted-oxide FinFETs. IEEE Trans Nanotechnol 16(2):209–216. https://doi.org/10.1109/TNANO.2017.2653099
9. Kumar K, Khiangte L, Dhar RS (2020) Design of DG MOSFET with tri-layered strained silicon channel. J Phys Conf Ser
10. Kumar K, Dhar RS (2020) Modelling and simulation of 10nm strained channel DG-nano-FET. In: 7th International conference on signal processing and integrated networks (SPIN), India
11. Singh N, Pandey R (2020) Design of 28 nm GAAFET and its digital applications. Int J Adv Sci Technol 29(3):14074–14088
12. Srivastava A, Nanda S, Aimol SE, Dhar RS (2021) Parametric analysis for varied gate work function in trigate n-channel FinFET. In: 2021 Devices for integrated circuit (DevIC). IEEE, India, pp. 359–362
13. Leobandung E, Chou SY (1996) Reduction of short channel effects in SOI MOSFETs with 35 nm channel width and 70 nm channel length. In: 54th Annual device research conference digest. IEEE, pp 110–111
14. Silvaco TCAD tool Atlas User's Manual
15. Karbalaei M, Dideban D, Heidari H (2020) A sectorial scheme ofgate-all-around field effect transistor with improved electrical characteristics. Ain Shams Eng J Electr Eng

Microstrip RFID Reader Antenna Analysis with Different Slot Configurations

Suraj Kumar, Priyadarshini, Neha Kumari, Sumanta Bhattacharyya, and Ravi Kumar Arya

Abstract The state-of-the-art designs are presented in this manuscript to improve the performance parameters of a standard patch antenna for the 2.45 GHz band. Several rectangular microstrip patch antennas with slots of different shapes and sizes have been proposed with inset-fed types of feed. Optimum values are selected through the simulation for different designs to increase their performance. The proposed antennas are designed and optimized using full-wave Ansys HFSS software. The antennas show resonance frequency near 2.45 GHz which can be used in Wireless Local Area Network (WLAN)/Radio Frequency Identification (RFID) applications. The substrate used for all the designs is economical FR-4 which has a permittivity of 4.4 and thickness of 1.6 mm. The overall dimension of all the antennas is 38.22 mm × 30.25 mm × 1.6 mm. The antennas can achieve peak gain up to 3.5 dBi and are suitable for 2.45 GHz WLAN/RFID applications.

Keywords Antenna · ISM band · RFID · Triangular slot · MSA · Slot · Microstrip antenna

1 Introduction

Radio Frequency Identification (RFID) is rising as an unrivaled contender for the automatic item identity [1]. RFID has evolved and has gone beyond the barcodes that have their limitations, such as limited range and the need for a near-perfect line of sight for communication. RFID has been growing in recent years and is becoming the de facto means of communication for carrier industries, logistics, and production companies [2]. A typical RFID system comprises a reader (interrogator) and tags

S. Kumar · Priyadarshini · N. Kumari
Cambridge Institute of Technology, Ranchi, Jharkhand, India
e-mail: priyadarshini.17ecl02@citranchi.in

S. Bhattacharyya
Greater Kolkata College of Engineering and Management, Kolkata, India

R. K. Arya (✉)
National Institute of Technology Delhi, New Delhi 110040, India

J. K. Mandal et al. (eds.), *Topical Drifts in Intelligent Computing*, Lecture Notes in Networks and Systems 426, https://doi.org/10.1007/978-981-19-0745-6_46

(transponders), and middleware which includes software programs and databases. Generally, the reader emits an electromagnetic frequency signal that is intercepted by the RFID tags in the way of the RFID signal. The antenna needed for the RFID reader should be compact so that the overall reader is compact. It leads to a smaller RFID reader that is easier to transport from one place to another.

Due to advancements in technologies, many researchers have developed devices that can operate in multiple frequency bands depending on the application requirements [2, 3]. Microstrip antennas (MSAs) being common and easier to design are one of the best contenders for RFID reader antennas. It has the advantage of simple design, low cost, mechanically durable, low profile, and lightweight. Due to all these properties, MSA has become a very common antenna and is used for multiple communication systems. Hence, MSAs are widely used in RFID, Satellite communication, direct broadcast service, mobile communication, radars, missile, and telemetry.

The manuscript is divided into multiple sections. Section 2 discusses different antenna designs with their design parameters. Section 3 compares all the designs. In the end, the manuscript concludes with Sect. 4.

2 Antenna Designs and Their Parameters

Microstrip antenna performance is determined by its size, operating frequency, radiation efficiency, directivity, and return loss. Microstrip antenna can be characterized by different parameters for a given substrate at the design frequency. There are closed-form equations that give the various size parameters of the antenna [4]. The following equations give the parameters of the conventional microstrip antenna (see Fig. 1a).

Width of the patch:

$$w = \frac{c}{2fr\sqrt{\frac{\varepsilon_r+1}{2}}} \tag{1}$$

where
c = speed of light.
ε_r = Dielectric constant.
Effective dielectric constant:

$$\varepsilon_{reff} = \frac{\varepsilon_r + 1}{2} + \frac{\varepsilon_r - 1}{2}\frac{1}{\sqrt{1 + 12\frac{h}{wp}}} \tag{2}$$

Length of the patch:

$$L_{\text{eff}} = \frac{c}{2fo\sqrt{E_{\text{eff}}}} \tag{3}$$

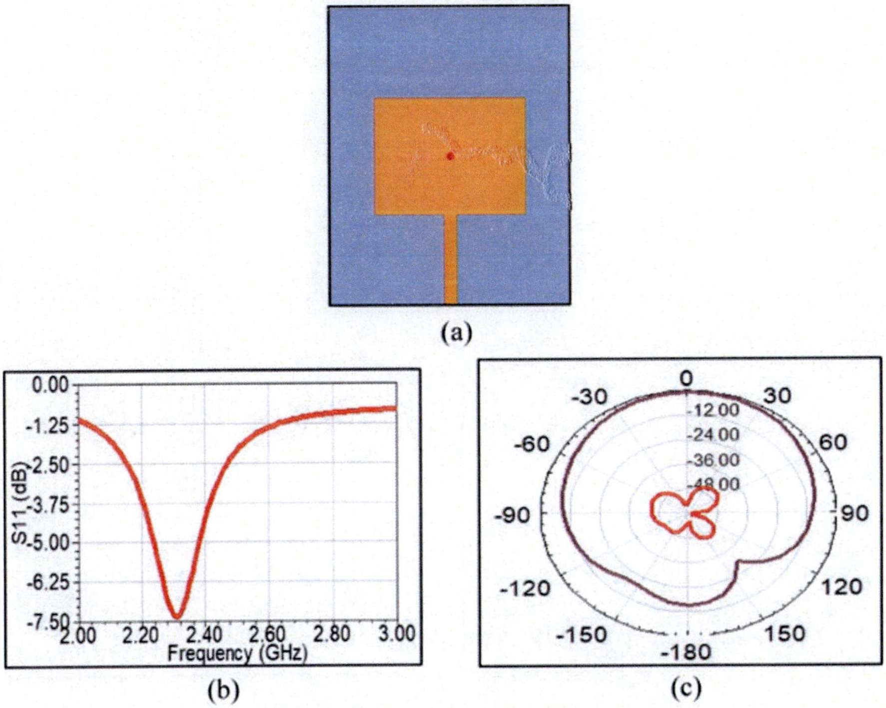

Fig. 1 Edge-fed conventional microstrip antenna (Design-1) **a** design; **b** S_{11}; **c** radiation gain pattern

Length extension:

$$\Delta L = 0.412h\frac{(\varepsilon_{\text{reff}} + 0.33)\left(\frac{w}{h} + 0.264\right)}{(\varepsilon_{\text{reff}} - 0.258)\left(\frac{w}{h} + 0.8\right)} \tag{4}$$

Figure 1a shows the conventional microstrip antenna fed by edge feed. In Fig. 1b, we see that the S_{11} of the microstrip antenna does not cover the desired 2.45 GHz band. Even the S_{11} does not reach a value of—10 dB signifying that the antenna is not matched well with 50 Ω impedance and most of the power is getting reflected back to the feed. To get the better matching design, the antenna is modified so that instead of edge feeding, the feed is changed to inset type as shown in Fig. 2a.

From Fig. 2b, it is evident that the S_{11} improves and shows a better impedance match. Next, we modify the antenna in Fig. 2a by etching out different slots [5, 6] to get the reconfigurability so that the antenna gives the desired bandwidth and resonant frequency. In Fig. 3, the design of the microstrip antenna with a single equilateral triangle slot is shown. Cutting slots in the antennas like this modifies the current distribution in the antenna which can contribute to frequency change or change in radiation patterns. It is visible in Fig. 3b that the resonant frequency of the slotted

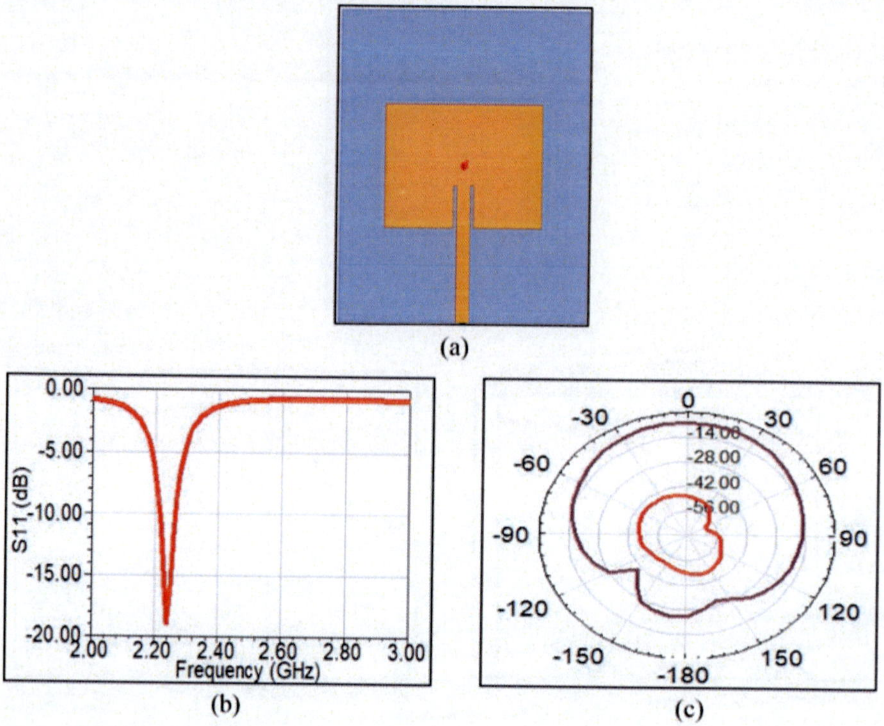

Fig. 2 Inset-fed conventional microstrip antenna (Design-2) **a** design, **b** S_{11}, **c** radiation gain pattern

antenna (Design-1) changes as compared to the antenna in Fig. 2. The antenna design parameters are given in Table 1.

Next, we show how the antenna response changes if three slots are cut in the microstrip patch (Design-4 in Fig. 4) instead of a single slot (Design-3 in Fig. 4). It is evident from Fig. 4b that the introduction of three slots instead of a single slot changes the resonance frequency. The antenna design parameters are given in Table 2.

The proposed antenna geometry (Design-5), which includes a dielectric substrate, patch length (PL), patch width (PW), microstrip feed line (Fl), feed width (Fw) after etching equilateral and rectangular slots is shown in Fig. 5a. The different design parameters of Design-5 are listed in Table 3. The S_{11} of Design-5 is shown in Fig. 5b.

3 Results and Discussion

In the previous section, we showed different configurations of the slotted microstrip antennas. Figure 6 shows the comparative S_{11} for different designs. Figure 7 shows the gain of the same antennas. From this analysis, we can see that cutting the slots in the

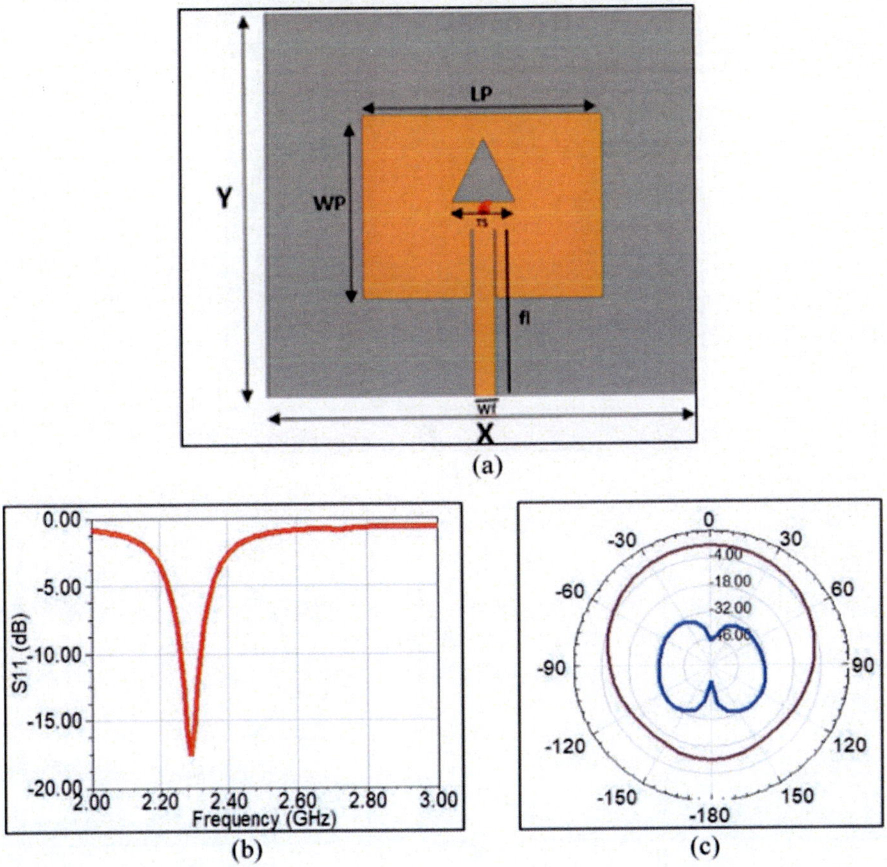

Fig. 3 Microstrip antenna with single triangular slot (Design-3) **a** design, **b** S_{11}, **c** radiation gain pattern

Table 1 Antenna parameters for Design-3

S. no.	Parameters	Dimension (mm)
1	Ground width (*X*)	68.1
2	Ground length (*Y*)	59.7
3	Height (*h*)	1.6
4	Patch width (WP)	37.5
5	Patch length (*w l*)	28.1
6	Feed length (*f l*)	16.2
7	Feed width (*w f*)	3.1
8	Triangle side (*T s*)	4

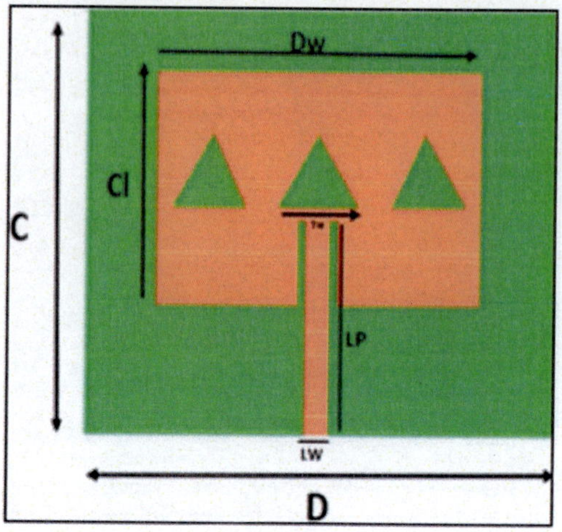

(a)

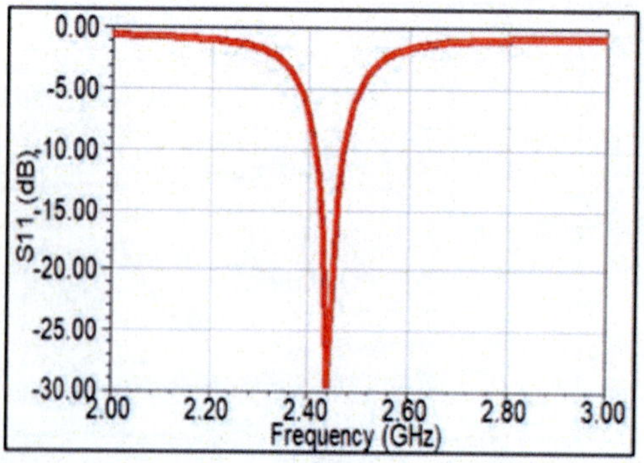

(b)

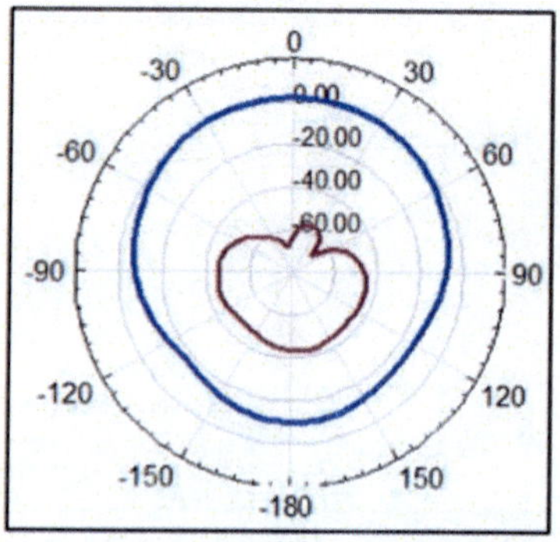

(c)

Fig. 4 Microstrip antenna with three triangular slots (Design-4) **a** design, **b** S_{11}, **c** radiation gain pattern

Table 2 Antenna design parameters for Design-4

S. no.	Parameters	Dimensions (mm)
1	Ground length (*C*)	92
2	Ground width (*D*)	58
3	Height (*h*)	1.6
4	Patch length (Cl)	26
5	Patch width (Dw)	40
6	Feed length (lp)	23
7	Feed width (lw)	3.1
8	Triangle sides (Tl)	5

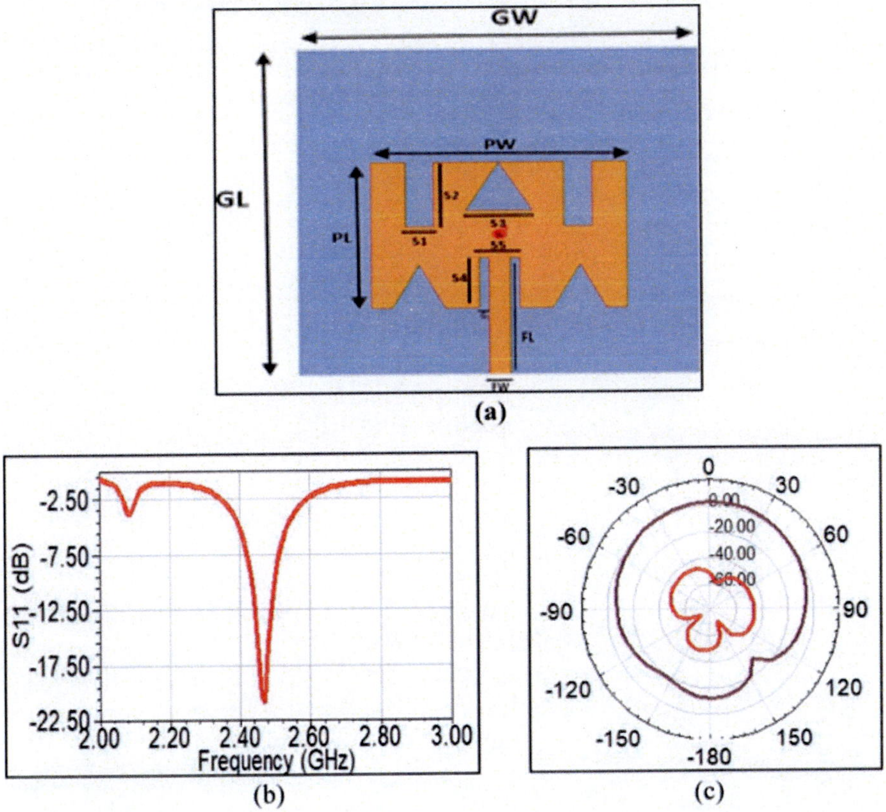

Fig. 5 Microstrip antenna with three triangular and two rectangular slots (Design-5) **a** design, **b** S_{11}, **c** radiation gain pattern

Table 3 Design parameters of Design-5

S. no.	Parameters	Dimensions (mm)
1	Patch length (Pl)	30.25
2	Patch width (Pw)	38.22
3	S1	4.3
4	S2	14.3
5	S3	5
6	S4	10.5
7	S5	6
8	Feed length (Fl)	19.5
9	Feed width (Fw)	3.1
10	G1	0.5
11	Ground length (Gl)	60.5
12	Ground width (Gw)	76.44

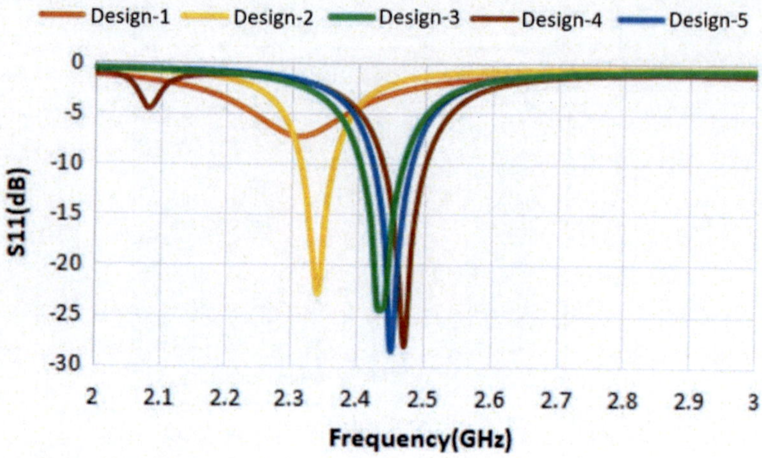

Fig. 6 S_{11} of different designs

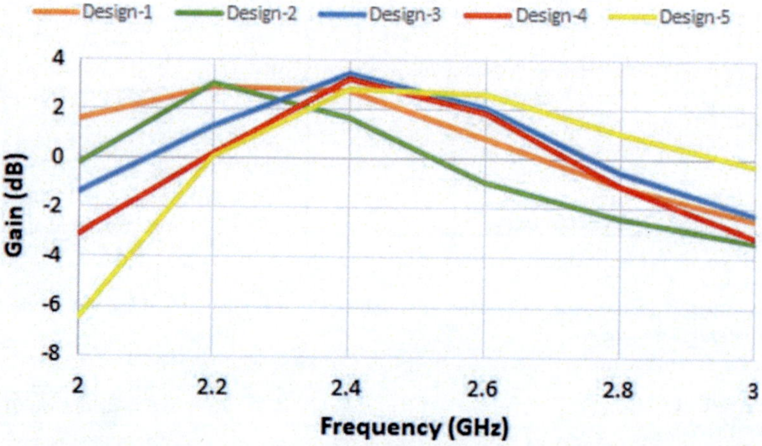

Fig. 7 Gain of different designs

microstrip antenna (of different sizes and shapes) can vary its behavior significantly. These antennas will be useful for 2.45 GHz band applications. Table 4 shows the properties of the different antennas discussed in this work.

4 Conclusion

In this work, we discussed different designs of microstrip antenna for RFID reader antenna. Three proposed designs use slots in the patch of the microstrip antennas.

Table 4 Comparison table for different shapes of microstrip antennas

S. no.	The shape of the antenna	Freq. (GHz)	Min. $\lvert S_{11} \rvert$ (dB)	Gain (diB)
1	Edge-fed conventional microstrip antenna	–	−7.5	2.8
2	Inset-fed conventional microstrip antenna	2.32–2.37	−22	3.1
3	Microstrip antenna with single triangular slot	2.4–2.47	−24	3.5
4	Microstrip antenna with three triangular slots	2.43–2.5	−28	3.3
5	Microstrip antenna with three triangular and two rectangular slots	2.42–2.48	−28.5	3.04

The behavior of different equilateral triangular slots was analyzed by simulations and then optimized dimensions of the slots were selected for the different designs. All the designs were compared in terms of return loss, gains, and radiation patterns. The antennas are resonant near 2.45 GHz and can be used for any application that works in the frequency range of 2.4–2.5 GHz. All the designs use FR-4 dielectric as substrate with permittivity of 4.4 and thickness of 1.6 mm. The overall dimension of all the antennas is 38.22 mm × 30.25 mm × 1.6 mm. The antennas can achieve peak gain up to 3.5 dBi. Such antennas can be used for WLAN/RFID working at 2.45 GHz.

References

1. Preradovic S, Karmakar NC, Balbin I (2008) RFID transponders. IEEE Microw Mag 9(5):90–103
2. Rasool M, Tahir N, Ijaz B, Alamgeer KS, Khan MS (2017) A compact tri-band microstrip slotted antenna with defected ground for wireless applications. In: 2017 11th International conference on signal processing and communication systems (ICSPCS)
3. Verma A, Arya RK, Bhattacharya R, Raghava SN (2021) Compact PIFA antenna with high gain and low SAR using AMC for WLAN/C-band/5G applications. IETE J Res. https://doi.org/10.1080/03772063.2021.1945958
4. Balanis CA (2005) Antenna theory analysis and design, 3rd edn
5. Moharamzadeh E, Javan AM (2013) Triple-band frequency-selective surfaces to enhance gain of X-band triangle slot antenna. IEEE Antennas Wirel Propag Lett 12:1145–1148
6. Song Y, Jiao YC, Zhao G, Zhang FS (2007) Multiband CPW-fed triangle-shaped monopole antenna for wireless applications. Progr Electromagn Res 70:329–336

Effect of High-K Dielectric Material of 14 nm Tri-Layered Strained Silicon Channel HOI FinFET on Short Channel Effect

Priyanka Saha and Rudra Sankar Dhar

Abstract In strain technology a twisting in the MOSFET by growing heterostructure bed connected with Si/SiGe/Si layers within the system is working broadly. Developing a tri-layered HOI n-channel FinFET devices at 14 nm gate length which have double strained Si layers and in between strained SiGe with the high-k dielectric material like HfO_2, ZrO_2 and Si_3N_4 as gate oxide. The short channel effects like DIBL, I_{on}/I_{off}, threshold voltage, etc. have been analyzed for better performance through changing the SiO_2 as gate oxide with the different high permittivity materials such as HfO_2, ZrO_2 and Si_3N_4. This paper explores it considering equivalent oxide thickness calculation and optimized using SILVACO TCAD software. Also compare it with SiO_2 SOI structure for both drain current and transconductance of FinFET and the result in both the drain current and transconductance are higher than SiO_2 SOI device when biased in the linear region. And DIBL, I_{on}/I_{off} are also developed by incorporating the high-k materials.

Keywords HOI · High-K dielectric · SILVACO TCAD · EOT · Strained silicon

1 Introduction

Enormous scaling down of the dimensions of conventional MOSFETs [1–4] to nanotechnology is taking immense presumes because of the actual narrowness of the conventional MOSFETs. A task to raise the representation of this device because of the impression of SCEs [5, 6] and abruptly increase in sub-threshold leakage current [7–9]. But the leakage current of gate reaches intensely for narrowing the gate oxide by the critical value and the role of new gate stack component to restore the setup structure in the formation have been immobile a very demanding research part. Instead of rising the doping of the channel demotes carrier mobility caused by enlarged scattering since the ionized atoms which is dopant and sub-threshold swing [10] degenerates as an outcome of former depletion capacitance owing to

P. Saha · R. S. Dhar (✉)
Department of Electronics and Communication Engineering, National Institute of Technology Mizoram, Aizawl, Mizoram 796012, India
e-mail: rdhar@uwaterloo.ca

J. K. Mandal et al. (eds.), *Topical Drifts in Intelligent Computing*, Lecture Notes in Networks and Systems 426, https://doi.org/10.1007/978-981-19-0745-6_47

reduced gate voltage effect. So, static leakage and band-to-band tunneling have been executed as meaningful factors for extremely elevated channel doping near the source/drain expansions of the design. In particularly scaled FinFETs, underlaid and high-k spacers process fin thickness to be unperturbed while hold down the same short channel performance, which can result in higher drive current per fin. This rise in leakage current causes addition in power dissipation because of DIBL [11]. To diminish this problem thickness of the oxide should be reduced and some techniques deployed are: (i) modification of effective oxide thickness, (ii) high-k dielectric material uses, (iii) increasing channel doping and (iv) introduces strained silicon technology in a device [12]. When the gate oxide layer thickness decreases day by day and sudden increase of gate leakage current due to 2 nm SiO_2 tunneling [13] then need to require a change of gate oxide, i.e., SiO_2 with the high-k materials to decrease the leakage gate current in Si/SiGe/Si channel heterostructure-on-insulator device. The tri-layered HOI device improves the drive current and along with high-k dielectric materials increases the I_{on} current as well as enhanced mobility of charge carriers without leakage effects [14]. The short channel effects like DIBL, I_{on}/I_{off}, threshold voltage, etc. have been analyzed and reduced in 14 nm tri-layered strained silicon channel HOI FinFET using various high-k materials [15, 16] such as HfO_2, ZrO_2 and Si_3N_4 namely, gate oxide.

Here, the inspiration is to ready up a much intending design to introducing a schematic where less leakage current [17] caused by various high-k dielectric material as gate oxide. The characteristics of HOI devices are likened sounding the 1 nm SiO_2 EOT on the equivalent TG FinFET schematic with the different high-k materials to give an advanced schematic of further devices.

2 Device Structure

Device fabrication is done on heterostructure-on-insulator with 14 nm gate extension of the part of the system with onward and backward gate through Si/SiGe/Si part as shown in Fig. 1. The tri-gate HOI FinFET structure is developed by height and length of fin with thickness of silicon. Employing the parameters is summarized in Table 1. and the invention of the modeling of the schematic is performed through SILVACO Atlas simulator [18].

Various high-k materials as a gate oxide is accustomed with FinFET-based design, where the dielectric constant are Si_3N_4 ($k = 7$), ZrO_2 ($k = 23$), HfO_2 ($k = 25$) and SiO_2 ($k = 3.9$) and to minimize the SCEs also those are reached into 1 nm of SiO_2 EOT. All parameters are compared under EOT consideration.

The EOT equalization, i.e., (1) pointed out [19]:

$$\mathrm{EOT} = t_{\mathrm{high}-k} \frac{k_{\mathrm{SiO_2}}}{k_{\mathrm{high}-k}} \tag{1}$$

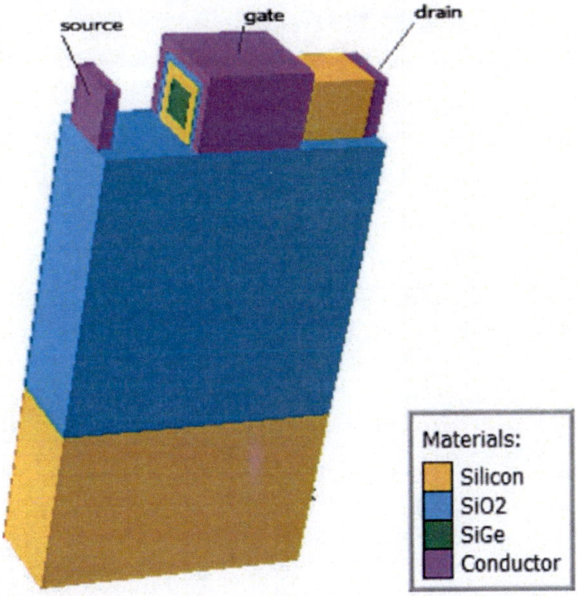

Fig. 1 Structure of the HOI TG FinFET

where t_{high-k} is the high-k material thickness, k_{SiO_2} is the SiO_2 dielectric constant and k_{high-k} is the high-k dielectric constant of materials.

3 Result and Discussion

The strained silicon channel HOI 3D n-FinFET is formed using SILVACO TCAD software. The narrow channel structure is formed with Si-SiGe/Si under consistency of 2 nm–6 nm–2 nm in that order [20]. The drain current—gate voltage graph are sketched in Fig. 2 using linear scale for the HOI TG FinFET design through different high-k dielectrics. Here, ZrO_2 high-k component requires small amount of improvement than SiO_2 in terms of advanced diffusion barrier. The HfO_2 graph line shows minimum leakage due to high barrier [21]. From Fig. 2, the V_{th} of the schematic is find out be present in 0.25 V for SiO_2 and the highest V_{th} is 0.32 V for HfO_2 with respect to drain to source voltage is 0.1 V, where the V_{th} of Si_3N_4 is 0.28 V and ZrO_2 is 0.24 V. Here, further showing a maximum leakage current for the schematic through SiO_2 namely, the gate oxide. The leakage current, I_{off} of HfO_2 is ascertained to be 1.33×10^{-11} µA at a $V_{GS} = 0.1$ V, where the I_{off} of SiO_2 is 1.15×10^{-10}µA. Hence, the HfO_2 as the dielectric implementation in HOI device maintains greater performance.

The V_{th} comparison is getting out of Fig. 3, where *y*-axis is taken as threshold voltage and *x*-axis is taken as different dielectric constant with various high-k materials.

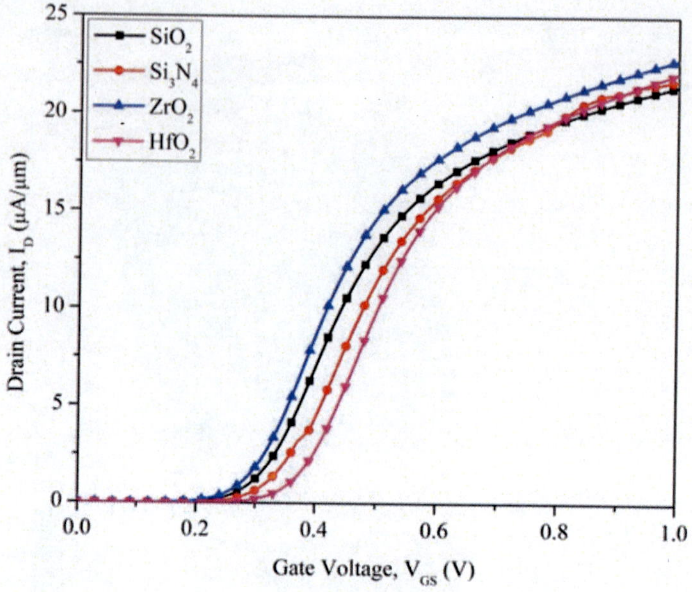

Fig. 2 Graphical range $I_D - V_{GS}$ line schematic of HOI design

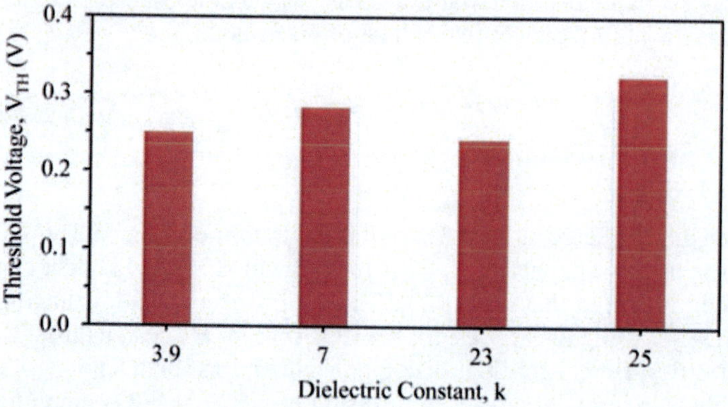

Fig. 3 Transformation of V_{th} of a HOI design

Depending upon the threshold voltage, introduces on current as the gate voltage increases and also introduces the leakage current. I_{off} calculation is (2) [22] following

$$I_{off}(nA) = 100\frac{W}{L}10^{\frac{-V_{Th}}{SS}} \tag{2}$$

where W and L are the channel width and length gradually, V_{TH} is the threshold voltage of the design and SS is the sub-threshold swing. Figure 4 shows the variation of off current through several dielectric constant for various high-k materials.

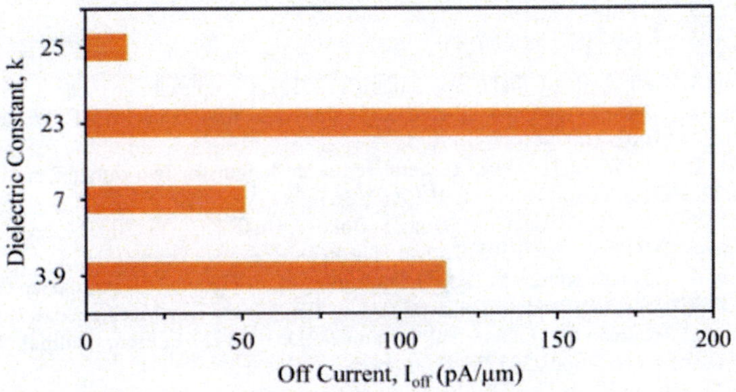

Fig. 4 Transformation of I_{off} of a HOI design

4 Conclusion

The analysis of 14 nm gate length tri-layered strained silicon channel HOI FinFET has been designed with SCEs parameters. Using high-k materials like HfO_2, ZrO_2 and Si_3N_4, the performance is compared with the nm EOT of SiO_2. An increased threshold voltage has been observed less leakage current. Also observed the improvement due to high-k materials where for HfO_2 very less leakage current introduces as compared with SiO_2 in HOI 3D FinFET devices using SILVACO TCAD.

Table 1 Parameters of the HOI device

Parameters	Dimensions
Drain/source length	10 nm
Channel length	14 nm
Lateral oxide thickness (SiO_2)	1 nm
Silicon fin thickness	10 nm
Silicon fin height	10 nm
Mole fraction of Ge	0.4
Si tier thickness	2 nm
SiGe tier thickness	6 nm
Buried oxide thickness	50 nm
Silicon substrate thickness	30 nm
Channel doping	10^{15} cm^{-3}
Drain/source doping	10^{18} cm^{-3}

References

1. Jandhyala S, Kashyap R, Anghel C, Mahapatra S (2012) A simple charge model for symmetric double-gate MOSFETs adapted to gate-oxide-thickness asymmetry. IEEE Trans Electron Devices 59(4):1002–1007
2. Frank D, Taur Y, Wong H (1998) Generalized scale length for two dimensional effects in MOSFETs. IEEE Electron Device Lett 19(10):365–387
3. Chang L, Choi YK, Ha D, Ranade P, Xiong S, Bokor J, Hu C, King TJ (2003) Extremely scaled silicon nano-CMOS devices. IEEE Trans Electron Devices 91(11):1860–1873
4. Chaudhry A, Jagadesh Kumar M (2004) Controlling short-channel effects in deep submicron SOI MOSFETs for improved reliability: a review. IEEE Trans Dev Mat Reliab 4(1):9–109
5. International Roadmap for Devices and Systems (IRDS™) 2017 Edition, [Online]. Avaliable: https://irds.ieee.org/editions/2017/more-moore
6. Veshala M, Jatooth R, Reddy KR (2013) Reduction of short-channel effects in FinFET. Int J Eng Innov Technol (IJEIT) 2(9):118–124
7. Kuhn KJ (2011) CMOS scaling for the 22nm node and beyond: device physics and technology. In: International symposium on VLSI technology, systems and applications (VLSI-TSA'11), pp 1–2
8. Roy K, Mukhopadhyay S, Mahmoodi-Meimand H (2003) Leakage current mechanisms and leakage reduction techniques in deep-submicrometer CMOS circuits. IEEE 91(2):305–327
9. Frank DJ, Dennard RH, Nowak E, Solomon PM, Taur Y, Wong HSP (2001) Device scaling limits of Si MOSFETs and their application dependencies. IEEE Volime 89(3):259–288
10. Tosaka Y, Suzuki K, Sugii T (1994) Scaling parameter-dependent model for subthreshold swing S in double-gate SOI MOSFET's. IEEE Electron Device Lett 15(11):466–468
11. Ghitani HE (1999) DIBL coefficient in short-channel NMOS transistors. In: Proceedings of the sixteenth national radio science conference, Egypt, pp D4/1–5
12. Mohsenifar S, Shahrokhabadi MH (2015) Gate stack high-κ materials for Si-based MOSFETs past, present, and futures. Microelectron Solid State Electron
13. Iwai H, Momose HS (1998) On-chip spiral inductors with diffused shields using channel-stop implant. In: International electron devices meeting 1998. Technical Digest (Cat. No.98CH36217)
14. Nirmal D, Shruti K, Thomas DM, Samuel PC, Kumar PV, Kumar NM (2011) Analysis of subthreshold characteristics for top and bottom flat-band voltages of junction less double gate MOSFET. In: 3rd International conference on electronics computer technology
15. Huff HR, Gilmer DC (2005) High dielectric constant materials. Springer, Berlin, Germany
16. Pradhan KP, Mohapatra SK, Sahu PK, Behera DK (2014) Impact of high-k gate dielectric on analog and RF performance of nanoscale DG-MOSFET. Microelectron J 45(2):144–151
17. Kerber P, Zhang Q, Koswatta S, Bryant A (2013) GIDL in doped and undoped FinFET devices for low-leakage applications. IEEE Electron Device Lett 34(1):6–8
18. International S (2016) Atlas user's manual device simulation software. Silvaco Int., Santa Clara
19. Wilk GD, Wallace RM, Anthony JM (2001) High-K gate dielectrics: Current status and materials properties considerations. J Appl Phys 89
20. Khiangte L, Dhar RS (2018) Development of Tri-Layered s-Si/s-SiGe/s-Si channel heterostructure-on-insulator MOSFET for enhanced drive current. Phys Status Solidi (B) 255:1800034
21. Wong H, Iwai H (2006) On the scaling issues and high-κ replacement of ultrathin gate dielectrics for nanoscale MOS transistors. Microelectronics Eng 83
22. Mohapatra SK, Pradhan KP, Sahu PK (2013) Some device design considerations to enhance the performance of DG-MOSFETs. Trans Electr Electron Mater 14(6)

A Center of Gravity-Based Novel Clustering Algorithm for Energy-Efficient Wireless Sensor Network

Deo Kumar and Sanjeet Kumar

Abstract Energy is a crucial constraint for wireless sensor networks (WSNs). Among the most fundamental approaches for prolonging the lifetime of a WSN is clustering. In this work, we present a Center of Gravity (COG)-based clustering algorithm for variable clustering. The formation of clusters uses the partitioning concept which is based on the node density for formation of optimum number of clusters. In the proposed algorithm, cluster head election is based on the concept of COG. The proposed algorithm is simulated and compared with standard algorithm like LEACH. The simulation results show that there is a significant improvement in all the lifetime metrics, i.e., First Node Die out (FND) and Half Node Die out (HND).

Keywords WSN · LEACH · COG

1 Introduction

Clustering is an important technique for wireless sensor networks (WSNs). Clustering allows enhancement of the network lifetime by preventing multiple long-distance transmissions and replacing them with multiple short-distance transmissions. In clustering, the entire network is divided into many clusters using some clustering algorithm. Each cluster is allocated a cluster head, which is chosen from among the cluster's nodes. Except for the cluster head, all nodes in the cluster sense their environment and provide data to their cluster head. The data from the normal nodes is received by the cluster head, which aggregates it before sending it to the sink.

Since WSNs are generally energy constrained and remotely located, so it is imperative to regulate the dissipation of energy. Cluster heads consume considerably more energy than normal nodes because they are responsible for gathering packets from the cluster's normal nodes, aggregating all this information and finally sending this

D. Kumar (✉) · S. Kumar
Department of Electronics and Communication Engineering, BIT Mesra, Ranchi, Jharkhand, India

S. Kumar
e-mail: sanjeet@bitmesra.ac.in

J. K. Mandal et al. (eds.), *Topical Drifts in Intelligent Computing*, Lecture Notes in Networks and Systems 426, https://doi.org/10.1007/978-981-19-0745-6_48

aggregated data to the sink [1, 2]. So, to improve the network's lifetime, it is necessary to ensure that optimal number of clusters should be formed. This is done using the optimal partitioning method described in this paper. Another important factor which affects the lifetime of a WSN is that the load of becoming the cluster head should be divided equally between all nodes within the network. This load balancing is achieved by changing the cluster head on round basis for each cluster. This ensures that the burden of being the cluster head falls equally on all nodes present in a cluster. LEACH was the first algorithm to provide an effective energy balancing approach by choosing the cluster heads on the basis of probability. In our algorithm, the cluster head election for each cluster is based on the Center of Gravity (COG) approach.

In [3], the detail survey of LEACH-based protocols is well defined. In IB-LEACH [9], relay nodes are introduced to send the aggregated data of cluster head to base station. In TL-LEACH [4], two cluster heads are used, namely primary and secondary cluster heads, within the cluster in dense network. In ETL-LEACH [5], the secondary cluster head acts as a relay node to transmit the aggregated data packet to base station.

2 LEACH and Its Shortcomings

LEACH is the abbreviation for low-energy adaptive clustering hierarchy. LEACH is a clustering algorithm which balances the load of becoming the cluster head almost uniformly among the nodes in the network. It ensures that a node which has not been a cluster head for the last $1/P$ rounds (P is the probability or the percentage of nodes that will become cluster heads in a particular round) [6, 7] gets a chance to become a cluster head. Each node evaluates the value of the threshold function $T(n)$ (n is the round number), and also, each node chooses a random number between 0 and 1. If the chosen number is less than the threshold, then it becomes the cluster head for the current round.

LEACH is very effective as a clustering algorithm. However, it is plagued by several shortcomings. In our proposed clustering approach, we have tried to overcome these shortcomings, and they are as follows.

- In LEACH, the cluster heads are elected on the basis of probability. This may result in a network with a non-uniform distribution of clusters. Also, the normal nodes are allocated to a cluster on the basis of RSSI of the cluster head. This may lead to the formation of non-uniform clusters (having different number of nodes), causing load imbalance among the cluster heads.
- In LEACH, the clusters are not formed effectively and the radio range of the cluster heads in two different clusters may overlap. So, LEACH can lead to the formation of redundant clusters.
- In LEACH, the numbers of cluster heads are determined prior to clustering by setting a particular value of P. However, this number may not lead to the most optimal lifetime.

- In LEACH, the cluster heads aggregate the data received from the normal nodes and then send this data directly to the sink (in a single hop). This leads to heavy load on the cluster head in terms of energy, especially when the sink and cluster head's distance are larger.

3 Proposed Clustering Algorithm

In our method of cluster clustering, the entire field area is first divided into partitions of equal area on the basis of the partition size (a). When the field is divided into partitions, the area of each region is less than or equal to the partition size. After the partitions are created, each partition is validated for being a cluster. If a partition has nodes at least equal to or greater than min_nodes (minimum number of nodes, a partition should have to qualify as a cluster), then it is declared as a valid cluster. The cluster head is then chosen using the concept of COG. The proposed algorithm is implemented at the base station, and the information of cluster head selection is relayed to the nodes. The operation is distributed into two phases, i.e., partitioning phase and cluster head election phase.

3.1 Partitioning Phase

In the beginning, a value of partition size (a) and min_nodes is selected so as to lead to the formation of optimal number of clusters. An analysis for choosing this optimal value of partition size and min_nodes is explained in the following section. After these values are selected, we take the value of subfield area (s) as the total field area (A). Subfield area is the area of each region formed after dividing the total field area. So, it is initially chosen as A. The subfield regions are recursively divided into two equal parts until the subfield area (s) becomes less than or equal to the partition size (a). Now, the number of nodes in each of these partitions is counted, and if this number is greater than or equal to min_nodes, then the partition is declared as a valid cluster. For the partitions which could not qualify as a cluster, their nodes are distributed to the neighboring clusters on the basis of the distance between the nodes and the COG of the neighboring clusters. Then, the value of COG is recalculated taking into consideration the reallocated nodes. The COG of a cluster is a point (X_o, Y_o) whose location is determined by taking the mean of the x- and y-coordinates of all the nodes in a cluster, respectively [8]. After this phase, the clusters are created and the cluster heads for these clusters are elected in the cluster head election phase.

3.2 Cluster Head Election

Center of Gravity (COG) of a cluster can be defined as a point (X_o, Y_o) which is obtained by taking the mean of the *x*-and *y*-coordinates of all the points present in the cluster. The X_o and Y_o can be calculated by using (1)–(2).

$$X_o = \frac{1}{n}\sum_{i=1}^{n} X_i \tag{1}$$

$$Y_o = \frac{1}{n}\sum_{i=1}^{n} Y_i \tag{2}$$

Here, n is the number of nodes in the cluster and X_i and Y_i are the coordinates of the ith node in the cluster. In our approach, the node that has the maximum residual energy and is closest to the COG is chosen as the cluster head for that cluster for the current round. This cluster head election is repeated in each round, so that the load of becoming the cluster head is evenly distributed throughout the network. In LEACH algorithm, the election of cluster head takes place on the basis of a distributed probabilistic approach. Due to this probability-based approach, the cluster heads may not be elected in the best possible position, leading to poor lifetime. In the case of COG-based approach, it is ensured that the cluster head is elected in a region of high node density. Also, in LEACH election of cluster head, the residual energy of the sensor nodes is not taken into account. Thus, load balancing is performed better in COG-based approach in comparison with LEACH. Figure 1 shows the flowchart for the proposed algorithm.

3.3 Determination of Optimal Value of Partition Size (a) and Min_Nodes

Partition size (a) and min_nodes are the variables that govern the number of clusters formed. It is important to select the most optimal value of these variables in order to achieve the best lifetime. In our work, we have empirically determined a plot between node density [total number of nodes (n)/field area (A)] and the ratio field area (A)/partition size (a) for random and Gaussian distribution of nodes in a sensor field. For a given distribution of n nodes in a sensor field of area A, the value of the optimal partition size can be determined using Fig. 2. The node density is calculated (n/A) first, and then, the corresponding value of a/A can be obtained from the plot. Since the value of A is known, thus we can determine the value of partition size (a). For min_nodes, it was empirically noted that the value of min_nodes does not have much influence on the lifetime. This is because in MATLAB the random function uses a PN sequence for generating random numbers. This leads to an almost equal

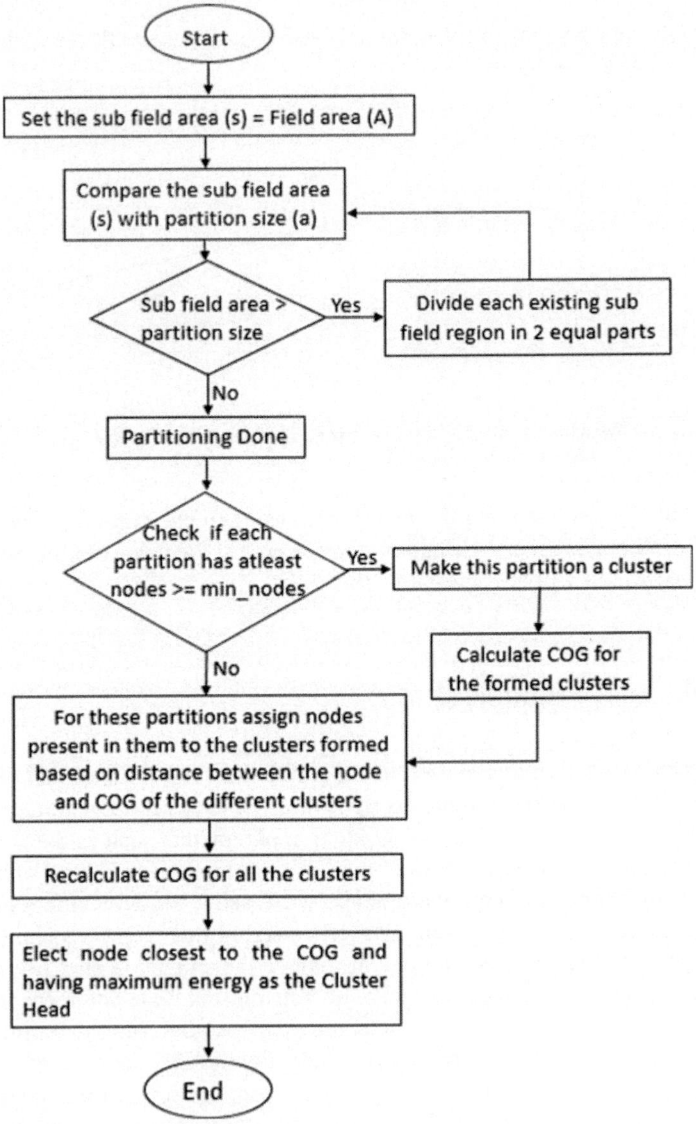

Fig. 1 Flowchart for the proposed COG-based clustering algorithm

distribution of nodes in all the regions. So, all the clusters have almost equal number of nodes and min_nodes does not have much significance. However, for a normal or Gaussian distribution of nodes, it was noted that the best lifetime is encountered at a value of min_nodes that is 30% of the nodes that would fall in a partition if nodes are uniformly distributed. For example, for field area $(A) = 100\text{ m} \times 100\text{ m}$ and partition size $(a) = 50\text{ m} \times 50\text{ m}$, four partitions will be formed. If 100 nodes were

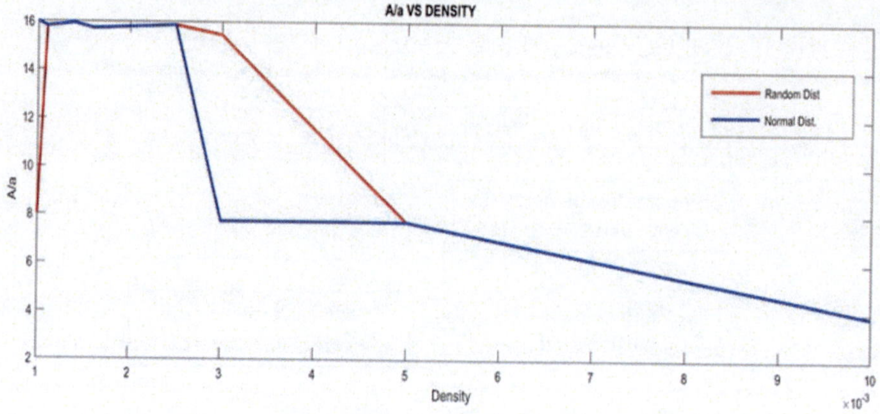

Fig. 2 Plot for *A/a* versus node density for determination of partition size (*a*)

to be distributed equally among the four partitions, then each partition will receive 25 nodes. So, for a Gaussian distribution of these nodes, the value of min_nodes will be 30% of 25, which will be approximately 8.

4 Simulation Parameters

The radio communication model used in our simulations is similar to that of [6]. To compare our results with some standard algorithm, we have set our parameters same as that of standard algorithms. All the simulation parameters such as the number of nodes, field area, initial energy of each node, energy consumed by transmitter and receiver circuitry, and data aggregation energy are same for achieving a common ground for comparison purpose. Energy model used in this work is a simple radio model for transmission and reception. In this model, the amount of energy consumption depends upon the distance d between the transmitting node and receiver. Here, parameter $E_{\text{Tx-elec}}$ and $E_{\text{Rx-elec}}$ means the energy dissipated to run the transmitter or receiver circuitry. E_{fsm} is the energy for the transmit amplifier for distances smaller than d_o. E_{fs} taken into account for the free space propagation without considering the multipath fading effect if the distance of transmission is less than d_o. E_{mp} is the energy for the transmit amplifier for distances greater than d_o. E_{mp} accounts for multipath fading; it also takes into account the fading exponent, and here, the value of fading exponent is set as 4. Below (3)–(5) are the equations of the energy model(Fig. 3).

$$E_T(k,\, d) = E_{\text{Tx - elec}}.\, k + E_{\text{fs}}.\, k.\, d^2 \quad \text{for}\, d \le d_o o \tag{3}$$

$$E_T(k,\, d) = E_{\text{Tx - elec}}.\, k + E_{\text{mp}}.\, k.\, d^4 \quad \text{for}\, d \le d_o o \tag{4}$$

$$E_R(k) = E_{\text{Rx - elec}}.kk \quad (5)$$

On equating Eqs. (1) and (2), we get (6) for the value of $d = d_o$.

$$d_o = \surd E_{\text{fs}} / \surd E_{\text{mp}} \quad (6)$$

The simulations were carried out in MATLAB, and the following simulation parameters were used.

(1) Total nodes = 100
(2) Random distribution of nodes
(3) Initial energy of each nodes = 0.5 J
(4) E_{elec} = 50nJ/bit
(5) E_{mp} = 0.0013 pJ/ bit / m^4
(6) E_{fs} = 10 pJ/bit/m^2
(7) Network area = 50 m x 50 m
(8) Sink location (25, 25).
(9) Every round, each node sends 2000 bits of data

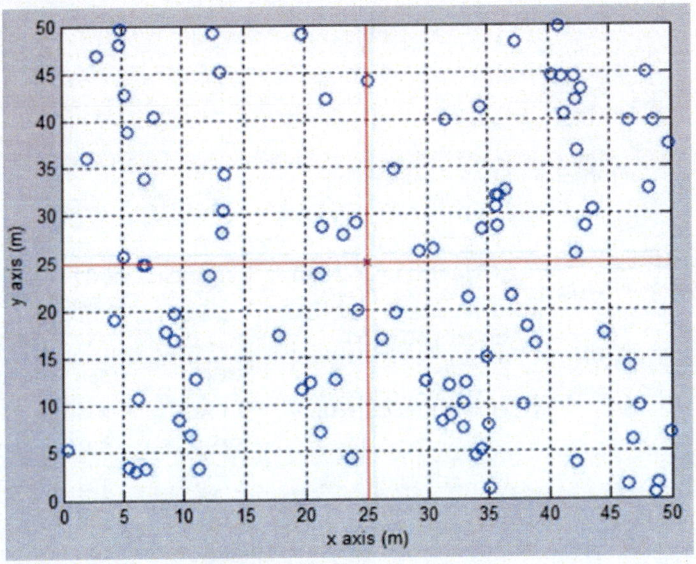

Fig. 3 Deployment of 100 nodes in a field of 50 m × 50 m

5 Simulation Results

The simulations for LEACH and the proposed algorithm were carried out in MATLAB for the simulation parameters mentioned in the last section. The comparison of results for the simulations is shown in Fig. 4. Comparisons for other scenarios are given in Table 1.

It is observed from Table 1 that the lifetime of the above five scenarios is increased about 200% using COG approach than LEACH protocol. It is also observed from Fig. 4 that FND and HND are increased about 205.5% and 197%, respectively, using COG approach than LEACH protocol in scenario 1.

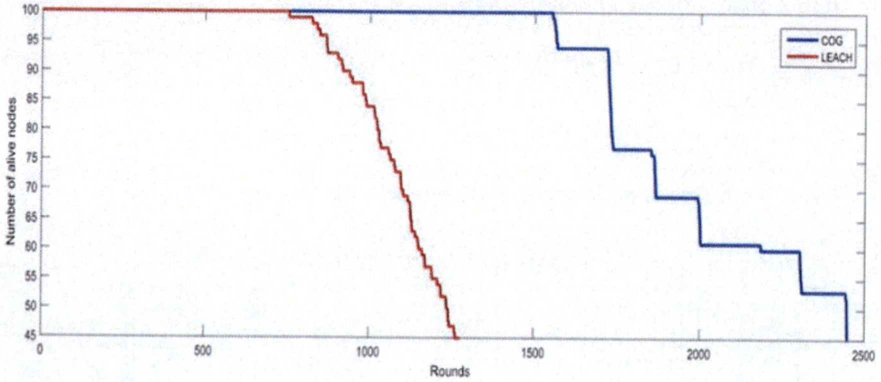

Fig. 4 Lifetime comparison for LEACH and COG approach

Table 1 Simulation results and comparison of different scenarios

(Field area, total nodes)	Algorithm	FND (rounds)	HND (rounds)
(50 m × 50 m, 50)	LEACH	754	1240
	COG	1550	2442
(100 m ×100m, 100)	LEACH	435	750
	COG	904	1303
(300 m × 300 m, 300)	LEACH	163	262
	COG	350	514
(500 m × 500 m, 500)	LEACH	94	151
	COG	168	283
(1000 m × 1000 m, 1000)	LEACH	44	62
	COG	81	114

6 Conclusion

From the lifetime plots given above, it can be concluded that the COG approach gives a much better energy performance and lifetime in comparison with the LEACH algorithm. This is mainly due fact that in COG approach cluster head is elected in a region of high node density and the residual energy of the nodes is also considered when choosing a cluster head. In LEACH, there is a formation of redundant clusters. However, in the COG-based clustering algorithm which we have proposed optimum, the numbers of clusters are formed leading to better lifetime. Further, COG approach can be useful for implementing energy-efficient routing protocol.

References

1. Akyildiz F, Su W, Sankarasubramaniam Y, Cayirci E (2002) A survey on sensor networks. IEEE Commun Mag 40(8):102–114
2. Yick J, Mukharjee B, Ghoshal D (2008) Wireless sensor network survey. Comput Netw 52(12):2292–2330
3. Singh SK, Kumar P, Singh JP (2017) A survey on successors of LEACH protocol. IEEE Access 5:4298–4328
4. Loscri V, Morabito G, Marano S (2005) A two-levels hierarchy for low-energy adaptive clustering hierarchy (TL-LEACH). In: 62nd IEEE vehicular technology conference, pp 1809–1813
5. Manzoor K, Jokhio SH, Khanzada TJ, Jokhio IA (2019) Enhanced TL-LEACH routing protocol for large-scale WSN applications. In: IEEE 2019 cybersecurity and cyberforensics conference (CCC), Melbourne, Australia, pp 35–39
6. Heinzelman WB, Chandrakasan AP, Balakrishnan H (2000) Energy-efficient communication protocol for wireless microsensor networks. In: Proceedings of 33rd annual Hawaii international conference on system sciences, Maui, HI, USA, vol 2, pp 1–10
7. Heinzelman WB, Chandrakasan AP, Balakrishnan H (2002) An application-specific protocol architecture for wireless microsensor networks. IEEE Trans Wirel Commun 1(4)
8. Bragingsky D, Estrin D (2002) Rumor routing algorithm for sensor networks. In: WSNA. Atlanta, Georgia, USA, pp 22–30
9. Salim A, Osamy W, Khedr MA (2014) IBLEACH: intra-balanced LEACH protocol for wireless sensor networks. Wireless Netw 20:1515–1525

Intelligent Computing in Electrical Power, Control Systems and Energy Technology

Model Coordinate System of Interval Regulation Train Traffic

Ravshan Aliev

Abstract The most limiting sections in terms of throughput are sections in areas of crossing sections, intermediate stations, station platforms, ascent and descent ferry sections, etc. The existing methods for determining the minimum passing interval cannot be applied with the coordinate system of interval regulation due to the fact that at small coordinates they do not give accurate results, since they do not take into account the dynamics of the approach of two trains. The latter circumstance requires an accurate mathematical study of the process of approaching trains, identifying the most difficult situation in terms of approach (critical moment of approach) and determining the value of the minimum interval of passing following from conditions of critical approach. For a quantitative analysis of the automatic control system of train traffic, analytical expression of the connections between the input and output characteristics of the system, to determine the degree of influence of one or another controlled characteristic on its operation, outlining the range of tasks that the system can solve, and establishing the redistribution of the system's capabilities, it is necessary to build a mathematical model of the system.

Keywords Railway system · Statistical · Distance · Coordinate · Interval

1 Main Part

A mathematical model of a system is an equation or multiple indicators of equations that reflect a quantitative relationship between the main task [1], the goal of the system, and the characteristics of the system [2]. The main task of the interval automatic control [3] of the movement behind the running train is T2, on the train T1 in front, ensuring the required throughput with complete traffic safety. Traffic regulation is expressed mainly in maintaining a safe interval between trains in all modes of movement [4, 5], i.e., an interval equal to or greater than a certain minimum safety interval corresponding [6] to a given system and being its parameters. Based on this,

R. Aliev (✉)
Department of Information Systems and Technologies, Tashkent State Transport University, Tashkent, Uzbekistan
e-mail: silara@mail.ru

J. K. Mandal et al. (eds.), *Topical Drifts in Intelligent Computing*, Lecture Notes in Networks and Systems 426, https://doi.org/10.1007/978-981-19-0745-6_49

on the left side of the expression of the mathematical model of the system, you can put the corresponding minimum spatial or time interval between trains S_{sm} (spatial min) or T_{bt} (between trains min), which it can create and maintain while maintaining complete traffic safety, and in the right parts—system characteristics [7–14].

Omitting the intermediate tabs, we present the final results of the derivation of the expression of the mathematical model of the coordinate system of the interval regulation of train traffic [15]. The expression of the mathematical model can be presented in a deterministic and statistical form.

The deterministic mathematical model of the automatic train control system has a simpler form [2], but is more approximate. It does not take into account the statistical nature of errors in measuring the initial characteristics in the system and the location of controlled objects within the quantization intervals in a discrete code system with time and level quantization.

Statistical mathematical model of the coordinate system of automatic interval control of train traffic, reflecting the connection of the minimum, for safety reasons, spatial Ssm and time Tbt intervals between trains following each other with the initial characteristics of the system, controlled by the system, has the form:

$$
\begin{gathered}
S_{sm} = l_{ti} + S_{t2} - S_{t1} + S_b + S_p = \\
l_{t1} + [0.278(v_2 + v_e + \delta v_{max} v_2)t_{t2} + \frac{(v_2+v_k+\delta_{vmax}v_2)^2}{\xi_2(v_{p2}\varphi_{cr2}+[\omega_0]_{x_2} \pm i_{c2}}]- \\
-[(v_1 - \delta_{vmax} v_1)t_{t1} + \frac{(v_1-\delta_{vmax}v_1)^2}{\xi_1(v_{p1}\varphi_{cr1}+[\omega_0]_{x2} \pm i_{c1}}] + (v_2 + v_k + \\
+ \delta v_{max} v_2) log_2[n^2 \left(\frac{L_{max}}{l_k}\right)^2 \left(\frac{v_{max}}{v_k}\right)^2 (\frac{i_{max}}{i_k})(\frac{R_{max}}{R_k})] \rightarrow \\
\rightarrow \frac{(n+1)}{\Delta f} + 2\delta s_{max} S_{\text{изм}1} + l_k + \Delta l_k \geq l_{t1};
\end{gathered} \tag{1}
$$

$$
\begin{aligned}
T_{\mathrm{bt}} = T\left(n_1 + T_{\tau 2} - T_{\tau 1} + T_b + T_p\right) = \frac{l_{t1}}{v_2 + v_k + \delta_{v\max} v_2} \\
+ \left[0.278 t_{t2} + \frac{500(v_2 + v_k + \delta_{v\max} v_2)^2}{\xi_2(1000 v_{p2}\varphi_{\mathrm{cr}2} + [\omega_0]_{x_2} \pm i_{c2}}\right] \\
- [\frac{0.278(v_1 - \delta_{v\max} v_1)t_{g1}}{v_2 + v_k + \delta_{v\max} v_2} + \frac{(v_1 - \delta_{v\max} v_1)^2}{(v_2 + v_k + \delta_{v\max} v_2)\xi_1(1000 v_{p1}\varphi_{\mathrm{cr}1} + [\omega_0]_x \pm i_{c1}} \\
- [\frac{0.278(v_1 - \delta_{v\max} v_1)t_{g1}}{v_2 + v_k + \delta_{v\max} v_2} + \frac{(v_1 - \delta_{v\max} v_1)^2}{(v_2 + v_k + \delta_{v\max} v_2)\xi_1(1000 v_{p1}\varphi_{\mathrm{cr}1} + [\omega_0]_x \pm i_{c1}} \\
+ \frac{2\delta_{s\max} S_{ch1}}{(v_2 + v_k + \delta_{v\max} v_2)} + \frac{l_k + \Delta l_k}{(v_2 + v_k + \delta_{v\max} v_2)} \geq \frac{l_{t1}}{(v_2 + v_k + \delta_{v\max} v_2)}
\end{aligned} \tag{2}
$$

Statistical mathematical del of the coordinate system of automatic interval control of train traffic, reflecting the connection of the minimum, for safety reasons, spatial S^{I}_{sm} and time T^{I}_{bt} intervals between trains following each other with the initial characteristics of the system, controlled by the system, has the form:

$$S^I_{sm} = l_{ti} + S_{t2} - S_{t1} + S_b + S_p =$$

$$= l_{t1} + [0.278(v_2 + v_e + \delta v_{max} v_2)t_{t2} + \frac{(v_2+v_k+\delta_{vmax}v_2)^2}{\xi_2(v_{p2}\varphi_{cr2}+[\omega_0]_{x_2}\pm i_{c2}}] -$$

$$-[0.278(v_1 - \delta_{vmax} v_1)t_{t1} + \frac{v_1-\delta_{vmax}v_1}{\xi_1(v_{p1}\varphi_{cr1}+[\omega_0]_{x1}\pm i_{c1}}] + (v_2 + v_k + \delta v_{max} v_2) *\rightarrow \quad (3)$$

$$\rightarrow* \frac{(n+1)}{\Delta f} log_2[n^2 \left(\frac{L_{max}}{l_k}\right)^2 \left(\frac{v_{max}}{v_k}\right)^2 (\frac{i_{max}}{i_k})(\frac{R_{max}}{R_k})] +$$

$$+2\delta s_{max} S_{ch1} + l_k + \Delta l_k + \delta_S S_{\text{изм}1} > l_{t1};$$

$$T^I_{\text{bt}} = (T_{t1} + T_{\tau 2} - T_{\tau 1} + T_b + T_p)$$

$$= \frac{l_{t1}}{v_2 + b_v v_2} + \left[\frac{0.278(v_2 + \delta_{v\max} v_2)t_{g1}}{v_2 + \delta_v v_2} + \frac{500(v_2 + v_k + \delta_{v\max} v_2)^2}{(v_2 + \delta_v v_2)\xi_2(1000 v_{p2}\varphi_{cr2} + [\omega_0]_{x_2} \pm i_{c2}}\right]$$

$$- [\frac{0.278(v_1 - \delta_{v\max} v_1)t_{g1}}{v_2 + \delta_v v_2} + \frac{500(v_1 - \delta_{v\max} v_1)^2}{(v_2 + v_k + \delta_{v\max} v_2)\xi_1(1000 v_{p1}\varphi_{cr1} + [\omega_0]_{x_1} \pm i_{c1}}$$

$$+ \frac{v_2 + v_k + \delta_{v\max} v_2}{(v_2 + \delta_v v_2)} \frac{n+1}{\Delta f} \log_2\left[n^2\left(\frac{L_{\max}}{l_k}\right)^2\left(\frac{v_{\max}}{v_k}\right)^2\left(\frac{i_{\max}}{i_k}\right)\left(\frac{R_{\max}}{R_k}\right)\right]]$$

$$+ \frac{2\delta_{S\max} S_{\text{ch}1}}{(v_1 + \delta_v v_2)} + \frac{l_k + \Delta l_k}{v_2 + \delta_v v_2} + \frac{\delta_S S_{\text{ch}1}}{v_2 + \delta_v v_2} \geq \frac{l_{t1}}{(v_2 + \delta_v v_2)} \quad (4)$$

In these expressions of the deterministic and statistical models of the system, the following designations are adopted:

l_{t1} is length of the train in front $T1$; v_1 and v_2 are measured speed ahead and behind the running trains, respectively; v_k is rate quantization step; $\delta_{v\max}$ is maximum relative error in measuring the speed;$\delta_{S\max}$ is maximum relative error in measuring the path; ξ_1, ξ_2 are deceleration in front and behind the running trains, respectively; v_{p1} and v_{p2} are calculated braking coefficient in front of and behind the running trains, respectively; $\varphi_{\text{cr}1}$ and $\varphi_{\text{cr}1}$ are the calculated coefficient of friction of the shoe on the tire at the highest speed for ahead and behind the running trains, respectively; $[\omega_0]_{x_1}$ and $[\omega_0]_{x_2}$ are force of the main specific resistance to movement in front and behind the running trains with the current turned off; i_{c1}. and i_{c2} are straightened path (in profile and plan); t_{g1} and t_{g2} are time to prepare the brakes for action; n is number of polled objects in the system; $L_{\max}$ is maximum distance measured by a distance meter without reset; l_k is path quantization step, coordinate segment; $v_{\max}$ is maximum possible train speed; v_k is rate quantization step; $i_{\max}$ is maximum rise (slope) of the track profile; i_k is quantization step of the path profile; $R_{\max}$ is maximum radius of the arc of the path in the plan of the path; R_k. is quantization step in terms of arc; Δf is bandwidth of the input filter of the system automation; $S_{\text{ch}1}$ is path traveled ahead of by the train in front; Δl_k is part of the coordinate segment that rounds the distance between trains to an integer number of coordinate segments; δ_s

is relative error in measuring the distance traveled, distributed according to a random law relative to the true value of the measured value;

δ_v is relative error in measuring the speed, distributed according to the normal law relative to the true value of the measured value.

1.1 Synthesis of the Optimal Algorithm for the Coordinate System of Automatic Control of Train Traffic

With the help of a mathematical model, you can solve a number of task interval automatic control of the movement of trains, among which one of the most interesting is the problem of synthesizing the optimal system algorithm. In particular, the derived mathematical model makes it possible to solve the problem of synthesizing the optimal algorithm for the coordinate system of automatic interval regulation of the movement of trains on railway transport. But in order to solve this and other particular problems, let us first consider in general terms the formulation and solution of synthesis problems the optimal algorithm of the system of interval regulation of train traffic.

Any task, the performed system of automatic regulation of train traffic can be solved in a way that is optimal in some sense (e.g., in terms of the throughput of railway lines, reliability, ease of execution, operation, economic efficiency of the system and several basic or some other criteria at once, or even all system criteria). Without the synthesis of an optimal algorithm, it is impossible to design optimal or close to them systems of interval automatic control of train traffic. On the other hand, it allows you to evaluate the end face, which can be achieved in the best optimal system, and compare it with the performance of existing systems. This comparison will make it possible to find out whether, in the case under consideration, it is necessary to develop an optimal system or it is possible to be satisfied with the existing one. Usually, the managed objects (trains) are defined and their properties cannot be changed. Meanwhile, the algorithm of the control device of the system interval regulation of train traffic is mostly not specified and it can be selected from a wide class of possible algorithms. A task of constructing an optimal system is reduced in this case to the development of such a control device that manages objects in the best way. The most difficult thing in creating an optimal control device is to determine the general form, the general structure of the algorithm that optimally fulfills the purpose of the control. In any case, the control goal can be considered as reaching an extremum of a certain value Q—an optimality criterion. Depending on the requirements, either a maximum or a minimum of Q is required. In the general case, the optimality criterion depends both on the setting program action $\overline{y}$ (where $\overline{y}$ is a complex vector parameter consisting of a number of simple actions, like other parameters with a bar) and on the output quantity $\overline{x}$. The optimality criterion can also depend on the control action on the object $\overline{u}$ and on the disturbing action, interference $\overline{Z}$, as well as on the time t. If for definiteness it is required that the value of Q be minimal, then in the general case the analytical formulation of the control circuit can be written as

$$Q(\overline{y,x,U,Z,t}) = \min,$$

where Q is a functional, i.e., a number depending on the form of the functions $\overline{y}$, $\overline{x}$, $\overline{u}$, $\overline{Z}$. In this case, the process of synthesizing the optimal algorithm for the coordinate system of the interval automatic control of train traffic will be reduced to finding the minimum optimality criterion Q on the system capacity, i.e., to finding the minimum of the temporary mathematical model T_{bt} of the system. With the help of variation methods, one can find the first and second derivatives of temporary mathematical model of the system. Then, equating the expression for the first derivative of temporary mathematical model of the system to zero ($T_{\mathrm{bt}} = 0$), we determine the minimum of temporary mathematical model of the system and express from it the speed of regulation $v_2(t)$ behind the running train through the initial primary parameters of the system, which are included in the right side of the mathematical model. The speed $v_2(t)$ will be optimal; $v_2(t)$ is optimal, since it will correspond to the minimum time interval between trains.

Let us illustrate the synthesis of the optimal algorithm for the coordinate system of the interval automatic control of train traffic using a simpler deterministic mathematical model of the system as an optimality criterion:

$$\begin{aligned}(T_e)^I = \Bigg\{ & \frac{l_{t1}}{v_2 + v_k + b_{v\max} v_2} + \left[0.278 t_{t2} + \frac{500(v_2 + v_k + \delta_{v\max} v_2)}{\xi_2(1000 v_{p2}\varphi_{\mathrm{cr}2} + [\omega_0]_{x_2} \pm i_{c2}}\right] \\ & - \frac{0.278(v_1 - \delta_{v\max} v_1)t_{t1}}{v_2 + v_k + \delta_{v\max} v_2} + \frac{500(v_1 - \delta_{v\max} v_1)^2}{(v_2 + v_k + \delta_{v\max} v_2)\xi_1(1000 v_{p1}\varphi_{\mathrm{cr}1} + [\omega_0]_{x_1} + i_{c1}} \\ & + \frac{(n+1)}{\Delta f}\log_2\left[n^2\left(\frac{L_{\max}}{l_k}\right)^2\left(\frac{v_{\max}}{v_k}\right)^2\left(\frac{i_{\max}}{i_k}\right)\left(\frac{R_{\max}}{R_k}\right)\right] + \frac{2\delta_{s\max} S_{\mathrm{ch}1}}{(v_2 + v_k + \delta_{v\max} v_2)} \\ & + \frac{l_k + \Delta l_k}{(v_2 + v_k + \delta_{v\max} v_2)}\Bigg\} = 0. \end{aligned} \tag{5}$$

Transforming the formulas, the final result synthesis of the optimal algorithm based on the deterministic mathematical model coordinate system for automatic control of train traffic, in which we denote the braking characteristics of trains by ȿ, will be as follows:

$$ȿ_1 = \frac{500}{\xi_2(1000\ \ _{p1}\varphi_{cr1} + [\omega_0]_{x_1} \pm i_{c1}};$$

$$ȿ_2 = \frac{500}{\xi_2(100\ \ _{p2}\varphi_{cr2} + [\omega_0]_{x_2} \pm i_{c2}}.$$

wherein

$$v_{2optimal} = \sqrt{\frac{\varsigma_2(1+\delta_{vmax})v_1^2}{\varsigma_1(1+\delta_{vmax})} - \frac{2\varsigma_2(l_t+l_k+\Delta l_k)}{1+\delta_{vmax}} - \frac{4\varsigma_2\delta_{Smax}S_{ch1}}{(1+\delta_{vmax})^2}} - \frac{v_k}{1+\delta_{vmax}}. \quad (6)$$

2 Result

The dynamics of the convergence of trains is illustrated in Fig. 1, which shows the following curves: OD—travel time of the tail of the train T1, and AT BC—the travel time of the head of the T2 train. Let us first assume that up to the breaking point T, the T2 train moves with a constant approach speed v_{as}. Then, the line MN, parallel to AT and spaced from it at a distance of the stopping distance from the speed v_{as} of the train T2, will represent the movement of the breaking point along the l-axis as a function of time. From the moment, the T2 train switches to the braking mode, the movement of the breaking point stops (line NB). The OL curve depicts the movement

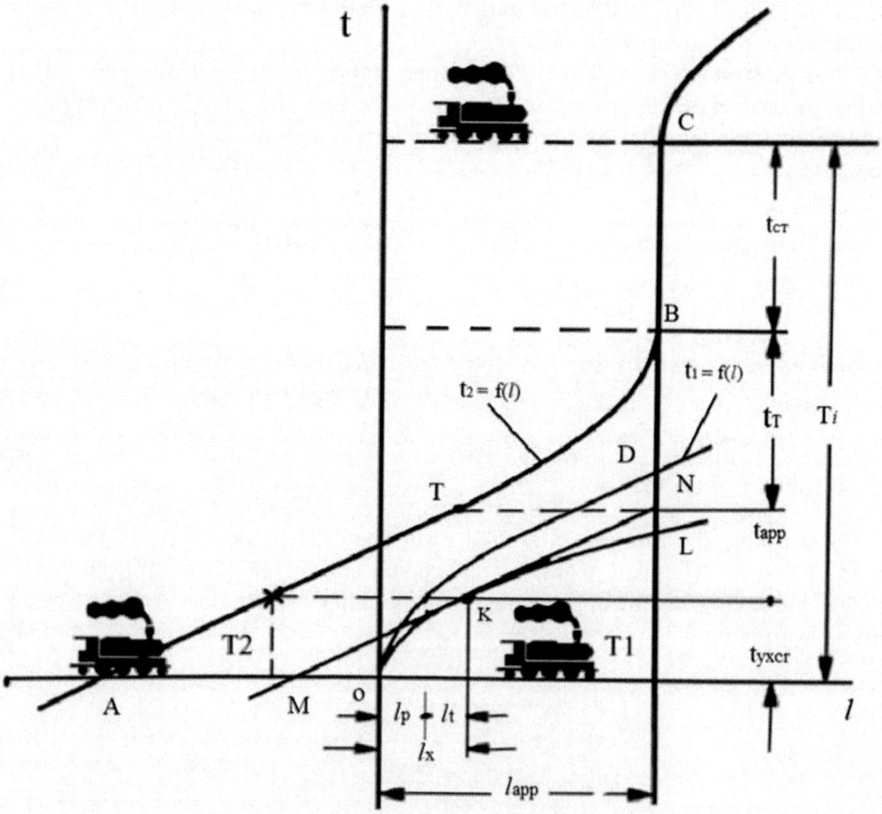

Fig. 1 Dynamics of the convergence of trains

of the point of aiming braking of the tail of the train T1 as a function of time. The equation for this curve will look like this:

$$t_x = t_p + t_\tau = \frac{a_1 t_{yx}^2}{2} + \frac{v_{yx}^2}{2b_1} = \frac{a_1}{2}\left(1 + \frac{a_1}{b_1}\right)t_{yx}^2 = c_1 t_{yx}^2. \quad (7)$$

In formula (8), the following designations are accepted: a_1 average acceleration of the train T1, and b_1—average deceleration of the most intensive type of braking for the T1 train;

at that

$$c_1 = \frac{a_1}{2}\left(1 + \frac{a_1}{b_1}\right). \quad (8)$$

The resulting expression (8) is the equation of a parabola with apex at the origin.

Taking into account the requirements of traffic safety, the sliding curve of the braking point of the T2 train should not cross the sliding curve of the aimed braking point. The extreme case, the most difficult to approach, is the moment when MN comes into contact with OL.

The minimum interval is determined by the moment of critical approach and is equal to the times of its components:

$$T_i = t_{yxcr} + t_{\text{approach}} + t_\tau + t_{\text{parking}}, \quad (9)$$

where t_{yxcr} is the departure time of the T1 train from the moment of starting to the moment of its critical location: t_{approach} is time of approach of T2 train with speed V_{as} until complete stop; t_τ is the braking time of the T2 train with a speed Vp until a complete stop; and t_{parking} is time of the train stay at the stopping platform.

Hence,

$$\frac{\mathrm{d}l_x}{\mathrm{d}t} = 2c_1 t_{yxcr} = v_t, \quad (10)$$

from where

$$t_{yxcr} = \frac{v_t}{2c_1} = \frac{v_t}{a_1\left(1 + \frac{a_1}{b_1}\right)}. \quad (11)$$

Braking time of the train T2

$$t_\tau = \frac{v_t}{b_2}, \quad (12)$$

where b_2 is the average deceleration of the service braking of the T2 train, taking into account the preparation time. Substituting expressions (10), (11), and (12) into expression (14), we obtain

$$T_i = \frac{2}{4c_1} v_t + \frac{l_t}{v_t} + \frac{v_t}{b_2} + t_{s\tau}. \tag{13}$$

Or after transformations

$$T_i = \frac{\left[2a_1\left(1 + \frac{a_1}{b_1}\right) + b_2\right] v_t^2 + 2a_1 b_2 t_{c\tau}\left(1 + \frac{a_1}{b_1}\right) v_t + 2a_1 b_2 l_t \left(1 + \frac{a_1}{b_1}\right)}{2a_1 b_2 \left(1 + \frac{a_1}{b_1}\right) v_t}. \tag{14}$$

3 Conclusion

The expressions obtained make it possible to take into account all the factors affecting the value of the passing interval. Since any system of interval regulation has the inertia and discreteness of measuring the location and speeds of approaching trains, the minimum interval value that the regulation system can practically implement will be determined by the proposed expressions.

References

1. Isaev DA, Stepin AV (2014) Review of modern systems of interval regulation of train traffic based on radio channel and coordinate principles. In: Materials of the international scientific-practical conference. Omsk State Transport University, pp 31–37
2. Aliev R (2021) Mathematical model of a tonal rail chain without insulating joints with current receiver in shunt mode. Transp Telecommun 22(3):312–320. https://doi.org/10.2478/ttj-2021-0024
3. Aliev R (2021) Trends in improving sensors for controlling the condition of track sections. In: CONMECHYDRO—2021 E3S web of conferences, vol 264, p 05045. https://doi.org/10.1051/e3sconf/202126405045
4. Rosenberg EN (2019) principles of construction of control systems and interval regulation of the movement of trains of the fourth generation, intelligent control systems in railway transport. In: Computer and mathematical modeling (isuzht-2019) proceedings of the eighth scientific and technical conference, pp 27–32
5. Sarbaev SSh (2019) Prospects for the use and implementation of DAS technology in railway automation and telemechanics systems. Tynyshbayev M, atyndagy Kazakh kulik zhune communicationalar academysynyk habarshysy. Bulletin of the Kazakh Academy of Transport and Communications named after M. Tynyshbayev. 1:153–157
6. Aliev R (2021) A Rail line model with distributed parameters of track circuit. In: 1st International Ninevah Conference on Engineering and Technology (INCET 2021), IOP Conf Ser Mater Sci Eng (1152):012018.https://doi.org/10.1088/1757-899X/1152/1/012018
7. Ananiev R (2020) Affordable and reliable/R. Ananiev // Railway worker of Belarus 8:1

8. Gapanovich VA (2019) Modern security complex. Gapanovich VA, Rumyantsev SV, Automation, communication, informatics 6:16
9. Digital technologies in the construction and operation of railway infrastructure. Transportnoe stroitel'stvo 2:9–12 (2019)
10. Brunner S (2020) GIS und BIM in Infrastrukturprojekten der Bahn / S. Brunner, J. Pfeifer. Der Eisenbahningenieur 7:20–23
11. Zhou YB, Jiajie L, Baohua M, Tang L (2018) Integrated optimization on train control and timetable to minimize net energy consumption of metro lines. Yuhe J Adv Transp (Article ID 7905820):1–19
12. Samà M, D'Ariano A, Pacciarelli D, Corman F (2017) A variable neighbourhood search for fast train scheduling and routing during disturbed railway traffic situations. Comput. Oper. Res. 78:480–499
13. Yang X, Chen A, Ning B, Tang T (2016) Astochastic model for the integrated optimization on metro timetable and speed profile with uncertain train mass. Transp Res Part B Methodological 91:424–445
14. Shrivastava P, Dhingra SL, Gundaliya PJ (2002) "Application of genetic algorithm for scheduling and schedule coordination problems. J Adv Transp 36(1):23–41
15. Aliev R et al (2020) Methodology for determining the optimal values of resistance at the ends of the jointless track circuit with considering twofold shunting. Int J Emerg Trends Eng Res 8(9):5048–5052

PSO and Firefly Algorithm Applied to EV-Based Hybrid Renewable Energy System for Load Frequency Control Considering Time-delay Effect

Hiramani Shukla and Siddhant Gudhe

Abstract The amalgamation of renewable sources into the preexisting grid system gives rise to stability, frequency deviation or power mismatch issues. Electrical vehicles (EVs) can supply power to the grid to solve these issues. EV is an energy storage system that is applied as a load and also as a source. To achieve high power quality, the management of automatic generation control (AGC) at the generation side is used. The system under investigation including thermal dish Stirling solar system (TDSSS) is used which emphasized the utilization of renewable into AGC. Along with TDSSS, distributed generators (DG) are also used consisting renewable sources like photovoltaic, aqua electrolyzes, fuel cell, wind, and diesel In this paper, two area system have been studied. Area 1 includes TDSSS with EV integration, and Area 2 has DG with EV penetration. An aggregate model of EV is used in both areas. The system is analyzed for step load and random load disruptions. To control the frequency, a proportional–integral–derivative controller with a derivative filer (PIDF) is used. Particle swarm optimization (PSO) and firefly algorithm (FF-A) optimization techniques are implemented to find the control variables, and these two techniques are compared. The time-delay effect is also taken into consideration, and the system under study is demonstrated in MATLAB/Simulink environment. The simulation results show that FA has low oscillation, very small settling time and have a considerable improvement in delay margin.

Keywords Load frequency control · Electric Vehicles · PID controller · Time-delay · PSO · Firefly algorithm

1 Introduction

Renewable energy systems have been integrating with the grid at a rapid rate with the advancements in power electronics technology. Electric vehicles (EV) as an energy storage system is used as a backup for the small-grid system. The penetration

H. Shukla (✉) · S. Gudhe
Electrical Engineering Department, Maulana Azad National Institute of Technology Bhopal, Bhopal, India

J. K. Mandal et al. (eds.), *Topical Drifts in Intelligent Computing*, Lecture Notes in Networks and Systems 426, https://doi.org/10.1007/978-981-19-0745-6_50

of EV along with the renewable sources possess more stability concerns. Electric vehicles in a large quantity connected to the grid affect the system stability, power, and frequency. EV can not only act as a load but also as an energy storage system. The variation of supply frequency and its stability along with its automatic control when the load varies is called automatic load frequency control (ALFC).

Stirling engine is the hot gas engine that runs on the heat energy collected at the solar dish. This system is called a thermal dish Stirling solar system (TDSSS). Such a system has been studied in the literature for multi-area multi-source systems [1–3].

PID controller along with derivative filter (PIDF) has been applied for a multi-area system for TDSSS using PSO [3]. Particle swarm optimization (PSO) is the most applied optimization method for the calculation of control parameters for the controller. Firefly algorithm (FF-A) optimization techniques have been proven efficient for the ALFC. In fact, FA outperforms PSO in all aspects of the control system.

A new model of EV has been proposed in [4]. The same aggregate model of EV is used along with the conventional sources [5]. But the renewable sources have not been discussed. The paper does not discuss the effect of random load disruption (RLD). The same EV model is used along with variable load deviation [6]. Model predictive control (MPC) controllers have been proposed to control frequency along with random loading [7]. A similar aggregate model with the conventional sources is also mentioned in [8]. But RLD discussion is lacking. Conventional energy sources have been discussed in [9]. Wind-energy-produced fluctuations in frequency have been mitigated with the help of EV [10]. The FF-A has been successfully implemented for the thermal power system [11], but the integration of EV and time-delay effects has not been considered.

The proposed model in the paper contains TDSSS and DG with renewable sources with the aggregated models of EV is under study. Section 2 describes the system under study, Section 3 discusses PSO and FF-A. Simulation Performance and conclusion are discussed in Sects. 4 and 5.

2 System Under Investigation

For the analysis of the system using controllers, two-area system model is considered. Two controllers are used, and their variables are calculated by using optimization technique. The electric vehicle model is introduced in both areas, and the analysis is completed. The system undergoes step load as well as random load variation. The time-delay effect is also taken into account.

2.1 time-delay Effect

Large-scale integration of conventional and renewable energy sources increase the inertia of the overall system along with low-frequency variation. There are delays in the power system which are not intentional, but the delay is provided to receive signals at the control unit. For the large power system, time-delays are usually not considered. This time-delay is usually taken into account for isolated smart grids. While considering the time-delay factor, the calculation of the time-delay margin is beyond the scope of this paper. A delayed linear system is marginally stable with all the roots lying on the left side of the s-plane. There is an infinite number of roots or time-delays available for the same system. Various methodologies are used in the literature to calculate the time-delay margin [12].

2.2 Model of Overall System

The overall system model consists of two area systems. The block diagram of the overall system under study is given in Fig. 1. Area 1 includes TDSSS with EV integration, and Area 2 has distributed generator (DG) with EV penetration. An aggregate model of EV is used in both areas. DG contains photovoltaic (PV), aqua electrolyzer (AE), fuel cell (FC), wind turbine system (WTS), and diesel engine generator (DEG). Both the areas are applied with a time-delay for communication with the command center. Both step load disruption (SLD) and random load disruption (RLD) are applied and tested in both areas. The system is studied for the change in frequency in both the areas and the change in tie-line power.

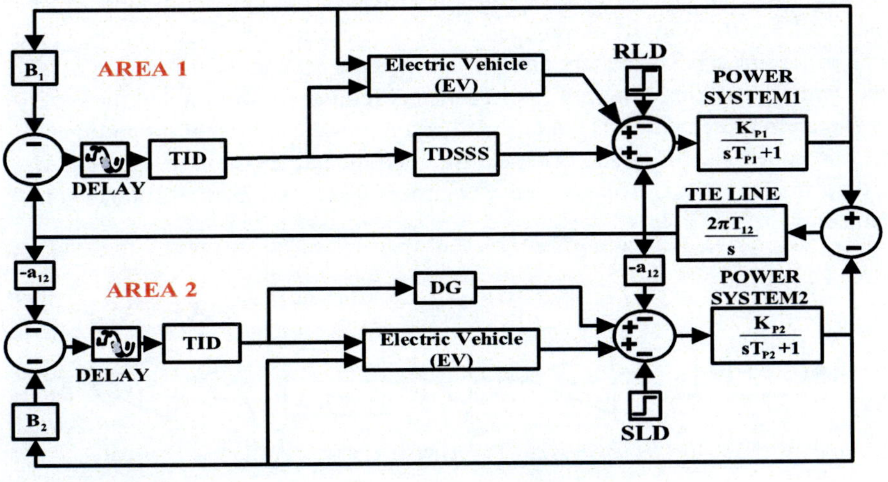

Fig. 1 Block diagram of the overall system under investigation

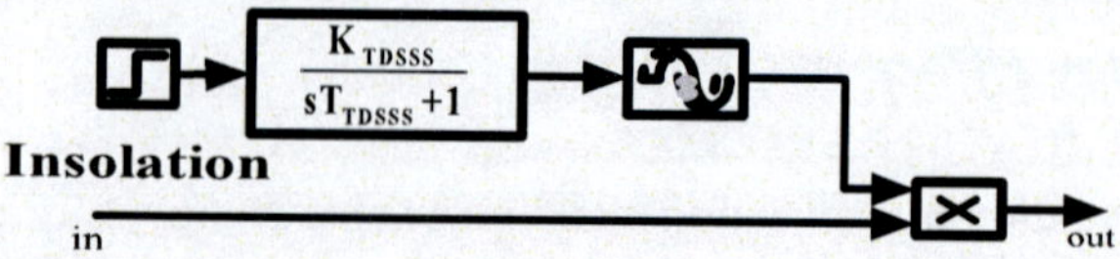

Fig. 2 Block diagram of TDSSS

2.3 Model of Thermal Dish Stirling Solar System (TDSSS)

The block diagram of TDSSS contains a solar-powered Stirling engine collector including a solar disc, hot gas engine, and electric generator which is used to generate electricity from the solar energy shown in Fig. 2. Incident sunrays are reflected by the disc to the receiver which is situated at the focus of the disc. Stirling engine is a hot gas engine that runs on the heat energy collected at the solar dish. The block diagram contains the insolation blocked coupled with the transfer function and the time-delay block. Insolation is the quantity of solar radiation for a certain area. This system is multiplied by controller parameters and penetrated the system. To obtain the dynamic performance of AGC using TDSSS, a time-delay of 1 s is provided. TDSSS has a gain of K_{TDSSS} and time constant T_{TDSSS} of its transfer function.

2.4 Model of Electric Vehicle (EV)

The aggregated model of EV gives the generalized model for the frequency control of the overall system which is shown in Fig. 3. The changed frequency is passed through the dead-band which limits the change in frequency within the particular band which is nearly +10 MHz and −10 MHz of the rated frequency. N_{EV} is the number of EVs connected. For the analysis purpose, all the EVs connected are assumed to be of the same battery rating. The battery charger is denoted by the transfer function of first order with a small-time constant T_{EV}. The upper limit and lower limit of the battery reserve are represented by (1) and (2). Charging power is directly proportional to the droop coefficient R_{AG} and is dependent upon the participation factor K_{EV}. The participation of every EV for frequency control is said to be a participation factor of that particular EV [13].

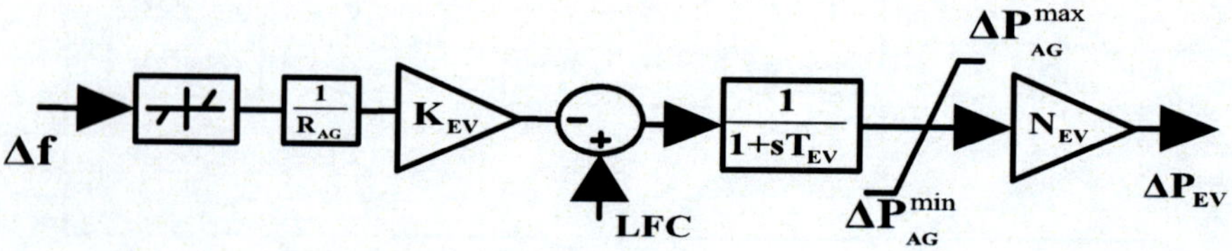

Fig. 3 Block diagram of an aggregated model of EV

$$\Delta P_{AG}^{max} = +[\Delta P_{EVi}/N_{EV}] \tag{1}$$

$$\Delta P_{AG}^{min} = -\left[\Delta P_{EVi}/N_{EV}\right] \tag{2}$$

where ΔP_{AG}^{max} is the maximum power output for a particular group of EV and ΔP_{AG}^{min} is the minimum power output for the same group of EVs. N_{EV} is the total number of EV which is assumed to be 1800 number of EV for Area 1 and 1600 number for Area 2 for the system under study.

2.5 Model of Distributed Generator (DG)

The term distributed generation (DG) is described as the generation of electricity for use on site instead of transmitting power over the electric grid. The model of DGm is shown in Fig. 4. The system is arranged in such a way that 60% PV power is directly supplying to the power system, and remaining 40% is input to the AE. Transfer function of the PV system contains K_{PV} and T_{PV} which are gain and time constant of the PV system. AE utilizes fraction of power (says 40%) from the PV system to make hydrogen (H_2) through water electrolysis using electricity, which is input to the FC. FC consumes H_2 to produce electricity. Transfer function of AE and FC has K_{AE} and K_{FC} terms which are gains, and T_{AE} and T_{FC} are time constant of the AE and FC, respectively.

WTS and DEG are also used in DG. WTS converts kinetic energy to mechanical which leads to produce electrical energy through induction generator. The relation between wind power (P_{WP}) and wind velocity (V_W) is $P_{WP} \propto V_W^3$. The first-order transfer function pf WTS has K_{WTS} which is the gain, and T_{WTS} is time constant

Fig. 4 Block diagram of model of DG

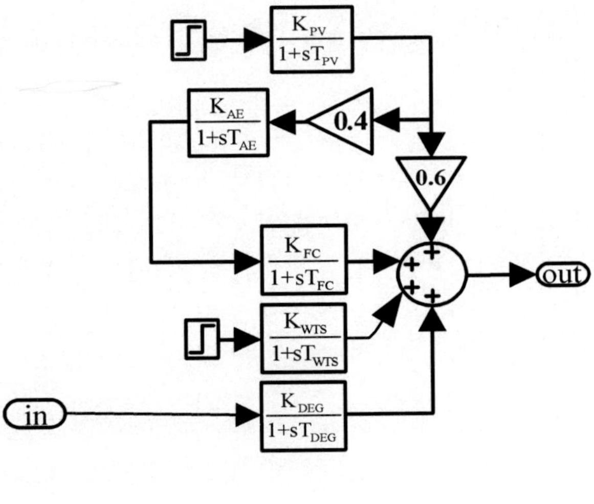

of the WTS. DEG uses liquid fuels as the primary fuel which works based on air compression. The transfer function of DEG contains K_{DEG} is the gain, and T_{DEG} is time constant of the WTS.

2.6 Controller Design

Equation of PIDF controller is given by $G(s)_{\text{PID}} = K_{Pk} + \frac{K_{Ik}}{s} + K_{Dk} \cdot \frac{sN}{s+N}$. By optimally tuning the control variables proportional gain K_{Pk}, integral gain K_{Ik}, and derivative gain K_{Dk} the performance is improved for the kth area. N is the coefficient filter for the controller. K_{Pk} controls the overshoots, rise time, and steady-state error with minimum effect on settling time. K_{Ik}affects the overshoots and rise time, but effect on settling time is still negligible. K_{Dk} is used to control settling time along with overshoot. The cost function to compare PSO and FF-A is shown in Eq. (3)

$$J = \int_0^T \left(\Delta f_{\text{area}-1}^2 + \Delta f_{\text{area}-2}^2 + \Delta P_{\text{tie}}^2\right)\text{d}t \tag{3}$$

3 Particle Swarm Optimization (PSO) and Firefly Algorithm Flowchart

The PSO technique is applied to extract the various controllers' parameters. The PSO parameters considered are: learning rates $c1 = c2 = 1.496$, damping factor (ωdamp) = 0.729, inertia weight $(\omega) = 1$, population size $(n) = 10$, and iterations (iter) = 100.

$$I = I_o \text{e}^{-\gamma \text{r}} \tag{4}$$

For FF-A, the motion of a firefly I is allured to other and enchanted firefly j [13, 14] is determined by Eq. (5)

$$x_i^{k+1} = x_i^k + \beta_o \text{e}^{-\gamma \text{r}_{\text{ij}}^2}(\text{x}_\text{j}^\text{k} - \text{x}_\text{i}^\text{k}) + \alpha \varepsilon_\text{i}^\text{k} \tag{5}$$

where α = randomization parameter and vector of random numbers is given by ε_i^k.

For FF-A optimization used in this study, tuned parameters: fireflies number = 10, total iterations = 100, $\beta = 0.2$, $\alpha = 0.5$, and $\gamma = 0.5$ (Figs. 5 and 6).

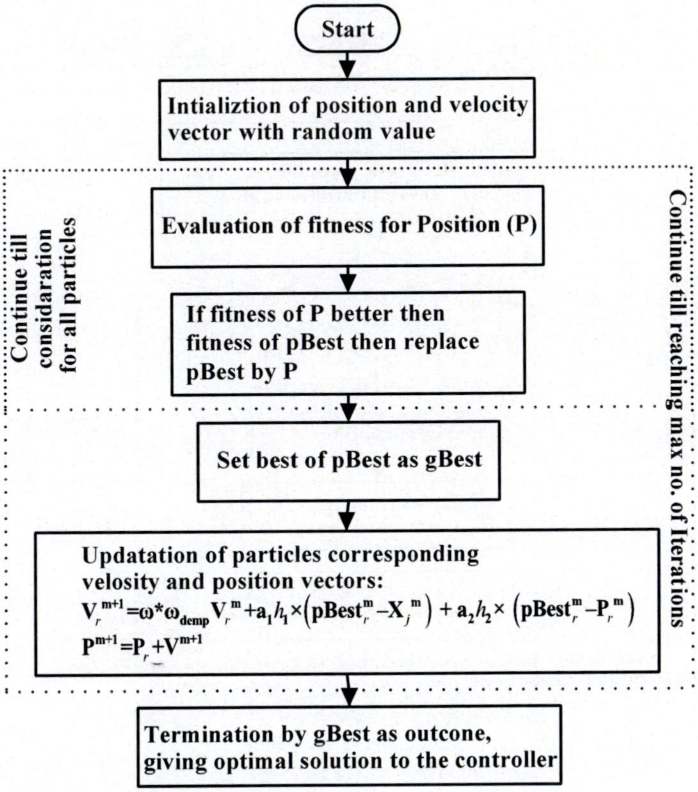

Fig. 5 Flowchart of particle swarm optimization

4 Simulation Performance

Simulation is performed in three parts. In the first part, the comparison of PID controller parameters using PSO and FF-A. In the second part, simulation results are shown for different loading, and in the last part, the system is tested for the time-delay. All the results are verified with the aggregate model of EV connected. The parameters for the simulation are shown in Table 1. The variable load applied to the system under investigation is shown in Fig. 7. The loading applied is in per unit (p.u).

The results for the PSO and FF-A comparison for step loading is shown in Figs. 8, and 9 shows the random loading disruption comparing PSO and FF-A.

Figure 10 shows the time-delay of 0.52 s applied to the system. At exact 0.52 s, PSO variables make the system marginally stable. Beyond 1.8 s, PSO gives control variables for PID which makes the system unstable. So, the time-delay margin for PSO-based controller is less than 0.52 s. The calculation of time-delay margin has tedious mathematical calculations depending on the system. The exact time-delay

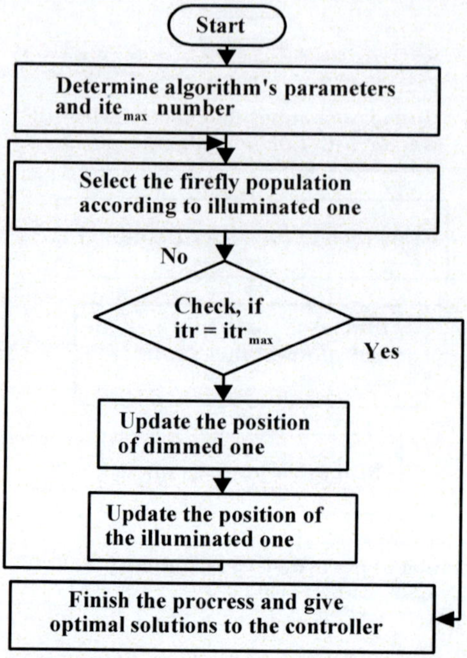

Fig. 6 Flowchart of firefly algorithm

Table 1 Simulation parameters for system under investigation

Parameters	Values
f (frequency), B_i (damping constant), R (regulation)	60 Hz, 0.425 pu/Hz, 2.4 Hz/pu
K_{TDSSS}, T_{TDSSS}, R_{AG}, K_{EV}, T_{EV}	1, 5 s, 2.4 Hz/pu, 1, 1 s
K_{PV}, T_{PV}, K_{AE}, T_{AE}, K_{FC}, T_{FC}, K_{WTS}, T_{WTS} K_{DEG}, T_{DEG}	1, 1.8 s, 1/500, 0.5 s, 1/100, 4 s, 1, 1.5 s, 3/1000, 2 s
K_{Pi}, T_{Pi}, T_{12}, a_{12}	120, 20 s, 0.0867, -1

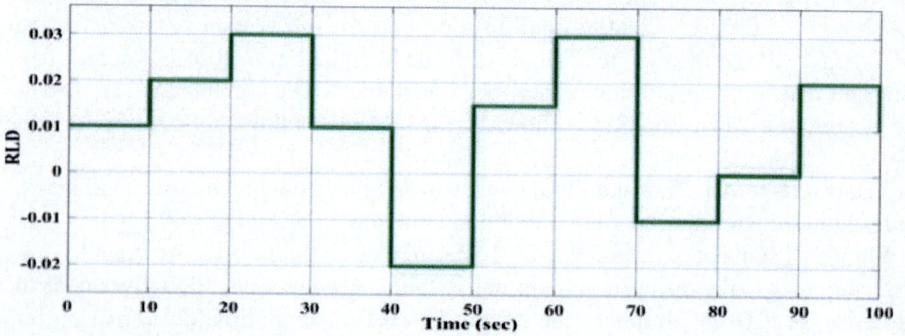

Fig. 7 Variable load applied to the system under study

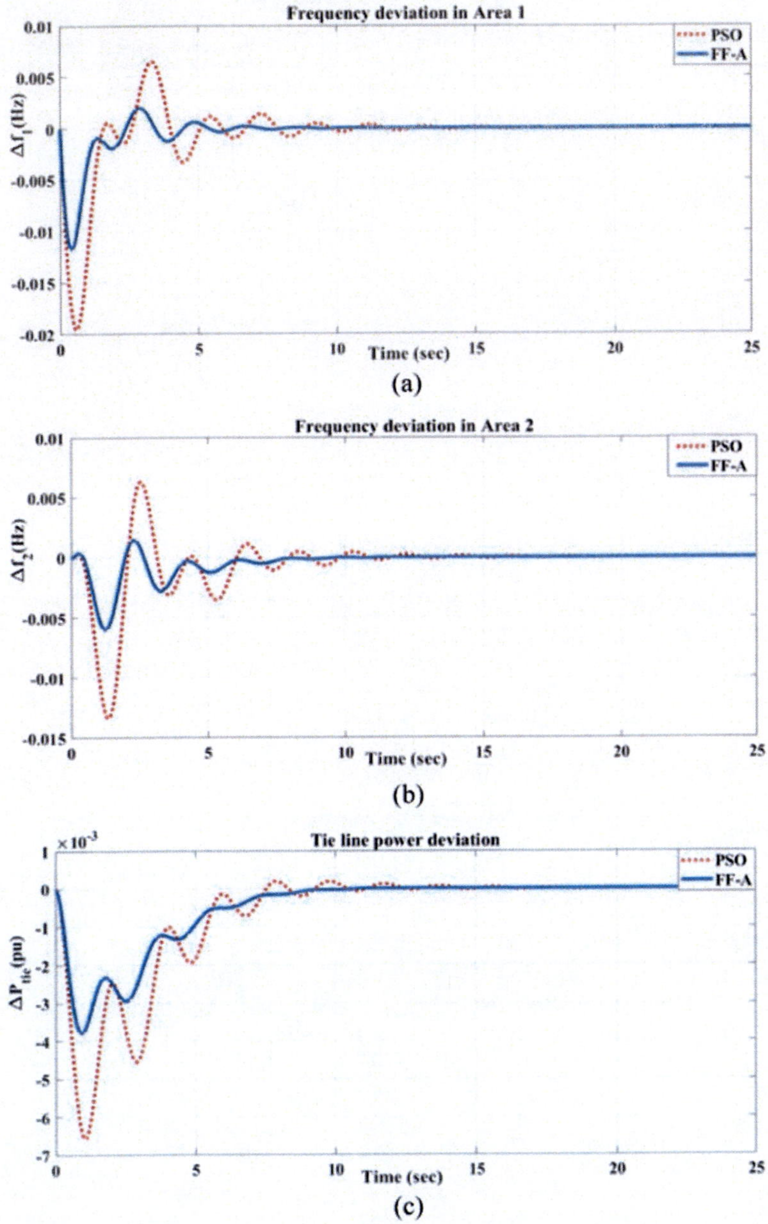

Fig. 8 Comparison of PSO and FF-A (SLD) **a** frequency fluctuations in area 1, **b** frequency fluctuations in area 2, **c** tie-line power variation

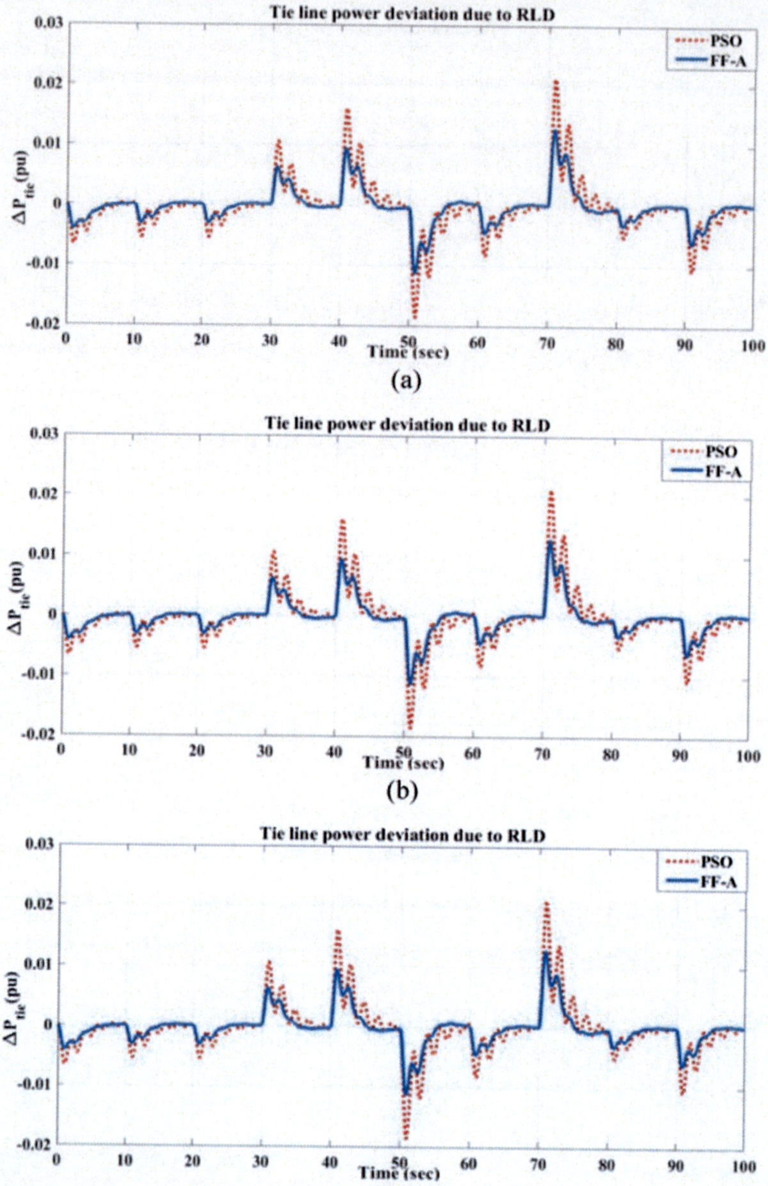

Fig. 9 Comparison of PSO and FF-A (RLD) **a** frequency fluctuations in area 1, **b** frequency fluctuations in area 2, **c** tie-line power variation

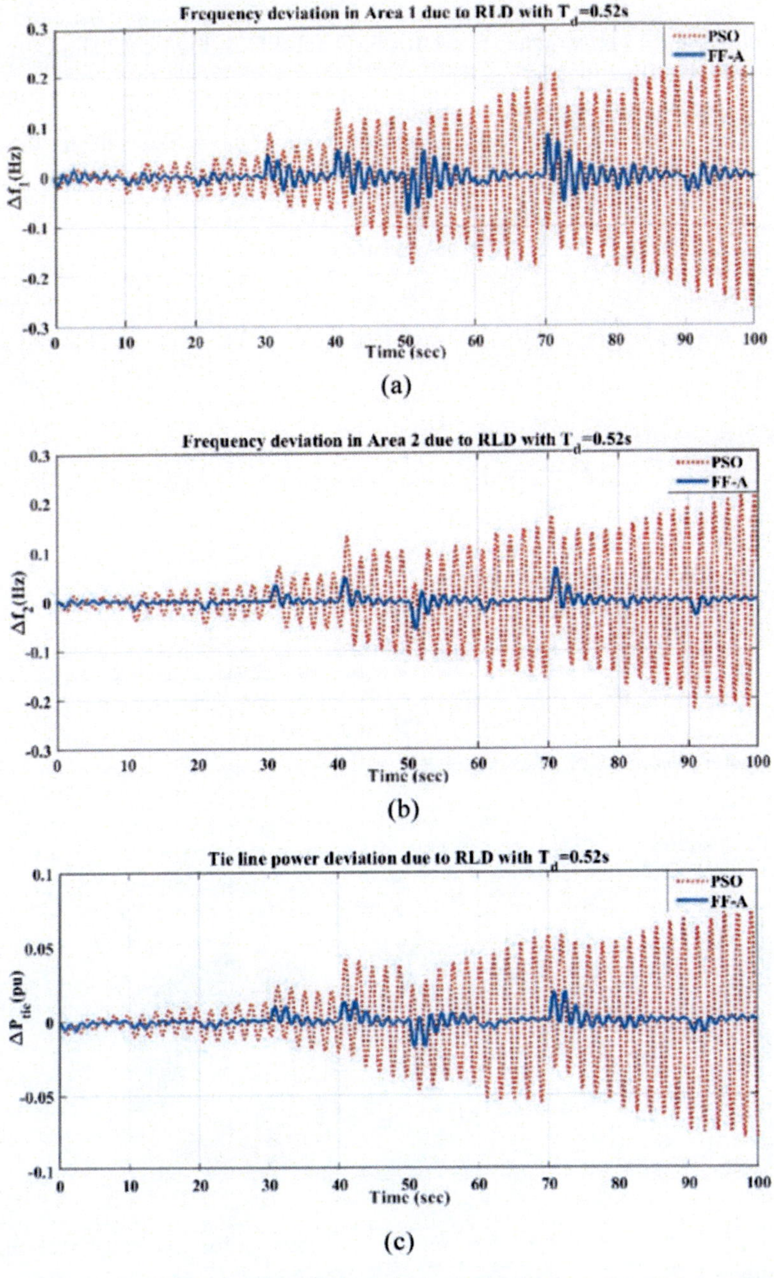

Fig. 10 Comparison of PSO and FF-A for time-delay margin of 0.52 s **a** frequency fluctuations in area 1, **b** frequency fluctuations in area 2, **c** tie-line power variation

margin is not shown as delay in simulation is based on Pade approximation [12]. Figure 11 gives the convergence curve for PSO and FF-A. Figure 12 shows the pie chart for the cost function for Eq. (3).

Tables 2 and 3 show the controller limitations found from both techniques. Table 4 shows the magnitude of peak overshoot, undershoot as well as settling time for PSO and FF-A for PIDF controller, respectively.

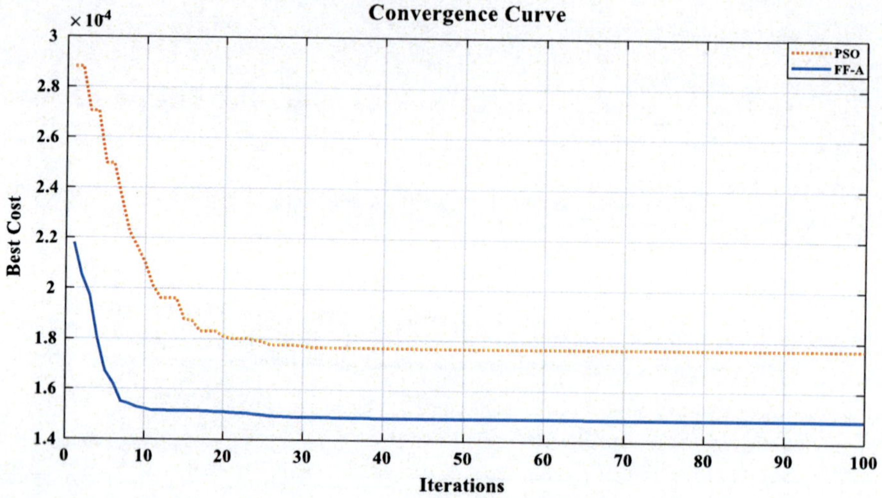

Fig. 11 Convergence curve for PSO and FF-A

Fig. 12 Pie chart for the cost function of Eq. (3)

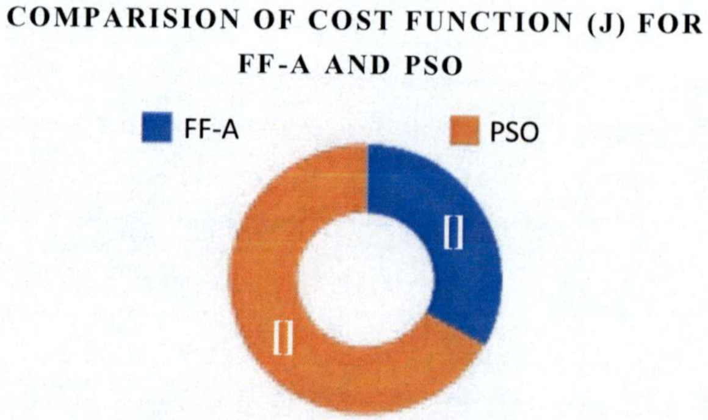

Table 2 PIDF control variables using PSO

Parameters (PSO)	K_P, K_I, K_D, N
Area 1	0.2271, 0.4409, 0.2614, 31.4114
Area 2	0.3569, 0.7162, 0.1684, 20.9199

Table 3 PIDF control variables using FF-A

Parameters (FF-A)	K_P, K_I, K_D, N
Area 1	1, 0.6303, 0.9989, 75.67
Area 2	0.9566, 0.6485, 0.2436, 96.71

Table 4 Comparison of simulation result parameters using PSO and FF-A

Parameters		Δf (Hz)	Δf (Hz)	ΔP (pu)
PSO	Peak overshoot (ms)	6.6	6.3	0.2
	Peak undershoot (ms)	19.6	13.4	6.6
	Settling time (s)	16.4	18.3	17.2
FF-A	Peak overshoot (ms)	1.9	1.4	–
	Peak undershoot (ms)	11.8	6	3.8
	Settling time (s)	10.2	10.8	10.4

5 Conclusion and Discussion

In this paper, a system having two areas is used to compare PSO and FF-A optimization techniques using the PIDF controller. Area 1 has TDSSS, and Area 2 contains DG. Aggregated model of EV is penetrated in both the areas. The simulation results show that FF-A control variables outperform PSO for settling time and oscillations. Both the techniques are tested for under step load and random load disruption. time-delay effects are considered, and the time-delay margin is approximated. For the system under study, the delay margin for FF-A obtained is 0.52 s, but the system becomes marginally stable at 0.52 s for PSO. More stability margin indicates that the system is robust and reliable. Also, the verification of FF-A superiority can be concluded from peak overshoot, undershoot as well as settling time. Peak overshoot for FF-A is considerably improved compared to PSO. The cost function using FF-A is lesser than PSO. The low value of the convergence curve and cost function gives optimized controller parameters. Therefore, it is easily concluded that FF-A gives very optimized parameters compared to the PSO algorithm.

Simulation results can be generalized for the multi-area multi-source system. In future research, the different controllers can be tested using both techniques. For the calculation of the time-delay margins, different methodologies for the large-scale power system can be explored.

References

1. Howard D, Harley RG (2010) Modeling of dish-Stirling solar thermal power generation. In: IEEE PES general meeting, pp 1–7
2. Rajbongshi R, Saikia LC (2018) Combined voltage and frequency control of a multi-area multisource system incorporating dish-Stirling solar thermal and HVDC link. IET Renew Power Gener 12(3):323–334
3. Panda A, Das A, Pati SS, Kumar Mishra S (2019) A PSO based PIDF controller for multiarea multisource system incorporating dish stirling solar system. In: 2019 International conference on intelligent computing and control systems (ICCS), pp 1458–1462
4. Izadkhast S, Garcia-Gonzalez P, Frías P (2015) An aggregate model of plug-in electric vehicles for primary frequency control. IEEE Trans Power Syst 30(3):1475–1482
5. Debbarma S, Dutta A (2017) Utilizing electric vehicles for LFC in restructured power systems using fractional order controller. IEEE Trans Smart Grid 8(6):2554–2564
6. Chorasiya G, Suhag S (2020) A comparative study of MVO and SSA optimized PID controller for LFC in EV integrated multi area network. In: 2020 11th International conference on computing, communication and networking technologies (ICCCNT), pp 1–7
7. Asghar R et al (2020) Load frequency control for EVs based smart grid system using PID and MPC. In: 2020 3rd International conference on computing, mathematics and engineering technologies (iCoMET), pp 1–6
8. Prabu RG (2018) Effect of plug-in electric vehicles on load frequency control. In: 2018 8th IEEE India international conference on power electronics (IICPE)
9. Kumar M, Agrawal S, Mohamed TH (2021) Application of AGPSO algorithm in frequency controller design for isolated microgrid. In: 2021 IEEE Texas power and energy conference (TPEC), pp 1–6
10. Casasola-Aignesberger L, Martinez S (2020) Electric vehicle recharge strategies for frequency control in electrical power systems with high wind power generation. In: 2020 IEEE international conference on environment and electrical engineering and 2020 IEEE industrial and commercial power systems Europe (EEEIC / I&CPS Europe), pp 1–5
11. Jagatheesan K, Anand B, Samanta S, Dey N, Ashour AS, Balas VE (2019) Design of a proportional-integral-derivative controller for an automatic generation control of multi-area power thermal systems using firefly algorithm. IEEE/CAA J Automatica Sinica 6(2):503–515
12. Macana CA, Mojica-Nava E, Quijano N (2013) time-delay effect on load frequency control for microgrids. In: 2013 10th IEEE international conference on networking, sensing and control (ICNSC), pp 544–549
13. Yang XS (2009) Firefly algorithms for multimodal optimization. In: Watanabe O, Zeugmann T (eds) Stochastic algorithms: foundations and applications. Lecture Notes in Computer Science, vol 5792. Springer, Berlin, Heidelberg, SAGA
14. Yanga X, Zhaob Y (2020) Firefly algorithm and flower pollination algorithm. In: Nature-inspired computation and swarm intelligence algorithms, theory and applications, pp 35–48

Incorporation of HVDC into Thermal-Gas-EV System for LFC Considering Time Delay Effect

Hiramani Shukla and Siddhant Gudhe

Abstract Electrical vehicle (EV) as an energy storage element plays a key role in stability, frequency deviation or power mismatch issues. HVDC links are acting as DC capacitors and its electrostatic energy is used as an energy storage device. To achieve high power quality, the management of automatic generation control (AGC) at the generation side is used. Along with the preexisting energy sources like thermal and gas, EV and HVDC are penetrated the system for load frequency control (LFC). The system under investigation includes thermal and gas with EV in both areas with HVDC link is connected between two areas. Area 1 includes thermal with EV integration and Area 2 has a gas system with EV penetration. An aggregate model of EV is used in both areas. The system is analyzed for step load and random load disruptions. To control the frequency, proportional-integral (PI) controller proportional-integral-derivative (PID) controller is used. Firefly algorithm (FF-A) optimization techniques are implemented to find the control variables. The time delay effect is also taken into consideration and the system under study is demonstrated in MATLAB/Simulink environment. The simulation results show that FA has low oscillation, very small settling time and have a huge improvement in delay margin.

Keywords Load frequency control · Electric vehicles · PID controller · HVDC link · Time delay · Firefly algorithm

1 Introduction

Electric vehicles (EVs) have also been increasing at an exponential rate. The penetration of EV along with the renewable sources possess more stability concerns. Electric vehicles in a large quantity connected to the grid affect the system frequency, power and stability. If such fleets of EV charging and discharging is not managed, which may lead to power quality issues and also the transformer may overload. There is

H. Shukla (✉) · S. Gudhe
Electrical Engineering Department, Maulana Azad National Institute of Technology Bhopal, Bhopal, India

J. K. Mandal et al. (eds.), *Topical Drifts in Intelligent Computing*, Lecture Notes in Networks and Systems 426, https://doi.org/10.1007/978-981-19-0745-6_51

rapid technological advancement in the battery. EV battery is used to drive the traction motor via inverter control. Therefore, the driving range for EVs depends on the stored energy of the battery. Efficient control of the EV is essential for efficiency improvement. The model developed for the analysis contains the conventional energy sources which include thermal and gas turbines. Gas turbine transfer function contains the effects of speed governor, valve position, combustion, fuel compressor and turbine. The thermal transfer function considers the effects of the steam turbine.

HVDC link connection makes system hybrid containing DC and AC energy sources. HVDC links have power electronics converters with considerable inertia. AGC is coordinated with the various system devices which include HVDC and FACTS devices. The effect of FACTS devices on LFC is beyond the scope of this paper. This paper focuses on the importance of HVDC electrostatic energy used for the frequency control of the system. HVDC integration for the multi-area system has been carried out by Rakhshani et al. [1] by solving the virtual inertia issue. The implementation of thermal and gas along with EV is studied by Debbarma et al. [2] without time delay effect consideration. The FF-A has been successfully implemented for the thermal power system by Jagatheesan et al. [3], but the integration of EV and time delay effects have not been considered. AGC of multi-area gas turbine and the thermal system has been carried out by Saikia et al. [4] with integral (I) controller and by Patel et al. [5] with PI controller. Implementation using PI controller has been also conducted by Bhadoria et al. [6] using the firefly algorithm (FF-A). A new model of EV has been proposed by Izadkhast et al. [7]. The same aggregate model of EV is used along with the conventional sources by Debbarma et al. [2]. The effect of the time delay on the load frequency control has been very beautifully studied by Macana et al. [8] and Hua et al. [9]. The decentralized optimal EV control approach is introduced by Kariminejad et al. [10]. The low value of the convergence curve and cost function gives optimized controller parameters [10].

The proposed model in the paper contains thermal and gas with the aggregated model of EV. Also, the interconnection of the HVDC link between Area 1 and Area 2 is under study. Section 2 describes the system under study, Sect. 3 discusses FF-A. Simulation performance and conclusion are discussed in Sects. 4 and 5.

2 System Under Investigation

For the analysis of the system using controllers, two area system model is considered. Two controllers are used and their variables are calculated by using optimization technique. The electric vehicle model is introduced in both areas and the analysis is completed. The system undergoes through step load as well as random load variation. The time delay effect is also taken into account.

2.1 Time Delay Effect

Large scale integration of conventional and renewable energy sources increases the inertia of the overall system and along with low-frequency variation. There are delays in the power system which are not intentional, but the delay is provided to receive signals at the control unit. For the large power system, time delays are usually not considered. This time delay is usually taken into account for isolated smart grids. While considering the time delay factor, the calculation of the time delay margin is beyond the scope of this paper. Various methodologies are used in the literature to calculate the time delay margin [8].

2.2 Model of Overall System

The overall system model consists of two area systems. The block diagram of the overall system under study is given in Fig. 1. Area 1 includes a gas turbine with EV integration and Area 2 has a gas turbine with EV penetration. An aggregate model of EV is used in both areas. Gas turbine contains governor, compressor, combustion and turbine. HVDC link as an energy storage device is connected between both areas. Both the areas are applied with a time delay for communication with the command center. Both step load disruption (SLD) and random load disruption (RLD)

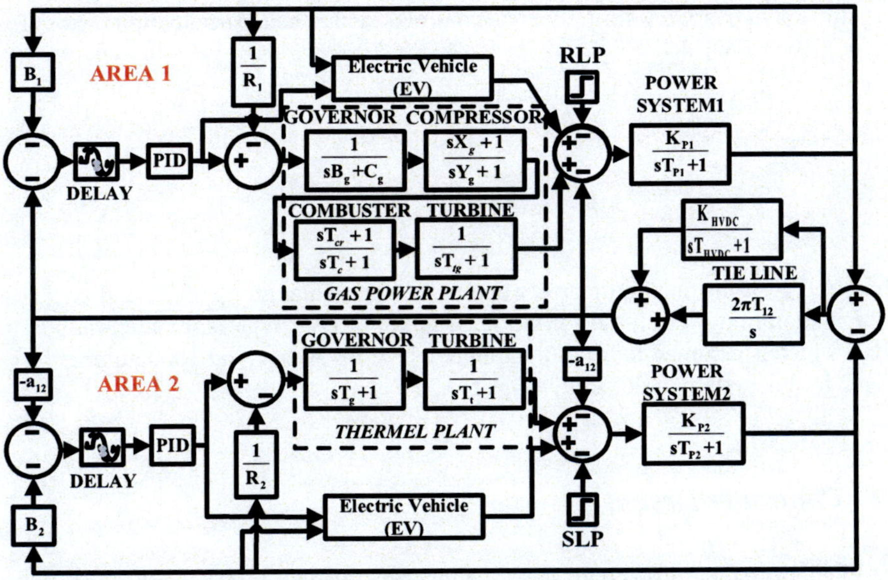

Fig. 1 Block diagram of the overall system under investigation

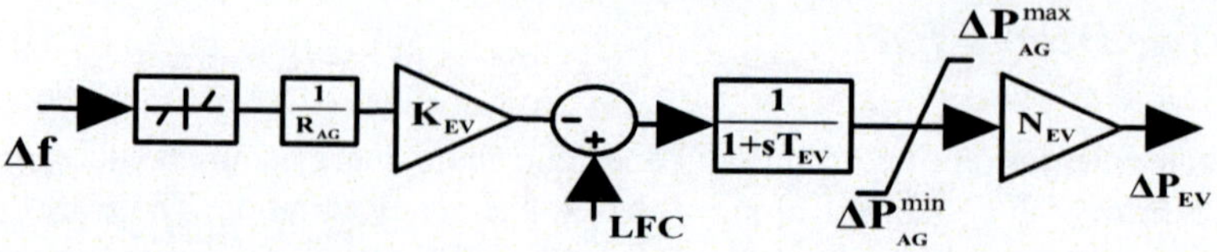

Fig. 2 Block diagram of aggregated model of EV

are applied and tested in both areas. The system is studied for the change in frequency in both the areas and the change in tie-line power.

2.3 Model of Electric Vehicle (EV)

The aggregated model of EV gives the generalized model for the frequency control of the overall system which is shown in Fig. 3. The changed frequency is passed through the dead-band which limits the change in frequency within the particular band which is nearly +10 MHz and −10 MHz of the rated frequency. N_{EV} is the number of EVs connected. For the analysis purpose, all the EVs connected are assumed to be of the same battery rating. The battery charger is denoted by the transfer function of first order with a small-time constant T_{EV}. The upper limit and lower limit of the battery reserve is represented by Eqs. (1) and (2). Charging power is directly proportional to the droop coefficient R_{AG}. Also is dependent upon the participation factor K_{EV}. The participation of every EV for frequency control is said to be a participation factor of that particular EV [2].

$$\Delta P_{\mathrm{AG}}^{\max} = +\left[{}^{\Delta P_{\mathrm{EVi}}}/_{N_{\mathrm{EV}}}\right] \tag{1}$$

$$\Delta P_{\mathrm{AG}}^{\min} = -\left[{}^{\Delta P_{\mathrm{EVi}}}/_{N_{\mathrm{EV}}}\right] \tag{2}$$

where $\Delta P_{\mathrm{AG}}^{\max}$ is the maximum power output for a particular group of EV and $\Delta P_{\mathrm{AG}}^{\min}$ is the minimum power output for the same group of EVs. N_{EV} is the total number of EV which is assumed to be 1800 number of EV for Area 1 and 1600 number for Area 2 for the system under study (Fig. 2).

2.4 Controller Design

The PI controller and PID controller equations are given by $G(s)_{\mathrm{PI}} = K_{\mathrm{Pk}} + K_{\mathrm{Ik}}/s$ and $G(s)_{\mathrm{PID}} = K_{\mathrm{Pk}} + K_{\mathrm{Ik}}/s + K_{\mathrm{Dk}} \cdot s$, respectively. By optimally tuning the control

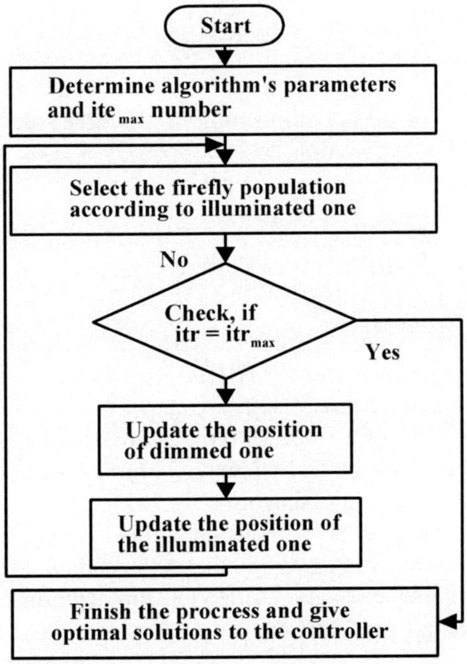

Fig. 3 Flowchart of firefly algorithm

variables proportional gain K_{Pk}, integral gain K_{Ik} and derivative gain K_{Dk} the performance is improved for the kth area. K_{Pk} controls the overshoots, rise time and steady-state error with minimum effect on settling time. K_{Ik} affects the overshoots and rise time, but effect on settling time is still negligible. K_{Dk} is used to control settling time along with overshoot. The cost function to compare the systems with and without HVDC is shown in Eq. (3)

$$J = \int_0^T \left(\Delta f_{\text{area}-1}^2 + \Delta f_{\text{area}-2}^2 + \Delta P_{\text{tie}}^2\right)\mathrm{d}t \tag{3}$$

3 Firefly Algorithm (FF-A) Flowchart

FF-A is applied to extract the various controllers' parameters.

$$\mathrm{I} = \mathrm{I_o}\mathrm{e}^{-\gamma \mathrm{r}} \tag{4}$$

For FF-A, the motion of a firefly i is allured to other and enchanted firefly j (15) is determined by Eq. (6)

$$x_i^{k+1} = x_i^k + \beta_o e^{-\gamma r_{ij}^2}(x_j^k - x_i^k) + \alpha \varepsilon_i^k \quad (5)$$

where α = Randomization parameter and vector of random numbers is given by ε_i^k.

For FF-A optimization used in this study, tuned parameters: fireflies number = 10, total iterations = 100, $\beta = 0.2$, $\alpha = 0.5$ and $\gamma = 0.5$.

4 Simulation Performance

Simulation is performed in three parts. In the first part, the system is simulated using HVDC and without HVDC with PI and PID controller using FF-A. In the second part, simulation results are shown for RPD and are tested for the time delay. All the results are verified with the aggregate model of EV connected. The parameters for the simulation are shown in Table 1. The variable load applied to the system under investigation is shown in Fig. 4.

The results for with and without HVDC link for step loading disruption (SLD) is shown in Figs. 5 and 6. shows the random loading disruption (RLD) results with and without application of HVDC link with time delay effect of 1 s. Results are verified using PI and PID controllers.

Table 1 Simulation parameters for system under investigation

Parameters	Values
f (frequency), B_i (damping constant), R (regulation)	60 Hz, 0.425 pu/Hz, 2.4 Hz/pu
B_g (valve position), C_g (valve position), X_g (least time constant), Y_g (lag time constant), T_{cr} (combustion reaction time delay), T_c (fuel time delay), T_{tg} (compressor discharge time constant)	0.05, 1, 0.6 s, 1 s, 0.01, 0.23, 0.2 s
K_g, T_g, K_t, T_t, K_{HVDC}, T_{HVDC}	1, 0.3 s, 1, 0.08 s, 1, 0.2 s
T_{12}, a_{12}	0.0867, −1

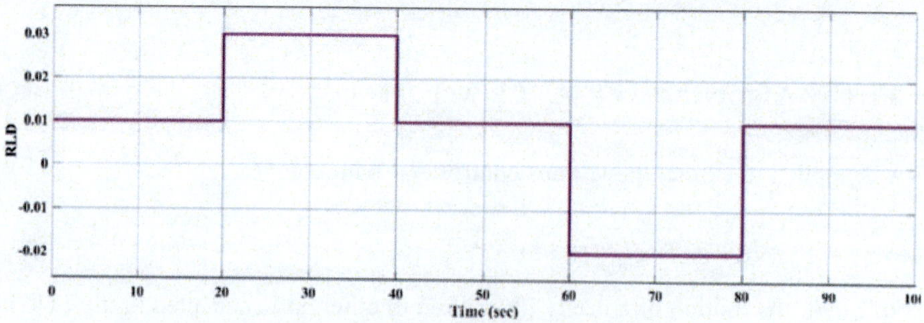

Fig. 4 Variable load applied to the system under study

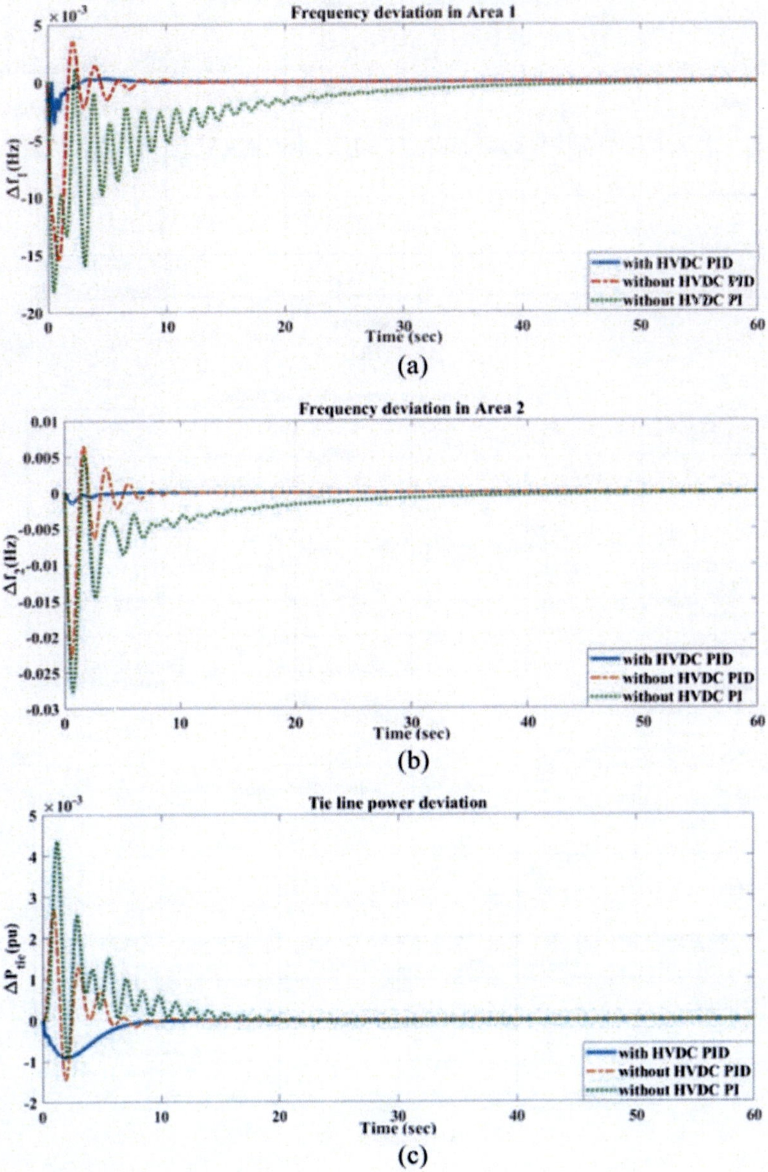

Fig. 5 SLD with and without HVDC link

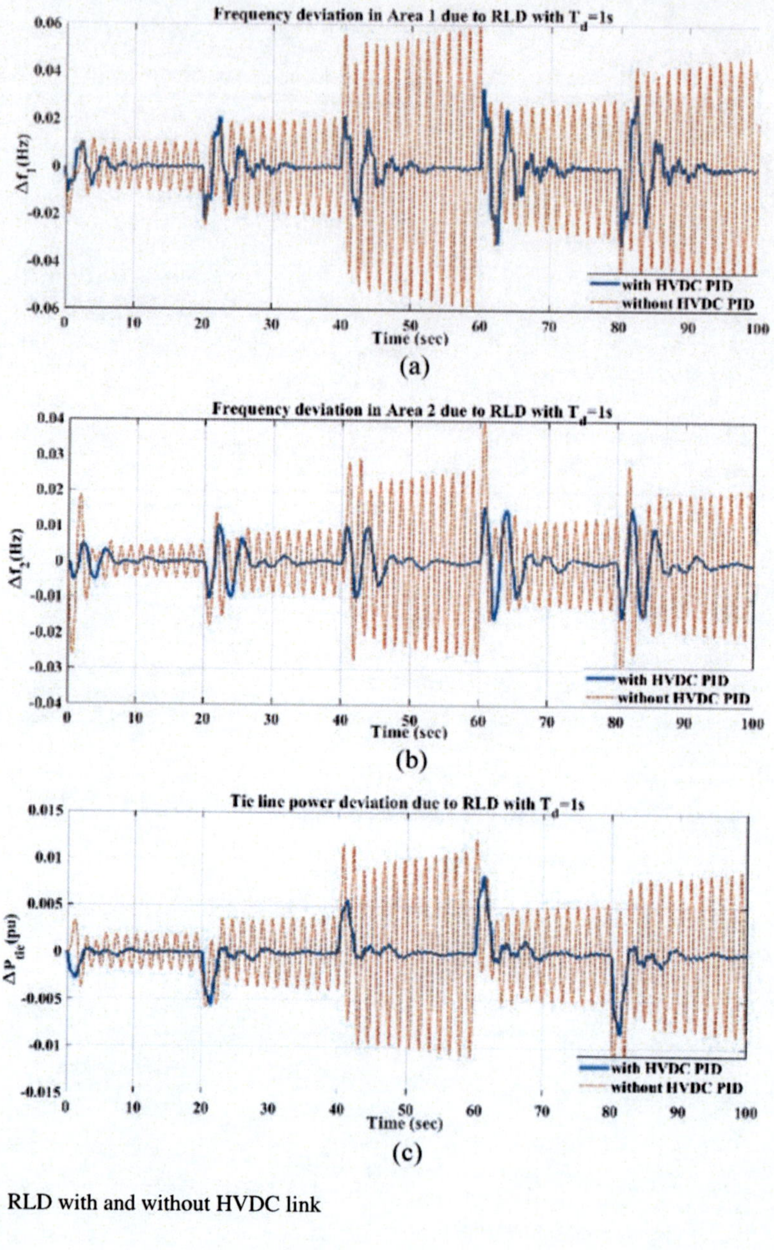

Fig. 6 RLD with and without HVDC link

At exact 1 s, FF-A variables make the system marginally stable. Beyond 1 s, FF-A gives control variables for PID which makes the system unstable. So, the time delay margin FF-A-based controller is less than 1 s. The calculation of time delay margin has tedious mathematical calculations depending on the system. The exact time delay margin is not shown as delay in simulation is based on Pade approximation

Table 2 PID control variables with HVDC link

Parameters (FF-A)	K_P, K_I, K_D
Area 1	1, 0.9987, 0.9523
Area 2	0.9987, 0.9646, 0.1691

Table 3 PI and PID control variables without HVDC link

Parameters (FF-A)	$K_P, K_I; K_P, K_I, K_D$
Area 1	0.746, 0.125; 0.0158, 1, 0.528
Area 2	0.0114, 0.0848; 0.1154, 0.4773, 0.0512

[8]. Figure 7 gives the convergence curve for PI and PID controller with HVDC link using Eq. (3).

Tables 2 and 3 shows the controller limitations found from both techniques. Table 4 shows the magnitude of peak overshoot, undershoot as well as settling time using PID controller without and with HVDC link, respectively (Fig. 7).

Table 4 Comparison of simulation result parameters for PID controller

Parameters		Δf_1 (Hz)	Δf_2 (Hz)	ΔP_{tie} (pu)
Without HVDC	Peak overshoot (ms)	3.45	6.5	2.7
	Peak undershoot (ms)	15.5	23	1.45
	Settling time (s)	10.2	11.5	13.2
With HVDC	Peak overshoot (ms)	0.018	–	–
	Peak undershoot (ms)	4.52	1.4	0.92
	Settling time (s)	7.9	6.3	13.1

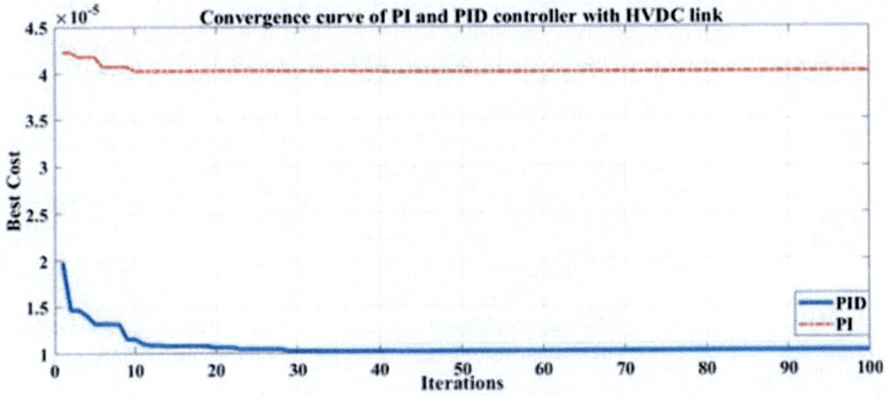

Fig. 7 Convergence curve with and without HVDC link

5 Conclusion and Discussion

In this paper, a system having two areas is used to compare system stability using the HVDC link. Area 1 has thermal and Area 2 contains gas turbine. Aggregated model of EV is penetrated in both areas. The simulation results show that the HVDC link with the PID controller outperforms the system without the HVDC link in terms of settling time and oscillations. Time delay effects are considered and the time delay margin is approximated. Even delayed system using HVDC link shows good performance as compared to the system not having HVDC link. For the system under study, the delay margin obtained is 1 s, but the system becomes marginally stable at 1 without an HVDC link. More stability margin indicates that the system is robust and reliable. Also, the verification of HVDC system superiority can be concluded from peak overshoot, undershoot as well as settling time. Cost function using HVDC is lesser than without an HVDC system. Therefore, it is concluded that a system having an HVDC link gives the optimized parameters.

Simulation results can be generalized for the multi-area multi-source system. In future research, the different controllers can be tested using both techniques. For the calculation of the time delay margins, different methodologies for the large scale power system can be explored.

References

1. Rakhshani E, Remon D, Cantarellas AM, Garcia JM, Rodriguez P (2017) Virtual synchronous power strategy for Multiple HVDC interconnections of multi-area AGC power systems. IEEE Trans Power Syst 32(3):1665–1677
2. Debbarma S, Dutta A (2017) Utilizing electric vehicles for LFC in restructured power systems using fractional order controller. IEEE Trans. Smart Grid 8(6):2554–2564
3. Jagatheesan K, Anand B, Samanta S, Dey N, Ashour AS, Balas VE (2019) Design of a proportional-integral-derivative controller for an automatic generation control of multi-area power thermal systems using firefly algorithm. IEEE/CAA J Autom Sinica 6(2):503–515
4. Saikia LC, Chowdhury A, Shakya N, Shukla S, Soni PK (2013) AGC of a multi area gas-thermal system using firefly optimized IDF controller. In: 2013 Annual IEEE India conference (INDICON), pp 1–6
5. Patel A, Bhongade S (2019) Designing of load frequency controller for two-area interconnected power system using grey wolf optimization. In: 2019 international conference on computing, power and communication technologies (GUCON), pp 318–325
6. Bhadoria VS, Jaiswal SP, Tayal A, Singh J (2019) Firefly algorithm for frequency controller of autonomous hybrid energy system. In: 2019 2nd international conference on power energy, environment and intelligent control (PEEIC), pp 150–155
7. Izadkhast S, Garcia-Gonzalez P, Frías P (2015) An aggregate model of plug-in electric vehicles for primary frequency control. IEEE Trans Power Syst 30(3):1475–1482
8. Macana CA, Mojica-Nava E, Quijano N (2013) Time-delay effect on load frequency control for microgrids. In: 2013 10th IEEE international conference on networking, sensing and control (ICNSC), pp 544–549

9. Hua C, Wang Y (2020) Delay-dependent stability for load frequency control system via linear operator inequality. In: IEEE transactions on cybernetics, pp 1–9
10. Kariminejad SAT, Khateri K, Pourgholi M (2020) Decentralized optimal control of electric vehicles for primary frequency control in power systems. In: 2020 28th Iranian conference on electrical engineering (ICEE), pp 1–8

Design of a Second Order System with Additional Actuating Signal for Desired Output

Bipa Datta, Arnab Das, Rajesh Dey, and Achintya Das

Abstract Due to the system nonlinearities, the expected response of a control system could not be achieved. The control function used by the controller is adaptive in nature, in order to deal with the significant changes of the parameters of the system due to environment or due to internal system disturbances. This adaptive control mechanism is dependent upon the value of adaptation gain. For higher order systems, the available range of adaptation gain becomes very less. A second order control mechanism, on the basis of the application of the MIT rule, is used in the paper. In order to achieve the desired response from the control system, introduction of a proposed compensating signal which is in addition to the normal actuating signal. This additional signal is applied to the controller and output response of system is maintained unaltered, even though there are changes in the system itself, input noise, ambiance or others likely to occur in any actual real life system. This signal, which may be called the additional actuating signal (AAS), helps in controlling output in the control system, without disturbing the standard actuating signal and the systems working environment. This technique would be useful in systems where the system parameters including the inputs cannot be fully controlled, like in an autonomous or independent control system or a physiological control system. Simulation is done in Simulink and MATLAB software.

Keywords Additional actuating signal (AAS) · MIT rule · Adaptation gain · MRAC · Expected outcome

B. Datta (✉) · A. Das
Brainware Group of Institutions-S.D.E.T, Kolkata, West Bengal, India

R. Dey
Brainware University, Kolkata, West Bengal, India

A. Das
Kalyani Government Engineering College, Kalyani, West Bengal, India

J. K. Mandal et al. (eds.), *Topical Drifts in Intelligent Computing*, Lecture Notes in Networks and Systems 426, https://doi.org/10.1007/978-981-19-0745-6_52

1 Introduction

In the case of real world systems, although they behave as linear when operating near a certain operating regions, under certain assumptions, but they are essentially nonlinear in nature at least when considered over broad operating range. A control system [1–3] is essentially a cluster of connected components that can be made to achieve an expected output while in the effect of external disturbances. The 'expected outcome' could be the trailing of a specified dynamic trajectory, in which case the control system acts as a servomechanism.

The growing interest in nonlinear control systems is due to the development of linear control systems, study of hard nonlinearities, design simplicity and the need to deal with system model uncertainties [4, 5]. Nonlinear strategies improve insignificant approaches by taking into accounts the active forces, which vary in proportion to the square of the speed. A simple nonlinear controller [6, 7] can reasonably balance the nonlinear forces thus achieving higher speeds in a sufficient working zone. Also, hard nonlinearities inherent in a control system, viz. backlash, dead-zones, Coulomb friction, hysteresis, stiction and saturation [8, 9] do not permit a real world system to achieve a linear approximation.

After taking into account, these nonlinearities, nonlinear approaches appropriately compensate these to achieve expected outcome [10, 11]. Due to sudden or slow change in the parameter values, real systems often exhibit uncertainties in the model parameters [12]. A nonlinear controller through strength or flexibility can handle the consequences due to system uncertainties. Also, the systems that are not linearly controllable may be controllable in a nonlinear [13, 14] way.

For the paper, MRAC under the MIT rule is used for achieving the adaptive [15] response. In most of the system environments, it is unavoidable for the existence of definite types of non-linearity in control [16] system. A static non-linearity system is where there exists a nonlinear relation between output and input, but there does not exist a differential equation. Whereas a dynamic non-linearity, has the output and input related through a nonlinear differential equation [16, 17].

2 Simulated Model

The system performance is described by a reference model, and the controller is such intended, so as to compel the system to perform closely to the reference model. The reference output is compared to the output of the plant, and the resultant error is utilized to regulate the controller [18], as shown in Fig. 1.

Reference Model: It reflects the ideal response of the control system in expected response of a control system. For this paper, the reference model used is an under damped second order system.

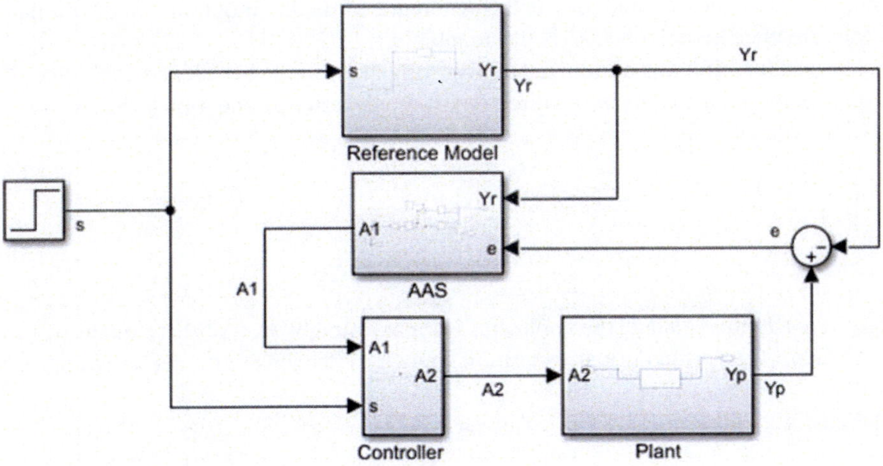

Fig. 1 Simulated model with AAS to maintain desired response

Controller: In this paper, the adjustable parameters $A_1$and A_2 are used to change the control algorithm of the adaptive law. The adaptation gain values in turn controls these parameters.

AAS: The introduction of an additional actuating signal (AAS) is proposed, in addition to the normal actuating signal, to control and stabilize the system for compensating the system to achieve the benchmark performance [13, 19]. This technique may find use in autonomous and independent process control systems and physiological control system, where the inputs and other system parameters cannot be fully controlled [19, 20]. Again, to an extent, the optimization hypothesis can be used to explain control strategies in a number of actual processes.

3 Discussion

Performance assessment of control system has been widely worked on. It is well known that the response of a system is likely to be deviated from desired one due to various reasons, like inherent parametric variations of the system, any changes occurring within the system due to arrival of any unwanted signal as may be due to noise, ambiance, etc. Other than the MIT rule, the design techniques in Lyapunov theory and MRAC system. In any practical system, MIT rule is used to apply the MRAC theory. In the MIT rule, the loss or cost function, is shown as

$$Z(A) = {e^2}/{2} \tag{1}$$

where output error, denoted as e, is the difference of the plant output and the output of the reference model, and 'A' is the adaptable.

The cost function was minimized by adjusting the variable 'A'. Hence, it is rational to modify the variable 'A' in the direction of the negative gradient of Z that is:

$$\mathrm{d}A/\mathrm{d}t = -\varsigma\,\partial Z/\partial \mathrm{A} \tag{2}$$

$$= -\varsigma e\,\partial e/\partial A \tag{3}$$

where the adaptive gain of the controller is represented by ς, a positive quantity.

The $\partial e/\partial A$, a partial derivative term, is known as the sensitivity derivative of the system.

Sign-sign algorithm:

$$\mathrm{d}A/\mathrm{d}t = -\varsigma\,\mathrm{sign}\left(\partial e/\partial A\right)\mathrm{sign}\,e \tag{4}$$

Or it can be selected as:

$$\mathrm{d}A/\mathrm{d}t = -\varsigma\left(\partial e/\partial A\right)\mathrm{sign}\,e \tag{5}$$

where

$$\mathrm{sign}\,e = \begin{cases} 0, & \text{for } e = 0 \\ -1, & e < 0 \\ 1, & \text{for } e < 0 \end{cases}$$

The option of adaptation gain is found to be critical in some applications and its value is dependent on the signal levels. So the rule might be changed as follows:

$$\mathrm{d}A/\mathrm{d}t = -\varsigma\,\xi e \tag{6}$$

where $\xi = \partial e/\partial A$

$$\mathrm{d}A/\mathrm{d}t = -\varsigma\xi\,e/(\beta + \xi^T\xi) \tag{7}$$

When $\xi^T\xi$ is small, to avoid division by zero, $\beta > 0$ is introduced. The rules as defined by Eqs. (1), (2) and (3) have been used in this paper for developing the control law.

4 Mathematical Analysis

In this work, MRAC using MIT rule [12], has been applied to a second order system. A second order under damped system gives an attribute with oscillations and this feature is close to the first order system. A second order under damped system has been used as a plant in this paper, which has a large settling time and high overshoot [13]. The performance of this system is to be improved by using adaptive control scheme. The second order system may be described by:

$$\mathrm{d}^2 Y_P/\mathrm{d}t^2 = -j\left(\mathrm{d}Y_P/\mathrm{d}t\right) - kY_P + kA_2 \quad (8)$$

Let $j = 6$ and $k = 400$.

Where the plant output (under damped second order system) Y_P and the output of the controller A_2 has been used.

The transfer function can be written as:

$$Y_P(s)/A_2(s) = 680/\left(s^2 + 6s + 400\right) \quad (9)$$

Likewise, the reference model can be described by:

$$\mathrm{d}^2 Y_R/\mathrm{d}t^2 = -j_r \mathrm{d}Y_R/\mathrm{d}t - k_r Y_R + k_r S \quad (10)$$

Take $j_r = 6$ and $k_r = 12$.

Where the reference input (unit step input) is S and the reference model output (under damped second order system) is Y_R. The transfer function can be written as:

$$Y_R(s)/S(s) = 16/\left(s^2 + 6s + 12\right) \quad (11)$$

The aim is to assess the error signal, which is the variation between the reference output (Y_R) and the actual output (Y_P), by applying MRAC, to refine the overall output. The controller may be shown as:

$$A_2(t) = A_1 S(t) - A_2 Y_P(t) \quad (12)$$

The selected controller parameters as follows:

$$A_1 = k_r/k \quad \text{and} \quad A_2 = (j_r - j)/k \quad (13)$$

The formula with MIT rule controller parameters can be described by:

$$\mathrm{d}A_1/\mathrm{d}t = -\varsigma\, \mathrm{e}\, \partial \mathrm{e}/\partial A_1$$

$$= -\varsigma\, e\, [kS/(p + j_r)]$$

$$= -\alpha\, e \left[{}^{j_r S}\!/_{p + k_r} \right] \tag{14}$$

$$ {}^{dA_2}\!/_{dt} = -\alpha\, e \left[{}^{j_r Y_P}\!/_{(p + j_r)} \right] \tag{15}$$

where error $e = Y_P - Y_R$ and adaptation gain $\alpha = \varsigma\, k/j_r$.

5 Simulation and Result

The simulation model as shown in Fig. 1 is simulated in SIMULINK and MATLAB. The performance track for different values of adaptation gain α are as shown in Figs. 2, 3, 4, 5, 6, 7, 8 and 9. It can be seen that the plant output Y_P is oscillatory in nature. In each case, it may be noticed from the figures that control parameters merge to a constant value. For different values of adaptation gain, the variation of AAS signal and error signal with time are shown in Figs. 10 and 11.

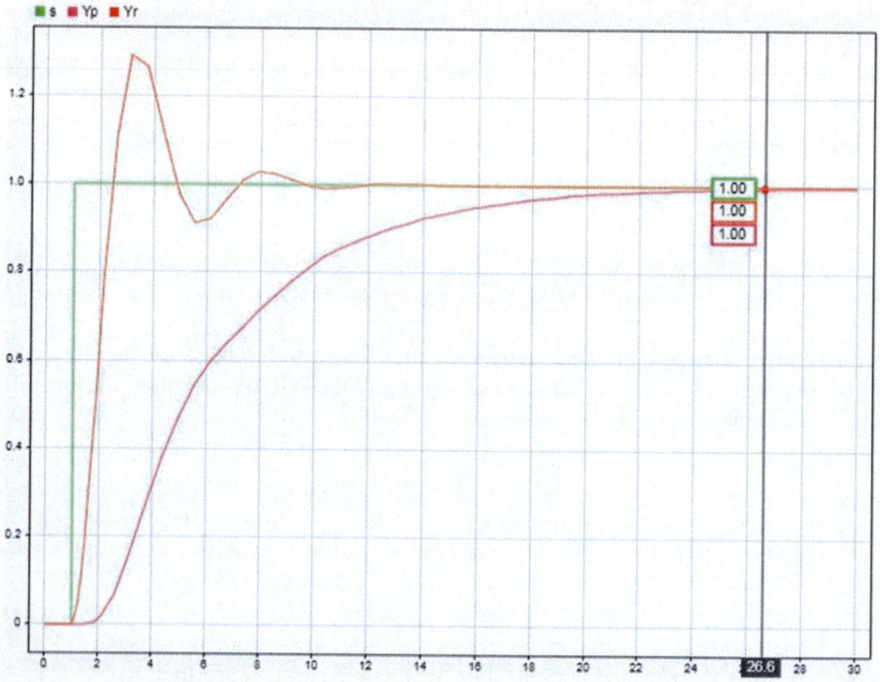

Fig. 2 Performance tracking for reference signal, $S\,(t) = 1$, adaption gain $\alpha = 1.5$

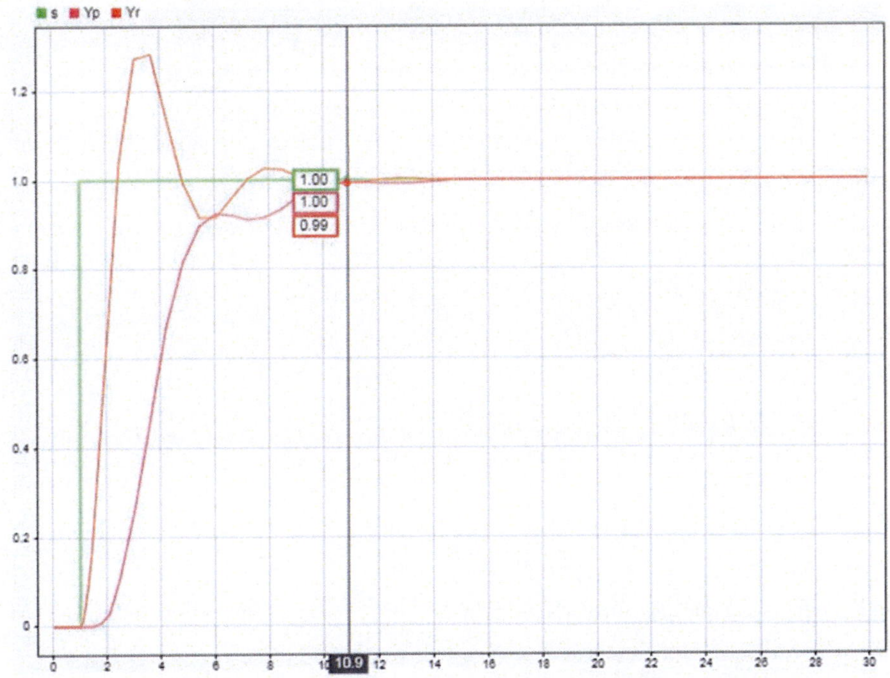

Fig. 3 Performance tracking for reference signal, $S(t) = 1$, adaption gain $\alpha = 2.0$

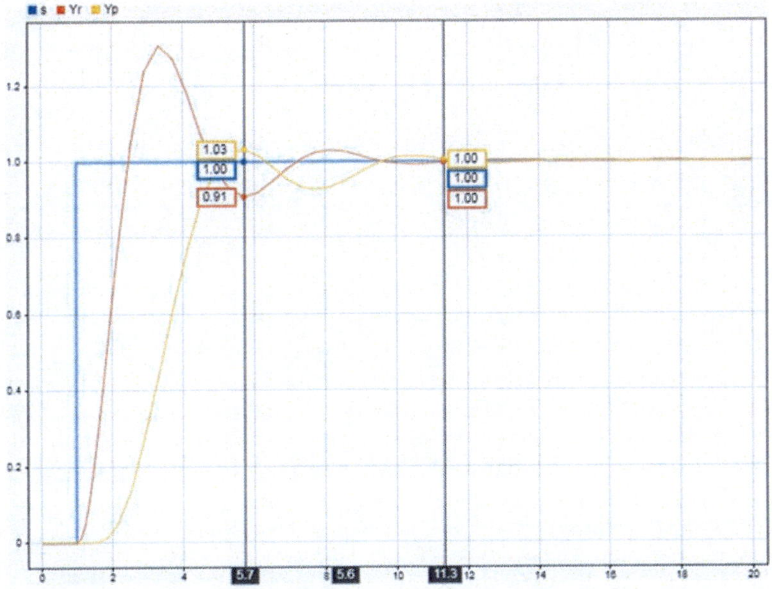

Fig. 4 Performance tracking for reference signal, $S(t) = 1$, adaption gain $\alpha = 2.5$

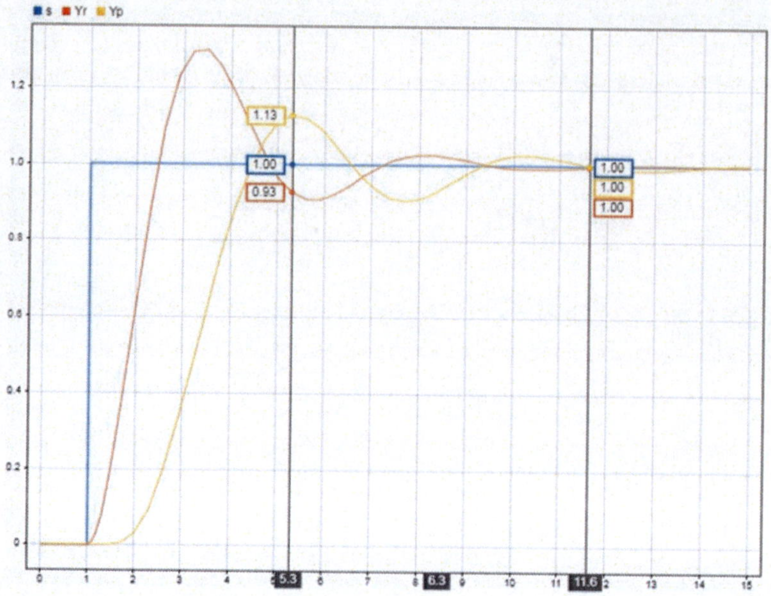

Fig. 5 Performance tracking for reference signal, $S(t) = 1$, adaption gain $\alpha = 3.0$

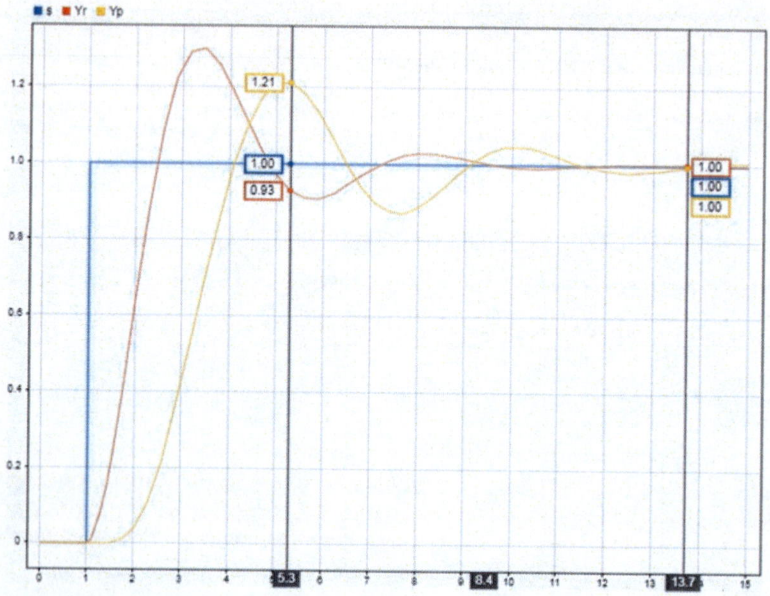

Fig. 6 Performance tracking for reference signal, $S(t) = 1$, adaption gain $\alpha = 3.5$

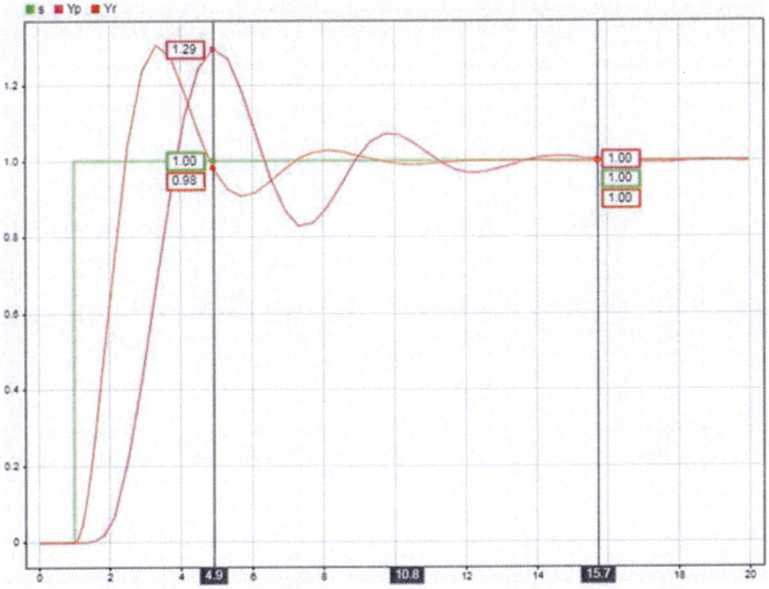

Fig. 7 Performance tracking for reference signal, $S(t) = 1$, adaption gain $\alpha = 4.0$

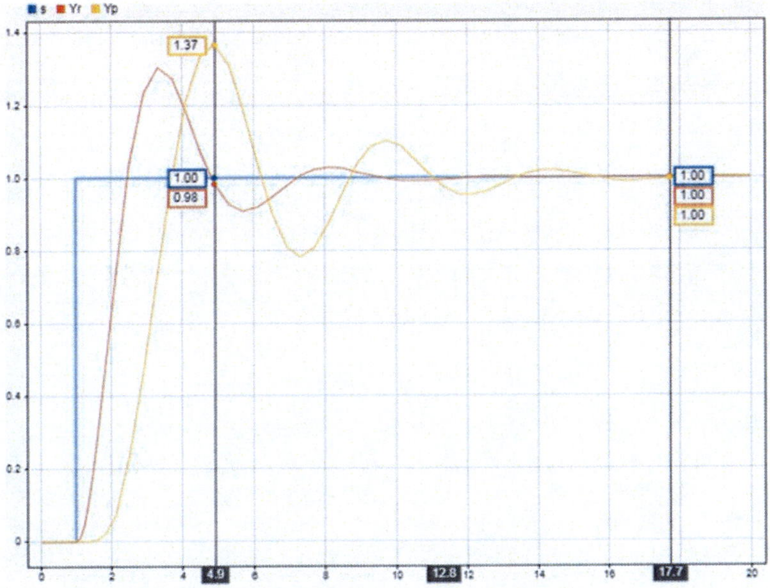

Fig. 8 Performance tracking for reference signal, $S(t) = 1$, adaption gain $\alpha = 4.5$

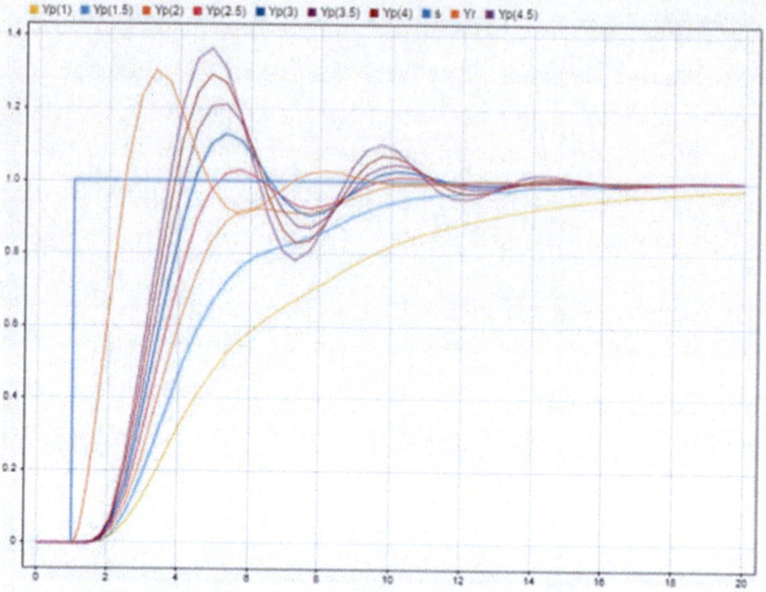

Fig. 9 Performance tracking for reference signal, $S\ (t) = 1$, adaption gain $\alpha = 1.5, 2.0, 2.5, 3.0, 3.5, 4.0, 4.5$

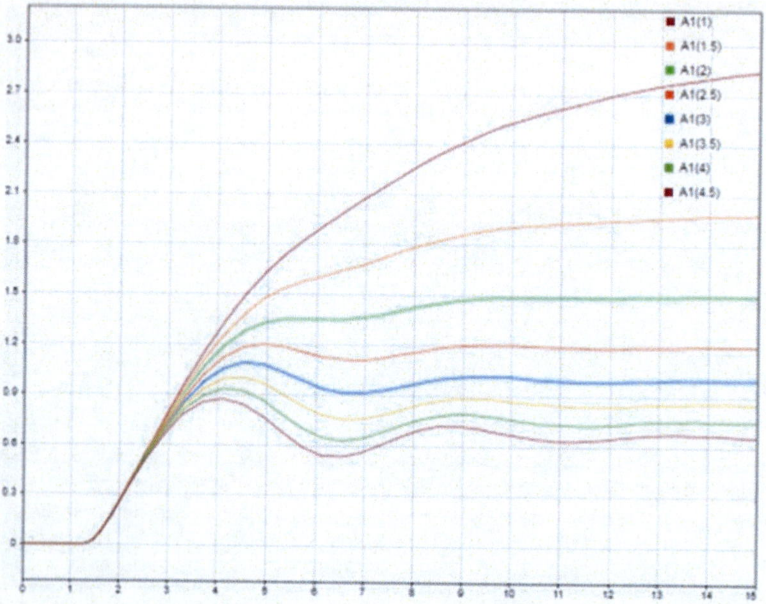

Fig. 10 Tracking AAS for reference signal, $S\ (t) = 1$, adaption gain $\alpha = 1.5, 2.0, 2.5, 3.0, 3.5, 4.0, 4.5$

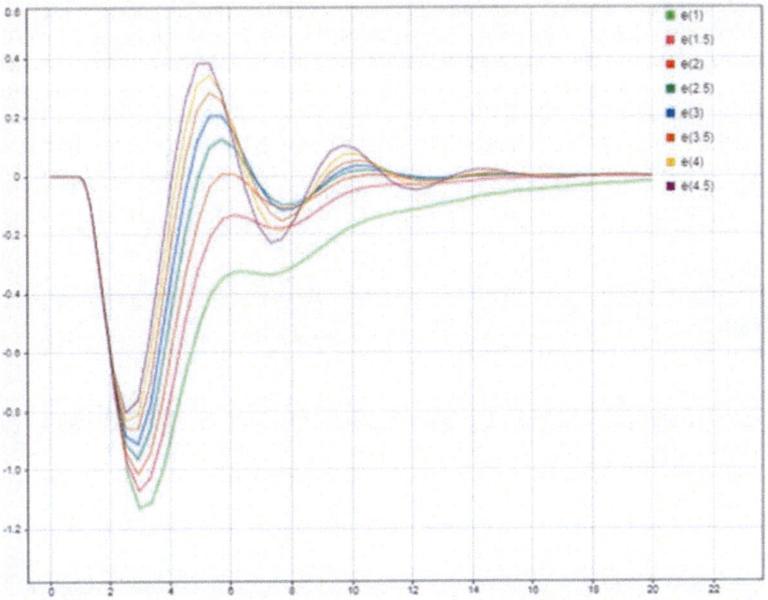

Fig. 11 Tracking error signal for reference signal, $S(t) = 1$, adaption gain $\alpha = 1.5, 2.0, 2.5, 3.0, 3.5, 4.0, 4.5$

6 Conclusion

This paper focuses on study the control systems ability to maintain expected outcome of a control system despite there being various changes in input or in the system. In the case of almost all real life examples of control systems, the system is inherent to disturbances which are nonlinear in nature. This problem aggravates more so, in the case of autonomous or independent systems, where the inputs to the system cannot be fully controlled. The introduction of an additional actuating signal (AAS) in addition to the normal actuating signal, helps in controlling output in the control system, without affecting the standard actuating signal and its working environment. It is clear from the results that the adaptive controllers are effective in handling the frequent parameter variations and environmental changes. The controller parameters were adjusted by changing the adaptation gain of the system, to get a preferred closed-loop performance of the system. As the output of the plant follows the output of the reference model, we can see from Fig. 11, the tracking error settles to zero. In Figs. 3, 4 and 5, it can been seen that corresponding to the value of adaptation gain the settling time is considerably reduced. System performance becomes poor beyond the range of adaptation gain ($2 < \alpha < 3$). This technique can be useful in any process control system, including nonlinear ones for maintaining the expected performance of the system. With the introduction of this additional actuating signal (AAS), it is expected to be suitable for control and compensation of a nonlinear system also.

Acknowledgements I cannot express enough thanks to my Ph.D. guide and supervisor Dr. Achintya Das. Without his guidance, continuous support and help, this paper and the research behind it would not have been possible. His valuable information and proper guidance to detail have been a motivation and kept my work on track. For this, I would like to be ever thankful to him. I also offer my sincere appreciation to M.A.K.A.U.T., West Bengal, for giving me an opportunity for the Ph.D. program and the learning opportunities extended to me during the course of this research work. Also I would express my sincere thanks to Brainware Group of Institutions-S.D.E.T., for providing me the time and space for the related works of this research work.

References

1. Spong MW, Vidyasagar M (2008) Robot dynamics and control. John Wiley & Sons
2. Ljung L (1999) System identification. Wiley encyclopedia of electrical and electronics engineering, pp 1–19
3. Poznyak AS (2009) Advanced mathematical tools for automatic control engineers: stochastic techniques. Elsevier Science
4. Bitmead R (1984) Persistence of excitation conditions and the convergence of adaptive schemes. IEEE Trans Inf Theory 30(2):183–191
5. Chen B, et al (2013) System parameter identification: information criteria and algorithms. Newnes
6. Åström KJ, Wittenmark B (1973) On self tuning regulators. Automatica 9(2):185–199
7. Tsakalis K, Ioannou P (1987) Adaptive control of linear time-varying plants. Automatica 23(4):459–468
8. Slotine JJE, Li W (1991) Applied nonlinear control, vol 199, No 1. Englewood Cliffs: Prentice hall
9. Ioannou PA, Sun J (1996) Robust adaptive control PTR Prentice-Hall Upper Saddle River
10. Narendra KS, Annaswamy AM (2012) Stable adaptive systems. Courier Corporation
11. Chen J, Chen W, Sun J (2017) Smooth controller design for non-linear systems using multiple fixed models. IET Control Theory Appl 11(9):1467–1473
12. Fan QY, Yang GH, Ye D (2018) Quantization-based adaptive actor-critic tracking control with tracking error constraints. IEEE Trans Neural Netw Learning Syst 29(4):970–980
13. Das A, Bag R, Nath NG (2006) A modification to realize dead-beat performance of control systems—signal correction technique. IEEE Trans Instrum Measur 55(5): 1546–1550
14. Qiao L, Zhang W (2018) Double-loop integral terminal sliding mode tracking control for UUVs with adaptive dynamic compensation of uncertainties and disturbances. IEEE J Oceanic Eng 99:1–25
15. Gaing Z-L (2004) A particle swarm optimization approach for optimum design of PID controller in AVR system. IEEE Trans Energy Convers 19(2):384–391
16. Liu T, García P, Chen Y, Ren X, Albertos P, Sanz R (2018) New predictor and 2dof control scheme for industrial processes with long time delay. IEEE Trans Industr Electron 65(5):4247–4256
17. Krstic M, Kokotovic PV, Kanellakopoulos I (1995) Nonlinear and adaptive control design. John Wiley & Sons, Inc.
18. Bobtsov AA, Pyrkin AA, Kolyubin SA (2014) Simple output feedback adaptive control based on passification principle. Int J Adapt Control Signal Process 28(7–8):620–632
19. Das A (2008) Advance control systems, 2nd ed. Matrix Educare, pp 286–287
20. Astrom KJ, Wittenmark B (2008) Adaptive control, Dover publications. INC. Mineola, New York

Design Approach for Online Parameter Estimators for Unknown Two-Parameter First-Order Scalar Plant

Arnab Das, Bipa Datta, Rajesh Dey, and Achintya Das

Abstract In practice, all the systems posses' some type of nonlinear dynamics for the intentional and inherent imperfection or properties of the system. Due to different dynamics, a nonlinear system exhibits some nonlinear phenomena, which are inevitable in any process. An approximate behavior of the nonlinear system can provide by the linearized system. But, often linearized models are inadequate or insufficient for analyzing the overall system behavior. To achieve the desired response, system parameter estimation is very much important in the face of external disturbances. For a specific value of initial conditions, structural model identification depends on the system's dynamical beaver, output, and input. The online parameter identification process is proposed in this paper for continuous-time plants to characterize the input required for structural identification. Generalization of nonlinear observability and incorporating extended lie is considered for structural identifiability derivation. The methodology evaluates structural identifiability for time-variant system inputs, and moreover, it can be used to limit the input profile that is required to design an expected system. MATLAB 'SIMULINK' software tools for model simulation and analysis are more effective for the proposed work.

Keywords Control systems · Online parameter identification · Parameter estimation · and Adaptation law

1 Introduction

To analyze dynamic system performance, the study of system trajectory performance is important. However, for any kind of practical or dynamic system, e.g., electrical, mechanical, electronic, or another, it is quite difficult to access or measured all

A. Das (✉) · B. Datta
Brainware Group of Institutions-S.D.E.T, Kolkata, West Bengal, India

R. Dey
Brainware University, Kolkata, West Bengal, India

A. Das
Kalyani Government Engineering College, Kalyani, West Bengal, India

J. K. Mandal et al. (eds.), *Topical Drifts in Intelligent Computing*, Lecture Notes in Networks and Systems 426, https://doi.org/10.1007/978-981-19-0745-6_53

information whose are used for identifying the actual estimated value [1–3]. So, for any control theory or a data processing system, successful parameters identification algorithms selection, in addition to trajectory estimation proceed of unmeasured state-space variables, play important roles [1–7]. So presently, the nonlinear systems modeling is an important research domain for researchers. An effective approach for nonlinear systems modeling the MRAC by Tsakalis et al. [3]. On another side, Narendra et al. planned [4] a steady adaptable system for dynamic systems. K. J. Astrom discusses the adaptation mechanism which follows the reference model output for analysis point of view [5]. A control system with adaptive-based feedback is proposed by Bobstov et al. [6]. For best execution and precision, one can utilize an adaptive control with suitable techniques to design a control plant [1–7]. MRAC is an instantaneous adaptable mechanism with some adapts controller parameters and changing elements to modify, to get desired response.

This paper discusses an adaptation mechanism applying the MRAC model by a typical adjustable adaptive controller. Presents design methodology consider a reference model for the proposed first-order system. The controller parameter enhances the system response by following the reference model. So, no alternation will occur in output response due to the reason for any disturbance, i.e., it will maintain desired output.

2 Discussion and Analysis

The abovementioned process is recognized as the closed-loop system, where by using a feedback mechanism as a reference linear time-invariant controller, a nonlinear system or plant is closed. The proposed paper approaches first by theoretical study and then analysis by MATLAB simulation for a controller-based two-parameter case first-order nonlinear system model. We considered the Lyapunov function for system stability analysis. Online parameter identification-based proposed work follows by the three procedure steps:

First Step: Signify a practical system or plant in the 'static' parametric model (SPM) form, i.e.,

$$z = \theta^{*^T} \emptyset,$$

where $z \varepsilon \Re$, $\emptyset \varepsilon \Re^n$ are accessible by measurement signals, and the vector contain $\theta^* \varepsilon \Re^n$ is with all the unknown parameters.

Second Step: Estimates θ of θ^* by proper estimation model. The difference between the estimation values is the estimation error, i.e., represented by the estimated parameters $\theta(t)$, and the parametric models output response where the estimated parameters $\theta(t)$ define how much distorted from the unknown parameters θ^* on the basis of some reference signal vector where by the adaptive law, online generates $\theta(t)$ is a function of the estimation error, represented as the follows:

$$\dot{\theta} = H(t)\varepsilon,$$

where $H(t)$ and ε are the time-varying measurement-based gain vector and the estimation error, respectively. Different adaptive laws with different ε and $H(t)$ may be designed by Lyapunov-type stability arguments and optimization techniques.

Third Step: Proposed to identify parameters and set up conditions to ensure that with time, $\theta(t)$ converges to θ^*. In parametric model, the input of the plant is defined as the signal vector $\emptyset(t)$ because it can observe or measure and the adaptive law which holds reasonable data with reference to the unknown parameters can be constrained by the observed or measured signals.

Now, we consider a two-parameter case first-order plant model, for signifying by the three design steps:

$$\left.\begin{array}{l} \frac{dy}{dt} = -ay + bu \\ y = \frac{b}{s+a}u \end{array}\right\}, \quad (1)$$

where a and b are two unknown constants, and assume that y, $\dot{y}$, u are obtainable by the measurement process. By creating a suitable online estimator, we can identify the unknown parameters a and b for the two-parameter case first-order plant model.

First Step: Approach for Parametric Model:

We can express (1) in the SPM form, as y, $\dot{y}$ are available, so

$$z = \theta^{*^T}\emptyset,$$

where $\emptyset$ and z are accessible for measurement and related as:

$$z = \dot{y}, \theta^* = [b, a]^T, \emptyset = [u, \ -y]^T.$$

Second Step: Parameter Identification (PI) Algorithm:

Representation of estimation model as

$$\hat{z} = \theta^T\emptyset,$$

where at time t, $\theta(t)$ is the estimate of θ^*.

Now, estimation error represent as

$$\varepsilon = \frac{z - \hat{z}}{m_s^2} = \frac{z - \theta^T\emptyset}{m_s^2}, \quad (2)$$

where m_s is the normalizing signal such that $\frac{\emptyset}{m_s} \in \mathcal{L}_\infty$. Simply, choice for m_s is $m_s^2 = 1 + \alpha\emptyset^T\emptyset$ for any $\alpha > 0$.

Adaptive Law: To minimize the cost, we use the gradient method

$$J(\theta) = \frac{\varepsilon^2 m_s^2}{2} = \frac{\left(z - \theta^T \emptyset\right)^2}{2m_s^2} = \frac{(z - \theta_1 \emptyset_1 - \theta_2 \emptyset_2)^2}{2m_s^2},$$

where $\emptyset_1 = u, \emptyset_2 = --- y$, and set

$$\dot{\theta} = -\Gamma \nabla J,$$

where

$$\nabla J = \left[\frac{\partial J}{\partial \theta_1}, \frac{\partial J}{\partial \theta_2}\right]^T,$$

where adaptive gain, $\Gamma = \Gamma^T > 0$, and θ_1, θ_2 are the elements of $\theta = [\theta_1, \theta_2]^T$. Since

$$\frac{\partial J}{\partial \theta_1} = -\frac{\left(z - \theta^T \emptyset\right)}{m_s^2} \emptyset_1 = -\varepsilon \emptyset_1,$$

$$\frac{\partial J}{\partial \theta_2} = -\frac{\left(z - \theta^T \emptyset\right)}{m_s^2} \emptyset_2 = -\varepsilon \emptyset_2,$$

we have

$$\dot{\theta} = \Gamma \varepsilon \phi, \theta(0) = \theta_0, \tag{3}$$

which define as the adaptive law for uploading $\theta(t)$ initiating from some initial condition $\theta(0) = \theta_0$.

Third Step: Stability Analysis and Parameter Convergence:

Now, the parameter error $\tilde{\theta} = \theta - \theta^*$ is calculated from (2) and (3) by noting that

$$\varepsilon = \frac{z - \theta^T \emptyset}{m_s^2} = \frac{\theta^{*^T} \phi - \theta^T \phi}{m_s^2} = -\frac{\tilde{\theta}^T \phi}{m_s^2} = -\frac{\phi^T \tilde{\theta}}{m_s^2} \tag{4}$$

and $\dot{\tilde{\theta}} = \dot{\theta}$, i.e.,

$$\dot{\tilde{\theta}} = \Gamma \phi \varepsilon = -\Gamma \frac{\phi \phi^T}{m_s^2} \tilde{\theta}. \tag{5}$$

So from (5), the stability of the equilibrium $\tilde{\theta}_e = 0$ will actually depend on the properties of the time-varying matrix $-\Gamma\frac{\phi\phi^T}{m_s^2}$, i.e., consecutively depends on the properties of ϕ. Now, assume that $a > 0$, i.e., the plant is stable. If we select $m_s^2 = 1, \Gamma = \gamma I$ for some $\gamma > 0$ and a constant input $u = c_0 > 0$, then at steady state $y = c_1 \triangleq \frac{c_0 b}{a} \neq 0$ and $\phi = [c_0, -c_1]^T$, giving

$$-\Gamma\frac{\phi\phi^T}{m_s^2} = -\gamma\begin{bmatrix} c_0^2 & -c_0c_1 \\ -c_0c_1 & c_1^2 \end{bmatrix} \triangleq A,$$

i.e.,

$$\dot{\tilde{\theta}} = A\tilde{\theta},$$

where constant matrix, A with eigenvalues 0 and $-\gamma\left(c_0^2 + c_1^2\right)$, which characterize that the equilibrium $\theta_e = 0$ system is only slightly stable. That is $\tilde{\theta}$ bounded but does not essentially converge to 0 as $t \to \infty$. So, plant input, u, will be chosen in such a way that at the equilibrium $\tilde{\theta}_e = 0$ the system will be guaranteed as exponential stable for certain properties of ϕ. Represent as

$$\phi = H(s)u,$$

where $H(s) = \left[1, -\frac{b}{s+a}\right]^T$ for this example. For the equilibrium $\tilde{\theta}_e = 0$ of (5), the appropriate choice of u can guarantee that implies exponential stability. For the equilibrium point, position exponential stability happens by $\theta(t)$ converges to θ^* exponentially fast. So, a stable input $u = c_0 > 0$ doesn't ensure exponential stability for the two-parameter case.

3 MATLAB Simulation Results

In this paper, a closed-loop control system proposed to control an unknown two-parameter first-order plant and simulated by MATLAB (Simulink) software. By comparing with (1), the first-order unstable plant is represented as:

$$\frac{\mathrm{d}y_p}{\mathrm{d}t} = -a_p y_p + b_p u$$

where u and y_p are defined as the system input and output, respectively, b_p and a_p are the unknown system or plant parameters. Consider, the unknown plant parameter, $a_p = 3$ and $b_p = 3.9$.

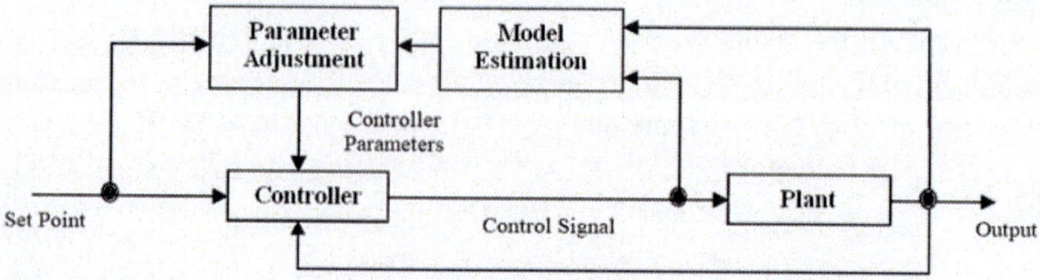

Fig. 1 Structure diagram for model reference adaptive controller

$$\frac{dy_p}{dt} = -3y_p + 3.9u$$

Now, a reference model can contrasted with (1), where y_m and r are the reference model output and input, respectively, and the plant reference parameter, $a_m = 5$ and $b_m = 5$:

$$\frac{dy_m}{dt} = -a_{\mathrm{m}}y_m + b_m r$$
$$\frac{dy_m}{dt} = -5y_m + 5r$$

The block diagram for an adaptive controller and adjustment mechanism-based control system is shown in Fig. 1.

For estimates, the unknown system parameter, with the help of the estimation model, the controller can develop an estimation error which is essential to confine the adaptive law. A step signal can consider as a reference signal or set point. We consider the control law as $u = \hat{a}_r r + \hat{a}_y y_p$, where $\hat{a}_r$ and $\hat{a}_y$ are variable feedback gain used for generating adaptation law. So, parameter estimation can be monitor by considering $\hat{a}_r$ and $\hat{a}_y$.

The simulated results, which are based on MATLAB, involve tracking errors between a two-parameter case of a first-order uncertain plant, and a reference model and continuous-time-domain response. From an observation point of view, the comparative time-domain response diagram is shown (see in Fig. 2). Figure 3 shows a defined comparative study for varying the adaption gain or reference signal. We can study the transient response of the system and observe system response by illustrating in Case 1, Case 2, and Case 3, respectively. Whereby, increasing the adaption gain value, overall response can be improved.

Case 1:

Initially consider the adaption gain, $\gamma = 0.1$, reference signal, $u(t) = 4$, and consider $y_p(0) = y_m(0) = 0$.

Case 2:

Let the adaption gain, $\gamma = 0.5$, reference signal, $u(t) = 4$ and consider $y_p(0) = y_m(0) = 0$.

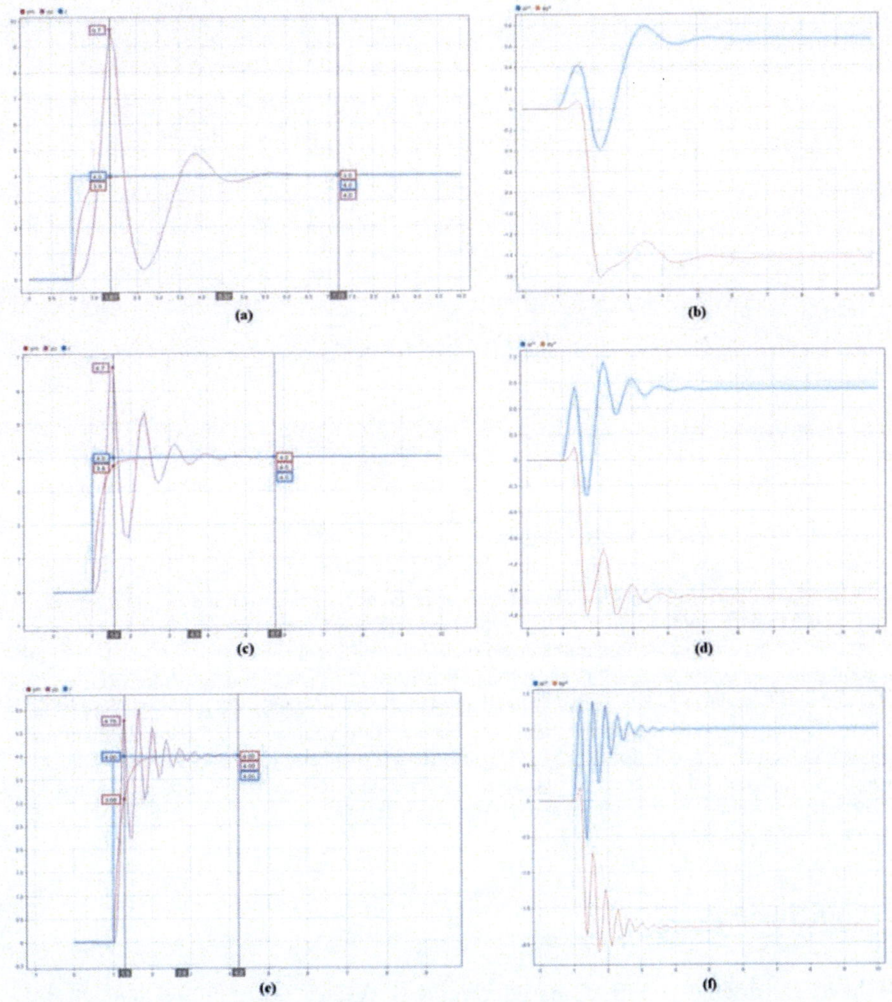

Fig. 2 **a** System performance and **b** system parameter estimation. **c** System performance and **d** system parameter estimation. **e** Tracking performance and **f** parameter estimation

Case 3:

Let the reference signal, $u(t) = 4$, adaption gain, $\gamma = 2.5$, and consider $y_p(0) = y_m(0) = 0$.

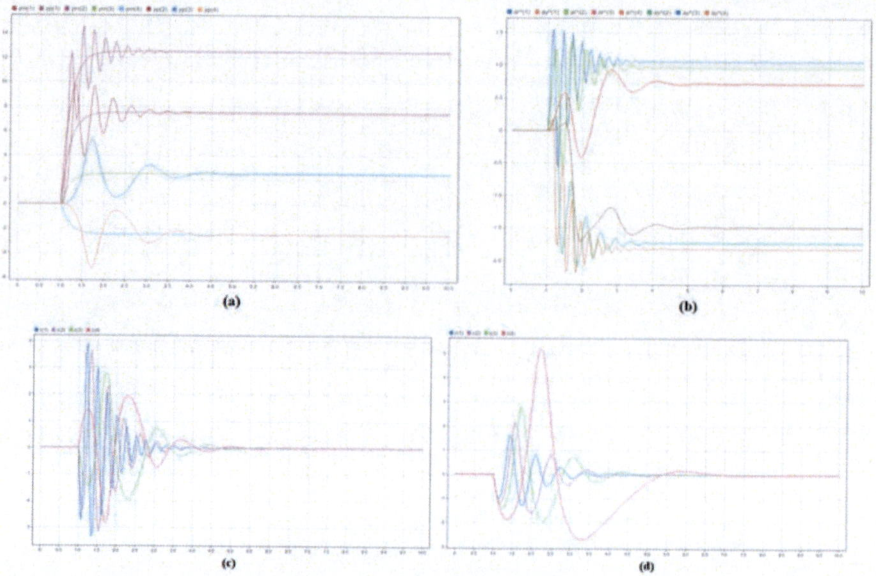

Fig. 3 **a** System performance for different step or reference signal, $u(t)$ and constant adaption gain, $\gamma = 0.5$ (where, yp(1) for $u(t) = 12.5$; yp(2) for $u(t) = 7.5$; yp(3) for $u(t) = 2.5$; and yp(4) for $u(t) = -2.5$.) and **b** system parameter estimation for different step signal, $u(t)$ and constant adaption gain, $\gamma = 0.5$ (where, ar^(1) and ay^(1) for $u(t) = 12.5$; ar^(2) and ay^(2) for $u(t) = 7.5$; ar^(3) and ay^(3) for $u(t) = 2.5$; and ar^(4) and ay^(4) for $u(t) = -2.5$.). **c** System tracking errors (ε) for different reference signal, $u(t)$ and constant adaption gain, $\gamma = 0.5$ between plant and reference model (where, $\varepsilon(1)$ for $u(t) = 12.5$; $\varepsilon(2)$ for $u(t) = 7.5$; $\varepsilon(2)$ for $u(t) = 2.5$; and $\varepsilon(3)$ for $u(t) = -2.5$.) and **d** system tracking errors (ε) constant reference signal, $u(t) = 2.5$ and for various adaption gain, γ between plant and reference model (where, $\varepsilon(1)$ for $\gamma = 2$; $\varepsilon(2)$ for $\gamma = 1$; $\varepsilon(3)$ for $\gamma = 0.5$ and $\varepsilon(4)$ for $\gamma = 0.1$).

4 Conclusion

We have considered a MRAC-based proposed system for stability analysis by Lyapunov approach, which can be appropriate for an unknown two-parameter case of first-order nonlinear systems in this paper. The adaptive law has to choose in such a way that both the speed of adaptation $\dot{\theta}$ and the parameter estimate $\theta(t)$ are bounded. And by this way, the estimation error ε can decreases with time. However, the parameter estimate $\theta(t)$ will get nearer to θ^* with time only for a few specific conditions on the vector $\emptyset(t)$. That particular vector $\emptyset(t)$ is specified as the regress or vector. For different adaptations gain, some specific numerical recreations can improve the planned control strategy.

It can observe by MATLAB simulated results for the proposed system that the simultaneous increasing of adaption gain value for a constant reference input can improve on the transient response. The gradient of the loss criterion relating to the estimated variables provides an improved system performance. Though, by this approach, system stability or convergence is not ensured.

The proposed MRAC system can be appropriate under upper norm bounded conditions for system uncertainties. This implies that for the equilibrium $\theta_e = 0$, the system will be only marginally stable, i.e., does not necessarily $\tilde{\theta} = 0$ as $t \to \infty$ for bounded values of $\tilde{\theta}$ where the appropriate choice of u is one of the important properties of ϕ, or it can guarantee the equilibrium $\tilde{\theta}_e = 0$ is exponentially stable. So, from the simulation result study, we can also conclude that for the unknown two-parameter first-order system, which does not assurance exponential stability for a constant input $u = c_0 > 0$. But, to a certain extent, these also make a sense that unknown parameters are maybe the reason for nonlinearity, e.g., inherent or purposefully introduced in the system, can be exponential stable and by proposed control law response will follow the reference signal.

Acknowledgments I offer my sincere appreciation to Maulana Abul Kalam Azad University of Technology (M.A.K.A.U.T.), West Bengal, for providing me an opportunity for the Ph.D. program. I would like to convey my special thanks of gratitude to my Ph.D. guide and supervisor Dr. Achintya Das. Without his guidance, continuous support, and help, this paper which is also an integral part of my research work would not have been possible. Our research work has been gradually realized thanks to the outstanding support of many esteemed individuals. One of them is Mr. Prasun Banerjee, Senior Account Manager of East at The MathWorks, and I am grateful for his unequivocal support to use MATLAB-based Simulink. Also, I would express my sincere thanks to Brainware Group of Institutions—S.D.E.T., for providing me the time and space for the related works of this research work.

References

1. Ljung L (1999) System identification. Wiley encyclopaedia of electrical and electronics engineering, pp 1–19
2. Das A, Datta B, Das A (2021) Design approach for parameter estimators for one-parameter first-order scalar plant. In: 2021 devices for integrated circuit (DevIC). IEEE
3. Tsakalis K, Ioannou P (1987) Adaptive control of linear time-varying plants. Automatica 23(4):459–468
4. Narendra KS, Annaswamy AM (2012) Stable adaptive systems. Courier Corporation
5. Astrom KJ, Wittenmark B (2008) Adaptive control, Dover publications. INC. Mineola, New York
6. Bobtsov AA, Pyrkin AA, Kolyubin SA (2014) Simple output feedback adaptive control based on passification principle. Int J Adapt Control Signal Process 28(7–8):620–632
7. Ordaz P et al (2021) Parameter identification based on nonlinear observer for mechanical systems. J Comput Nonlinear Dyn 16(2)

Performance Assessment of Hybrid Triple-Tied BIPV Array Configurations for Maximising Power Output Under Patterns of Partial Shading

Debayan Sarkar and Pradip Kumar Sadhu

Abstract Building integrated photovoltaic (BIPV) installations in modern buildings necessitate a large BIPV array (LBiAr) for both off-grid and on-grid system applications. The problem of LBiAr is partial shading (PS) that reduces considerable power output (PO), which happens due to various reasons. To mitigate PS and to maximise PO of LBiAr, one of the effective solutions is the arrangement of fixed BIPV array configurations (BIPV-ArCns). This research paper proposes four different 5×6 fixed hybrid (Hbr) triple-tied (TrTd) BIPV array configurations (BIPV-ArCns), such as series–parallel triple-tied (Hbr-SePl TrTd), bridge-linked triple-tied (Hbr-BdLk TrTd), honey-combed triple-tied (Hbr-HnCb TrTd), and ladder triple-tied (Hbr-Ld TrTd) to improve the PO. The design, simulation and analysis of the proposed Hbr TrTd BIPV-ArCns under four diverse patterns of PS are effectively executed in Matlab/Simulink. Finally, the evaluation of performance for the Hbr TrTd BIPV-ArCns is compared and judged in respect of global maximum-power (Gl-MxPo), mismatch-loss (MsLo), fill-factor (FlFc) and efficiency (Efcy).

Keywords Large BIPV array (LBiAr) · Partial shading (PS) · Power output (PO) · BIPV array configurations (BIPV-ArCns) · Hybrid (Hbr) · Triple-tied (TrTd)

1 Introduction

Presently, building integrated photovoltaic (BIPV)-based solar power generation systems are becoming extensively popular amongst photovoltaic (PV)-based technologies [1]. The essential components of BIPV systems are BIPV modules. The BIPV modules are incorporated mainly in building structures such as (rooftops, facades, skylights, balconies, windows, etc.) [2]. The power rating (PRt) of an individual BIPV module is significantly less. Hence, the BIPV modules are interconnected to form a large BIPV array (LBiAr) with sufficient power output (PO) that can be utilised for both off-grid and on-grid system applications. The problem of

D. Sarkar (✉) · P. K. Sadhu
Department of Electrical Engineering, Indian Institute of Technology (ISM), Dhanbad 826004, India

J. K. Mandal et al. (eds.), *Topical Drifts in Intelligent Computing*, Lecture Notes in Networks and Systems 426, https://doi.org/10.1007/978-981-19-0745-6_54

LBiAr is partial shading (PS) that reduces considerable PO, which happens due to various reasons (neighbouring or same building, building structure, trees, clouds, electric pole, bird droppings, dust, etc.) [3]. To mitigate PS and to maximise PO of LBiAr, one of the effective solutions is the arrangement of fixed BIPV array configurations (BIPV-ArCns) or PV array configurations (PV-ArCns) [4]. Different conventional (Convnl) fixed PV-ArCns are available in the literature. The Convnl PV-ArCns include series–parallel (SePl), bridge-linked (BdLk), honey-combed (HnCb), ladder (Ld), triple-tied (TrTd) and total-cross-tied (TtCrTd) [5–8]. Almost all researchers proved that TtCrTd is the best PV-ArCn in terms of maximum PO during PS, but it has some disadvantages like the requirement of extensive wiring resulting in more wiring cost [8]. To overcome this problem, TrTd PV-ArCn is proposed by [9]. TrTd PV-ArCn is the second-best amongst the Convnl PV-ArCns. The other Convnl PV-ArCns like SePl, BdLk, HnCb and Ld gives less PO compared to TrTd and TtCrTd in conditions of PS. Hence, to maximise the PO as well as reducing the wiring cost, hybrid (Hbr) triple-tied (TrTd) BIPV array configurations (BIPV-ArCns) are proposed. The main contributions of this research paper are design, simulation and analysis of four different 5 × 6 fixed hybrid (Hbr) triple-tied (TrTd) BIPV array configurations (BIPV-ArCns), such as series–parallel triple-tied (Hbr-SePl TrTd), bridge-linked triple-tied (Hbr-BdLk TrTd), honey-combed triple-tied (Hbr-HnCb TrTd), and ladder triple-tied (Hbr-Ld TrTd) to improve the PO under four diverse patterns of PS. Finally, the evaluation of performance for the Hbr TrTd BIPV-ArCns is compared and judged in respect of global maximum-power (Gl-MxPo), mismatch-loss (MsLo), fill-factor (FlFc) and efficiency (Efcy).

2 Modelling of 5 × 6 Hybrid Triple-Tied BIPV Array Configurations

In this research paper, 54 W Sun Energy tile (poly-crystalline BIPV module) with model number BIPV054-T86 [10] is utilised for modelling the proposed 5 × 6 fixed Hybrid (Hbr) Triple-Tied (TrTd) BIPV array configurations (BIPV-ArCns). The electrical specifications of 54 W BIPV modules are (P_{peak} = 53.976 W, V_{peak} = 6.92 V, I_{peak} = 7.8 A, V_{opckt} = 8.87 V and I_{shckt} = 8.22 A). The mechanical specifications of 54 W BIPV modules are (L = 1160 mm, W = 336 mm and A = 0.39 m^2). Modelling and explanation of four different proposed 5 × 6 fixed Hbr TrTd BIPV-ArCns are performed in this section. The large BIPV array (LBiAr) comprises thirty (5 × 6 = 30) number of 54 W Sun Energy tile BIPV modules. The thirty different 54 W BIPV modules are arranged in a 5 × 6 structure of a matrix with five rows and six columns. The Hbr TrTd BIPV-ArCns that are modelled and explained include series–parallel triple-tied (Hbr-SePl TrTd), bridge-linked triple-tied (Hbr-BdLk TrTd), honey-combed triple-tied (Hbr-HnCb TrTd) and ladder triple-tied (Hbr-Ld TrTd). The diagram of connection for four different Hbr TrTd BIPV-ArCns is showcased in Fig. 1.

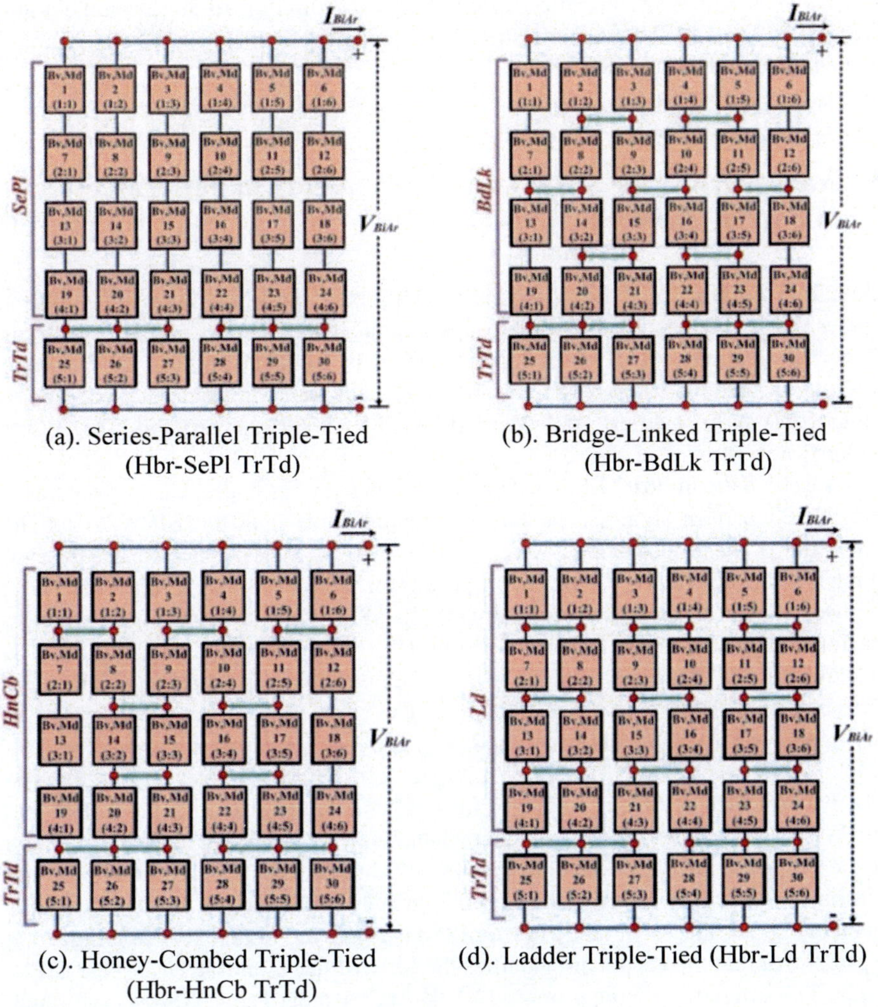

Fig. 1 Diagram of connection for four different Hbr TrTd BIPV-ArCns

The diagram of the connection for 5 × 6 fixed Hbr-SePl TrTd BIPV-ArCn is showcased in Fig. 1a. From Fig. 1a, it is prominent that the Hbr-SePl TrTd BIPV-ArCn is designed and modelled by merging the conventional (Convnl) SePl and TrTd BIPV-ArCns. The diagram of the connection for 5 × 6 fixed Hbr-BdLk TrTd BIPV-ArCn is showcased in Fig. 1b. From Fig. 1b, it is prominent that the Hbr-BdLk TrTd BIPV-ArCn is designed and modelled by merging the conventional (Convnl) BdLk and TrTd BIPV-ArCns. The diagram of the connection for 5 × 6 fixed Hbr-HnCb TrTd BIPV-ArCn is showcased in Fig. 1c. From Fig. 1c, it is prominent that the Hbr-HnCb TrTd BIPV-ArCn is designed and modelled by merging the conventional (Convnl) HnCb and TrTd BIPV-ArCns. The diagram of the connection for 5 × 6 fixed

Hbr-Ld TrTd BIPV-ArCn is showcased in Fig. 1d. From Fig. 1d, it is prominent that the Hbr-Ld TrTd BIPV-ArCn is designed and modelled by merging the conventional (Convnl) Ld and TrTd BIPV-ArCns.

3 Analysis of Proposed Hybrid BIPV Array in Patterns of Partial Shading

Modelling of Hbr TrTd BIPV-ArCns is completed; now, the succeeding objective is the performance assessment of the proposed BIPV-ArCns under patterns of partial shading (PS). Hence, for simulation in Matlab/Simulink, four diverse patterns of partial shading (PS) are modelled with four different solar irradiation levels {250 W/m^2 (violet), 450 W/m^2 (blue), 650 W/m^2 (green) and 1000 W/m^2 (white)} as portrayed in Fig. 2.

The schematic of Ptrn1:ESSd is displayed in Fig. 2a. From Fig. 2a, it is prominent that the PS pattern signifies the form of capital letter E in the BIPV-ArCn. The simulation of the proposed BIPV-ArCns under Ptrn1:ESSd is performed, and the *P–V* graph is achieved. The *P–V* graph of proposed Hbr TrTd BIPV-ArCns for Ptrn1:ESSd is given in Fig. 3a. F-shaped shading (Ptrn2:FSSd) is the second pattern amongst the four diverse patterns of partial shading (PS). The schematic of Ptrn2:FSSd is displayed in Fig. 2b. From Fig. 2b, it is prominent that the PS pattern signifies the form of capital letter F in the BIPV-ArCn. The simulation of the proposed BIPV-ArCns under Ptrn2:FSSd is performed, and the *P–V* graph is achieved. The *P–V* graph of proposed Hbr TrTd BIPV-ArCns for Ptrn2:FSSd is given in Fig. 3b. The schematic of Ptrn3:HSSd is displayed in Fig. 2c. From Fig. 2c, it is prominent that the PS pattern signifies the form of capital letter H in the BIPV-ArCn. The simulation of the proposed BIPV-ArCns under Ptrn3:HSSd is performed, and the *P–V* graph is achieved. The *P–V* graph of proposed Hbr TrTd BIPV-ArCns for Ptrn3:HSSd is given in Fig. 3c. The schematic of Ptrn4:ISSd is displayed in Fig. 2d. From Fig. 2d, it is prominent that the PS pattern signifies the form of the capital letter I in the BIPV-ArCn. The simulation of the proposed BIPV-ArCns under Ptrn4:ISSd is performed, and the *P–V* graph is achieved. The *P–V* graph of proposed Hbr TrTd BIPV-ArCns for Ptrn4:ISSd is given in Fig. 3d.

4 Performance Assessment Results and Discussions

The performance of Hbr-SePl TrTd, Hbr-BdLk TrTd, Hbr-HnCb TrTd and Hbr-Ld TrTd BIPV-ArCns is assessed in this section in respect of global maximum-power (Gl-MxPo), mismatch-loss (MsLo), fill-factor (FlFc) and efficiency (Efcy). Performance assessment is required to select the best BIPV-ArCn amongst the proposed Hbr BIPV-ArCns for maximising the power output (PO) under conditions of partial

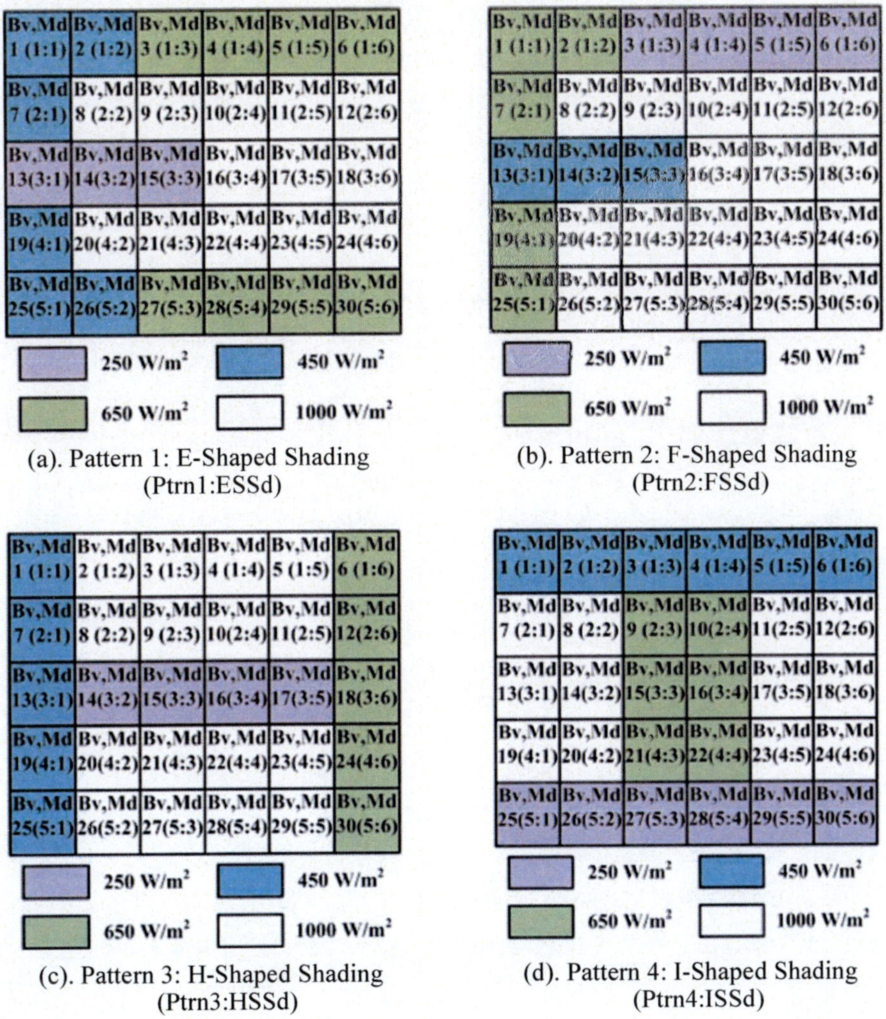

Fig. 2 Four diverse patterns of partial shading (PS)

shading (PS). Mismatch-loss (MsLo) % is computed by Eq. (1), as conveyed below.

$$P_{\text{MsLo}}(\%) = \frac{P_{\text{M\_Unfm}} - P_{\text{M\_Gl,MxPo}}}{P_{\text{M\_Unfm}}} \times 100 \tag{1}$$

Here, MsLo is implied by (P_{MsLo}); maximum power (MP) developed by BIPV modules under uniform irradiation (UI) is implied by ($P_{\text{M\_Unfm}}$); and global maximum-power (Gl-MxPo) under partial shading (PS) is implied by ($P_{\text{M\_Gl,MxPo}}$). The Gl-MxPo (W) and MsLo (%) for four different proposed Hbr TrTd BIPV-ArCns under four diverse patterns of PS are exhibited in Fig. 4a, b. Fill-factor (FlFc) % is

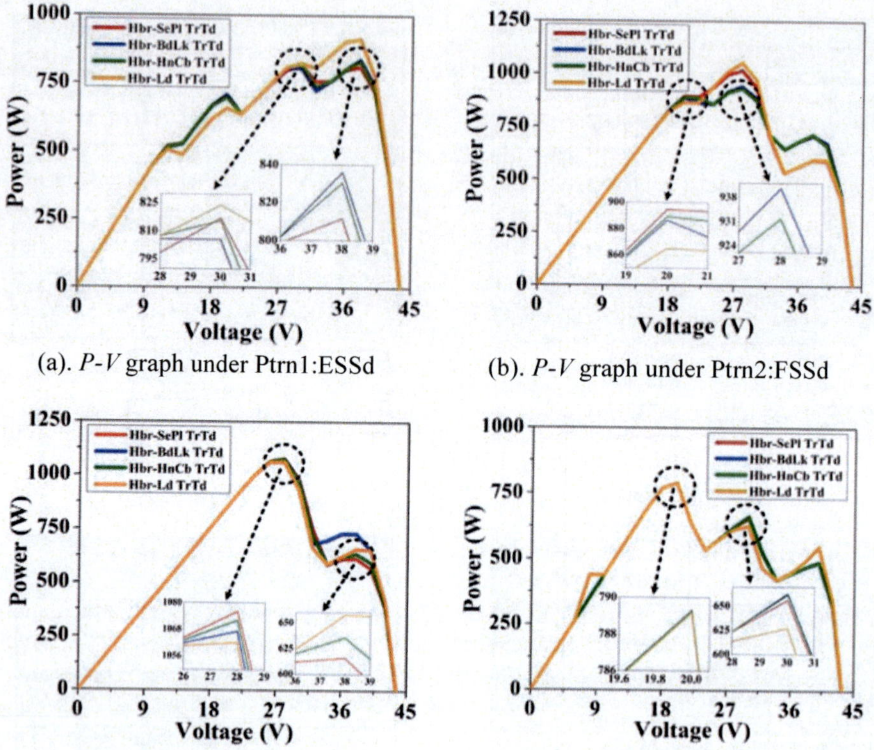

(a). *P-V* graph under Ptrn1:ESSd

(b). *P-V* graph under Ptrn2:FSSd

(c). *P-V* graph under Ptrn3:HSSd

(d). *P-V* graph under Ptrn4:ISSd

Fig. 3 *P–V* graph of proposed Hbr TrTd BIPV-ArCns

computed by Eq. (2), as conveyed below.

$$\text{FlFc}(\%) = \frac{P_{\text{M\_Gl,MxPo}}}{V_{\text{Op\_Ckt}} \times I_{\text{Sh\_Ckt}}} \times 100 \tag{2}$$

Here, global maximum-power (Gl-MxPo) under partial shading (PS) is implied by ($P_{\text{M\_Gl,MxPo}}$); open (Op) circuit (Ckt) voltage is implied by ($V_{\text{Op\_Ckt}}$); and short (Sh) circuit (Ckt) current is implied by ($I_{\text{Sh\_Ckt}}$). The FlFc (%) for four different proposed Hbr TrTd BIPV-ArCns under four diverse patterns of PS are exhibited in Fig. 4c. Efficiency (Efcy) % is computed by Eq. (3), as conveyed below.

$$\text{Efcy}(\%) = \frac{P_{\text{M\_Gl,MxPo}}}{IR_{\text{BiMd}} \times A_{\text{BiMd}}} \times 100 \tag{3}$$

Here, global maximum-power (Gl-MxPo) under partial shading (PS) is implied by ($P_{\text{M\_Gl,MxPo}}$); solar irradiation received by the BIPV module per m^2 is implied by (IR_{BiMd}); and area of the BIPV module in m^2 is implied by (A_{BiMd}). The Efcy (%)

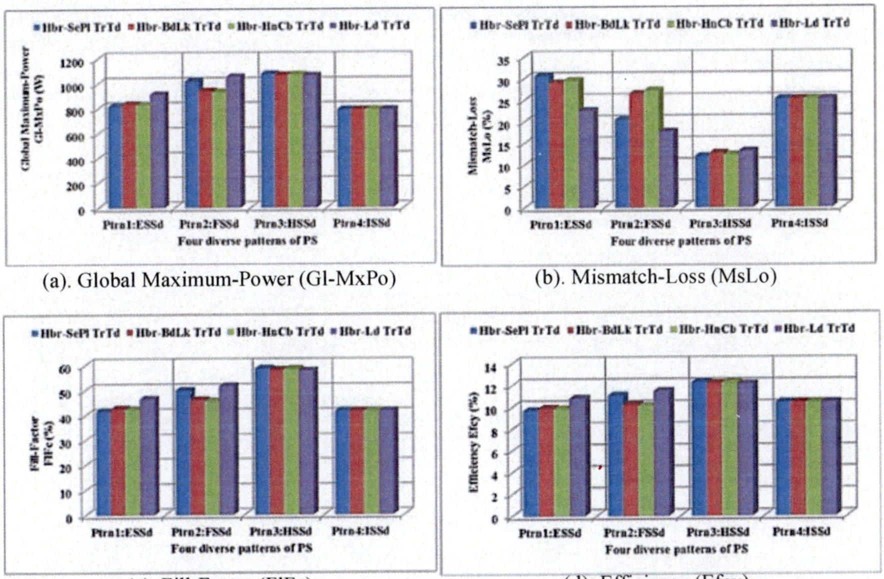

(a). Global Maximum-Power (Gl-MxPo)

(b). Mismatch-Loss (MsLo)

(c). Fill-Factor (FlFc)

(d). Efficiency (Efcy)

Fig. 4 Performance of four different proposed Hbr TrTd BIPV-ArCns under four diverse patterns of PS

for four different proposed Hbr TrTd BIPV-ArCns under four diverse patterns of PS are exhibited in Fig. 4d.

Table 1 provides the performance assessment results for eight different conventional and proposed Hbr TrTd BIPV-ArCns under four diverse patterns of PS. From Table 1, it is noticed that power is maximised in hybrid configurations as compared to conventional configurations. Under Ptrn1:ESSd; the performance of Hbr-BdLk TrTd and Hbr-HnCb TrTd is average, while the performance of Hbr-SePl TrTd is poor. Under Ptrn2:FSSd; the performance of Hbr-SePl TrTd and Hbr-BdLk TrTd is average, while the performance of Hbr-HnCb TrTd is poor. Under Ptrn3:HSSd; the performance of Hbr-HnCb TrTd and Hbr-BdLk TrTd is average, while the performance of Hbr-Ld TrTd is poor. Under Ptrn4:ISSd; all the BIPV-ArCns gives almost similar results.

5 Conclusion

This research paper proposes four different 5 × 6 fixed hybrid (Hbr) triple-tied (TrTd) BIPV array configurations (BIPV-ArCns) for both off-grid and on-grid system applications. The proposed BIPV-ArCns maximise the power output (PO) as well as reduces the wiring cost (by decreasing the wiring requirement) of a large BIPV array (LBiAr). The performances of the proposed BIPV-ArCns are assessed under

Table 1 Performance assessment results for eight different conventional and proposed Hbr TrTd BIPV-ArCns under four diverse patterns of PS

Patterns of PS	Conventional and proposed Hbr BIPV-ArCns	Performance assessment results			
		Gl-MxPo (W)	MsLo (%)	FlFc (%)	Efcy (%)
Ptrn1:ESSd	SePl	792.107	34.712	36.243	8.862
	BdLk	796.673	33.891	37.892	8.991
	HnCb	804.742	32.167	38.432	9.126
	Ld	812.714	31.234	39.721	9.372
	Hbr-SePl TrTd	817.018	30.655	41.583	9.676
	Hbr-BdLk TrTd	835.955	29.048	42.541	9.900
	Hbr-HnCb TrTd	830.305	29.527	42.257	9.833
	Hbr-Ld TrTd	912.204	22.576	46.402	10.803
Ptrn2:FSSd	SePl	904.678	29.567	44.678	9.834
	BdLk	912.056	28.452	43.234	9.906
	HnCb	916.142	28.124	42.986	10.124
	Ld	952.987	25.356	47.687	10.208
	Hbr-SePl TrTd	1017.490	20.506	49.840	11.054
	Hbr-BdLk TrTd	940.609	26.512	46.048	10.219
	Hbr-HnCb TrTd	931.778	27.202	45.626	10.123
	Hbr-Ld TrTd	1053.501	17.692	51.596	11.446
Ptrn3:HSSd	SePl	1032.786	15.987	53.556	10.843
	BdLk	1042.574	15.465	54.876	11.542
	HnCb	1048.453	14.879	54.672	11.765
	Ld	1052.567	14.342	54.783	11.945
	Hbr-SePl TrTd	1076.167	11.919	58.530	12.264
	Hbr-BdLk TrTd	1067.132	12.659	58.001	12.161
	Hbr-HnCb TrTd	1072.074	12.254	58.295	12.217
	Hbr-Ld TrTd	1062.124	13.068	57.709	12.103
Ptrn4:ISSd	SePl	753.989	27.967	39.656	9.345
	BdLk	754.621	27.763	39.564	9.568
	HnCb	756.973	27.432	40.125	9.724
	Ld	757.982	27.134	41.653	9.983
	Hbr-SePl TrTd	789.282	25.214	41.566	10.431
	Hbr-BdLk TrTd	789.297	25.212	41.564	10.432
	Hbr-HnCb TrTd	789.282	25.214	41.565	10.431
	Hbr-Ld TrTd	789.181	25.223	41.517	10.430

four diverse patterns of partial shading (PS). The Hbr-Ld TrTd BIPV-ArCn gives the best performance, while the Hbr-SePl TrTd BIPV-ArCn gives poor performance. In one pattern case, Hbr-SePl TrTd BIPV-ArCn gives the best performance, while Hbr-Ld TrTd BIPV-ArCn gives poor performance. Hence, in the majority of cases, Hbr-Ld TrTd is the best BIPV-ArCn while Hbr-SePl TrTd is the poor BIPV-ArCn. The performance of Hbr-BdLk TrTd and Hbr-HnCb TrTd BIPV-ArCns is average. Critical analysis in more extreme cases of PS patterns is necessary that will serve as future work.

References

1. Zhang T, Wang M, Yang H (2018) A review of the energy performance and life-cycle assessment of building-integrated photovoltaic (BIPV) systems. Energies 11
2. Sarkar D, Kumar A, Sadhu PK (2020) A survey on development and recent trends of renewable energy generation from BIPV systems. IETE Tech Rev 37:258–280
3. Al-Janahi SA, Ellabban O, Al-Ghamdi SG (2020) A novel BIPV reconfiguration algorithm for maximum power generation under partial shading. Energies 13
4. Pendem SR, Mikkili S (2018) Modeling, simulation, and performance analysis of PV array configurations (series, series-parallel, bridge-linked, and honey-comb) to harvest maximum power under various partial shading conditions. Int J Green Energy 15:795–812
5. Pendem SR, Mikkili S (2018) Modelling and performance assessment of PV array topologies under partial shading conditions to mitigate the mismatching power losses. Sol Energy 160:303–321
6. Jha V, Triar US (2019) A detailed comparative analysis of different photovoltaic array configurations under partial shading conditions. Int Trans Electr Energy Syst 29:e12020
7. Premkumar M, Subramaniam U, Babu TS, Elavarasan RM, Mihet-Popa L (2020) Evaluation of mathematical model to characterise the performance of conventional and hybrid PV array topologies under static and dynamic shading patterns. Energies 13
8. Bonthagorla PK, Mikkili S (2020) Performance analysis of PV array configurations (SP, BL, HC and TT) to enhance maximum power under non-uniform shading conditions. Eng Rep 2:e12214
9. Bonthagorla PK, Mikkili S (2021) A novel fixed PV array configuration for harvesting maximum power from shaded modules by reducing the number of cross-ties. IEEE J Emerg Sel Top Power Electron 9:2109–2121
10. BIPV054-T86 54W Sun Energy Tile BIPV modules, https://solarhub.com/product-catalog/pv-modules/27017-BIPV054-T86-BIPV. Accessed 23 June 2021

Electricity Generation Using Soil and Living PKL Tree

Salman Rahman Rasel, K. A. Khan, and Sumanta Bhattacharyya

Abstract Zn/Cu electrodes-based electrochemical cell has been designed and developed using soil of a pot and living *Bryophillum pinnatum* tree for cultivation of electricity. V_{oc}, I_{sc}, P_{max}, and r_{in} have been studied. Firstly, Zn plate was placed in the soil of the pot, and Cu plate was placed onto the living PKL tree. Secondly, Cu plate was placed in the soil of the pot, and Zn plate was placed onto the living PKL tree. Different soil pots and different living PKL trees have been used for getting V_{oc}, I_{sc}, P_{max}, and r_{in} of the electrochemical cell. It is shown that the performance of the second condition is better than the first condition. This work is very new and innovative. This work can help to light the LED bulb.

Keywords Soil · Living PKL · Zn/Cu electrodes · Electrochemical cell · Internal resistance

1 Introduction

It has been conducted research on living PKL trees using Zn/Cu-based electrodes before this research work [1–6]. This has also been conducted—voltage cultivation from fresh leaves [7–12]. But this has not ever been done—Zn/Cu electrodes-based electrochemical cell using soil of a pot and living PKL tree for electricity cultivation [13–19]. It is mentioned that the PKL tree grows itself in the same soil pot, where Zn and Cu plates have been immersed it to the soil [20–24]. Electricity can be produced during the PKL gardening surrounding the house or home or terrace. Any person can cultivate electricity by this research work [25, 26]. Even a handicapped person can cultivate electricity through this research work. It will help the people to cultivate electricity when the traditional fuels like oil, gas, and coal will be finished within

S. R. Rasel
Local Government Engineering Department (LGED), Fulbaria, Mymensingh, Bangladesh

K. A. Khan
Department of Physics, Jagannath University, Dhaka-1100, Bangladesh

S. Bhattacharyya (✉)
Department of ECE, Greater Kolkata College of Engineering and Management, Kolkata, India

J. K. Mandal et al. (eds.), *Topical Drifts in Intelligent Computing*, Lecture Notes in Networks and Systems 426, https://doi.org/10.1007/978-981-19-0745-6_55

Table 1 Length and surface area of Cu and Zn electrodes for two cases

Name of the plate	Condition of the plate	Length (cm)	Breadth (cm)	Area (cm^2)
Zn	Zn plate immersed in the soil of the pot	4.50	4.40	19.80
Zn	Zn plate occupied length embedded by the PKL	8.50	4.40	37.40
Cu	Cu plate immersed in the soil of the pot	7.80	5.80	45.24
Cu	Cu plate occupied length embedded by the PKL	11.00	5.80	63.80

twenty-first century [27]. The open-circuit voltage was 1.01 V for Cu/Zn electrodes-based electrochemical cell using soil of a pot and living PKL tree, whereas it was 0.99 V for Zn/Cu electrodes-based electrochemical cell using soil of a pot and living PKL tree.

Objectives:

(i) To generate electricity using soil and living PKL tree.
(ii) To study the internal resistance of this system.

2 Methodology

2.1 *Materials Used for This Research Work*

The specifications for both Zn and Cu plates are given by the following and has been summarized in Table 1:

For Zn Plate: Zn plate immersed in the soil of the pot = 4.50 cm, breadth of the Zn plate = 4.40 cm, area of the Zn plate = 19.80 cm^2, Zn plate occupied length embedded by the PKL = 11.0 cm, breadth of the Zn plate = 4.40 cm, and area of the Zn plate embedded by the PKL = 48.40 cm^2.

For Cu Plate: Cu plate immersed in the soil of the pot = 7.8 cm, breadth of the Cu plate = 5.8 cm, area of the Cu plate = 45.25 cm^2, Cu plate occupied length embedded by the PKL = 8.5 cm, breadth of the Cu plate = 5.8 cm, and area of the Cu plate embedded by the PKL = 49.3 cm^2.

Table 1 shows the length and surface area of Cu and Zn electrodes for two cases. The first case is Zn and Cu plates immersed in the soil of the pot, and the second case is Zn and Cu plate occupied length embedded by the PKL.

It is shown from Fig. 1 that the voltage measurement for Zn in soil and Cu is above leaf, and it is also shown from Fig. 2 that the voltage measurement for Cu in soil and Zn is above leaf. The living PKL tree lives on the soil of the pot. The leaf of the living tree and the soil of that tree makes an electrochemical cell. The soil of

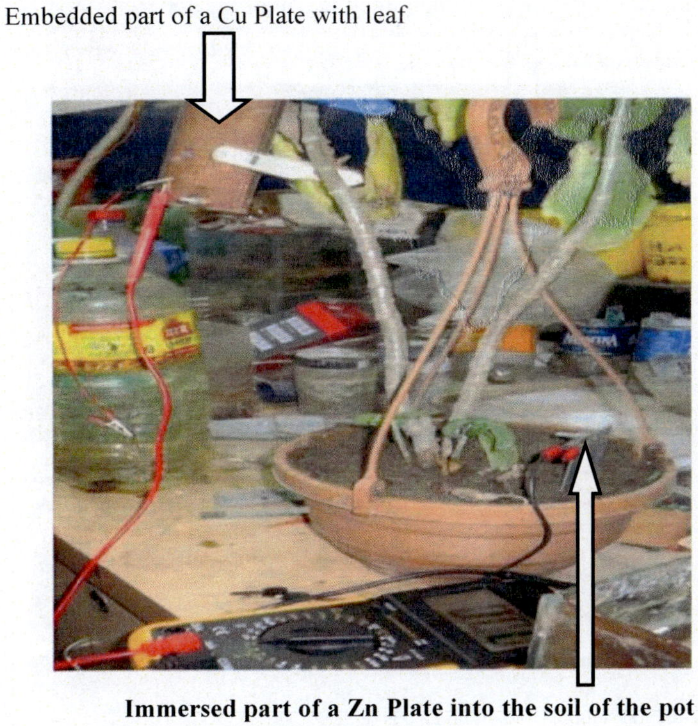

Fig. 1 Experimental setup Zn in soil and Cu is on leaf

the pot and the leaf works together as an electrolyte. Zn and Cu plates work as an electrolytes. The two different mediums—soil and living leaf—are in the same pot and apart by 0.6 m. The electrochemical cell was made by a single Cu plate and a single Zn plate. A part of the Cu plate and Zn plate was immersed in the soil of the pot and was embedded with the leaf. It has been used plastic clip for tightening the embedded living PKL leaf (Table 2).

3 Results and Discussion

It is shown that (in Fig. 3) V_{oc} versus T.D. (h) for Cu plate is embedded with leaf and Zn plate is in the soil. It is mentioned that the observation was for 13 h. The surface area of the Cu plate embedded with leaf is 63.80 cm^2 and the surface of the Zn plate is in the soil is 19.80 cm^2. From this observation, it can be concluded that the self-discharge characteristics for such type of soil–leaf electrochemical cell is viable and feasible compare to other electrochemical cells/battery system (Table 3).

It is shown (Fig. 4) that the variation of I_{sc} with the variation of T.D. (h) for Cu plate is embedded with leaf and Zn plate is in the soil. It is mentioned that the

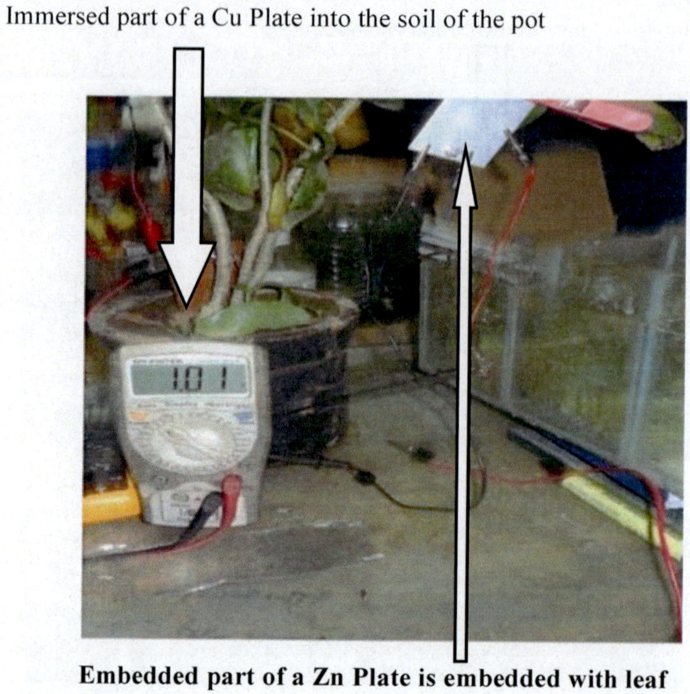

Fig. 2 Experimental setup Cu in soil and Zn is on leaf

Table 2 Data collection for V_{oc}, I_{sc}, internal resistance (r_{in}), and P_{max} during Cu plate is embedded with leaf and Zn plate is in the soil

T.D. (h)	V_{oc} (V)	I_{sc}(A)	$P_{max} = V_{oc} * I_{sc}$(W)	$r_{in} = V_{oc}/I_{sc}$ (Ω)
00	1.00	0.002	0.205	487.80
01	1.00	0.002	0.205	487.80
02	1.00	0.002	0.205	487.80
03	1.00	0.002	0.205	487.80
04	1.00	0.002	0.205	487.80
05	1.00	0.002	0.205	487.80
06	1.00	0.002	0.205	487.80
07	1.01	0.002	0.206	495.10
08	1.01	0.002	0.206	495.10
09	1.01	0.002	0.206	495.10
10	1.01	0.002	0.206	495.10
11	1.01	0.002	0.206	495.10
12	1.01	0.002	0.206	495.10
13	1.01	0.002	0.206	495.10

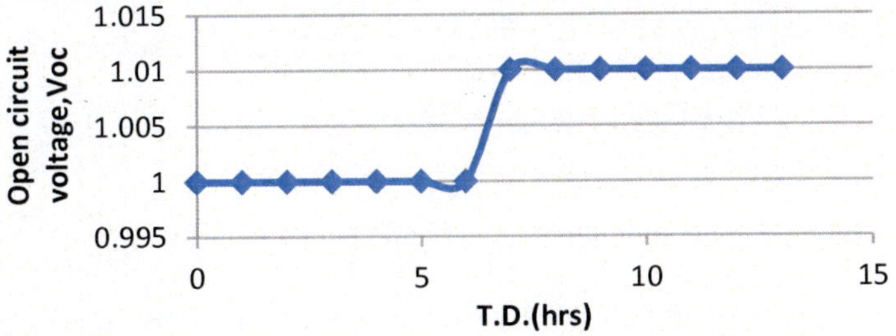

Fig. 3 *V*oc—T.D. graph during Cu plate is embedded with leaf and Zn plate is in the soil

Table 3 Data collection for V_{oc}, I_{sc}, internal resistance (R_{in}), and P_{max} during Cu plate into the soil and Zn plate is embedded with leaf

T.D. (h)	V_{oc}(V)	I_{sc} (A)	$P_{max} = V_{oc} * I_{sc}$ (W)	$R_{in} = V_{oc}/I_{sc}$ (Ω)
00	0.75	0.006	0.45	126.48
01	0.75	0.006	0.48	116.28
02	0.75	0.006	0.46	121.56
03	0.75	0.006	0.48	117.37
04	0.75	0.006	0.46	121.95
05	0.75	0.006	0.47	118.86
06	0.75	0.006	0.46	122.35
07	0.74	0.007	0.49	112.29
08	0.74	0.006	0.46	118.21
09	0.74	0.006	0.48	115.26
10	0.74	0.007	0.48	113.85
11	0.74	0.006	0.48	114.20
12	0.74	0.007	0.48	113.50
13	0.74	0.007	0.48	113.67

Fig. 4 *I*sc—T.D. (h) graph during Cu plate is embedded with leaf and Zn plate is in the soil

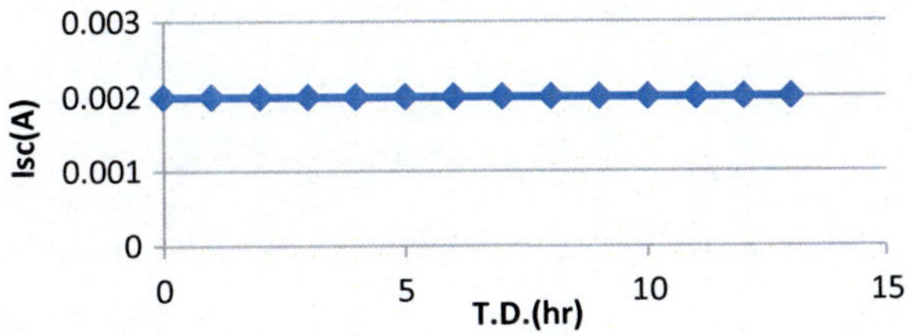

observation was for 13 h. It shows that the variation of $I_{sc} = 0$ for 13 h. The surface area of the Cu plate embedded with leaf is 63.80 cm^2, and the surface of the Zn plate is in the soil = 19.80 cm^2.

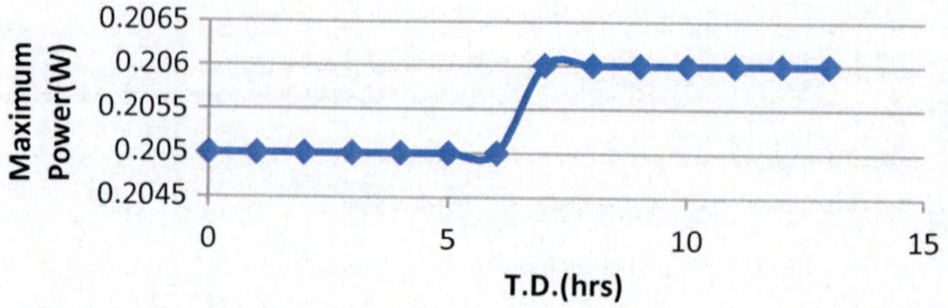

Fig. 5 P_{max}—T.D. graph during Cu plate is embedded with leaf and Zn plate is in the soil

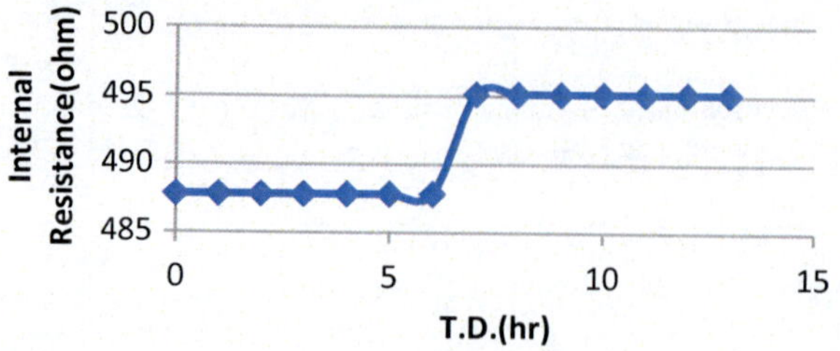

Fig. 6 R_{in}—T.D. (h) curve during Cu plate is embedded with leaf and Zn plate is in the soil

It is mentioned that the observation was for 13 h. It is shown that the difference of maximum power is 0.01 W for 13 h. The surface area of the Cu plate embedded with leaf is 63.80 cm^2, and the surface of the Zn plate is in the soil is 19.80 cm^2. From this observation, it can be concluded that the self-discharge characteristics for such type of soil–leaf electrochemical cell is viable and feasible compared to other electrochemical cells/battery system.

Figure 7 shows the open-circuit voltage versus time duration for Cu plate in the soil and Zn plate is embedded with PKL leaf. It is mentioned that the observation was for 13 h. It shows that the difference of open V_{oc} is almost 0.01 V for 13 h. The surface area of the copper plate immersed in the soil was 45.24 cm^2, and the surface of the Zn plate embedded with leaf was 37.40 cm^2. Finally, it is concluded that the

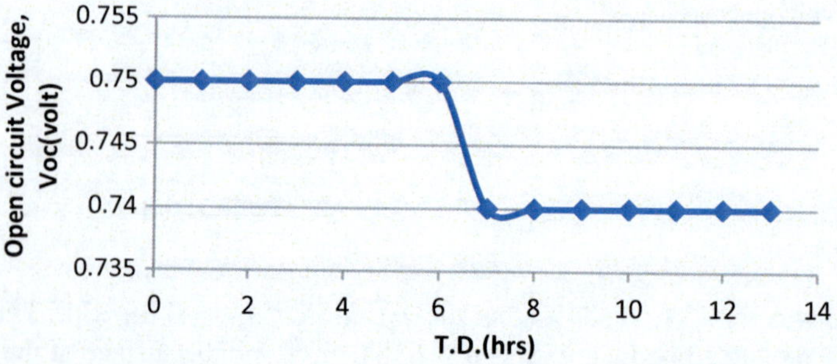

Fig. 7 Voc—T.D. graph during Cu plate immersed in the soil and Zn plate is embedded with leaf

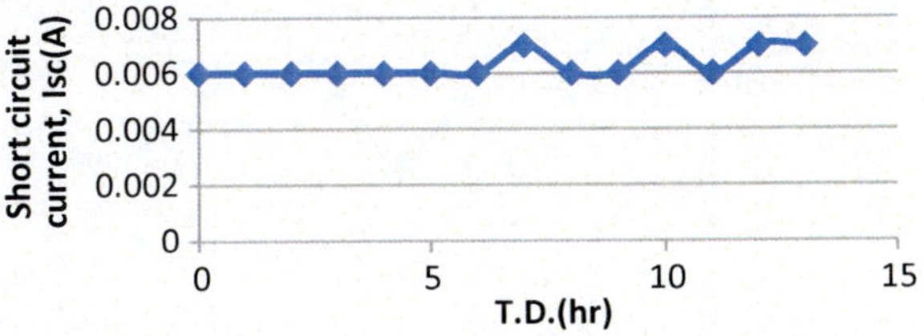

Fig. 8 I_{sc}—T.D. (h) curve during Cu plate into the soil and Zn plate is embedded with leaf

discharge characteristics of the soil–leaf electrochemical cell is better than the other electrochemical cell comparison to other electrochemical cell.

It is shown from (Fig. 8) that the graph of short I_{sc} versus the T.D. (h) for Cu plate into the soil, and Zn plate is embedded with leaf. It is mentioned that the observation was for 13 h. It is shown that the variation of I_{sc} is 0.001 A for 13 h. The surface area of the Cu plate immersed in the soil was 45.24 cm^2, and the surface of the Zn plate embedded with leaf was 37.40 cm^2. It is shown that the I_{sc} was greater for Cu plate in the soil, and Zn plate is embedded with leaf than for Zn plate into the soil, and Cu plate is embedded with leaf comparison with Figs. 4 and 8.

It is shown from Fig. 9 that the change of Pmax versus T.D. (hr) for Cu plate in the soil, and Zn plate is embedded with leaf. It is mentioned that the observation was for 13 h. It is shown that the variation of maximum power is 0.03 W for 13 h. The variation of maximum power was not constant, but it was fluctuated up to 13 h from starting. The surface area of the Cu plate immersed in the soil was 45.24 cm^2, and the surface of the Zn plate embedded with leaf was 37.40 cm^2.

Figure 10 shows the variation of internal resistance with the variation of time duration for Cu plate into the soil, and Zn plate is embedded with leaf. It is mentioned that the observation was for 13 h. It shows the variation of R_{in} is 14.19 Ω for 13 h. The variation of internal resistance was not constant but it decreases, and it was fluctuated up to 13 h from starting. The surface area of the Cu plate immersed in the soil was 45.24 cm^2, and the surface of the Zn plate embedded with leaf was 37.40 cm^2.

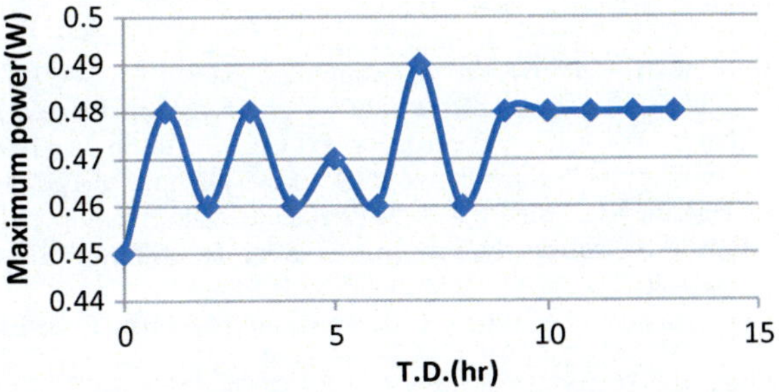

Fig. 9 P_{max}—T.D. (time duration) curve during Cu plate into the soil and Zn plate is embedded with leaf

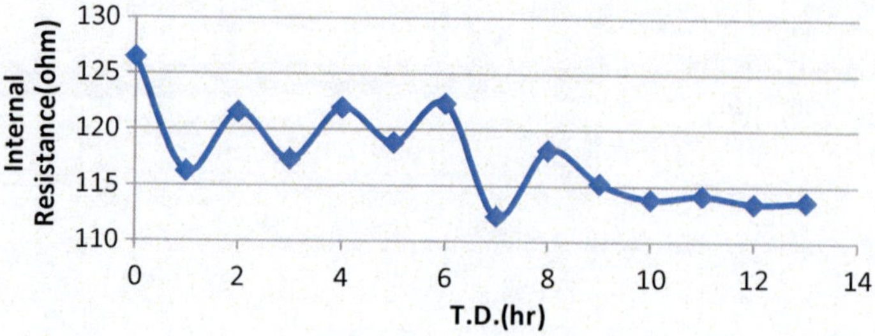

Fig. 10 R_{in}—T.D. (h) curve during Cu plate into the soil and Zn plate is embedded with leaf

4 Conclusions

It is found that for the Zn plate was placed in the soil of the pot and the Cu plate was embedded on to the living PKL tree, the maximum and minimum V_{oc}, I_{sc}, P_{max}, and internal resistance (r_{in}) were 1.01 V and 1.00 V, 0.002 and 0.002 A, 0.206 W and 0.205 W,, and 495.5 Ω and 487.5 Ω, respectively. It is also found that for the Cu plate was placed in the soil of the pot and the Zn plate was embedded on to the living PKL tree, the maximum and minimum V_{oc}, I_{sc}, P_{max}, and internal resistance (r_{in}) were 0.75 V and 0.74 V, 0.007 A and 0.006 A, 0.49 W and 0.45 W, and 126 Ω and 112 Ω, respectively. By comparing the above two conditions, it can be concluded that when the Cu plate was placed in the soil of the pot, and the Zn plate was embedded on to the living PKL tree is better than the Zn plate was placed in the soil of the pot and the Cu plate was embedded on to the living PKL tree.

Acknowledgements The authors are grateful to the Ministry of Education for financing during the research work (Project/User ID: PS2019949).

References

1. Saad B, Abdelmoneim I, Adam G, Elghazali L (2006) Traditional Arab herbal medicine, evidence-based complementary and alternative medicine. J Ethnopharmacol 90:625–627
2. Larrey DA, Chan JE, Ngogang JY, Cai X, Vinson JA (1994) Antioxidant capacity of some herbs/spices: a comparative study of two methods. J Agric Food Chem 53:6819–6824
3. Jagessar RC, Allen R (2012) Phytochemical screening and atomic absorption spectroscopic studies of solvent type extract from leaves of Terminalia catappa, (Almond). Acad Res In 3:17–26
4. Riazul Hamid M (2013) Characterization of a battery cell fueled by Bryophyllum Pinnatum sap. Int J Sci Eng Res 9(3). ISSN 2229-5518
5. Riazul Hamid M, Yusuf A, Abdul Wadud AM, Mosfiqur Rahaman M (2016) Design and performance test of a prototype of a 12 Volt DC battery Fueled by Bryophyllum Pinnatum Sap and improvement of its characteristics, department of electrical and electronic engineering, Ahsanullah University of Science and Technology, Dhaka, Bangladesh, Email: {drhamidbd, shohan933}@gmail.com, {akib147, sshaon95}@yahoo.com. Int J Electron Electr Eng 4(5)

6. Pal S, Chaudhuri AKN (1991) Studies on the anti-ulcer activity of a Bryophyllum pinnatum leaf extract in experimental animals. J Ethanopharmacol 33:97–102
7. Supratman U, Fujita T, Akiyama K, Hayashi H (2000) New insecticidal bufadienolide, bryophyllin C, from Kalanchoe pinnata. Biosci Biotechnol Biochem 64:1310–1312
8. Ojewole JAO (2005) Antinociceptive, anti-inflammatory and antidiabetic effects of Bryophyllum pinnatum (Crassulaceae) leaf aqueous extract. J Ethnopharmacol 99:13–19
9. Asiedu-Gyekye IJ, Antwi DA, Bugyei KA, Awortwe C (2012) Comparative study of two kalanchoe species: total flavonoid, phenolic contents and antioxidant properties. Afr J Pure Appl Chem 6:65–73
10. Khan KA (1999) Copper oxide coatings for use in a linear solar Fresnel reflecting concentrating collector, Published in the Journal of Elsevier, Renewable Energy, An International Journal, WREN(World Renewable Energy Network), UK, RE: 12.97/859,1998, Publication date 1999/8/1. J Renew Energy 17(4):603–608. Publisher—Pergamon, 1999
11. Ruhane TA, Tauhidul Islam M, Saifur Rahman M, Bhuiyah MMH, Islam JMM, Bhuiyah TI, Khan KA, Khan MA (2017) Impact of photo electrode thickness annealing temperature on natural dye sensitized solar cell. Sustain Energy Technol Assess Elsevier. http://dx.doi.org/https://doi.org/10.1016/j.seta.2017.01.012
12. Ruhane TA, Tauhidul Islam M, Saifur Rahaman M, Bhuiyan MMH, Islam JMM, Newaz MK, Khan KA, Khan MA (2017) Photo current enhancement of natural dye sensitized solar cell by optimizing dye extraction and its loading period. Optik—International Journal for Light and Electron Optics, Elsevier
13. Hasan M, Khan KA (2018) Dynamic model of Bryophyllum pinnatum leaf fueled BPL cell: a possible alternate source of electricity at the off-grid region in Bangladesh, microsystem technologies micro—and nanosystems information storage and processing systems, Springer, ISSN 0946–7076. Microsyst Technol. https://doi.org/10.1007/s00542-018-4149-y
14. Khan KA, Hazrat Ali M, Obaydullah AKM, Wadud MA (2019) Production of candle using solar thermal technology. In: Microsystem technologies micro- and nanosystems information storage and processing systems, Springer, ISSN 0946-7076. Microsyst Technol 25(12). doi:https://doi.org/10.1007/s00542-019-04390-7
15. Khan KA, Rasel SR, Ohiduzzaman M (2019) Homemade PKL electricity generation for use in DC fan at remote areas. In: Microsystem technologies micro- and nanosystems information storage and processing systems, ISSN 0946-7076. Microsyst Technol 25(12). Doi:https://doi.org/10.1007/s00542-019-04422-2
16. Hasan M, Khan KA (2019) Experimental characterization and identification of cell parameters in a BPL electrochemical device, Springer. SN Appl Sci 1:1008. https://doi.org/10.1007/s42452-019-1045-8
17. Hassan L, Khan KA (2019) A study on harvesting of PKL electricity, Springer Journal. Microsyst Technol 26:1031–1041 (2020). doi:https://doi.org/10.1007/s00542-019-04625-7, 26(3):1032–1041
18. Khan KA, Mamun MA, Ibrahim M, Hasan M, Ohiduzzaman M, Obaydullah AKM, Wadud MA, Shajahan M (2019) PKL electrochemical cell: physics and chemistry, Springer Journal. SN Appl Sci 1:1335. https://doi.org/10.1007/s42452-019-1363-x
19. Hazrat Ali M, Chakma U, Howlader D, Tawhidul Islam M, Khan KA (2019) Studies on performance parameters of a practical transformer for various utilizations, microsystem technologies, Springer. Accepted 03 Dec 2019. doi:https://doi.org/10.1007/s00542-019-04711-w
20. Khan KA, Hassan L, Obaydullah AKM, Azharul Islam SM, Mamun MA, Akter T, Hasan M, Shamsul Alam M, Ibrahim M, Mizanur Rahman M, Shahjahan M (2018) Bioelectricity: a new approach to provide the electrical power from vegetative and fruits at off-grid region. Microsystem Technologies, Springer. https://doi.org/10.1007/s00542-018-3808-3
21. Khan KA, Bhuyan MS, Mamun MA, Ibrahim M, Hasan L, Wadud MA (2018) Organic electricity from Zn/Cu-PKL electrochemical cell. In: JK Mandal et al (eds) Contemporary advances in innovative and applicable information technology, advances in intelligent systems and computing, vol 812, Chapter 9, pp 75–90. © Springer Nature Singapore Pvt. Ltd.

22. Khan KA, Hazrat Ali M, Mamun MA, Mahbubul Haque M, Atique Ullah AKM, Islam Khan MN, Lovelu Hassan AKM, Obaydullah MAW (2020) Bioelectrical characterization and production of nanoparticles (NPs) Using PKL extract for electricity generation, received: July 2018/Accepted: 4 February 2020. Microsyst Technol Springer J. https://doi.org/10.1007/s00542-020-04774-0
23. Khan KA, Rahman A, Rahman MS, Tahsin A, Jubyer KM, Paul S (2016) Performance analysis of electrical parameters of PKL electricity (an experimental analysis on discharge rates, capacity and discharge time, pulse performance and cycle life and deep discharge of Pathor Kuchi Leaf (PKL) electricity cell). In: Innovative smart grid technologies-Asia (ISGT-Asia), 2016 IEEE, pp 540–544. IEEE
24. Khan MKA, Paul S, Rahman MS, Kundu RK, Hasan MM, Moniruzzaman M, Mamun MA (2016) A study of performance analysis of PKL electricity generation parameters (an experimental analysis on voltage regulation, capacity and energy efficiency of pathorkuchi leaf (PKL) electricity cell). In: IEEE 7th power India international Conference (PIICON), 2016, pp 1–6. IEEE
25. Khan MKA, Rahman MS, Das T, Ahmed MN, Saha KN, Paul S (2017) Investigation on parameters performance of Zn/Cu electrodes of PKL, AVL, tomato and lemon juice based electrochemical cells: a comparative study. In: 3rd International Conference on electrical information and communication technology (EICT), 2015, pp 1–6. IEEE
26. Hasan M, Khan KA (2017), MA Mamun, "an estimation of the extractable electrical energy from Bryophyllum pinnatum leaf. Am Int J Res Sci Technol Eng Math (AIJRSTEM) 01(19):100–106
27. Khan MKA, Rahman MS, Das T, Ahmed MN, Saha KN, Paul S (2017). Investigation on parameters performance of Zn/Cu electrodes of PKL, AVL, tomato and lemon juice based electrochemical cells: a comparative study. In: 3rd International conference on electrical information and communication technology (EICT), 2017, pp 1–6. IEEE, 2017. Doi:https://doi.org/10.1109/EICT.2017.8275150, IEEE, Khulna, Bangladesh, Bangladesh, 7–9 Dec 2017

Comparative Studies of V_L, I_L, and P_L from Different Vegetative and Fruits Electrochemical Cells

Kamrul Alam Khan, Md. Sayed Hossain, Salman Rahman Rasel, and Sumanta Bhattacharyya

Abstract The experiment is carried out to study the electrical energy harvested from the four types of living plant extracts like PKL, pandan leaf, red spinach, and green chili based on electrochemical cells and to utilize the generated electricity for lighting systems. Plant extracts were used as electrolytes where Zn and Cu plates were used as a cathode and anode in this investigation. This research work aims to find out the more sustainable energy source by comparing the sources of PKL, pandan leaf, red spinach, and green chili extracts. Here, six electrochemical cells were used and connected in a series combination. LED bulb was used as a load. From experiments, three circuit parameters (load voltage, load current, and load power) have been compared for the PKL, pandan leaf, red spinach and green chili electrochemical cells such as (i) the maximum and minimum load voltages are 5.52 V, 3.75 V, 3.74 V, and 3.73 V, respectively; and 5.18 V, 3.73 V, 3.72 V and 3.69 V, respectively; (ii) the maximum and minimum load currents are 760 mA, 10.10 mA, 7.50 mA, and 0.60 mA; and 440 mA, 9.40 mA, 7.10 mA, and 0.30 mA, respectively, and (iii) the maximum and minimum load powers are 420 mW, 26.26 mW, 18.00 mW and 0.90 mW, respectively, and 230 mW, 24.94 mW, 16.04 mW, and 0.45 mW, respectively. Electrochemistry is responsible for electricity generation. Finally, it is concluded that the electrochemical cell for the PKL extract is better than the other cells. This work is innovative.

Keywords Plant energy harvest · Comparative studies · Electrochemical cell · VL · I_L and P_L

K. A. Khan
Department of Physics, Jagannath University, Dhaka-1100, Bangladesh

Md. Sayed Hossain
Center for Research Reactor, Bangladesh Atomic Energy Commission, Dhaka, Bangladesh

S. R. Rasel
Local Government Engineering Department, Sherpur Sadar, Sherpur, Bangladesh

S. Bhattacharyya (✉)
Department of ECE, Greater Kolkata College of Engineering and Management, Kolkata, India

J. K. Mandal et al. (eds.), *Topical Drifts in Intelligent Computing*, Lecture Notes in Networks and Systems 426, https://doi.org/10.1007/978-981-19-0745-6_56

1 Introduction

The fossil fuels are the mineral resources, and they will be finished due to the increased usages of these fossil fuels in near future [1–5]. Since it will become extinct after some period and then, the more sustainable form of different sources of energy like solar, wind, geothermal, biogas, biomass, water, wave, tidal, and ocean thermal energy conversion (OTEC) energy will be needed. Furthermore, the burning of fossil fuels is not environment-friendly and eco-friendly also [6]. The world became polluted over the past centuries due to the use of fossil fuels like oil, gas, and coal [7, 8]. The main cause of greenhouse effect is the use of fossil fuels [9–13]. Several studies have been conducted to determine the amount of electrical potential generated by living plants [14]. A PKL electric converter has been designed and fabricated for using different electric appliances like LED lamp, radio, TV, fan which was invented in the Physics department of Jagannath University of Bangladesh [15–18]. This technology is based on a natural process that can save the plants and environment all over the world [19, 20].

2 Methodology

2.1 *Materials Used for This Research Work*

Six unit cells are used for this research work. Copper is used as the cathode, and zinc is used as the anode. The electrodes are summerset into the cell and connected in a series combination to flow the electrons to generate electricity. Experiments have been done at room temperature (about 25 °C), and the average humidity was about 66%. The composition of electrolytes is 60% extracts of PKL, pandan leaf, red spinach, and green chili, 5% $CuSO_4$ and 35% H_2O. The circuit diagram is shown in Fig. 1.

Figure 1 showed the six electrochemical cells for PKL, pandan leaf, red spinach, and green chili electrochemical cells. Zinc and copper plates are used as electrodes. Zn and Cu plates are connected in series combinations to flow electrons. This flow of electrons produces electricity. PKL, pandan leaf, red spinach, and green chili extract works as an electrolyte. Electrochemistry is responsible for electricity generation for PKL, pandan leaf, red spinach, and green chili electrochemical cells.

It is shown the PKL extracts preparation in Fig. 2. The PKL was blended by a blender machine. It is also shown that the bi-product was solid wastage.

Figure 3 shows the PKL unit cell for electricity generation. The Zn and Cu plates were immersed into the PKL electrolyte.

Figure 4 shows the cycle from the production of pandan leaf extraction to the electricity generation. It is also seen that the LED bulb is illuminated as final product using pandan leaf electrochemical cell.

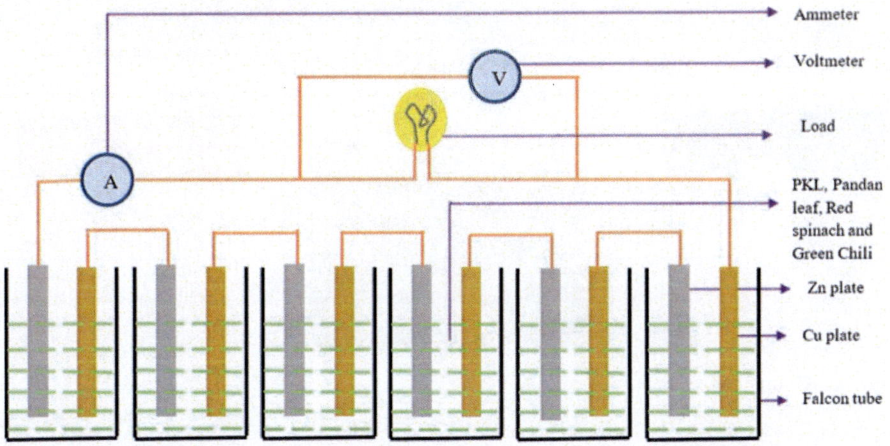

Fig. 1 Experimental setup of an LED bulb lighting system using PKL, pandan leaf, red spinach, and green chili electrochemical cell

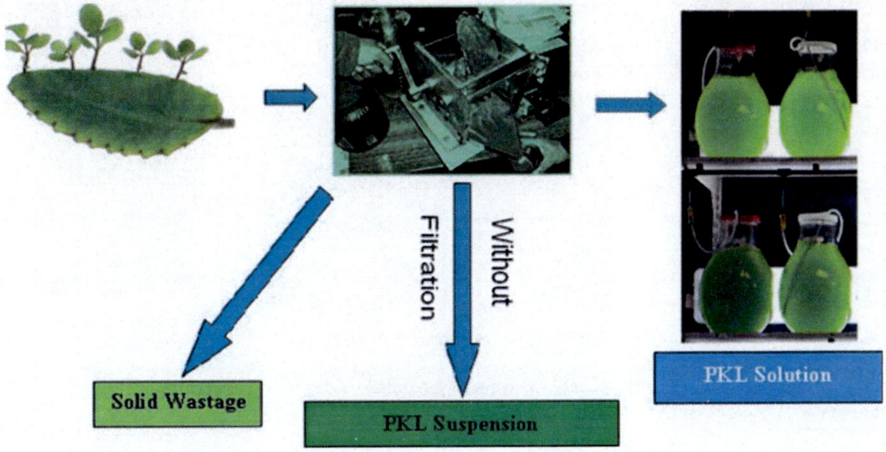

Fig. 2 PKL extracts preparation

Figure 5 shows the workflow of the red spinach electricity generation system. It is also seen that the LED bulb is illuminated as final product using red spinach electrochemical cell.

Figure 6 shows the graphical abstract of the green chili electricity generation system. It is also seen that the LED bulb is illuminated as final product using green chili electrochemical cell.

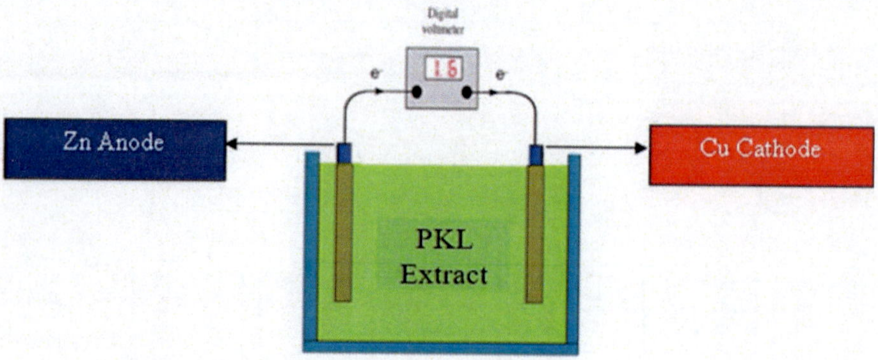

Fig. 3 PKL unit cell

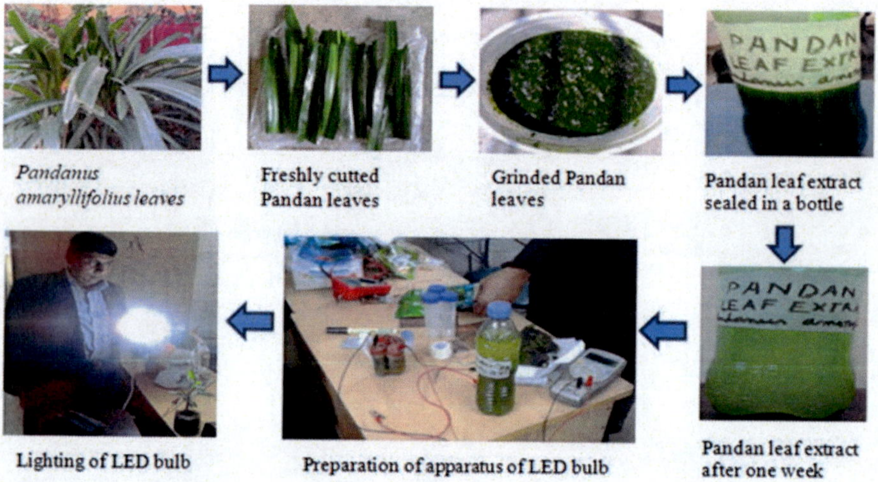

Fig. 4 Graphical abstract of the pandan leaf electricity

2.2 *Chemical Reactions*

The oxidation has occurred at anode (Zn-plate), and reduction has occurred at cathode (Cu-plate). This creates a movement of electrons flow from anode (Zn-plate) to the cathode (Cu-plate) which generates current in the circuit. The total cell reaction is given by the following:

$$Zn + Cu^{2+} + H^{+} \rightarrow Zn^{2+} + Cu + H_2 \uparrow \tag{1}$$

Here, Cu^{2+} = Reactant ions, H^{+} = Reactant ions, and Zn^{2+} = Product ions.

The loss of electron occurs at anode, and the gain of electrons occurs at cathode. Furthermore, the metal at anode dissolved, and the metal at cathode grows.

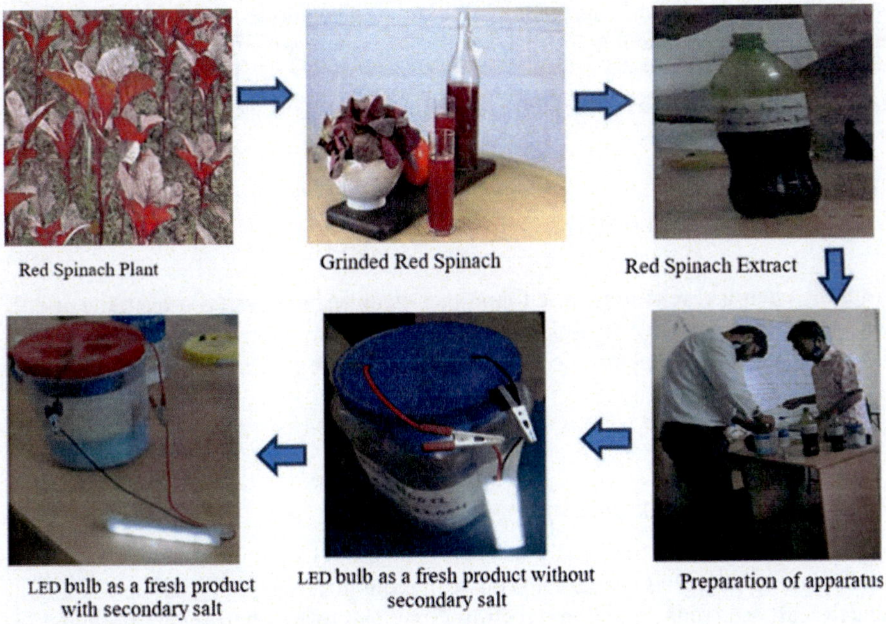

Fig. 5 Graphical abstract of the Red spinach electricity

Fig. 6 Graphical abstract of the green chili electricity

2.3 Energy Source

The energy for the PKL, pandan leaf, red spinach, and green chili electrochemical cells comes from the chemical change in anode (Zn-plate) when it dissolves in PKL, pandan leaf, red spinach, and green chili extract. The anode (Zn-plate) is oxidized inside the electrochemical cell. It has been also exchanged some electrons with the organic acid for electricity generation. As the cell does not require any external means to harvest the green energy, therefore, this kind of cell is worked as primary cell.

2.4 *Flow of Electrons*

From the chemical reactions, it is found that electrons always flow from the anode (Zn-plate) to the cathode (Zn-plate). The cell potential at standard state condition (E^o_{cell}) indicates that electron flows from the more −ve half reaction to the more + ve half reaction.

3 Results and Discussion

A comparative study has been carried out using four different plant extracts like PKL, pandan leaf, red spinach, and green chili. Graphical results have been discussed in the following.

It is shown (in Fig. 7) that the V_L harvested from PKL, pandan leaf, red spinach, and green chili extract over time. It is seen that the load voltage was maximum (5.52 V) for PKL, and it reached to 5.18 V after 120 h. It is found that for all cases except PKL, the harvested load voltage was almost stable up to 120 h. From Fig. 7, it is also observed that the change of load voltages with time for pandan leaf, R spinach, and green chili electrochemical cells was less than the load voltage of the cell.

Figure 8 shows the change of load current versus time duration for PKL, pandan leaf, red spinach, and green chili extract electrochemical cells. It is shown that that the load current (760 mA) was maximum for PKL. It shows that the change of load current with time for pandan leaf, red spinach, and green chili electrochemical cells was less than the I_L of PKL cell.

Figure 9 shows the change of load power with the variation of time duration for PKL, pandan leaf, Red spinach, and green chili extract electrochemical Cell. It is shown that that the load power (420 mW) was maximum for PKL. It shows that the change of load power with time for pandan leaf, red spinach, and green chili electrochemical cells was less than the load power of electrochemical cell based on PKL.

From the above all comparative study, it is found that the V_L, I_L, and hence P_L were higher for the PKL extract than pandan leaf, red spinach, and green chili electrochemical cells.

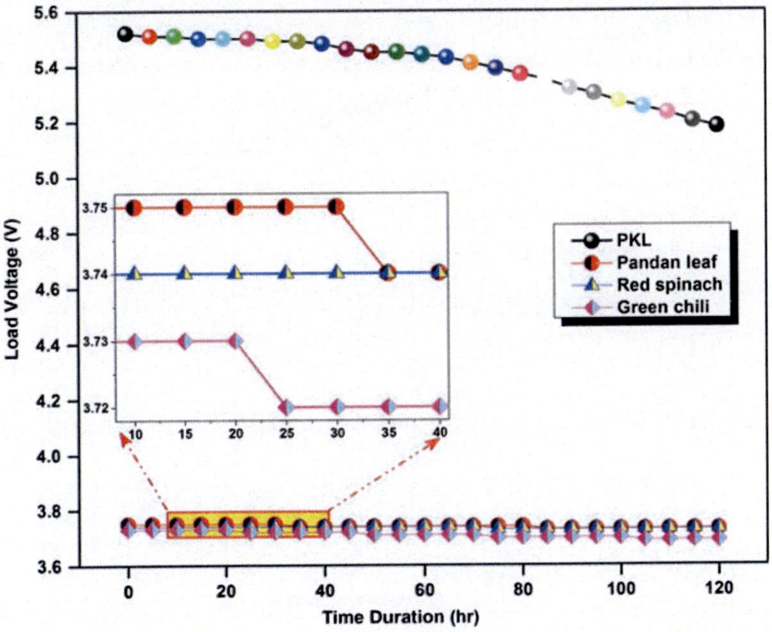

Fig. 7 V_L versus T.D. (h) curve

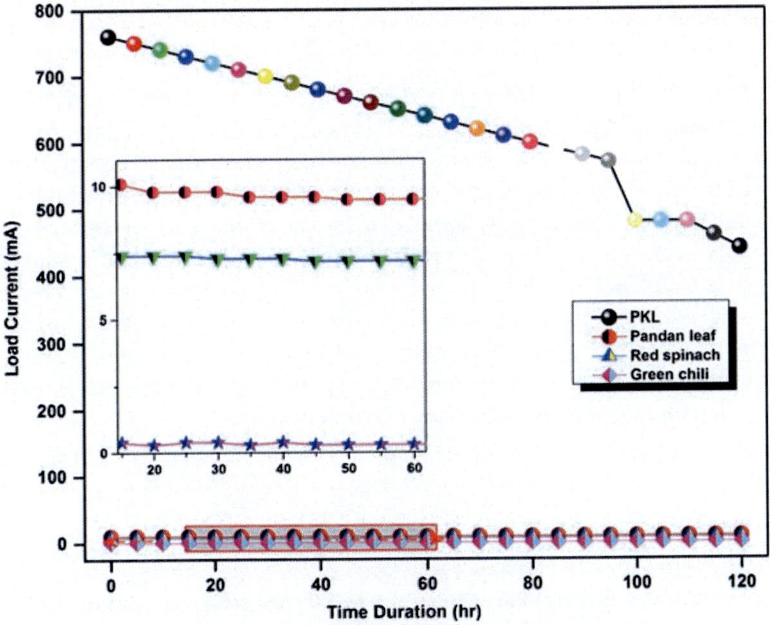

Fig. 8 Load current versus time duration curve

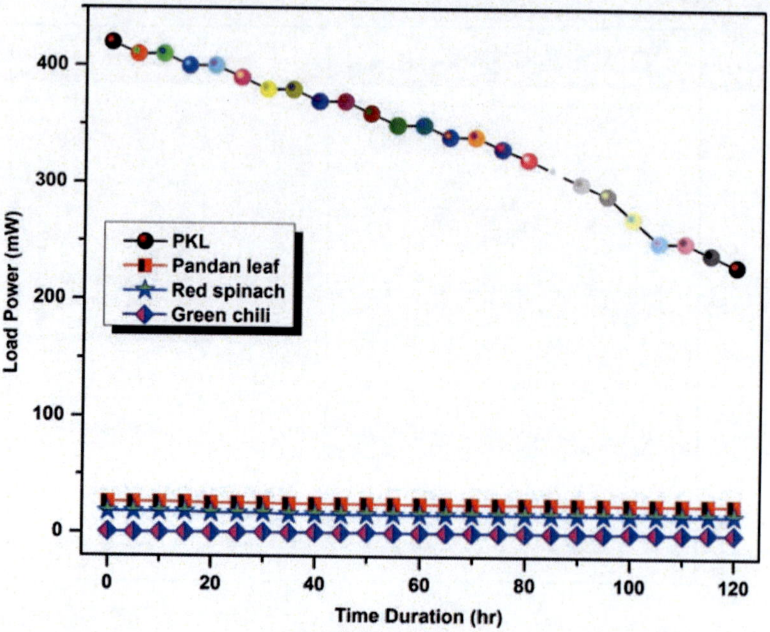

Fig. 9 Load power versus time duration curve

4 Conclusions

Electricity generation using different types of fruits and vegetables like PKL, pandan leaf, red spinach, and green chili extract has been discussed. Vegetative and fruits store chemical energy in their leaves and fruits in the form of chemical bonds. When electrodes are immersed into the electrolytes, ions move toward the electrodes which produce electricity. Finally, it is found that PKL is better than the other pandan leaf, red spinach, and green chili extract-based electrochemical cells. This work is beneficial to reduce the use of conventional energy sources, and it can reduce the greenhouse gasses released during power generation. This research area on harvesting green energy can play an instrumental role in developing a system for electronic/electrical appliances and create a new horizon in green energy research for scientific communities through electrochemical cells.

References

1. Jagessar RC, Allen R (2012) Phytochemical screening and atomic absorption spectroscopic studies of solvent type extract from leaves of Terminalia catappa, (almond). Acad Res Int 3(3):17–26
2. Khan MKA, Paul S, Rahman MS, Kundu RK, Hasan MM, Moniruzzaman M, Al Mamun M

(2016) A study of performance analysis of PKL electricity generation parameters (an experimental analysis on voltage regulation, capacity and energy efficiency of pathor kuchi leaf (PKL) electricity cell). In: IEEE 7th power india international conference (PIICON) 2016, pp 1–6. IEEE
3. Khan MKA, Rahman MS, Das T, Ahmed MN, Saha KN, Paul S (2017) Investigation on parameters performance of Zn/Cu electrodes of PKL, AVL, tomato and lemon juice based electrochemical cells: a comparative study. In: 3rd International conference on electrical information and communication technology (EICT) 2017, pp 1–6. IEEE
4. Khan KA, Rahman A, Rahman MS, Tahsin A, Jubyer KM, Paul S (2016) Performance analysis of electrical parameters of PKL electricity (an experimental analysis on discharge rates, capacity and discharge time, pulse performance and cycle life and deep discharge of Pathor Kuchi Leaf (PKL) electricity cell). In: IEEE Innovative smart grid technologies-Asia (ISGT-Asia) 2016, pp 540–544. IEEE
5. Paul S, Khan KA, Islam KA, Islam B, Reza MA (2012) Modeling of a biomass energy based (BPL) Generating power plant and its features in comparison with other generating plants. IPCBEE 44(3):12–16
6. Pal S, Chaudhuri AN (1991) Studies on the anti-ulcer activity of a Bryophyllum pinnatum leaf extract in experimental animals. J Ethnopharmacol 33(1–2):97–102
7. Khan KA, Bhuyan MS, Mamun MA, Ibrahim M, Hassan L, Wadud MA (2019) Organic electricity from Zn/Cu-PKL electrochemical cell. In: Mandal J, Sinha D, Bandopadhyay J (eds) Contemporary advances in innovative and applicable information technology. Advances in intelligent systems and computing, vol 812, pp 75–90. Springer, Singapore (2019). https://doi.org/10.1007/978-981-13-1540-4_9
8. Khan KA, Ali MH, Mamun MA, Haque MM, Ullah AA, Khan MI, Hassan L, Obaydullah AKM, Wadud MA (2020) Bioelectrical characterization and production of nanoparticles (NPs) using PKL extract for electricity generation. Microsyst Technol. https://doi.org/10.1007/s00542-020-04774-0
9. Khan MKA, Paul S (2013) A analytical study on electrochemistry for PKL (Pathor Kuchi Leaf) electricity generation system. In: IEEE Energytech 2013, pp 1–6. IEEE. doi: https://doi.org/10.1109/EnergyTech.2013.6645296
10. Hasan M, Alam K (2016) Bryophyllum pinnatum leaf fueled cell: an alternate way of supplying electricity at the off-grid areas in Bangladesh. In: 4th International conference on the development in the in renewable energy technology (ICDRET) 2016, pp 1–5. IEEE. doi: https://doi.org/10.1109/ICDRET.2016.7421522
11. Hasan MEHEDI, Khan KA, Mamun MA (2017) An estimation of the extractable electrical energy from Bryophyllum pinnatum leaf. Am Int J Res Sci Technol Eng Math (AIJRSTEM) 1(19):100–106
12. Ullah AA, Haque MM, Akter M, Hossain A, Tamanna AN, Hosen MM, Khan MKA (2020) Green synthesis of Bryophyllum pinnatum aqueous leaf extract mediated bio-molecule capped dilute ferromagnetic α-$MnO2$ nanoparticles. Mater Res Express 7(1):015088
13. Prakash R, Bhat IK (2009) Energy, economics and environmental impacts of renewable energy systems. Renew Sustain Energy Rev 13(9):2716–2721
14. Choo YY, Dayou J (2013) A method to harvest electrical energy from living plants. J Sci Technol 5(1):79–90
15. Ojewole JA (2005) Antinociceptive, anti-inflammatory and antidiabetic effects of Bryophyllum pinnatum (Crassulaceae) leaf aqueous extract. J Ethnopharmacol 99(1):13–19
16. Khan KA (1999) Copper oxide coatings for use in a linear solar Fresnel reflecting concentrating collector. Renewable Energy 17(4):603–608. https://doi.org/10.1016/S0960-1481(98)00023-8
17. Ruhane TA, Islam MT, Rahaman MS, Bhuiyan MMH, Islam JM, Bhuiyan TI, Khan KA, Khan MA (2017) Impact of photo electrode thickness and annealing temperature on natural dye sensitized solar cell. Sustain Energy Technol Assess 20:72–77. https://doi.org/10.1016/j.seta.2017.01.012
18. Ruhane TA, Islam MT, Rahaman MS, Bhuiyan MMH, Islam JM, Newaz MK, Khan KA, Khan MA (2017) Photo current enhancement of natural dye sensitized solar cell by optimizing

dye extraction and its loading period. Optik 149:174–183. https://doi.org/10.1016/j.ijleo.2017.09.024

19. Khan KA, Ali MH, Obaydullah AKM, Wadud MA (2019) Production of candle using solar thermal technology. Microsyst Technol 25(12):4505–4515. https://doi.org/10.1007/s00542-019-04390-7
20. Khan KA, Rasel SR, Ohiduzzaman M (2019) Homemade PKL electricity generation for use in DC fan at remote areas. Microsyst Technol 25(12):4529–4536. https://doi.org/10.1007/s00542-019-04422-2

Fractional Order Modified AWPI Based DC-DC Converter Controlled SEDC Motor

Rimi Paul

Abstract In this paper, a fractional order modified Anti Windup (AW) proportional–integral (PI) scheme is designed to generate the switching gate pulse of buck converter to control the speed of separately excited DC (SEDC) motor for low speed applications. Input saturation is a constraint on all real-time motors. Furthermore, for DC-DC power converters, the duty cycle is the natural control input. Therefore, to keep off the wind up phenomena, the output of a new modified fractional order back calculation-based AW scheme with PI controller is compared to the repetitive sawtooth waveform for switching pulse generation in closed loop speed tracking system. Here, fractional tuning gain of integrator contributes additional flexibility. To ameliorate the steady state error as well as tracking speed performance compared to the AWPI and fractional order AWPI controller, this new controller is formulated by taking the difference between AW tracking time constant and proportional gain connected to the output capacitor voltage signal. It can be observed that the new controller, with its two configurable parameters, outperforms in terms of speed tracking. Also, it can minimize the integral time absolute error (ITAE) and integral squared error (ISE). In presence of noise, both fractional order controllers perform well compared to the integer order controller by showing their potency for speed tracking as well as disturbance rejection.

Keywords Fractional order modified AWPI controller · DC-DC Buck converter · SEDC motor

1 Introduction

A DC-DC converter, an electromechanical device, turns the unregulated DC input into a controlled DC output at a preferred voltage level by adjusting the duty ratio for different DC motor drive applications such as fuel cell vehicles [1], in PV system [4], battery operated system [6], etc. DC-DC buck converter [2–4] is addressed by employing an anti windup PI approach in [3]. Here, a perturbed buck converter is

R. Paul (✉)
Aliah University, New Town, Action Area II, A/27, Kolkata, West Bengal 700160, India

J. K. Mandal et al. (eds.), *Topical Drifts in Intelligent Computing*, Lecture Notes in Networks and Systems 426, https://doi.org/10.1007/978-981-19-0745-6_57

designed using a Lyapunov based constructive approach for output feedback stabilization. Valenzuela [4] introduced a Lyapunov based PI with back calculation AW controller for input saturation. To optimize the performance during saturation and to restrict inductor current within a specific range, an anti windup strategy is used in [5–8]. Because of the superior performance, dependability, and changeable speed control, direct current (DC) motors play a large part in modern industrial drives such as battery operated [17] vehicles, electric trains, and so on [9]. Researchers have used a variety of control measures [9, 10] for controlling the motor's speed to avert machine damage, slow rise time and high overshoot. Khubalkar et al. used an embedded platform to construct the digital fractional order proportional integral derivative (FOPID) controller for speed control of buck converter fed DC motor [11]. In simulation environment of the buck converter driven DC motor in [12], two feedback control techniques (LQR and PI controller) are constructed and tested. A solar panel [13] is linked to the DC-DC converter's input for controlling speed of separately excited DC motor [15, 16] for getting below rated speed [13, 14]. To drive the gate pulse of DC-DC converter fed DC motor's speed control in closed loop approach, PI controller is compared with repetitive sawtooth waveform to generate pulse width modulation [18]. However, traditional controllers have constraints in optimizing between contradicting objectives such as fast reaction, small overshoots, and zero steady state errors as well as being insensitive to load-torque disturbances [10]. To obtain more precise and robust control performances, the fractional order (FO) [19] PID controller [20] is also studied in speed control of DC motor [10, 11, 15]. In practice, it is often observed that the performance of a PID controller [4] or FOPID controller [8, 9, 15] can be severely restricted due to the saturation of the actuators [7, 8], converter driven DC motor drives [11, 15, 18]. Therefore, consideration of actuator saturation is necessary to avoid poor performance of the controller and the instability [11]. Here, an effective methodology for speed control of buck converter driven SEDC motor model is taken and a fractional order modified anti windup PI controller is designed and compared to the repetitive wave to generate PWM signal for the converter. The performance of this new controller is compared with FO-AWPI and IO-AWPI controller. This paper has been articulated as follows: in the next Sect. 2, the description of converter with SEDC motor is demonstrated. The tuning parameter of derivative gain is kept zero. A fractional order modified AWPI controller is manifested in Sect. 3. Section 4 delineates the performance study and the comparison of the three controllers.

2 Converter Design for Separately Excited DC (SEDC) Motor

One of the main performance features for industrial applications in electric drives systems is an efficient dynamic speed command tracking response [18]. A separately

Table 1 Parameters of plant with converter

Plant with converter	Specifications
SEDC motor	5 hp, 240 v, 1750 rpm; Armature resistance (r_a) = 0.78Ω, inductance (L_a) = 0.0161 H, Field armature mutual inductance (L_{af}) = 1.234 H, Inertia (J) = 0.05 Kg-m^2, Viscous friction coefficient (B_m) = 0.01 N-m-S, coulomb friction torque = 0
DC-DC Buck converter	Input voltage (V_{Hv}) = 500 V, Output voltage (V_{Lv}) = 120 V-160 V, switching frequency = 16 kHz, 0.5%voltage-ripple ratio on the capacitors

excited DC motor (SEDC) is utilized as a plant in simulation instead of permanent-magnet DC motors due to constrained ratings such as few horsepower and a maximum speed constraint. This paper addresses armature voltage control approach to obtain the speed below rated speed of the SEDC motor using DC-DC Buck converter. The parameters of the plant (SEDC motor) with the converter are shown in Table 1.

The speed (ω_m) in rpm can be controlled [18] using (1)

$$\omega_m = \frac{1}{k_a\varphi_f}\left(V_t - \frac{R_a}{T}T_{em}\right) \tag{1}$$

Here, k_a = Armature constant; φ_f = Constant field flux in Weber; V_t = Buck converter output in volt; T_{em} = Electromagnetic Torque in N-m.

3 Fractional Order Modified AWPI Controller

The FO-PI controller with anti windup technique [8] is introduced here to ameliorate the controller performance by addressing the problem of actuator saturation [4, 8], i.e. a long time is required to reach the output in steady state [5]. However, inclusion of the fractional parameter in this method, the integral action for zero actuating signals is not fixed. In addition, by choosing the right fractional parameter value, it is possible to get a faster time response and better denoising. The AWPI controller [8] output signal is

$$u(t) = K_p e(t) + (K_I + T_I(u_{sat} - u))\int e(t)d(t) \tag{2}$$

In this AW scheme, controlling the tracking constant (T_I), the integral value can be reduced during integral saturation. Here, modified back calculation-based anti windup scheme [8] is used and $\frac{1}{T_I}$ is chosen as half of $\sqrt{\frac{K_p}{K_i}}$. In the proposed fractional order AW strategy, integer order integrator in Eq. (2) is substituted by a fractional order integrator to obtain switching frequency of the gate terminal of buck converter.

Here, the Riemann–Liouville definition of fractional integration is employed [19]. To generate FO-AWPI control output ($u_{fc}(t)$), Eq. (2) is modified:

$$u_{fc}(t) = K_P e(t) + (K_i + T_i(u_{sat} - u))D^{\mu} \int e(t)d(t) \tag{3}$$

where K_P and K_I are proportional and integral gain value. Here, $\mu \in \Re; \Re < 0$. D^{μ} can be written as $D^{-\mu}$. The term fractional integration $D^{-\mu}\ e(t)$ [19] is defined as

$$D^{-\mu}e(t) = \frac{1}{\Gamma(\mu)} \int_0^t (t - \tau)^{\mu-1} e(\tau)d\tau \tag{4}$$

Here, control laws are taken as (i) overshoot should be less than 20%; (ii) 2% settling time. Based on this control law, the FO-AWPI scheme for SEDC motor control furnishes large steady state error (e_{ss}) though it can reduce the settling time and percentage (%) overshoot for tracking speed performance compared to the IO-AWPI controller. But in both cases, e_{ss} is much higher. In this paper [20], a new integer order PID control scheme is proposed to control the gate pulse of the boost converter. Motivated from the given idea in [20], this new PID control structure was incorporated initially with a FO-AWPI controller for buck converter-based SEDC motor control. But implementing this scheme for FO controller does not help to reduce the steady state error. Therefore, a modified scheme is introduced with a FO-AWPI controller. The output voltage of capacitor is fed back and the modified control (u_{mfAW}(t)) signal is shown below:

$$u_{mfAW}(t) = K_P e(t) + (K_i + T_I(u_{sat} - u) - K_{P1} * V_c)D^{\mu} \int e(t)d(t) \tag{5}$$

Here, the feedback capacitor voltage is multiplied with proportional gain K_{P1} only. Initially, the term ($K_{P1} * V_c$) was added to the tracking time constant (T_I). If K_{P1} increases, the integration of three added gain values increase the overshoot and settling time with reduction of steady state error. But low value of K_{P1} provides longer time to settle. Therefore, the control signal in Eq. (5) is designed to improve the tracking performance. Here, Z-N tuning method is utilized for K_P and K_i value. These values are kept constant for all control structures. For modified fractional order AWPI controller, K_{P1} value is used by taking $\frac{K_P}{T_I}$ and regulated to study its performance. Fractional order modified AWPI controller is implemented using MATLAB SIMULINK which is shown in Fig. 1.

4 Experimental Result

Here, a SEDC motor is controlled by a DC-DC step down converter or buck converter whose input DC voltage is 120 V. The motor is considered for low speed application

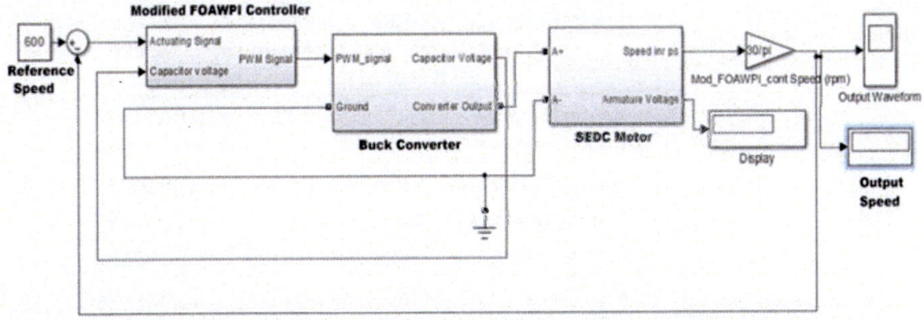

Fig. 1 Fractional order modified AWPI controller for buck converter controlled SEDC motor

with 2% pulsating torque at 600 rpm. Here, 300 V NiMH battery is considered and a battery source is used as input of the converter in the MATLAB SIMULINK block. The design specifications are shown in Table 1. From Table 1 data, for 5 hp motor with converter, the calculations are shown below:

2% basis permissible pulsating torque = 0.4069 N-m;

Rated torque = 20.3427 N-m.

The fundamental current to produce 2% pulsating torque = 0.3272 A;

To meet the fundamental current, the required switching frequency is 15.826 kHz. Therefore, in constant torque region, pulse width modulation (PWM) technique with switching frequency $\left(f_s = \frac{1}{T_s}\right)$ of 16 kHz is applied to the buck converter. Here, closed loop feedback speed control of SEDC motor is designed to keep constant the speed during any load disturbances.

Here, duty cycle is $D = \frac{T_{On}}{T_s}$; T_{On} = On time of the pulse in ms; T_s = Switching time. During T_{ON}, inductor current is calculated.

$$\Delta I_L = \frac{1}{f_s L}(V_s - V_o)D = 49.7067 \text{ Amp}$$

The average inductor current (I_{Lavg}) or output current = 31.0667 Amp. For continuous inductor current operation, minimum inductance 72.425 μH is considered. Here, in an open loop mode, the motor provides sluggish output speed response.

To generate the switching pulse to trigger the power switch, PI controller with anti windup strategy is used and compared to the sawtooth signal in [4, 5, 8]. Here, to enhance the performance of AWPI controller, integer order integral controller is replaced by fractional parameter. The additional tuning knob provides extra flexibility without disturbing the proportional and integral gain value in presence of noise [19]. This fractional order AWPI controller is then modified by connecting the gain K_{P1} to add an extra tuning knob to achieve more flexibility as well as to improve tracking performance and denoising performance without disturbing the other three tuning knobs with or without noisy system. The MATLAB SIMULINK block with new

Table 2 Performances of controllers

Specifications	AWPI controller	FO-AWPI controller	FO modified AWPI controller
% Maximum overshoot	2.2347	1.2876	0.8887
Rise time in s	0.0607	0.0632	0.0641
Settling time in s	0.0819	0.0725	0.0668
Peak time in s	0.0778	0.0789	0.0789
Steady state error	2.2480	2.2798	0.1049
% Initial undershoot (%IU)	2.5780	2.5760	2.5761

control scheme-based converter controlled motor shown in Fig. 1 is used to control the speed in 600 r.p.m. The performances of three controllers, i.e. IO-AWPI, FO-AWPI, and fractional order modified AWPI controller are shown in Table 2. The output speed responses are shown in Fig. 2. From Table 2, it is seen that by keeping the gain values K_P, K_i, T_I, and μ constant and by varying only K_{P1} value, the tracking performances of new controller such as percentage maximum overshoot (%MO), settling time (T_s), and steady state error (e_{ss}) are reduced. In comparison with IO and FO controller, it is found from Table 2 that fractional parameter helps to improve tracking performance by reducing %MO, T_s, and %IU. But it leaves a higher e_{ss} value for the FO-AWPI controller-based motor speed tracking case. New modified controller improves tracking performance but it increases rise time which indicates slightly sluggish response compared to other two controllers.

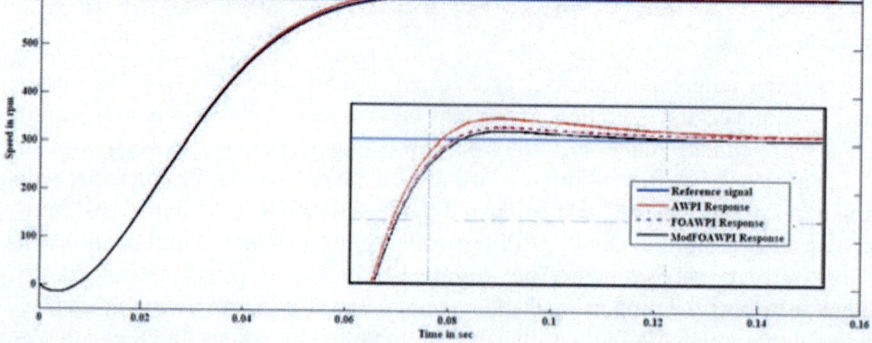

Fig. 2 The output speed tracking response of three controllers based buck converter controlled SEDC motor

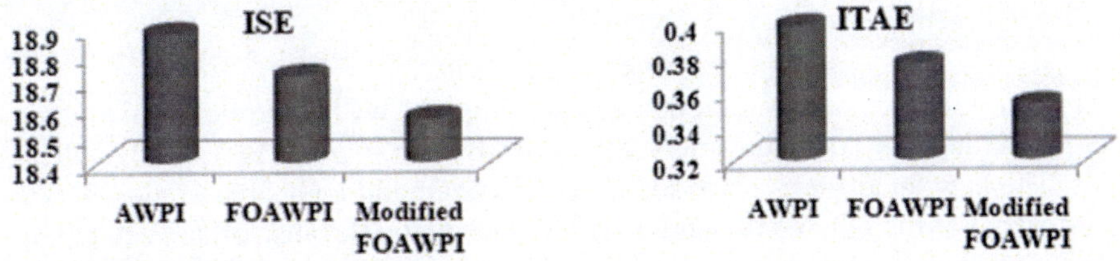

Fig. 3 Performance indices of three controllers

Here, the performance of the error signal, i.e. ISE or e_2 and ITAE or time multiplied e_1 is also measured and compared for three controllers shown in Fig. 3. It is observed that the new controller minimizes both ISE and ITAE. Also, both FO controllers perform well by minimizing integral time squared error (ITSE) compared to the integer one in this case.

To showcase the efficacy of the new controller in presence of noise, the band limited white noise with noise power 0.0005 W/Hz is taken. Compared to the performances of IO version shown in Table 3, FO-AWPI controller reduces rise time and steady state error. It also improves disturbance rejection performance, i.e. increases signal to noise ratio (SNR) by lowering mean squared error (MSE) [19]. In this case, IO controller slightly reduces %MO and settling time (T_s). Due to one extra knob, new controller yields better reduction of %MO and T_s compared to others. Therefore, by adjusting two flexible parameters of fractional order modified AWPI controller, both tracking and denoising performances can be ameliorated compared to the IO-AWPI controller-based converter controlled motor speed tracking system. It is also observed that both FO and new controller reduce the ISE and ITAE compared to integer order controller. The similar tracking and denoising performances are also observed for the magnitude of 0.001 W/Hz process noise with higher sampling rate case.

Table 3 Performances of controllers in presence of noise

Specifications	AWPI controller	FO-AWPI controller	FO modified AWPI controller
% Maximum overshoot	8.6333	8.8845	8.0734
Rise time in s	0.1156	0.1153	0.1163
Settling time in s	0.2657	0.2689	0.2586
Steady state error	13.2391	12.2657	14.0928
SNR in dB	13.6192	13.79	13.74

5 Conclusion

This paper introduces a new fractional order modified AWPI controller for switching pulse generation of buck converter controlled SEDC motor for low speed applications. The new controller's performance is measured and compared in closed loop mode to that of the FO-AWPI and IO-AWPI controllers. Because of its extra tuning value, this new controller furnishes lower maximum overshoot, settling time, and steady state error than the FO and IO versions. In addition, for noisy case, when compared to an IO controller, both FO and modified FO controllers perform well in terms of tracking the reference speed and improving denoising performance, i.e. increasing SNR and lowering MSE.

References

1. Zhang Y, Liu H, Li J, Sumner M, Xia C (2019) A DC DC-DC Boost Converter with a wide input range and high voltage gain for fuel cell vehicles. IEEE Trans on Power Electron 34(5):4100–4111
2. Garg MM, Hote YV, Pathak MK (2015) Design and performance analysis of a PWM dc–dc Buck converter using PI–lead compensator. Arab J Sci Eng 40(12):3607–3626
3. Ibanez CA, Valenzuela JM, Alarcón OG, Lopez MM, Acosta JA, Castanon MSS (2021) PI-type controllers and Σ-Δ modulation for saturated DC-DC Buck power converters. IEEE Access 9:20346–20357
4. Valenzuela JM (2020) A class of proportional-integral with anti-windup controllers for DC–DC Buck power converters with saturating input. IEEE Trans Circuits Syst II Express Briefs 67(1):157–161
5. Xiao W, Wen H, Zeineldin HH (2012) Affine parameterization and anti-windup approaches for controlling DC-DC converters. In: IEEE international symposium on industrial electronics
6. Cheng Y, Yang C, Wen G, He Y (2017) Adaptive saturated finite-time control algorithm for buck-type DC–DC converter systems. Int J Adapt Control Signal Process 31(10):1428–1436
7. Yang SK (2012) A new anti-windup strategy for PID controllers with derivative filters. J Control 14(2):564–571
8. Paul R, Afroz N (2018) Anti-windup FOPI controller for step motor. In: IEEE, International conference on electronics, materials engineering and nano-technology (IEMENTech), (2018).
9. Viola J, Angel L, Sebastian JM (2017) Design and Robust Performance evaluation of a fractional order PID controller applied to a DC motor. J IEEE/CAA Autom Sincia 4(2):304–314
10. Hsu YY, Chan WC (1984) Optimal variable-structure controller for DC motor speed control. IEEE Proc Control Theory and Appl 131(6):233–237
11. Khubalkar S, Chopade A, Junghare A, Aware M, Das S (2017) Design and realization of stand-alone digital fractional order PID controller for buck converter fed DC motor. Turk J Electr Eng Comput Sci 35:2189–2211
12. Ismail RMTR, Ahmad MA, Ramli MS (2009) Speed control of Buck-converter Driven Dc motor using LQR and PI: a comparative assessment. In: International conference on information management and engineering
13. Jayaprakash S, Ramakrishnan V (2014) Simulation of solar based DC-DC converter for armature voltage controlled separately excited motor. In: International conference on advances in electrical engineering (ICAEE)
14. Singh SN, Kohli DR (1986) Performance determination of a chopper controlled separately excited DC motor. IEEE Trans Ind Electron 31(1)

15. Seo S, Choi HH (2019) Digital implementation of fractional order PID-type controller for boost DC–DC converter. IEEE Access 7:142652–142662
16. Blasko V (1985) Model of chopper-controlled DC series motor. IEEE Trans Ind Appl IA-21(1)
17. Arbetter B, Erickson R, Maksimovic D (1995) DC-DC converter design for battery-operated systems. In: Proceedings of PESC '95—power electronics specialist conference, vol 1, pp 103–109
18. Mohan N, Undeland TM, Robbins WP (2002) Power electronics: converters, applications, and design. 3rd Bk&Cdr edn. Wiley
19. Paul R, Sengupta A (2021) Fractional order intelligent controller for single tank liquid level system. In: IEEE Second international conference on control, measurement and instrumentation (CMI), pp 24–29
20. Adnan MF, Oninda MAMM, Nishat M, Islam N (2017) Design and simulation of a DC-DC boost converter with PID controller for enhanced performance. Int J Eng Res Technol (IJERT) 6(09)

A Non-Linear Modeling Towards the Pade Approximated Electric Ventricular Assist Device Using Describing Function Technique

Tanmoy Singha, Soumyendu Bhattacharjee, Rudra Sankar Dhar, Arindam Biswas, and Joydeep Dutta

Abstract One of the most remarkable menaces that attract attention in this ".com" era is cardiac surgery. To prevent this, the electric ventricular assist device (EVAD) has widely been put into practice, especially during the execution of the cardiac surgery. Previous research works present function with the exponential time delay, but this function has widely been accepted as it has proved to be helpful in creating not just polynomial form but exponential forms as well through the "Pade approximation method." This paper aims to unveil a nonlinear analysis through the EVAD system by considering dead zone combined with saturated nonlinearity, which has resulted in the creation of a converging stable limit cycle through analysis. Ultimately, by the application of the Lyapunov stability theorem on the proposed nonlinear model, it is vivid that due to the semi-definite energy function, Sylvester's criteria are used, and proper energy function for our proposed design has also been calculated.

Keywords Electric ventricular assist device (EVAD) · Non-linearity · Lyapnuov stability

1 Introduction

In 1964, prosthetic research was initially started at the national level. The specific part of the human physique comprising with heart, veins, capillaries and arteries, the components of the desktop responsible (circulation of fluids, blood) for short-term,

T. Singha · R. S. Dhar (✉)
Department of ECE, National Institute of Technology Mizoram, Aizawl, Mizoram, India
e-mail: rdhar@uwaterloo.ca

S. Bhattacharjee
Department of Electronics, Barasat State University, Kolkata, West Bengal, India
e-mail: principal@rpcollege.ind.in

A. Biswas
School of Mines and Metallurgy, Kazi Nazrul University, Asansol, West Bengal, India

J. Dutta
Department of CS, Kazi Nazrul University, Asansol, West Bengal, India

J. K. Mandal et al. (eds.), *Topical Drifts in Intelligent Computing*, Lecture Notes in Networks and Systems 426, https://doi.org/10.1007/978-981-19-0745-6_58

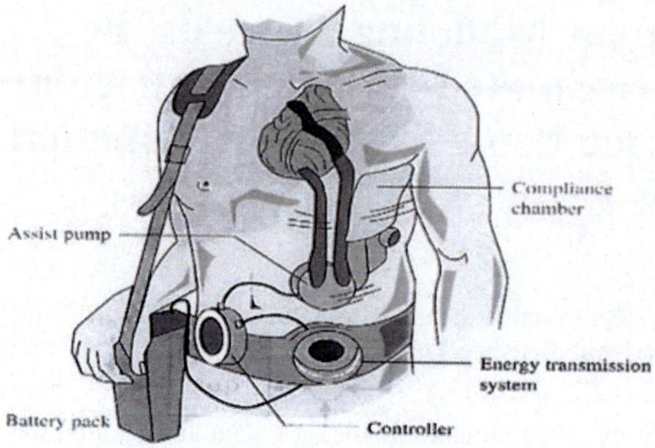

Fig. 1 Standard model approach for EVAD

is used in end-stage coronary heart failure [1]. EVAD is a device that mostly aids the medical clinical activities and is usually used for the mobility and compression of liquid and particularly for those who have feeble hearts. Battery-operated electric ventricular assist device that is implanted by the body is depicted by Fig. 1.

In the previous years, EVDA was mostly used as a device or technique for objective therapy, especially with patients in the end stage of heart failure.

2 Linear Control Model Analogy of EVAD System

Primarily, with the help of EVDA, two predominant jobs are executed. It powers the pusher plate by adjusting the motor voltage. It fulfills the demand of the body cardiac system by modifying the flow of blood circulation. As per the EVAD beat rate, the flow of blood is accordingly adjusted with the aid of the blood drift controller [2, 3].Figure 2 describes the control system-based analogy of EVAD system. $G_c(s)$ represents the controller transfer function which is connected in series to improve the response.

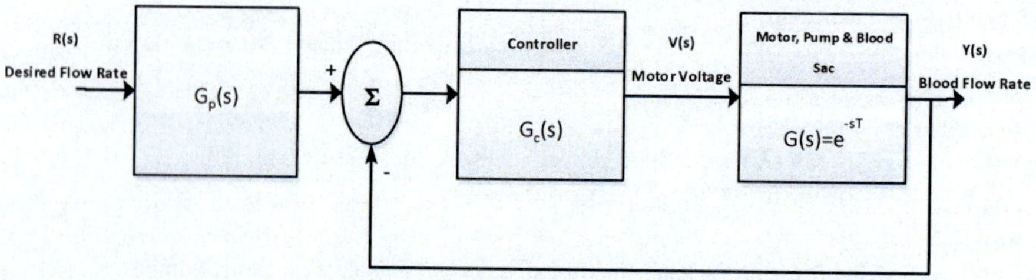

Fig. 2 Linear control system-based model of EVAD

3 PADE Approximation Technique Toward EVAD System

Electric ventricular assist device transfer function is depicted as $G(s) = e^{-s\tau}$. In the field of engineering of control system, the frequency responses study should be assessed as a logical transfer function, or we can say it is a rational transfer function. In Control system engineering, the frequency response analysis of as a time delay function must be approximated in a form of the rational transfer function, otherwise the improvement of the system response will be a difficult task for a researcher. The method that is most widely utilized and applied that has now been adapted is Pade approximation. This function helps for approximation over a defined region inside its arena in order to minimize errors. On $[a, b]$, the rational approximation is $f(x)$. The most general wide accepted useful practice is the Pade Approximation. This function helps to approximate over a small region of its domain to reduce the errors as low as possible. The rational approximation of $f(x)$ on $[a, b]$ is quotient of two different polynomials $p_u(x)$ and $q_v(x)$ which are having degree u and v, usually $S_{u,v}$ (x).

$$S_{u,v}(x) = \frac{p_u(x)}{q_v(x)} \quad \text{for } a \le x \le b \tag{1}$$

Using computational algorithm, a rational approximation can easily be produced which has a small error on $[a, b]$ than an ordinary approximated polynomial [4, 5]. At $x = 0$, the functions $f(x)$ and $f(x)$ have to be continuous, according to the Padé approximation approach. If $(m + n)$ is true, and only if $(m + n)$ is true, in terms of $f(s)$ and $S_{u,v}$ (x), power series expansions are equal; the rational function $S_{m,n}$ (x) of factors of S as u x v is also known as Padé approximantion of the function $f(s)$. The following Padé approximant can be used to approximate the function $f(s)$.

$$S_{u,v}(x) = \frac{p_u(x)}{q_v(x)} = \frac{a_0 + a_1 s + a_2 s^2 + \cdots + a_{n-1} s^{v-1}}{b_0 + b_1 s + b_2 s^2 + \cdots + b_{n-1} s^{v-1} + s^v} \tag{2}$$

The set of relation can be established as given below:

$$a_0 = b_0 c_0 \tag{3}$$

$$a_1 = b_0 c_1 + b_1 c_0 \tag{4}$$

$$a_{n-1} = b_0 c_{n-1} + b_1 c_{n-2} + \cdots + b_{n-1} c_0 \tag{5}$$

Once the co-efficient c_j, $j = 0, 1, 2 \ldots$ is found and $c_j = (-1)^j a_{j+2,1}$, the total model can be represented as follows:

$$G(s) = \frac{d_0 + d_1 s + d_2 s^2 + \cdots + d_{n-1} s^{v-1}}{e_0 + e_1 s + e_2 s^2 + \cdots + e_{n-1} s^{n-1} + e_n s^v} \tag{6}$$

The deviations from the logical approximation of the delay function $e(-s)$, as well as the truncated errors within a finite series expansion, are steadily reduced using this approximation method. The function $e(-s)$ can be written in the following way.

$$e^{-\tau s} \approx \sum_{i=0}^{u+v} (-1)^i \frac{(\tau s)^i}{i!} = \frac{\sum_{i=0}^{u} p_i (\tau s)^i}{\sum_{i=0}^{v} q_i (\tau s)^i} \tag{7}$$

The polynomial approximations must be built in such a way that the transfer function $e^{-\tau s}$ is easy to accomplish, robust, and optimal. The polynomials for S (u, v) (x) from Eq. (7) can be estimated as follows:

$$e^{-\tau s} = \frac{1 - k_1 s + k_2 s^2 - k_3 s^3 + \cdots \pm k_n s^v}{1 + k_1 s + k_2 s^2 + k_3 s^3 + \cdots + k_n s^v} \tag{8}$$

The order of the approximation is represented by v, and the components $k\ i$ are functions of v. The values collated for K_i for $v = 1$ and $v = 2$ have been given in Table 2.

The second order Padé approximation ($v = 2$) is used in Eq. (8), and $H_{\text{pade}}(s)$ has been formulated with the purpose of obtaining a simple and optimal transfer function choice:

$$H_{\text{pade}}(s) = \frac{1 - k_1 s + k_2 s^2}{1 + k_1 s + k_2 s^2} = \frac{1 - \frac{\tau}{2}s + \frac{\tau^2}{12}s^2}{1 + \frac{\tau}{2}s + \frac{\tau^2}{12}s^2} = \frac{\tau^2 s^2 - 6\tau s + 12}{\tau^2 s^2 + 6\tau s + 12} \tag{9}$$

As shown in Table 3, numerous values of settling time have been recorded and tabulated.

Table 2 Values of k for the order of v

When $v = 1$	When $v = 2$
$k_1 = \frac{\tau}{2}$	$k_1 = \frac{\tau}{2}$, $k_2 = \frac{\tau^2}{12}$

Table 3 Settling time with different τ value

τ	Settling time (s)
2	3.11
1.5	2.34
1	1.56
0.6	0.934
0.2	0.311
0.1	0.156

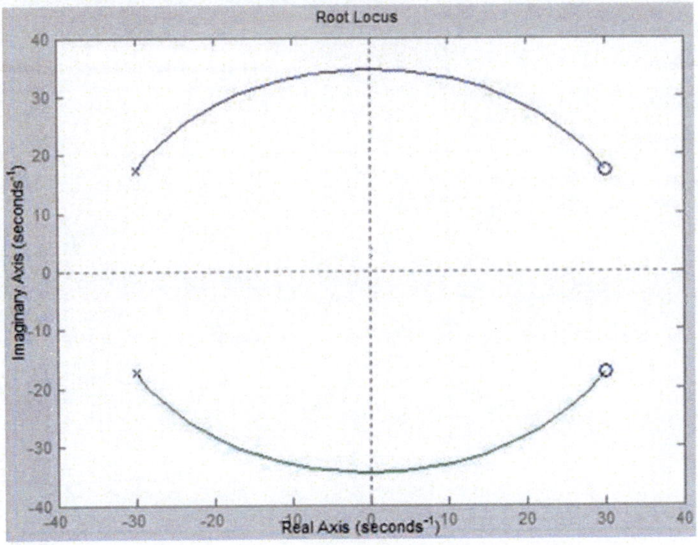

Fig. 3 Representing the root locus of EVAD system

From the above table it is clear that when $\tau = 0.1$, the approximated system has the least settling time of 0.156 second for different Settling Time. So,

$$G_{\text{pade}}(s) = \frac{0.01S^2 - 0.6S + 12}{0.01S^2 + 0.6S + 12} \tag{10}$$

The above equation represents the Padé estimated transfer function for EVAD system in linear control system-based approach. The root locus and Bode plot of the system are given in Figs. 3 and 4, respectively.

From Figs. 3 and 4, it can be observed that the Pade approximated model of EVAD system considering transportation delay is stable in time domain. Now, in the next section, we will analyze the system and check the stability considering all types of nonlinearity present in the proposed system.

4 Designing of PID Controller Using Tuning Process for Beneficial Performance of EVAD System

Ziegler Nichols tuning rule generally refers to the tuning method for the basic PID controller. A computational approach helps to find out optimal sets of parameter values of PID controllers. K_p is the proportionate gain. These variables are the integral time T and the derivative time T_d together are expressed through the transfer function.

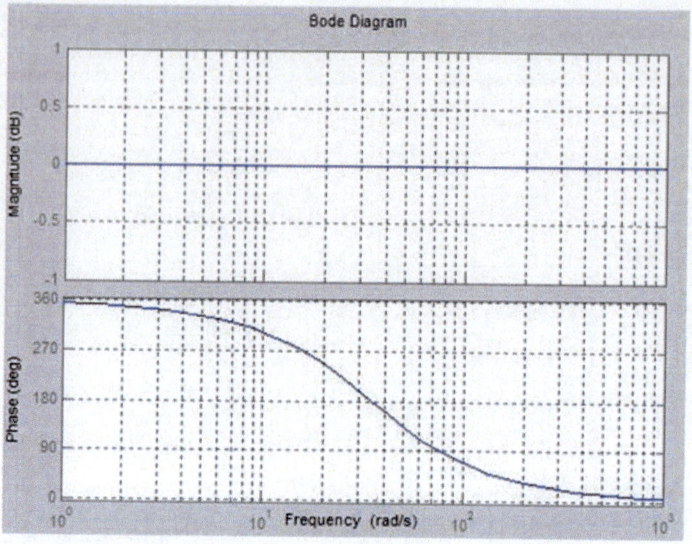

Fig. 4 Representing the bode plot

$$G_c(s) = K_p\left(1 + \frac{1}{T_i s} + T_d s\right) \tag{11}$$

Initially, $T_i = \infty$ and $T_d = 0$ are set by the usage of proportional control action only, whereas there is a tendency of k_p The value of k_p should be increased until it reaches the critical value K_{cr}, when the output tends to show sustained oscillations. The abovementioned parameters are reported in Table 4 according to the directed formula.

Now, PID controller is fine-tuned and designed for EVAD system as per the Ziegler–Nichols rules for Eq. (11) displays:

$$= 0.6K_{cr}\left(1 + \frac{1}{0.5P_{cr}s} + 0.125P_{cr}s\right) \tag{12}$$

If the frequency of sustained oscillation is ω_{cr}, the R–H criterion is used to determine the value of the critical gain K_{cr}; then, the system can be said to have a mathematical model, where $\frac{2\pi}{\omega_{cr}} = P_{cr}$.

Table 4 Considered setting PID values for Z–N tuning

Type of controller	K_p	T_i	T_d
P	0.5 *K*cr	0	0
PI	0.45 *K*cr	1/1.2 *P*cr	0
PID	0.6 *K*cr	0.5 *P*cr	0.125 *P*cr

The open-loop transfer function (approximated) for EVAD system considering least settling time and without tuning is referred to as the Eq. (12) is given below.

$$G_{\text{pade}}(s) = \frac{s^2 - 60s + 1200}{s^2 + 60s + 1200} \tag{13}$$

In closed-loop transfer function in Ziegler–Nichols tuning method, with tuning is exhibited as follows:

$$G(s) = \frac{\mathrm{K}_P}{s^3 + 60s^2 + 1200s + \mathrm{K}_P} \tag{14}$$

[Setting $T_{\rm i} = T_{\rm d} = 0$] Now, a Routh stability criterion is the primary section which determines the value $k_{\rm p}$, and the device becomes slightly stable as a result of the $k_{\rm p}$. The characteristics of the polynomial for the closed loop can be represented as follows:

$s^3 + 60s^2 + 1200s + K_{\rm P} = 0.$

The Routh table has been formed and given below.

$$\begin{array}{lcc} s^3 & 1 & 1200 \\ s^2 & 60 & K_{\rm P} \\ s^1 & \frac{[60\times(1200)-k_{\rm p})]}{60} & 0 \\ s^0 & K_{\rm P} & 0 \end{array}$$

In order to get the value of critical gain ($K_{\rm cr}$) of the system, the s1 row is to be made to zero. Thus, $\frac{72000-60K_{\rm P}}{60} \geq 0$.

From the above condition, the value of has been calculated as $K_{\rm p} = 1200$. Hence, $K_{\rm p} = K_{\rm cr} =$ critical gain $= 1200$. Thus, maintaining that $K_{\rm P}$ equal to $K_{\rm cr}$, the attribute equation is stated as.

The R–H array can be used to write the auxiliary equation as.

$$60s^2 + K_{\rm p} = 0 \tag{15}$$

Using the value of $K_{\rm p} = 1200$, we get $s^2 = -60/1200 = -0.05$ in Eq. (18).

By inserting $s = j$, the equation can also be substituted with the value of s. As a result,

$$\omega = 0.23 \text{ is the equation} \tag{16}$$

This value of ω is called $\omega_{\rm cr}$ or critical frequency for PID controller.

Now, calculating $P_{\rm cr} = \frac{2\pi}{\omega_{\rm cr}} = \frac{2\pi}{0.23} = 27$. We may also calculate the values of $K_{\rm p}$, using the Z–N tuning rule. $T_{\rm i}$, $T_{\rm d}$ using the values of $K_{\rm cr}$ and $P_{\rm cr}$.

$$K_{\rm P} = 0.6K_{\rm cr} = 720 \tag{17}$$

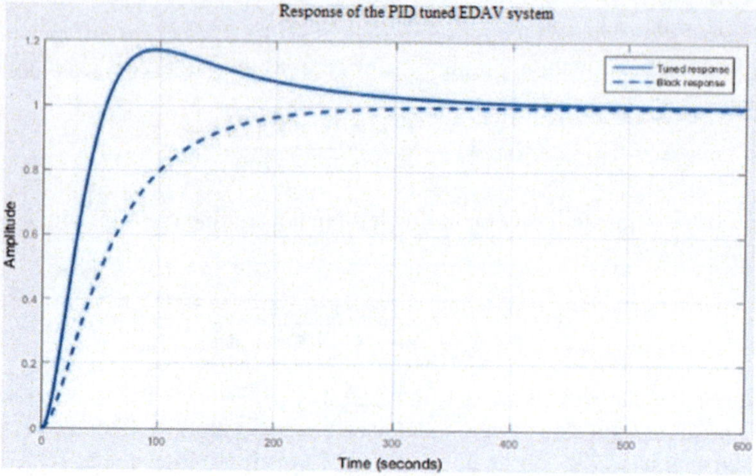

Fig. 5 The PID-tuned EDAV system's response with $\tau = 0.1$

$$T_{\mathrm{i}} = 0.5P_{\mathrm{cr}} = 13.5 \tag{18}$$

$$\mathrm{T_d} = 0.125\mathrm{P_{cr}} = 6.75 \tag{19}$$

As the value of K_{p}, T_{i}, T_{d}, the transfer function of the PID controller was also calculated using the equations from Eqs. (18) and (19).

$$Gc(s) = 1200\left(1 + \frac{1}{0.15s} + 0.07s\right) \tag{20}$$

The simulation result of EVAD system considering tuning is given below in Fig. 5. IIt is observed after studying the response of Fig. 5, steady state error has been minimized and stability criterion of the system that after tuning, i.e., the system practically gives all of the required performance and stability, owing to the closed-loop design or the 2nd order tuned by the EVAD system.

5 Type of Nonlinearity Present in EVAD System

To trace a nonlinear system's function approach, it is important to define it as the ratio of amplitudes (both complex and real parts) plus the phase angle between the fundamental harmonic components in the output to input sinusoid. Sinusoidal description function [6] is another name for it. $V(X, \omega) = \frac{Y_1}{X}\angle\Phi$ is the mathematical expression. The phase shift Φ inside the sinusoidal component that is fundamental component at output is depicted in the diagram below.

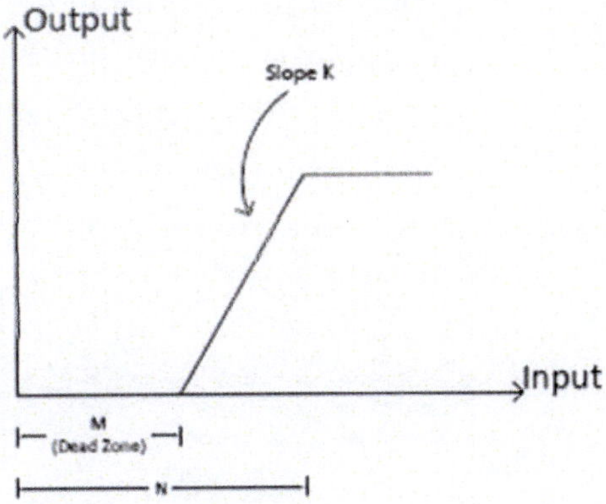

Fig. 6 Overall nonlinearity's input–output properties

The combination of dead zone and saturation, according to the results obtained above, is the nonlinearity present in heart muscle.

Figure 6 represents the input verses output features of overall nonlinearity, whereas Figs. 7 and 8 show the input and output waveform individually in the diagram below.

Let us consider $x = X_m \sin \omega t$, then the corresponding output is given by the expression $y = 0$, for $0 \le \omega t \le \theta 1$, $= K.(X_m - \frac{M}{2})$, for $\theta 1 \le \omega t \le \theta 2$, $= 0$, for $(\pi - \theta 1) \le \theta 1$,

$$A_1 = \frac{2}{\pi}\left[\int_{\theta 1}^{\theta 2} k\left(X_m . \sin \omega t - \frac{M}{2}\right) \cos \omega t d(\omega t)\right.$$

$$+ \int_{\theta 2}^{\pi - \theta 2} k\left(s - \frac{M}{2}\right) \cos \omega t d(\omega t)$$

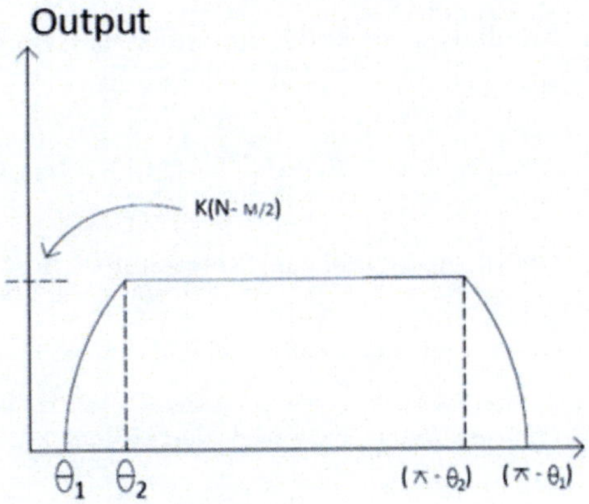

Fig. 7 Input waveform

Fig. 8 Output waveform

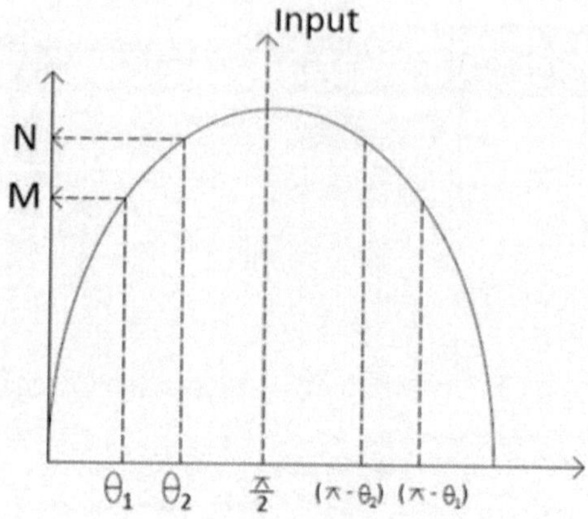

$$+\int_{\pi-\theta 2}^{\pi-\theta 1} k\left(X_m.\sin\omega t-\frac{M}{2}\right)\cos\omega t\, d(\omega t)\Bigg] \tag{21}$$

$$A_1 = 0 \tag{22}$$

After this, in order to calculate the value of B_1, we must follow the necessary steps:

$$B_1 = \frac{2K}{\pi}\Bigg[X_m.\theta 2 - X.\theta 1 - \frac{X_m}{2}\sin 2\theta 2 + \frac{X_m}{2}\sin 2\theta 1 + 2N\cos\theta 2 - M\cos\theta 1\Bigg] \tag{23}$$

As input is $x = X_m.\ \sin\omega t$, and at $\omega t = \theta 1$, amplitude is $\frac{M}{2}.M = 2X_m \sin\theta 1$. So, $\sin\theta 1 = \frac{M}{2X_m}$.

Similarly at $\omega t = \theta 2$, amplitude is s. So $N = X\sin\theta 2$. So, $\sin\theta 2 = \frac{N}{X_m}$. Putting these values

$$B_1 = \frac{KX_m}{\pi}[2(\theta 2 - \theta 1) - \sin 2\theta 1 + \sin 2\theta 2] \tag{24}$$

Hence, mathematical expression of describing function for the nonlinearity is expressed by

$$V(X_m, \omega) = \frac{\sqrt{A_1^2 + B_1^2}}{X}\angle\tan^{-1}\frac{A_1}{B_1} = \frac{B_1}{X}\angle 0 \tag{25}$$

$$= 1 - \frac{2}{\pi}[2(\theta 2 - \theta 1) - \sin 2\theta 1 + \sin 2\theta 2] \quad \text{for} \quad N > X_m > \frac{M}{2} \tag{26}$$

The required description function for the suggested nonlinear architecture is shown in Eq. (26).

6 Approach for Analyzing of Nonlinearity with the Use of Describing Function and Simulation Result

We can keep track of all the different types of nonlinearity in the EVAD system by putting them all in one block that represents the output attributes of the non-input linearity, with the representing function already evaluated in the prior section, and all the different types of linearity's in another block. Through the use of a block diagram, the following figure depicts the system with all of its linearity's and many nonlinearity's (Fig. 9).

Now, the polar plot of the transfer function of EVAD system is simulation showing in MATLAB and explained in Fig. 10.

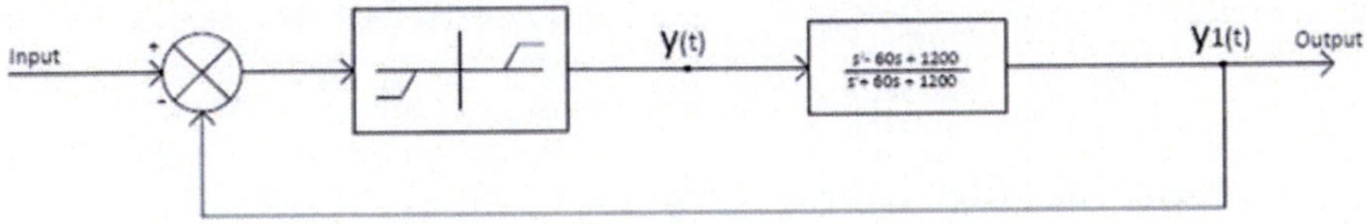

Fig. 9 Represents the type of nonlinearity along with linear section

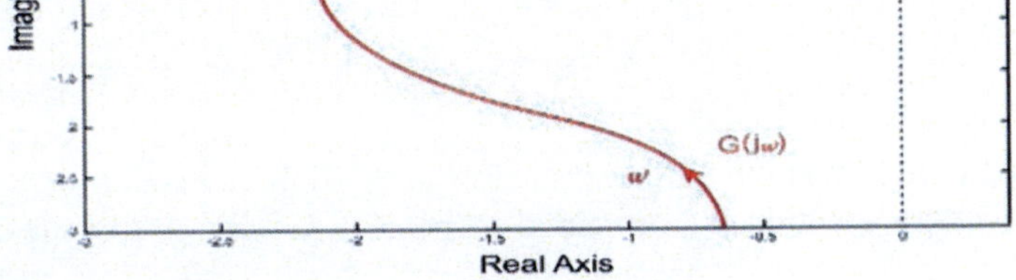

Fig. 10 Represents the polar plot of EVAD system

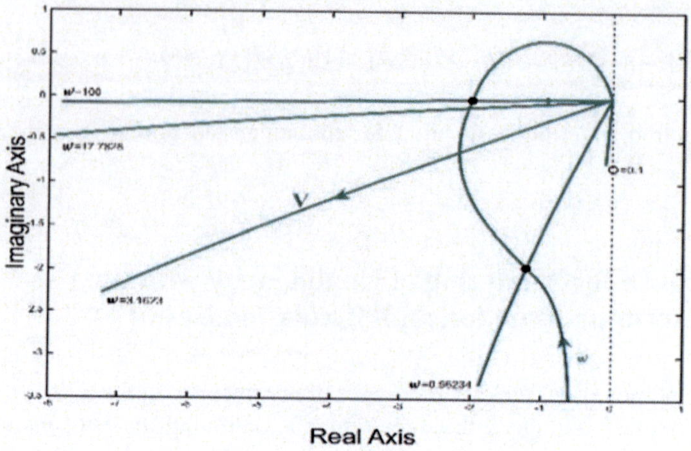

Fig. 11 Represents the polar plot describing function of EVAD system

The frequency is changed in the MATLAB simulation according to the description function shown in Fig. 11, and it is obvious that a robust limit cycle is constructed for the suggested nonlinear EVAD system configuration based on simulation results. As a result, it is reasonable to infer that the design will work in practice, opening the way for implementation.

7 Lyapnuov Stability

In order to execute this, Lyapnuov's direct method is applied on our preferred nonlinear Pade approximated EVAD system to find the system stability. External input is regarded to be zero for an autonomous system [6]. The dead zone is considered based on transportation delay. It has been observed that for $T = 0.1$ the system passes the approximate value, the shortest time for settling 0.156 s.$G_{\text{pade}}(s) = \frac{S^2-60S+1200}{S^2+60S+1200}$. The state equations are given by

$$\dot{x}_1 = -60x_1 - 1200x_2 \tag{27}$$

$$\dot{x}_2 = x_1 \tag{28}$$

where x_1 and x_2 represent the different states of the system. In case of our proposed model, the energy function is considered to be $v(t) = ax_1^2 + bx_2^2$. Taking the derivative,

$$\dot{v}(t) = \frac{\delta v}{\delta x_1} : \dot{x}_1 + \frac{\delta v}{\delta x_2}.\dot{x}_2 \tag{29}$$

Using Eqs. (34) and (35),

$$\dot{v}(t) = -120ax_1^2 - (2400ax_1x_2) + 2b \tag{30}$$

Now, for $a = 1$, $b = 1200$, our energy function for EVAD system will be $v(t) = x_1^2 + 1200x_2^2$. As a result, Eq. (30) can be expressed as follows:

$$-\dot{v}(t) = 1200x_1^2 \tag{31}$$

Equation (31) represents clearly that the time derivative is negative for our energy function which indicate the stability of the system. Though Eq. (31) is a positive quantity, yet, it is a semi-definite function as the function is totally independent of x_1. To find the trajectory, Jacobian of the function has been calculated and given below. Equations (30) and (31) are considered to be the function "f" and "g," respectively.

$$J = \begin{vmatrix} \frac{\delta f}{\delta x_1} & \frac{\delta f}{\delta x_2} \\ \frac{\delta g}{\delta x_1} & \frac{\delta g}{\delta x_2} \end{vmatrix} = \begin{vmatrix} -60 & -1200 \\ 1 & 0 \end{vmatrix} \tag{32}$$

The characteristics equation $|\lambda I - A| = 0$ is solved to find the roots λ_1 and λ_2 which gives

$$\lambda^2 + 60\lambda - 1200 = 0 \tag{33}$$

So, $\lambda_1 = -30 + j60$ and $\lambda_2 = -30 - j60$ shows the system worked in an asymptotically steady manner.

In our approached design, a proper energy function has also been calculated using Sylvester's criteria as the system's energy function shows negative semi-definiteness. Le $V(x) = X^T.P.X$, i is a energy function where "X" is the state vector, and "P" represents a two-dimensional matrix. By considering the derivative of the above-mentioned energy function with respect to time, verification of the stability of the proposed system can be given as below:

$$\dot{V}(X) = X^T\left(A^T.P + P.A\right)X \tag{34}$$

Solving Eqs. (33)–(36), the values of P_{11}, P_{22}, P_{21}, P_{12} are evaluated.

$$P_{11} = \frac{1}{40}, \; P_{22} = 90 \quad \text{and} \quad P_{21} = P_{12} = 1.$$

Hence, $P = \begin{vmatrix} \frac{1}{40} & 1 \\ 1 & 90 \end{vmatrix} > 0$, which indicates P is positive definite, and as a result, $V(x)$ is also positive definite. According to Sylvester's criteria,

$$V(X) = X^T.P.X = \frac{1}{40}x_1^2 + 2x_1x_2 + 90x_2^2 \tag{35}$$

Equation (35) represents the proper energy function that are used in our proposed nonlinear design of EVAD system.

8 Conclusion

The transfer function of an electric ventricular assist device (EVAD) is turned into a useful polynomial function using the Pade approximation approach in this work, and then, a nonlinear analysis of the model is shown because the EVAD system is not linear but rather nonlinear. The system exhibits nonlinearity as a result of the dead zone and saturation. The descriptive function technique is used to analyze the nonlinearities. We are using descriptive function method which is used to determine the nonlinearity's. After significant research and observation, a convergent limit cycle has been established, which is stable. The application of Lyapunov stability to our proposed nonlinear model reveals that the technique is actually asymptomatically stable in the form of Lyapunov since it has a semi-definite energy function. Finally, using Sylvester's criteria the proper energy function for our proposed design has also been calculated which will be very much useful for future research work.

References

1. Sharp J, Henry R (1979) Designing policies the Ziegler Nichols way. Dynamica 5(2):79–94
2. Bahrami V, Mansouri M, Teshnehlab M (2016) Designing robust model reference hybrid fuzzy controller based on LYAPUNOV for a class of nonlinear systems. J Intell Fuzzy Syst 31(3):1545–1564
3. Cook AM, Simes JG (1972) A simple heart model designed to demonstrate biological stem simulation. IEEE Trans Biomed Eng Rme 19(2)
4. Guo B-Z, Zhao Z-L (2013) On convergence of the nonlinear active disturbance rejection control for MIMO systems. SIAM J Control Optim 51:1727–1757
5. Ang KH, Chong G, Li Y (2005) PID control system analysis, design, and technology. IEEE Trans Control Syst Technol 13(4):559–576
6. Silva GJ, Datta A, Bhattacharyya SP (2007) PID controllers for time delaysystems. Springer Science & Business Media
7. Das S, Bhattacharjee S, Bhattacharyya S, Neogi B (2016) Study on systemic modeling for cardiac and respiratory system towards support on prosthesis facets. Comput Sci Eng CRC Press, p 81
8. Diaz-Insua M, Delgado M, Modeling and simulation of the human cardiovascular system with bond graph: a basic development. Comp Cardiol IEEE, 393–396
9. Biswas DG (2006) Mathematical model of cardiovascular system by transfer function method. Calcutta Med J 103(4):15–17
10. Rosenberg M, Kung RT (1998) Extra cardiac ventricular assist device 3. US Patent 5,713,954
11. Vajta, M.: Some remarks on pad´e-approximations (2000).
12. Salem FA (2013) New efficient model-based PID design method. European Scientific J ESJ 9(15)
13. di Bernardo D, Signorini MG, Cerutti S (1998) A model of two nonlinear coupled oscillators for the study of heartbeat dynamics. Int J Bifurcation Chaos Appl Sci Eng 8

14. Gospodinov E, Gospodinov M, Dey N, Domuschiev A, Ashour S, Sifaki-Pistolla D (2015) Analysis of heart rate variability by applying nonlinear methods with different approaches for graphical representation of results. Analysis 6(8)
15. Atassi AN, Khalil HK (1999) A separation principle for the stabilization of a class of nonlinear systems. IEEE Trans Autom Control 44:1672–1687
16. Ziegler JG, Nichols NB (1942) Optimum settings for automatic controllers. ASME 64(11)
17. Wang X, Zhang Y, Zhang X (2019) AVR fuzzy PID control system based on MCU. J Comput Methods Sci Eng, 1–10

MVO-Optimized Linear Quadratic Regulator for Automatic Voltage Controller System

Vineet Kumar, Veena Sharma, R. Naresh, and V. Kumar

Abstract The automatic voltage regulator (AVR) is regarded as a critical component of an electrical power network, and its proper operation is essential for the network's efficient and protected process. The focus of this research is on developing an inexpensive and reliable control mechanism for AVR systems. As a result, the multi-verse optimizer (MVO)-based linear quadratic regulator (LQR) scheme has been explored and realized using MATLAB 2018a software, with the transient response stipulations equated to standard PID and two-degree of freedom (2-dof) PID control provisions. The robustness of the proposed methodology has also been explored in terms of parameter fluctuations caused due to external factors. The obtained results have demonstrated that the proposed LQR-MVO-linked AVR system has yielded better transient resolution as well as robustness when compared with existing methodologies.

Keywords Linear quadratic controller · Optimal controller · Multi-verse algorithm · Robust controller

1 Introduction

The AVR is required in any energy production network around the world to ensure the voltage constancy all through their functioning zone. Voltage steadiness is disrupted for a variety of reasons, including fluctuating load demand, terminal disturbances, reactive power demand and supply mismatches, and so on. To enable uninterrupted

V. Kumar (✉) · V. Sharma · R. Naresh · V. Kumar
Electrical Engineering Department, NIT Hamirpur, Hamirpur, Himachal Pradesh, India
e-mail: vineet.kumar@nith.ac.in

V. Sharma
e-mail: veena@nith.ac.in

R. Naresh
e-mail: rnaresh@nith.ac.in

V. Kumar
e-mail: vineet@nith.ac.in

J. K. Mandal et al. (eds.), *Topical Drifts in Intelligent Computing*, Lecture Notes in Networks and Systems 426, https://doi.org/10.1007/978-981-19-0745-6_59

power apportionment with steadfast and innocuous practice, any energy-producing network must operate at a predefined value of terminal voltage. Voltage variations that exceed specified tolerated limitations can cause substantial destruction to apparatus on both the user and supplier sides, compromising the power system network's reliability and security [1].

Traditionally, AVR structures used an amplifier to standardize the rotor field excitation of synchronous generators' rotor, but in order to achieve more specific and rapid reaction, a separate control tool is required. The functions of traditional controllers like PI and PID controllers have been developed by considerable research on AVR control over the last few decades [2]. Later on, the researchers experimented with several PID controller and its other modifications (FOPID, PID-N) [3]. According to the literature study, analytical gain factor setting tactics suffer from several bottlenecks such as poor transient performance, inability to address system's changing parameter uncertainty, model configuration complexities cannot be handled with the inclusion of the particle swarm optimization, an innovative tactic of fine-tuning PID gains had been investigated initially in [4]. In the sphere of voltage regulation issues, the PSO technique is the highly investigated meta-heuristic optimization procedure. The researchers also attempted to improve the quality of the AVR system's time domain response by using a PID controller that was tweaked using a variety of meta-heuristic approaches [5]. Based on the findings of the literature review, optimal control methods such as LQR can improve the operational efficiency of AVR. As a result, additional research is required in this region. The major contribution of this work is given as:

1. An AVR system has been contemplated in this work, and it is regulated with the advantage of LQR controller.
2. In a novel attempt, the cost function parameters of LQR have been tweaked with the backing of multi-verse optimization (MVO) approach.
3. The LQR-MVO method's efficacy was assessed in terminologies of transient response specification, and it was equated to meta-heuristically restrained traditional methods of controller design.

2 System Considered

A straightforward AVR structure with controller composition and four essential elements, amplifier, exciter, generator, and sensor, has been explored in this study. A first-order transfer function was also used to represent these components.

The parameters of AVR system are provided in reference [6]. Since this study involves state feedback controller, it is necessary to acquire state-space modelling of this particular system given in Fig. 1. The required state-space model for this fourth-order system has been provided in this section.

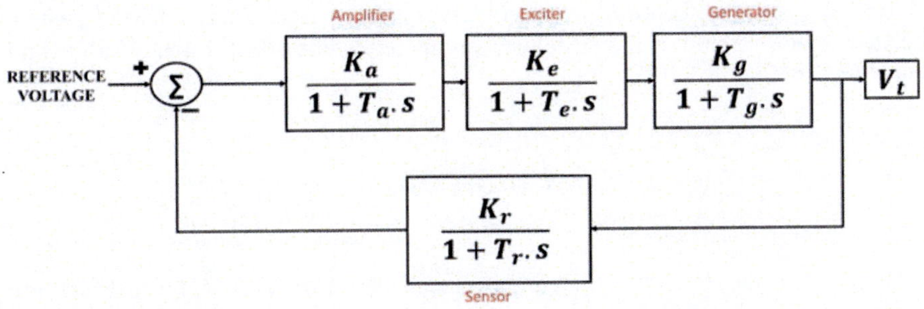

Fig. 1 Schematic for AVR model

$$\dot{X}(t) = \begin{bmatrix} 0 & 1 & 0 & 0 \\ 0 & 0 & 1 & 0 \\ 0 & 0 & 0 & 1 \\ -27500 & -3775 & -1388 & -113.5 \end{bmatrix} X(t) + \begin{bmatrix} 1 \\ 0 \\ 0 \\ 0 \end{bmatrix} u(t) \text{ and},$$

$$Y(t) = \begin{bmatrix} 0 \; 0 \; 250 \; -3375 \end{bmatrix} X(t)$$

3 Linear Quadratic Regulator

For state feedback control design, the LQR approach is the most prevalent control approach (Fig. 2). To find the best controller input, the LQR technique uses a linear or linearized plant model and a quadratic performance criteria [7]. The LQR control design is straightforward and easy to formulate. The fundamental goal of LQR control design is to determine the value of control action that allows the system's states to be driven from an initial to a final state while meeting a set of performance criteria. The ideal state feedback gain matrix *(K)* will be used to determine the value of LQR control action, and the best state feedback gain is achieved by solving the algebraic matrix Riccati equation (AMRE). The basic Lagrangian principle of

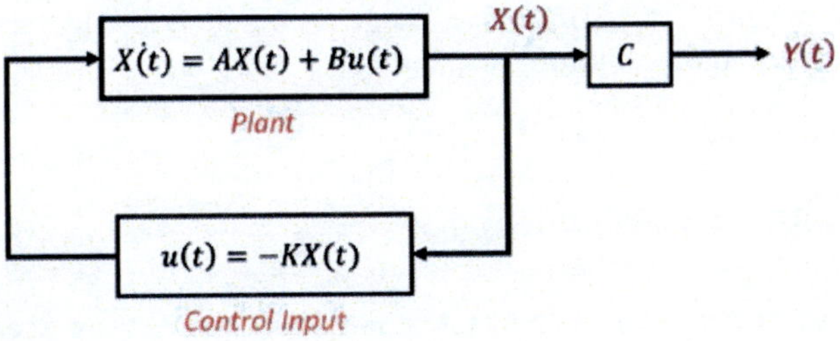

Fig. 2 Simplified pictorial interpretation of LQR

addressing constrained optimization problems has been used to derive DMRE, where the state-space version of the structure is regarded as the constriction. The general objective function is defined as follows [7].

$$\text{Cost Function}\,(J) = \int_{t_0}^{t_f} [x'(t)Q(t)x(t) + u^t(t)R(t)u(t)]\text{dt}$$

Here, $x(t)$ is the state trajectory, and $u(t)$ is the input vector [8]. The cost function weight matrices $Q(t)$ and $R(t)$ can be tweaked to get the LQ controller to perform as needed [7]. These weights, in technical terms, describe the degree of freedom associated with a specific state or input. Assume that if a given input's weight is high, the designer is attempting to minimize the cost function (or stabilize the system) with a limited amount of that input. With any input, a lower weight indicates that the input can be varied from its nominal value. Typically, these matrices are chosen for design purposes using a trial-and-error method, and the proper tuning of these weight matrices is very important for optimal calculation of control input (u). Therefore, multi-verse optimization algorithm has been selected to properly tune the cost function weights of the LQR controller design. In this study, the diagonal matrix has been considered for simplicity, and off-diagonal elements of $Q(t)$ will be considered as zero [9].

4 Multi-verse Optimization Algorithm

MVO is a meta-heuristic technique inspired by the nature that strikes a good balance between the exploration and exploitation phases. As a result, the chances of a solution becoming trapped in local optimum are much reduced (Fig. 3). This allows the operator to obtain an ideal solution in the fewest number of iterations possible, as well as a quality resolution in a solitary run. The MVO algorithm has very limited factors to stipulate, and henceforth, it has also diminished algorithm intricacy [6, 10–12].

The MVO algorithm is divided into three holes. Every single resolution is compared to a universe, and each alterable component in the result is linked to an item in that universe. In addition, we assign an inflation rate to each resolution that is proportionate to the solution's capability.

5 Results and Discussion

After running the simulation in MATLAB/Simulink, a variety of scenarios were considered in order to evaluate the suggested methodology's performance.

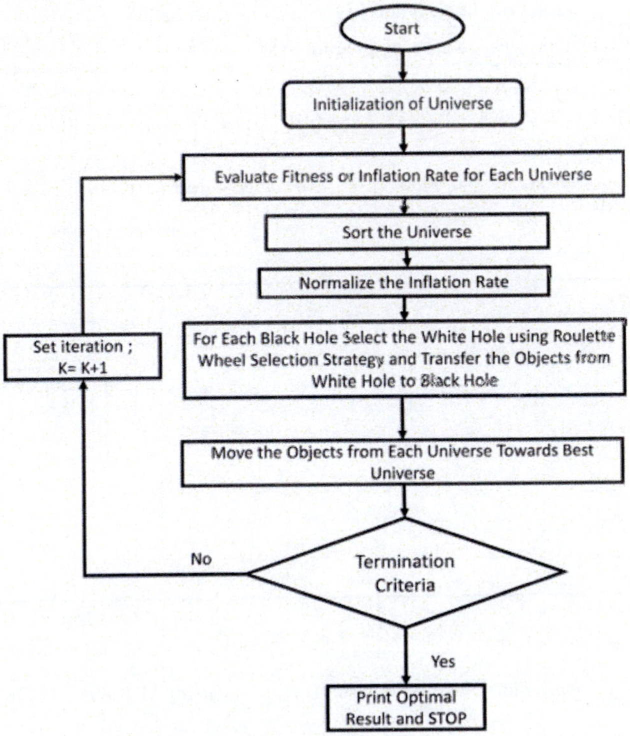

Fig. 3 Algorithmic steps followed by MVO algorithm

5.1 General Assessment Case

Under this case, the MVO algorithm has been employed to optimally tune the parameters of LQR cost function. After using these optimized cost function weights, the LQR method has been implemented to find best possible controller for AVR arrangement in concern. Here, the diagonal elements of *Q(t)* matrix ($q1$, $q2$, $q3$, and $q4$), along with a single element of *R(t)* matrix ($r1$), have been tuned using MVO algorithm. To evaluate the performance of LQR controlled system, the transient response plot has been obtained and is compared with the response from existing PID and 2-dof PID controlled systems. Table 1 presents the MVO-tuned augmented parameters of PID, 2-dof PID, and LQR controllers. And the expression for integral squared error (ISE) for this problem is given as:

$$\mathrm{ISE} = \int (V_{\mathrm{Reference}} - V_t)^2 \mathrm{dt}$$

From Fig. 4, it may be observed that the suggested MVO-based LQR controller postulates enhanced transient functioning as compared to the conventional methods.

Table 1 Numerical values of classical and LQ regulators' gain factors

Controller	Parameters tweaked after employing MVO	Fitness value (ISE)
PID	$Kp = 0.6256$; $Kd = 0.0622$; $Ki = 0.3086$	0.3259
2-dof PID	$Kp = 0.6975$; $Kd = 0.0832$; $Ki = 0.3071$; $b = 0.9984$; $c = 0.2865$	0.3036
LQR	$q1 = 0.0011$; $q2 = 0.0083$; $q3 = 1.1812$; $q4 = 1.6873$; $r1 = 4.4647$	0.2739

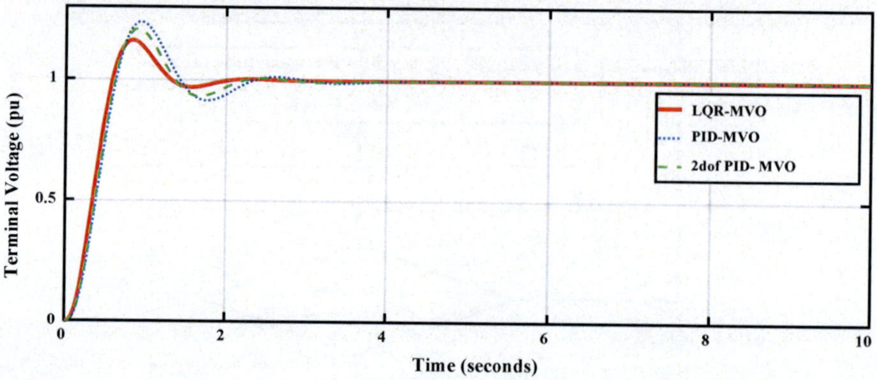

Fig. 4 Transient response plots of the AVR system controlled using MVO-tuned LQR and classical controllers

Table 2 Specifications for transient reaction acquired from the standardized AVR

Controller	RT (s)	ST (s)	PO (%)
PID	0.405	2.215	23.9
2-dof PID	0.378	2.07	20.8
LQR	0.357	1.783	

Table 2 compares the transient response specifications obtained from different controllers and establishes that the proposed method (LQR-MVO) gives better settling time (ST), rise time (RT), and peak overshoot (PO) as matched to the classical control means.

5.2 *Sensitivity Analysis*

For this case study, the system parameter values of LQR-MVO-controlled AVR prototype have been deviated from their rated value and the subsequent response has been recorded while keeping the same control system parameters.

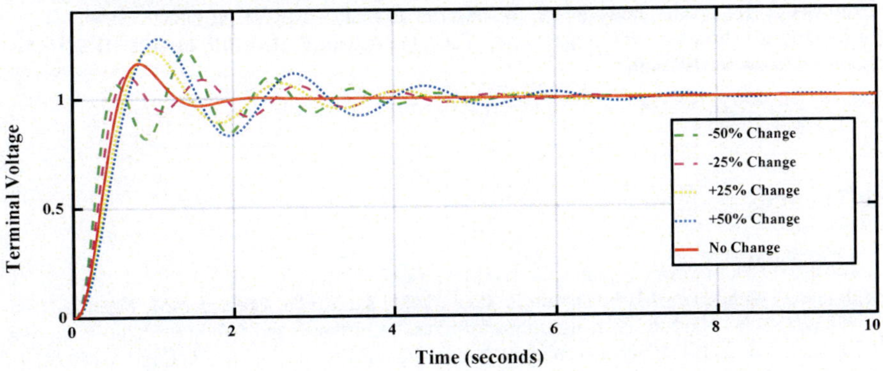

Fig. 5 Transient graphs of terminal voltage after adjusting the exciter time constant

Table 3 Performance assessment after enumerating uncertainty

Change in T_e (%)	Settling time (s)	Overshoot (%)
−50	4.17	21.5
−25	4.70	10.9
+25	5.02	22.1
+50	6.12	

After considering plots in Fig. 5, it can be validated that the obtained transient characteristic plots for parameter variation case are stable in nature. Also, Table 3 has confirmed that the LQR-MVO-controlled AVR system delivers good quality dynamical effect even when system boundaries are disturbed from their fundamental standards.

6 Conclusion

In this manuscript, an optimal control strategy (LQR) has been discussed and implemented for the efficacy enhancement of AVR system in the power system network. The linear quadratic regulator strategy has been embedded with multi-verse optimization approach which has allowed the users to optimally tune the LQR cost function weights. The proposed LQR-MVO methodology has provided better transient results when equated together with the existing traditional controllers. Moreover, the proposed algorithm has provided good robustness against the parameter sensitivity case when system parameters were perturbed from their original value. The presented work can be expanded after considering the generator field circuit dynamics in the AVR system modelling.

Acknowledgements The study in this manuscript is financed by HIMCOSTE, SCST, and E (R&D)/2019-20, beneath project grant no. STC/F(8)-6/2019(R&D 2019-20)-408, H.P., India, sanctioned to the second author.

References

1. Saadat H (2004) Power system analysis. McGraw-Hill
2. Jumani TA, Mustafa MW, Hussain Z, Rasid MM, Saeed MS, Memon MM, Khan I, Nisar KS (2020) Jaya optimization algorithm for transient response and stability enhancement of a fractional-order PID based automatic voltage regulator system. Alexandria Eng J 59(4):2429–2440
3. Sikander A, Thakur P, Bansal RC, Rajasekar S (2018) A novel technique to design cuckoo search based FOPID controller for AVR in power systems. Comput Electr Eng 70:261–274
4. Gaing ZL (2004) A particle swarm optimization approach for optimum design of PID controller in AVR system. IEEE Trans Energy Convers 19(2):384–391
5. Bingul Z, Karahan O (2018) A novel performance criterion approach to optimum design of PID controller using cuckoo search algorithm for AVR system. J Franklin Inst 355(13):5534–5559
6. Kumar V, Sharma V, Kumar V (2022) performance evaluation of hho optimized model predictive controller for AVR system and its comparison with conventional controllers. In: Tomar A, Malik H, Kumar P, Iqbal A (eds) Machine learning, advances in computing, renewable energy and communication. Lecture Notes in Electrical Engineering, vol 768. Springer, pp 151–159. https://doi.org/10.1007/978-981-16-2354-7_14
7. Kumar V, Sharma V, Rahi OP, Kumar U (2019) Optimal position tracking for an AC servomotor using linear quadratic and model predictive control. In: 4th international conference on information systems and computer networks (ISCON), Mathura
8. Prasad LB, Tyagi B, Gupta HO (2014) Optimal control of nonlinear inverted pendulum system using PID controller and LQR: performance analysis without and with disturbance input. Int J Autom Comput 11:661–670. https://doi.org/10.1007/s11633-014-0818-1
9. Saleem O, Mahmood-ul-Hasan K (2021) Hierarchical adaptive control of self-stabilizing electromechanical systems using artificial-immune self-tuning mechanism for state weighting-factors. J Mech Sci Technol 35:1235–1250. https://doi.org/10.1007/s12206-021-0237-5
10. Mirjalili S, Mirjalili SM, Hatamlou A (2016) Multi-verse optimizer: a nature-inspired algorithm for global optimization. Neural Comput Appl 27:495–513. https://doi.org/10.1007/s00521-015-1870-7
11. Kumar V, Naresh R (2020) Monarch butterfly optimization-based computational methodology for unit commitment problem. Electr Power Comp Syst 48(19–20):2181–2194
12. Kumar V, Sharma V, Naresh R (2021) HHO-based model predictive controller for combined voltage and frequency control problem including SMES. IETE J Res. https://doi.org/10.1080/03772063.2021.1908180

An Observation of Energy Density for PKL, Aloe Vera, Myrobalan, Lemon, and Tomato Electrochemical Cell

Md. Abdul Wadud, Kamrul Alam Khan, Md. Sayed Hossain, Salman Rahman Rasel, and Sumanta Bhattacharyya

Abstract It has been discussed about an observation of energy density of a PKL, aloe vera, myrobalan, lemon, and tomato electrochemical cell. The PKL, aloe vera, myrobalan, lemon, and tomato electrochemical cells are the primary cell. It can generate the electrical power from chemical reaction. Everywhere of every country needs electricity for their development. It is now proved that world needs renewable energy sources to safeguard the future earth. It can be used 24 h unlike solar energy. The each of the electro chemical cells were fabricated 2 half cells. It is obtained that the energy density of PKL electrochemical cell is the best among all other electrochemical cells.

Keywords Electrochemical cell · Leaves · Vegetative · Fruits · Energy density · Electrodes · Electrolytes

1 Introduction

Green technology is the most important technology for nowadays [1]. It is minimized the adverse effects of the environment [2]. It is using as a safeguard of the future earth. For any development, electricity is the utmost important for our daily life [3–5]. Renewable energy is also called alternative energy. When non-renewable energy sources is not possible to provide electricity, then renewable energy will provide electricity. Because renewable energy does not have a limited source. It can

Md. Abdul Wadud
Department of Chemistry, Dhaka Cantonment, BAF Shaheen College, Dhaka, Bangladesh

K. A. Khan
Department of Physics, Jagannath University, Dhaka 1100, Bangladesh

Md. Sayed Hossain
Center for Research Reactor, Bangladesh Atomic Energy Commission, Dhaka, Bangladesh

S. R. Rasel
Local Government Engineering Department, Sherpur Sadar, Sherpur, Bangladesh

S. Bhattacharyya (✉)
Department of ECE, Greater Kolkata College of Engineering and Management, Kolkata, India

J. K. Mandal et al. (eds.), *Topical Drifts in Intelligent Computing*, Lecture Notes in Networks and Systems 426, https://doi.org/10.1007/978-981-19-0745-6_60

be generated again and again and will never run out. For the growing development of any country, it is needed the uninterrupted electricity generation system [6–8]. In Bangladesh, around 79% electricity comes from the gas-based power system [9–12]. The renewable energy sources are the ecofriendly [9–13]. To keep it in mind, the energy density of the PKL, aloe vera, myrobalan, lemon, and tomato electrochemical cell has been studied in this research experiment.

2 Materials and Methodology

2.1 Methodology

It is shown in Fig. 1a, the experimental technique of basic chemical reactions to generate electricity. Copper is called cathode or it is also called collector electrode. The zinc electrode is called sacrificial electrode. When is dissolved into the leaves extract, then Zn^{2+} flows from zinc plate to copper plate as an ionic wind. Figure 1b, and c are for PKL and aloe vera cell, respectively.

Figure 2a shows an experimental set up for myrobalan cell. Figure 2b shows an experimental set up for lemon cell, and Fig. 2c shows an experimental set up for tomato cell.

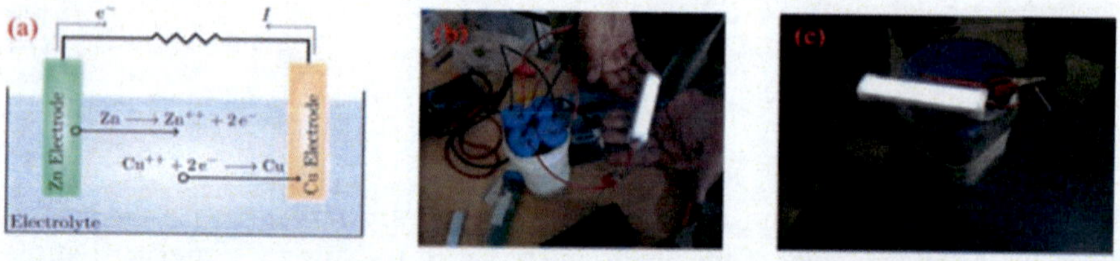

Fig. 1 **a** Experimental technique of chemical reaction. **b** An experimental set up for PKL cell. **c** An experimental set up for aloe vera cell

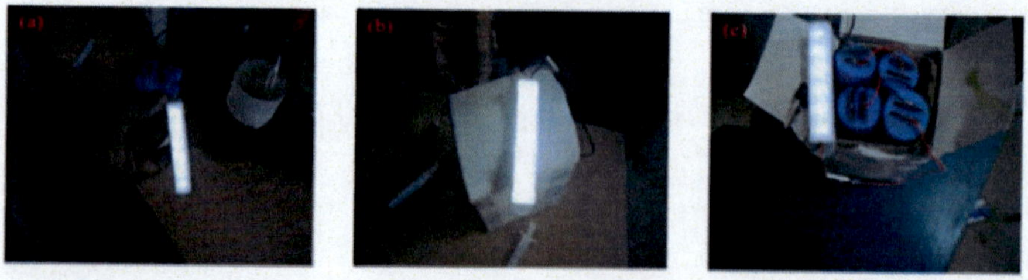

Fig. 2 **a** An experimental set up for myrobalan cell. **b** An experimental set up for lemon cell. **c** An experimental set up for tomato cell

2.2 Chemical Reactions

The oxidation has occurred at anode (Zn-plate), and reduction has occurred at cathode (Cu-plate). This creates a movement of electrons flow from anode (Zn-plate) to the cathode (Cu-plate) which generates current in the circuit. The total cell reaction is given by the following

$$Zn + Cu^{2+} + H^{+} \rightarrow Zn^{2+} + Cu + H_2 \uparrow \quad (1)$$

Here, Cu^{2+} = Reactant ions, H^{+} = Reactant ions, and Zn^{2+} = Product ions.

The loss of electron occurs at anode, and the gain of electrons occurs at cathode. Furthermore, the metal at anode dissolved and the metal at cathode grows.

2.3 Flow of Electrons

From the chemical reactions, it is found that electrons always flow from the anode (Zn-plate) to the cathode (Zn-plate). In other words, it can be said that electrons always flow from the oxidation half-cell to the reduction half-cell. The cell potential at standard state condition (Eocell) indicates that electron flows from the more −ve half-reaction to the more +ve half-reaction.

2.4 Electrochemistry

Zinc (anode) is oxidized:

$$Zn_{(s)} \rightarrow Zn^{2+}_{(aq)} + 2e^{-} \quad (2)$$

(Standard electrode potential −0.7618 V).

Copper (cathode) is reduced:

$$Cu^{2+}_{(aq)} + 2e^{-} \rightarrow Cu_{(s)} \quad (3)$$

(Standard electrode potential +0.340 V).

Hydrogen (cathode) is reduced:

$$H^{+}_{(aq)} + 2e^{-} \rightarrow H_2 \uparrow \quad (4)$$

(Standard electrode potential +0.340 V).

The net reaction is

$$Zn_{(s)} + Cu^{2+}_{(aq)} \rightarrow Zn^{2+}_{(aq)} + Cu_{(s)} \quad (5)$$

(Open-circuit voltage 1.1018 V).

3 Results and Discussion

Figure 3 shows the curve of energy density (WH/Kg) for PKL cell with T.D. (min).This finds that energy density for PKL cell is 6.995 × 10^{-3} WH/Kg, and the minimum energy density for PKL cell is 0 WH/Kg. Therefore, the distinction of energy density = 6.995 × 10^{-3} WH/Kg. It is also shown that energy density increases linearly for 150 min and then it remains same for 570 min. Then, this increases linearly for 1050 min and then after, this increases slowly for 1950 min. Again, this follows zig-zag pattern and lastly, this has decreed exponentially till 23,491 min.

Figure 4 shows the curve of energy density (Wh/Kg) for aloe vera cell with T.D. (min). This shows that the higher energy density for aloe vera cell is 3.3 × 10^{-3} WH/Kg, and the minimum energy density for aloe vera cell is 0 WH/Kg. Therefore, the distinction of energy density = 3.3 × 10^{-3} Wh/Kg.

Figure 5 shows the curve of energy density (WH/Kg) for arum leaf cell with T.D. (min). This found that the energy density for arum leaf cell = 8.49 × 10^{-3} Wh/Kg and the minimum energy density for arum leaf cell = 0 Wh/Kg. Therefore, the distinction of energy density = 8.49 × 10^{-3} Wh/Kg.

Figure 6 shows the curve of energy density (Wh/Kg) for lemon cell with T.D. (min). This found that the large energy density for lemon cell is 4.56 × 10^{-3} WH/Kg,

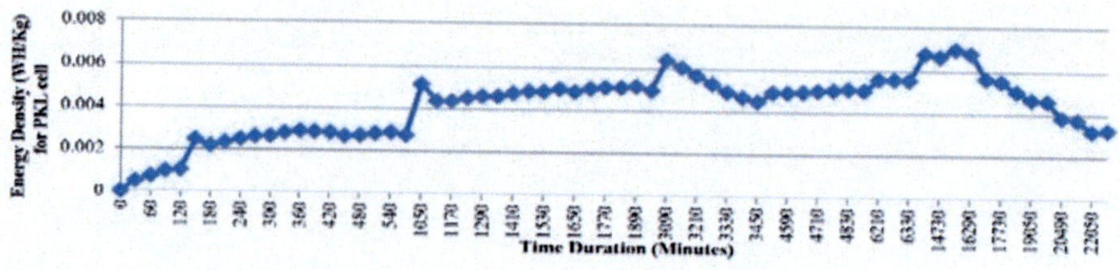

Fig. 3 Energy density (WH/Kg) for PKL cell with time duration (min)

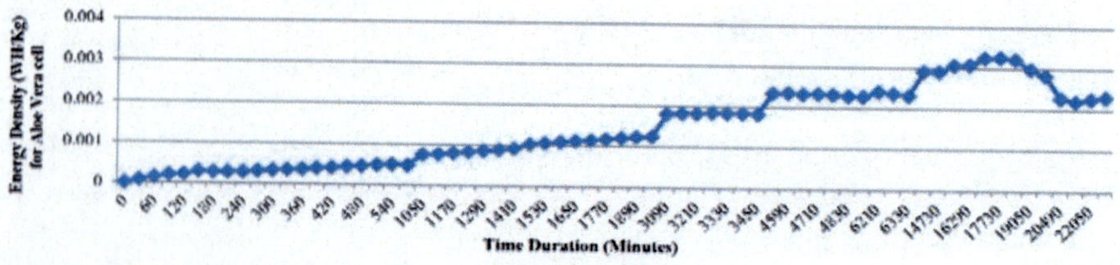

Fig. 4 Energy density (WH/Kg) for aloe vera cell with time duration (min)

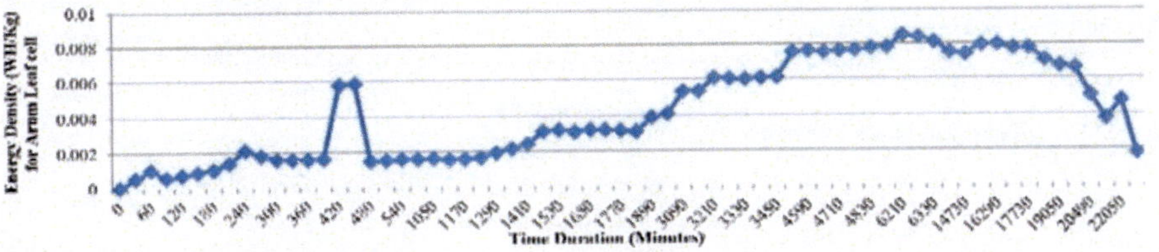

Fig. 5 Energy density (WH/Kg) for arum leaf cell with time duration (min)

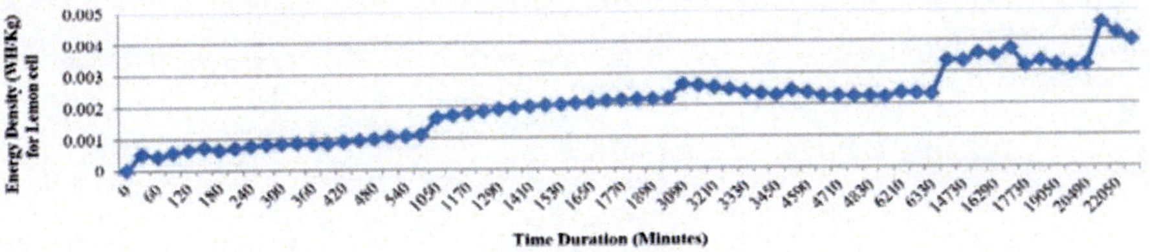

Fig. 6 Energy density (WH/Kg) for lemon cell with time duration (min)

and the minimum energy density for lemon cell is 0 WH/Kg. Therefore, the distinction of energy density is 4.56×10^{-3} WH/Kg. It is also shown that energy density increases for 20,550 min and this has decreased for 23,490 min.

Figure 7 shows the curve of energy density (WH/Kg) for myrobalan cell with T.D. (min). This found the specific energy density for myrobalan cell is 2.65×10^{-2} WH/Kg and, the minimum energy density for myrobalan cell is 0 WH/Kg. Therefore, the distinction of energy density is 2.65×10^{-2} Wh/Kg. This found that energy density has increased in a zig-zag way for 3270 min, and thereafter, this has decreased for 23,490 min.

Figure 8 shows the curve of energy density (WH/Kg) for tomato cell with T.D. (min). This has found that the large specific energy density for tomato cell is 4.01×10^{-3} Wh/Kg, and the minimum specific energy density for tomato cell is 0 Wh/Kg. Therefore, the distinction of energy density = 4.01×10^{-3} Wh/Kg.

Figure 9 shows the curve of energy density (WH/Kg) for PKL cell with time of the day (26.08.2017). It is shown that the maximum energy density for PKL cell with time of the day (26.08.2017) is 2.9×10^{-3} WH/Kg, and the minimum energy density for PKL cell with time of the day (26.08.2017) is 0 WH/Kg. So the difference

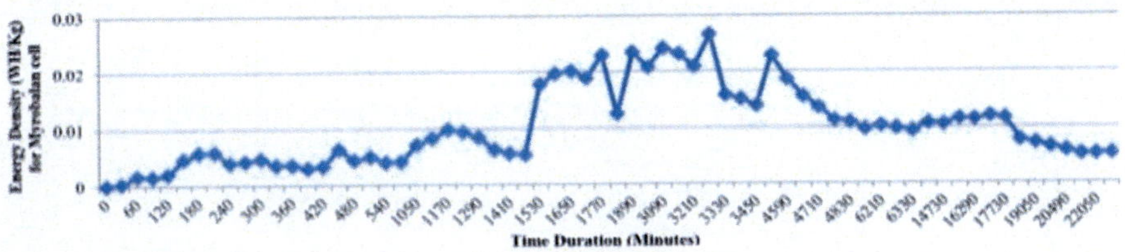

Fig. 7 Energy density (WH/Kg) for myrobalan cell with time duration (min)

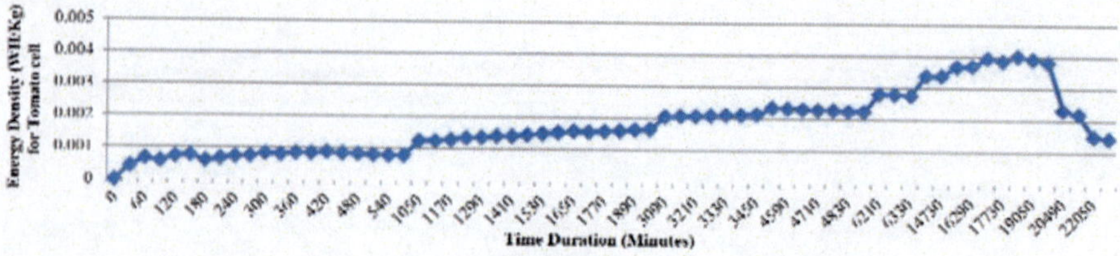

Fig. 8 Energy density (WH/Kg) for tomato cell with time duration (min)

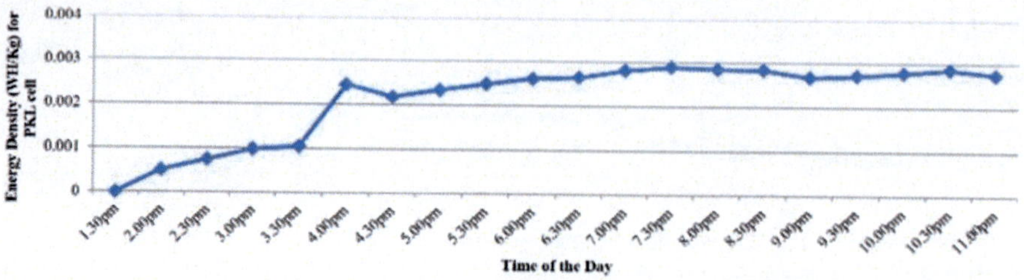

Fig. 9 Energy density (WH/Kg) for PKL cell with time of the day (26.08.2017)

between the maximum and minimum energy density is 2.9×10^{-3} WH/Kg. It is also shown that energy density increases linearly up to 7:30 PM, and then, it maintains almost constant pattern up to 11:00 PM.

Figure 10 shows the curve of energy density (Wh/Kg) for PKL cell with time of the day (27.08.2017). It is shown that the maximum energy density for PKL cell with time of the day (27.08.2017) is 5.1×10^{-3} WH/Kg, and the minimum energy density for PKL cell with time of the day (27.08.2017) is 4.35×10^{-3} WH/Kg. Therefore, the distinction of energy density $= 0.75 \times 10^{-3}$ Wh/Kg. It is also shown that energy density decreases linearly up to 8:00 AM, and then, it increases gradually up to 9:00 PM. Finally, it started to decrease up to 10:00 PM.

Figure 11 shows the curve of energy density (WH/Kg) for PKL cell with time of the day (28.08.2017). It is shown that the maximum specific energy density for PKL cell with time of the day (28.08.2017) is 6.3×10^{-3} WH/Kg, and the minimum energy density for PKL cell with time of the day (28.08.2017) is 4.4×10^{-3} WH/Kg.

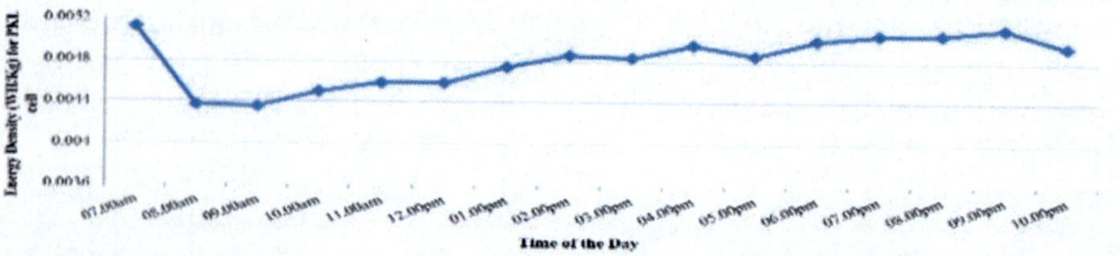

Fig. 10 Energy density (WH/Kg) for PKL cell with time of the day (27.08.2017)

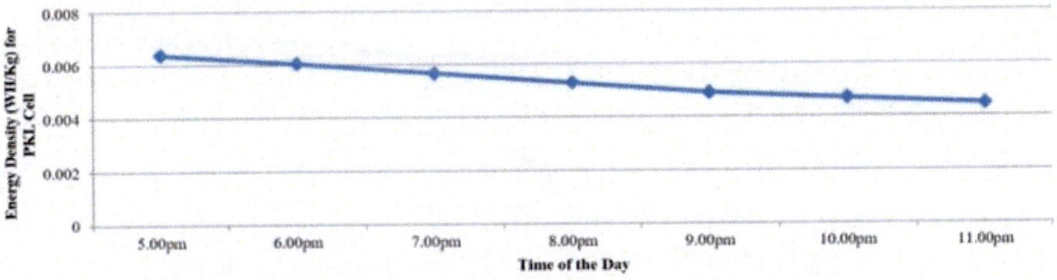

Fig. 11 Energy density (WH/Kg) for PKL cell with time of the day (28.08.2017)

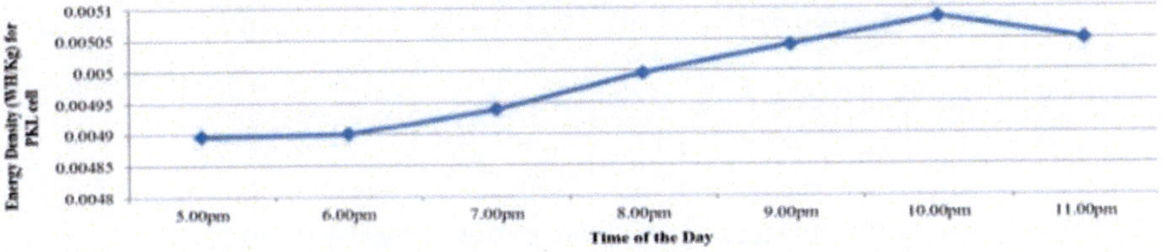

Fig. 12 Energy density (WH/Kg) for PKL cell with time of the day (29.08.2017)

Therefore, the distinction of energy density = 1.9×10^{-3} WH/Kg. It is also shown that specific energy density decreases exponentially up to 11:00 PM.

Figure 12 shows the curve of energy density (WH/Kg) for PKL cell with time of the day (29.08.2017). It is shown that the maximum specific energy density for PKL cell with time of the day (29.08.2017) is 5.08×10^{-3} WH/Kg, and the minimum energy density for PKL cell with time of the day (29.08.2017) is 4.89×10^{-3} Wh/Kg. Therefore, the distinction of energy density = 0.19×10^{-3} Wh/Kg.

Figure 13 shows the curve of energy density (WH/Kg) for PKL cell with time of the day (30.08.2017). It is shown that the maximum energy density for PKL cell with time of the day (30.08.2017) is 5.56×10^{-3} WH/Kg, and the minimum energy density for PKL cell with time of the day (30.08.2017) is 5.54×10^{-3} WH/Kg. So the difference between the maximum and minimum specific energy density is 0.02 $\times 10^{-3}$ WH/Kg.

Figure 14 shows the curve of energy density (Wh/Kg) for aloe vera cell with time of the day (26.08.2017). It is shown that the maximum energy density for aloe

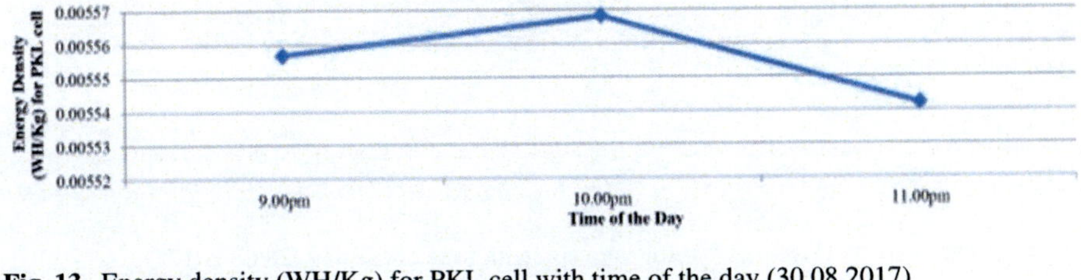

Fig. 13 Energy density (WH/Kg) for PKL cell with time of the day (30.08.2017)

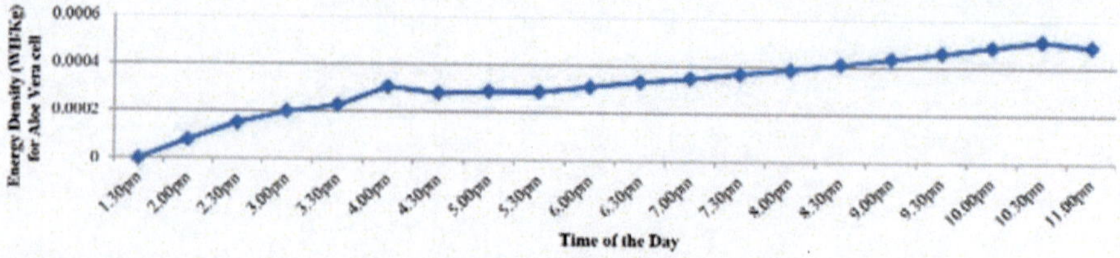

Fig. 14 Energy density (Wh/Kg) for aloe vera cell with time of the day (26.08.2017)

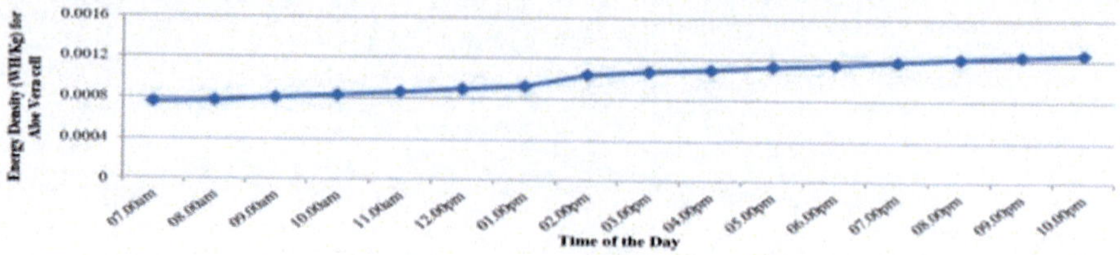

Fig. 15 Energy density (Wh/Kg) for aloe vera cell with time of the day (27.08.2017)

vera cell with time of the day (26.08.2017) is 5.1×10^{-4} WH/Kg, and the minimum specific energy density for aloe vera cell with time of the day (26.08.2017) is 0.792×10^{-4} Wh/Kg. Therefore, the distinction of energy density = 4.308×10^{-4} Wh/Kg. It is also shown that energy density increases gradually up to 10:30 PM and then decreases up to 11:00 PM.

Figure 15 shows the curve of energy density (Wh/Kg) for aloe vera cell with time of the day (27.08.2017). It is shown that the maximum energy density for aloe vera cell with time of the day (27.08.2017) is 1.25×10^{-3} Wh/Kg, and the minimum energy density for aloe vera cell with time of the day (27.08.2017) is 0.75×10^{-3} Wh/Kg. Therefore, the distinction of energy density = 0.5×10^{-3} Wh/Kg. It is also shown that energy density increases slowly up to 10:00 PM from its starting.

Figure 16 shows the curve of energy density (Wh/Kg) for aloe vera cell with time of the day (28.08.2017). It is shown that the maximum energy density for aloe vera cell with time of the day (28.08.2017) is 1.85 × 10–3 Wh/Kg, and the minimum energy density for aloe vera cell with time of the day (28.08.2017) is 1.81×10^{-3} Wh/Kg. Therefore, the distinction of energy density = 0.04×10^{-3} Wh/Kg.

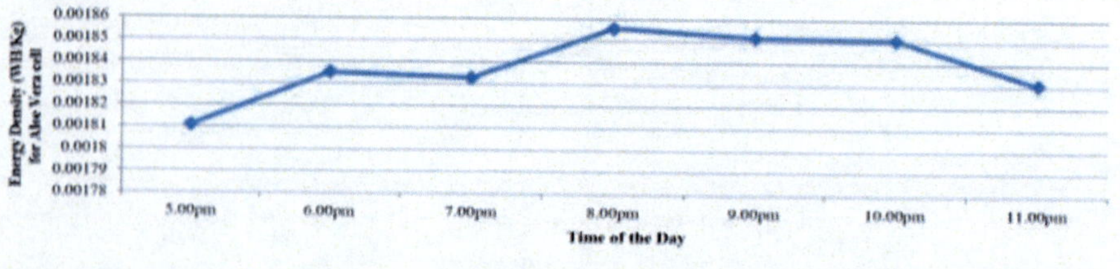

Fig. 16 Energy density (Wh/Kg) for aloe vera cell with time of the day (28.08.2017)

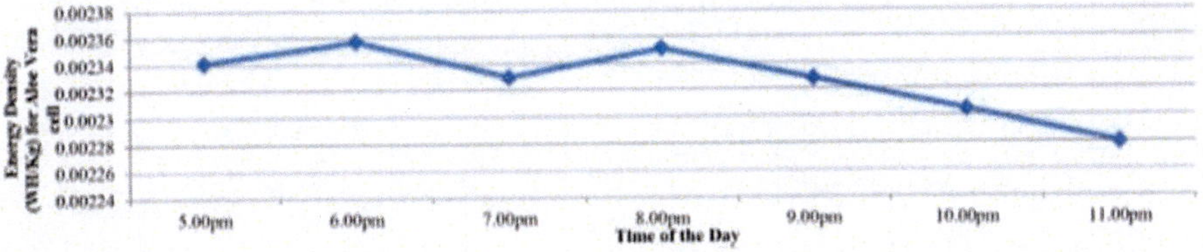

Fig. 17 Energy density (WH/Kg) for aloe vera cell with time of the day (29.08.2017)

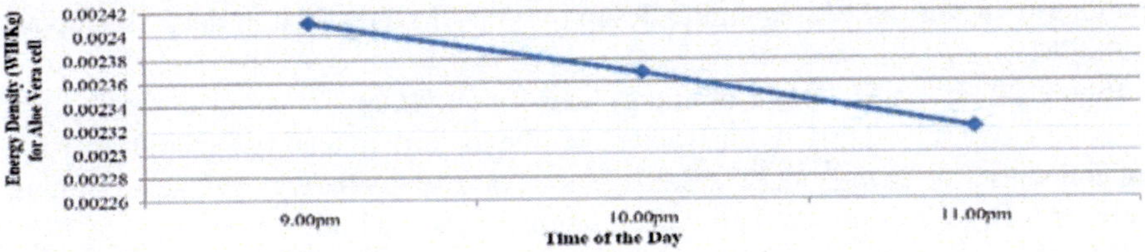

Fig. 18 Energy density (Wh/Kg) for aloe vera cell with time of the day (30.08.2017)

Figure 17 shows the curve of energy density (WH/Kg) for aloe vera cell with time of the day (29.08.2017). It is shown that the maximum energy density for aloe vera cell with time of the day (29.08.2017) is 2.35×10^{-3} WH/Kg, and the minimum energy density for aloe vera cell with time of the day (29.08.2017) is 2.28×10^{-3} WH/Kg. Therefore, the distinction of energy density is 0.07×10^{-3} WH/Kg.

Figure 18 shows the curve of energy density (Wh/Kg) for aloe vera cell with time of the day (30.08.2017). It is shown that the maximum energy density for aloe vera cell with time of the day (30.08.2017) is 2.41×10^{-3} Wh/Kg, and the minimum energy density for aloe vera cell with time of the day (30.08.2017) is 2.32×10^{-3} Wh/Kg. Therefore, the distinction of energy density = 0.09×10^{-3} Wh/Kg. It is also shown that energy density decreases linearly up to 11:00 PM from its starting.

Figure 19 shows the curve of energy density (Wh/Kg) for arum leaf cell with time of the day (26.08.2017). It is shown that the maximum energy density for arum leaf cell with time of the day (26.08.2017) is 5.86×10^{-3} Wh/Kg, and the minimum energy density for arum leaf cell with time of the day (26.08.2017) is 0 Wh/Kg. Therefore, the distinction of energy density = 5.86×10^{-3} Wh/Kg. It is also shown

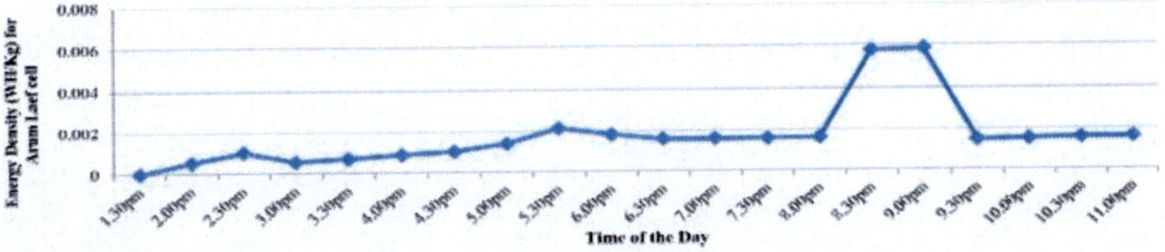

Fig. 19 Energy density (Wh/Kg) for arum leaf cell with time of the day (26.08.2017)

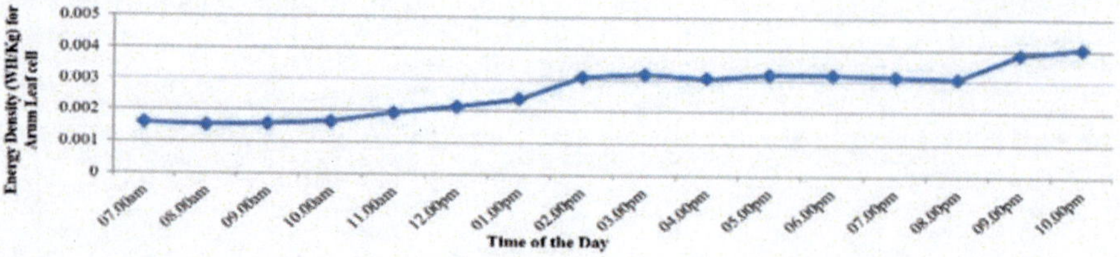

Fig. 20 Energy density (WH/Kg) for arum leaf cell with time of the day (27.08.2017)

that energy density increases gradually up to 9:00 PM, and then, it decreases up to 11:00 PM.

Figure 20 shows the curve of energy density (WH/Kg) for arum leaf cell with time of the day (27.08.2017). It is shown that the maximum energy density for arum leaf cell with time of the day (27.08.2017) is 4.06×10^{-3} WH/Kg, and the minimum energy density for arum leaf cell with time of the day (27.08.2017) is 1.53×10^{-3} WH/Kg. Therefore, the distinction of energy density is 2.53×10^{-3} Wh/Kg. It is also shown that energy density increases gradually up to 10:00 PM from its starting.

Figure 21 shows the curve of energy density (WH/Kg) for arum leaf cell with time of the day (28.08.2017). It is shown that the maximum energy density for arum leaf cell with time of the day (28.08.2017) is 6.13×10^{-3} WH/Kg, and the minimum energy density for arum leaf cell with time of the day (28.08.2017) is 5.33×10^{-3} WH/Kg. Therefore, the distinction of energy density $= 0.8 \times 10^{-3}$ WH/Kg. It is also shown that energy density increases linearly up to 7:00 PM from, and then, it decreases up to 9:00 PM and finally increasing linearly up to 11:00 PM.

Figure 22 shows the curve of energy density (Wh/Kg) for arum leaf cell with time of the day (29.08.2017). It is shown that the maximum energy density for arum

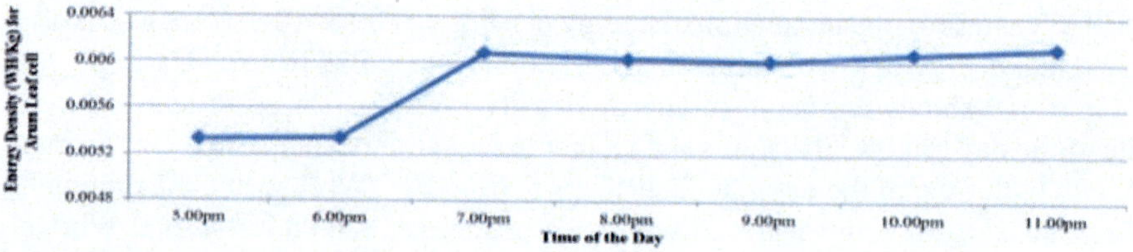

Fig. 21 Energy density (WH/Kg) for arum leaf cell with time of the day (28.08.2017)

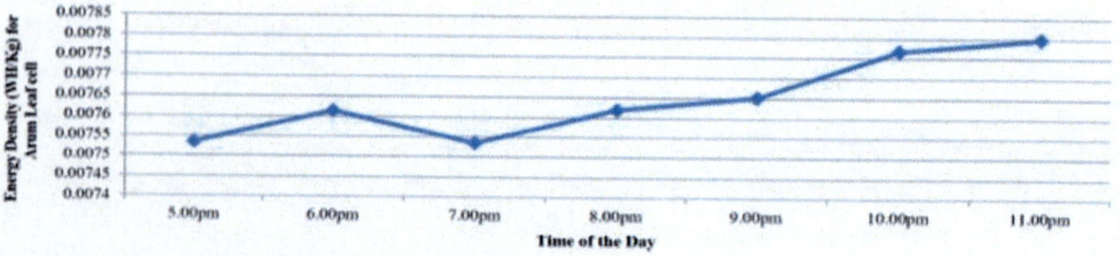

Fig. 22 Energy density (Wh/Kg) for arum leaf cell with time of the day (29.08.2017)

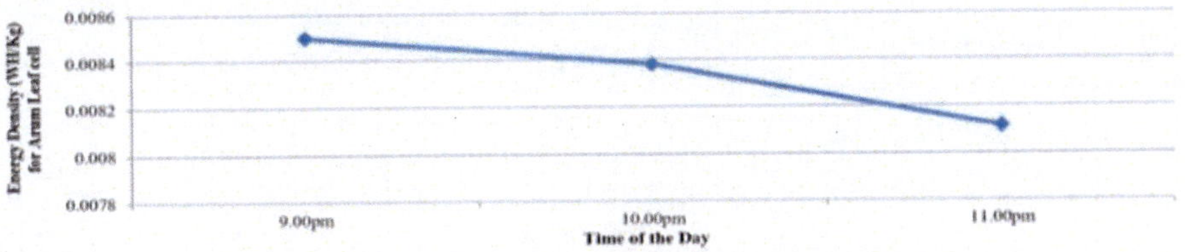

Fig. 23 Energy density (Wh/Kg) for arum leaf cell with time of the day (30.08.2017)

leaf cell with time of the day (29.08.2017) is 7.78 × 10^{-3} Wh/Kg, and the minimum energy density for arum leaf cell with time of the day (29.08.2017) is 7.53 × 10^{-3} Wh/Kg. Therefore, the distinction of energy density = 0.25 × 10^{-3} Wh/Kg. It is also shown that energy density increases linearly up to 6:00 PM from, and then, it decreases up to 7:00 PM and finally increasing linearly up to 11:00 PM.

Figure 23 shows the curve of energy density (Wh/Kg) for arum leaf cell with time of the day (30.08.2017). It is shown that the maximum energy density for arum leaf cell with time of the day (30.08.2017) is 8.49 × 10^{-3} Wh/Kg, and the minimum energy density for arum leaf cell with time of the day (30.08.2017) is 8.11 × 10^{-3} Wh/Kg. Therefore, the distinction of energy density = 0.38 × 10^{-3} Wh/Kg. It is also shown that energy density decreases linearly up to 11:00 PM from its starting.

Figure 24 shows the curve of energy density (Wh/Kg) for lemon cell with time of the day (26.08.2017). It is shown that the maximum energy density for lemon cell with time of the day (26.08.2017) is 1.08 × 10^{-3} Wh/Kg, and the minimum energy density for lemon cell with time of the day (26.08.2017) is 0 Wh/Kg. So the difference between the maximum and minimum energy density is 1.08 × 10^{-3} Wh/Kg. It is also shown that energy density increases linearly up to 11:00 PM, but it decreases at few points like 2:30 PM and 4:30 PM.

Figure 25 shows the curve of energy density (Wh/Kg) for lemon cell with time of the day (27.08.2017). It is shown that the maximum energy density for lemon cell with time of the day (27.08.2017) is 2.20 × 10^{-3} Wh/Kg, and the minimum energy density for lemon cell with time of the day (27.08.2017) is 1.63 × 10^{-3} Wh/Kg. Therefore, the distinction of energy density = 0.57 × 10–3 Wh/Kg. It is also shown that specific energy density increases slowly up to 11:00 PM from its starting.

Figure 26 shows the curve of energy density (WH/Kg) for lemon cell with time of the day (28.08.2017). It is shown that the maximum energy density for lemon cell

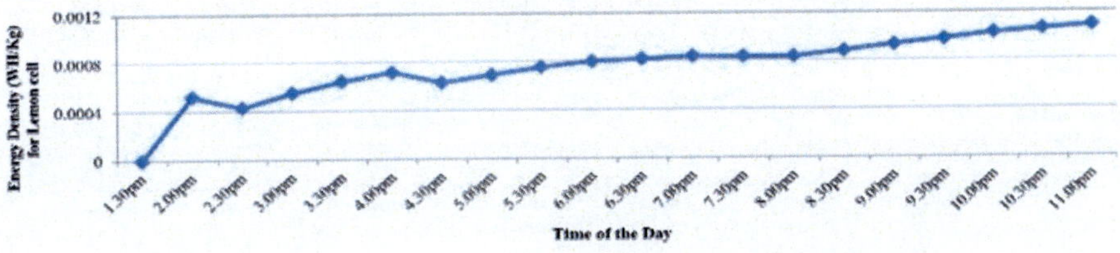

Fig. 24 Energy density (Wh/Kg) for lemon cell with time of the day (26.08.2017)

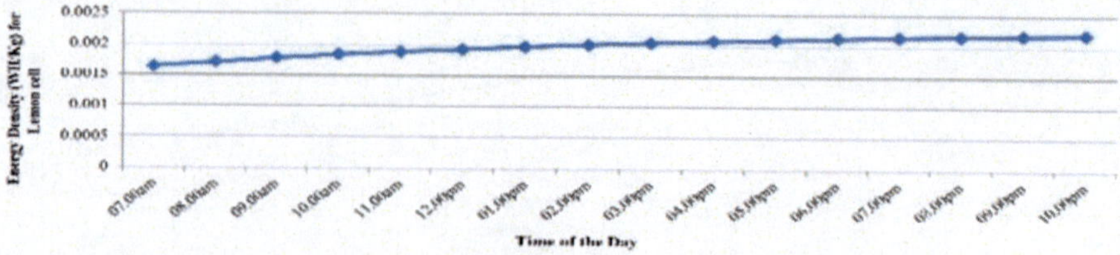

Fig. 25 Energy density (Wh/Kg) for lemon cell with time of the day (27.08.2017)

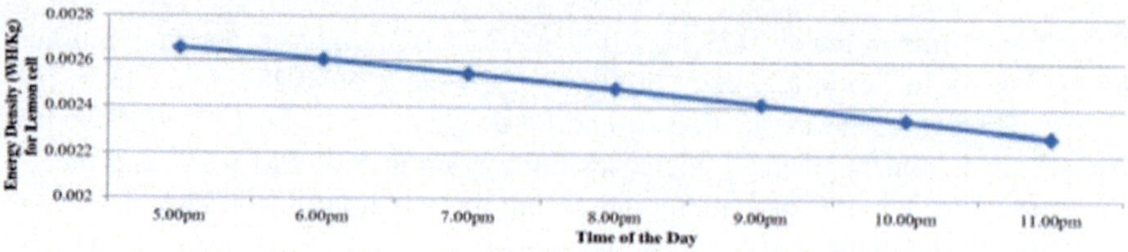

Fig. 26 Energy density (WH/Kg) for lemon cell with time of the day (28.08.2017)

with time of the day (28.08.2017) is 2.65 × 10^{-3} WH/Kg, and the minimum energy density for lemon cell with time of the day (28.08.2017) is 2.27 × 10^{-3} WH/Kg. Therefore, the distinction of energy density = 0.38 × 10^{-3} WH/Kg. It is also shown that energy density decreases linearly up to 11:00 PM from its starting.

Figure 27 shows the curve of energy density (Wh/Kg) for lemon cell with time of the day (29.08.2017). It is shown that the maximum energy density for lemon cell with time of the day (29.08.2017) is 2.45 × 10^{-3} Wh/Kg, and the minimum specific energy density for lemon cell with time of the day (29.08.2017) is 2.22 × 10^{-3} Wh/Kg. So the difference between the maximum and minimum energy density is 0.23 × 10^{-3} Wh/Kg. It is also shown that energy density decreases linearly up to 11:00 PM from its starting.

Figure 28 shows the curve of energy density (WH/Kg) for lemon cell with time of the day (30.08.2017). It is shown that the maximum energy density for lemon cell with time of the day (30.08.2017) is 2.34 × 10^{-3} WH/Kg, and the minimum specific energy density for lemon cell with time of the day (30.08.2017) is 2.29 × 10^{-3} WH/Kg. Therefore, the distinction of energy density = 0.05 × 10^{-3} WH/Kg. It is also shown that energy density decreases linearly up to 11:00 PM from its starting.

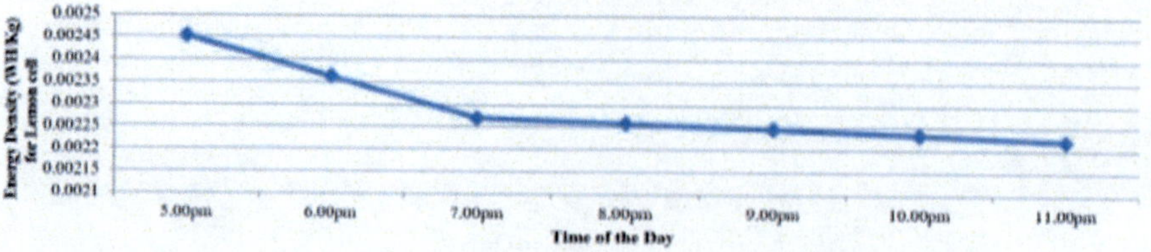

Fig. 27 Energy density (Wh/Kg) for lemon cell with time of the day (29.08.2017)

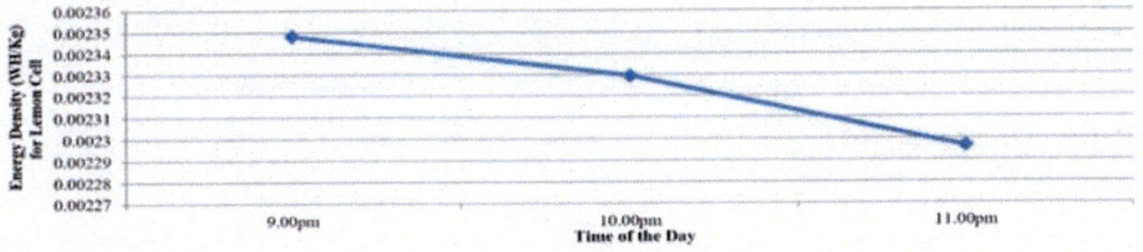

Fig. 28 Energy density (WH/Kg) for lemon cell with time of the day (30.08.2017)

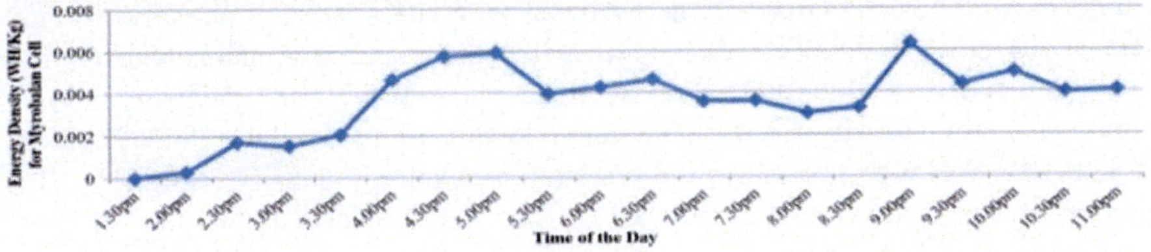

Fig. 29 Energy density (WH/Kg) for myrobalan cell with time of the day (26.08.2017)

Figure 29 shows the curve of energy density (WH/Kg) for myrobalan cell with time of the day (26.08.2017). It is shown that the maximum energy density for myrobalan cell with time of the day (26.08.2017) is 6.31×10^{-3} WH/Kg, and the minimum energy density for myrobalan cell with time of the day (26.08.2017) is 0 WH/Kg. Therefore, the distinction of energy density = 6.31×10^{-3} WH/Kg. It is also shown that energy density increases up to 5:00 PM, then it decreases up to 8:00 PM. Then, again, it increases linearly up to 9:00 PM then it decreases up to 11:00 PM.

Figure 30 shows the curve of energy density (WH/Kg) for myrobalan cell with time of the day (27.08.2017). It is shown that the maximum energy density for myrobalan cell with time of the day (27.08.2017) is 2.34×10^{-2} WH/Kg, and the minimum energy density for myrobalan cell with time of the day (27.08.2017) is 0.53×10^{-2} WH/Kg. Therefore, the distinction of energy density = 1.81×10^{-2} WH/Kg. It is also seen that energy density decreases linearly up to 2:00 PM, then it increases up to 7:00 PM. Then, again, it decreases linearly up to 8:00 PM then it increases up to 10:00 PM.

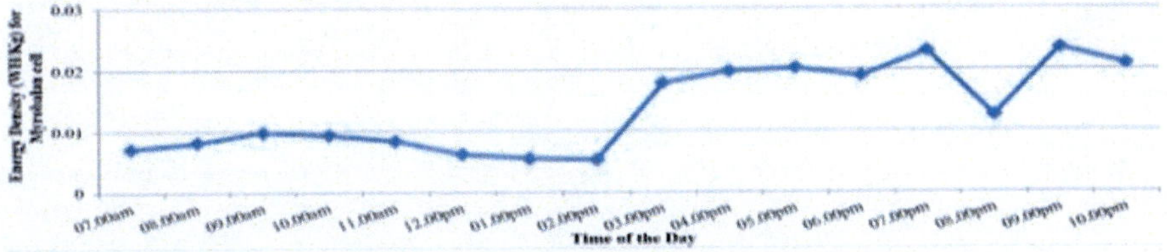

Fig. 30 Energy density (WH/Kg) for myrobalan cell with time of the day (27.08.2017)

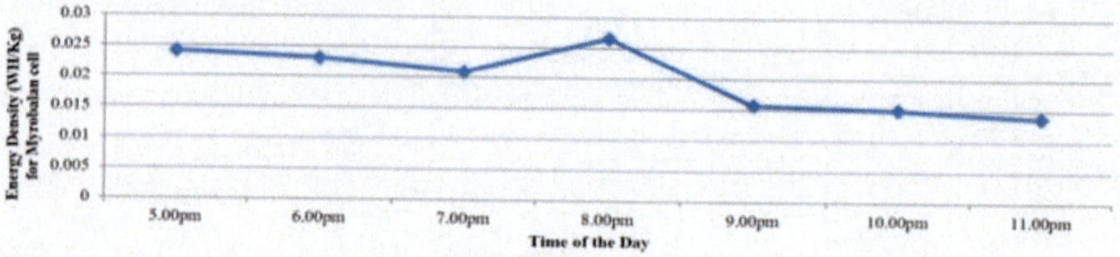

Fig. 31 Energy density (WH/Kg) for myrobalan cell with time of the day (28.08.2017)

Figure 31 shows the curve of energy density (WH/Kg) for myrobalan cell with time of the day (28.08.2017). It is shown that the maximum energy density for myrobalan cell with time of the day (28.08.2017) is 2.65×10^{-2} WH/Kg, and the minimum energy density for myrobalan cell with time of the day (28.08.2017) is 1.38×10^{-2} WH/Kg. Therefore, the distinction of energy density = 1.27×10^{-2} WH/Kg. It is also shown that energy density decreases linearly up to 7:00 PM, then it increases up to 8:00 PM. Finally, it decreases linearly up to 11:00 PM.

Figure 32 shows the curve of energy density (WH/Kg) for myrobalan cell with time of the day (29.08.2017). It is shown that the maximum energy density for myrobalan cell with time of the day (29.08.2017) is 2.28×10^{-2} WH/Kg, and the minimum energy density for myrobalan cell with time of the day (29.08.2017) is 0.97×10^{-2} WH/Kg. Therefore, the distinction of energy density = 1.31×10^{-2} WH/Kg. It is also shown that energy density decreases linearly up to 11:00 PM since its starting.

Figure 33 shows the curve of energy density (WH/Kg) for myrobalan cell with time of the day (30.08.2017). It is shown that the maximum energy density for myrobalan cell with time of the day (30.08.2017) is 1.02×10^{-2} WH/Kg, and the

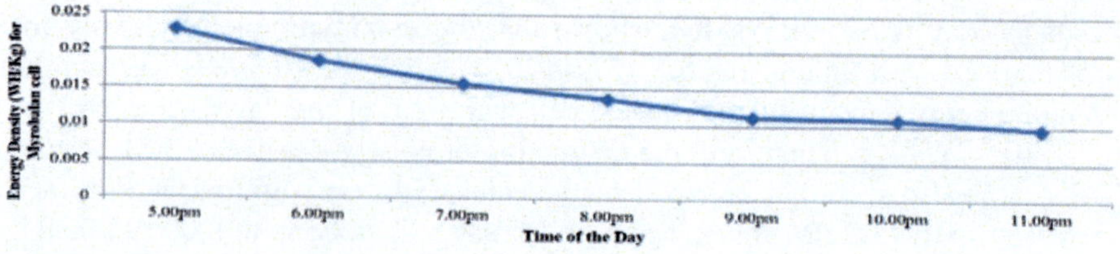

Fig. 32 Energy density (WH/Kg) for myrobalan cell with time of the day (29.08.2017)

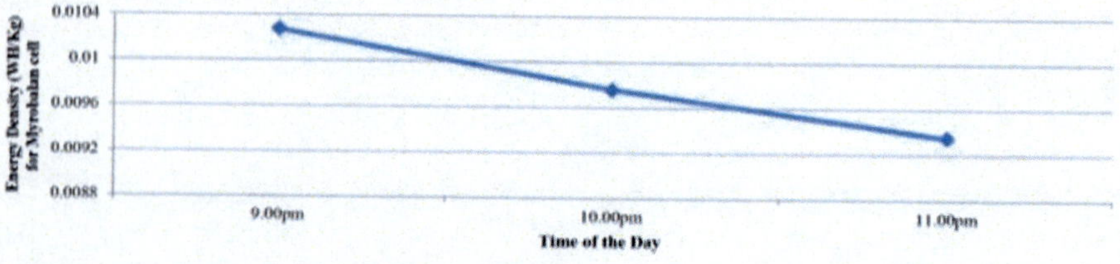

Fig. 33 Energy density (WH/Kg) for myrobalan cell with time of the day (30.08.2017)

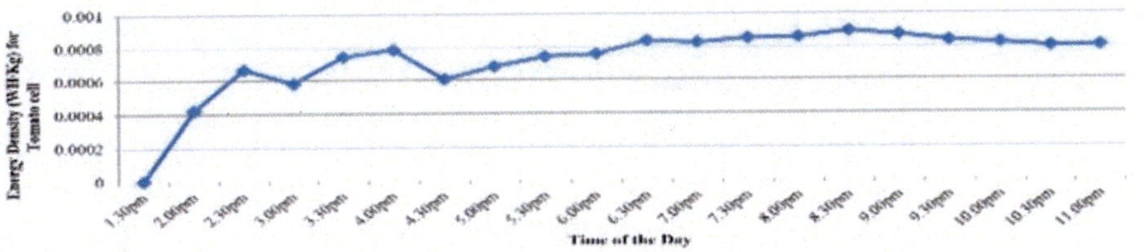

Fig. 34 Energy density (WH/Kg) for tomato cell with time of the day (26.08.2017)

minimum energy density for myrobalan cell with time of the day (30.08.2017) is 0.93×10^{-2} WH/Kg. Therefore, the distinction of energy density = 0.09×10^{-2} WH/Kg. It is also shown that energy density decreases linearly up to 11:00 PM since its starting.

Figure 34 shows the curve of energy density (WH/Kg) for tomato cell with time of the day (26.08.2017). It is shown that the maximum energy density for tomato cell with time of the day (26.08.2017) is 0.89×10^{-3} WH/Kg, and the minimum energy density for tomato cell with time of the day (26.08.2017) is 0 WH/Kg. Therefore, the distinction of density = 0.89×10^{-3} WH/Kg. It is also shown that energy density increases linearly up to 6:30 PM, but it decreases at few points like 3:00 PM and 4:30 PM. Then, it follows almost constant pattern up to 11:00 PM.

Figure 35 shows the curve of energy density (WH/Kg) for tomato cell with time of the day (27.08.2017). It is shown that the maximum energy density for tomato cell with time of the day (27.08.2017) is 1.66×10^{-3} WH/Kg, and the minimum energy density for tomato cell with time of the day (27.08.2017) is 1.25×10^{-3} WH/Kg. Therefore, the distinction of energy density = 0.41×10^{-3} WH/Kg. It is also shown that energy density increases linearly up to 11:00 PM since its beginning.

Figure 36 shows the curve of energy density (WH/Kg) for tomato cell with time of the day (28.08.2017). It is shown that the maximum energy density for tomato cell with time of the day (28.08.2017) is 2.1×10^{-3} WH/Kg, and the minimum energy density for tomato cell with time of the day (28.08.2017) is 2.05×10^{-3} WH/Kg. Therefore, the distinction of energy density = 0.05×10^{-3} WH/Kg. It is also shown that energy density increases linearly up to 11:00 PM since its beginning.

Figure 37 shows the curve of energy density (WH/Kg) for tomato cell with time of the day (29.08.2017). It is shown that the maximum energy density for tomato cell with time of the day (29.08.2017) is 2.33×10^{-3} WH/Kg, and the minimum energy

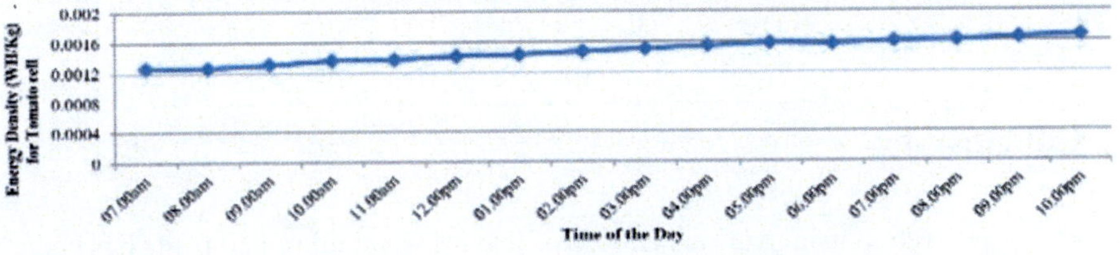

Fig. 35 Energy density (WH/Kg) for tomato cell with time of the day (27.08.2017)

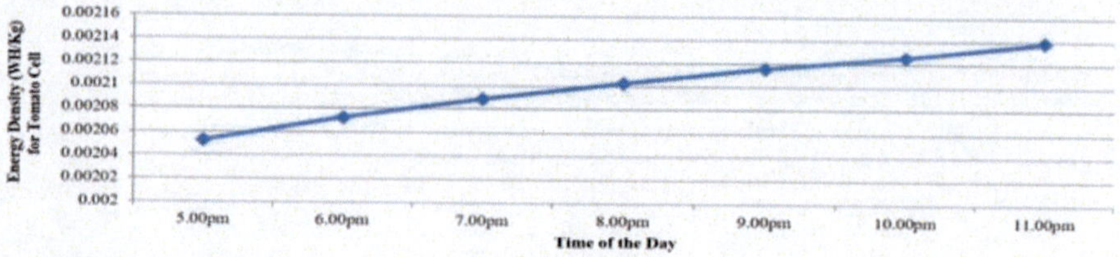

Fig. 36 Energy density (WH/Kg) for tomato cell with time of the day (28.08.2017)

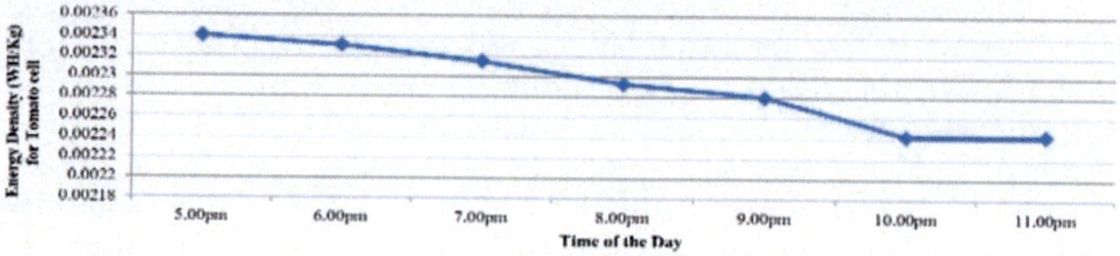

Fig. 37 Energy density (WH/Kg) for tomato cell with time of the day (29.08.2017)

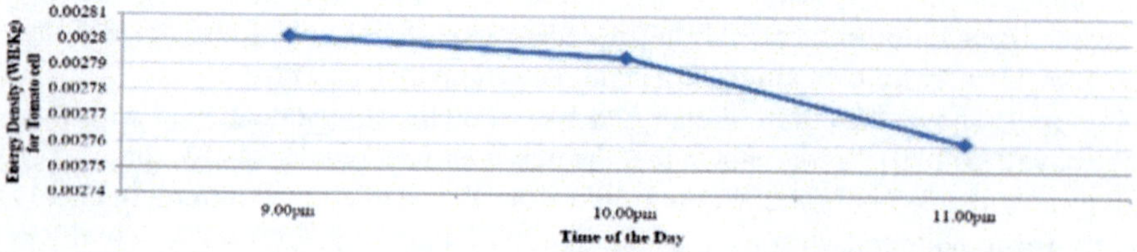

Fig. 38 Energy density (WH/Kg) for tomato cell with time of the day (30.08.2017)

density for tomato cell with time of the day (29.08.2017) is 2.24 × 10^{-3} WH/Kg. Therefore, the distinction of energy density = 0.09 × 10^{-3} WH/Kg. It is also shown that energy density decreases linearly up to 10:00 PM since its beginning. Then, energy density was almost constant up to 11:00 PM.

Figure 38 shows the curve of energy density (WH/Kg) for tomato cell with time of the day (30.08.2017). It is shown that the maximum energy density for tomato cell with time of the day (30.08.2017) is 2.80 × 10^{-3} WH/Kg, and the minimum energy density for tomato cell with time of the day (30.08.2017) is 2.76 × 10^{-3} WH/Kg. Therefore, the distinction of energy density = 0.04 × 10^{-3} WH/Kg. It is also shown that energy density decreases linearly up to 11:00 PM since its beginning.

4 Conclusions

Electricity generation using various classes of leaves, vegetables, and fruits has been discussed. It is concluded that the electrochemical cells were designed and fabricated for 2 half cells. The energy density depends on different parameters like concentration of the juice, temperature of the electrolyte, distance between two plates, etc. From the experiment, it is observed that the energy density of PKL electrochemical cell is the best among all other electrochemical cells. The findings of the system showed that it is possible to generate electricity using different leaves, vegetative, and fruits extract.

References

1. Timurkutluk B, Timurkutluk C, Mat MD, Kaplan Y (2011) Anode-supported solid oxide fuel cells with ion conductor infiltration. Int J Energy Res 35(12):1048–1055
2. Bokhari M, Abdullah AN, Hassan SNS, Jano Z, Saadan R (2014) Relationship between understanding, awareness and the practice of green technology based on ethnicity perspectives in Melaka. J Human Capital Devel (JHCD) 7(2):33–46
3. Lee MF, Zain MM, Lai CS (2018) Lighting system design using green energy from living plants. J Phys Conf Ser 1019(1):012019. IOP Publishing
4. Howells M, Alexander R (2012) Perspectives on sustainable energy for the 21. Retrieved from https://sustainabledevelopment.un.org/content/documents/1131Energy_SD21.pdf
5. Timmers RA (2012) Electricity generation by living plants in a plant microbial fuel cell
6. Choo YY, Dayou J (2013) A method to harvest electrical energy from living plants. J Sci Technol 5(1)
7. Bhardwaj M, Neelam K (2015) The advantages and disadvantages of green technology. J Basic Appl Eng Res 2(22):1957–1960
8. Hassan L, Khan KA (2021) Applications of PKL electricity for use in DC instruments. In: Microelectronics, circuits and systems. Springer, pp 191–202
9. Khan KA, Shaiful Islam M, Awal A, Khan MN, Ullah AKM (2021) Studies on performances of copper oxide nanoparticles from catharanthus roseus leaf extract. In: Microelectronics, circuits and systems. Springer, pp 179–190
10. Alam Khan K, Rahman Rasel S, Zian Reza SM, Yesmin F (2020) Energy efficiency and sustainability in outdoor lighting—a bet for the future. https://doi.org/10.5772/intechopen.89413
11. Khan KA, Awal MA (2020) A study on connection between chemistry and electricity. IJARIIEISSN (O)-2395–4396, 6
12. Khan KA, Yesmin F, Wadud MA, Obaydullah AKM (2019) Performance of PKL electricity for use in television. In: International conference on recent trends in electronics & computer science-2019, Venue: NIT Silchar, Assam, India, Conference date: 18th and 19th of March, p 69
13. Rab MNF, Khan KA, Rasel SR, Ali MH, Hassan L, Salek MA, Ohiduzzaman M (2020) Voltage cultivation from fresh leaves of air plant, climbing spinach, mint, spinach and Indian pennywort for practical utilization. In: Energy systems, drives and automations. Springer, pp 161–172

Review on Various Techniques for Load Frequency Control in Deregulated Power Structures

Veena Sharma, Ayushi Dogra, R. Naresh, and Vineet Kumar

Abstract The power system has expanded immensely during the past few years due to continuous emerging innovations and changing ideas for improvement in the system stability. Hence, it becomes very necessary to maintain different parameters of the system as per the norms without disturbing the system. For fulfilling the load demand of the consumers, most reliably and economically power systems are interconnected to each other forming multi-area control systems. Also, the conventional power system structure is getting replaced by deregulated structures for the betterment of the power quality and increasing the competition in the power sector, which is also immensely impacting the way of how the whole power sector works. The main area impacted due to these changes is load frequency control. This paper highlights the recent work done on load frequency regulation techniques most recently applied in a deregulated environment. A brief review of various control methodologies based on robust control and soft computing control techniques are discussed.

Keywords Deregulated power system · Load frequency control · Automatic generation control · Integrated power system

1 Introduction

Nowadays, smart grids are being installed in power systems for efficient working. In an interconnected power system, different types of power plants are being connected for fulfilling the load demand [1–5]. Due to the interconnection of the power system, the complexity increases, and it becomes highly prone to tie-line power and frequency

V. Sharma · A. Dogra (✉) · R. Naresh · V. Kumar
Electrical Engineering Department, NIT Hamirpur, Hamirpur, H.P., India

V. Sharma
e-mail: veena@nith.ac.in

R. Naresh
e-mail: rnaresh@nith.ac.in

V. Kumar
e-mail: vineet.kumar@nith.ac.in

J. K. Mandal et al. (eds.), *Topical Drifts in Intelligent Computing*, Lecture Notes in Networks and Systems 426, https://doi.org/10.1007/978-981-19-0745-6_61

deviations. The smooth operation of the modern power system is very complex and depends on the frequency of the system, hence making load frequency control an essential part of the AGC (automatic generation control). Sudden load change can give rise to frequency deviation which in turn can effectively disturb the whole power system. Load frequency control is the most researched part of automatic generation control, in the last few years; this area has immensely got new techniques and improvements. Also, due to changing trends in power supply, i.e., deregulation of the power sector, it becomes an urgent need to make power systems highly robust against sensitivity toward frequency fluctuations [6–12]. The detailed review on existing methodologies available for the AGC under deregulation case is necessary for new researchers to find a proper direction and tools for the thorough research.

This survey briefly describes all the recent methods applied in deregulated power system (DPS) for load frequency control. Section 2 explains the deregulated supply system. Section 3 describes some of the soft computing techniques proposed in the recent past. Section 4 introduces some robust control techniques [13–18]. Section 5 gives highlights on MPC controller, and Section 6 is a survey on various combinations of different techniques. A brief of the techniques surveyed is shown in Table 2, and some of the observations are concluded by authors on basis of the survey.

2 Deregulation of Power Supply

During the past few years, power sector is going through various changes and one of which is the deregulation of the power supply. The old way of supplying electric power to consumers under government policies and norms is getting replaced by deregulated system. Deregulation is unbundling of the power system component both vertically and horizontally. New rules are formed for selling the power and its regulation. Hence, increasing the competition in the power sector and further resulting in innovations and enhancement in power quality. In India, electricity reforms are in process and will take much time to grow all over, but it is a big step toward the transformation of the power sector [19–21].

As shown in Fig. 1, deregulated power sector consists of different unbundled parts.

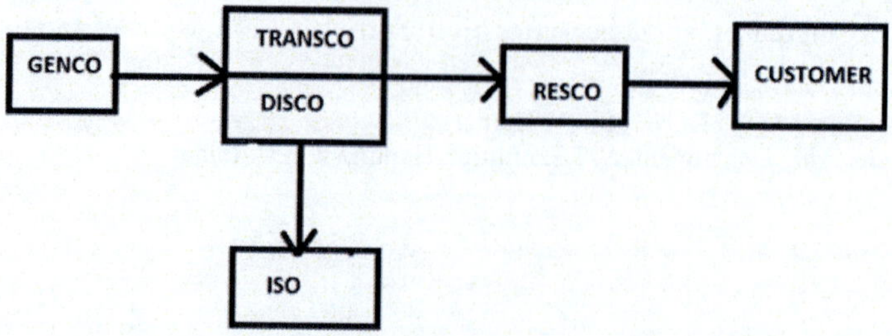

Fig. 1 Different entities in deregulated power supply

Table 1 Merits and demerits of DPS

Advantage	Disadvantage
Increased reliability	Increase in network congestion
More options for buying electricity	High complexity
Reduced electricity prices	Hedging and risk management

1. GENCO (GENERATION COMPANY)—Generating power and selling.
2. TRANSCO (TRANSMISSION COMPANY)—It transfers power from GENCO to the delivery point.
3. DISCO (DISTRIBUTION COMPANY)—Power is distributed to customers.
4. RESCO (RETAIL ENERGY SERVICE)—Purchase energy from the GENCO's/spot market and sell.
5. ISO (INDEPENDENT SYSTEM OPERATOR)—Entire control of the entire system is done by this unit.
6. Customers—End receiver of power.

Many countries have already adapted the deregulated power system (DPS) such as Peru, the USA, England, Columbia, and Argentina. In India, in 2003, Government of India notified the electricity act. Under this act, accelerating the power system reforms was the main objective [22–30]. The government of India is taking many initiatives in this field. Many states have initialized the power sector reforms, starting with the restructuring of the power system and tariff rationalization. High accountability can be achieved practically by power sector reform [31–35] (Table 1 and Fig. 2).

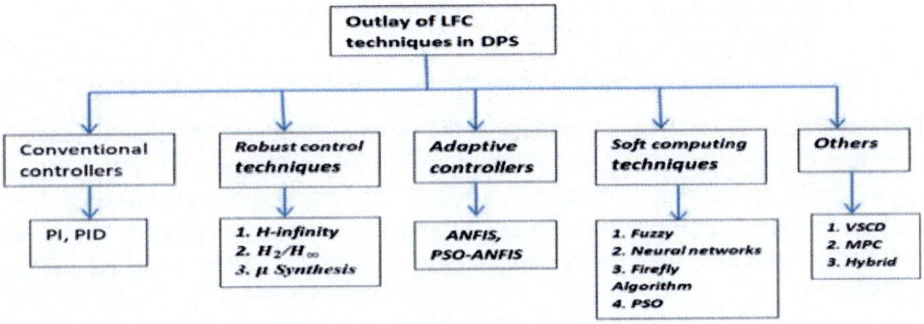

Fig. 2 Various techniques of LFC in deregulated environment

Table 2 Highlight on success of the reviewed techniques

Techniques	Advantages	Disadvantages
Fuzzy, fuzzy PID controller with MBA, ANN	Better performance than that of conventional PID, less error, less overshoot, and less settling time	Not suitable for large power systems
PSO, HCPSO, FFA	Dynamic performance of the system improves	Not tested on real-time basis. (With effect of time delay, GRC, and dead band.)
QOHS-PI controller	Fast convergence speed Low computational cost	Limited iterations
PSO based ANFIS	High robustness, Easy implementation	Less flexible Not practical for real-time applications
H-infinity control	Suitable for large power system disturbances	Higher-order control
μ-synthesis	Effective in disturbance rejection	Shows low-frequency regulation under perturbations
H_2/H_∞	Robust for bilateral contracts, good performance	Unknown attacks and inputs cannot be investigated
Variable structures control/sliding mode control	Low computational cost, good results shown for model-based schemes	Not suitable for large power systems
MPC/DMPC	Closed-loop performance is enhanced	Online optimization needs expensive computation power

3 Various Soft Computing Techniques for Deregulated Power System

The changing structure and complexity of the power system are on updating every single day. Renewable sources when introduced to the interconnected power structure, a very high-frequency deviations due to load perturbation is experienced due to the inconsistent nature of RES, creating large blackouts in the system. In these scenarios, the old LFC may not be successful in the modernized deregulated power environment [36–42].

Hence, to cope with this problem, the intelligent control scheme along with soft computing techniques such as an artificial neural network (ANN), fuzzy logic controllers, genetic algorithm (GA), bacterial foraging optimization algorithm (BFOA), particle swarm optimization (PSO) algorithm, and firefly algorithm are explored [5–8].

3.1 Fuzzy Controllers

Fuzzy logic controllers work on a knowledge-based learning system and can be successfully applied for LFC in the power system. Fuzzy control is implemented for load frequency regulation of an isolated interconnected power system for both single areas and two-area control system by Sambriya and Viveknath [5]. The results are compared with the PID controller and found better in fuzzy-based LFC system. LFC for multi-interconnected reheat thermal system is proposed using optimized fuzzy PID incorporated with mine blast algorithm (MBA) improving the system performance by Ahmad Fathy [6]. For better performance in deregulated power system, a fuzzy PID is studied, and the performance found is better than that of PID and FPID. Results were found better than that of the classical integral controllers when FLC is designed using Sugeno and Mamdani as membership functions by Ram et al. [7]. The author designed FLC-based load frequency control system for a four-area interconnected control system. Comparison with other conventional controllers has shown improved results. Fuzzy sliding mode controller is integrated with SEMS to improve the dynamic response of IPS. PID is tuned by the multi-objective optimization algorithm. This design has improved the system rejections and keeps control ranges high in operating ranges. Robust design to uncertainties in the system is proposed by Khasraviani [8].

3.2 Artificial Neural Networks

Artificial neural networks are human brain-inspired intelligence controllers. ANN consists of interconnected processing units called neurons. ANN's through learning can perform tasks like pattern recognition and classification. Danwei Qian et al. have proposed a control scheme using integral sliding mode. RBF networks are utilized to suppress the nonlinearities in generation rate constraint (GRC). Results have shown superiority in NN-based I-SMC method, and effects of wind turbines have also been demonstrated [8]. Kassem has discussed the LFC of two-area interconnected systems using the neural predictive method. Simulation results show the improvement of NN-based MPC than that of fuzzy-based MPC [24]. Mosaad proposed a design based on ANFS, and ANN. GA is being utilized for training of both ANFS and ANN. Over the full loading range, both controllers are showing great performance [26]. Design for the deregulated power system is proposed by Heidar Ali Shyantar et al. by using ANN-based controller (decentralized). Performance evaluated shows good results during parameter uncertainties. The author defined the LFC problem when RES is connected to the power system. RBF NNs are designed for decreasing the system uncertainties. D. K Chaturvedi et al. worked on GNN controllers with backpropagation time. Application of generalized NN for LFC has been successfully implemented.

3.3 *Firefly Algorithm*

The firefly algorithm (FA) is inspired by nature, i.e., by the flashing of fireflies (social behavior) in tropical areas. It is metaheuristic; its main advantage is easy implementation. Chander Sekhar et al. [34] have designed (FA) algorithm optimized FPID controller (FUZZY PID) along with the derivative filter. This proposed algorithm has shown good performance under different load perturbations. GDB and GRC are considered as constraints, and the algorithm is tested on the deregulated environment. The firefly algorithm is applied for the tuning of the PI controllers for load frequency regulation of the hybrid system (comprising of the PV system and thermal generators) by Abd-Elazim et al. [3]. An optimal solution is achieved and can be successfully applied for the large-scale power system. O. Abeinia et al. has proposed an FPID-based FA for the deregulated environment. It is found that under different load conditions, it gives better results than the CPID and PID. An attempt was made by Dilip Khamari et al. for the designing firefly algorithm-based PID controller, and results are found better than GSA tuned PID controllers.

4 Robust Control Techniques for Deregulated Power Structure

Robust load frequency control techniques are the center of attraction nowadays due to their robustness against uncertainties of the power plant, load perturbation in DPS.

Arlene Davidson et al. [10] proposed a decentralized loop shaping H-infinity controller for a deregulated non reheat Thermal Power System. Dynamic performance is improved concerning disturbance at low frequencies and robustness against uncertainty at higher frequencies. The author has also proposed an H-infinity event triggered multi-area power system concerning a three-area control system. He Zhang et al. [12] have solved H-infinity control and stability analysis by the Lyapunov method. The method proposed has shown good performance. For reducing the communication bandwidth network burden an adaptive control scheme is framed. Combination of a new intelligent agent-based control scheme, using Bayesian networks is proposed for decentralized AGC by Bevrani et al. [43]. Results are compared with controllers and improvement is shown. Based on modern control tools, a robust PI controller is designed for LFC, and a new decentralized technique in DPS has been utilized by several authors. High robustness is achieved by this combination. For robustness of load frequency control in DPS, author has also presented synthesis (singular value) for decentralized AGC. A two-area thermal control system (deregulated) loop shaping controller is designed by Arlene et al. [10].

5 Model Predictive Controller

MPC is the advanced control of the process working under a certain set of constraints. Model predictive controllers are dependent on the dynamic models of the process.

Jizhen Liu et al. [11] proposed an LFC scheme for a hybrid power system comprising of wind farms and thermal power plants. The model predictive controller is used for the calculation of optimization parameters. It is being observed that frequency can be successfully regulated using the MPC-based controller in a hybrid system. DMPC scheme is being proposed by the author for four areas IPS. When a wind turbine is introduced to the system, DMPC has shown good results. Results are compared between centralized and decentralized MPC. Anne Mai Ersdal et al. [20] analyzed MPC-based load frequency regulation. Improved performance is seen, and cost reduction is being observed by using MPC-based AGC. Xilin Zhao et al. [21] have given model predictive-based AGC method for multi-area IPS. Random time delay and penetration of photovoltaic systems are taken into consideration. Reduction in negative effects of time delay is seen; results are satisfactory. Problems related to load change are mitigated by adding MPC by the author. Results are tested on a real-time basis. MPC is found robust, and optimal values are calculated irrespective of the load changes. Guo-Qiang-Zeng et al. [23] have designed MPC-based LFC for IPS with PV generators. Results show superiority in performance than GA-PI, FA-PI, and PEO-PI.

6 Hybrid Techniques for LFC

Techniques based on combination of both algorithm and controllers are utilized for getting better version of load frequency regulation system. Most work is based on these hybrid techniques as robustness of the system against variation in power system parameters is seen.

Elsisi et al. [14] have proposed a model predictive controlled (MPC) LFC with the BAT algorithm. This model is successfully implemented for coping up with the nonlinearities of the system such as GRC and GDB's. For minimizing the overshoot and settling time, domain-based function is incorporated. Comparison with GA-based PI controllers and BA-based MPLC has shown improved results. A LUS-TLBO hybrid algorithm is combined with the PID and fuzzy PID controller for optimization of gains. Local unimodal sampling algorithm (LUS) and teaching learning-based optimization (TLBO) algorithm. This hybrid algorithm has shown improvement in undershoot, overshoot, and the settling time. For wide variations in system parameters, it has shown robustness. Sukhwinder Singh Dhillon et al. [16] have given a model tuning the PI parameters with BFOA-PSO technique. Effectiveness of both with and without DFIG generators is tested on three-area control network. Design has shown that system performance is stable during the system parameter variations. G Shankar et al. [29] have proposed a model which uses concept of quasi-oppositional-based

(QOHSA) learning to accelerate the exploring capacity of harmony search algorithm (HSA). Results have shown improvement compared to that of metaheuristic techniques. Chandan Kumar Shiva et al. [31] have attempted to work on large overshoots of hydro–hydropower system. QOHS is applied to optimize the controller gains. Quality solutions are found by this approach, better results are achieved compared to MRPS and IMC methods. Table 2 presents merits and demerits of various control methodologies available in the existing literature.

7 Conclusions

In this paper, different techniques applied for load frequency regulation in the deregulated environment are reviewed. The review has emphasized the importance of latest control techniques on the deregulated load frequency control in recent scenario where renewable integration has increased. Based on the authors' best knowledge, following aspects need more consideration:

1. Issues related to cyber-attack must be taken care of in reregulated AGC.
2. Both power production and system parameters should be efficiently handled through optimization techniques.
3. Algorithms proposed should be tested more practically.
4. Robust controller tools like MPC and SMC need more investigation for LFC problems under deregulated case.

This paper covers few recently applied techniques for LFC in the deregulated environment and will provide necessary information to researchers for further investigation.

Acknowledgements This work was supported by H.P Council for Science, Technology and Environment, Shimla, India, under HIMCOSTE Project grant STC/F (8)-6/2019(R&D2019-20)-408, H.P., India, sanctioned to the first author.

References

1. Esmail M et al (2017) Review of automatic generation control in deregulated environment. IFAC-PapersOnLine 50(2):88–93
2. Tungadio DH, Sun Y (2019) Load frequency controllers considering renewable energy integration in power system. Energy Reports 5:436–453
3. Abd-Elazim SM, Ali ES (2018) Load frequency controller design of a two-area system composing of PV grid and thermal generator via firefly algorithm. Neural Comput Appl 30(2):607–616
4. Sekhar G, Chandra T et al (2016) Load frequency control of power system under deregulated environment using optimal firefly algorithm. Int J Electric Power Energy Syst 74:195–211
5. Sambariya DK, Nath V (2015) Load frequency control using fuzzy logic based controller for multi-area power system. Brit J Math Comput Sci 13:1–19

6. Fathy A et al (2020) Optimal design of fuzzy PID controller for deregulated LFC of multi-area power system via mine blast algorithm. Neural Comput Appl 32(9):4531–4551
7. Ram P, Jha AN (2010) Automatic generation control of interconnected hydro-thermal system in deregulated environment considering generation rate constraints. In: International conference on industrial electronics, control and robotics, pp 148–159
8. Khosraviani M, Mehrshad et al (2018) Load-frequency control using multi-objective genetic algorithm and hybrid sliding mode control-based SMES. Int J Fuzzy Syst 20(1):280–294
9. Aldi M, Karapici G (2019) Load frequency control in four areas by using fuzzy logic controller. Int J Eng Res 8(10)
10. Arlene D, Ushakumari S (2016) H-infinity loop-shaping controller for load frequency control of a deregulated power system. Procedia Technol 25:775–784
11. Liu J et al (2019) Event-triggered H load frequency control for multiarea power systems under hybrid cyber attacks. IEEE Trans Syst Man Cybern 49(8):1665–1678
12. Zhang H et al (2018) H-infinity load frequency control of networked power systems via an event-triggered scheme. IEEE Trans Ind Electron 67(8):7104–7113
13. Kumar N, Indragandhi V (2018) Analysis on various optimization techniques used for load frequency control in power system. Serbian J Electric Eng 15(3):249–273
14. Elsisi M et al (2016) Bat inspired algorithm based optimal design of model predictive load frequency control. Int J Electr Power Energy Syst 83:426–433
15. Sahu BK et al (2016) A novel hybrid LUS–TLBO optimized fuzzy-PID controller for load frequency control of multi-source power system. Int J Electric Power Energy Syst 74:58–69
16. Dhillon SS et al (2016) Multi objective load frequency control using hybrid bacterial foraging and particle swarm optimized PI controller. Int J Electric Power Energy Syst 79:196–209
17. Kayalvizhi S, Vinod Kumar DM (2017) Load frequency control of an isolated micro grid using fuzzy adaptive model predictive control. IEEE Access 5:16241–16251
18. Liu J et al (2019) Model predictive control for load frequency of hybrid power system with wind power and thermal power. Energy 172:555–565
19. Zhang Y et al (2015) Model predictive control for load frequency control with wind turbines. J Control Sci Eng 1–17
20. Ersdal AM et al (2016) Model predictive load-frequency control. IEEE Trans Power Syst 31(1):777–785
21. Zhao X et al (2018) A system compensation based model predictive AGC method for multiarea interconnected power systems with high penetration of PV system and random time delay between different areas. Math Probl Eng 1–10
22. Gulzar MM et al (2020) Mitigating the load frequency fluctuations of interconnected power systems using model predictive controller. Electronics 8(2):156
23. Zeng G-Q et al (2017) An adaptive model predictive load frequency control method for multi-area interconnected power systems with photovoltaic generations. Energies 10(11):1840
24. Qian D et al (2016) Load frequency control by neural-network-based integral sliding mode for nonlinear power systems with wind turbines. Neuro Comput 173:875–885
25. Kassem AM (2010) Neural predictive controller of a two-area load frequency control for interconnected power system. Ain Shams Eng J 1(1):49–58
26. Mossad MI, Salem F (2014) LFC based adaptive PID controller using ANN and ANFIS techniques. J Electric Syst Inf Technol 1(3):212–222
27. Dianwei Q, Fan G (2018) Neural-network-based terminal sliding mode control for frequency stabilization of renewable power systems. IEEE/CAA J Autom Sinica 5(3):706–717
28. Ramachandran R et al (2018) Load frequency control of a dynamic interconnected power system using generalised hopfield neural network based self-adaptive PID controller. Iet Gen Transm Distrib 12(21):5713–5722
29. Shankar G, Mukherjee V (2016) Load frequency control of an autonomous hybrid power system by quasi-oppositional harmony search algorithm. Int J Electr Power Energy Syst 78:715–734
30. Guha D et al (2020) Quasi-oppositional backtracking search algorithm to solve load frequency control problem of interconnected power system. Iran J Sci Technol Trans Electric Eng 44(2):781–804

31. Shiva CK, Mukherjee V (2016) Automatic generation control of hydropower systems using a novel quasi-oppositional harmony search algorithm. Electric Power Comp Syst 44(13):1478–1491
32. Kunya AB et al (2020) Improved model predictive load frequency control of interconnected power system with synchronized automatic generation control loops. Beni-Suef Univ J Basic Appl Sci 9(1):1–13
33. Pappachen A, Peer Fathima A (2017) Critical research areas on load frequency control issues in a deregulated power system. A state-of-the-art-of-review. Renew Sustain Energy Rev 72:163–177
34. Gorripotu TS et al AGC (2015) of a multi-area power system under deregulated environment using redox flow batteries and interline power flow controller. Eng Sci Technol Int J 18(4):555–578
35. Verma YP, Kumar A (2013) Load frequency control in deregulated power system with wind integrated system using fuzzy controller. Front Energy 7(2):245–254
36. Davidson R, Arlene, Ushakumari S (2016) H-infinity loop-shaping controller for load frequency control of an uncertain deregulated power system. Int Conf Electric Electron Opt Tech (ICEEOT), 2185–2191
37. Lakshmi R, Zahira (2019) Load frequency control in deregulated power system. Int J Res 5:124–133
38. Evans E, Ejegi et al (2015) Model predictive load frequency control of a two-area deregulated power system. Eur Control Conf (ECC), 1044–1049
39. Anitha P (2016) An adaptive load frequency control using soft computing techniques in deregulated power system. University
40. Kumar VK et al (2020) Load frequency control of three area multi-unit deregulated power system with FOSMC and performance analysis using regulation constant. Int J Eng Technol 389–406
41. Gorripotu TS, Sahu RK, Panda S (2015) AGC of a multi-area power system under deregulated environment using redox flow batteries and interline power flow controller. Eng Sci Technol Int J 18(4):555–578
42. Mishra RN, Chaturvedi DK, Kumar P (2020) Recent philosophies of AGC techniques in deregulated power environment. J Inst Eng India Ser B 101:417–433. https://doi.org/10.1007/s40031-020-00463-8
43. Bevrani H, Daneshfar F, Hiyama T (2012) A new intelligent agent-based AGC design with real-time application. IEEE Trans Syst Man Cybern Part C (Applications and Reviews) 42(6):994–1002

Investigations on the Impact of Soiling on Bifacial Gain

Gautam Raina, Shubham Sharma, and Sunanda Sinha

Abstract Accumulation of dust in conjugation with various environmental adversities over the surface of PV modules causes soiling phenomenon, thereby generating shading scenarios and leading to reduced irradiance available to the module. Soiling has been recorded as one of the most common detrimental factors to module health and energy output. An overview on the soiling phenomena has been discussed in this manuscript, with a brief discussion regarding the soiling of bifacial PV modules. Furthermore, a case study investigating the impact of bifacial gain of a 90° bifacial PV module installed over two different orientations East–West (E–W) and North–South (N–S) has been conducted. The results show that the performance ratio (PR) of soiled system installed at E–W orientation (1.37) is greater than PR of N–S-oriented system (0.99). The bifacial gain of a soiled system under E–W orientation (0.81) is greater than the bifacial gain of soiled system (0.34) under N–S orientation due to higher irradiance collection throughout the day.

Keywords Bifacial gain · Soiling · Performance ratio · Orientation

1 Introduction

The phenomenon of accumulation and deposition of certain contaminants on the top surface of a PV module glass cover, which limits the transmittance of incident solar radiation received by the PV module, is known as soiling [1]. The examples of such contaminants include mineral dirt/dust particles, salt particulates (near coastal areas), bird droppings, algae, fungi, bacterial films in wet climates, pollen, agricultural emission (rice husk, feed dust, etc.), and industrial emission (engine exhaust, fly ash, cement, charcoal, etc.) [2, 3]. Soiling is considered as one of the most dominant environmental factors which causes the degradation of power in PV modules. A layer of dirt/dust deposited on PV glass surface causes a reduction in the transmittance of solar radiation to the PV solar cells, which in turn leads to decreased generation. It

G. Raina · S. Sharma · S. Sinha (✉)
Centre for Energy and Environment, Malaviya National Institute of Technology, Jaipur, Rajasthan, India

J. K. Mandal et al. (eds.), *Topical Drifts in Intelligent Computing*, Lecture Notes in Networks and Systems 426, https://doi.org/10.1007/978-981-19-0745-6_62

also introduces additional expenditure in terms of operation and maintenance cost. Soiling is a site-specific phenomenon which increases the uncertainty in power output prediction and forecasting of PV plants production capacity, which is incurred in the form of financial losses.

As the worldwide PV installations are increasing day by day, it has been anticipated that global losses in PV power generation may rise drastically in between 4 and 7% annually causing global financial losses of almost 7 billion Euros by 2023 [3, 4]. Unlike traditional solar modules, bifacial PV modules generate electricity from both front and rear sides and this ability to increase electric output has attracted the industry.

In this paper, an overview of factors affecting soiling has been discussed in brief along with literature analysis based upon soiling impact investigations for bifacial PV modules worldwide. A simulation-based case study using PVsyst, for two different bifacial PV module orientations, East–West (E–W) and North–South (N–S), has been conducted. Performance of bifacial modules in terms of performance ratio and bifacial gain has been carried out for first time in Indian composite climate.

Section 2 of this paper gives an overview of various factors which affect the phenomenon of soiling on solar PV modules. Section 3 presents the literature review of recent studies which highlighted the influence of soiling on the performance of bifacial modules. Section 4 describes a case study focused on determining the dependency of soiling on bifacial gain of PV modules. Finally, the noteworthy outcomes of this study have been concluded in Sect. 5.

2 Factors Affecting the Soiling Process

Soiling is a highly site-specific phenomenon which is influenced by several environmental factors. In arid regions, soiling occurs due to electrostatically active inorganic materials (mostly comprising of desert); in coastal areas, salt accumulation on the surface is quite common; while colder and industrial zones mostly suffer soiling issues from organic dust, carbon emissions from industries, etc. This uniform and non-uniform coverage cause more obstruction to light reaching the solar cell and thus cause additional optical losses [5, 6]. Wind and rainfall are considered as natural cleaning sources, but sometimes, even heavy rainfall and high-speed winds are not able to completely remove soiling layers of fine dust particles. In the presence of moisture, the particles tend to exert cohesive attractive forces on each other causing cementation. It has been observed that a change in humidity from 40% to 80% can enhance the adhesion between particulate matters by 80%. Few important factors are the covering materials, height, orientation, and tilt angles of PV module [7, 8] which have an implication on the dust effect. Figure 1 describes the most important factors which influence the dust accumulation on the surfaces of PV modules.

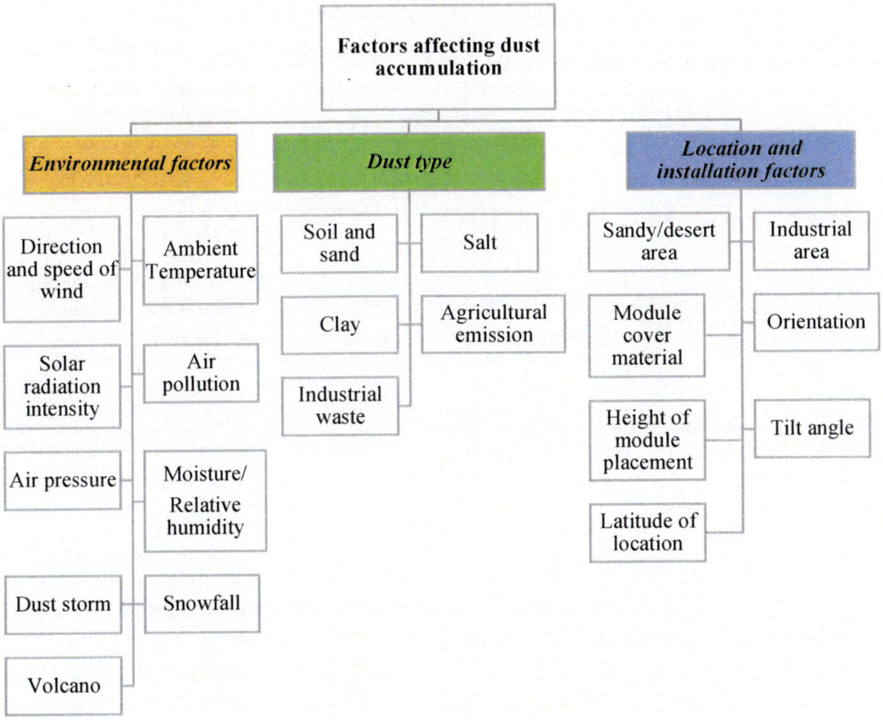

Fig. 1 Factors affecting the dust accumulation on PV modules [7]

3 Effect of Soiling on Bifacial Modules

Bifacial modules are capable of absorbing solar radiation from the front as well as rear side which makes their operation quite efficient and offer 10%–15% more electrical yield than their monofacial counterparts [9, 10]. As rear side of bifacial module also contributes in power generation, the decrease in total current caused by reduced irradiance due to dust accumulation will also be from both sides of the module. Bifacial technology has gained quite popularity in recent days, and adequate research on impact of soiling on the performance of this technology is lacking. Some recent studies on soiling of bifacial modules, conducted in different parts of the world, have been summarized in Table 1.

Albeit bifacial technology has been prevalent for quite some time, adequate research into different aspects of bifacial nature, and performance of bifacial PV modules is not available in abundance. Among them, the effects of soiling on bifacial performance and bifacial gain need to be examined. It can be assumed that understanding the soiling effect in bifacial modules, and thereby studying the soiling rates and soiling losses become of vital importance, to further enhance the acceptance and understanding of this technology.

Table 1 Literature review of recent studies on soiling of bifacial modules

Reference	Reference no	Duration of study	Location	Major findings
Singh et al. (2021)	[10]	6 months	Utah	A method obtained to separately study the contributions of rear and back side of bifacial modules to power output
Baloch et al. (2016)	[11]	60 days	Qatar	Study highlights the real-time performance of bifacial modules during winters and summers under soiling conditions. Cleaning improved the performance of modules by 29%
Cabrera et al. (2018)	[12]	45 days	Chile	If soiling losses are maintained at 0.10% per day, then vertically mounted bifacial modules perform better than traditionally tilted bifacial modules
Bahaduri and Kottantharayil (2018)	[13]	120 days	Mumbai	Vertical mounted module shows fewer soiling losses and low temperatures as compared to the modules tilted at latitude angle
Ayala et al. (2018)	[14]	8 months	Chile	Soiling rate less than 2.5% was obtained during the starting phase of study. Later events of rain caused natural cleaning of the accumulated dust
Luque et al. (2018)	[15]	70 days	Chile	Soiling rate of 0.236%/day was observed for bifacial modules Soiling rate of rear side was almost 8.8 times smaller than the rate obtained at front side
Ullah et al. (2020)	[16]	120 days	Lahore	Soiling losses decreased from 1.11% to 0.11% when the tilt angle changed from 0° to 90°

4 Dependency of Bifacial Gain on Soiling of Module: A Case Study

The energy generated (kWh) by a PV system annually, known as annual energy yield (AEY), is a critical parameter used in estimating the performance of any PV system [17]. This performance metric is further utilized to evaluate the bifacial gain possible by using a PV system comprising of bifacial PV modules in place of conventional monofacial modules. Bifacial gain (BG) is the dimensionless gain in produced AEY realized by using a bifacial PV module [18]. To estimate the BG of a given system, Eq. (1) is used.

$$BG = \frac{AEY_{Bi} - AEY_{Mono}}{AEY_{Mono}} \tag{1}$$

In Eq. (1), subscripts "bi" and "mono" refer to a bifacial and monofacial PV system, respectively.

For the purpose of this research work, a case study has been carried out for Jaipur, India (26.9124° N, 75.78° E) through simulations in PVsyst, which enables calculation of system generation characteristics for a pre-defined time scale. The simulations are conducted for a PV system with a bifacial module of specification given in Table 2, and a monofacial module of same specifications is built in the software.

To determine the impact of soiling on the bifacial gain of a PV system, soiling loss of 5% is assumed, which is kept constant for both bifacial and monofacial PV systems. It should be noted that although various research works have established a higher soiling loss and soiling rate for monofacial PV systems [12, 13, 15], an equal value for soiling loss for both monofacial and bifacial PV systems is assumed. This is done to ensure any discrepancy in final results of BG is not attributed to higher soiling loss of monofacial systems. In other words, the only variable upon which the final results will be dependent shall be the type of PV system. Bifacial PV also presents an opportunity to install the system in a vertical E–W facing orientation instead of the conventional

Table 2 Specifications of test bifacial PV module

Specification	Ground reflectance				
	0%	15%	20%	25%	30%
Peak power (φ) (W_p)	355	400	415	430	446
Maximum voltage V_{mpp} (V)	37.9	38	38.1	38.2	38.3
Maximum current I_{mpp} (A)	9.37	10.53	10.90	11.27	11.63
Open circuit voltage V_{oc} (V)	46.4	46.7	46.9	47.0	47.1
Short circuit current I_{sc} (A)	9.72	10.95	11.34	11.72	12.10

* All data measured to STC (irradiance 1000 W/m^2, 25 °C and AM 1.5G according to EN-60904-3)
**Power gain from rear side depends upon bifaciality factor ($\varphi_{bi} = 0.85$) and albedo (ρ_g).

N–S, optimally tilted module orientation [19]. This enables the power generation peak to be realized in the morning as well as evening, which in turn provides certain advantages [20]. Bifacial modules installed in vertical orientation have also reported lower soiling losses as compared to optimally tilted modules. Keeping this in mind, the simulations have been carried out for two orientations, i.e., tilt = 90° azimuth ($\gamma°$) = 0° (N–S) and 90° (E–W). Figure 2a, b present a graphical representation of the results obtained upon simulation conducted for defined parameters.

Occurrence of soiling on both bifacial and monofacial PV systems presents an opportunity for losses to be incurred by the system. These losses are quantified as irradiance loss in PVsyst software. These irradiance losses subsequently lead to power output loss from a PV system. This phenomenon is evident in Fig. 2a, wherein for every orientation under study, there is a decline in the daily energy yield from both mono and bifacial PV systems. However, the characteristic of bifacial technology to generate greater output due to rear side absorption is the reason that under similar ambient and soiling conditions and the daily average yield of a bifacial

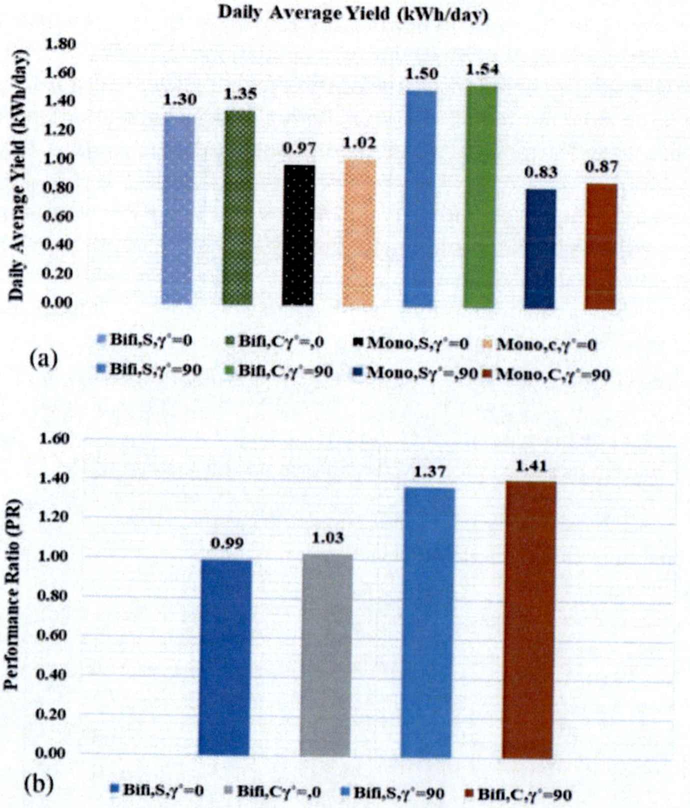

Fig. 2 a Daily average yield for soiled and clean bifacial PV systems at **$\beta = 90°$ $\gamma = 0°$ and $\beta = 90°$ $\gamma = 90°$**. b PR for soiled and clean bifacial PV systems at **$\beta = 90°$ $\gamma = 0°$ and $\beta = 90°$ $\gamma = 90°$**

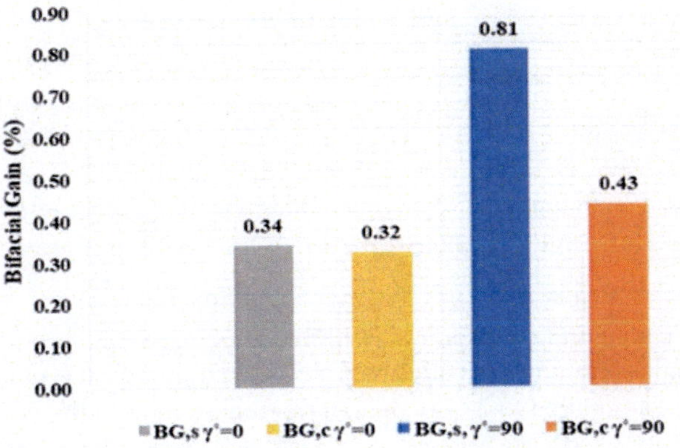

Fig. 3 Comparative representation of BG of soiled and clean bifacial PV module

system is greater than a monofacial PV system under soiled state. In every case under investigation, a soiled bifacial system performs imperfectly when compared to a clean PV system, however, due to lower collection losses in case of a bifacial PV system on account of extra rear side absorption, even a soiled bifacial PV system shows better PR and yield than a monofacial system. From Fig. 2b, it can be observed that system installed in E–W orientation realizes a better PR than a N–S-oriented system. This is due to greater collection of diffuse irradiance by a E–W-oriented bifacial PV system throughout the day, which leads to reduced irradiance collection loss. These collection losses occur due to factors such as lower absorption of irradiance due to obstruction in form of dust particles, reduced beam irradiance absorption [21, 22], higher reflection of incident light from the dust particles [23, 24], and zero rear side irradiance in case of monofacial PV system. Subsequently, BG is calculated from Eq. (1) and the results are plotted in Fig. 3.

In Fig. 3, it is interesting to note that BG for a soiled bifacial system is greater than clean system for both orientations under investigation. This can be attributed to the fact that while the yield of a clean system is relatively higher than a soiled PV system, similar difference in yield of bifacial and monofacial systems (numerator in Eq. 1) under soiled and clean state is offset by the divergence of yield of monofacial system. The yield of monofacial system is far lower under soiled state, and this discrepancy allows for greater BG to be realized under soiled state of the modules. It is safe to say that while the yield of bifacial system under soiled state will be considerably lower than from a clean system, BG is governed at a greater extent by the yield from a monofacial PV system under comparison with a bifacial PV system. The orientation of the PV system also has an effect on the BG. A bifacial PV system installed facing east will generate a higher gain due to higher irradiance collection throughout the day.

5 Conclusions

Dust accumulation on the surface of PV module leads to reduction in irradiance for absorption by the module. Lower irradiance directly relates to reduced power generation. Along with power losses, soiling also causes additional expenditure in terms of operation and maintenance costs to be incurred. An elaborate discussion about the impact of soiling on PV modules' performance has been presented.

- Case study presented wherein simulations conducted to highlight the impact of soiling on the bifacial gain of a bifacial PV module.
- System with modules oriented in E–W direction shows higher PR (1.37 for soiled and 1.41 for clean) compared to N–S-oriented modules (soiled-0.99, clean-1.03) due to greater diffuse irradiance collection in case of E–W-oriented modules.
- BG of soiled system (0.81) greater than BG of clean system (0.43), mainly due to lower yield from monofacial module under soiled state, which leads to higher gain value for bifacial system.

While soiling considerably impacts the yield from the bifacial module, the performance of a bifacial module is considerably better than monofacial modules under soiled state. Also, E–W-oriented modules show better results than N–S-oriented modules due to greater irradiance collection throughout the day.

References

1. Ghosh A (2020) Soiling losses: a barrier for india's energy security dependency from photovoltaic power. Challenges 11:9
2. Chanchangi YN, Ghosh A, Sundaram S, Mallick TK (2020) An analytical indoor experimental study on the effect of soiling on PV, focusing on dust properties and PV surface material. Sol Energy 203:46–68
3. Ilse K, Micheli L, Figgis BW, Lange K, Daßler D, Hanifi H, Wolfertstetter F, Naumann V, Hagendorf C, Gottschalg R, Bagdahn J (2019) Techno-economic assessment of soiling losses and mitigation strategies for solar power generation. Joule 3(10):2303–2321
4. Li X, Mauzerall DL, Bergin MH (2020) Global reduction of solar power generation efficiency due to aerosols and panel soiling. Nature Sustainability 3(9):720–727
5. Ilse K, Figgis B, Khan MZ, Naumann V, Hagendorf C (2018) Dew as a detrimental influencing factor for soiling of PV modules. IEEE J Photovoltaics 9(1):287–294
6. Laarabi B, El Baqqal Y, Rajasekar N, Barhdadi A (2021) Updated review on soiling of solar photovoltaicsystems Morocco and India contributions. J Clean Prod 127608
7. Maghami MR, Hizam H, Gomes C, Radzi MA, Rezadad MI, Hajighorbani S (2016) Power loss due to soiling on solar panel: a review. Renew Sustain Energy Rev 59:1307–1316
8. Elminir HK, Ghitas AE, Hamid RH, El-Hussainy F, Beheary MM, Abdel-Moneim KM (2006) Effect of dust on the transparent cover of solar collectors. Energy Convers Manage 47(18–19):3192–3203
9. Raina G, Sinha S (2021) A simulation study to evaluate and compare monofacial Vs bifacial PERC PV cells and the effect of albedo on bifacial performance. Mater Today Proc 46:5242–5247
10. Singh A, Perry M, Jones D (2021) Measurement of soiling loss on bifacial PV modules. In: 2021 IEEE 48th photovoltaic specialists conference (PVSC). IEEE, pp 1125–1127

11. Baloch AA, Armoush M, Hindi B, Bousselham A, Tabet N (2017) Performance assessment of stand alone bifacial solar panel under real time conditions. In: 2017 IEEE 44th photovoltaic specialist conference (PVSC). IEEE, pp 1058–1060
12. Cabrera E, Araya F, Schneider A, Marzo A, Roescu R, Rabanal J, Ferrada P, Ayala P, Berrian D, Libal J, Fuentealba E (2018) Atamos tec project: Soiling impact on bifacial modules with different mounting geometry in the atacama desert in chile. In: 2018 IEEE 7th world conference on photovoltaic energy conversion (WCPEC) (A Joint Conference of 45th IEEE PVSC, 28th PVSEC & 34th EU PVSEC). IEEE, pp. 3653–3659
13. Bhaduri S, Kottantharayil A (2018) Mitigation of soiling by vertical mounting of bifacial modules. IEEE J Photovoltaics 9(1):240–244
14. Ayala P, Muñoz C, Osorio N, Hernández C, Zurita F, Gutierrez V, Ramirez G, Mancilla F, Valdivia P, Cuevas F, Ferrada P (2018) Bifacial technology performance compared with three commercial monofacial PV technologies under outdoor high irradiance conditions at the Atacama Desert. In 2018 IEEE 7th world conference on photovoltaic energy conversion (WCPEC) (A Joint Conference of 45th IEEE PVSC, 28th PVSEC & 34th EU PVSEC). IEEE, pp 0672–0675
15. Luque EG, Antonanzas-Torres F, Escobar R (2018) Effect of soiling in bifacial PV modules and cleaning schedule optimization. Energy Convers Manage 174:615–625
16. Ullah A, Amin A, Haider T, Saleem M, Butt NZ (2020) Investigation of soiling effects, dust chemistry and optimum cleaning schedule for PV modules in Lahore, Pakistan. Renew Energy 150:456–468
17. Janssen GJ, Van Aken BB, Carr AJ, Mewe AA (2015) Outdoor performance of bifacial modules by measurements and modelling. Energy Procedia 77:364–373
18. Yusufoglu UA, Pletzer TM, Koduvelikulathu LJ, Comparotto C, Kopecek R, Kurz H (2014) Analysis of the annual performance of bifacial modules and optimization methods. IEEE J Photovoltaics 5(1):320–328
19. Ledesma JR, Almeida RH, Martinez-Moreno F, Rossa C, Martín-Rueda J, Narvarte L, Lorenzo E (2020) A simulation model of the irradiation and energy yield of large bifacial photovoltaic plants. Sol Energy 206:522–538
20. Deline CA, Ayala Pelaez S, Marion WF, Sekulic WR, Woodhouse MA, Stein J (2019) Bifacial PV system performance: separating fact from fiction (No. NREL/PR-5K00–74090). National Renewable Energy Lab (NREL), Golden, CO (United States)
21. Smestad GP, Germer TA, Alrashidi H, Fernández EF, Dey S, Brahma H, Sarmah N, Ghosh A, Sellami N, Hassan IA, Desouky M (2020) Modelling photovoltaic soiling losses through optical characterization. Sci Rep 10(1):1–13
22. Caron JR, Littmann B (2012) Direct monitoring of energy lost due to soiling on first solar modules in California. IEEE J Photovoltaics 3(1):336–340
23. John JJ, Rajasekar V, Boppana S, Chattopadhyay S, Kottantharayil A, TamizhMani G (2015) Quantification and modeling of spectral and angular losses of naturally soiled PV modules. IEEE J Photovoltaics 5(6):1727–1734
24. Green M, Dunlop E, Hohl-Ebinger J, Yoshita M, Kopidakis N, Hao X (2021) Solar cell efficiency tables (version 57). Prog Photovoltaics Res Appl 29(1):3–15

Classification and Area Computation Modelling of Remote Sensing Images Using Histogram and Convolutional Neural Network

Swarna Kamal Pradhan, Dipon Das, and Ujjwal Mondal

Abstract Remote sensing is an important field in science and technology and consists of the images of the Earth taken by the means of artificial satellites or aircraft. Satellite images or high-resolution aerial images are flexible to work with and easy to monitor. Since the total area of the earth is so large, high-resolution remote sensing images produce vast amount of data, even image processing is time consuming. This project represents a combination of unsupervised and supervised process to classify high spatial resolution satellite images so that minimal human intervention is needed. For this purpose, histogram peak-based classification approach is used to classify remote sensing image into subcategories like urban land, vegetation land, water body, etc. To detect different objects, present in the image, convolutional neural network-based approach is used. The neural network model is trained using custom dataset. Then, object localization operation is performed to get the coordinates of the object present in the image. Then, histogram-based segmentation operation is performed to compute the area of different objects present in the image. After that 3D model is constructed using the coordinates obtained. Georeferencing technique is used to calculate the area of different classes observed.

Keywords Remote sensing · Image classification · Convolutional neural network · 3D modelling

1 Introduction

The use of image signal is huge, and until today, we need to keep on innovating in the field of remote sensing for the satisfaction of human need. Field-based measurements and 3-dimentional modelling give results at small areas with high temporal data, usually require a lot of measuring and hard work. Additionally, this process is time consuming, may be less efficient, and varies from person to person. Satellite images or high-resolution aerial images, on the other hand, are flexible to work with and easy to monitor. Colour and texture features in colour image provide better result [1].

S. K. Pradhan (✉) · D. Das · U. Mondal
Department of Applied Physics, University of Calcutta, Kolkata, West Bengal 700009, India

J. K. Mandal et al. (eds.), *Topical Drifts in Intelligent Computing*, Lecture Notes in Networks and Systems 426, https://doi.org/10.1007/978-981-19-0745-6_63

This information can be well observed by plotting colour histogram of the image. Colour histogram is related closely to scene properties, and colour histogram has features that are identifiable and relates in a mathematical way to scene properties precisely [2]. When histogram is plotted for an image, it represents different colour classes present on that image and cluster formed on that histogram. These clusters together form different type of shapes, height, and width, which represent distinct feature. Though histogram plotting is a powerful process for image information retrieval, it can be time consuming for a large image dataset. There are several methods proposed for calculating histogram in real time [3]. Histogram peak-based approach is proposed by Strelkov in 2008 [4]. In a sampled data, a peak is a value, which has higher amplitude than two direct neighbouring samples. In this approach, a similarity between ordered histogram is used. Here, similarity is observed between positions and shapes of peaks in the histograms. Histogram peak-based approach also used for histogram stretching for contrast enhancement purpose using plateau equalization technique. Here a scene-adaptive plateau threshold is used to correct the raw image [5]. The histogram peaks that are above average value represent number of clusters. A cluster is a small group of objects having similar property. Here, a cluster represents a group of histogram peaks [6]. There are basically two types of clustering of data: (1) unsupervised classification and (2) supervised classification. Unsupervised classification does not require human involvement, and it is a faster process on the other hand supervised classification needs large human intervention [7]. There are several unsupervised clustering techniques are available. *K*-means is a very popular unsupervised algorithm and widely used. It is used for partitioning *N*-dimensional points into *k* sets on the basis of a sample [8]. In supervised image classification, an input image is assigned with labels to form a dataset. Since 2012 machine learning methods are used in various fields for classification purposes such as medical imaging [9], hand written letter recognition [10], face recognition, security systems, and automatic zone detection of UAVs [11]. Machine learning in the field geographical image classification is relatively new and has provided better result in the area of vegetation detection [12]. Using machine learning method, we obtained features from the image and train a neural network model using those features so that neural network model can predict the class of an unknown image. Followed by we proposed a technique for generating similarly looking 3-dimentional model of the area given on the image. We successfully used machine learning techniques for the classification of a remote sensing image. Used CNN-based deep learning method for object detection and *K*-mean-based unsupervised classification technique to classify the urban land. Have successfully managed to classify image and calculate their area using georeferencing technique. Have successfully built a 3-dimensional model of the target image using the object coordinates.

2 Proposed Method

Here, the steps required for image classification and 3D modelling are described. This approach exploits a combination of unsupervised and supervised image classification technique to broadly classify the urban land. All high-resolution images are taken from United States Geological Survey (USGS) official Website [13] (Fig. 1).

2.1 Image Classification Using Histogram and K-Mean Clustering

Different types of class present in a terrain image impact highly the shape of the histogram of the image. Depending on the classes present on the image, the histogram will produce unimodal, bimodal, or multimodal distributions. These peaks of histogram distribution represent one, two, and multiple classes (depending on colour) present in the image. Depending on the location of the peak and width, images can be classified into colour-based subcategories. To plot the image histogram, at first, the RGB model image is converted to HSV model which is a colour model where RGB pixel values are represented in cylindrical coordinate system. Then, the position of the peaks in the histogram has to be calculated. Then, histogram peaks are detected. The detected peaks are plotted with respected to their position. Then, the peaks are classified using one-dimensional K-mean clustering. K-means clustering is a method of vector quantization (Fig. 2).

Fig. 1 Original image of an area taken from USGS Earth Explorer Website

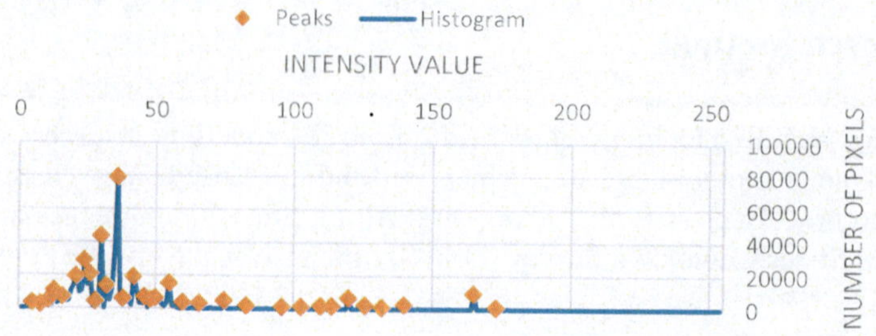

Fig. 2 Peak detected of a histogram

2.2 *Waterbody Detection Using Edge-Based Operator*

Some part of water body cannot be detected properly by only looking at the histogram as the colour of water has wide range of variety. So, frequency-based analysis is used. Water body region on the image has very low-frequency noise compared to urban land. This is done using Canny edge detection. It has basically four stages. They are noise reduction with a Gaussian or median filter, finding intensity gradient non-maximum suppression, hysteresis thresholding.

2.3 *Area Computation*

To calculate the area of the highlighted part of the images, we had to georeference the image. The purpose is to transform image from geometric domain to geographic domain using base map or base image. To georeference an image, we followed certain steps. Ground control points (GCPs) are point in terms of latitude and longitude on surface of earth of known location used to georeference map or remote sensing images of a region on earth. Also, this is necessary to determine the pixel values of the above-mentioned places on the raw image that we are going to georeference, so that we can determine latitude and longitude variation per pixel.

2.4 *Object Detection Using Convolutional Neural Network*

Structure of Convolutional Neural Network

To build the structure of convolutional neural network, we had to construct it layer by layer. They are convolutional layer, pooling layer, flatten layer, and fully connected layer.

Convolutional Layer: The convolutional layers are used for feature extraction. Using this layer at first, features, like colours and edges, are detected. Later, layers recognize larger structures or shapes of the object until whole object features are covered. The convolution operation is performed by sliding a filter matrix w, with the input image x, then the dot product between them is calculated. Here, square filters are used to detect object, meaning that they have a size (200, 200) pixels and (100, 100) pixels. Though various combinations of height and width can be used. Mathematically, convolution of a matrix, x, with a filter, w, for all valid positions of an input image can be given by

$$(w*x)_{i.j} = \sum_{l=0}^{k-1}\sum_{m=0}^{k-1} w_{l,m}x_{i-l,j-m} \tag{1}$$

Pooling Layer: Pooling layers are used to reduce the dimension with the help of down sampling operation such that number of parameters can be reduced. Pooling layers also use sliding window technique to perform the whole operation. Here, only max pooling is used.

Flatten Layer: In this layer, two-dimensional data are converted to one-dimensional array so that it can be passed through the fully connected layer.

Fully Connected Layer: Fully connected layers are compared to neurons in human brain. It has multiple number of layers, and all former layer neurons are connected to the later layer with synaptic weights. Fully connected layers perform the following operation which can be mathematically expressed as

$$\text{Output} = \text{activation}((\text{input} * \text{weights}) + \text{bias}) \tag{2}$$

Evaluating the Loss.

Using the loss function, we evaluate performance of neural network. This function creates relation between predictions and network parameters. Here, we used binary cross entropy (BCE) loss function. BCE loss function is used to convert the predicted probabilities into logarithmic scale. The BCE function can be expressed by the equation given below (Fig. 3):

$$B\left(y, y^{\text{true}}\right) = \sum_i \left[-y_i^{\text{true}} \log(y_i) + \left(1 - y_i^{\text{true}}\right) \log(1 - y_i)\right] \tag{3}$$

Training the Convolutional Neural Network Model and Object Detection.

To train the model, a Python programme is written and images are imported into an array. Images are imported into an array with their corresponding labels. For an example, house images are labelled as 1 and not house is labelled as 0, this is done to create house detection dataset. Similar operation is also done to create tree detection model. After importing the image data, they are converted to grayscale image for 2-dimensional convolution. Then, the image pixel values are normalized

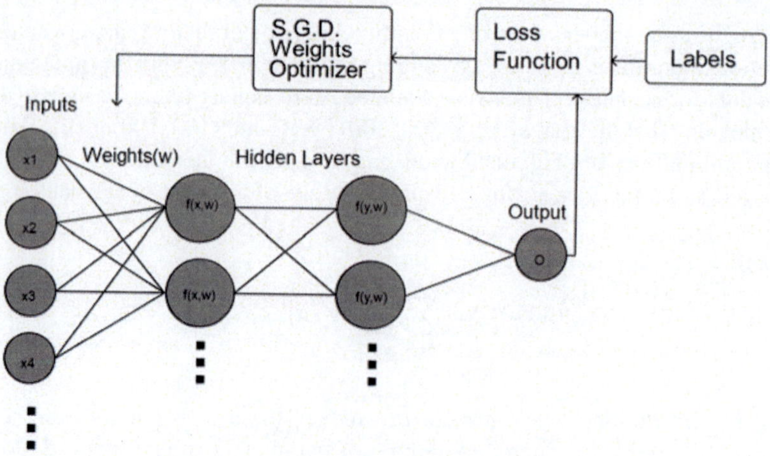

Fig. 3 Structural model of convolutional neural network

from 0–255 to 0–1.0 value for the ease of calculation. After that the whole dataset is divided into training and testing dataset so that we can test the model after training. After training, the model is saved for further use and also validation accuracy is also observed for both house and tree model. The generated model basically works as a filter. We used sliding window method to detect the objects present in the image. There was multiple object detection of the same object. So, we applied non-maximum suppression operation to remove the repetitive detection of the neural network. To remove repetitive detection, we took all the probability values for a single object and performed non-maximum suppression operation so that only the probability value that represents the object most accurately remains. Then, the coordinates of the final result are stored in a list so that these coordinates can be used for 3-dimensional modelling purposes.

2.5 3-Dimensional Modelling

Once the location coordinates of the object are found, the 3-dimensional modelling is done using Ursina library and using those coordinates (Fig. 5). For that purpose, we had built a model for each class in Blender software. The 3-dimensional shape file is created for those two classes and stored into '. Blend' file format. Also, colour texture files are created using blender software and stored into the 'png' file format. Then, those 3-dimensional model files and texture files are imported into Ursina model using Python. The ground colour is set us set using *k*-means clustering. The final output is given in the Fig. 4.

Fig. 4 Side view of generated 3-dimensional model

3 Conclusion and Scope for Future Work

We have successfully managed to classify image and calculate their area by histogram. Though we have faced certain difficulties and errors whilst performing this project, some of them are mainly atmospheric errors caused due to pollution, smog, dust particles present in the atmosphere and cloud cover. The histogram provides a complete solution for colour-based classification problem. Although, we have managed to classify images using convolutional neural network, which helped us to find objects present in the image, such as house and tree. Using those object coordinates, we have successfully built a 3-dimensional model of the target image. Though rendering synthetic 3-dimensional models has enabled a variety of computer vision applications, it is, however, not the perfect remedy for data scarcity. Gathering hundreds of images for each new element to recognize is costly, time consuming, and sometimes completely impractical, for instance, when the target objects are not produced or are only available at some remote location. We have only used grayscale images to train and test the neural network. So, in case of image classification, colour-based analysis can be used for further study and for better accuracy. This project includes limited data (one thousand images for each class). So, in the future, the image classification technique can be improved by using augmentation techniques and deep learning methods [14]. We have successfully completed the classification of images based on their class and 3-dimensional structural modelling using convolutional neural network which can be used in various applications such as construction surveying, virtual tour, and industrial structural planning. Till now, in 3-dimensional modelling, we have successfully managed to define the outline

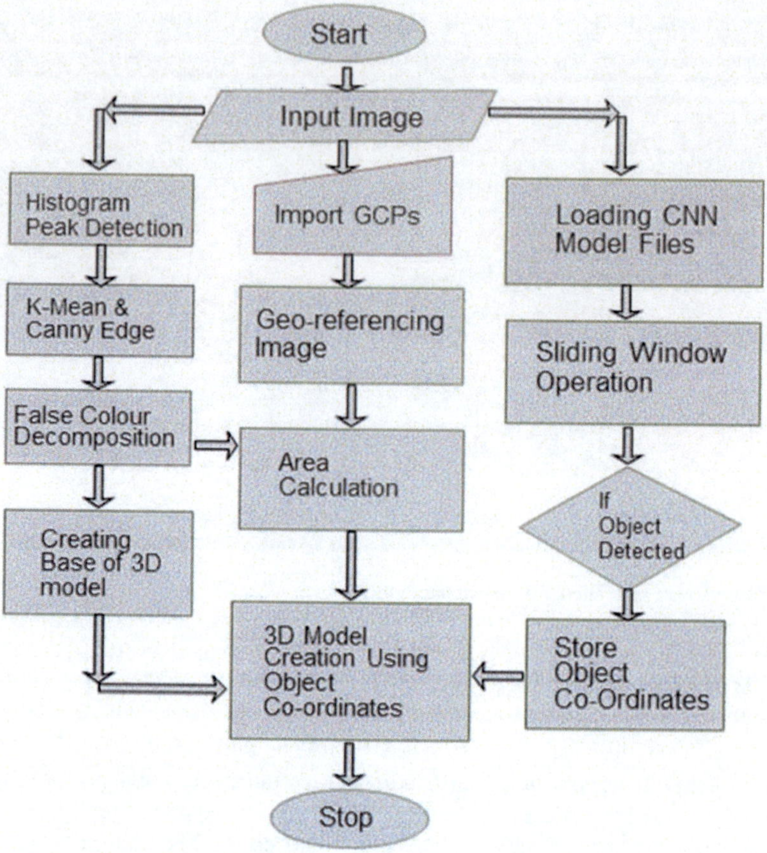

Fig. 5 Flowchart of the complete process

(specific location and area) of a particular object, but we could not define the exact structure and condition of the objects. So, semantic segmentation can be used for defining the precise contours of the object for object classification.

References

1. Color texture based image retrieval system (2011) Int J Comput Appl. https://doi.org/10.5120/2958-3910
2. Novak C, Shafer SA (1992) Anatomy of a color histogram. CVPR
3. Porikl F (2005) Integral histogram: a fast way to extract histograms in cartesian spaces. https://doi.org/10.1109/Cvpr.2005.188
4. Strelkov V (2008) A new similarity measure for histogram comparison and its application in time series analysis. https://doi.org/10.1016/J.Patrec.2008.05.002

5. Lai R, Yang Y-T, Wang B-J, Zhou H-X A quantitative measure based infrared image enhancement algorithm using plateau histogram
6. Chowdhury S, Roy S, Mitra A, Das P (2016) Automated numbers of cluster determination using the combination of entropy and histogram peaks from multiple images. https://doi.org/10.14257/Ijsip.2016.9.7.10
7. Image classification in remote sensing (2013) J Environ Earth Sci 3(10)
8. Lloyd SP (1982) Least squares quantization in Pcm (Pdf). IEEE Trans Inf Theorhttps://doi.org/10.1109/Tit.1982.1056489. Retrieved 5 Apr 2009
9. Erickson B, Korfiatis P, Akkus Z, Kline T (2017) Machine learning for medical imaging. Radiographics
10. Handwritten digit recognition using convolutional neural networks (2016). IJIRCCE 4(2) https://doi.org/10.15680/Ijircce.2016.0402001
11. An automatic zone detection system for safe landing of UAVs (2019). Expert Syst Appl 122. https://doi.org/10.1016/J.Eswa.2019.01.024
12. Vegetation detection through smoke-filled AVIRIS images: an assessment using modis band passe. J Geophys Res 103(D24):32,001–32,011
13. https://www.usgs.gov/
14. Vegetation detection using deep learning and conventional methods. MDPI, Remote Sensing (2020). https://doi.org/10.3390/Rs12152502

Performance Analysis of Lead-Free Perovskite Solar Cells

Riya Sen and Menka Yadav

Abstract The evolution of perovskite-based solar cells has recently generated a huge attention, with the goal of removing harmful lead from perovskite materials. Important goal of this work is to enhance current research by numerically simulating various lead-free perovskite solar cells (PSCs) with the SCAPS-1D software. Device simulation is carried out for five different lead-free perovskite materials in n-i-p configuration of FTO/TiO_2/perovskite layer/Spiro-OMeTAD/Au and analyzed. The impacts of various perovskite material layers on solar cell performance, such as hole and electron transport layer thickness and doping concentration, have been thoroughly investigated and optimized. Among lead-free perovskite-based devices, the $CsSnI_3$-based PSC has the highest power conversion efficiency of 32.11%. This suggests that lead-free PSCs with comparable performance could be produced experimentally in the future.

Keywords Perovskite solar cell · Non-toxic · SCAPS-1D · Simulation · Performance analysis · Optimization

1 Introduction

The energy demands of today's society will continue to rise as a result of increased industrialization and a growing population [1]. Due to the great demand for energy to promote a country's growth, researchers are always working to replace fossil fuels with renewable energy sources. Crystalline silicon has held a large percentage of the market for the past few decades, but it has reached its maximum efficiency, forcing researchers to focus on new photovoltaic technologies, such as perovskite solar cells, which have substantially improved in a brief span of time [2]. Perovskite material can absorb a wide range of wavelengths of light and has a direct bandgap, which conventional silicon does not have. There are numerous advantages, such as decreased exciton binding energy, outstanding photoelectric property, carrier life

R. Sen (✉) · M. Yadav
Department of ECE, Malviya National Institute of Technology, Jaipur, India
e-mail: 2019rec9508@mnit.ac.in

J. K. Mandal et al. (eds.), *Topical Drifts in Intelligent Computing*, Lecture Notes in Networks and Systems 426, https://doi.org/10.1007/978-981-19-0745-6_64

time, low recombination, large dielectric constant, and long transmission distance, all of which add to high open-circuit voltage and efficiency [3].

Miyasaka et al. used these materials in solar cells for the first time in 2009, yielding the highest efficiencies of 3.81% [4]. These efficiencies were increased to 6.54% in 2011 by Lee et al. [5]. The created devices, on the other hand, decayed in minutes. As a result, the researchers were forced to focus on all solid-state perovskite devices and reported with solid electrolyte by Spiro-OMeTAD with more than nine percent efficiency in 2012 [6]. Since then, the efficiency has risen dramatically, from 3.81 to 25.2%, in 2021. [2].

FTO, electron transport layer (ETL), absorber, hole transport layer (HTL), and metal back contact are the five pillars that constitute a perovskite solar cell. The sun rays pass through the FTO layer and reach the absorber, where it generates a free charge carrier whose movements are harnessed to generate current [4]. Methyl ammonium lead iodide, or $MAPbI_3$, is the most commonly used absorber layer to strengthen power conversion efficiency. Since lead is present in methyl ammonium lead iodide, researchers are encouraged to focus on creating new PSC. As a result, new non-toxic or low-toxic materials are regularly studied. In PSC, the ions $Sn^{(2+)}$ and $Ge^{(2+)}$ are the most viable substitutes for Pb [7, 8]. Wu et al. suggested a double perovskite solar cell, the most popular structure with $Cs_2AgBiBr_6$ as the absorber [9]. Kang et al. demonstrated a $CsSnX_3$ perovskite-based lead-free solar cell [10]. Liu et al. investigated the structure of $CsBi_3I_{10}$ in order to develop a Pb-free perovskite [11]. In fact, in recent times, a new promising material compound, Cs_2TiBr_6, has also been reported to obtain photovoltaic material with good physical, electrical, and optical properties [12, 13].

Simulations, in addition to experiments, are the most essential tool for analyzing the performance of diverse materials. One of the most widely used open-source tools for this simulation is Solar Cell Capacitance Simulator—1 Dimension (SCAPS-1D) [14]. The remaining paper is structured as: Sect. 2 discusses the designed architecture of various PSCs without lead, Sect. 3 presents all simulation results and discusses the effect of different layers and doping concentration variations on the performance of lead-free PSCs, and Sect. 4 concludes the paper.

2 System Model

The device architecture of a PSC is depicted in Fig. 1, with the absorber layer sandwiched between two electrodes. HTL and ETL play a critical role in developing the perovskite's working principle, such as collecting electrons/holes from the absorber, restricting opposing charge carriers to eliminate recombination in the HTL/ETL layer, and guiding charge carriers to their respective electrode terminals.

There have been discoveries of PEDOT:PSS, CuI, CuSCN, NiO, spirobifluorenes, thiophenes, triphenylamines, Spiro-MeOTAD, P3HT, PTAA, as well as other inorganic and organic HTL materials. Spiro-MeOTAD is indicated as best-suited HTL for perovskite solar cells in terms of efficiency [40]. As an n-type semiconducting

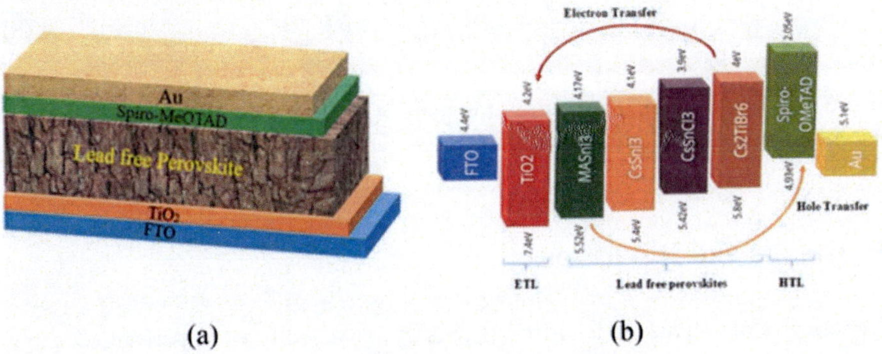

Fig. 1 **a** Schematic view of the proposed Au/Spiro-MeOTAD/lead-free perovskite/TiO_2/FTO photovoltaic device for simulation and **b** device energy band diagram

material, TiO_2 enables electron extraction from the perovskite absorber layer and functions as an efficient ETL in perovskite structure. So, in this work, we have taken Spiro-MeOTAD as HTL and TiO_2 as ETL for simulation.

The prime objective of this work is to determine how different lead-free halide perovskites behaved in the identical device arrangement. To determine which lead-free perovskite performs best in this setting, the attributes of each perovskite is compared and analyzed. The structure configuration, i.e., FTO/TiO_2/absorber layer/Spiro-MeOTAD/Au is used for device simulation in this work, as shown in Fig. 1. As the principal absorber layer, various lead-free perovskite layers such as $CsSnI_3$ (1.27 $_{eV}$), Cs_2AgBiI_6 (1.6 $_{eV}$), $CsSnCl_3$ (1.52 $_{eV}$), Cs_2TiBr_6 (1.8 $_{eV}$), and $MASnI_3$ (1.9 eV) are chosen [15–20]. After that, the optimization is carried out by varying the thickness of HTLs and ETLs. The device is furthermore optimized on the basis of the thickness variation of the perovskite absorber layer and by considering the defect density.

The equations that play a vital part in the simulation study are listed below. For a semiconductor, Poisson's equation is expressed as follows continuity equation accordingly,

$$\frac{d^2\Psi(x)}{dx^2} = \frac{q}{\varepsilon}(n - p + N_A - N_D) \tag{1}$$

where ε is the semiconductor's permittivity, N_A stands for acceptor concentration, N_D stands for donor concentration, and Ψ represents the electrostatic potential.

The electron and hole continuity equations are:

Continuity equation for electron:

$$\frac{\partial J_n(x)}{\partial x} - q\frac{\partial \text{n}}{\partial \text{t}} = +qR \tag{2}$$

Continuity equation for hole:

$$\frac{\partial J_p(x)}{\partial x} + q\frac{\partial \mathrm{p}}{\partial \mathrm{t}} = -qR \tag{3}$$

where J_n denotes the electron current density, J_P denotes the hole current density, and R is the carrier recombination rate.

3 Results and Discussion

Numerical simulations of lead-free perovskite solar cells were performed using tabular parameters obtained by considering a variety of theoretical and practical literature. This section describes in depth the calibration and improvement of the lead-free perovskite solar cell (PSC) structure. To begin, a PSC with a lead-free absorber layer is calibrated using ETL (TiO_2) and HTL (Spiro-MeOTAD). In the device, configuration of FTO/TiO_2/perovskite layer/Spiro-MeOTAD/Au, different lead-free perovskite materials have been employed as an absorber layer while keeping all other parameters same, except for the light-absorbing layer, which has been altered among many lead-free perovskite materials.

HTLs, ETLs, and absorber layer thickness changes are also taken into account while calibrating the device. The device models have been simulated using this configuration to achieve a precise condition for each case. Table 1 contains all of the material properties required for simulation work.

Impact of Absorber Layer Variation

The active absorber layers' thickness was adjusted to achieve optimal device performance from 0.2 to 1.0 μm while all other parameters, including temperature, were maintained constant. For majority of the perovskites presented, the open-circuit voltage (V_{OC}) declines or almost constant. The nature of the curves produced for short-circuit current density (J_{SC}) and efficiency (η) versus thickness is almost identical to that shown in Fig. 2. For all perovskite materials, the J_{SC} and η value increases as the thickness of the absorber grows and reaches the optimal value, and then, it is gradually constant as the layer thickness increases in Fig. 2. This trajectory could be explained by the fact that as the active layer thickness increases, the device absorbs more light, leading to increased charge carrier growth and better light-induced current, resulting in higher J_{SC}, and efficiency. Further increase in thickness may exceed the diffusion length of these materials, creating an increase in charge carrier recombination, resulting in higher saturation current, lower V_{OC}, and lower J_{SC}, all of which diminish efficiency. The fill factor plot Fig. 2c demonstrates that as perovskite thickness grows the value is almost constant or decreases, which can be explained by expressing the fill factor (FF) dependence on perovskite layer thickness. FF denotes the ease with which electron–hole pairs can pass through the device without recombination. So, charge route resistance should increase with increased perovskite layer thickness, resulting in a drop in FF.

Table 1 Input electrical properties of solar cell layers

	ETL		Perovskite absorber layer					HTL
	FTO	TiO_2	$MASnI_3$	$CsSnI_3$	Cs_2BiAgI_6	$CsSnCl_3$	Cs_2TiBr_6	Spiro-OMeTAD
Thickness (μm)	0.050	Varied	Varied	Varied	Varied	Varied	Varied	Varied
E_g (eV)	3.5	3.2	1.9	1.27	1.6	1.52	1.8	3
χ (eV)	4	3.9	3.98	4.47	3.93	3.9	3.9	2.45
ε_r	9	9	10	10.59	6.5	29.4	4	3
Nc (cm^{-3})	2.2×10^{18}	1.0×10^{19}	1×10^{16}	1.58×10^{19}	1×10^{19}	1×10^{19}	6×10^{19}	2.8×10^{19}
Nv (cm^{-3})	1.8×10^{19}	1.0×10^{19}	1×10^{15}	1.47×10^{18}	1×10^{19}	1×10^{19}	2.14×10^{19}	1.0×10^{19}
Electron thermal velocity (cm/s)	1×10^7	1×10^7	1×10^7	1×10^7	1×10^7	1×10^7	1×10^7	1×10^7
Hole thermal velocity (cm/s)	1×10^7	1×10^7	1×10^7	1×10^7	1×10^7	1×10^7	1×10^7	1×10^7
μ_n (cm^2/Vs)	20	20	16.2	4.37	2	2	0.236	2.0×10^{-4}
μ_p (cm^2/Vs)	10	10	10.1	4.37	2	2	0.171	2.0×10^{-4}
N_A (cm^{-3})	–	–	Varied	Varied	Varied	Varied	Varied	-
N_D (cm^{-3})	Varied	Varied	Varied	Varied	Varied	Varied	Varied	Varied
References	[33]	[36]	[38]	[35]	[37]	[37]	[34]	[33]

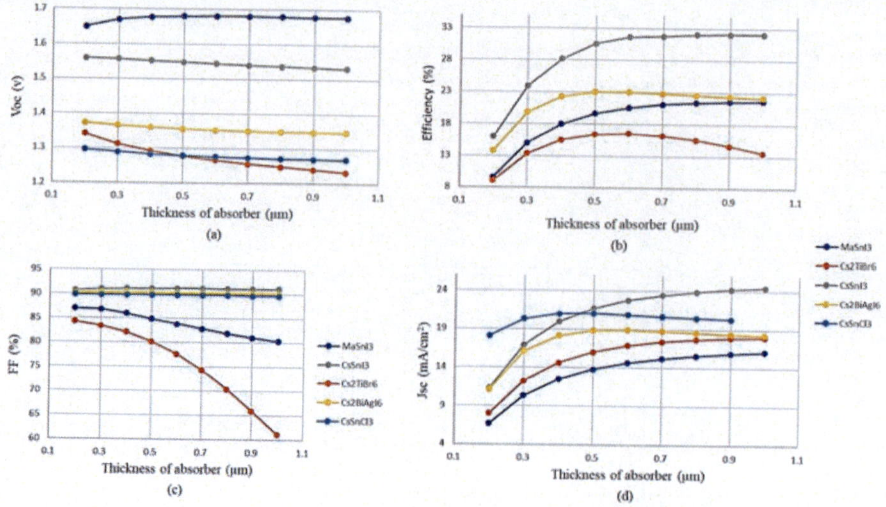

Fig. 2 Effects of absorber layer thickness on **a** Voc, **b** efficiency, **c** FF, and **d** Jsc

Impact of HTL Variation

In order to create efficient solar cells, the HTL properties must be carefully chosen and Spiro-MeOTAD has an extremely efficient charge transport technique as HTL. As a result, many interface defects are incorporated to simulate high-resistive Spiro-MeOTAD, as listed in Table 2.

To improve the HTL layer for this research, we first tuned the thickness of the Spiro-MeOTAD layer and then the acceptor doping density. To begin, we used the parameters in Table 2 and ran the software for the proposed device. Interaction between the absorber layer and the anode material allows it to handle simultaneously the processes of collecting and recombining charges as well as hopping charge transport. Figure 3a demonstrates that as HTL thickness grows the value is almost constant or decreases.

Table 2 Defect impact parameter [17]

Parameters	Unit	Values
Defect type	–	Neutral
Capture cross section for electron	cm^{-3}	1.00E + 14
Capture cross section for hole	cm^{-3}	1.00E + 14
Energetic Distribution	–	Single
Energy level with respect to E_V	eV	6.00E-01
Characteristic energy	eV	approx. 0.1
Defect density	cm^{-3}	4.50E + 17

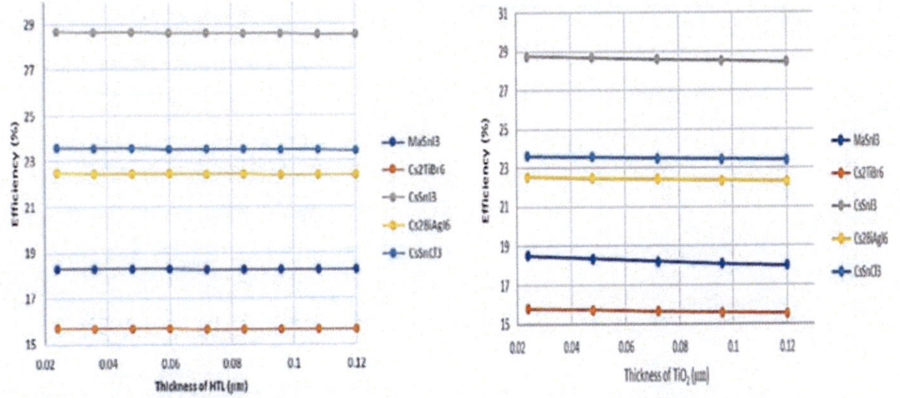

(a) (b)

Fig. 3 **a** Effects of HTL thickness on efficiency and **b** effects of ETL thickness on efficiency

Effect of Variation in ETL

The ETL properties must be carefully chosen in order to create efficient solar cells. At the absorber-ETL interface recombination reduction is crucial to achieving efficient carrier extraction. TiO_2's interaction with perovskite material is characterized by high chemical reactivity, unexpected charge buildup, inordinate recombination, and unpredictability under ultraviolet light, as well as low mobility, non-uniform distribution, and a high density of electronic trap states.

So, for the aforementioned device setup, the thickness of ETM and TiO_2 is changed from 0.02 μm to 0.120 μm. Variations in the ETL layer, i.e., TiO_2, have a negligible effect on the performance parameters of all perovskites. The optimum performance characteristics are found for an electron transport layer (ETL) thickness of 0.05 μm, as shown in Fig. 3b. Also, very similar results are found for V_{oc}, J_{sc}, FF, and PCE.

Effect of Variation in Doping Concentration

Perovskites' electrical behavior is determined by the concentration of dopants, which has a significant impact on the solar cell's performance. Figure 4 shows the variance in doping for all lead-free perovskites as a function of the solar cell performance parameters. The doping of absorber layer, HTL and ETL has been adjusted between 5×10^{13} and 5×10^{21} cm^{-3} keeping all other properties constant, and the influence is investigated by contrasting the behavior of various materials.

Figure 4 depicts the effect of varying doping concentrations in different absorber materials, HTL and ETL on various solar cell properties. The efficiency of most materials increases with increasing doping values as shown in Fig. 4.

The J–V characteristics and quantum efficiency (QE) of all of the devices are excellent as shown in Fig. 5. The absorption spectrum of $CsSnI_3$ with lowest band gap is the broadest and that of Cs_2TiBr_6 is the narrowest of all, with the rest of the materials descending in order. Table 3 shows the performance parameters for five

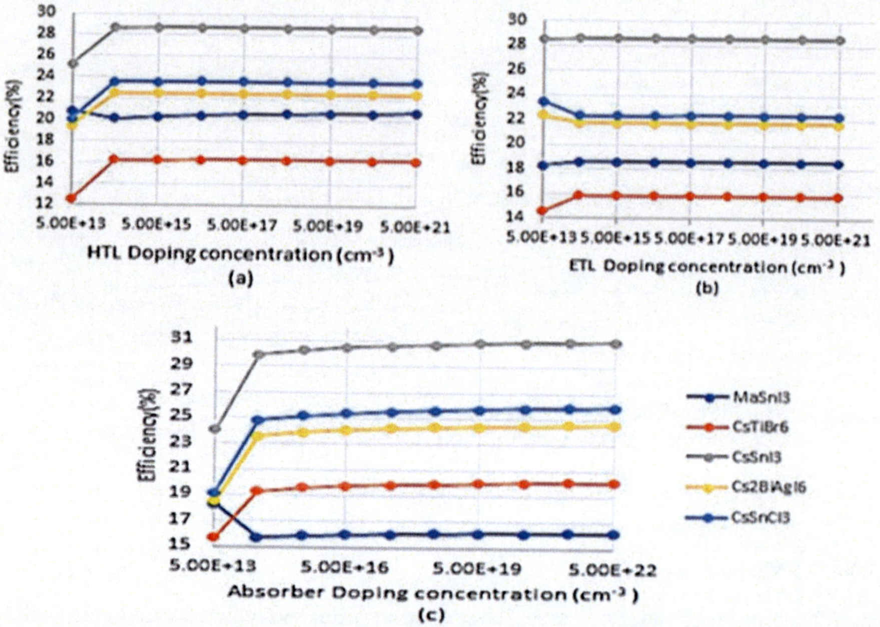

Fig. 4 Effects of **a** HTL, **b** ETL, and **c** absorber doping on PSC performance

distinct Pb-free perovskites-based solar cells when simulated under air mass (AM) 1.5G spectra, 1 sun standard, 300 K temperature, and all other working circumstances remaining the same. $MASnI_3$ has the greatest V_{OC} of 1.6775 $_V$ of all the optimal setup, while $CsSnCl_3$ has the lowest V_{OC} of 1.391 $_V$. With a J_{SC} of 24.446291 mA/cm^2, the device with $CsSnI_3$ has the highest short-circuit current density and Cs_2TiBr_6 has the lowest J_{SC} of 13.179807.

When compared to other devices, $CsSnI_3$- and Cs_2BiAgI_6- based devices have a greater FF. Among these various perovskite-based solar cells, the $CsSnI_3$-based device has the highest PCE of 34.11% with V_{OC} of 1.5312 V, J_{SC} of 24.446291 mA/cm^2 and FF of 91.11% which are listed below in Table 3.

4 Conclusion

The overall performance of many lead-free perovskite-based solar cells is assessed using numerical simulation on SCAPS-1D. The impact of varied perovskite material characteristics, doping, thickness, and their impact on the PCE on the same device configuration with HTL as Spiro-MeOTAD and ETL as TiO_2 has been studied in detail. We improved many parameters including as doping density, active material thickness, and hole and electron transport layers to boost device performance even

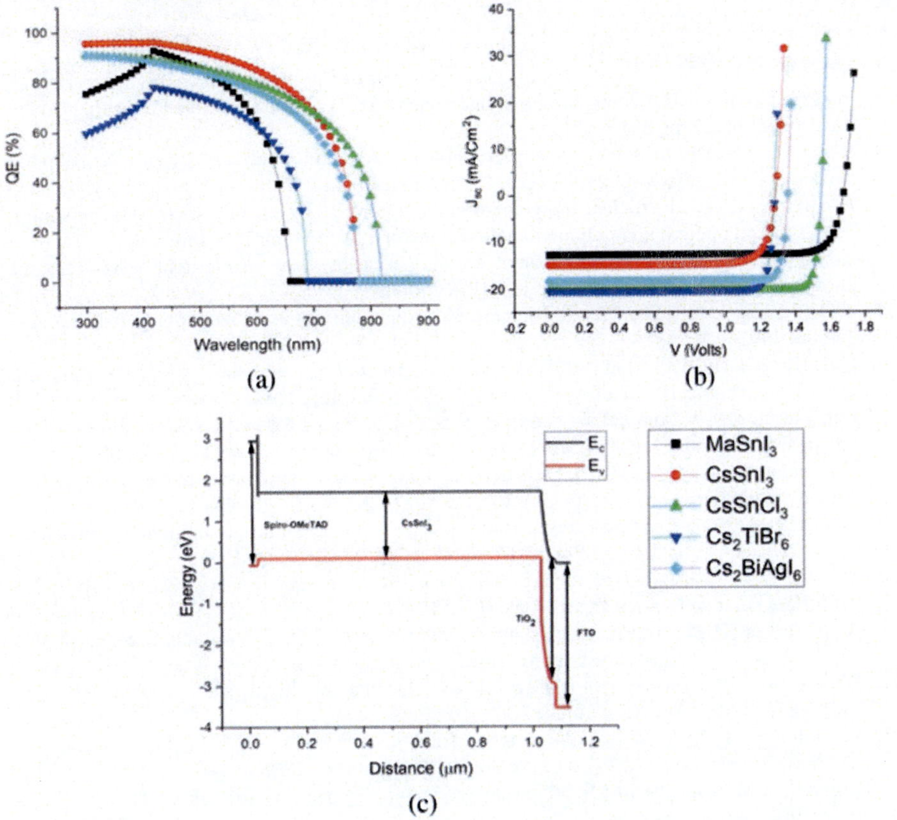

Fig. 5 **a** *JV* characteristics of various Pb-free perovskite solar cells, **b** corresponding QE spectra of PSCs with different Pb-free perovskites, and **c** energy band diagram of $CsSnI_3$-based perovskite solar cell

Table 3 Performance parameters for various Pb-free perovskites in optimal setups

Perovskite	V_{OC} (V)	J_{SC} (mA/cm^2)	FF (%)	η (%)
$CsSnI_3$	1.5312	24.446291	91.11	32.11
$CsSnCl_3$	1.391	20.440362	90.59	25.83
Cs_2BiAgI_6	1.4789	18.293236	90.95	24.6
$MASnI_3$	1.6775	16.092058	80.32	21.68
Cs_2TiBr_6	1.6669	13.179807	91.59	20.12

more. The $CsSnI_3$-based device has the maximum PCE of 32.11% among all lead-free perovskite-based devices, with an Voc of 1.5312 V, Jsc of 24 mA/cm^2, and FF of 91.11%. As a result of our simulation analysis, it is apparent that lead-free perovskite-based solar cells will undoubtedly push the scientific community toward deeper exploration and experimental realization in the future.

References

1. Irama R et al (2021) Prioritization of renewable solar energy to prevent energy insecurity: an integrated role. In: The Singapore economic review (SER), World Scientific Publishing Co Pte Ltd, vol 66, no 02, pp 391–412
2. Best research-cell efficiencies: emerging photovoltaics. https://www.nrel.gov/pv/assets/pdfs/cell-pv-eff-emergingpv-rev210726.pdf. Last accessed 9 Aug 2021
3. Heo JH et al (2013) Efficient inorganic-organic hybrid heterojunction solar cells containing perovskite compound and polymeric hole conductors. Nat Photon 7(6):486–491
4. Kojima A et al (2009) Organometal halide perovskites as visible-light sensitizers for photovoltaic cells. J Am Chem Soc 131(17):6050–6051
5. Im J-H, Lee et al (2011) 6.5% efficient perovskite quantum-dot-sensitized solar cell. Nanoscale 3:4088–4093
6. Kim HS et al (2012) Lead iodide perovskite sensitized all-solid-state submicron thin film mesoscopic solar cell with efficiency exceeding 9%. Sci Rep 2:591
7. Hao F et al (2014) Anomalous band gap behavior in mixed Sn and Pb perovskites enables broadening of absorption spectrum in solar cells. J Am Chem Soc 136(22):8094–8099
8. Krishnamoorthy T et al (2015) Lead-free germanium iodide perovskite materials for photovoltaic applications. J Mater Chem A 3(47):2050–7488
9. Wang H et al (2018) Enhanced crystallization and performance of formamidinium lead triiodide perovskite solar cells through PbI2-SrCl2 modulation. Mater Today Energy 239
10. Zhang J et al (2016) n-type doping and energy states tuning in CH3NH3Pb1−xSb2x/3I3 perovskite solar cells. ACS Energy Lett 1(3):535–541
11. Kang C et al (2021) Antioxidative stannous oxalate derived lead-free stable CsSnX3 (X=Cl, Br, and I) perovskite nanocrystals. Angew Chem Int Ed Engl 60(2):660–665
12. Wu C et al (2017) The dawn of lead-free perovskite solar cell: highly stable double perovskite Cs2AgBiBr6 film. Adv Sci (Weinh) 5(3)
13. Locardi F et al (2018) Colloidal synthesis of double perovskite Cs2AgInCl6 and Mn-doped Cs2AgInCl6 nanocrystals. J Am Chem Soc 140(40):12989–12995
14. Burgelman M et al (2000) Modelling polycrystalline semiconductor solar cells. Thin Solid Films 361–362:527–532
15. Roy P et al (2021) An investigation on the influence of temperature variation on the performance of tin (Sn) based perovskite solar cells using various transport layers and absorber layers. Optics 4
16. Khattak YH et al (2020) Numerical analysis guidelines for the design of efficient novel nip structures for perovskite solar cell. Sol Energy 207:579–591
17. Singh AK (2021) Performance optimization of lead free-MASnI3 based solar cell with 27% efficiency by numerical simulation. Opt Mater 117:111193. ISSN 0925-3467
18. Ahmed S et al (2020) A study of Cesium Titanium Bromide based perovskite solar cell with different Hole and Electron transport materials, ICAICT, pp 297–301
19. Rahman MS et al (2020) Numerical simulation of CsSnI3-based perovskite solar cells: influence of doped-ITO front contact, TENCON, pp 140–145
20. Smith B (2018) Efficient lead-free perovskite solar cell

Author Index

J. K. Mandal et al. (eds.), *Topical Drifts in Intelligent Computing*, Lecture Notes
in Networks and Systems 426, https://doi.org/10.1007/978-981-19-0745-6